NEW
COMPUTING TECHNIQUES
IN PHYSICS
RESEARCH III

Proceedings of the Third International Workshop on Software Engineering, Artificial Intelligence and Expert Systems for High Energy and Nuclear Physics

October 4 – 8, 1993
Oberammergau, Germany

NEW COMPUTING TECHNIQUES IN PHYSICS RESEARCH III

edited by

K.-H. Becks
University of Wuppertal

D. Perret-Gallix
LAPP-IN^{2}P^3 (Annecy) CNRS

World Scientific
Singapore • New Jersey • London • Hong Kong

Published by

World Scientific Publishing Co. Pte. Ltd.
P O Box 128, Farrer Road, Singapore 9128
USA office: Suite 1B, 1060 Main Street, River Edge, NJ 07661
UK office: 73 Lynton Mead, Totteridge, London N20 8DH

Cover picture: "Aztec feather shield or how to get out of this maze without re-inventing the wheel". (Cliché Württermbergisches Landesmuseum, Stuttgart)

NEW COMPUTING TECHNIQUES IN PHYSICS RESEARCH III

ISBN 981-02-1699-8

Printed in Singapore by JBW Printers & Binders Pte. Ltd.

WORKSHOP ORGANIZING COMMITTEE

G. Auger
GANIL, CAEN

K.H. Becks
BERGISCHE UNIV., WUPPERTAL

O. Benhar
INFN, ROME

R. Brun
CERN CN, GENEVA

B. Denby
FNAL. CHICAGO

F. Etienne
CPPM, MARSEILLE

R. Gatto
GENEVA UNIV., GENEVA

G. Gonnet
ETHZ, ZURICH

M.Green
ROYAL HOLLOWAY COLL., EGHAM SURRAY

V.Ilyn
MOSCOW UNIV., MOSCOW

A. Kataev
INR, MOSCOW

C. Kiesling
MPI, MUNICH

P. Kunz
SLAC, STANFORD

M. Kunze
RUHR UNIV., BOCHUM

J. Lefoll
DPHPE, SACLAY

V. Matveev
INR, MOSCOW

K. McFarlane
SSC, DALLAS

M. Metcalf
CERN CN, GENEVA

T. Nash
FNAL. CHICAGO

D. Perret-Gallix
LAPP, ANNECY

C. Peterson
LUND UNIV., LUND

P. Ribarics
MPI, MUNICH

M. Sendall
CERN ECP, GENEVA

Y. Shimizu
KEK, TSUKUBA

D. Shirkov
JINR, DUBNA

J. Vermaseren
NIKHEF-H, AMSTERDAM

C. Vogel
CISI, PARIS

E. Wildner
CERN PS, GENEVA

Conference Secretaries

Ulrike Becks-Malorny
WUPPERTAL

Verena Offermann
WUPPERTAL UNIVERSITY

Foreword

Under very tight budget conditions in fundamental research all over the world, the "Third International Workshop on Software Engineering, Artificial Intelligence and Expert Systems for High Energy and Nuclear Physics" took place in the nice ambiance of the village of Oberammergau in Bavaria, Germany, from October 4 to 8, 1993. It was the third workshop in a series; the first was held in Lyon in 1990[1] and the second at France-Télécom site near La Londe les Maures in 1992[2]. Although the weather was very kind to us, our mood was to some extend shadowed by the uncertain future of the SSC Laboratory. When writing these lines, we got aware of the decision of the U.S. governing bodies to terminate the SSC project. We are all deeply depressed by this defeat for science and the anti-science tendency of people in times of recession. We very much feel with our colleagues from SSCL and hope that they will find ways to overcome this disaster and remain the traditionally strong competitors and partners in our discipline. To overcome these difficulties, High Energy Physics should become a true worldwide enterprise, and topical workshops where common scientific projects can be discussed and developed in a nice social atmosphere will play a major role to achieve this goal.

Taking all this into account we were quite happy to welcome about 120 participants from many countries, most of them stayed for the whole week. It seems clear to us that this relatively small number has only to do with the present travel budgets of the national and international research centers and universities and by no means with the actuality of the topics of the workshop. This can be seen from the large number of people who made a preliminary registration at the beginning of this year, and from the large number of contributions by people who finally could not participate. This was also demonstrated at the vivid discussions during the entire workshop.

This series of workshops offers an important platform for discussions of a broad spectrum of problems. New, highly sophisticated experiments demand new techniques in computing, in hardware as well as in software. An ever growing number of people develops software packages for their experiments. This needs a high degree of planning and coordination. Software Engineering Techniques are discussed which could in principle satisfy the needs for forthcoming accelerator experiments in the next century. The growing complexity of detector systems demands new techniques in experimental error diagnosis and repair suggestions; Expert Systems seem to offer an adequate way of assisting the experimental crew during data-taking. Evolutionary Algorithms offer the possibility to resolve hard optimization problems. Artificial Neural Networks are on the way to be used in the present generation of experiments. Special hardware is being implemented in trigger systems, software simulated networks are used for event classification in data analysis. The higher statistics of experimental data ask for higher precision computations in theoretical physics; systematic procedures are being developed to compute higher order Feynman diagrams with the Techniques of Algebraic Manipulation.

---

[1] The proceedings were published by 'Les Presses du CNRS', Paris, ISBN 2-222-04514-2, 1990.

[2] The proceedings were published by 'World Scientific Publishing', Singapore, ISBN 981-02-1122-8, 1992.

Also experts from outside physics research took part in the workshop: B. Botella (Constraint Logic Programming), O. Krämer-Fuhrmann (Software Engineering), F. Kursawe (Evolutionary Algorithms), M. Lades (Neural Networks), E. Littmann (Neural Networks). We highly appreciate their active and enthusiastic participation!

The workshop was held under the auspices of the European Physical Society and the German Physical Society. We are very grateful for financial support by the

- Bundesminister für Forschung und Technologie (BMFT), Bonn
- Centre National de la Recherche Scientifique (CNRS), Paris
- Deutsche Forschungsgemeinschaft (DFG), Bonn
- International Science Foundation, Washington DC
- Minister für Wissenschaft und Forschung NRW, Düsseldorf
- Village of Oberammergau
- University of Wuppertal.

The smooth running of the workshop was mainly due to the continuous action of Frau I. Lang from the Oberammergau Tourist Office. She was responsible for the local organization and resolved all problems online — by day and by night. On behalf of the entire workshop we would like to thank her and U. Becks-Malorny, H.-W. Markus, V. Offermann, M. Schmidt, and B. Voswinkel for the efficient organization as well as the hardworking staff of the Ammergauer Haus managed by Dr. Bauer.

On behalf of the organizing committee,
Karl-Heinz Becks, *University of Wuppertal*
Denis Perret-Gallix, *LAPP Annecy*

Table of Contents

Artificial Intelligence

Symbolic AI

Evolutionary Algorithms

Symbolic Manipulation

Auomatic/Semi-Automatic Computations

Miscellaneous

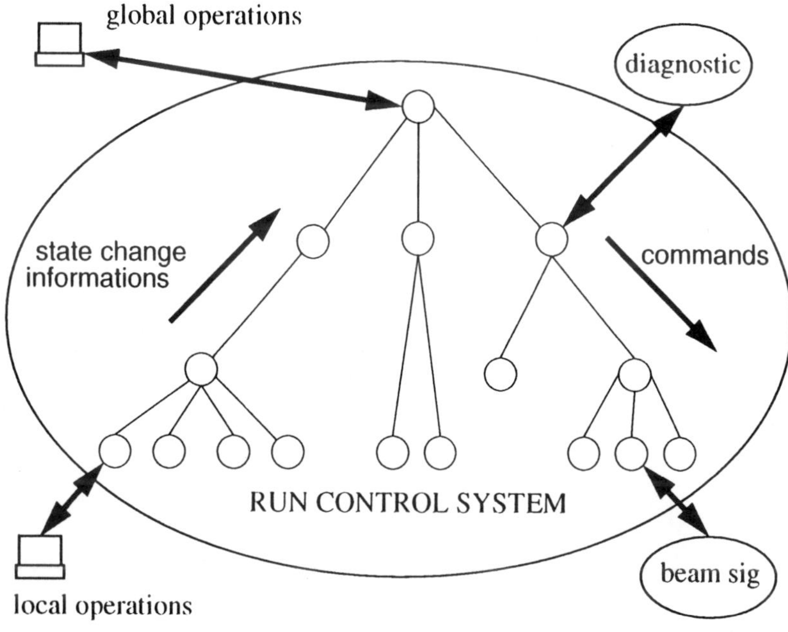
global operations
diagnostic
state change
informations
commands
RUN CONTROL SYSTEM
local operations
beam sig

— TRAPPER —

A GRAPHICAL PROGRAMMING ENVIRONMENT FOR PARALLEL SYSTEMS

O. KRÄMER-FUHRMANN
German National Research Center for Computer Science (GMD)
Schloss Birlinghoven, D-53757 Sankt Augustin, Germany

L. SCHÄFERS, C. SCHEIDLER
Daimler-Benz Research, Alt Moabit 91b, D-10559 Berlin, Germany

ABSTRACT

TRAPPER is a graphical programming environment which supports the development of parallel applications. The programming environment is based on the programming model of communicating sequential processes.

TRAPPER contains tools for the design, mapping and visualization of parallel systems. The Designtool supports a hybrid program development, where the parallel process structure is described using a graphical representation and the sequential behavior is described by sequential program code. The configuration of the target hardware and the mapping of the application onto the hardware is supported by the Configtool. During run-time, the monitoring-system records events which can be animated by the Vistool and the Perftool which visualize the behavior of the hard- and software components.

This paper describes the support of two machine independent message passing interfaces: PVM (Parallel Virtual Machine) and PARMACS (Parallel Macros).

1. Introduction

Parallel computing is accepted as the only technology which offers a long term chance of improving the performance of computer systems. Parallel processing has been widely accepted in the field of numerical computing, but will also have a great impact on technical systems when the software problem will be solved.

In this paper we describe TRAPPER, a graphical programming environment that assists the programmer in developing application software for systems which use parallel processing as a key technology for high computing power. The TRAPPER philosophy is to have the programmer explicitly specify the parallel structure of the application and to aid the programmer as much as possible in the various phases of the development cycle. TRAPPER is based on the programming model of communicating sequential processes, which is suited for a large class of applications.

TRAPPER has been under development in the GMD since 1991 [14]. The programming environment supports different target systems like PVM [5] and the PARMACS [7]. In cooperation with Daimler-Benz, embedded industrial real-time systems based on the Transputer technology are supported [20]. A first release has been delivered to different research groups of the GMD, the Mercedes-Benz vehicle research center and other members of the Daimler-Benz corporation.

2. Related Work

Ever since parallel computers have existed, researchers have investigated the difficulties associated with programming them. Tools for parallel MIMD computers (multiple instruction multiple data) can be classified into systems which support the data-parallel approach or explicit message passing. In the data-parallel concept, the programmer deals with arrays of data, getting distributed over the parallel architecture by compiler technology [8]. In the message passing paradigm, the parallel program consists of an ensemble of communicating processes [12]. The benefit of the data-parallel programming model is the absence of multiple control flows, which leads to an easy but inflexible programming model. The process-parallel programming model is more flexible, but the existence of multiple control flows raises new problems in the various program development tasks. Several research activities are dealing with design, mapping, debugging, animation and optimization of parallel programs. A good overview over those tools is given in [13].

During the design phase, the programmer has to describe the parallelism of the application. Text-oriented programming languages reflect the parallel structure of a program very poorly. Therefore various tools make use of graphical representations instead of textual representations to describe the parallelism. HeNCE [3], Code [18] and Paralex [2] follow the data-flow approach, where nodes represent subroutines and directed arcs represent data-dependencies between subroutines. Millipede [1], MP [16] and TRAPPER use process graphs, where nodes represent processes and arcs represent communication channels.

The mapping of the application onto the target hardware is a task that does not exist in the traditional development cycle of sequential programs. A lot of research has been done in this area, a survey is given in [6]. TRAPPER integrates a semi-automatic mapping algorithm into a graphical programming environment.

Animation is crucial for understanding the run-time behavior of a parallel program. Parallel activities are better understandable when they are represented graphically. ParaGraph [11] is a very popular visualization tool, which can display trace data in a lot of different graphical views. Unfortunately ParaGraph works only in offline mode, has no backstepping facilities and cannot display user defined data. PATOP [4] is part of the Topsys project and enables very detailed observation of hardware activity including statistical evaluation. Pablo [19] is a very flexible animation tool, where the user can define and configure the views of his data interactively. Unfortunately Pablo has no stepping/backstepping facilities and offers no time diagrams. Furthermore there exist many graphical visualization tools for special systems. Xab is a monitor and visualizer for PVM applications. Express has its own monitoring and visualizing toolset. Prism [10] is a graphical tool for the connection machine, which is a SIMD architecture. Prism gives excellent debugging support and is well integrated into the system software, but an extension for message passing systems and the portation to other MIMD architectures is difficult.

3. TRAPPER Components

TRAPPER comprises different tools for all phases of the software development cycle. Figure 1 gives an overview of the TRAPPER toolset.

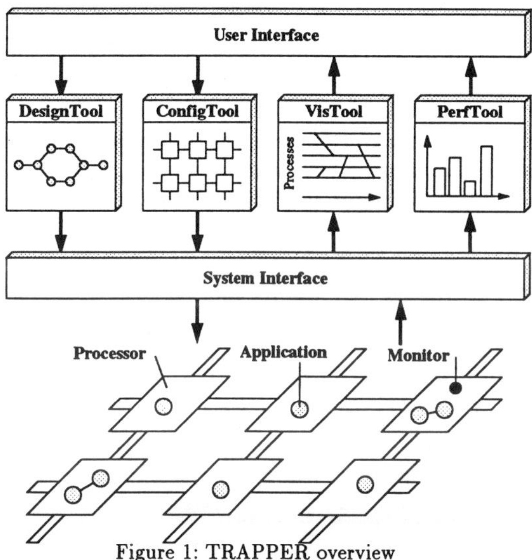

Figure 1: TRAPPER overview

With the `Designtool` the programmer specifies the process graph, where the nodes in the graph represent processes and the edges denote communication channels. The process graph describes the parallel structure of the application, independently of the target hardware. Each process is a sequential task with access to local memory only. Processes can communicate with each other by message passing constructs.

The `Configtool` allows the user to specify the configuration of the hardware system and determines the mapping of the process graph onto the hardware.

The `Monitor` collects run-time information about software-events like process invocation, interprocess-communication, computation and communication loads of the target hardware and of user defined events. All events are time stamped automatically and are stored in an animation file.

Two different tools allow the visualization of the program execution. The `Vistool` enables program animation, i.e. the graphical display of the software behavior, like execution phases, variable contents and application specific information. The `Perftool` displays information about the hardware, i.e. load characteristics and scheduling information. Both animation tools offer not only online animation but also offline animation, including stepping and backtracing facilities. TRAPPER supports the snapshot concept, which displays the system states of all components at a given point of time. Furthermore TRAPPER can visualize the dependencies between different events on a time scale and draws various statistics.

TRAPPER resides on the host (currently a Sun) and is implemented in C++, using the InterViews graphics library [15], which is based on the X window system.

3.1. Designtool

The Designtool supports a hybrid program development. The parallel structure of the application is described by a graphical representation, the process graph. The sequential components are described by textual representations, the traditional program code. The process graph consists of nodes and edges, where nodes represent processes and edges represent communication channels. Each process consists of a unique process identifier, a process type denoted by the process name and dedicated communication interfaces called ports.

Large process graphs can be designed hierarchically as a composition of subsystems. A subsystem is a graphical entity and can be considered as a black box, which contains a subgraph of the process graph. A subsystem itself can contain other subsystems. TRAPPER can create standard topologies (rings, grids, tori and trees) by a built-in net generator.

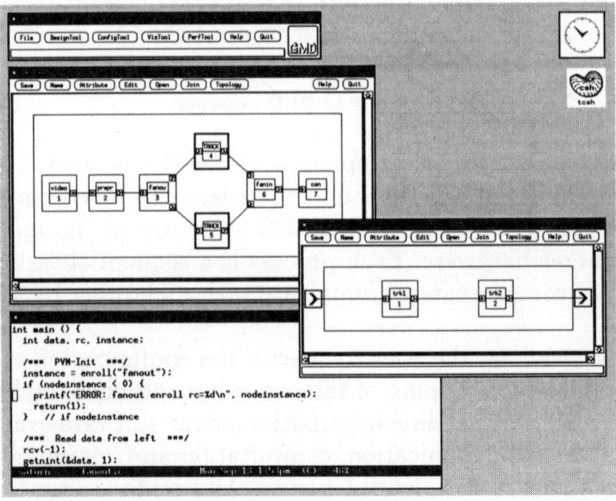

Figure 2: TRAPPER Designtool

A screen dump of the Designtool is given in figure 2. The upper left window shows the main TRAPPER control panel. The window below the control panel shows the activated Designtool with the (simplified) process graph of an application "autonomous vehicle guidance" being developed by Daimler-Benz. In this example data in the process graph flow from left to right. Normal processes are represented by boxes with one frame while subsystems are represented by boxes with double frames. The right window shows an opened subsystem consisting of two processes. Ports are represented by squares located at the frame of the subsystems.

The behavior of a process is described textually by the program code. The programmer selects a process in the process graph and activates a text editor with the associated C or Fortran code file. Such a program code is shown in the lower left window.

The process code is associated with the process type, not with the process itself. In other words, processes having the same name share the same process code.

3.2. Configuration Tool

With the aid of the Configtool the application is mapped onto the target hardware. The mapping is done in two steps.

First the programmer has to specify the target hardware by a special designtool. With the aid of a graphic editor the user draws the configuration of the target hardware. Nodes represent processors and edges connecting nodes represent communication links. These links are not necessary for workstation clusters, because usually all processors are connected to a shared bus. Different processor types can be introduced by using different node names. Each type takes into account a relative speed in order to model different computation speed of the various workstations.

In the second step the application is mapped onto the hardware. This can be done either automatically or manually. In the automatic mode TRAPPER computes a mapping of the process graph. The TRAPPER mapping algorithm searches a partitioning of the process graph with a well distributed computation load and a small communication load between partitions. It takes into account the following criteria:

- computation time of each process,
- communication amount between the processes and
- speed of the processors.

These input data for the mapping algorithm can be defined in two ways. Either the programmer adds weights to the nodes and edges of the process graphs, or these data are extracted from a real test-run. During program execution the monitor extracts the start/stop and communication events, which can be gathered by the TRAPPER utility `trace2load`. This program extracts the computation and communication amount and adds these data to the software and hardware graphs.

In general, the mapping algorithm can find only sub-optimal solutions, because the underlying optimization problems (e.g. graph partitioning) belong to the class of NP-hard problems. Therefore a heuristic algorithm — called iterated 2-Opt [17] — is used to determine a good solution in short time. In the first phase a valid solution is constructed at random. Then the 2-Opt algorithm tries to improve the solution by pairwise exchanges until no more improvements can be found. The solution is disturbed by few random exchanges before the 2-Opt tries to improve this solution again. After every 2-Opt phase the actual solution is compared to the best solution so far in order to keep the best mapping. This iteration is repeated until a time limit (e.g. 5 seconds) is reached.

The process graphs build the user-interface of the Configtool. The mapping solution is displayed by coloring the nodes of the software graph. Processes mapped onto the same processor have the same color. The mapping computed by TRAPPER can be modified by the user. The programmer can select nodes interactively in the process graph and specify the desired CPU number.

After finishing the previous steps, TRAPPER generates the configuration file needed by the TRAPPER utility `startPvm` to start the PVM application. This will be explained in section 4.

3.3. Monitoring System

The monitoring-system provides run-time information as input to the Vistool and the Perftool. Vistool needs information about the the application and Perftool needs information about the underlying hardware. The monitoring-system is the only component running on the target hardware and is therefore not as portable as the other TRAPPER components. In following sections we describe two different monitoring systems. Section 4 describes a monitor for PVM applications and in section 5 an offline monitoring for PARMACS applications is presented.

3.4. Visualization Tool

The Vistool aids the observation of the run-time behavior of the application. It is integrated with the Perftool in such a way that a consistent view on the states of hardware and software components is offered. The Vistool supports the programmer in the analysis of the parallel algorithm by displaying run-time data of the application. This helps the user to understand the dynamics of distributed systems, gives debugging information, allows the detection of errors (i.e. deadlocks by the visualization of cycles of incomplete communication requests) and gives important information for code optimization.

The animation tools have an online and an offline mode. In short applications, the offline mode is to be preferred. The events are collected in a monitoring-file which is read by the animation system. This decoupling of the animation from the execution allows the observation of the software events in an individual speed. Additional features like single stepping and backtracing are offered by TRAPPER, to enable the programmer to lead the animation to the interesting program phases.

The Vistool has three different views: one is based on the process graph, a second is based on the time scale, and at last a kiviat diagram.

The process graph view is based on the graphical representation developed with the Designtool. The programmer can select between a variety of different animation features like coloring of nodes or edges, textures on nodes or edges, changing the line width or drawing arrows on the edges, displaying plots, histograms or rastered squares in the process boxes. The process graph animation can be used to display the process state, variable values or interprocess-communication.

The time scale animation shows the events on a time scale. This animation visualizes communication operations as arrows between processes. This function allows a detailed insight in the cooperation of distributed processes and therefore helps the programmer to debug and optimize his program. Statistics and a critical path analysis can be supplied displaying the dependencies in the application, which determine the overall execution time.

The kiviat diagram displays load distribution.

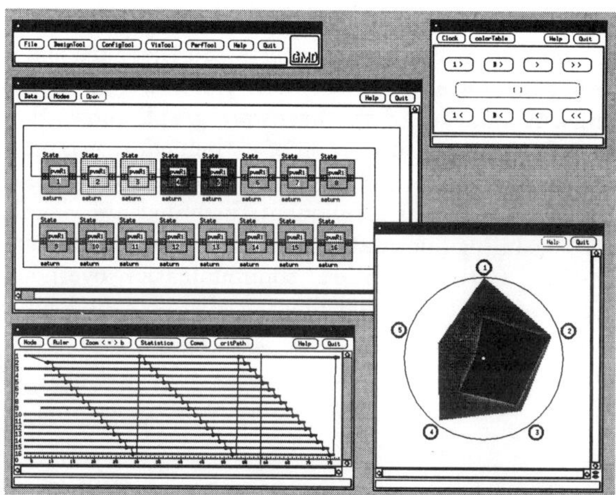

Figure 3: TRAPPER animation tools

Figure 3 shows a typical session with the TRAPPER animation tools. The upper right window shows the animation controller with an interface similar to a tape recorder. Each animation display is controlled by this component. The animation controller supports single-stepping, normal play and fast forward in both directions.

The middle left window shows the software graph. The colors of the nodes indicate the state of a process: active, waiting, communicating or idle. This diagram shows a snapshot of the application, i.e. the state at a given time. The kind of data displayed in each node can be changed interactively by the user such that a comprehensive analysis of the interresting points in the application is possible.

The lower window shows a time diagram, where each process is represented by a horizontal line. State changes of the processes are displayed by different colors, communication is indicated by a black arrow between the time lines of the partners. With this diagram the analysis of the dynamic behavior of an application is possible.

The lower right window is a Kiviat diagram showing the load distribution of the processors. Each CPU is represented by a spoke of the wheel which gets colored corresponding to the current load. A so-called high water mark indicates the maximal value ever reached.

3.5. Performance Analysis Tool

The Perftool supports the user in the optimization phase, which completes the software development after the design and debugging phases. The software developer receives hints on possible bottlenecks that are due to load imbalances. Use of the Perftool is tightly coupled to the use of the Vistool. An important purpose of the Perftool is to find a relation between the behavior of the hardware and the software, e.g. to relate a non-satisfactory CPU load and the specific code segment which causes it.

The Perftool offers three different views: the first is based on the hardware graph, the second is based on time scaled charts and the third is a Kiviat diagram.

Within the hardware graph the CPU load and link load are shown by coloring the nodes and links. Animation of the hardware graph shows the whole parallel machine and gives a first, rough impression of its behavior. A more detailed insight can be gained with animations based on time scaled charts. Each chart shows the temporal behavior of the selected component. Included are visualization of CPU load, link load and scheduling information. Performance statistics are also provided. They include run-time, CPU utilization, communication overhead, speedup and efficiency of the application. The Kiviat diagram displays the load of all processors including with a so-called high water mark. This Kiviat diagram is useful for the analysis of load balance.

4. PVM

This section describes the cooperation between PVM [5] applications and the TRAPPER programming environment.

The PVM support consists of two independent functionalities. In the program design phase, PVM applications can be specified and mapped by TRAPPER. Running PVM applications can be monitored and their dynamic behavior can be visualized by the TRAPPER animation tools. Figure 4 gives an overview of the two mechanisms which will be explained in the following.

Parallel applications being designed and mapped by TRAPPER can be started on PVM via the `startPvm` utility. For this purpose TRAPPER generates a configuration file (*.pvm) which describes the application in terms of processes and their target processors. With the TRAPPER design tool only static process nets can be specified.

For the definition of interprocess communication two different paradigms are supported by TRAPPER. PVM and the PARMACS allow communication between arbitrary process pairs. Addressing is done by global unique process identifiers, such that an explicit connection between communicating partners is not necessary. The second model is based on port communication, as it is used in the Transputer world. Here processes must be connected explicitly by a communication channel. For this paradigm, the process topology can be defined by the TRAPPER Designtool which

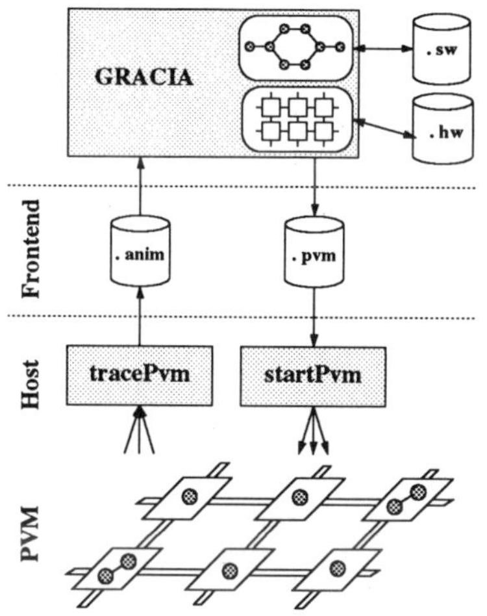

Figure 4: PVM integration

is instantiated by the `startPvm` utility.

Monitoring under PVM is implemented as a special process, called `tracePvm`. This process is started automatically by `startPvm` and runs additionally to the application. Its purpose is twofold. First the hardware load has to be recorded by polling the PVM processors via the Unix command `rstat`. This is done cyclic in a given time interval (e.g. 5 seconds). Second the monitoring process receives messages being sent by the application processes, like

task activation creation and termination of processes,

communication events begin and end of a communication and

user events color and write textures to nodes or edges; display application data in the nodes.

Each event must carry an individual time stamp, which can either be inserted by the application or, if this is impossible, automatically by the monitoring process.

TRAPPER animation can be either online or offline. In the online case the monitor writes the events to a named pipe which is being read simultaneously by TRAPPER. This enables the observation of the distributed application during run-time. In the offline mode the events are written to a file (*.anim). After sorting them by using the time stamps, this file can be read and visualized by the TRAPPER visualization tools.

5. PARMACS

This section describes the cooperation between PARMACS [7] applications and the TRAPPER programming environment. We describe the offline version, but online animation is possible by introducing a tracer process similar to the PVM implementation in section 4.

PARMACS applications use a set of machine independent functions to express parallelism in sequential C or Fortran programs. Most important are functions that allow process creation (CREATE), synchronization (BARRIER) and communication (SEND, RECV). Process nets are mapped automatically to the actual hardware and communication partners are addressed by global process identifiers. This implies that PARMACS applications cannot make direct use of design and mapping functionality of TRAPPER. Therefore our work concentrates on an automatic instrumentation of PARMACS programs and their visualization. An automatic detection of process creation and termination, barriers and process communication is realized.

The instrumentation of a PARMACS program is done by preprocessing the program code, similar to the instrumentation for Paragraph implemented by the Pallas GmbH. This preprocessing is activated by an additional flag to the `parmacs` call and therefore no changes in the source code are needed. The preprocessing expands the PARMACS macros such that before and after a PARMACS call a library routine is called. The following program segment clarifies this "sandwich" technique for the example of an asynchronous send operation.

```
                                        sendBegin(...)
SEND(target,data,length,type)    ->     SEND(target,data,length,type)
                                        sendEnd(...)
```

These newly included functions collect the event together with the actual time stamp in an internal event queue. Each PARMACS program executes an `ENDNODE` of `ENDHOST` command before termination. These macros have also been modified such that the internal event queues get sent to the host program, where the events get written to the animation file. TRAPPER can sort and read this event file and allows the visualization with the animation tools.

To instrument a PARMACS program only the makefile has to be changed in two places: `parmacs` must be called with an additional flag (-patools) and the linking step must include the `trapperlib-offline`, which implements the monitoring functionality.

6. Conclusion

We presented TRAPPER, a graphical programming environment for parallel applications. The TRAPPER philosophy is to have the programmer explicitly specify the parallel structure of the application. TRAPPER supports a hybrid program development, where the process structure is described using a graphical representation and the sequential behavior is described using textual representations. TRAPPER consists of tools which support the design, mapping, monitoring and animation of parallel applications.

With the aid of the Designtool the programmer specifies the process graph. The Configtool allows the user to specify the configuration of the hardware system and determines the mapping of the process graph onto the hardware. The monitoring-system collects run-time information. The Vistool enables program animation, i.e. the graphical display of execution phases, variable contents and application specific information. The Perftool displays information about the hardware, i.e. load characteristics and scheduling information.

In this paper we described two mechanisms for automatic program instrumentation, one for online monitoring of PVM programs and another for offline monitoring of PARMACS programs. The aim was to collect runtime information about a program run automatically, without the need of source code mainpulation. Important information about the program behaviour, like process creation and termination, and interprocess communication can be monitored and displayed by the TRAPPER animation tools.

6.1. Project Status

TRAPPER is an active research project. Its first releases are in use in different research projects of the GMD: in the PEGASUS project a parallel genetic algorithm toolbox is being developed, and in the ROTOR project the air flow in jet-propulsion is simulated. A first release has also been delivered to the Mercedes-Benz vehicle research center and other members of the Daimler-Benz corporation.

6.2. Future Work

Future activities may concern all parts of the programming environment:

- Other message passing interfaces like the MPI will be supported by TRAPPER. The extensions concern mainly the monitoring-system.
- Currently the graphical representation only describes the static process structure, extensions to support dynamic nets are planned.
- The debugging features of the visualization can be extended by integrating existing debuggers or by extending the online features of the Vistool.
- It is planned to use TRAPPER as an user interface for interactions during runtime. This can be useful for interactively steered simulation processes.

7. Acknowledgements

We thank the TRAPPER team, namely Beatrix Hornef, Hans-Christof Lenhard, Josef Roggenbuck and Angelika Weihermüller, which did an excellent work implementing the TRAPPER toolset.

8. References

1. M. Aspnäs, B. R. J. R., and T. Langbacka. *Millipede - A Programming Environment Providing Visual Support for Parallel Programming.* In European Workshop on Parallel Computing, Barcelona, Spain, 92.
2. Ö. Babaoğlu, L. Alvisi, A. Amoroso, R. Davoli, and L. A. Giachini. *Paralex: An Environment for Parallel Programming Distributed Systems.* In International Conference on Supercomputing, pages 178–187, Washington, USA, July 92.
3. A. Beguelin, J. J. Dongarra, A. Geist, R. Manchek, and V. Sunderam. *HeNCE: Graphical Development Tool for Network-Based Concurrent Supercomputing.* In Proc. of Supercomputing, Albuquerque, 91.
4. T. Bemmerl and A. Bode. *An Integrated Environment for Programming Distributed Memory Multiprocessors.* In Second European Distributed Memory Conference, Munich, Apr. 82.
5. A. Benguelin, J. Dongarra, A. Geist, R. Manchek, and V. Sunderam. *A Users Guide to PVM - Parallel Virtual Machine.* Technical Report ORNL/TM-11826, Oak Ridge National Laboratory, Sept. 91.
6. F. Berman. *Experience with an Automatic Solution to the Mapping Problem.* In L. H. Lamieson, D. Gannon, and R. J. Douglas, editors, The Characteristics of Parallel Algorithms, Series in Scientific Computation — The Characteristics of Parallel Algorithms, pages 307–334. MIT Press, 87.
7. L. Bomans, D. Roose, and R. Hempel. *The Argonne/GMD Macros in FORTRAN for Portable Parallel Programming and their Implementation on the Intel iPSC/2.* Parallel Computing, 15:119–132, 90.
8. T. Brandes. *Compiling Data Parallel Programs to Message Passing Programs for Massively Parallel MIMD Systems.* In Working Conference on Massively Parallel Programming Models, Berlin, Sept. 93.
9. T. Bräunl. *Structured SIMD Programming in Parallaxis.* Structured Programming, 10(3):121–132, 89.
10. T. M. Corporation. *Prism User's Guide.* Version 1.2, Thinking Machines Corporation, März 93.
11. M. T. Heath and J. A. Etheridge. *ParaGraph: Visualizing the Performance of Parallel Programs.* IEEE Software, 8(5):29–39, Sept. 91.

12. C. A. R. Hoare. *Communicating Sequential Processes.* Commun. ACM, 21(8):666–677, Aug. 78.

13. H.-C. Hoppe, T. Kentemich, O. Krämer-Fuhrmann, and W. Krotz-Vogel. *Evaluation of Graphical Performance Analysis Tools for Local Memory Parallel Computers.* Technical Report D6.2.b, Esprit Project PPPE: Portable Parallel Programming Environments, July 93.

14. O. Krämer-Fuhrmann and T. Brandes. *GRACIA: A Software Environment for Graphical Specification, Automatic Configuration and Animation of Parallel Programs.* International Conference on Supercomputing, pages 67–74, June 91.

15. M. A. Linton, J. M. Vlissides, and P. R. Calder. *Composing User Interfaces with InterViews.* Computer, 22(2):8–22, Feb. 90.

16. J. Mangee and N. Dulay. *MP: A Programming Environment for Multicomputers.* In Proc. of the IFIP Working Group on programming Environments for Parallel Computers, Edinburgh, Scotland, Apr. 92.

17. C. S. R. Murthy and V. Rajaraman. *Task Assignment in a Multiprocessor System.* Microprocessing and Microprogramming, 26:63–71, 89.

18. P. Newton and J. C. Browne. *The CODE 2.0 Graphical Programming Environment.* In International Conference on Supercomputing, pages 167–177, Washington, USA, July 92.

19. D. A. Reed, R. A. Aydt, T. M. Madhyasta, R. J. Noe, K. A. Shields, and B. W. Schwartz. *An Overview of the Pablo Performance Analysis Environment.* Technical Report, University of Illinois, Department of Computer Science, Nov. 92.

20. L. Schäfers, C. Scheidler, and O. Krämer-Fuhrmann. *TRAPPER: A Graphical Programming Environment for Industrial High-Performance Applications.* In Parle, Parallel Architectures and Languages Europe, pages 403–413, Munich, June 93.

PHYSICS RESEARCH INTEGRATED DEVELOPMENT ENVIRONMENT (PRIDE)

Jackie Burton, Larry Cormell
Superconducting Super Collider
2550 Beckleymeade Avenue
Dallas, TX 75237-3997

ABSTRACT

Past efforts to implement a Software Engineering approach to High Energy Physics computing have been met with significant resistance and have been, in many cases, only marginally successful. At least a portion of the problem has been the lack of an integrated development environment, tailored to High Energy Physics and incorporating a suite of Computer Aided Software Engineering tools. The Superconducting Super Collider Physics Research Division Computing Department is implementing pilot projects to develop just such an environment.

1. Introduction

Software Engineering (SE) technology continues to evolve in the direction of a coherent, integrated environment based on the complementary principles of Object Orientation (OO) and the Open Systems Environment (OSE). The unique teaming of physicists from both the Solenoidal Detector Collaboration (SDC) and the Gammas Electrons and Muons (GEM) collaboration with the Physicists and Software Engineers of the Physics Research Division (PRD) Computing Department (PRCD) at the Superconducting Super Collider Laboratory (SSCL) to develop such an environment, the Physics Research integrated development environment (PRide), portends a remarkable new phase in software development for High Energy Physics (HEP). It promises: 1) improved productivity in the development of software; 2) greater accuracy and reliability of the software; and 3) quicker turn-around time for both the results of analysis and for the implementation of new software tools/versions.

2. An Historical Perspective

Previous efforts to introduce structured SE techniques into the realm of HEP have met with, at best, mixed results. The adoption of formal SE methods by experimental high energy physicists has proceeded very slowly. Several recent articles by Loken[1], Knobloch[2], and Nash[3] provide interesting discussions of this lack of formal SE in HEP. It does seem strange that scientists in a field that constantly pushes the envelope by demanding the state of the art in cryogenics, electro-optics, structural materials, electronics and certainly computing hardware would not demand the same of SE. There are a number of historical, philosophical, and even emotional reasons for this bias against, or at least reluctance to SE methods. While a discussion of this reluctance would probably be an excellent topic for a panel discussion, it is beyond the scope of this paper. We simply note in passing that structured analysis/structured design (SA/SD) methodologies were introduced in the early '70s and have been used successfully throughout the software industry since that time to reduce development time and costs and to improve product quality.

Despite this apparent bias against SE, several HEP experiments have recently made use of SE methodologies. The most notable are D0 and ALEPH. Both of these experiments have had considerable benefit from their use of SE methods and tools. The software developers in these experiments basically followed the classical life cycle or waterfall development plan[4] employed SA/SD formalism[5], and embodied their design in data flow diagrams (DFD) ala DeMarco[6].

The D0 collaboration used SA/SD methodology in the initial design and implementation of their software. DFDs and their associated data dictionaries were the primary tools used to develop the D0 software models. These along with structure charts and state transition diagrams provide the basic design, description and documentation of the D0 software system[7]. At the time that D0 began developing their software they were not able to obtain adequate computer-aided software engineering (CASE) tools to maintain their design and diagrams.

ALEPH has made extensive and effective use of DFDs throughout the life cycle of their software development effort[8]. In addition the ALEPH collaboration has introduced data modeling and data base techniques to provide organization for their analysis tasks. ADAMO (Aleph DAta MOdel) was developed by the Aleph collaboration where it is used in several programs including detector description, data acquisition, event reconstruction and display, data analysis and group administration[9]. ADAMO includes the elements of an entity-relationship (ER) model in a form suitable for HEP computing. Entity-relationship diagrams were drawn to provide graphical descriptions of the data structures and the relationships among them. Data dictionaries as derived from the DFDs were generated and maintained. A CASE tool from Tektronix was used to automate the process of drawing and updating the diagrams.

The ALEPH collaboration found several advantages to using SA/SD methods and CASE tools:

- There is a common/uniform style of work throughout the development effort.
- The extended analysis phase had a very positive aspect because it led to the clarification, documentation and easy communication of concepts and ideas to other colleagues.
- The design documentation was available before coding began, providing a powerful means to describe abstract details.
- The top level breakdown and organization of the design permitted partitioning of the work among several independent teams.
- Frequent walk-throughs and reviews for verification and validation provided increased quality assurance in the final products.

Thus far, those high energy physicists who have used SE methods and CASE tools have primarily followed waterfall development efforts with some rapid proto-typing and feedback (modified waterfall). The waterfall method can be quite effective, especially for fairly well defined, closed-end applications. However, for many problems, such as those typically encountered in HEP, the solutions can not been written in closed form at the beginning of the project. The input requirements and desired output may change during the development cycle. In these cases, the formalism of the waterfall method may be too rigid. Newer methods have attempted to address the need for dynamic flexibility within the development effort. Several of these include Dearnley and Mayhew's model[10], Boehm's spiral development model[11], and other variants of iterative prototyping methods[12 13 14 15]. There is no single silver bullet[16]. Each software project is unique with its own set of user requirements, desired level of quality assurance, available resources, deliverables and schedules. The development approach, planning, and design methodology must be tailored accordingly. In the end there is still no substitute for careful planning, lots of thinking, and plenty of hard work.

3. On-Going Efforts

Despite the limited success of previous efforts, there is still an awareness of the need for a better way to develop software. Numerous efforts are under way to incorporate sound software engineering techniques into HEP software development efforts in order to take advantage of rapidly changing hardware and software technologies. Several papers were presented at this conference last year on this topic including: 1) a proposal by F. Bruyant to design a coherent software framework for future HEP experiments; 2) a report by P. Kunz on SLAC B efforts to utilize Object Oriented Programming (OOP), C, C++ and UNIX; 3) a report by Durr and Lourens on a formal specification language for Object Oriented Design (OOD); 4) a report by A. Nilsson on an event generation toolkit in C++; 5) a report by C. Arnault on the use of OOA for the Delphi Online Event Display; and 6) a report by J. D. Shiers on HEPDB[17]. There are additional examples, all pointing to the conclusion that HEP software development is moving inexorably to an integrated, software engineering approach to solving the problems of HEP data capture, storage, filtering, reconstruction and analysis.

4. The SSCL Computing Environment

The SSCL PRCD, SDC and the GEM collaboration are in a unique position to take advantage of the leap in computer engineering efficiency and potential for improvement in software development cost-to-benefit ratios which presents itself with the implementation of an integrated software engineering development environment based on OO and OSE concepts. The lack of legacy constraints of outdated hardware, uncontrolled software and immovable bureaucracy provides the opportunity to plan and implement the most cost effective, scientifically feasible solution for the long term. Cost, resource and schedule constraints add to the attractiveness of alternative solutions

The viability of the utilization of scalable, loosely coupled networks of workstations as proven by the SSCL Physics Detector Simulation Facility (PDSF) argues successfully for the implementation of an open systems concept[18]. The need for the integration of a variety of analysis tools developed and used by physicists around the world with the latest commercial off the shelf (COTS) SE tools points out the need for an applications programmer interface (API) within some integrated environment.

In the past, physicists have been able to rely on "free" manpower resources in the form of post-docs and graduate students. The scale and longevity of the efforts at the SSCL limit the effectiveness of such resources. The limited budget and resources allocated to PRCD and the collaborations for computer and software engineering makes it an imperative that alternatives be identified. The extensive requirements by both of the collaborations coupled with these limited resources begs for the implementation of reusable code and the sharing of resources to the maximum extent possible without endangering the integrity of either of the collaboration's experiments.

The long time scale for completion of the SSCL as well as the anticipated duration of each of the detector experiments requires the ability to provide technology forecasts beyond reasonable expectations of accuracy. As such, the implementation of interim software solutions must maximize the ease of modification, expansion and seamless replacement of the components of the environment. That is, software tools or sets of tools should be readily replaceable by newer versions or even new tools with minimum impact on the end user, the physicist.

5. The Beginning Of PRide

The unique environment at the SSCL set the stage for an unprecedented combination of physics collaborations and laboratory computing department resources. Recognizing the need for an innovative solution to the problems of insufficient funding, limited personnel resources and tight schedule constraints, PRCD has long championed a confederated approach to maximizing commonalty in the solutions to the computer engineering and software development problems facing the collaborations[18]. The SDC Computing subgroup took the first steps towards such an environment by inviting members of the GEM collaboration computing subgroup to an SDC computing subgroup workshop at Fermi National Accelerator Laboratory (FNAL). Attendees left with a conviction that there was some common ground.

In May, 1993 an SDC computing workshop hosted by Rice University[19] featured full participation by both GEM collaboration computing subgroup members and members of PRCD. An effort was made, at that time, to establish and prioritize software development requirements shared by the two collaborations. Among those prioritized requirements were a requirement for a software framework environment and for an integrated Computer Aided Software Engineering (CASE) environment.

Subsequently, the first joint computing subgroup workshop was held at the SSCL[20]. At that time, a brief definition of each of the proposed joint projects was provided by the PRCD Computer Engineering Group. In addition, a detailed explanation of CASE and the integrated software engineering environment was presented. The presentation included a brief proposal to develop the PRide. More importantly, PRCD, the SDC and the GEM computing subgroups agreed to develop a Memorandum of Understanding for management of joint projects between PRCD and the two collaborations[21]. The Experimental Computing Management Committee (ECMC) was established to oversee initiation, assignment prioritization and allocation of resources to projects. Alternative approaches for the prioritization and resource allocation process were discussed and initial drafts were developed for the procedures by which joint projects would be initiated, approved, prioritized, managed and have resources assigned from PRCD and the collaborations[22 23 24]. In addition to the need to identify projects, priorities and how resources were to be shared, the intent of this effort was to prevent the development of a client server relationship in which PRCD received requirements from the collaborations and returned with a complete system some time later that might or might not fulfill those requirements.

6. Framework Software

Among the first projects initiated by the ECMC was an effort to develop a standardized environment for joint off-line computing applications software to be known as the FRAMEWORK software. The concept of the FRAMEWORK software had grown out of the previous SDC core software efforts[25]. The FRAMEWORK software was seen as necessary to provide support to the various tools, utilities, application programs and subroutines that will be developed by the two collaborations.

7. PRide Reference Model Selection

Integral to the ECMC FRAMEWORK software development project was the requirement to develop an integrated software development environment, the PRide. In fact, the FRAMEWORK software can be seen as a component of PRide itself. It should be noted that the purpose of the reference model is to describe environments that support projects that engineer, develop or maintain computer based systems. It is explicitly aimed at establishing a conceptual basis for an environment, not at establishing any particular environment product. There is no part of the reference model that involves choosing a standard toolset. That is the purpose of the PRide development effort.

Evaluations of various reference models for integrated environments, including the Portable Common Tools Environment (PCTE)[26], the Reference Model for Frameworks of Software Engineering Environments[27], and the POSIX Open Systems Environments. POSIX model seeks to define the open systems environment. The NIST/ECMA reference model domain is the PSE frameworks that support software engineering.

These reviews led to the evaluation selection of the Next Generation Computer Resources (NGCR) program Project Support Environment (PSE) reference model for the initial attempts to define and model the PRide. The domain of the NGCR model encompasses each of the other standards and more in an attempt to establish a conceptual base for an environment.

8. NGCR PSE

The NGCR PSE reflects the aspects of a large collection of existing environment efforts and models. These include the Software Technology for Adaptable and Reliable Systems (STARS) program, the National Institute of Standards and Technology (NIST) Integrated Software Engineering Environments (ISEE) working group, the European Computer Manufacturers Association (ECMA) TC33 Task Group on the Reference Model, the United States Air Force Software Life Cycle Support Environment (SLCSE) project, Honeywell's Engineering Information System (EIS) program, the Conceptual Environment Architecture Reference Model (CEARM) effort, and the standardization committees within the Institute for Electrical and Electronics Engineers (IEEE) and American National Standards Institute (ANSI) for POSIX and for CASE Tool Integration Models (CTIM)[28].

The approach of the model is comparable to the POSIX OSE and the NIST/ECMA Reference Model for Frameworks of Software Engineering Environments. The model incorporates the two different domains of interest of these environments: 1) the Open Systems Environment of Posix; and 2) the NIST/ECMA PSE frameworks that support software engineering.

Since both of these efforts developed independently, there are some minor discrepancies between them that are being resolved within the two affected agencies[29]. For example, the Network services described in POSIX are included in the Object Management Services Distribution and Location service and in the replication and synchronization service. Additional areas of interest are incorporated as well. For instance, it has been proposed that the NIST/ECMA Framework's Presentation service be revised to accommodate some of the emerging technologies in the area of Audio and Video Processing Service.

9. Case Tools

Before we look at the PRide, it is worthwhile to look at what CASE is in order to relate it to the PRide. We will discuss CASE in terms of the standard industry classification of Upper, Lower and ICASE or Integrated CASE. Upper CASE tools are those which support the initial phases of a software development project, to include systems analysis and trade-off studies. Upper CASE tools are generally selected to support the software development methodology selected by the organization such as SA/SD or one of the varieties of Object Oriented (OO) Analysis (OOA) and Design (OOD) such as Shlaer/Mellor[30] or Rumbaugh[31]. Lower CASE traditionally includes more programming specific tools such as editors, compilers, linker/loader, debuggers, restructures and other programming aides. ICASE tries to integrate all of the CASE tools into a coherent whole.

Examples of CASE tools may include products as simple as code editors like TECO, Vi, eve ,emacs, xedit, etc., debuggers or syntax checkers such as dbx, gdb or Kadb, or life cycle support products such as Cadre Technology's Teamwork or Interactive Development Environment's Software through Pictures which may provide some automated project management, analysis and design, quality assurance, configuration management and testing tools as well as code generation support. Generally speaking, CASE tools provide automated support for some or all of the following activities:

- REQUIREMENTS MODELING USING DIAGRAMS - CASE tools in this area generally provide for initial graphic representation of the system, subsequent changes to and output of in-progress or completed diagrams. Object input validation and diagram manipulation rules are provided. It should be noted that graphic representations have not yet become standardized within the industry.

- MODEL VALIDATION - Tools in this area check for model completeness, and identify inconsistencies or errors. Validation may be interactive, batch or both.

- SPECIFICATION DEVELOPMENT - Tools in this category provide for the conversion of the initial model into preliminary and detailed design specifications.

- DESIGN VALIDATION - Tools in this area check for design completeness, and identify inconsistencies or errors. At this time, most such tools are platform dependent.

- CODE GENERATION - Tools in this category support the transformation of detailed design specifications into executable code. Again, code generators are typically specific to a single platform.

- TESTING - Automated testing support includes both the automated generation of test plans and procedures and the exercising of the target system either directly or through the use of a simulation of the target environment.

- PROJECT MANAGEMENT - These tools provide a framework for the recording of the work planned, the actual work done, and an assessment of the work to be completed.

- DOCUMENTATION PRODUCTION - These tools provide for the automated generation of documentation, in a format based on site specific templates, from information provided by the specification models for both the requirements and the design.

- RE-ENGINEERING OF EXISTING APPLICATIONS - These tools provide support for the automated documentation of existing code, the analysis of existing systems and the subsequent redesign/optimization of the existing systems either manually or automatically[32].

10. Description Of The Pride Model

In the development of the PRide, we have chosen a course which seeks to optimize the advantages of CASE tool utilization by integrating them into a single, cohesive environment. Among the advantages identified for this type of integrated environment are:

- The maintenance of a single set of object definitions, particularly data definitions, in a variety of development efforts.

- The ability to maintain project continuity via the access of the same information from all phases in the development process.

- The ability to identify targets for re-use and to re-use objects in different application development projects.

- The ability to integrate project management, analysis, design, and development efforts to maximize the opportunities for the use of automated tools and to prevent the disconnects prevalent in other development environments which result in incorrect or missed requirements in the target implementation[33].

The PRide will provide services which will span the functionality of a populated environment. We define an environment to be a collection of software and hardware components, with a degree of compatibility sufficient to render these components harmonious.

Services are defined as an environment's capabilities. Those capabilities which directly support an end-user are referred to as tools. The components that comprise an actual infrastructure are referred to as the framework.

The Pride services will be grouped in different categories reflecting degrees of abstraction, granularity, or functionality. The highest level division classifies services either as end-user or framework services. End user services will be further subdivided into Technical Engineering, Technical Management, Project Management, and Support Services. The first three of these groups partition the execution of a project into engineering, management, and engineering management. The fourth group , Support Services, is, per the NGCR reference model, orthogonal to the other three, since it includes capabilities potentially used by all other users[34]. The logical relationship of these service groups is illustrated in Figure 1. Framework services form a central core with a potential relationship to all other services in the environment. Support services underlie all the other end-user services. The remaining groups provide an envelope around the framework services and utilize the support services. The boundaries do not explicitly indicate interfaces, as the representations are at the service groups level, not the services level. To emphasize this point, the service groups might be represented as in Figure 2.

Figure 1

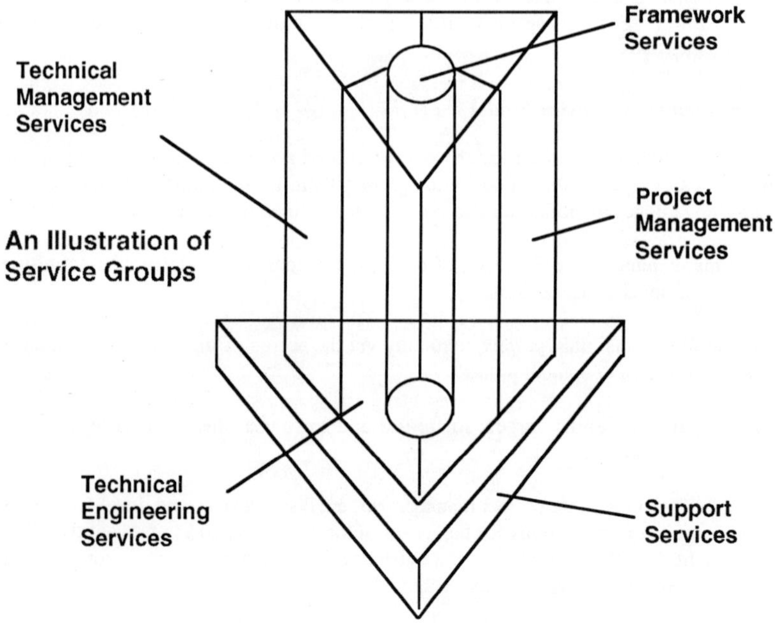

Framework Services
Technical Management Services
Project Management Services
An Illustration of Service Groups
Technical Engineering Services
Support Services

Figure 2

Support Services		
Project Management Services	Technical Management Services	Technical Engineering Services
Framework Services		

Another Illustration of Service Groups

Each of the end user service categories is further subdivided by engineering domain, user role, or life cycle phase. Technical engineering services are directed toward the technical aspects of project development. These services are organized by specific engineering domain. Tasks typically include designing and coding which require services such as compilation and testing. Technical management provides services that are closely related to engineering activities; these include services which provide a managerial complement to engineering activities in the areas of configuration management, reuse, and metrics. Project management services are relevant to the overall success of the SSCL. They include such things as scheduling, planning, and tracking overall progress of a project. Support services focus on tasks and activities common among all users of the PRide. They include a group of common services for information processing, as well as publishing, user communication, presentation, and administration services.

The framework service categories include Operating System, Object Management, Process Management, Policy Enforcement, User Interface, Communication, Network, and User Command Interface services.

Real software components span various service groups with many components considered to be end-user tools also providing capabilities properly regarded by the NGCR reference model as framework services. The model provides a common conceptual basis against which to examine these environmental implementations. The boundary between service groups, particularly the boundary between end-user and framework services is a dynamic one that changes over time.

11. Development Of PRide

The development of PRide is following an iterative multi-stage development path. Parallel efforts are underway at the SSCL to: 1) identify the current physics computing environment both within PRCD and the collaborations; 2) to establish an approach to developing framework services within the ECMC FRAMEWORK project[35]; and 3) to acquire, build and integrate components of the PRide through innovative acquisition techniques, including the utilization of

Creative Research and Development Agreements (CRADAs) which would allow software developed by teams of PRCD engineers and commercial vendors to be made available at little or no cost throughout the collaborations.

12. References

1. S. C. Loken, Software Engineering: What Do Experiments Need?, Proceedings of the International Conference on Computing in High Energy Physics '92 (1992) 87.
2. J. Knobloch, Reality of Software Engineering in High Energy Physics, Proceedings of the International Conference on Computing in High Energy Physics '91 (1991) 291.
3. T. Nash, High Energy Physics Experiment Triggers and the Trustworthiness of Software, Fermilab-CONF-91/270, August, 1991.
4. W. W. Royce, Managing the Development of Large Software Systems: Concepts and Techniques, in the Proceedings of WESCON (1970)
5. B. W. Boehm, Software Engineering, IEEE Transactions on Computing (1976) 1226.
6. M. Paige-Jones, The Practical Guide to Structured Systems Design, Yourdon Press (1980).
7. T. DeMarco, Structured Analysis and System Specification, Yourdon Press (1978).
8. J. Linneman, et al., The Use of SA/SD Methods in D0 Software, Computer Physics Communications 45 (1987) 245.
9. G. Kellner, Development of Software for ALEPH Using Structured Techniques, Computer Physics Communications 45 (1987) 229.
10. S. M. Fisher and P. Palazzi, Using a Data Model from Software Design to Data Analysis: What Have We Learned? , Computer Physics Communications 57 (1989) 169.
11. P. A. Dearnley and P. J. Mayhew, In Favour of System Prototypes and Their Integration into the Systems Development Cycle, The Computer Journal 26 (1) (1983) 36.
12. B. W. Boehm, IEEE Computer, 21 (1988) 5.
13. B. Boehm, et al., Prototyping Versus Specifying: A Multiproject Experiment, IEEE Transactions on Software Engineering SE-10 (1984) 290.
14. F.P. Brooks, No Silver Bullet: Essence and Accidents of Software Engineering, IEEE Computer, 20 (4) April 1987.
15. K. S. Lantz, The Prototyping Methodology, Prentice-Hall (1986)
16. J. L. Connell and L. B. Shafer, Structured Rapid Prototyping: An Evolutionary Approach to Software Development, Yourdon Press (1989).
17. In the Proceedings of the Second International Workshop on Software Engineering, Artificial Intelligence and Neural Networks in High Energy Physics.
18. B. Scipioni, et.al., Physics Detector Simulation Facility Phase II System Software Description, Proceedings of the Conference on Computing in HEP, Annecy (1992) 597.

19. L. Cormell, Experimental Systems Computing, SSCL-N-814, February, 1993.
20. I. Gaines ed., Proceedings, SDC Computing Workshop, Rice University, May 17-19, 1993.
21. I. Gaines ed.l Proceedings, Joint GEM/SDC/PRCD Computing Workshop, Superconducting Super Collider Lbaoratory, July 1993.
22. Memorandum Of Understanding, R40-000006
23. Project Initiation, R40-000007
24. SSCL PRCD PRACTICE #R40-000008 Project Assignment Plan.
25. SSCL PRCD PRACTICE #R40-00009Project Execution Plan.
26. S. Frederiksen, I. Gaines, L. Price editors, SDC Off-Line Software Concepts, version 1, December 5, 1992.
27. ECMA, Technical Report TR/55, European Computer Manufacturer's Association December 1991.
28. NIST, Reference Model for Frameworks of Software Engineering Environments, NIST Special Publication 500-201, December, 1991.
29. PSESWG, Reference Model for Project Support Environments, Next Generation Computer Resources Program, Project Support Environment Standards Working Group, United States Navy, March 1, 1993, pp2.
30. S. Shlaer, S.J. Mellor, Object Oriented Systems Analysis, Modeling the World in Data, Yourdon PRess Computing Series, Prentice-Hall, 1988.
31. J. Rumbaugh et.al., Object Oriented Modeling and Design, Prentice-Hall, 1991.
32. PSESWG, Reference Model for Project Support Environments, Next Generation Computer Resources Program, Project Support Environment Standards Working Group, United States Navy, March 1, 1993, pp2.
33. John Parkinson, Making CASE Work, NCC Blackwell Limited, 1991, pp37-38.
34. ibid pg39.
35. PSESWG, Reference Model for Project Support Environments, Next Generation Computer Resources Program, Project Support Environment Standards Working Group, United States Navy, March 1, 1993, pp7-8.
36. Frederiksen, et.al., FRAMEWORK Software Project Assignment Plan, SSCL, September 1993.

Software Engineering Techniques and CASE Tools in RD13

AGUER M[4], AMBROSINI G[1], BUONO S[2], DUVAL P.Y. [3], FERRATO D[3], FUMAGALLI[1], GAPONENKO I[1], HUET M[4], JONES R[2], KHODABANDEH A[2], MAPELLI L[2], LE VAN SUU A[3], MORNACCHI G[2], POLESELLO G[1], PRIGENT D[2], QIAN Z[3], RONDOT C[3], SANCHEZ-CORAL E[1], SKIADELLI M[1], TOPPERS A[1]

1: Dipartimento di Fisica dell'Universita'e Sezione INFN di Pavia, Italy

2: CERN, Geneva, Switzerland

3: Centre de Physique des Particules de Marseille, IN2P3, France

4: Deptartement de Physique Nucleaire - STEN, C.E. Saclay, France

ABSTRACT

The RD13 project was approved in April 1991 for the development of a scalable data taking system suitable to host various LHC studies. One of its goals is the exploitation of software engineering techniques, in order to indicate their overall suitability for DAQ software design and implementation. This paper describes how such techniques have been applied to the development of the RD13 DAQ used in test-beam runs at CERN. In particular, we describe our experience with various software methodologies (e.g. SA/RT, Protob) and their associated CASE tools (StP, Artifex) as well as the software development environment and highlight the limitations of using code generated by CASE tools in real-time environments. The areas addressed by CASE tools include dbms, user-interface, code management, documentation, s/w metrics and project management.

1. Introduction

The evaluation of software engineering technology is one of the major areas for study indicated in the RD13 proposal [1] and has been performed throughout the activity of the project. We have investigated the exploitation of software engineering techniques to establish their suitability for DAQ software design and developments. We describe our experiences and efforts in the areas of methods and tools, database systems, user interfaces, information diffusion, code management and software quality.

2. Methods and Tools

2.1. The Data Flow Protocol

We selected the RD13 data flow protocol (DFP) as an initial task to study the application of software engineering technology. We aimed to cover the full life-cycle of the project: from requirement analysis through to final implementation and maintenance.

2.1.1. Developing the Data Flow Protocol with SA/SD

In the first phase of the project we applied to the analysis phase the real-time Structure Analysis method (as described in Ward/Mellor [2] and Hatley/Pirbhai [3]) and used the StP (Software through Pictures) CASE tool which supports, among several diagramming techniques, structured methods with real-time extensions. On completion of the exercise we concluded that

- StP has many interesting and important features (data flow decomposition, consistency checking, openness, document generation) but the lack of state transition decomposition, simulation and code generation makes it weak for protocol design.
- Structured methods, and the waterfall like life-cycle which goes with them, seriously lack continuity between different phases (e.g. from analysis to design) requiring error prone manual transformations from the language of a phase to that of the next one. Also, there is no guaranteed consistency between the different phases (nothing prevents the modification of a design while the requirements are left untouched and therefore out of phase) and they offer poor support for code generation.

For of the above reasons and while the DFP analysis was taking place with SART and StP, we selected an alternative methodology and supporting tool which we considered the most mature at that time, Artifex (Artis srl, Italy).

2.1.2. Artifex methodology and CASE tool

Artifex consists of several tools supporting modelling, simulation and code generation activities within the framework of the operational software life-cycle methodology [4]. This latter idea is that of providing a formal language suitable to the specification of the requirements of a selected problem domain. The model can then be simulated, thus providing a rapid validation of the requirements, and automatically translated into code for production. This kind of methodology does guarantee continuity and enforces consistency throughout the full software life-cycle. As its major application area Artifex addresses discrete event dynamical systems, such as real-time applications, distributed systems, communication protocols and manufacturing control systems.

2.1.3. Developing the Data Flow Protocol with Artifex

To model the protocol we have used Artifex High Level Colored Concurrent Petri Nets (HLCCPN) starting from the analysis work performed with SART and StP. We defined the system structure and behavior in terms of HLCCPN, capturing the same information that was modelled in StP with Data Flow, Control Flow and State Transition diagrams.

2.1.4. The test-beam Artifex based DAQ System

Having completed the definition of the DFP, we decided to go to a full application which we could test under real conditions. We built a data acquisition system by composing the objects developed with Artifex that could run in a test-beam environment. Because of its exploratory nature and the uncertainties of the outcome, this was an activity independent of the development of the production version of the RD13 acquisition system. Control and ancillary processes (initialization, message logging) were developed as well with Artifex, running on the front-end RAID and the back-end SUN (control, initializes, message logger). The inter-process communication (via tcp/ip) is managed transparently by Artifex.

2.1.5. Running experience and conclusions

The Artifex based data acquisition system was used part time during the November 92 RD2 test-beam period. It ran reliably and a few data tapes were written. Performance wise, it does not compare favorably with the (hand coded) RD13 acquisition. We found some points where performance could certainly be improved but it is clear that the generated code has an inherent performance penalty due to the dynamic management of the concurrent behavior of the petri nets. The full (from analysis to code generation) applicability of such a tool is, therefore, unsuitable for a very demanding environment such as that of the acquisition in a test-beam but it is certainly suitable for other less demanding applications such as the case of large events and/or low rates and for experiment controls.

2.2. Comparison of Methods and Tools

Comparing Artifex and StP, it appears that both tools are very good in terms of support of their corresponding methodologies. Artifex was preferred because of its simulation and code generation features that we believe are very important for protocol design. The Artifex drawbacks, beyond the performance issue already discussed, lie mainly in the "static" nature of the modelling; every class instance which will participate in the model has to be drawn on the screen and connected to other instances. While this may be acceptable during the initial modelling and validation, it becomes cumbersome when used in the final application where a certain degree of dynamism is necessary.

Encouraged by the results of our work with Artifex, we have extended its use to two other areas of the DAQ where the run-time requirements are not so stringent as the data flow protocol: the run-control system and the development of the second level trigger algorithm.

3. Databases

The data acquisition system needs a large number of parameters to describe its hardware and software components. We have considered using a database management system (DBMS) for organizing and storing such parameters and determined the following distinct databases:

- Hardware configuration: describes the layout of the hardware in terms of crates, modules, processors and interconnects
- Software configuration: describes the layout of the software in terms of processes, services provided, connections (network and shared memory) and host machines

- Run parameters: information pertaining to the current DAQ run (run number, recording device, Level 2 trigger state etc.)
- Detector parameters: information pertaining to the detector (defined by the detector group themselves within a fixed framework)

3.1. ORACLE and StP implementation

The initial implementation of the software configuration data base was built with a combination of tools provided by the Oracle relational DBMS and StP CASE toolset:

- StP Entity-Relationship (E-R) diagram editor to draw the diagram implementing the data base schema.
- StP to generate code from the E-R diagram

  SQL statements to create Oracle tables, C type definitions for the table rows, C arrays implementing the tables in memory and C functions to load/dump a data base from memory from/to disk files or Oracle.
- Hand-coded Data Access Library (DAL)

  written in C as a set of functions working directly on the arrays in memory that are called by the application programs which wish to access the data in the DBMS.
- Each application reads from disk (or Oracle) a copy of the data base for its personal use There was no provision for distributed access to the data.
- Versioning (i.e. the maintenance of different configurations) was handled off-line by having different files. There is currently no provision for multiple versions on Oracle.

The major drawback of this scheme comes from the need to hand code the data access library directly in C, in particular for what concerns the navigation of the data base.

3.2. QUID implementation

The current implementation of the databases listed above has been developed with a commercial system called QUID (Artis srl, Italy). QUID is a data base development environment targeted to real time applications where performance is needed but the full functionality of a DBMS is not required. A QUID data base is kept in the application's memory space and its contents can be saved/loaded to/from a disk file. Multiple versions of a data base, e.g. different system configurations, may be maintained by keeping different files. The QUID system provides most of the features we require from an environment for developing real time data bases:

- A data model and CASE tools supporting its development: this is provided by the QUID extended E-R model and the QUID editor.
- Code generation: C code supporting the management of the data base is produced by the QUID code generator taking as input the data base description created by the editor.

- Data Access Library: the production of a data access library is greatly simplified by the availability of the QUID language which mixes fairly well with C. The code generator translates a QUID module into a C module.
- Real time access: the data are kept in the application memory, hence high performance is guaranteed. However, this also limits the size of the database to that of the machine's memory.
- Versioning: Backup copies on files can easily be created and accessed
- Distributed aspects: there is no provision for distributed transactions, nor for the distribution of data among many processes. For read-only data bases this restriction is easily overcome by having all participating processes access the file in read-only mode.
- User interface (modifying and browsing a data base): a monitor tool produces an application to navigate and modify the contents of a data base given the schema definition. This is important since it provides a way to interactively populate the database without developing our own graphical front-ends.

In short the QUID environment meets most of our requirements for a data base framework, with the major exception of the distributed aspects.

3.3. Object Oriented DataBase Management Systems

To overcome the limitations of distributed access and size we are currently investigating a number of Object-Oriented DBMS (OODBMS). OODBMS also offer other advantages such as the ability to store multi-media data, better support for versioning and consistency constraints. We made a first investigation into the use of Ontos OODBMS (ONTOS Inc., USA) but it was not possible to introduce such a system during the evaluation (early 1992) because the current version only ran on a collection of homogenous machines which meant the database could not be shared between our back-end workstations and front-end processors. Also it only offered a C++ programming interface while, at the time, there was no suitable C++ compiler available on the front-end processors.

We plan to continue our investigation of the usefulness of OODBMS in our problem domain and compare them to relational database management systems. To this end, we are currently evaluating GemStone (Servio Corp, USA) and Itasca (Itasca Systems Inc., USA).

4. User Interfaces

We have chosen Motif as the standard graphical user interface (GUI) toolkit because of its general availability and wide acceptance by workstation vendors and software developers. We have also chosen a commercial GUI builder tool, X-Designer (Imperial Software Technology, UK), as a means of developing Motif based interfaces to our applications. This tool generates portable C code using the Motif application programming interface. Such a tool has saved many hours of laborious hand coding, to the extent that all the major components of the DAQ are now equipped with user-friendly GUIs.

For interfaces that require more powerful graphics capabilities than are provided by the Motif toolkit, we used the DataViews commercial tool. While we were successful in producing a status display for the RD2 test-beam run last year, we found a number features of the tool which discouraged us from continuing it use in its present form:

- Graphical editor was difficult to use and rather complex.
- The editor had to be relinked with our code for capturing the data in order to test the interface being constructed.
- The proprietary application library used to pass data to the graph widgets is very large (several hundred routines) and required a lot of training in order to use it successfully
- It is difficult to modify the graph widget parameters at run-time.
- Individual graphical objects of the same class cannot be distinguished between at run-time.

In light of these difficulties we decided to use another approach for the latest version of the status display. The DataViews widgets that are used to display data are available separately without the editor and programming library. The X-Designer tool offers a means of incorporating these widgets and allowing them to be used along side the standard Motif widgets. This offers the advantages of using only one editor to produce the complete status display and at the same time giving a means to display data in graphical form and providing data for the graphs in a more common manner (via widget resources). We have used this combination of tools to produce a version of the status display for the RD6 test-beam run.

5. Information Diffusion

We considered it important to have a well organized documentation system for the project. To this end we chose the commercial FrameMaker (Frame Corp, USA) interactive WYSIWYG document preparation system as our primary authoring tool. To date we have used FrameMaker to produce:

- Technical notes (more than 80) covering all aspects of the DAQ development and representing around 1000 pages of information
- Reports and Conference papers (including this document)
- Hypertext based online help for the DAQ system
- Letters, faxes and slides

5.1. Document Preparation

We are very satisfied with FrameMaker and appreciate the WYSIWYG style of authoring compared to mark-up languages such as LaTex and HTML. We have not chosen to write our documents in HTML directly because, as can be seen from the list above, not all of the documents are destined for online viewing. Presentation on paper is very important, especially for Conference papers, and we don't know of any HTML interpreters which can give a good and totally controllable output on paper in the same manner as FrameMaker or LaTex. A LaTex to HTML converter exists but we prefer FrameMaker because it is a WYSIWYG system - with LaTex one has to print to postscript then view the output on paper or use a postscript previewer in order to see the results. Also a separate drawing tool (such as xfig) is needed to make figures while FrameMaker incorporates such a tool and other facilities (e.g. table and equation handler, spelling checker) that make writing documents simpler. Finally, it is easier for people to only have to learn one tool rather than many and we have found FrameMaker addresses the needs of all the document types listed above.

5.2. Making RD13 documentation available via WWW

In order to make these documents available to people outside the project we have setup a WorldWideWeb [5] server for RD13 documentation. This required the conversion of the documents from FrameMaker's own format to HTML format suitable for use with WWW. Rather than performing the conversion by hand we have used a publicly available package that performs the conversion automatically. Unfortunately, the converter does not cover all of the features in FrameMaker that we use and so we are obliged to make available the postscript version of some of the documents. We expect this situation to improve as a new version of the WWW protocol allows more formats to be used and the next version of FrameMaker will allow the output of documents in SGML/CALS like format which is closer to HTML.

5.3. On-line Help

It is important to provide on-line information to users and operators of the DAQ system. One of the reasons we chose FrameMaker was its ability to be used as a document viewer and to provide hypertext capabilities. We have used FrameMaker to produce on-line help for all the components of the DAQ system which is accessible from the run-control user interface. When the user presses a help button on a window a FrameMaker document is displayed. The user can navigate through the document and access others using FrameMaker's hypertext capabilities.

This facility is integrated with our GUI builder, X-Designer, which generates code to automatically open a FrameMaker document when the "help" button is pressed on the interface being developed. If a more complete FrameMaker to HTML converter existed we could use a WWW browser, such as XMosaic, to display the online help documents after they have been converted. X-Designer is configurable so it could also generate code that works with XMosaic instead of FrameMaker. Also, WWW's hypertext references work over the network whereas all the documents referenced by a FrameMaker document must be on the same file system. In such a configuration, we would not need to distribute the online information but they could be keep on a central server and retrieved by WWW on the user's request.

6. Code Management

The management of source code to track software modifications, releases and configurations is an important activity in a software project of any size. Originally RD13 members used the SCCS UNIX tool, which is normally available on any UNIX system, on a personal basis. The size of the RD13 developed software, as well as the interaction between different developers, is such that we have decided to define a general scheme for the maintenance of the source code.

While SCCS is probably satisfactory for a single user, a more sophisticated system was needed for the management of a complex software system made of different products and developed by several people. Convinced by the positive experience reported by SLAC [6], we have adopted the Concurrent Version System, CVS [7]. The source code of the different products is now organized into a repository managed by CVS. Each product is maintained in several releases and the full RD13 data acquisition system releases are superimposed on to the structure and are maintained by CVS.

7. Software Quality

In order to continue to improve the quality of our software we have been concerned with the following issues:

- Minimizing the number of defects in software.
- Creating mechanisms for controlling software development and maintenance so that schedules and costs are not impaired.
- Making certain that the software can be easily used and fulfils the necessary functions.
- Improving the quality of future releases.

We have adopted a number of steps that have a minimal impact on our day-to-day working but which help us to produce high quality software by addressing the above issues:

7.1. Measure the amount of effort used in software development

In order to determine the effects of any change to the way we produce software (e.g. use a new CASE tool or apply a new methodology) we need to know how much effort is put in to its production. Developers record their activity on a half-day basis (mornings and afternoons) using the X calendar tool installed on the workstations. A script file, periodically run in batch, collects all the activity entries and totals the time used for each phase of each project. This information is then recorded in a spreadsheet (Microsoft Excel on a Macintosh) to provide a perform statistical analysis and output in graphical form.

7.2. Record and track defects

An important aspect of improving software quality is reducing the defects is the software produced. Before we can reduce the number of defects we must measure how many defects are detected. We have developed a simple database application, based on the

Quid commercial database and a graphical user interface developed with X-Designer, which records defects. Details such as date found, project name (e.g. run-control), phase (e.g. design), defect category (e.g. serious or moderate) are recorded as well as the cause of the defect when determined. This information allows us to estimate the number of defects attributed to each activity of the development and indicates where we should concentrate our efforts so as to reap greater quality improvements.

As a means of providing better support for the tracking of software defects, we intend to gain experience with a more sophisticated bug tracking system. Gnats [8] is a bug-tracking system based upon the paradigm of a central site or organization which receives problem reports and negotiates their resolution by electronic mail.

7.3. Collect and record metrics

To measure the complexity of the software we have applied the McCabe [9] metric. This metric gives a numerical value, on a per-routine basis, of source code. We have installed a public domain package that implements the McCabe metric and written some script files that give a size, in source code lines, of the software measured. As a second step we have started to use the Logiscope commercial tool (Verilog, France) to provide more software metrics and an overall quality assessment in a graphical format.

7.4. Use metrics as an indication of the software quality

We have applied the above metrics to the version of the RD13 DAQ that was used in the RD2 test-beam and the results are used as input to our software production cycle. We have already introduced software reviews and so we are able to use the results from our metrics as input to these meetings. That is to say that the review panel examines the metric values and only pass the software if these measurements fall within accepted levels. As the amount of data collected from these metrics increases, we will be in a position to cross check them against the defects recorded in the problem-database as a means of estimating the accuracy of the metrics themselves.

7.5. Testing software

We have tried to take a more organized approach to testing the software that has been written by the project. To this end we have used lint which is a generally available UNIX package to detect features of C code that are likely to be defects, non-portable or wasteful.

We have also started to use another facility offered by the Logiscope tool to perform code coverage analysis. The DAQ source code is modified (instrumented) by Logiscope so that the paths taken through the code (e.g. routine entries and exits), branches and loops are recorded in a log file during execution. The tool then analyses the results and displays them in a graphical format to indicate which parts of the code have been exercised during the execution. Based on this information, the developer can produce new tests that exercise the sections of the code not previously covered. This facility is very useful when combined with the CVS repository since the source code can be checked-out, parsed to calculate the metrics values, instrumented, executed and the results analyzed from a batch job that can be run over-night so that the information is available to the developer the following morning.

Another area of testing we intend to investigate is the exercising of GUI interfaces. Normally, the developer is required to test such interfaces by using the mouse and keyboard to select buttons and enter text whenever a new version is made. Commercial tools exist that can record this activity and then play-back the mouse and keyboard strokes without human intervention.

8. Conclusions

Our experience has been very positive and we believe that there is a great future in software engineering and CASE technology for HEP, especially for large software production. The benefits in communications between the developer and the user, with provisions for project management and support for the full life-cycle including code reliability and maintainability are tremendous. The quality and quantity of commercial CASE tools is increasing rapidly and are sure to offer the most cost-effective and organized manner for HEP projects to develop their software. The software engineering field, and the CASE market in particular, are in rapid evolution and, although industry is investing effort in this direction, standardization is still far away. We plan to continue our emphasis on software engineering by extending the use of CASE tools, addressing software quality issues and the introduction of object-oriented technology.

9. References

1. L.Mapelli et al., A Scalable Data Taking System At A Test Beam for LHC, (CERN/DRDC/93-25).

2 Ward & Mellor, Structured development for Real-Time systems, Prentice Hall, 1986.

3 Hatley & Pirbhai, Strategies for Real-Time system specification, Dorset House, 1987.

4 P. Zave, An Operational Approach to Requirements Specification for Embedded Systems, IEEE Trans. Softw.Eng, Vol. SE-8, No3, May 1982

5 T.Berners-Lee et al. World Wide Web; An Architecture for Wide-Area Hypertext. CERN, 1991.

6 T. Hung, P.F. Kunz, UNIX Code Management and Distribution, Proceedings of the International Conference on Computing in High Energy Physics '92, CERN 92-07, 1992

7 B. Belliner, CVS II:parallelizing software development, Proceedings of the Winter 1990 USENIX Conference, Washington, DC, January 1990. USENIX, 1990.

8 Bug Management With GNATS: (gnats). GNU Problem Report Management System, Free Software Foundation, Inc., 1993

9 T.J. McCabe, A Complexity Measure, IEEE Transactions on Software Engineering, VOL. SE-2, NO. 4, Dec. 1976

COMPUTING IN THE ATLAS EXPERIMENT AT LHC

KORS BOS
National Institute for Nuclear Physics and High Energy Physics, NIKHEF
PO Box 41882, 1009 DB Amsterdam, The Netherlands

and

STEPHEN W. O'NEALE
School of Physics and Space Research, University of Birmingham
Birmingham B15 2TT, United Kingdom

ABSTRACT

We present the computing needs of the ATLAS experiment at the LHC. The current simulation and reconstruction software used for the optimisation of the detector design is described. The activities of the future computing group are reviewed and some areas are shown where computer assisted software engineering products are (yet) inadequate. The replacement of the CERN central mainframe system with workstation group services for ATLAS is discussed.

1. Introduction

In this paper we give a brief description of the data handling needs of the general purpose ATLAS detector [1] proposed for operation at the Large Hadron Collider (LHC) at CERN. The software effort now is primarily concerned with the simulation and reconstruction of events in ATLAS to be used for the selection and refinement of the detector design options – to be presented in a detailed Technical Proposal in May 1994. This work, the *current* software system, relies on the mature HEP products of the LEP era and is reviewed in the third section. Then we give an overview of the software activities of the *future computing* group. Here modern software engineering techniques, in particular object oriented analysis, design and implementation problems, are being studied. This work has not reached the stage in which results would normally be presented to a conference but the flavour of the problems arising may be useful input for further work in, and arising from, this workshop. Lastly, we address the computing infrastructure needs and the working environment for ATLAS which is migrating to workstation services from a central mainframe facility.

2. The ATLAS Experiment

The ATLAS collaboration, comprising a thousand physicists from a hundred universities and research laboratories, proposes a general purpose pp experiment for operation at the LHC. The main focus of interest is the discovery of the Higgs particle which is hypothesised as the origin of mass on the electroweak scale. ATLAS is designed to be sensitive to many processes in order to cover the full mass range above the LEP discovery limit. The detector is also able to cope with studies of top quark decays, supersymmetry searches and has some potential to deal with any unexpected new physics.

The LHC is planned for installation in the LEP ring so a major investment in the civil engineering has already been made. The current design has 8 TeV colliding beams in the proton proton mode. The luminosity is 1.7 10^{34} cm^{-2} s^{-1} and the bunch crossing time is 25 ns. For an inelastic cross-section of 80 mb this leads to an average pileup of about 30 overlapping events per beam crossing. The cross sections for Higgs production and other interesting processes such at $t\bar{t}$ pair production may be ten or more orders of magnitude below the total cross section at LHC energies so a large data rejection power will be needed.

The ATLAS detector has four major subsystems built around an interaction point of the LHC beam line. The inner detector records points on the trajectory of electrically charged particles moving in a 2 Tesla toroidal magnetic field. An analysis of these points (reconstruction) leads to a measurement of the momentum of the particles and an identification of the decay topology of hadrons containing b c and s quarks and of tau leptons. The electromagnetic calorimeter provides a large signal for incident electrons and positrons which measures their energy. Incident photons yield a similar energy signal as they also convert into a shower of electrons, positrons and photons. The granularity of the calorimeter permits some resolution of the two photons produced by the decay of a neutral pion in the intersection region. Iron absorbers are used for the next layer, a calorimeter to record the passage of particles emerging from electromagnetic calorimeter and to measure the energy of the hadrons. The iron may also form part of the return yoke of the magnetic field. Charged particles which penetrate the iron are identified as muons and their momentum and charge may be measured with a stand-alone muon system in the toriodal field outside the calorimeters.

The data flow from the detector will be dominated by the inner tracking system which may yield 1 MByte every 25 ns. The ATLAS trigger and data acquisition (DAQ) model has three levels. Level-1 may trigger at 10^4 - 10^5 Hz and deal with 10^{10} - 10^{11} Bytes/s. Level-2 decisions may require the transfer of 10^8 - 10^9 Bytes/s from level-1 and run at 10^2 - 10^3 Hz. At level-3 we wish to run event reconstruction to make further data reductions yielding a data recording rate of 10-100 MBytes/s - possibly data suitable for physics analysis. The software running in Level-3 may be the traditional offline reconstruction. We estimate we will need 10^6 Mips for the level-3 filter and reconstruction and 10^7 Mips for simulation. The computer assisted software engineering (CASE) tools and techniques used to develop and document the DAQ system in the context of the RD13 project are presented at this workshop and the parallel sessions have talks on formal methods and simulation for the level-2 trigger and DAQ system.

An extrapolation of today's technologies for cpu power, data storage and data access to the LHC era [2] suggests we may obtain 10^7 Mips in 10^4 workstations in a few hundred boxes for 10-20 M$. Data recording and robotic manipulation do not present major problems, 10 PByte systems seem quite feasible. However data access for analysis is less clear, a few percent of recorded physics data may be placed on 10 TByte of random access (disk) devices for 5M$. At this stage we must defer the complexity of detailed design and note the cost estimates in the budgets for the construction and operation of our experiment.

We anticipate a need to go from 1 Mbit/s wide area networks of today to 1 Gbit/s for the preparation and analysis of data at the LHC. Some new applications include audio-visual mail and tele-conferencing sessions and the use of ASF mounted disks to avoid tedious and repetitive file transfers. Many physicists have just realised that there is a crisis looming in HEP networking as EBONE has just gone down and alternative routes have much reduced bandwidths. Links to DESY have been reduced and the transatlantic T1 link is not funded for operation beyond January 1994. The situation in Europe is particularly bad when compared with de-regulated commercial systems and data highways proposed in the USA. It is difficult to avoid the conclusion that European monopoly holders enjoy inflated profits and/or shameful inefficiencies.

We expect to use our *current* software for the Technical Proposal up to May 1994 and hope for formal approval of the LHC project and the ATLAS experiment by the end of 1994. Design refinements and test-beam studies are likely to use this software through 1995-96. *Future* computing studies should come to fruition during 1995-96 with decisions on methodologies, tools, data models, milestones and quality control techniques. This leaves more than six years for an incremental implementation of the ATLAS computing model ready for beam in 2003.

3. Detailed Studies for the Technical Proposal

The ATLAS collaboration is engaged in a detailed study of detector options which are to be presented in the form of a Technical Proposal in May 1994. This proposal needs software to simulate interactions, the response of the ATLAS detector and some reconstruction of events to optimise the detector design. One disappointment is that the CASE vendors appear to have little to offer such as industrial strength tools which might help this effort, on this time scale, today.

The current ATLAS software is written in Fortran 77, a draft style guide and some verification with packages like FLOPPY [3] are close to formal approval within the collaboration. The code management is performed with CMZ [4]. The ZEBRA [5] package is used as a memory manager and for the organisation of the data structures. It was first proposed to write code using "bare" ZEBRA but several authors have felt the need to use access routines to avoid coding the navigation of the banks themselves.

The strongest support of ATLAS software is on the VM/CMS and HP-UX platforms. This is simply because the entire collaboration has network access to the high quality service available on the CERNVM mainframe machine and because the CERN simulation facility (CSF) and the ATLAS cluster of workstations at CERN are HP 9000/700 series workstations. There is no enthusiasm to support ATLAS code on VAX/VMS or its more recent open variety within the central group. Investment in the infrastructure on CERNVM is kept to a minimum and it is expected that reference systems on several Unix platforms will be available in 1994.

The simulation code is based on GEANT version 3.16 [6]. The geometry description and the response of the detectors to particles are handled by the DICE [7] package. DICE provides the CONS, GEOM, HITS and DIGI user functions in the GEANT framework and will also facilitate the use of other packages in reconstruction and other codes.

The SLUG [8] package is an interface between GEANT and DICE which controls input/output and also acts as an interface to physics generators like PYTHIA [10]. For a project of 15 to 20 years duration the "expert down the corridor" approach will clearly fail and already the principle author of SLUG, and a generator package GENZ [9], has left the ATLAS collaboration to enhance British Industry. Fortunately he has written excellent documentation and implemented the features needed for current ATLAS work.

Work has started on a reconstruction package ATRECON [11] and an event viewing system is operational but not released for public usage yet.

There are 100 institutes in the ATLAS collaboration of which 45 have designated a contact person who contributes to the overall computing effort or simply represents the needs and interests of his home laboratory. The ATLAS automated code update system has only 10 subscribers off the CERN site. Although the code is available to all, this raises the possibility that a complete software development environment will not be needed in each and every laboratory. Given the trend towards commercial tools for software development and the reluctance of many physicist-managers to fund a proper working environment, this may help to raise the lowest common level which constrains offline software development.

The quality and detail of the current software was demonstrated with a few pictures from the ATLAS collection produced through the GEANT graphics interface which can not be reproduced in this publication. They may be obtained from the ATLAS Secretariat at CERN (email atlassec@cernvm.cern.ch). A display from the current software applied to the evaluation of the ATLAS potential for B physics at low luminosity is presented in figure 1.

4. The Future Software Working Group

ATLAS has a future computing working group which is concerned with all aspects of computing between now and the start of ATLAS running. The membership is the same as the object oriented working group. A software effort of the scale of ATLAS needs effective planning with a strategy, tools and a budget for project management. We expect to find some specialisation of people working as software engineers, electronics engineers and physicists and may go as far as allowing physicists to specify their requirements to another group who build the software based upon the analysis of these requirements.

Formal software engineering methodologies have not been particularly successful in particle physics. The SA/SD methodology was in vogue in the planning stages of the LEP experiments and there is a fairly broad concensus that the dataflow diagrams were a very useful aid to discussion. However the transition from analysis to design was clumsy and the modifications introduced in the design and implementation stages were rarely fed back into the documentation. Extension for real time systems came rather late. The tools available for SA/SD, and indeed any other variant examined by LEP experiments, were rather poor. Implementations running on the PCs of the day were very slow once a project of a moderate size had to be handled and in some cases there was no access to the "internal" tables. The efforts of the ALEPH collaboration in the use of SA/SD and the entity relationship model are well publicised [12]. We might now be helped by better tools but it is difficult to find a consultant in software engineering who would recommend the use of these techniques today [13].

Object Oriented methodologies are the current fashion and ATLAS is currently studing the Object Modeling Technique (OMT) [14]. The methodology covers analysis, design and coding with no awkward transitions in notation. The Objecteering CASE tool is being used for pilot projects. Objecteering covers a subset of the OMT methodology and generates code in C++ [15]. Code management (of the input to objecteering) is migrating to CVS [16].

The first OO prototype model was a naive detector which gave the group some experience of OO analysis and design. Effort was transferred to a more realistic project - prototype II - an OO approach to the Atlas Inner Tracker reconstruction and electron identification. A C program has been written to read (construct objects) from Zebra banks for the GEANT GEOM, HITS and KINE banks. The pattern recognition for the tracker hits uses an external point from a calorimeter cluster to start the search for tracks in the barrel region. For a project of this type a lot of time is spent installing tools, learning how to use them and reporting errors, learning the methodology and developing suitable class libraries. Nevertheless it was decided to deal with the algorithmic complexity of tracking in a magnetic field before presenting the results to the collaboration. There are now several projects, notably GISMO [17], which demonstrate the feasability and advantages of the OO paradigm. They do not (yet) cover the same functionality and performance of existing software running with procedural languages like Fortran. Prototype II should be able to provide a direct comparison with an ATLAS Fortran 77 implementation with the same functionality.

Writing objects (data and methods) to external files for later use in analysis or for reprocessing on other machines with new algorithms is still considered to be a problem area (associated with object persistence and/or object control). A short term solution has been found via Lonnblad's scheme in the CLHEP C++ class libraries. For Eiffel I/O a few changes have been made to the format to make it fully self describing (removing the markers at the beginning and end of subobjects and adding a marker at the beginning of every complete object). The data is written in ascii format for convenience in program development. Reading data / objects is less satisfactory and the schemes we have devised so far require a dictionary - hardly a satisfactory technique for eventual use in a DAQ system passing data at a few 100 Mbytes per second. We would particularly like to raise this topic to the attention of the experts and commercial software vendors attending this workshop.

Several members of the oo-atlas group feel that Eiffel is a cleaner language than C++ and a better choice for physicists moving from Fortran to a high level language supporting object oriented work. The prototype model has been translated to Eiffel and an ATLAS Eiffel style guide has just been completed.

Moving away from Fortran 77 will be a major problem for many physicists. Fortran 90 has many innovations which are particularly suitable for high performance scientific computing but the language is very different from its Fortran 77 subset. Classic works on numerical methods are now available in C versions but the language seems most popular with people wanting to exploit system features in an online environment. If one uses an object oriented methodology then it seems sensible to use a language which supports object oriented techniques. The C++ language is readily available in varied implementations on many platforms. The commercial success of the language leads us to expect there will be many software tools which first appear in C++ implementations. However C++ (C with Objects) has been under development for many years and development continues after the ANSI base document has been produced. There are many other object oriented languages now receiving serious consideration [19]: C+@, Eiffel, Sather, Beta, Liana, Paradox and lots more. Although we now place more emphasis on analysis and design and we hope to find tools and working environments with reliable code generation it will still be neccessary to provide code for the methods and algorithms. It will be difficult to persuade experienced physicists–programmers to migrate to a language with a lifetime of a few months.

Studies of Object Oriented Data Base Management Systems (OODBMS) have begun with GEMSTONE and ITASCA. OnX [20] and TkTcl [21] are candidates for programming a production environment on unix platforms.

Documentation, agenda for meetings and so on are made available through the World Wide Web [22] (WWW). This has proved to be very popular, far more so than early conferencing systems, but the need to bring new items to the browser's attention has often been expressed. There is a large amount of HEP documentation written in LateX and a converter to the HTML language of WWW is a useful tool. A desktop publishing package, for example Framemaker, may also be extended and then used to collect information from many sources and give output for publication on paper and in HTML.

5. Computing Infrastructure and the Working Environment

The ATLAS computing model follows the approach of OPAL in the exploita-

7. K. Bos *A Subroutine Package for the Atlas Detector Simulation* Atlas Note (under WWW, unpublished).
8. R. DeWolf *Simulation for LHC Using Geant* Atlas Note (under WWW, unpublished).
9. R. DeWolf *Generated Event Handling Using Zebra: User's Guide version 2.0* Atlas Note (under WWW, unpublished).
10. H. Bengtsson and T. Sjostrand *Pythia version 5.3 - Hadron Hadron high-Pt Scattering* Reprinted in The LUND Monte Carlo Programs long writeup.
11. M. Dodgson *The ATLAS Atrecon Manual* (under WWW, unpublished)
12. J. Knobloch *Reality of Software Engineering in High Energy Physics* in Computing in High Energy Physics '91 (Universal Academic Press, Inc., Tokyo 1991).
13. E. Yourdon *Decline and Fall of the American Programmer* (Yourdon Press, Englewood Cliffs 1992).
14. J. Rumbaugh et al *Object-Oriented Modeling and Design.* (Prentice Hall, Englewood Cliffs 1991).
15. M. Ellis and B. Stroustrup *The Annotated C++ Reference Manual* (Addison-Wesley, Reading 1991).
16. D. Grune et al, *cvs - Concurrent Versions System* Unix man pages.
17. W. Atwood and T. Burnett *Gismo: C++ Classes for HEP* in Computing in High Energy Physics '92 ,CHEP92 (CERN 92-07, Geneva 1992).
18. B. Meyer *Eiffel: the Language* (Prentice Hall 1992)
19. Dr Dobb's Journal *Annual Object Oriented Edition* October 1993
20. G. Barrand *The OnX Manual* LAL Orsay (unpublished)
21. J. Ousterhout *An Introduction to Tcl and Tk* (Addison Wesley, Reading 1994)
22. T. Berners-Lee *World-Wide Web* Proceedings CHEP92, CERN 92-07
23. C. Andrews et al *A Guide to Personal Computer Networks at CERN* The NICE Working Group
24. P. Defert and I. Reguero *ASIS: The Installer's Guide* CERN CN/CO/152

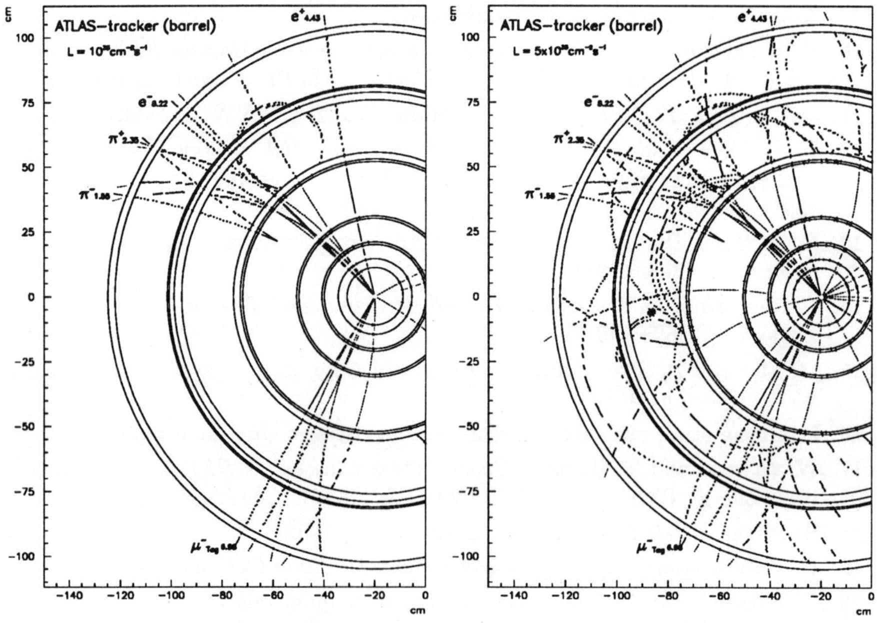

Figure 1: *Display of one half of ATLAS barrel tracker in the transverse plane for a signal event from* $B_d^0 \rightarrow J/\psi K_S^0$ *decay at a luminosity* $L = 10^{33}\ cm^{-2}s^{-1}$ *(left), where two minimum bias events have been added on top of the signal event, and at a luminosity* $L = 5 \times 10^{33}\ cm^{-2}s^{-1}$ *(right), where ten minimum bias events have been added on top of the signal event. The straw tracker hits are shown as large dots and the precision hits as dashes. The reconstructed tracks with* $p_T > 0.5$ *GeV/c and* $0 < \eta < 0.8$ *are shown as dotted lines between the detector planes and the interesting ones corresponding to the electrons from the* J/ψ *decay, the charged pions from the* K_S^0 *decay and the muon tag as full lines.*

COMPUTING FOR THE CMS EXPERIMENT

V. Karimäki
CERN ECP/SA
1211 Geneva 23, Switzerland

ABSTRACT

The Compact Muon Solenoid apparatus proposed as a general purpose detector at the future Large Hadron Collider will comprise some 20 million readout channels. The anticipated readout rate after level 1 trigger selection is about 25 GB/s and about 25 MB/s at the filter farm output. This very large data flow poses a computing challenge without a precedence in the history of particle physics. Present and future computing aspects of the CMS data handling are discussed including on-line and off-line data reduction.

1. Introduction

The Compact Muon Solenoid[1] (CMS) experiment is being designed as one of the two general purpose detectors for the CERN Large Hadron Collider[2] (LHC) scheduled to start running towards the change of the next millenium. The LHC will collide protons at a centre-of-mass energy of $\sqrt{s} = 15$ TeV every 25 ns with the intensity of a few thousands of million pp interactions per second. The experiments will address key questions in particle physics such as the existence of Higgs bosons and super-symmetric particles.

The very large pp interaction rates are needed in order to produce the predicted particles copiously enough for observation. Large rates and high particle energies imply that all the detector design and construction aspects as well as the related computing must be pushed to a limit.

A large number of scientists and other personnel are involved. The present member institutes of the CMS collaboration come from some 25 different countries and 75 institutes with a total of 600-700 people.

2. Detector

The design goal of the CMS is to build a cost/performance optimized detector with emphasis on the muon trigger and measurement. Large super-conducting solenoid will provide a 4 Tesla magnetic field for a precise momentum measurement (1% precision at 100 GeV/c) in a relatively compact inner tracker (radius 1.3 m). Four layers of muon stations are built within the flux return yoke so as to optimize between the size of the apparatus and performance.

Figure 1 shows a 3-dimensional view of the CMS. The overall dimensions of the detector are: a length of 20 m, a diameter of 14 m and a total weight of 12 000 tons. The inner tracker consists of a silicon detector with some 20 m^2 total area of silicon wafers. The outer layers of the inner tracker will be built of micro strip/micro gap gas counters. Crystal or shashlik designs are the options for the e.m. calorimeter and a copper-scintillator design that of the hadronic calorimeter. The muon stations will cover a total area of some 3000 m^2. The geometric acceptance of the central apparatus will be about ±2.5 units in pseudo rapidity.

The inelastic pp interaction rate at the LHC with the luminosity $L=10^{34}$ $cm^{-2}s^{-1}$ will be about 0.8 GHz per interaction point assuming 80 mb for the total inelastic cross section. The pp collisions take place in bunches which collide each other in 25 ns intervals. Hence the mean number of the inelastic pp collisions per one bunch-bunch collision will be about 20 at this luminosity. One pp interaction will yield 7 charged tracks on the average per unit of rapidity so that one trigger event will contain some 20x5x7 = 700 particles.

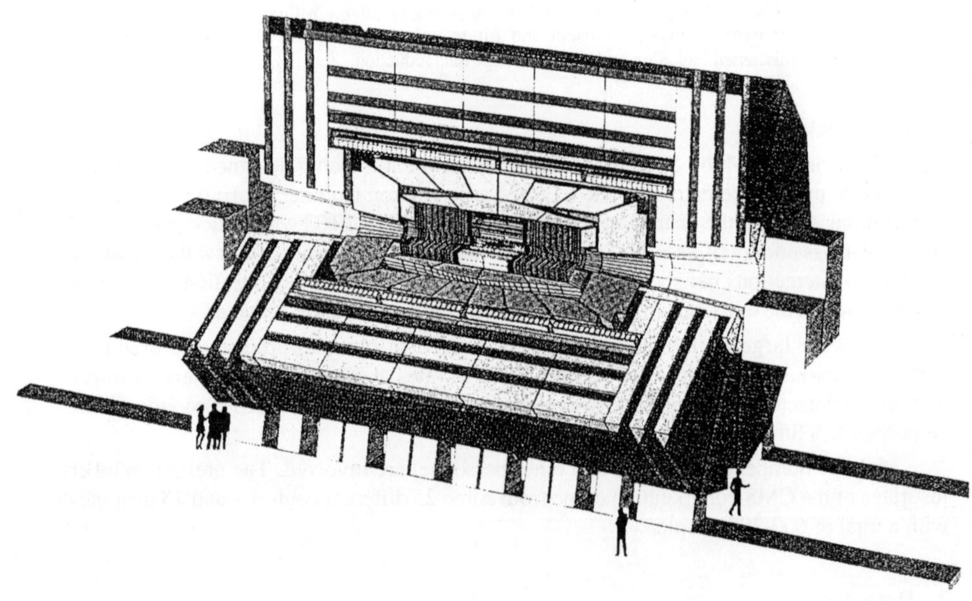

Fig. 1: Three-dimensional view of the CMS detector

3. On-line Computing

The data acquisition must be able to function at the bunch crossing rate of 40 MHz. The detector consists of some 17 million readout channels (Table 1) from which about 1% get active after one bunch crossing. The information flow from the detector is then on the average 170 000 signals every 25 ns which implies 7 THz signal frequency. In digitized form (30 bits per signal) the information flow is about 200 Tbits/s. The on-line system should be able to cope with this tremendous information flow.

TABLE 1

CMS data acquisition main parameters

Detector	Tracker	Calorimeter	Muon
No. Channels	1.6×10^7	2×10^5	10^6
Occupancy (%)	1	3	0.01
Hit information (bits)	30	64	28
Average event size (Mbit)	4.8	0.38	0.0013
at level 1 ≈ 4 0 kHz	*Readout network throughput ≈ 200 Gbit/s*		
Data rate (Mbit/s)	192 000	16 000	100

Only general principles of the on-line system are described in the following. Solutions to the still open details are being looked for in the R&D programme being carried out by the community. In general terms the on-line data processing is proposed to go as follows:

- analogue response from each channels is amplified, shaped and digitized
- fast Digital Signal Processing is performed to filter the signals out of noise
- digital signals are fed into the trigger processors and pipeline buffers
- trigger processors make the decision and tag the selected events
- selected data fragments are synchronized and collected to form a full event
- events are fed into processor farms to perform further selection

The following features of this design are to be noted. The trigger processing is proposed be performed in digital form with the benefit that the decision algorithms will be programmable. Only the signals must be filtered out of noise before feeding into the trigger. The technology of fast *digital signal processing* (DSP) is going to be adequate for this task[3]. In this design the very triggering will be made on *level 1 only* (Fig. 2). Further selection will be made by powerful *event filtering farms* built out of commercially available processors.

The design goal for the level 1 trigger is to gain a rejection power of 10^{-3} or better for the global trigger. The trigger decision will take some 2 μs during which time the digitized events must be kept in buffers. A large number of pipelines ($\sim 10^5$) will be used for this purpose. At the level of event builder the event rate will be about 50 kHz or less and the data flow about 25 GB/s.

The filtering farm must be able to cope with this rate. It is estimated that the selection algorithm will take some 100 million instructions per event. Hence the total processing power of the filtering farm at 50 kHz input event rate must be of the order five million MIPS (5 TIPS). This processing power can be obtained e.g. by 100 farms of 50 CPU's each capable of 1000 MIPS. Extrapolation from the present day perfomance suggests that 1000 MIPS CPU's will be commercially available at a reasonable price in a few years time.

It should be mentioned here that due to the high event frequency no other parallelism is needed but the one at the event level.

The rejection power of the filtering farm should be of the order 10^{-3} or better in order to reduce the event rate at the on-line output to a level of 40 Hz or less. At this rate the flow

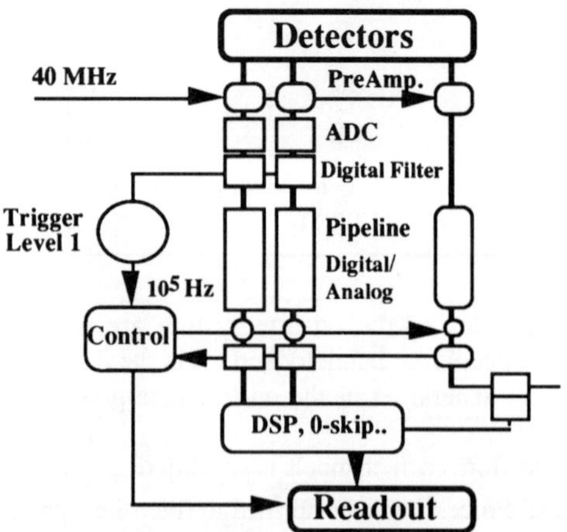

Fig. 2: CMS readout frontend

to the data storage would be about 25 MB/s (5 Mbit/ev). Assuming that the effective collider time will be about 20% of the elapsed time, the total amount of data recorded in one year will amount to about 150 TB. Supposing a technology of 10 GB cassettes will be available in 5-10 years, then more than 10 000 data cassettes a year would be needed.

The filtering farm can also be used for full event reconstruction for the selected events. This would be feasible in case the full reconstruction algorithm would require less than 10 Giga instructions per event which would bring some $5 \cdot 10^5$ MIPS additional load to the farm.

4. Off-line Computing

4.1. Environment

CMS computing environment at present will be briefly described. At CERN the VM mainframe system (IBM9000 machines) is used as a reference machine for the time being. It is mainly used for the code development whereas the CSF workstation cluster (HP9000/735) is used for production.

CMZ code manager is used as a principal programming environment. Patchy was used as a program maintenance and development tool in the beginning, but gradually almost all the users have moved to the CMZ.

At present all the off-line coding is made in Fortran 77. Recommendations are given to the programming team members to avoid statements obsolete in Fortran 90 so as to facilitate conversion to Fortran 90, when compilers will be more widely available.

The GEANT package[4] is the basis of the detector simulation whereas the physics simulation packages Pythia[6] and Jetset[7] are used for kinematics generation. BOS system[5] is being used for memory management and simulation output data are stored in BOS bank format.

No choice of database system has yet been made. The TZ title package of the Zebra system[8] is provisionally used as a kind of intermediate data storage to keep constants for detector materials and media as well as geometry volume descriptions.

4.2. CMS simulation system

A GEANT based simulation package is under development in CMS by a team of some 20 physicists. About half of the team is concentrated on physics aspects of the simulations and the other half on the detector simulation. The CMS specific simulation code consists of some 40 000 Fortran lines at the present stage.

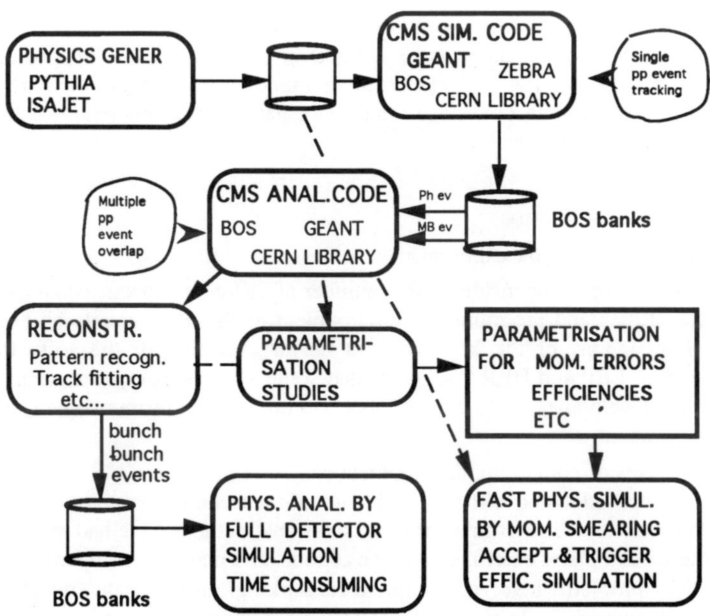

Fig. 3: Schematic view of the CMS simulation model

The overall view of the package is shown in Fig. 3. Generated events are normally written on files which then provide kinematics input for the detector simulation. The option to generate kinematics at run time of the detector simulation is also available.

One of the basic features of the package is to simulate and store the hit and energy deposition information of single pp interactions only at the GEANT tracking stage of the detector simulation. A separate job ('analysis') reads the detector simulated pp-events and builds up full bunch-bunch collision events by summing up hits and energy depositions of individual pp-events thus forming a 'trigger' event. For each trigger a Poisson generated number of minimum bias (ordinary 'uninteresting' pp collisions) events are read from one file and one 'physics' event from another file. Hits and energy depositions from individual pp collisions are summed detector element by element and digitised after event overlap..

The minimum bias (MB) event file is read cyclically so that when an EOF is detected the input will start again from the beginning of the file. In order to minimize correlations between different bunch-bunch events a random event skipping procedure is used for the MB file.

The simulation package is used mainly for the following 3 purposes: 1) optimisation of the detector, 2) investigation of the feasibility to observe various interesting physical phenomena and 3) development and verification of the event reconstruction algorithms. Most urgently one needs simulation results for the items 1 and 2. Since full simulation is very time consuming, parametrisation methods are used at present.

Momentum error and acceptance tables as a function of momentum and rapidity have been created for particles like muon and K^0. The error estimates are then applied directly to the generated events (see Fig. 3) by performing a random smearing of the particle four-momenta. This provides a fast means to simulate the particle measurent in the detector thus facilitating physics simulations.

5. Future Trends and Plans

5.1. Computing aspects of the experiment

One has to take into consideration a number of different aspects, when developing computing facilities and programming environment of the experiment. An example is shown in Fig. 4. One of the key issues is the organisation of the software work. Experience from LEP and HERA suggests that some 1000 person years are needed for LHC software development i.e. a team of about 50 full time software people is needed in one LHC experiment.

Three phases of computing development can be seen in the CMS experiment. The first phase is supposed to end with the completion of the technical proposal next year. Due to limited manpower and strict time table old tools and methods have had to be accepted. Towards the second phase careful evaluation of tools and plans for computing model are necessary. The problems involved comprise networking, mass storage, documentation, programming languages, code managers, memory management, training issues etc.

During the third phase the computing environment should be fairly steady and well established so as to allow productive work towards the start of the experiments.

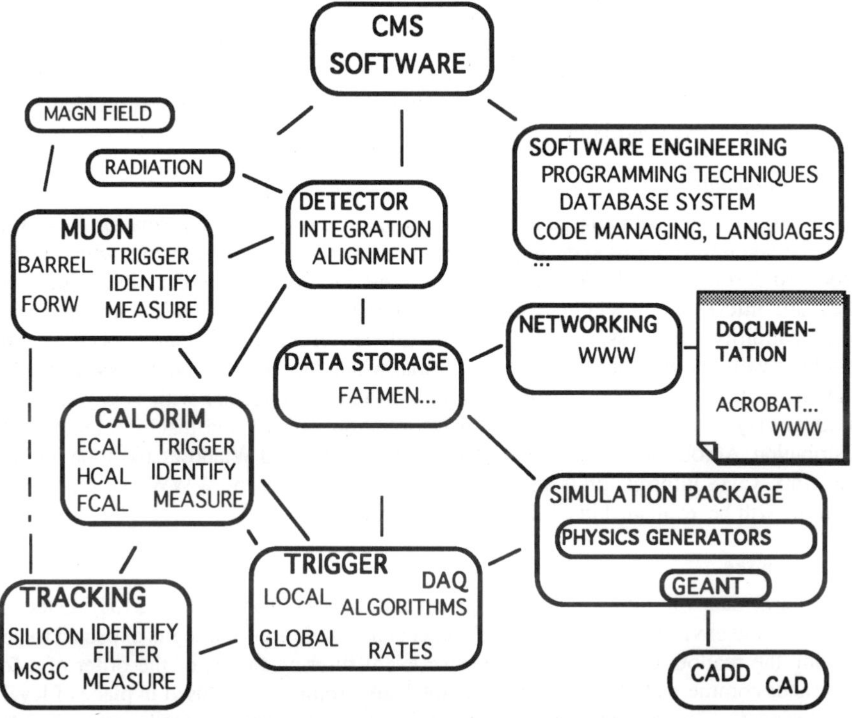

Fig. 4: Basic computing requirements of a large experiment

5.2. Code management

A key issue in the code management is to synchronize work of a large number of team members. Patchy proved to be very effective and successful in this sense in the large experiments like UA1 during the last decade. CERN continues to support CMZ and Patchy so as to keep them mutually compatible. In addition CERN has invited a user group (Code Manager User's Committee) to help coordinating the development of these two products. Hence either of these products seems like a possible solution. On the other hand there are commercial code managers on the market which are claimed to be efficient in the hands of big software houses. The same is not necessarily true, however, amongst a team of scientists where more room may be left to certain individualism.

5.3. Massively Parallel Processing

Solutions to problems such as powerful networking may be provided by industry independent of the particle physics community. Many problems, however, are HE physics specific and contribution to the industrial projects may prove to be vital. One such example is the problem of massively parallel processing, where CERN plays an active role e.g. in the project GPMIMD 2 (General Purpose Multiple Instructions Multiple Data) for which two MPP machines CS-2 will be installed at CERN[9]. MPP may prove to be a solution in long term to problems like ever increasing CPU time consumption of HE physics simulations. In CMS these activities are closely followed though manpower problem has so far prevented active contribution.

5.4. Documentation

Large project with a large number of people implies a large information flow. Care must be taken in order to keep some order in the thousands of documents which will be produced during the work. World Wide Web is one solution to provide people information they are interested in via networks. Still remains a problem of document format incompatibility. One possible solution is the Adobe Acrobat product announced recently. Adobe is well known from it's PostScipt format which has become more a less a standard and Acrobat is it's extended form (Portable PostScript). Together with WWW this new product may be a powerful solution for information storage and machine independent distribution. Acrobat software is already available for Mac and Windows and soon for DOS and Unix. CMS information is already being distributed over the WWW and the Acrobat software will be evaluated in CMS in a near future.

6. Summary

Large data volumes with high frequency are foreseen from the future LHC detectors. In the on-line environment trecent developments in fast digital filtering is expected to help solving the readout problem. Large on-line event filtering farms, of the order of a few thousands commercially available units in total, are proposed to be used in place of level 2 and 3 triggers. The amount of data to be handled off-line is estimated to be more than 100 TB per year. CMS simulation procedures are described in some detail and future trends and plans discussed.

7. Acknowledgements

I would like to thank Sergio Cittolin, Fabrizio Gagliardi, Sami J. Inkinen, Norbert Neumeister, Martti Pimiä, Jean-Pierre Porte, Nick Sinanis and other colleagues who contributed to the material presented.

8. References

1. M.Markytan et al., The Compact Muon Solenoid, Letter of Intent, CERN/LHCC 92-3.
2. Design study of the Large Hadron Collider, CERN 91-03.
3. S.J.Inkinen, IInd International Workshop on Software Engineering, January 13-18,92.
4. GEANT User's Guide, CERN Computing and Networks division.
5. V.Blobel, The BOS System, Dynamic memory management, Universität Hamburg.
6. H-U. Bengtson and T. Sjöstrand, Computer Physics Commun. 46 (1987) 43.
7. T. Sjöstrand, Computer Physics Commun. 39 (1986) 347.
8. Zebra memory management system, CERN program library.
9. Espirit project P7255 GPMIMD 2.

BRIDGING FORTRAN TO OBJECT ORIENTED PARADIGM FOR HEP DATA MODELING TASK

Jean Huang
Superconducting Super Collider Laboratory,
2550 Beckleymeade Avenue,
Dallas, Texas 75237, USA
jhuang@ssc.gov

ABSTRACT

It seems to be appropriate to seriously consider using the object-oriented (oo) technology, if the tangible benefit offered by the computer industry is desired. Yet, many physicists view this newest software development trend with much reservation and reluctance. Facing the reality of having to support the existing physics applications, which are written in FORTRAN, the software engineers in the Computer Engineering Group (CEG) of the Physics Research Division (PRD) at the Superconducting Super Collider Laboratory (SSCL) have accepted the challenge of mixing an old language with the new technology. This paper describes the experience and the techniques devised to fit Fortran into the OO paradigm (OOP).

1. Background

Traditionally, physicists who design and implement particle physics experiment software use the Fortran language. While the Fortran language is well-suited to writing algorithms, its homogeneous element of ARRAY and COMMON BLOCK are very inadequate for complex data structure design. To facilitate the data structuring deficiency in the Fortran language, the high energy physics community has developed with several solutions of their own. ZEBRA[7], the hierarchical structure; YBOS[11], the list of banks; and ADAMO[8], the relational model, are among the most commonly used data management systems.

Since the 80s, different approaches has been taken to implement the complex algorithm and data abstraction for the physics application software. Instead of hanging on to the Fortran language with the constraint of old technology, groups of physicists have adapted the newest trend in the computer industry, the object oriented approach. The PION project[1] at CERN, using OOP techniques for a specialized graphics application, then the SLAC Reason project[2,3], that implemented a graphical interactive analysis environment, and recently the Gismo project[4,5,6], which was originally developed on a NeXT computer using Objective-C. These projects were not only successful in testing the oo technology, their works are having a significant influence in changing the way physics applications are designed and developed.

From the computing point-of-view, it is inevitable to seriously consider using the object-oriented technology if the tangible benefit offered by the computer industry is desired. The physics application maps quite well to object-oriented paradigm, since the original concepts are of an abstract nature, however, technology migration of any sort is a gradual process. It takes years to complete such a transition. Many physicists will continue to devote their energy in analyzing the HEP data using the Fortran language and view this newest software development trend with reservation and reluctance. Facing the reality of having to support the FORTRAN applications, the physicists and software engineers in the CEG of the PRD at the SSCL have accepted the challenge and have done some experiments that mix the old language with the new technology.

The following sections describe various approaches of bridging the language to the object-oriented paradigm, as well as lessons learned from the experiments. The experience will be applied to the data modeling project, of which the objectives are to provide a data modeling environment to establish a standardized data model and its repository systems for supporting physics computing activities.

2. Generic Interface Approach

The first experiment done by the PRD computer engineering group was the Object Class Library for the SZ++ Package by Dr. B. Traversat. This package is an initial C++ version for hierarchical data structure memory management services that was originally provided by the ZEBRA Management System. It organized ZEBRA banks into a double-linked list class, and further defined bank attributes through derived classes. The initial work started with six classes. Figure 1 shows the inheritance of the classes and the following list is a brief description of each class:

Double-linked List Base Class implements a generic double-linked list.

Zebra Bank Base Class derived from the dblink. It adds basic Zebra bank information.

bank_head Class derived from the bank class. It serves as a link point.

bank_fix Class derived from the bank class. This implements a fixed size bank.

bank_var Class derived from the bank class. It implements variable size banks.

SZ Class manages the entire event structure and implements the SZ API to Fortran.

The organization of the data bank is separated from the design of the application programming interface. This approach leads to the design of a platform that spares the user from worrying about the underlying details of data management systems. The following is the application interface subroutines of the SZ++ package for the Fortran language:

szbini() initializes the SZ control structure

szfill() fills a bank

szget() gets bank value

szdrop() drops a bank

szsurv() displays structure of a bank

szshow() displays content of a bank

szexist() tests to see if a bank tree name exists

sznum() returns the max bank number

szwipe() wipes out a bank

Although the SZ++ package provides a generic Fortran interface to the object-oriented realm, due to its close relationship with ZEBRA package, the design is very much bank oriented.

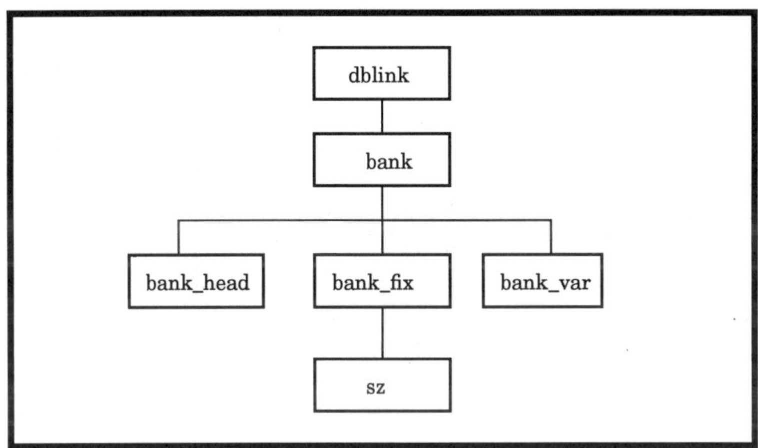

Figure 1: SZ++ Classes

3. Ad Hoc Approach

There are several projects in PRD that involve programs written in mixed languages of Fortran, C, and C++. They come in two different forms: the C or C++ main program calls Fortran subroutines, or visa versa. These projects are usually concentrated on the database and graphical user interface designed for storing and retrieving data with a limited set of functionality.

These projects were able to take advantage of the most updated computer technology of the databases and graphical user interface, but none of them put much consideration in providing a unified platform that allow Fortran applications to work freely. Most of the time, a small modification in the header file structure requires changes to be made in the Fortran subroutine to reflect the mapping. This nature of volatility makes it impossible for large scale projects that require services for the general applications.

Figure 2 shows the call pattern of the ad hoc approach:

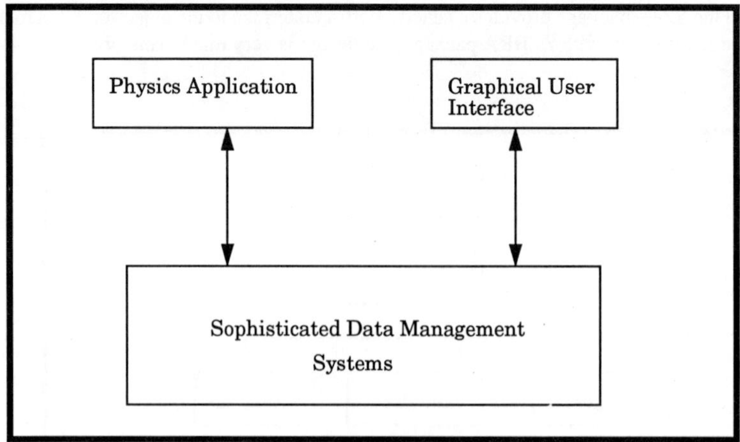

Figure 2: Ad Hoc Approach

4. Bridging Concerns

When bridging the Fortran language to the oo paradigm, the major concern is how to communicate with the more sophisticated data structuring facility of a oo-suitable language. Both generic and ad hoc approaches encounter problems related to naming of routines, parameter passing and call positioning:

- ***Naming***

  It usually is solved by adding an underscore to the subroutine names on the Fortran side.

- ***Calling***

  The concern is which language should be used for writing the main program, and its associated linking considerations.

- ***Parameter passing***

  It deals with the matching between the scattered parameters of the Fortran and the attributes defined in a more sophisticated language.

All these issues stated above are compiler dependant. Porting software between systems may require some attention to the differences between compilers. According to the experience of experimental projects done in the PRCD, it is not difficult to mix the Fortran language with C++ or C.

5. Conclusion

Lessons learned from these oo experiments will be applied to the development tasks of the joint Data Modeling (DM) Project effort of the Gamma, Electrons, and Muons (GEM) Collaboration, Solenoidal Detector Collaboration (SDC) and Physics Research Computing Department (PRCD). The purpose of the DM Project is to develop the standardized data modeling environment and High Energy Physics (HEP) data models necessary to support computing activities of GEM and SDC.

Since mixing the programming languages is not a difficult task, the future DM effort will be focused on the HEP data model design and a generic platform that provides data access services to software developers. Eventually, this data modeling environment, which also includes tools for data analysis, data design, and data definition, will be integrated with the SSCL Framework[9] Object Management Services environment and the Physics Research Integrated Computing Development Environment (PRIDE)[10].

With the increasing complexity and advancement pace of computer technology, it is obvious that the tools and services provided by the DM environment will be acquired mostly from the computer industry, research and academic institutions. Evidently, the object-oriented technology continues to gain momentum in setting the direction of today's computer software development. In order to take advantage of the technology offered by the industry, the object-oriented approach has been chosen for development work of the Data Modeling project.

The C++ language will most likely be used due to its close relationship with the oo technology and its wide acceptance by the software community. The development of computer technology in the physics community, however, is perceived in different shades of light. Changing from one programming language to the other in a community with diversified awareness is a slow process. To best serve the GEM/SDC collaboration, an access mechanism that supports mapping of multiple programming languages, especially the Fortran language, is necessary until the majority has made the transition.

6. References

1. J. Bettels and D. Myers, "The PIONS Graphics System", IEEE computer graphics and applications, Vol 6, No 7 (July 1986).
2. W.B. Atwood, et al., "The Reason Project", SLAC-PUB-5242(april 1990) 11p.
3. Paul F. Kunz, "Physics analysis tools", SLAC-PUB-5520(April 1990) 11p.
4. W.B. Atwood, et al., "GISMO: An object-oriented program for high-energy physics event simulation and reconstruction", Cern Preprint ECP 91-15
5. W.B. Atwood, et al., "Gismo: C++ Classes for HEP", C++ Report March-April 1993
6. W.B. Atwood, et al., "The Gismo Project - C++ at work in high energy physics"
7. R. Brun, et al., "ZEBRA - kernel Data Structure Management System", Cern Computer Center Program Library Long Write-Up, Q100
8. S.M. Fisher, "The ADAMO data system", Cern/ECP/PT xx/yy Real..., November 1991
9. S. Frederiksen, et al., "Framework Software Project Assignment Plan" and "Framework Software Project Execution Plan", SSCL, PRCD R40-000010.
10. J. Burton, et al, "Physics Research Integrated Development Environment (PRIDE)", Presented at the Third International Workshop on Software Engineering, Artificial Intelligence and Expert Systems for High Energy and Nuclear Physics.
11. David Quarrie, et al., "YBOS programmers reference manual", Fermilab, CDF Note No. 156.

REAL TIME, DATA DRIVEN ARCHITECTURES SIMULATED IN CONCURRENT C++ FOR THE LHC SECOND-LEVEL TRIGGER

R.K. Bock, J. Carter, I.C. Legrand

CERN-ECP, Geneva 23, Switzerland

ABSTRACT

For the future CERN Large Hadron Collider (LHC) experiments, real-time algorithms capable of selecting interesting physics events at 100 KHz frequency in the second-level trigger are central to the data acquisition scheme. The parallel decomposition of the problem has been motivated by the necessity to cope with an expected input data rate of several Gbytes/s, and an assumed maximum decision latency of the order of 1ms. Due to the complexity of the physics classification and the potential for unprecedented new physics, consideration has also been given to the flexibility of the decision and to the scalability of the system. A non-farming model for the global second-level trigger, able to keep up with the input rate, requires massively parallel and pipeline decomposition of the data processing problem. The study of such systems motivates the design in a concurrent object-oriented language (μC++), of a simulator for networks of DSP-like processing units able to handle computing and multiple communications in parallel. An object-oriented graphic interface allows an easier way to optimize the system and to study features such as latency and scalability. Results using a TMS320C40 network systolically connected are presented.

1. Introduction

The aim of this paper is to present candidate algorithms for the second-level trigger for LHC type experiments, and one possible implementation of them in a DSP based architecture capable of keeping up with 100 KHz input rate of events passing the first level trigger. Our general trigger scheme [1], presented in figure 1, assumes that the first level trigger guides the second-level trigger by indicating "regions of interest" (RoIs) in the relevant subdetectors. This reduces (by a factor of ~20) the input bandwidth and data processing required for the second-level trigger system. In this scheme the RoIs have to be selected from the raw data and sent to the feature extraction units. This routing problem, which is detector read-out dependent, is one of the key parts in performing a low latency decision [2]. The feature extraction units use detector dependent algorithms and produce a small number of features describing the detected phenomena.

The global second level trigger [3] combines the features coming from different subdetectors from the same geometrical region in order to obtain better particle identification and measurement (RoI task). By combining all the RoIs from each event and computing kinematical variables (such as the invariant masses of pairs) the final "physics decision" is performed by comparing the resulting set of parameters with all possible physics hypotheses (i.e. different interesting physics channels). For the development of the algorithms we have used simulations of typical detectors which are expected to be used in LHC experiments [4]. Production of the appropriate data sets, filtered by the first level trigger algorithms, requires considerable computational resources.

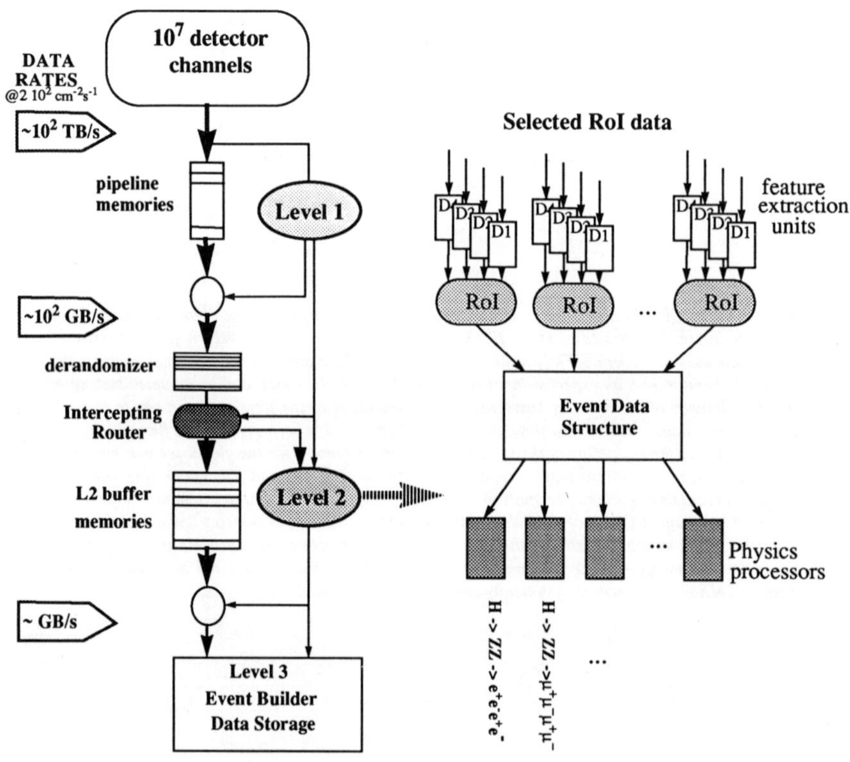

Figure 1. Overview of the data flow in the DAQ system and functional representation of the second level trigger system.

In the following we present the study of an TSM320C40 [5] based network, running in data driven mode, to perform the RoI task and the physics decision task in a non-farming scheme. We assume for the moment that the feature extraction architectures are able to keep up with the input data rate [6], and that the features are available almost in phase to the global structure. The structure of the global decision engine is based on parallel, functional and pipelined decomposition of the algorithms in order to minimise the decision latency. We have developed a μC++[7] simulator for DSP like processing units able to perform computation and communication operations concurrently. Based on these processing units, complex networks running different programs on each node can be simulated and optimised. Graphical X window based tools have also been developed for the optimization of the overall performance, and to provide debugging tools of the programs "running" in the network nodes.

2. Second level trigger algorithms

Using Monte Carlo simulated data for detectors such as a calorimeter and a transition radiation detector (TRD), we have optimised the algorithms in a general scheme of data processing which is hierarchical and at each stage reduces the data flow bandwidth. The natural parallelism in decomposing the problem is always used, and the different hierarchical steps are implemented as a pipeline. The three main steps for this trigger model are briefly presented below.

2.1. Feature extraction

The model assumes that upstream of the feature extraction units a router intercepts the regions of interest (RoIs) pointed out by the first level trigger, and sends the raw RoI data for subdetectors from the same geometrical region in proper format to the feature extractors. The objective of the feature extraction algorithms is to find a few decision variables, suitable for describing the detected phenomena in each type of detector, and provide a bandwidth reduction of more than an order of magnitude over the raw RoI data.

The calorimeter algorithm [8] is based on fine granularity 20X20 cell windows for two calorimeter layers (electromagnetic and hadronic). The algorithm finds the centre of the energy deposition of the cluster, computes the total energy in both electromagnetic and hadronic regions and cluster shape estimators. These features are likely to contain the relevant information for electron/pion/jet discrimination in the calorimeter.

The TRD algorithm [9] consists of finding a track and its parameters (slope, number of low/high pulse hits) in a projected window in the Z ϕ plane indicated by the first level trigger. In this case the size of the window is rapidity dependent. The algorithm recognizes patterns of high momentum tracks and, if found, discriminates between tracks produced by electromagnetic and hadronic particles.

A detailed study implementing such algorithms for two detectors (calorimeter and TRD) in commercially available high performance architectures and custom-design systems [6] has shown that with the currently available technology the required input rate of 100 KHz in performing these tasks can be achieved.

2.2. RoI task

The RoI task [10] combines all the features coming from different sub-detectors in the same geometrical region in order to define the identity and kinematical properties of the detected phenomena. It is an artificial neural network (ANN) inspired algorithm, developed and optimised in connection with the physics decision task, and is based on simulated data [4] for several interesting channels versus QCD jets as background. The fully connected feed-forward "analogic" ANNs take at their input layer the features coming from different subdetectors in 12 neurons, use 6 neurons in a hidden layer, and provide an analogic output in a 4 neuron layer (figure 2).

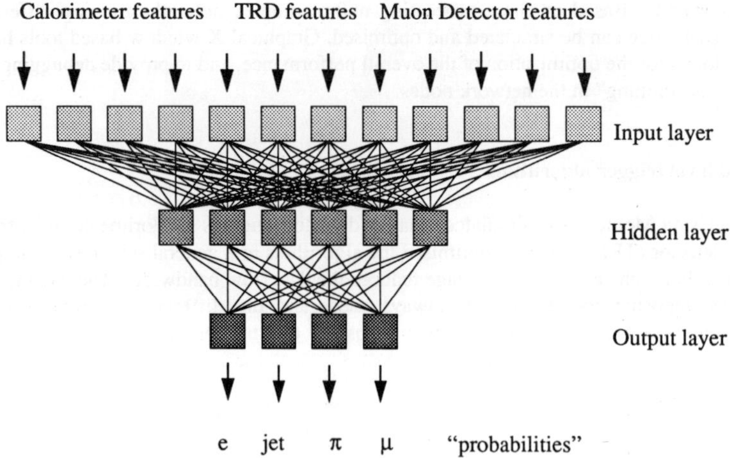

Figure 2. Schematic view of the ANN-like algorithm used to combine features from different subdetectors in order to obtain probabilities for each type of particles.

The new output features obtained by training the network in the appropriate way can be interpreted as probabilities that the detected particle is an electron/muon/pion/jet. By sending these four "probabilities" to the final Physics Decision task together with the kinematical variables for each RoI the event classification is more flexible, and provides a better quality decision than by uniquely identifying the particle at this level. This ANN algorithm is based on accumulated multiplication, well suited to a DSP like TMS320C40, which can perform such operations in one instruction.

2.3. Physics Decision

The final trigger decision [3], [10] is required to be flexible, user programmable and scalable in the number of different interesting channels considered. It combines all the data coming from the RoI tasks by generating an event data structure, and computes the invariant masses and "pair probabilities" for each combination of RoI pairs, quantities which we have found important in performing an efficient decision. Testing the different physics hypotheses in parallel is a natural way to perform the classification task, and this requires the distribution of the full event data structure to a scalable system in which flexible decision criteria are applied. The event classification procedure is data dependent and is performed by testing against conditions which characterise the predefined interesting channels. These conditions are typically thresholds for particle and pair "probabilities", invariant masses, energies and conditions on the number of RoIs of a given class. Results of the algorithms in selecting six test channels vs QCD jets as background at high luminosity ($<n>=20$ collision per bunch crossing) are presented in Table 1.

Note that the simulation data used in these calculations is provisional and model dependent, and was not intended to represent in full detail the detector behaviour. Our aim was

Physics Channels	Accepted Signal %	Accepted Background (QCD jets) %
H->ZZ->4e	100	0
H->ZZ->4μ	99.5	0
H->ZZ->2e2μ	99.6	0
H->ZZ->2e2j	96	0
tt->e3j	94.6	5.6
tt->μ3j	100	0.4

Table 1. The global second level discrimination results for six channels vs. QCD jets as background. For each channels and the background we used 500 events filtered by the fist level trigger.

to develop algorithms of representative computational complexity for benchmarking purposes, and the results in table 1 are intended to show only that the algorithms are appropriate for the task, and not to accurately represent physics performance. Further simulation work is underway to address in the proper detail the physics performance of this approach.

3. DSP simulation

μC++ is a concurrent object-oriented language [7] designed to support parallel programming models on shared-memory uniprocessor and multiprocessor computers. Inheriting elementary objects such as coroutine, monitor, coroutine-monitor and task, one can describe simple components such as communication links, buffers and "software objects" (C-like programs running on different "hardware objects"). The real-time behaviour of more complex objects like the DSP may be modelled using composite objects. The real time simulation of a multiprocessor environment requires a unique time structure and scheduler for all the objects in the system. μC++ provides these features and in the case of simulating parallel systems each task controls its own time behaviour through delay (n - clock cycles) statements. Each such delay statement specifies how many clock cycles are necessary for each sequence of computing.

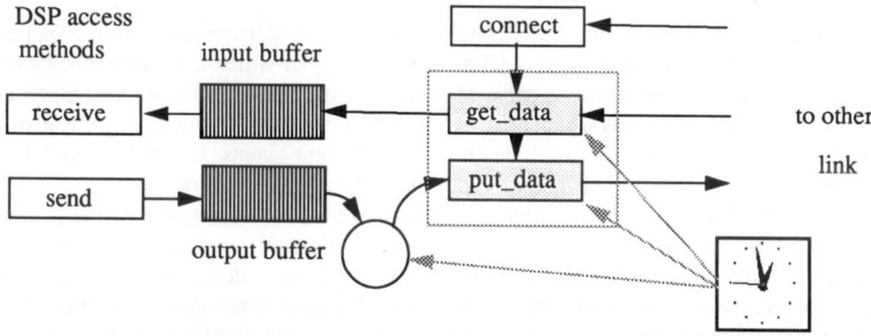

Figure 3. Schematic structure of the task used to simulate one link, and the main methods used to access it.

The TMS320C40 has six 8-bit parallel links which can send or receive data concurrently in either CPU controlled or DMA modes. Such links are an integral part of the TMS320C40 processor, and the task which simulates the behaviour of the links is schematically presented in figure 3. Each link has an input and an output buffer (32 bytes for each buffer), and a number of public methods are defined for using this object. One method provides for connecting the link uniquely to another by simply passing the pointer of the new link. Once a link is configured its methods can be accessed by the connected DSP. The DSP which owns the link can read and write to its input and output buffers respectively. Two mutual exclusion methods allow access to these buffers in order to perform data transfer to/from a connected DSP. The mutual exclusion attribute ensures that the maximum bandwidth of the link is not exceeded in duplex mode. A small circular buffer is used to simulate the link latency in data transfer.

The DSP is a composite object which when created generates six new link tasks. The DSP object allows access to the links through its own methods in a simple way. The DSP task can simulate for example the DMA coprocessor of the TMS320C40 or a communication kernel used in long distance message passing through predefined channels. The "software objects" are C-like programs which can run in each DSP object, and may have communication and computational functions. Each such software object requires a pointer to the DSP task in which it is running. Based on this model the main simulation program initially creates all the DSP objects, configures them, and then creates all the required software objects. In this way the main program is small and straightforward to adapt to a new architecture. The same structure was used with the proper modifications to simulate networks based on T9000 transputers.

In order to see the state of each DSP and its links in time, a X window graphical part was added. Each DSP is responsible for plotting itself (when the object is created the position on the screen is specified) and the image is updated when its state is changed. In addition we have developed a "logic analyser", for which each software object simply indicates when a predefined event of interest occurs, and an interactive graphical display shows the time structure of these events for all concurrent objects. We have found the latter particularly useful in load balancing the network under study. We have used this simulation program to define and optimize a data driven architecture for the global decision part of the second level trigger.

4. Implementation of the global decision

The global decision consists of RoI tasks and physics decision tasks. Two broad architectural distinction can be made: systolic data driven solutions in which the data for one event is processed in phase in an architecture capable of sustaining the operational frequency; and asynchronous processor farm solutions which need not operate at the input frequency by virtue of having many more processing elements each dealing with different events. Our aim of studying the use of a DSP network is to provide a data driven solution for global decision.

The TMS320C40 is a very attractive architecture for performing the RoI task since its links can be used to collect data from different feature extraction units concurrently with the computation of the RoI parameters. In order to perform different physics decision tasks in parallel the event data structure has to be duplicated to many nodes at the same time. To achieve this we chose to use communication performed by the CPU, since in this way the latency is smaller than using DMA controller, and the internal parallel instructions allow computing while communication is also handled by the CPU. Morever, the data which has to be copied is of short length (few words), and it seems that in this case the synchronous communication is more efficient. The architecture simulated to perform these tasks is schematically presented in figure 4.

The first layer receives the features coming from different subdetector feature extractors and performs the RoI tasks. For simplicity this scheme assumes that these features arrive at this layer effectively in phase. If the latency in the feature extraction units is different or part of the feature extraction is performed in farms of asynchronous processors, one more layer of DSPs is necessary to receive the data and to re-synchronize the data flow. The RoI task as it has been defined can be performed in about 10μs on a TMS320C40, with full optimization at assembler level by using parallel instructions.

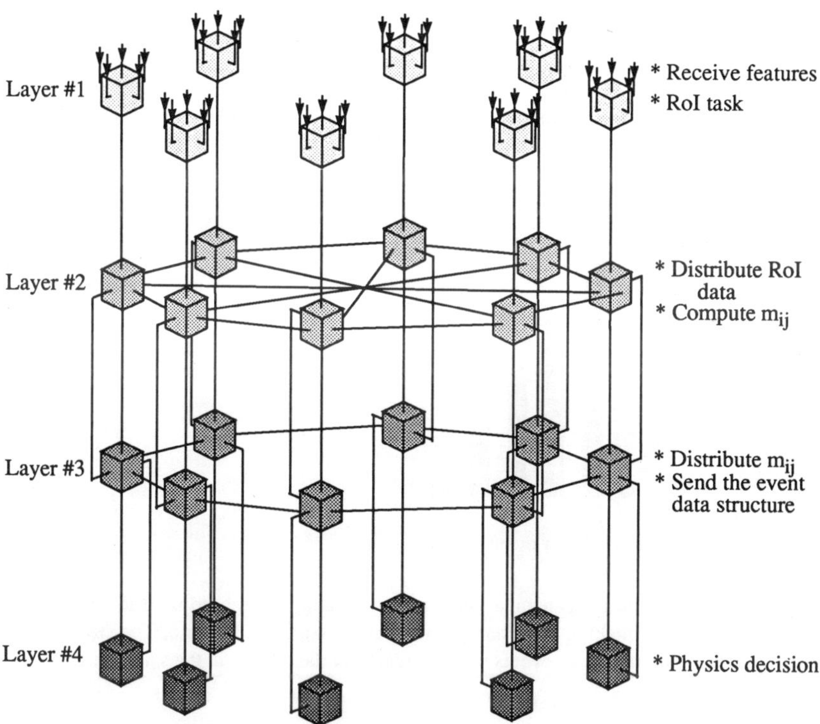

Figure 4. The TMS320C40 based global decision architecture (m_{ij} is the invariant mass of the i,j th pair).

The second layer is the key part in generating the event data structure. The output from each RoI task is received through a link and sent to the nearest neighbours and also across the diameter of the ring. Once a new RoI data stream arrives in a node it is sent down to the third layer. For this two links on each DSP are used because the bandwidth is higher. Layer two is also used to perform the invariant mass calculation, one column of the invariant mass matrix being computed in each node. The invariant mass is computed using look-up tables for the trigonometric functions and may be concurrent with data transfer for each subsequent RoI. In this way each invariant mass is computed twice, but this is preferable to having different programs for different numbers of RoIs per event. Simplicity of the structure is considered a more important ingredient in parallelism than efficient use of all resources.

The third layer distributes all the invariant masses by copying the data to nearest neighbours in the ring and passes down to the next layer the complete event data structure. This part is optimised for passing each invariant mass through the shortest way as long as it was computed in two nodes.

Each node in the lowest layer performs different physics classifications. The number of physics processors may be scaled by adding further layers to form a pipeline.

5. Results

In the simulation program we use values for the latency, communication speed and program execution time measured on real TMS320C40 systems. The number of RoIs per event was generated randomly from a distribution obtained by physics simulation [4] and is presented in figure 5.This distribution is dependent on the thresholds used by the first level trigger in selecting RoIs.

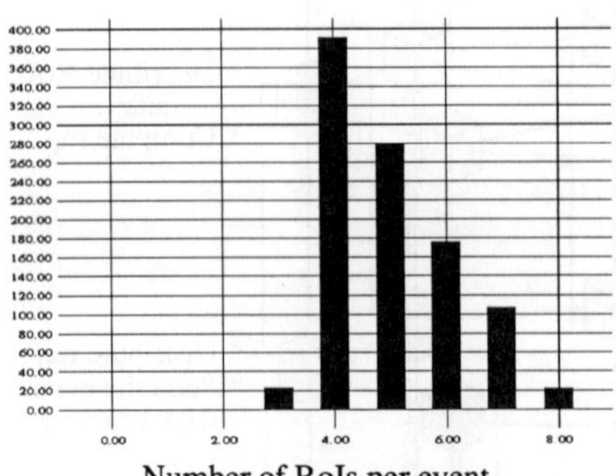

Number of RoIs per event

The simulation indicates that the structure is able to keep up with the 100KHz input rate, and the latency in making the global decision is presented in figure 6. The structure is running as a pipeline, so up to three events may be treated in the system in the same time.

6. Conclusion

We have successfully simulated a complex multi-processor network used as a trigger architecture at 100 KHz. For our R&D environment, we have found the concurrency and the object-orientedness of μC++ of great help. We added a graphical user interface via X windows, which resulted in an adequate development tool, allowing us to quickly design and optimize different architectures.

We have been able to show that a high performance DSP-based mesh connected architecture can cope with our target using the benchmark algorithms of our model. This first attempt in performing the second-level task without any farming in a pipelined data driven mode, seems, based on this work, a possible low latency option.

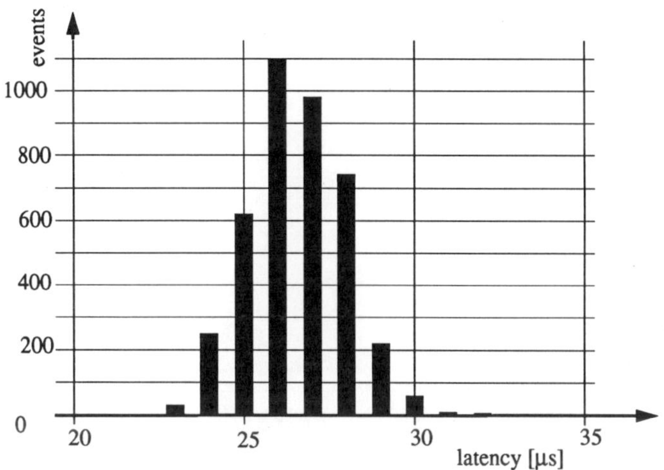

Figure 5. The latency [μs] distribution for the global decision.

Acknowledgment

We express our gratitude to Jos Vermeulen (NIKHE -Amsterdam), Barry Green (RHBNC -London), Peter Wegner (Zeuthen-Berlin), Adriaan Borgers and Frank Globach (Utrecht University) for useful discussions and helping us in measuring real features on TMS320C40.

REFERENCES

1. EAST(RD-11), Proposal and Status Report: CERN/DRDC 90-56, (1990)
2. P. Kammel et al, CERN/EAST Note 93-13 (1993)
3. R.K.Bock, J. Carter, E. Denes, I.C. Legrand, M. Novac, J. Varela, Second-level trigger: global decision structure CERN/EAST Note 92-06 (1992).
4. R.K. Bock, J. Carter, I.C. Legrand, J. Varela, Test data for the global second level trigger CERN/EAST Note 93-01 (1993)
5. Texas Instruments, TMS320C4x, User's Manual (1992)
6. J. Badier et al, Evaluating Parallel Architectures for Two Real-Time Applications with 100kHz Repetition Rate, IEEE Transactions on Nuclear Science, vol 40, No. 1, (1993).
7. P.A. Buhr and R.A. Stroobosscher, μC++ Annotated Reference Manual, version 3.2.5 (1992)
8. J.Badier, R.K. Bock, C. Charlot and I.C. Legrand, Benchmarking architectures with Spacal data, CERN/EAST Note 91-10, (1991)
9. P. Bialas, J. Chwastowski, P. Malecki and A. Sobala, Benchmarking with data from the transition radiation detector, CERN/EAST Note 91-11,(1991).
10. R.K.Bock, J. Carter, I.C. Legrand, Modelling of global decision structure, CERN/EAST Note 93-03 (1993).

THE SSCL FRAMEWORK SOFTWARE PLANS

Soren G. Frederiksen
Superconducting Super Collider
Dallas, TX U.S.A.

abstract

In about ten years the Superconducting Super Collider Laboratory (SSCL) will be producing 40 TeV proton-proton interactions. The size and scale of the effort demands new approaches to design and develop software used by the experimental collaborations. The Physics Research Division Computing Department (PRCD) of the SSCL is developing (in collaboration with the Solenoidal Detector Collaboration (SDC) and Gamma, Electron and Muon (GEM) collaboration) a support system which will be used to build and run the collaboration software. It will be used for simulating the events needed for detector development and for the analysis of these complicated events. The plans and status of this program will be discussed.

1. Introduction

A standardized environment or set of framework services is required to support the various tools, utilities, application programs and subroutines that will be developed by the GEM and SDC experimental collaborations. The Reference Models developed by the U.S. Navy's Project Support Environment Standards Working Group (PSESWG) (Reference 1), a working group in the Next Generation Computing Resources (NGCR) program, and the National Institute of Standards and Technology (NIST) (Reference 2) were heavily referenced when developing the plans for the Framework Software project. These Reference Models define the Framework services as the services which comprise the underlying infrastructure of an environment. In this context Services refer to an environment's capabilities or is an abstract description of the work being done. The purpose of the Framework Software project is to develop the Framework Software services that will be used by the SSC collaborations.

2. Framework Software

The Framework Software will be developed to run on the PRCD supported platforms which currently include most major UNIX vendors (DEC, HP, IBM, SGI and SUN). The software will always reflect an open system concept, to ensure that the software will be able to migrate as the technology changes. The distributed nature of SSCL computing requires that this software be able to run in a distributed computing environment, and will also allow for a parallel processing environment. The software will be written in a modular and expandable manner using Object Oriented Analysis, Object Oriented Design, and Object Oriented Programing.

The PSESWG reference model has two basic categories of services, the Framework Services and the End-User Services. The End-User Services are further subdivided into Technical Engineering, Technical Management, Project Management, and Support Services. Figure 1 shows the relationships among these 5 service groups. The Framework Software forms the central core with relationships to all the other services. The Support Services underlie and support all the other services.

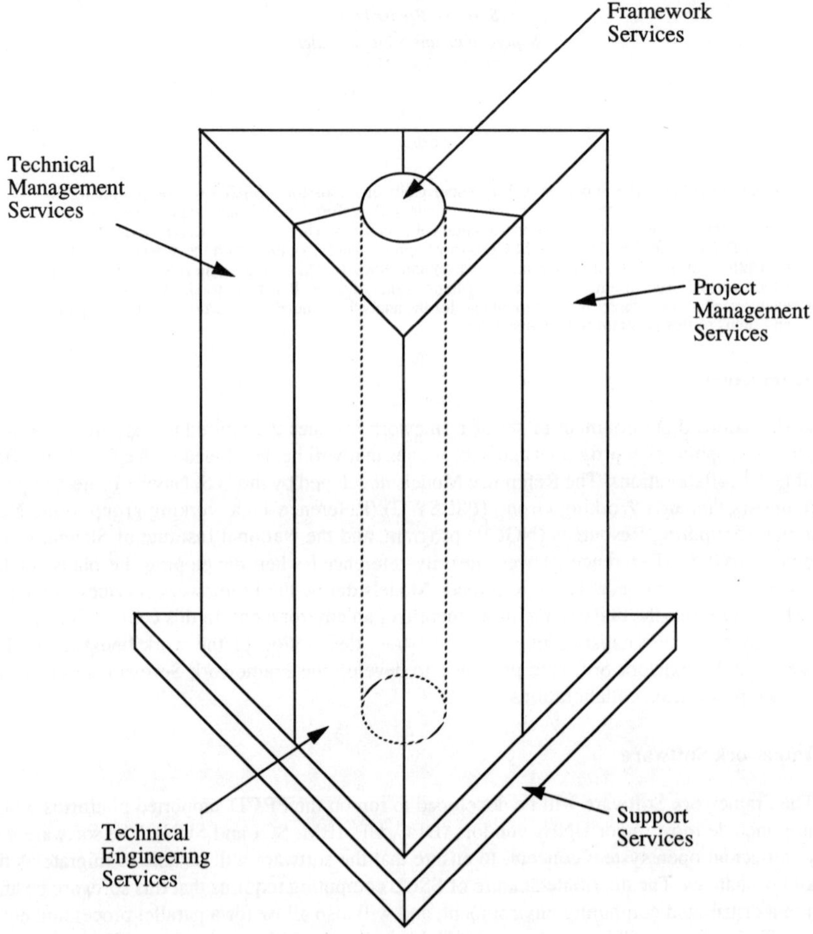

Figure 1. Service Groups relationships.

The standardized Framework services and tools that need to be developed are:

- Operating System services: The services normally associated with the operating system.
- Object Management services: The services responsible for the definition, storage, maintenance, management, and access of object entities and the relationships among them.
- Policy Enforcement services: The services which take care of security enforcement, integrity monitoring, and various object management functions.
- Process Management services: The services responsible for the definition and the computer-assisted management of project development activities.
- Communication services: The services which provides a standard communication mechanism that is used for inter-tool and inter-service communication.
- User Interface services: The services which provide for the integration of the object management system, the process management services and the tools.
- Framework Administration services: The services responsible for configuring and administering the Framework Environment.
- A C++ Class Library for use by the Framework Software. It is assumed that the object oriented programming will be done using C++.
- An Interface Definition Language (IDL).
- A Data Definition Language (DDL).

In order to facilitate the development of the Framework Software a complete Software Development Environment (SDE) will also be required. This SDE is referred to as the Physics Research Integrated Development Environment (PRIDE) and is described in more detail in Reference 3. The scope of this project will also include the development of the PRIDE.

3. Plans

The Framework Software project is a joint GEM/SDC/PRCD project, and requires the cooperation of Physicists and Computer Specialists. The final integrated computing environment that this project will develop will be used by both collaborations.

The Framework Software development will involve the following steps:

1 A review of the Framework Reference Models (Reference 1 and 2) to establish their relevance for the Framework Software.

2 Establish Functional Requirements. This will require interviews of the collaboration physicists in order to determine their requirements. It is anticipated that this will only be the preliminary requirements which will be refined during follow-on activities of the project and as new versions of the Framework Software are developed.

3 Develop system design. The system design step will also require Pilot Projects and Prototypes to help refine the Function Requirements and the System Design. These Pilot Projects will be used by the collaboration physicists to give them a feel for what the final production system will be like. It is anticipated that the System Design will also be refined during follow-on steps.

4 Develop the Framework Software development plan.

5 Develop the requirements specifications.
6 Develop the test plans specifications for the Framework Software. Pilot projects may also be required as part of the Framework software development plan.
7 Develop the software design.
8 Code development and/or integration of software.
9 Final integration and testing of the production system.

These steps will result in a final work Framework Software program. The procedure will then repeat itself in order to develop evolutionary versions of the Framework Software program.

In order to compress the development time and to implement a workable environment for the physicists as soon as possible, the first few steps will also involve the use and testing of IBM's SEE Pathfinder. SEE Pathfinder is a fully integrated Software Development Environment developed by IBM's Software Engineering Technology (SET) group. The SEE Pathfinder will be evaluated as a candidate for the PRIDE, and will be used to help in the first phase of the Framework Software development. SEE Pathfinder is the integration of a number of programs in an IBM Workbench (or HP Softbench) environment for use in the development of software. It has integrated both upper and lower CASE tools.

The PRIDE development will involve the following steps:

1 Establish Functional Requirements. After establishing the Functional Requirements the SEE Pathfinder will be evaluated to see if it is a candidate solution for the PRIDE. If it is determined that the SEE Pathfinder is not the appropriate candidate for the PRIDE then the remaining steps will be followed.
2 Develop system design. It is anticipated that the System Design will also be refined during follow-on steps.
3 Develop the PRIDE development plan.
4 Develop the requirements specifications.
5 Develop the test plans specifications for the PRIDE. Pilot projects may also be required as part of the PRIDE development plan.
6 Develop the software design.
7 Code development and/or integration of software.
8 Final integration and testing of the production system.

The Framework Software project has just started down the trail of developing the software that will be used by the SSCL collaborations. This software should meet and exceed the requirements of the collaborations and will continue to evolve and be used well into the 21st century.

4. References

1. PSESWG, Next Generation Computing Resources, *Reference Model for Project Support Environment*, Version 1.0, March 1, 1993.
2. NIST & ECMA, *Reference Model for Frameworks of Software Engineering Environments*, NIST Special Publication 500-211, August, 1993.
3. Jackie Burton & Laird Cormell, *Physics Research Integrated Development Environment (PRIDE)*, These Proceedings.

A HIERARCHICAL AND DISTRIBUTED CONTROL ARCHITECTURE FOR HEP BIG DATA ACQUISITION SYSTEMS

S. BUONO, I. GAPONENKO, R. JONES, L. MAPELLI, G. MORNACCHI, D. PRIGENT, E. SANCHEZ-CORRAL, M. SKIADELLI, A. TOPPER
CERN, Geneva, Switzerland

P-Y. DUVAL, A. LEVANSUU, R. NACASCH, Z. QIAN, C. RONDOT
CPPM, , IN2P3, France

G. AMBROSINI, G. FUMAGALLI
INFN, sezione Pavia, Italy

M. AGUER, M. HUET
STEN, C.E. Saclay, France

ABSTRACT

One of the purposes of the RD13 project [6] is to evaluate software technologies that could be used to fulfill the requirements of the LHC DAQ systems. One issue concerns the ability to keep an efficient control over an enormous number of processes, supported by many processors, distributed over a wide range of hardware elements. This paper presents a software architecture, based on a network of tree connected reactive control agents which implement a global DAQ run control system.

1. Introduction

A recent paper [7] makes the following statement about the ZEUS Run Control System:" ... However, sequencing a series of run commands precisely, ensuring that a large number of components (~30) running in different systems are placed into the correct state at each step of the run cycle, proves to be a challenging task. ...". This has definitely convinced us that things needs to be thought about seriously, well in advance for LHC.

The run control (RCL) architectures rely upon several software technologies among which two can be considered as "core technologies":

- control description models and languages
- distributed computing.

Today state-of-the-art software offers various description models. Some of which are associated to software environment and tools, to cope with control system design and distributed implementations. Two are currently evaluated within the RD13 project. One uses the Petri net modelling and the ARTIFEX environment [8]. This paper presents the second one which is based upon finite state machine modelling and the ISIS distributed computing toolkit [5].

2. The RCL system model

To be manageable, complex systems composed of a large number of components, are usually split-up into a tree of sub-systems. Each level presents a local perception over one of the system elements. It is associated to a certain level of functionalities inside a technological point of view. The same decomposition rules are adopted for the software (hiearchical design), the functional point of view (functional decomposition) and hardware point of view (assembly and maintenance levels). The following RCL architecture applies to systems that are structured according to this tree model.

Our RCL system is designed as a network of tree connected DAQ sub-systems controllers called RCL modules. In the DAQ system design, one RCL module is associated to each one of the DAQ sub-systems.

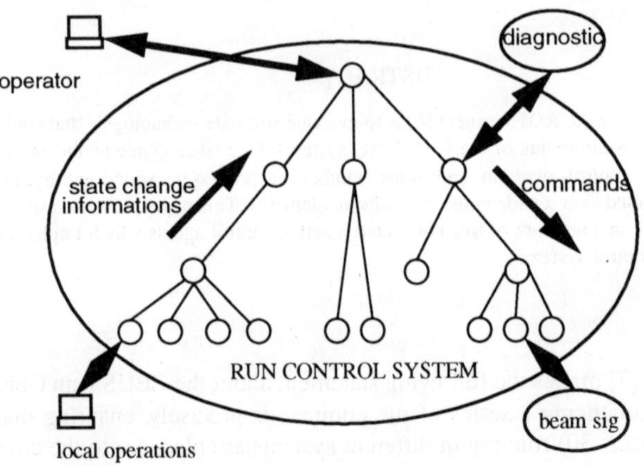

FIGURE 1. The run control system

The control organization is inspired by a "goal oriented" type of organization management. Each sub-system controller has its own local level of responsibilities and its own resources to achieve the global RCL system goals which are:

1. to put the controlled DAQ system in the working mode the user wants it to be,
2. to maintain and publish information about the state of its controlled subsystems,

3. assume the states coherency between the various subsystems states.

RCL modules resources are either programs, they can directly execute, or lower levels RCL modules they can command.

RCL modules execute commands received from their father RCL module. This father is only interested by the commands results, not the way they were executed. RCL modules also execute RCL commands they received from external DAQ services. If an external command changes the controlled subsystem's state, they have to advise thier father module to let it make the related actions that fit with the new situation.

The consequence of this organization is that two types of data flow along the control tree. A downward flow of commands from the higher controllers to their child modules, and an upward flow of states information from the lower level controllers to their fathers.

Run control modules are pure reactive agents that produce no activity by themselves. They only react to RCL messages received from other RCL modules or external DAQ processing entities. Each RCL module is implemented as a separate process. Those processes run upon different machines and use an ISIS based communication service.

2.1. The RCL system and its DAQ environment

The RCL system provides a service to manage the DAQ system states in a coherent way but it does not decide in which state the data acquisition system should be. Those decisions are taken by other DAQ system entities which uses the RCL system to have their change of state decision executed coherently all over the DAQ system.

The main run control system clients are the DAQ systems operators. But other DAQ system components can use the RCL services. Such components may be responsible for some system monitoring and decision taking like: on-line failure detection and diagnostic systems, or machine signal and sensors monitoring entities. They interact with the RCL system by querying it about some sub-systems states or sending it change state commands to execute.

3. The reactive control agent model

3.1. The agent structural model

The RCL modules are modelled as finite state machines (FSM) . Their behaviour is described by a set of states which are connected by arcs called transitions. At least one transition starts from each state. When the RCL module is created it is in a special state called the initial state.

A list of actions is associated to each transition. These actions are executed when the transition is fired to put the controlled sub-system in the configuration corresponding to the new state.

Agent receive messages that can produce reaction. RCL messages can be of two types: commands messages and state change information messages. One command and a set of transition rules are associated to each transition. RCL agent receptivities in a certain state are:

1. the set of commands associated to each transitions starting from this state,
2. the state change informations that can validate a transition rule.

The RCL system coherency rules are coded in two ways. They are directly coded into the FSM transitions actions list executed when a state change commands is received or they are also coded into the FSM transitions rules evaluated each time a new state information is received from the underlying RCL modules. In the present implementation, those transitions rules are expressed as a C logical expressions whose members are others RCL modules states.

The FSM descriptions are stored in the RCL database and are loaded at run time and interpreted by the RCL module.

3.2. The agent activities description

Agent activities executed in transitions are described in actions lists called scripts. One script is associated to each transition. Local activities are C functions that can be called by the RCL module to be executed by its local machine. In the script language those functions are called by their symbolic names. To let the RCL modules use their sons as resources, the script language provides script commands to send RCL commands to others RCL modules. Those remote commands can be executed with 3 synchronization modes:

1. Synchronous remote command execution. In this mode the script execution in the sending RCL module is suspended until the command result is received. For exemple a command to send the Stop command to all the child modules is:

   send_sons_sync(Stop);

2. Asynchronous remote command execution. In this mode the script execution go on as soon as the command has been successfully received by the target but no result or execution acknowledgment is expected by the sending RCL module. For exemple a command to send the Run sommand to the DETECTORS module is:

   send_async(Run,DETECTORS);

3. Parallel remote command execution. In this mode the script execution go on as soon as the command has been successfully received by the target. But the script interpreter will wait for the results of the command execution at the end of the script list execution. For example a command to send the Configure command to the DEC_UNIT group is:

   send_group_par(Configure,DEC_UNIT);

The command target can be specified as an individual module, either by its module name or by the constant name “father”. The target can be a set of modules which are specified either by their group name or by the constatnt name “sons”.

Combining local actions and remote commands in scripts executed within each level, the RCL system enables the DAQ system builder to have synchronized actions executed throughout the DAQ system with different sequencing politics. Actions can be executed starting with the higher subsystems levels and then downtree, level by level. This is typical of initialization sequences. They can be executed starting with the lower levels and then propagated upward to the higher level. This is typical of ordered stop sequences. They can be done in parallel in all subsystems, which is typical of local testing or parameters setting functions.

4. The control system creation and supervision

The RCL system is created by a RCL supervisor. This process creates all the RCL modules on the various DAQ system machines. Once created, the processes connect themselves together to form the RCL control tree. This RCL supervisor is based upon the ISIS Resource Manager (RM)[3]. It provides a way to remotely monitor the running RCL processes from any machine running ISIS in the network. The supervisor was written using the ISIS RM application programming interface. In case any RCL process crashes, the supervisor is advised by an ISIS message. The RCL supervisor picks up the necessary informations to create all the RCL modules in the RCL database.

5. Database description

The RCL system is completely described in a database. Adding, removing, changing the behaviour of any RCL modules is made by changing the database description. This database contains several DAQ configurations from which the user choose when starting the DAQ system.This database contains the configuration data, used by the RCL supervisor and the FSM data used by each module. Each RCL module retrieves its FSM definition, its transition scripts and coherency rules at run time to interpret them.

The database can be populated by two types of interfaces. Data that can be more easily described with graphical diagram: RCL connections tree and modules FSM, are inserted using a graphical interface. Others uses a MOTIF based textual interface.

The database is implemented with QUID[2], a real time, memory oriented, database management system. The database interface uses the ARTIFEX[1] petri net editor for the graphical parts.

6. Project status and conclusions

This work is the second step in the achievement of a distributed control architecture for DAQ systems. The first version was used last year in a test beam to control the RD2 [9] project detector test. The present more sophisticated version will be used in October 1993 in a test beam with the RD6 [10] detector group.The work will go on using alternatives languages, tools and methods to re-implement it or part of it. Functionalities will be increased for efficiency.

The run control area is a DAQ key point seriously explored in RD13 project. Since 1991 several approaches have been implemented. In the limit of tests that are within our reach today, ISIS proves an efficient tool to implement distributed control. We have also shown that a fully database described RCL system is workable and easy to use.

7. References

1. ARTIS, *Artifex Environment User Guide*,(Corso Cairoli, 8 Torino, Italy).
2. ARTIS, *Quid Development Environment*, (Corso Cairoli, 8 Torino, Italy).
3. T.Clark,K.Birman, *Using the ISIS Resource Manager for Distributed, Fault-Tolerant Computing*,(Dept. of Computer Science, Cornell Univ. USA)
4. Imperial Software Technology, *The Xdesigner User's guide*,(95 London Street, Reading RG1 4Qa)
5. ISIS Distributed System Inc, *The ISIS distributed toolkit*, (111 South Cayuga street, Ithaca, New York 14850).
6. L.Mapelli & al., *A Scalable Data Taking System At A Test Beam for LHC*,(CERN/DRDC/90-64/P16,CERN/DRDC/90-64/P16 Add 1,CERN/DRDC/90-64/P16 Add 2, 1990).
7. I.H. Park, in *The on-line control data acquisition system of ZEUS*,(Real-time Conference, Vancouver Canada).
8. E. Sanchez-Corral & al., in *Software engineering Techniques and CASE tools in RD13*, (ICALEPCS'93 Berlin).
9. RD2 Collaboration, *A 66Mhz, 32-channel analog memory circuit with data selection for fast silicon detectors*, Nuc. Inst. & Meth. Phys Research A326 (1993) p100-111.
10. C.W. Fabjan & al. *Integrated high rate transition radiation detector and tracking chamber for the LHC*, (CREN/DRDC/90-38 DRDC/P8 ,1990).

EXPERIENCE WITH A RUN FILE ARCHIVE USING DATABASE TECHNOLOGY

Ute Nixdorf

Superconducting Super Collider
Physics Research Division
Computing Department
2550 Beckleymeade Avenue
Dallas, Texas 75237-3997

ABSTRACT

High Energy Physics experiments are known for their production of large amounts of data. Even small projects may have to manage several Giga Byte of event information. One possible solution for the management of this data is to use today's technology to archive the raw data files in tertiary storage and build on-line catalogs which reference interesting data. This approach has been taken by the Gammas, Electrons and Muons (GEM) Collaboration for their evaluation of muon chamber technologies at the Superconducting Super Collider Laboratory (SSCL). Several technologies were installed and tested during a 6 month period. Events produced were first recorded in the UNIX filesystem of the data acquisition system and then migrated to the Physics Detector Simulation Facility (PDSF) for long term storage. The software system makes use of a commercial relational database management system (SYBASE) and the Data Management System (DMS), a tape archival system developed at the SSCL. The components are distributed among several machines inside and outside PDSF. A Motif-based graphical user interface (GUI) enables physicists to retrieve interesting runs from the archive using the on-line database catalog.

1. Introduction

The GEM Collaboration designed a test stand to evaluate various muon chamber technologies for their future detector. A test rig was built, the chambers installed and the electronic data acquisition system (DAQ) set up to register cosmic muons passing through the detector prototype. Data was initially registered during a period of 6 months. This produced several Giga Byte of event data, which was first recorded on the disk and tape of the DAQ system. The data was then migrated to PDSF for long term storage. The purpose of this setup was the comparison of the quality of different types of muon chambers. The amount of data anticipated drove the requirement to be able to select interesting runs without having to retrieve and scan through all stored run files. Retrieval of run files from tape and sequential analysis of all run files would consume a lot of resources. A list of criteria was defined by which the collaboration members are able to decide, up front, which files they need to access. The criteria catalog is implemented using the relational database management system (RDBMS) SYBASE. The data itself resides in its original form in the tape archival system, DMS, of the PDSF. The project used commercial and in-house products and integrated them into an archival system that provides a fast and efficient way to access the run file archive. In order to ensure continuous operations even for several consecutive days of data recording, the software is designed to need as little attention as possible and is able to restart itself, in most cases.

2. System Architecture

The Texas Test Rig (TTR) archival system manages the run file archive. It is implemented in a client server environment. The archival system reads and analyzes the run files from the disk of the DAQ system to build the criteria catalog. The run files are then moved from the DAQ disk to the DMS of PDSF. Users select files, using the Motif-based graphical user interface to the database, and request associated runs to be restored from tape. Figure 1 gives an overview of the hardware and software components of the archival system. Catalog data is transferred, from the DAQ System to the DBMS, and retrieved by a user on the physics workstation. The files containing the run data are moved, from the DAQ system to DMS, and loaded to a physics workstation, on request.

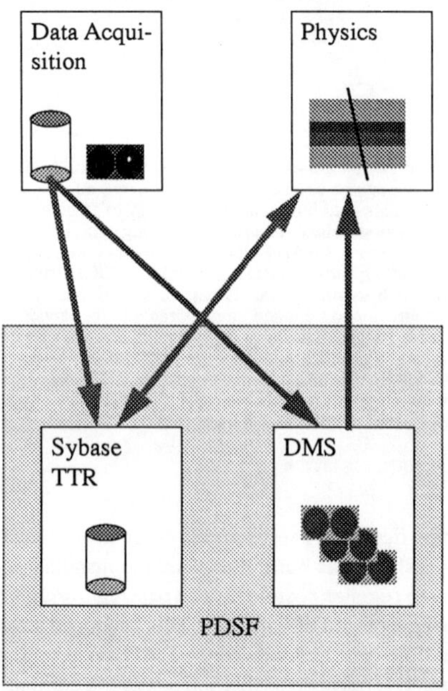

Figure 1: System Overview of the TTR Archival and Catalog System

The TTR software system is implemented in a client-server fashion. Different machines execute parts of the overall task. Authorized users can access the catalog and the DMS from any PDSF workstation. The database management system is installed on one of the workstations of the PDSF. DMS uses the batch ranches of PDSF and drives the two tape robots. Each robot contains a carousel which can hold up to 54 8mm ANSI-labelled tapes. Part of the DMS software is a SYBASE database application.

The TTR System is a complex network of machines and software which will usually fail if one of the components does not function properly. The SYBASE software for the TTR is installed on participating physics workstations. The database server, the database application programs and query tools are installed on the TTR database server machine, as well as on the physics workstations. It has to be ensured that the DAQ system is not affected by these off-line activities. It will continue to store recorded data as long as sufficient space is available. This guarantees that the events can be recorded, even if the long term archival is not available.

3. TTR Catalog Database

For the TTR, we used SYBASE a relational, client-server based database management system. The TTR database management system provides concurrent access to the stored data through a standard interface. The TTR software was implemented using the C Library interface of SYBASE. The DBMS allows storage of data in a central place, so applications can access it from any workstation that has a network connection to the database machine and the appropriate permissions. Consistency and safety of the stored data are basic features of database management systems. To be able to build an efficient relational database, certain rules have to be applied. These rules determine the way data is stored during the design of the database. The purpose of the system is to select run files based on the technology used during the run, environmental conditions, gas mixture used for the chambers and other technical information. The graphical database design tool ERdraw[1] and the associated schema translation tool SDT[2] were used for database design and implementation. ERdraw is based on an extended entity relationship model and generates SYBASE SQL code for database creation and consistency control. It proved to be a very useful tool for SYBASE database development. Figure 2 shows an overview of the TTR database design using ERdraw notation.

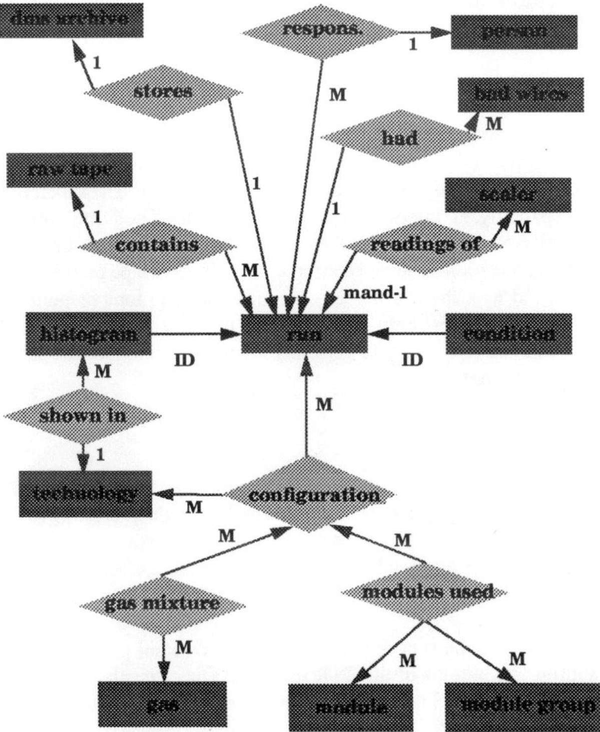

Figure 2: TTR Database Design (ERDraw notation) [1]

A graphical user interface was built for database access. It ties predefined database queries, programmed using SQL, to buttons and menus in the window. Physicists can build queries by the "point and click" method and review the results on-line. The query results can be saved in files for later reference. Additionally selected run file identifications are used directly to invoke other tools supported by the TTR window environment. Generated histograms are also managed and stored on PDSF disks by the TTR software. This allows physicists to view results immediately after selecting the interesting runs without having to restore any file from the archive.

4. Integration with Data Acquisition

The database applications are integrated with the DAQ only through the data itself. At the end of a run, the file containing the events is written to a specific directory on one of the disks of the DAQ system. The archival software periodically checks this directory, analyzes the files for catalog data and sends it to DMS for long term storage. The file is then removed from the DAQ disk. The run files contain "special events" and "special comments" which, together with other information, is used to build the database catalog. This causes the archival software to be sensitive to changes in the run file structure. The mechanism used for periodic checking is a UNIX "cron" table invoking a "fresh" copy of the program every time. This enables the software to recover automatically from most system problems.

5. DMS

DMS, the Data Management System of the PDSF[3,4,5], provides tools to support users in the off-line tape storage of files. DMS shields users from having to know anything about the particular storage medium. The storage is addressed like a standard UNIX file. The actual transfer of the data to tape is done transparently by internal mechanisms of DMS. The DMS software uses the four batch ranches, currently part of PDSF. DMS also provides a catalog about files and storage locations via a SYBASE application. The tapes are managed by two tape carousels with the least recently used tapes moved to a shelf. This can sometimes lead to long response times. The TTR applications follow a policy of restoring the second file requested while the first one is being processed. The transfer of a file to/from DMS currently involves an intermediate staging of the file on one of the PDSF batch ranches.

6. Graphical User Interface

The TTR catalog provides on-line information about runs and associated technologies. Users select runs using a MOTIF-based graphical interface to determine interesting run files before they request them from the DMS. The graphical user interface provides the user with a set of predefined buttons and menus to compose a query by the "point and click" method. It retrieves the data from the database using SQL, the standard query language for RDBMS, and displays the results on the screen. Users can then detail the query or use the resulting file name to retrieve the run from the tape storage. The run file can be sent to the event display or other physics analysis programs by clicking on the appropriate buttons of the TTR interface. Query results can also be saved in files for future reference. The graphical user interface also allows changes to the database, provided the user has the necessary permissions.

The TTR GUI currently supports a predefined set of queries. The system is designed to extend this set of queries as required. Any query to the database can also be generated using the SYBASE toolset.

Figure 3 shows a snapshot of a session using the TTR GUI.

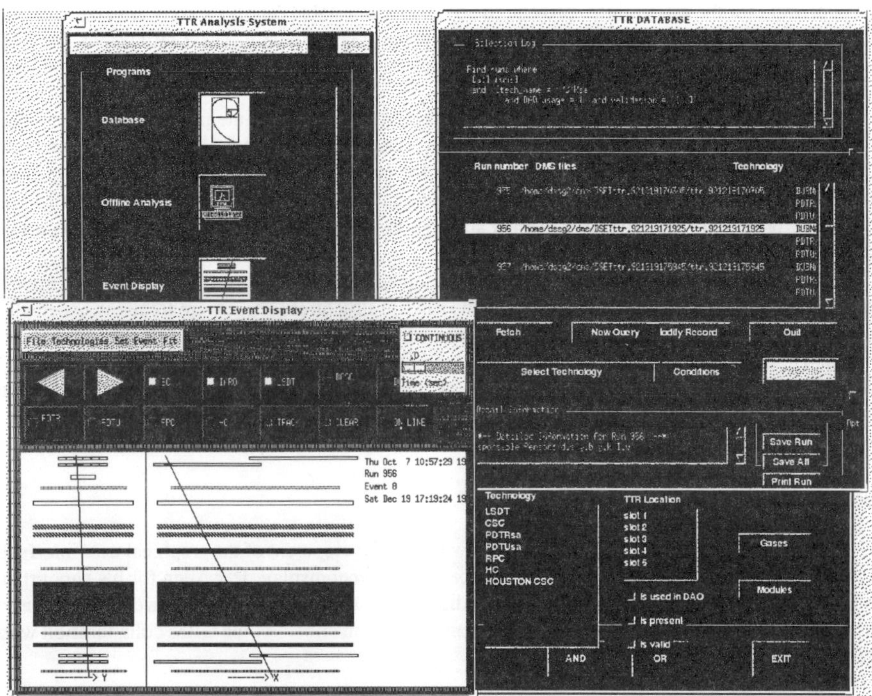

Figure 3: Graphical User Interface

7. Integration with Physics Analysis Programs

The graphical user interface integrates the database query application with physics analysis programs. It is possible to determine the name of a file, which contains interesting events, by applying the query mechanisms. The user can then immediately apply the result of the database query, - usually a run file name - to retrieve this file from DMS. It can then be input into other physics analysis programs or displayed directly with the on-line event display.

The results of run file analysis frequently are kept in histograms files, which are also archived in PDSF and entered into the TTR catalog for future reference. This allows users to reference the associated histograms immediately after selecting a run.

8. Status and Conclusions

Between October 1992 and March 1993 approximately 560 runs were recorded and stored in DMS using the TTR catalog software. DMS tapes currently store 7 Giga Byte of run files, more than 12 million events. Run file sizes vary between a few Kilo Byte and 50 MB. Sybase and DMS have proven, for the scope of this project, to be reliable and stable. The TTR software and database maintenance is fully automated and needs attention only occasionally. The TTR system, for long term storage and selective retrieval for run files provides tool to select interesting files without having to retrieve and analyze all of them.

The database catalog is generated automatically. This requires the presence of the special events and special comments in the run file. The Entry was encouraged but not enforced. This caused one third of run files not to have any catalog entry besides their name.The lesson learned here is that as much as possible has to be done automatically. If someone has to remember to add certain information to run files and to enter some keystrokes to start a catalog entry, at least part of the files will not use this feature.

The interface between the DAQ and the catalog application is the run file itself, which makes the catalog builder dependent on changes in the run file structures. These were frequent, especially in the beginning of the project. The dependencies between run file and catalog software are not obvious. The advantage of this approach is that the DAQ system can run completely independent from the database system. On the other hand, the integration of the DAQ software and the catalog builder would make things faster and less vulnerable to changes in the run file. Future projects may be able to solve this trade off.

9. Acknowledgments

The author would like to thank Ivan Chow for his development of the graphical user interface for the TTR software package, Luis Villasenor for the event display, Jeff Allen for his support as a DMS expert and George Yost for the many discussions and suggestions. She would also like to acknowledge the contributions from our students and the members of the GEM collaboration who participated in this project.

10. References

1. E. Szeto, V. Markowitz, *"ERDRAW A Graphical Schema Specification Tool"*, Lawrence Berkeley Laboratory 1990
2. V. Markowitz, Weiping Fang, *"SDT A Database Schema Design and Translation Tool"*, Lawrence Berkeley Laboratory 1990
3. Jeffrey Allen, *"SSCL-PDSF Data Management System"*, presented at the CHEP 92
4. Jeffrey Allen, *"SSCANSI, SSC Laboratory's Implementation of ANSI Labelled Tapes"*, 1992
5. Ch. Chang, Cathy Nelson, *"Computer System Operator's Manual for the Data Management System (DMS)"*, SSCL Documentation, September 1993

COMPARATIVE PERFORMANCE MEASURES OF RELATIONAL AND OBJECT-ORIENTED DATABASES USING HIGH ENERGY PHYSICS DATA

J.Marstaller
Superconducting Super Collider
2550 Beckleymead Avenue
Dallas, TX 75237-3997

ABSTRACT

The major experiments at the SSC are expected to produce up to 1 Petabyte of data per year. The use of database techniques can significantly reduce the time it takes to access data. The goal of this project was to test which underlying data model, the relational or the object-oriented, would be better suited for archival and accessing high energy data. We describe the relational and the object-oriented data model and their implementation in commercial database management systems. To determine scalability we tested both implementations for 10MB and 100MB databases using storage and timing criteria.

1. Introduction

Historically, analysis of High Energy Physics (HEP) event data from detectors has been done using a serial, tape-based approach. Database computing offers several advantages to the traditional approach in physics analysis: 1) It may not be necessary to read an entire event into the computer, when only a few quantities are needed for a given analysis, thus reducing the I/O factor considerably. 2) Databases allow simultaneous access of the data. 3) Analysis of the data is immediate and interactive. 4) Data can be selected via simple queries using the standard query languages.

The basis for the project described here is within the framework of the Petabyte Access and Storage project (PASS)[1]. The extent to which HEP can benefit from database technology is to be evaluated. A HEP database, using existing data from the Collider Detector at Fermilab (CDF), was designed to make initial performance measurements of a relational database management system (RDBMS) versus an object-oriented database management system (OODBMS). Approximately 1GB of CDF data was divided into 100 datasets to be stored on disk. A database requires the data stored to be put into some kind of underlying data model, which provides in turn some form of stable structures. The goal is to determine the data model that produces better results.

2. Conceptual Design

When designing a database, the first step is to collect user requirements. These requirements can be put into a high-level construct that provides a preliminary view of how the user perceives the database. This high-level model, the conceptual model, describes how the user, i.e. a physicist, conceives the domain area (HEP data). A semantic data model, the entity-relationship(ER) model, was chosen to illustrate the conceptual design. An entity-relationship diagram is a graphical technique that depicts the enterprise being modeled as a collection of sets, relationships between the sets, and attributes associated with both entity and relationship sets. The most common extension to the ER diagram includes the abstraction concepts of aggregation and generalization. Aggregation abstracts a collection in such a way that objects (entities) are viewed as a single higher-level object. Generalization is a form of abstraction which captures the commonalties of objects (entities), while at the same time ignoring the differences. The extended entity-relationship model, (EER) model, is fairly simple to construct and has a reasonably unique interpretation. This model produced a diagram (Figure 1) which served as a means of communication among users and designers. The conceptual design ensured that no loss of information occurred, i.e. all entities mentioned in the data analysis tool[2] were captured.

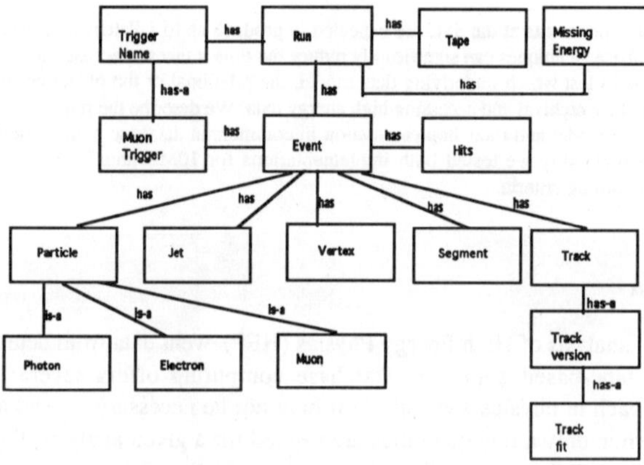

Figure 1. Conceptual Design

3. Relational Design

The next step in designing the data model is to create a logical data model. The logical model has constructs that are easy for the users to follow and yet still avoids physical details of implementation. It results, however, in a direct computer implementation, in this case in an implementation in the relational database management system.

The EER model also depicts the logical data model. Figure 2 shows the partial model. The conceptual data model was used as a means to produce a logical data model specific to the relational and the object-oriented implementation. The nature of the objects have an inherent hierarchical structure based upon level of details. To store this data in a traditional relational database, links need to be established between the objects to represent this hierarchy. Each run and event is given a unique identifier. Each child relation (such as "Tack", "Vertex", etc.) is augmented to include the primary key of its parent. In other words, any relation in the hierarchy must include the primary-key of its parent relation[3]. This compares to data from computer-aided design application. This hierarchy was modeled using weak entities. Weak entities depend on the existence of other entities for their identification[4].The CASE tool we used, ERDRAW[5], uses non-directed edges, labeled ID, to show the dependency. The entity "Track" was modeled to reflect the CDF data. The entity "Track" is the generalization of the two homogenous entities[6] "Defaulttrack" and "Otherversiontrack". The homogenous entities "Defaulttrack" and "Otherversiontrack" only contain the common attributes, whereas the generalization "Track" has additional attributes. ERDRAW provides the directed edges labeled "ISA" for illustration.

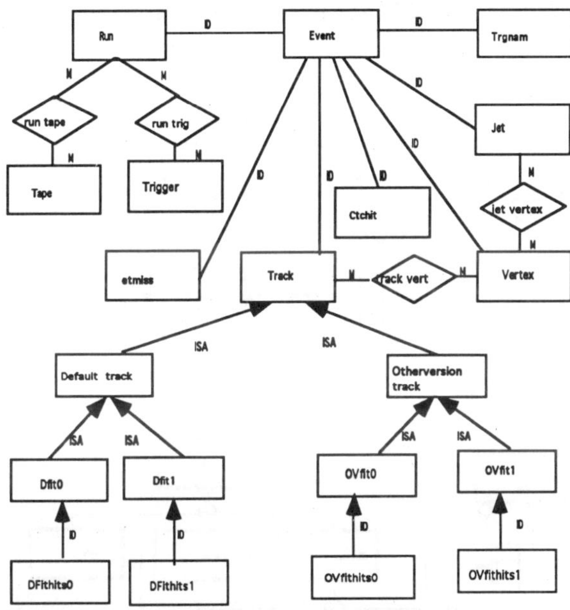

Figure 2. Part One of the logical design for RDBMS using ERDRAW notation.

4. Object-Oriented Design

The ultimate goal of the object-oriented data model design is to produce a system that consists of a collection of objects. The focal point of the design task is to determine and specify those classes that govern the structure and functionality of these objects. As in the relational design method, the key is to identify the abstractions that occur in the problem domain.

Objects are concepts, abstractions, or things with crisp boundaries within the problem domain (HEP). An object class describes a group of objects with similar properties (attributes), common behavior (methods), common relationships to other objects and common semantics. A class can be viewed as a template, which specifies the intended purpose of the objects or as an abstraction mechanism. Once the object classes were established, we identified and recognized relationships that exist between the classes within the application domain (HEP). The main goal is to recognize patterns for aggregation, generalization, specialization or inheritance on both the object and the class level. This object-oriented logical data model was produced using OMTool[7]. OMTool follows the basic conventions for object model notation as described in Rumbaugh[8]. Figure 3 shows the logical design for the OODBMS. The class "Event" is the top of the hierarchy and depending on "Event" the corresponding instansciations of the other classes can be found. "Hit", "Track", etc, have a relationship to "Event" so each of these classes is-part-of an "Event". The class "Particle" is modeled as an abstract class (one that has no direct instanciations). The subclasses of "Particle", the different types of particles, inherit all attributes from their superclass.

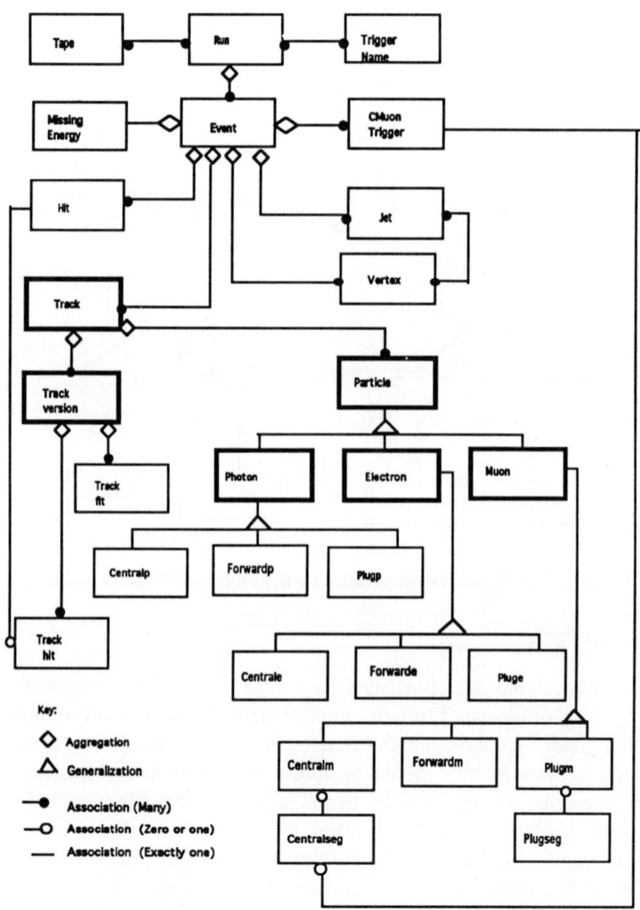

Figure 3. Logical Design for the OODBMS using OMTool notation.

5. Testing Criteria

For the purpose of this project, we wanted to measure how well the underlying data model reflects and enhances the performance of the system. Our test measures response time, i.e., the real time elapsed, from the point a program calls the database system with a particular query, until the results of the query, if any, have been placed into the program's variables.

There are two broad categories of performance measures: *time* and *storage*. We wanted to determine the storage efficiency of the database management systems; i.e. determine what fraction of a database is actual source data and what is database management overhead. We had a 10MB and a 100MB database and checked the size of the database dumpfiles on disk.

We looked at only 3 time measurements: 1) the time it takes to create a database, 2) time it takes to archive a database and 3) the time it takes to complete the queries. All tests are run on disk resident databases.

6. Results

The goal of our test was to model a specific environment, the high energy physics analysis environment, and the aspects of how the two database systems are performing in this environment; therefore, a lot of time was spent in modeling the HEP data and reflecting this environment in the relational and object-oriented data model. All measurements are summarized in Figure 4.

	RDBMS		OODBMS	
	10MB	**100MB**	**10MB**	**100MB**
Storage:				
	8.4MB	**84MB**	**6.4MB**	**64MB**
Time:				
Populate	**3:47.0**	**28:18.0**	**2:54.5**	**32:18.9**
Dump db	**7.2**	**1:37.6**	**22.7**	**4:12.1**
Load db	**27.8**	**5:33.4**	**22.7**	**4:12.1**
Query:vfit	**24.4**	**36.0**	**15.5**	**44.9**

Figure 4. Time Measurements in Seconds.

Storage:

To measure the overhead of each database management system, we determined the size of the files to be archived. The relational database dumpfile is larger because it stores a self-contained database that includes all of its pertaining information. The object-oriented database schema is not part of the size of the database file size recorded. For both database systems, the schema remains static throughout analysis. The schema must, however, exist prior to loading a database file into an existing healthy database. For the OODBMS, the schema files are of insignificant size, so we did not archive them with the database file. The OODBMS database file is approximately 25% smaller than the RDBMS database file.

Time:

1. The first time measurent determined how long it takes to create and fill (populate) a database for both database systems and both sizes (10MB and 100MB). The RDBMS showed much more consistent results in creating and loading both database sizes. When loading a relational database, only the actual data is loaded, whereas when loading an object-oriented database the pointers, or the paths, have to be loaded also. This reflects the time increase for the OODBMS. For 10MB the OODBMS performed approximately 25% better than the RDBMS, however, for 100MB the performance deteriorates and becomes approximately 12% worse than the RDBMS.

2. Archiving the database files was the next time measurement recorded. For both DBMS we used vendor-provided utilities to archive the datafiles. Archiving included loading a database file from disk and dumping a database file for storage. Both measurements are important when 1GB of data is divided into separate datasets so that archiving can eventually be done in a parallel distributed fashion. The RDBMS performed consistently better than the OODBMS. To retrieve a database file, the RDBMS requires approximately 18% less time for the 10MB and 24% less for 100MB. To store 10MB for the RDBMS it requires approximately 31% and for 100MB 38% less time than for the OODBMS.

3. For the query used as a test measurent, we selected a "track refitting" query. There are other physics query results[8], but the "refitting" query reflects a standard traversal over three large tables, five levels deep. The RDBMS scales up better, without optimization from the designer (25%). In the OODBMS, where the user has to take care of all query optimization, using such features as "path of", which will store a predefined indexed path for a query, the performance can be increased dramatically by implementing such changes.

Both relational and object-oriented databases have their own inherent problems for the physics research community. Relational databases provide a simple way of extracting data with SQL, but all analyses must be done outside the database. In other words, for the algorithms there still needs to be a layer of application programs on top of the database. SQL does not have sufficient capabilities, such as recursion. This causes problems, such as the cost of writing application programs, the granularity of the data present in the programs and the impedance mismatch, however, this approach seems to be rather attractive to the physicists since most of the algorithms are already implemented in FORTRAN. Object-oriented databases, on the other hand, are less mature than relational systems and are more difficult to fine tune for performance.

The grand picture is, of course, to develop a system that is very flexible, modular, scalable and distributed. It should, essentially, take data and process it into information, preferably visual information, so that the user can extract facts and make meaningful conclusions from that. . Use of databases is essential for these projects, however, the design of such a database is not yet fully understood. The next-generation database application has little in common with current business data processing. Since a need for much larger, more complex datasets exists, new capabilities in areas such as type extensions, complex objects, rule processing and archival storage are required.

This project was not designed to be "another" benchmark in the area of database comparison, therefore, traditional database benchmarks, such as "insert" were not measured. We wanted to show the performance characteristics of the database management systems in the application domain, i.e. high energy physics applications. The benchmarks designed for generic use and overall performance were applicable, but less desirable than designing specific tests.

7. Acknowledgments

This paper has been produced within the framework of the PASS project [1] at the SSCL. The author has participated in this project, designing and implementing the databases and creating the performance measures. She would like to thank Alain Gauthier for developing the "track refitting" query, incorporating existing FORTRAN code. She would like to express special thanks to Ed May and Ute Nixdorf from the SSCL who were crucial to the success of this exercise.

8. References

1. E. May, D. Lifka,R.Lusk, L.Price, D.Baden, R. Grossman, C.T.Day, S. Loken, J.F. Mcfarlane,L.Cormell, P.Leibold, D.Liu, M.Marquez, U.Nixdorf, B.Scipioni, T.Song,*The PASS Project: Database Computing for the SSC,* Proposal to the DOE HPCC Initiative, Argonne National Lab, 1991, unpublished
2. J. Marstaller, *Comparative Performance Measures of Relational and Object-Oriented Databases for High Energy Physics Data,* Baylor University, 1993, unpublished.
3. Codd, E.F. *Extending the Database Relational Model to Capture More Meaning* ACM TODS (December) 1979.
4. Markowitz, V.M.and Shoshanie, A, *Representing Extended Entity-Relationship Structures in Relational Databases: A Modular Approach,* ACM Transaction on DatabaseSystems 1992.
5. Szeto, E, V.M. Markowitz, *ERDRAW A Graphical Schema Specifcation Tool,* Lawrence Berkley Laboratory, 1990..
6. Markowitz, V.M.and Shoshanie, A, *Representing Extended Entity-Relationship Structures in Relational Databases: A Modular Approach,* ACM Transaction on Database Systems 1992.
7. General Electric Company, *OMTool. User Guide for Sun Microsystem Workstations.*, Pennsylvania, General Electric Company, 1992.
8. Rumbaugh, J. et al, *Object-Oriented Modeling and Design*, New Jersey, Prentice Hall, 1991.

SOFTWARE AND COMPUTING IN THE OPAL EXPERIMENT AT LEP

STEPHEN W. O'NEALE
School of Physics and Space Research, University of Birmingham
Birmingham B15 2TT, United Kingdom

ABSTRACT

The OPAL experiment has overcome the complexity of software development at LEP through a well disciplined organisation of its software and data management. To assist in the explotation of new techniques we give a frank description of the problems we have faced and some comments on the solutions we employed.

1. Introduction

The LEP experiments have all been a great success and OPAL's record of physics output is second to none. However, for this workshop, I give as some details of the problems encountered as well as the successes[1]. First I present the software development in OPAL and then I give a review of data management techniques where the LEP experiments needs were ahead of the enabling technologies. In the fourth section I describe the user services provided to OPAL physicists. The evolution of the production system for reconstruction of OPAL data, the Monte Carlo simulation service and data import and export facilities are covered in section five. The use of a central data repository, primarily for DST analysis – the SHIFT project is described in section six. The manpower and resources needed to provide these facilities are reviewed in the seventh section and then conclusions are presented.

2. Programs Development

When I came to CERN, to coordinate the OPAL offline software, I found CERN expected to contribute of 3 (or at most 6) CERN units (12 Mips) of "mainframe" cpu power against an estimated need of 15 units. It seemed likely that we would need to exploit the cpu power within the online system and export data to several institutes for even the first pass of reconstruction and analysis.

Early experience showed moving code from say a VAX to another machine was sufficient to expose errors. The ZBOOK memory manager returned control to the user with a return code set (but rarely checked) when an error occurred.

These two experiences lead me to steer OPAL towards strict adherence to the Fortran standard, with an extension to strong typing, so that programs could, if neccessary, be run on online and offline systems throughout the collaboration. A clean, modular, separation of machine dependent code for file handling and error detection and recovery was essential. Data formats were chosen to be machine independent (free with Zebra) and a need for efficient data access on many media and for a data management system which could run anywhere in the collaboration was clear. I was also very positive towards establishing OPAL's production code on new systems when joint projects were offered by vendors donating hardware and technical expertise.

A few OPAL people attended software engineering courses in SASD. These were physicist-programmers who were the analysts and designers for the needs of their sub-detector, and were responsible for the implementation and testing of the code and at the same time the customers who needed the code. In HEP this whole range of activities is often called programming but care should be taken when talking to a CASE guru, as each role above is considered to be separate activity in an a lifecycle model. The SASD data flow diagrams (maintained by the PROSA package) proved to be a valuable aid in analysis meetings but the general application of the methodology was inhibited by the poor performance of the support tools and the consequent lack of funding to continue the training.

The general framework for OPAL's (and any other particle physics experiment's) code provides facilities for memory management, for I/O of event data, control information and databases for calibration and book-keeping; for monitoring of cpu, memory and I/O resources, errors detected by runtime libraries and reported by user code; for debugging ranging from simply printing extra information to memory shaking of check-summed data structures at each data access. The user can run physics four-vector generation, detailed or partly parameterised simulation (GOPAL) to produce raw-data, reconstruction (ROPE) of simulated or real raw-data to produce dsts, fast simulation from four-vector level to dst, and dst analysis to produce histograms, ntuples and other mini-dsts of selected events. This may be done starting with a random number going to final analysis with each event or by reading and writing event streams at any intermediate level. Interactive graphics interfaces are provided for the simulation (GEANX) reconstruction (GROPE) and analysis (PAW) stages.

Within this framework OPAL "detector" code is modular, this is good coding practise which has been followed in HEP for many years. It is also facilitates the development of code in many remote institutes in a structured (top-down) style. Data from each detector is processed and reduced as much as is reasonably possible before the event is built in the DAQ system. In the "offline" code the data from detectors are processed and combined so that, for example, the muon endcap and muon barrel build a structure of muon segments. Similarly tracks and clusters are merged for central tracking, electromagnetic calorimetry and hadronic calorimetry followed by vertex fitting, extrapolation of tracks to outer detectors and particle identification. The results are presented to the physicist in a DST. The information retained and easily available on DST depends on the understanding of the detector and the need to consider alternative interpretations of detector response.

Each subdetector or major analysis process (*xx*) has its own file, organized in a standard way, for its production, analysis, and interactive code and documentation of the process and its data structures. The high-level routines are defined so that the GOPAL and ROPE steerings may call *zz*INIT, *zz*SETR, *zz*IKU, *zz*CALR, *zz*GEOM, *zz*HITS, *zz*DIGI, *zz*BAT, *zz*DROP, *zz*FIN, and so on, for each process *zz*. The ZEBRA data structures are addressed via links and offsets (Fortran parameters) and each major process has a link in the high-level tree or fanout banks from which its data structures are supported.

PATCHY is used for code management, one of the few formal requests I received as offline co-ordinator was to guarantee that no OPAL institute should have to pay money to get access to the code they had developed (i.e. not use Historian). Each ROPE processor, such as CE311 for track extrapolation, has a version number, its own cradle, pam file and library. There is rarely more than one person contributing to the development of one pam file and when there is, we prefer human control to tools for concurrent potentially inconsistent updates from several authors. The pam file for each library also contains a LaTeX description of the processor, its input and output and the parameters which may be changed by the user and a DZDOC description of the banks. The writeups are extracted and made available on paper and in WWW. The target audience is the physicist whose analysis needs more than blind trust in the contents of databanks on the DST. There are some facilities to write coding stubs from the bank descriptions but their use is not compulsory and not easily integrated with code management tools; the accuracy of the documentation therefore relies on verification processes.

The production PRO version of a library may be modified only via cradle corrections but a new public development pam DEV may be introduced and later modified with the author's favourite editor. This involves some loss of the history of modifications but avoids the complexity of applying corrections to corrections to generate new releases and in the Patchy framework allows the new code to be editted directly. Reference versions of production and development libraries are maintained on many platforms and are updated automatically in any institute subscribing to the distribution systems. Modules are built for production running to ensure a uniform treatment of large samples of data. They are not changed when small corrections are made to libraries.

OPAL has made great use of and many contributions to the packages in the CERN Program Library. Such scientific program libraries are a fine example of software re-use – one of today's fashions. Many of the speakers in this workshop have told us that their project follows the OO paradigm and may be re-used; none have reported using any existing class libraries.

A milestone for OPAL set by the LEPC requested a study of the hadron calorimeter which needed a simulation of several components of the detector. This gave some leverage to request that manpower was directed to the production of a reliable detailed montecarlo package for the whole OPAL detector, rather than approximate stand-alone simulations which would have no lasting value.

I found it difficult to get information on support structures as the group constructing a detector subsystem were primarily interested in the response of the active components of their detector, and in the scattering properties of the support structures between them and the interaction region. A tendency to run simulation development with only those parts of OPAL which were needed for code development (possibly because the initialisation required a few minutes on the computers of the day), or to test with a limited range of particle types, resulted in code updates being submitted which failed when running with the full, standard OPAL setup.

The simulation code was checked against test-beam studies of detector components or prototypes. An exercise to analyse simulated montecarlo data samples of the first running of LEP started in 1988 gave a deadline for offline code development. A pilot run of LEP in 1989 gave useful experience of running the detector and the whole analysis chain.

There is now an annual cycle in OPAL activities. From December to April upgrades may be made to the detector and data handling systems and corresponding code is produced. New algorithms and calibrations may become available for recorded data. Data taking starts in May and the first few weeks of data are used for alignment and calibration measurements. After a few months a second pass at the reconstruction of recorded data is prepared and the entire sample is re-processed. The montecarlo program is modified to handle new detector configurations and tuned to the resolutions achieved by the re-processing. The production of standard montecarlo samples begins, typically in mid-October, and exciting new physics results are presented at the winter conferences.

It is difficult to know if reconstruction programs are producing the correct answer (althought sometimes it is easy to see they are wrong). Formal verification procedures and code generators are not sufficiently mature to be of any value and are, of course, limited to demonstrating the software is built to the physicist's specification - if he has confused a non-linear effect in the drift velocity with a cross-talk effect on his pedestals then no CASE tool or expert system on the market today is going to help. Attempts to (use manpower to) improve the understanding and precision of the detector are moderated or motivated by a corresponding need in physics analysis. Testing procedures include the use of an interactive graphic event viewing system to look at the results of the code and a comparison of results with expected distributions and resolutions. Data sets are prepared to verify new versions of code and improvement in algorithms. Muon pairs and tau pairs are easy to understand and the D^* signal is a useful benchmark channel in multihadron events. A file of all events which failed to pass through the standard online reconstruction program in production are kept and passed through all new versions.

3. Data Management

The ZEBRA memory management package was developed for use in the LEP era and was introduced into GEANT version 3. It seemed a natural choice to use the same package for the other production programs of a CERN based experiment. The ZEBRA documentation system was written as an OPAL facility. The bank descriptions may be viewed in the OPAL WWW files and are distributed on paper at the release of a new production version of the code. ZEBRA provides machine independent sequential and direct access I/O facilities and many debugging and error handling facilities. The OPCAL package is used to maintain and distribute OPAL's hierarchical RZ calibration database. It has been adopted by other experiments and incorporated into HEPDB.

Production control packages on the central mainframe were written in REXX. Commercial database products (ORACLE) were found to be unsuitable for use in a central role. Run control information from the experiment were logged in ORACLE tables on VAXEN and APOLLO machines at the experiment. These tables can be interrogated from CERNVM and UNIX systems and are accessible to OPAL physicists. The results of the most common enquiries, (a status report on each run) are maintained independently to maintain the response of the machine updating the database. There is some use of spreadsheet packages. A nanodst is written at the filter stage in the online system and transmitted to CERNVM. This may be compared with the event header records on SHIFT to verify the integrity of the data handling chain.

I presented OPAL's ideas for file and tape management to a MEDDLE meeting in December 1988, in particular generic names to access data, direct access to physical records on all media and job submission facilities working across many platforms. It became clear that many of the tools we needed would not be ready for the start of data taking. The FATMEN task force was setup early in 1989 and developed a layered model. All four LEP experiments contributed manpower for the implementation of the upper layers with co-ordination from CN division – I thought this was a window of opportunity to exploit commercial software engineering techniques which OPAL had abandoned some years early but there was no enthusiam for this from any of the other members of the group. Many parts of production control systems were implemented privately and required modification as facilities like TMS, SYSREQ, SMCF and STAGE evolved. There was also, later, considerable resistance introducing the FATMEN system into production code as we already had something which worked and somewhat less than full co-operation from the systems side of many of the computing facilities used at CERN.

This is clearly an area where LHC might do better, perhaps at LEP we were too concerned with cpu power and started too late, but it should be noted that many of the packages eventually used were not available very much earlier.

4. User Services

The VM/CMS system was introduced on the CERN central computers while the LEP experiments were setting up and many contributions were made by OPAL physicists with several years use of this system elsewhere. A move towards personal workstation systems came two years later. The same work environment at CERN and in the home laboratory was generally consider more important than using the machine with the cheapest cpu power or CERN support for system management. Most OPAL physicists now have access to a workstation, there are seventy nodes in the Vax cluster and sixty in the OPAL unix systems at CERN. Terminal access to CERNVM computer is still significant.

CERNVM is the primary system for OPAL's communication via electronic mail, news, and online help information, as well as holding the reference version of code and calibration, and file and tape management databases. Papers, in LaTeX, and PostScript figures may be dispatched to each OPAL institute for review before publication or acceptance as an internal physics note. WWW is reported to be very popular and some OPAL documents have been entered with a LaTeX to HTML converter. I suggest that the discipline and tools to enter information (and the ability to get printed copy) are as important as the quality of the browser tools. Packages which require the user to search out the information he needs, rather than having it drawn to his attention (via mail or news) have not so far been a widespread success in HEP. A conferencing system and a documentation database with search facilities for five thousand OPAL documents are now rarely used.

5. Production Computing Resources

We define production computing to be the reconstruction of events recorded by the OPAL detector, the generation of corresponding montecarlo samples and the distribution of these samples within the collaboration. In 1989 data cartridges were transported from the experimental pit for the first, and subsequent, passes of ROPE on central computers. Processing yielded a six to one reduction in data storage but despite the use of multiple files on cartridges we were often waiting for mounts. An urgent request for a buffer disk space of 500 Mb made in September was not answered until Easter 1990.

The installed mainframe computing capacity was saturated by early 1990 and we were saved by the loan of an Apollo DN10K from the Microcosm project. Cartridges were read and written with STK4280 drives attached to a spare VMEbus-based OS-9 system. The emulator buffer manager was used to pipe data in and out of the DN10K. Shifts of physicists mounted cartridges (one every 10 minutes) with an efficiency equivalent to the loss of half of the processing capacity of the system. A data compression package for raw data (DD) was written during 1990.

In 1991 new collaborators from West Canada joined OPAL and undertook the installation and operation of a ROPE farm at the pit. This facility now provides monitoring, calibration and DST production services for online data taking and a re-processing of accumulated data. Compressed raw data is stored on an optical disk library and output dsts are read across a network to be available on shift.

OPAL physicists write four-vectors and receive DSTs from the Monte Carlo production service. The system is controlled and managed on CERNVM, files are cataloged in FATMEN and information on each run is in the XFIND system (and is being ported to WWW). Event simulation was setup to absorb capacity on IBM mainframe machines and was OPAL's only use of central CRAY supercomputers. The HOPE project demonstrated this work can be performed in a straightforward way on workstations and lead on to the CERN simulation facility (CSF). It was first proposed that LEP experiments could purchase a share of CSF when HP offered a project to use clusters of 9000/700 series workstations for compute intensive work at one per cent of the cost of the Cray (assuming no benefit from vectorisation). A new OPAL group (UCR) was prepared to take this responsibility and later setup an independent facility (RSF) at CERN when CSF received central funding. The UCR group also provide system support for hp-ux workstations used by physicists at CERN. At times of peak demand we also use ODF: running many short processes (3 hrs each) via remote shell on physicist's workstations and collecting event files on a large work disk. The files are combined and written to cartridge re-using the HOPE scheme. The physicist donates about 50Mb of disk space and is able to signal the process to stop and start if it interferes with his use of the workstation. This simple but effective scheme was setup in a few weeks while production was running and its general acceptance encourages us to persue more general packages like Condor or LOADLEVELER from IBM when they are available on HP-UX.

The import and export of data is managed on CERNVM with FATMEN as a catalog and TMS used to manage the media. Data may be copied to EXABYTE 8200 or 8500 cassettes, IBM 3480 cartridges and 60m or 90m DAT cassettes and exported. DAT writing is done on HP workstations using CLIO to copy from CERNVM disks. Many physicists in OPAL institutes analyse real data on SHIFT and retrieve n-tuples over networks for local manipulation with PAW. There is a growing demand to export Monte Carlo datasets and selected OPAL data as local facilities become more powerful and the volume of accumulated data grows.

6. DST Analysis and Data Access

The general features of OPAL data access were established before the start of data-taking[1]. OPAL uses (FATMEN) generic names so that the same data may be accessed by the same name on several systems and media. The ZEBRA FX format is used for almost all data and lists of pointers to give direct access to event records on disk and cartridge. DADlists, giving direct access to OPAL data on a disk via a list of events and file offsets, are now implemented via the CLibrary I/O in ZEBRA.

OPAL's share of central computers was hardly enough to cope with data collected in 1989 and MAW, a package for running parallel VMS batch jobs which read only the data on local low cost SCSI disks, made a significant contribution to the analysis power of OPAL. In November 1990, Indiana undertook a major upgrade of the OPAL DST farm. The HOPE project had been a major success with batch computing and led on to CSF for low I/O applications. CERN CN division proposed the SHIFT system of compute, disk and tape staging servers connected to a high speed network with an appropriate remote file access system. We considered simply expanding the Vax based farm (and convertion unix based workstations which were about a factor two cheaper) but the management of hundreds of workstations with local disks was a major disadvantage. The UltraNet gigabit network gave access to computer centre data-handling services and the throughput permitted greater flexibility in reading data. After benchmarking studies with our own code we pre-

[1] See, for example, CERN DD 88-1 or CERN Meddle meeting December 6, 1988

ferred Silicon Graphics multiprocessor machines for cpu and disk servers (to use both in the farm mode if UltraNet failed). SG had a working UltraNet interface and demonstrated they could reformat SCSI disks for use on VaxStations so we had a fall back to extending MAW. CN efforts were focussed on the needs of the dst access facility and OPAL developed its code, library and data management facilities on shift and job submission facilities from VMS, VM/CMS and UNIX systems. The pilot system was operational for the start of 1991 data-taking.

The choice of SG gave some problems as this was not an acceptable product to CERN's management - this was circumvented via Indiana University purchasing all of the computer systems. The acceptable products included the Apollo DN10K with cpu upgrade, DN10K VME-UltraNet Interface, SCSI disks attached to a DN10K by Rorke Data, UltraNet attached to an Ultrix machine via a VME-Turbo Channel Interface – none of which were ready on time and some were never made to work. I would like to make two general points from this example. Firstly, salesmen are often optimistic to the point of dishonesty, if you really need to install something ready for next years data taking then ask to see a working product. Secondly, HEP should be justified on its physics merit, not spin-off, and purchases should not be restricted to those assembled in member state countries when no really suitable product is available.

A central repository requires that storage space increases as the accumulated data sample grows, particularly as event data is often analysed using DADlists. As we could demonstrate that in 1992 cpu utilisation on shift was shared by many physicists widely distributed throughout member and non-member state institutes we were able to persuade our Financial Review Committee to let us buy some disks from the media budget. Unfortunately centralisation of computing activities at CERN is still regarded unfavourably by many funding bodies even though most of them are do not provide the manpower and facilities needed just to copy a cartridge for their domestic OPAL physicists and jobs can be submitted to SHIFT from institutes just as simply as submitting them from a computer on the CERN site.

In order to ensure the continued support for the project we were also persuaded to donate a significant part of the cpu resources for a pilot project to help Aleph get their system running on SHIFT and to reduce our use of CERNVM. The share of mainframe quality computing (i.e. excluding cheap, low I/O mc production) given to the LEP experiments in 92,93 is shown below. The cost is the lowest that could reasonable be used to keep the machine on the site and excludes all staffing and running costs. The savings which have been made thanks OPAL's efforts now amount to a few MSF per year, or rather physicists enjoy 4 to 5 times the analysis power they would have had without SHIFT. The SHIFT system was recently upgraded to a SG Challenge system for OPAL and Aleph and Delphi are now running SHIFT facilities funded through CN Division and L3 are next in line.

Cost of computing resources allocated to LEP Experiments from COCOTime.

	IBM VM/CMS		CRAY	VAX/VMS	
Cost kchf	6000.	6000.	2000.	1200.	1200.
chf/168hr	7.7	6.5	10.1	16.7	17.4

Experiment	VM92	VM93	CR92	VX92	VX93	Total kchf 92-93
ALEPH XU	542.	978.	747.	133.	139.	2540.
DELPHI XX	828.	1043.	121.	533.	330.	2857.
LEP3 XV	681.	685.	30.	50.	52.	1499.
OPAL WS	774.	652.	10.	33.	35.	1505.

As we record more events, direct-access data storage is becoming a limiting factor. We might expect to get five times the current accumulated data sample before abandoning the OPAL detector. So we need 5 times the current disk storage space for event data and a large staging space for DSTS from simulated events. We may tighten the physics cuts and remove the DD record from most events in the next ROPE pass since the understanding of the detector has improved. The use of classical data compression techniques gain a factor 2.7 on the DST at some cost in cpu time. A major use of CERNVM is for access to the cartridge library for data which are not on the SHIFT disks - primarily Monte Carlo samples - which approach twice the volume of the accumulated data from OPAL. This work can easily be transferred to an equivalent SHIFT facility, funded from the savings obtained by closing down CERNVM. The CN shift team are now concentrating on improving the reliability of cartridge reading and writing.

7. Manpower and Management

The responsibility for the offline code and calibration for twenty OPAL subdetector clearly lies with the institutes engaged in their construction and maintenance. Following this model the responsibilities for new facilities, the ROPE farm at the pit, the SHIFT DST analysis facility, and the UC Riverside simulation facility RSF are well defined. Twenty-four central offline tasks are reviewed annually. They include management of workstations, computers and networks, the coordination of the code for GOPAL, ROPE and, GROPE and for central processors like the generator library, particle identification, data compression, data base access, and the DST processor. Data-handling tasks include Monte Carlo and ROPE production and the import and export of data. Two full-time people are employed for workstation operations via service contracts and the collaboration provides two (half-time) people via a scheme of shifts, each session lasting for two months. For seven of the tasks the responsible person has changed annually. As OPAL has had some technically interesting projects we have, so far, been able to attract fellows and associates to take on these responsibilities. The success of OPAL analysis is largely due to the members of the offline team. Each person has taken well defined responsibilities and performed well. Their contribution is clearly visible to the entire collaboration.

A very simple view of most LEP experiments shows many institutes undertaking to design, build, run and maintain major detector components, provide the corresponding computer codes and to assist in their exploitation in physics analysis. The host laboratory promises to try to provide beam time as well as other involvement in the detector subsystems. The responsibility for the central offline tasks above is simply not defined, perhaps it was assumed to be the responsibility of CERN data handling divisions but the manpower was not formally requested and has not been provided from CERN staff. The funding for the project falls considerably when construction is complete but this is the time when resources for analysis are needed. OPAL were fortunate in attracting new members, with funds, to provide major components of our production and analysis facilities.

8. Conclusion

The success of OPAL's offline system is evident from our published physics results and demonstrates that it is possible to deal with the complexity of LEP and LHC analysis with discipline and good organisation. However, it would be foolish to ignore the advances of the CASE industry, so I have presented frankly many of the problems we have faced and overcome in OPAL. It seems to me that many wheels remain to be re-invented to exploit new technologies, for example error handling, machine independent I/O and RISC compiler performance on Fortran code. I hope to see some progress at the next conference in this series. The workstation systems developed in joint projects with OPAL may also be of lasting value to the HEP community. We recommend a central DST analysis facility or data repository to new experiments as, with current technologies and HEP budgets, it is much easier to install cpu power close to the data than it is to disperse bulk data throughout a collaboration. The planning, funding, and manpower for the analysis system should be given proper consideration at the proposal stage of an experiment.

In my presentation I closed with a reference to "the Man with the Golden Gun" who will assist the coordinater or class librarian for new experiments to follow the rules when persuasion fails. I was surprised to find afterwards that only one person present had read "No Silver Bullet" by Fred Brooks in IEEE computer, April 1987 or his book The Mythical Man Month. I can also recommend as Decline and Fall of the American Programmer by E Yourdon as a review of software engineering principle although it is dissappointing that he, also, can not find a single magic solution to our needs.

9. Reference

1. S.W. O'Neale, *The Offline Analysis System of the OPAL Experiment at LEP* in Computing in High Energy Physics '92 p.336 (CERN 92-07, Geneva 1992)

Simulation of Data-Acquisition and Trigger Systems in C++

J. C. VERMEULEN
NIKHEF-H, PO BOX 41882,
1009DB Amsterdam, The Netherlands

ABSTRACT

Modelling in C++ is discussed. Next is shown how support for discrete event simulation can be implemented in C++. A comparison with MODSIM II is made and the experience gained with applying the approach outlined in this paper is discussed.

1. Modelling

1.1. Objects

For simulation a model is needed of the system to be simulated. The latter will consist of interacting objects. Therefore object-oriented programming provides a natural way to construct software models. In the model all entities in the system to be simulated are represented by objects. In many cases modelling can be done on basis of a hierarchy of objects. For example for modelling a system consisting of CPU boards, each consisting of a CPU, a normal random access memory and FIFO memories for communication between the boards, one could construct an object modelling the CPU board from an object modelling the CPU, an object modelling the random access memory and objects modelling the FIFO's. Objects assembled from other objects, "component objects", are referred to as "aggregate objects"[1]. Component objects may in turn be aggregate objects. Non-aggregate objects are referred to as "basic objects".

1.2. Structure of basic objects

A basic object is modelled as a state machine. Events (cf. 2.1) concerning the object cause transitions in the state machine, may generate new events for the object or for other objects, and may lead to invocation of member functions of other objects. Each basic object should have a function for handling events (the "event handler"), that will be invoked using the information in the event data structure when the event is retrieved from the event list.

1.3. Instantiation of objects

Objects are coming into existence by instantiation. Instantiation of an aggregate object should automatically cause instantiation of the component objects. In the example of 1.1 instantiation of the object modelling the CPU board should lead to instantiation of the objects modelling the CPU, the random-access memory and the FIFO's. When an object is instantiated, its constructor function, if defined, is called. In the constructor other objects can be instantiated. This is the key for instantiating an aggregate object : in its constructor the component objects should be instantiated. In their constructors component objects should be instantiated, if these objects are also composite, etc. . It may be necessary for a component object to access data members or to invoke function members of its parent object. This can be made possible by means of a data member of the component object pointing to the parent object, which can be set in the constructor of the aggregate object after instantiation of the component object.

1.4. Interactions between the objects

The interaction between two objects is modelled on basis of calling a member function of one of the objects from a member function of the other object, and vice versa. It is desirable to have the possibility to specify the interactions at the level of the aggregate objects, without the need to code explicitly of which object certain member functions should be invoked. For example for modelling the system mentioned in 1.1 at the level of the CPU boards only specification of the connections between the boards should be necessary. An event causing a change of state in a FIFO of one of the boards may lead to an event for a FIFO of one of the other boards, so at the level of the FIFO objects it should be known which of these objects interact with each other.

The interaction pattern can be established on basis of pointers to the basic objects of which member functions should be invoked. With a procedure like outlined in fig. 1 for obtaining pointers and a similar procedure for setting pointers to component objects, the interaction pattern can be completely defined at the level of the aggregate objects without explicitly referencing component objects.

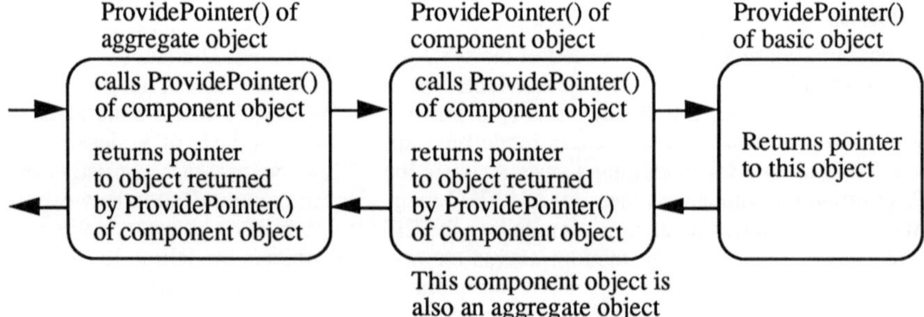

Fig. 1. Procedure for obtaining a pointer to a basic object.

2. Discrete Event Simulation

2.1. Principle

Simulation of a data-acquisition and trigger system can be done on basis of the generation of events (not to be confused with events due to particle interactions, observed in an experiment) to occur at certain simulation times and of determining of the response of the simulated system to each event. The response may consist of the generation of new events at the same simulation time as the original event occurred or at future simulation times. This type of simulation is called discrete event simulation. It can be efficiently implemented on basis of a so-called "event list". Data structures representing the events are stored in the list in an order determined by the (simulation) times at which they should occur. They are inserted in the list by the actions of the simulated objects. The data structure will contain (i) the simulation time at which the event should occur, (ii) an identifier from which the action to be caused at the occurrence of the event can be derived (a pointer to an object), and (iii) additional information. It should be noted that events caused by one object may lead to events for and state transitions in another object. Hence, when for each entity to be simulated a separate event list is maintained it will be necessary (for a meaningful simulation) to scan all event lists for the earliest event. Therefore the use of a single global event list is to be preferred, while parallel implementations are difficult to realise.

```
Event *next;

for (;;)
{
      next = List->RetrieveEvent(); // get event from list
      // call event handler of object to which next->objectPointer points
      (next->objectPointer)->HandleEvent(next);
      delete next; // discard event
}
```

Fig. 2. Basic structure of the main loop of a discrete event simulation program. The event list is implemented as an object with name List. RetrieveEvent() is a member function of this object.

2.2. Implementation

In its simplest form, a single global event list can be maintained in C++ as a linked list of events, hence of pointers to data structures representing the events (cf. 2.1). The events in the list are ordered with respect to the times at which they should occur. Events are retrieved from the head of the queue. The memory space used for the event data structure should be released after returning from the event handler. New events are inserted in the queue by calling a function with as arguments at least the time of occurrence and a pointer to the object that should handle the event. The function allocates memory for a new event data structure and determines after which entry it should be inserted in the list. The use of pointers to objects requires that either all objects belong to the same class, or that their classes are derived from the same class. The first requirement in general cannot be satisfied, hence all classes have to be derived from a single class. The main loop of a simulation program will have the structure shown in fig. 2.

2.3. Efficiency considerations

The implementation described in 2.2 provides the functionality needed. However, for large simulations the computation time can be reduced considerably by changing the implementation of the event list. There are three different possibilities for optimization :

(i) For each event the memory needed for the event data structure corresponding to it is dynamically allocated and deallocated. By pre-allocating memory for a number of events and maintaining a linked list of free event data structures, this can be avoided. The number of pre-allocated event data structures can easily be increased with the normal memory allocation mechanism, if necessary,

(ii) When using a double linked list for the event list, with for each node a pointer to the next node and a pointer to the previous node, the list can be searched from the back, while it is still possible to take the earliest event from the head of the list. This may result in shorter times for searching for the point at which the event should be inserted in the event list, as it is likely that for most events less event list nodes have to be traversed when searching starting from the end of the list,

(iii) The search for the insertion point can in addition be made more efficient by splitting the event list in an array of fixed size (the "event wheel") and in a number of linked lists[2]. The elements of the array are pointing to the linked lists, except for one linked list, the so-called "overflow list". The array corresponds to a certain range in time (see fig. 3). Thus, searching of a single long list is avoided. In stead, a (fast) determination of which list has to be searched is followed by a search of a shorter list. Each list can be a double linked list, as described before.

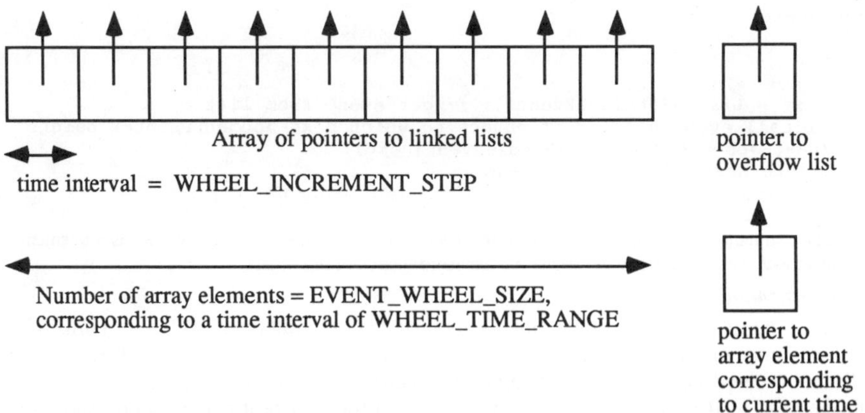

Fig. 3. Splitting a long event list into shorter lists. For finding the list in which an event to occur at time T should be inserted the following rule is applied : if T < current time or T > WHEEL_TIME_RANGE the event goes to the overflow list, otherwise the list in which the event should be inserted is pointed to by the array element with index : index = (T / WHEEL_INCREMENT_STEP) % EVENT_WHEEL_SIZE

The methods described also cause additional overhead. For large event lists the gain however may be substantial. Inspection of the times associated with the events in the list (e.g. by histogramming the difference between the time associated with an event and the current simulation time, when inserting an event in the list) and of the length of the list can be very helpful to judge the possible effectiveness of these methods.

2.4. Complications

An event may cause new events to be generated. Until now it has been ignored that one or more of these events may invalidate one or more events already stored in the event list, or one or more future events not yet stored in the list. As an example, consider the simulation of a transputer link that can transport data in two directions simultaneously. The transport speed per data stream will be lower for bi- than for uni-directional transfers. An event may signal the end of a uni-directional transfer. If during the uni-directional transfer also a transfer in the other direction starts, this event will be invalid : it has to be delayed because the transport speed decreases. On the other hand, when the transfer is started while already a transfer was in progress in the opposite direction, and which finishes during the other transfer the event will arrive too late, as the transfer speed is increasing : the event has to be cancelled. In both cases searching the event list for events to be delayed or to be cancelled can be avoided. First we look at events to be delayed. Objects that need to deal with possible delayed events maintain a flag indicating whether the next event for that object has to be delayed. When the event handler of the object is called, the flag is inspected and when the event has to be delayed a new event is put into the event list to occur at the correct time, while the flag is reset. This approach requires that for any object at most a single event can be present in the event list. This can be arranged for by having each object only generating events for itself, and only when it is not waiting for an event. The requirement can always be fulfilled, except when an event has to be replaced by an earlier event, i.e. the event has to be cancelled. This can be done by maintaining a counter in the object, which is incremented each time an event is placed in the event list. The contents of the counter is stored in the data structure associated with the event. Events received with a stored value of the counter lower than the current value have to be ignored, hence they are effectively cancelled.

2.5. The data structure representing an event

The discussion in 2.4 shows that an event can be represented by three quantities : (i) the simulation time at which the event should occur, (ii) a pointer to the object that will handle the event and (iii) the contents of a counter maintained by the object that will handle the event, so that an event can be replaced by an earlier event. The data structure representing an event will also contain one or two (in case of a double linked event list) pointers needed for the event list.

3. Comparison with MODSIM II

MODSIM II is an object-oriented simulation language, available from CACI[3]. It directly supports discrete event simulation with several language primitives.

3.1. Modelling

The same approach of building aggregate objects from basic objects, each of which is implementing a state machine and which are interacting on basis of pointers ("reference variables" for MODSIM II), as outlined in 1. , can also be applied in MODSIM II.

3.2. Discrete event simulation

This is an area where MODSIM II supports several approaches. With `WAIT` a method of an object can wait during the time specified, which is similar to the waiting for the occurrence of an event in the event handler of a basic object. It is also possible to wait for the invocation of the `trigger` method of an object of type `TriggerObj`. Another type of wait is waiting for the completion of an invoked method, where this method can terminate the wait with the `TERMINATE` statement. All types of waiting can be interrupted. A so-called `TELL` method of an object can be invoked with a specified delay.

In the C++ framework described in this paper the same functionality is possible. However, the programmer has to implement e.g. the procedures described in 2.4 and has to invoke function members of basic objects that cause events for those objects at later times and that also cause appropriate state transitions. This is less straightforward than what is possible in MODSIM, but the state transitions that occur are explicitly exposed, which may be an advantage and may not be the case in a MODSIM program.

3.3. Libraries, user interface

MODSIM offers support for gathering and reporting statistics, support for graphics and animation, and a set of libraries useful for simulation. When using C++, these facilities have to be supplied by external packages, which is not a serious problem with the packages available in the high-energy physics community.

4. Application

A system built of transputer-like processors, with their communication links connected in a tree configuration, has been simulated. The simulation was done as part of the study of second-level trigger architectures by the RD-11 collaboration[4]. The purpose of the system is to concentrate input data streams from a large number of front-end processors into a few output data streams. Each transputer-like processor was modelled as consisting of 4 communication link engines and a CPU running a program consisting of several processes and a multi-priority task scheduler. A processor was modelled as an aggregate object, assembled from 4 objects for the 4 link engines, an object representing the scheduler and an object for the program. The latter is also an aggregate object, consisting of objects representing the processes. Most objects are derived from a single class (see 2.2), objects representing the processes are however derived from a class which takes care of the influence of the scheduler and of the scheduling state (e.g. waiting for scheduling or blocked until communication in progress is finished) on the processes. In this way the behaviour of a process can be described without taking into account the scheduling state, as this is already taking care of.

Using the approach as outlined did not show problem areas, although a few enhancements for debugging have been found to be useful. First each object has a function member for printing information on the state of the project and of the simulation time. These functions can be turned on and off by conditional compilation or with logical variables. They can be very useful for following the actions of the system. Second when a function member of an object is invoked, it is always checked whether the invocation of it is consistent with the state of the object. An error message is printed or the simulation is halted when this is not the case. The same type of check is done when an event is received. Also the time order of the events is checked. Bugs are quickly detected by these checks, while they do have only a minor influence on the computation time.

User interfacing and presentation of results was done with the SUIT package[5] and with histogram display routines, originally written in FORTRAN by C. de Laat[6] and translated into C with AT&T's f2c translator. SUIT provides an interactive graphical user interface, and runs on Unix workstations, PC's and on Apple Macintosh computers. For program development Symantec's ThinkC integrated development environment for the Apple Macintosh was used, which offers a compiler for a subset of C++ (the newest version also has a compiler for full C++), a graphical browser and an interactive debugger. Production running of the simulation was done on HP9000/700 series workstations, where AT&T's C++ compiler was used.

5. Conclusion

A method for modelling and performing discrete event simulation in C++ has been presented. Simulation languages like MODSIM II offer direct support for discrete event simulation. Although programming in a general purpose language like C++ is less straightforward in this respect, the explicit exposure of the state transitions in the program code may be an advantage. Another advantage of using C++ is the possibility to optimize the run-time efficiency with the methods described in section 2. The approach followed was validated by means of the simulation described in section 4.

6. References

1. J. Rumbaugh et al., *Object-Oriented Modeling and Design*, Prentice-Hall International, Inc., 1991.
2. H. H. Desai, *International Journal in Computer Simulation* **1** (1991) 99.
3. CACI Products Company, 3344 North Torrey Pines Court, La Jolla, California 92037, USA.
4. RD-11, Proposal : CERN/DRDC 90-56 (1990), Status Reports : CERN/DRDC 92-11 (1992), CERN/DRDC 93-12 (1993).
5. SUIT, the "Simple User Interace Toolkit" is available via anonymous ftp from uvacs.cs.virginia.edu.
6. C. Th. de Laat et al., in *New Computing Techniques in Physics Research II*, ed. D. Perret-Gallix (World Scientific, Singapore, 1992), p. 215.

FORMAL METHODS IN THE DESIGN OF A SECOND-LEVEL TRIGGER

J. Haveman, A.C. Balke, W. Lourens
Department of Physics and Astronomy, Utrecht University
PO Box 80.000, 3508 TA Utrecht, The Netherlands
haaf@fys.ruu.nl

ABSTRACT

Due to the complexity of large data extraction and analysis systems it is essential that they be designed using a well-structured development method. The applicability of VDM^{++}, an object oriented extension of the well-established formal method VDM, is being investigated. In its minimal form VDM^{++} includes encapsulation of data in objects (with methods) and an inheritance scheme, with full formal reasoning. In its full form the language also deals with synchronisation and parallelism, however the formal semantics of these constructs is still under development.

We will give examples of our work on both local triggers (feature extractors) and the global second level trigger. In the former case a solution using essentially sequential techniques is possible when one uses a synchronous / pipelined architecture. In the latter case concurrency and real-time properties are essential parts of the description of the system.

1. The Afrodite project and VDM^{++}

Afrodite is an ESPRIT III project with as main objective to make formal methods more easy for industrial use in technical environments. Utrecht and CERN members of the EAST collaboration are 'application partners' in this project, with as aim the specification of part of a data acquisition system, emphasising on the second level trigger.

1.1. Formal methods

Formal methods provide implicit, property-oriented specification in addition to constructive, algorithm-oriented specification. This makes it possible to specify a system at a high level of abstraction in a mathematical framework[1], without the ambiguities of a natural language or diagrammatic representation. The absence of unnecessary, and often unwanted, detail from such a specification makes it possible to independently develop different parts of the specification towards an implementation. In each development step one should prove that the new specification satisfies the more abstract one. Although it is usually impractical to rigorously construct all these proofs, the formal method does provide the mechanism to do so if the proofs are non-trivial or if the reliability of the system under construction is of major importance.

[1] The construction of this abstract specification from the demands of the future users of the system, often vague and evolving with time, is a non-trivial task in itself.

The existing formal method VDM[1], on which the development in the Afrodite project is based, distinguishes a number of proof obligations. First there are those dealing with the internal consistency of one level of specification: *syntax-correctness*, *type-correctness* and *satisfiability* of the specification. Second are the proof obligations dealing with refinement relations between components of an abstract specification and components of a refined specification. For *data reification*, the transformation of (abstract) data structures to a more concrete (implementation-oriented) representation, the proof obligations centre around the retrieve function that maps the concrete data structure onto the abstract one. For *operation decomposition*, the introduction or detailing of an algorithm to accomplish a certain abstract task, there are a number of proof rules dealing with the introduction of for instance if-statements, for-loops and while-loops.

1.2. From VDM to VDM++

VDM++[2] is an extension to VDM on a number of points. Through the introduction of object oriented concepts both top-down development (hiding of implementation details of a certain class from its clients) and bottom-up development (reuse of already written class libraries or hardware specifications) are supported. In the specification of one class one cannot freely use information about another class's properties, because this would break encapsulation. Relations between objects should be modelled through 'family' or 'container' classes. Only in the reasoning about the total system one can freely use all private information of the objects.

A system is thought to be constructed from independently operating objects. The concurrency between the objects can be limited by two techniques: *trace* statements to limit the allowed sequences of method invocations, and *permission* statements to allow or disallow invocation of methods depending on the current state of the object. Objects can be made active by giving them a *thread.* This introduces a new proof obligation: the satisfiability of the combined concurrency constraints, first on the class level but also on the global specification level.

Certainly for the EAST applications the specification of real-time behaviour is an essential part of any system description. Unfortunately the development of real-time constructs is only in a preliminary stage. Therefore the only way to put real-time constraints on our specification is through adding comments. Note that the satisfiability of real-time constraints cannot usually be checked on the abstract level of a top-down specification, because the timing properties of the implementing hard- and software are not available at this level.

The use of the language VDM++ is made easier by an integrated tool set consisting of a syntax checker, a type checker and a proof assistant. The tools should keep track of dependencies between class definitions and include a version control mechanism.

2. Specification of a feature extractor

To gain experience in the use of VDM++ we have chosen to specify an already existing system, the MaxVideo implementation[3] of the feature extractor for the transition radiation detector[4]. A region of interest for this detector consists of two rectangular pixel images. For each image a histogram of the number of '1' pixels in each row should be made. The position of the maximum of one of these histograms and the histogram values at this position form the required feature. The full VDM++ specification of this system is described in elsewhere[5].

2.1. The abstract specification

The following specification should be seen as part of the specification of a larger system, provided by the designers of this larger system as the interface which our implementation should satisfy. Since the format of real-time obligations in VDM++ has not yet been determined we were forced to include this obligation in the form of a comment. The most important part of the resulting single-class specification is:

```
class TrdMaxVideo;
  values
    nrows: ℕ = 25;
    ncolumns: ℕ = 240
  types
  -- definitions of various types are left out
    PixelImage = map (RowIdx * ColumnIdx) to Bit
    inv p == dom p = { (x, y) | x ∈ {1,...,nrows} ∧ y ∈ {1,...,ncolumns} }
  functions
  -- definitions of histogram and positionOfMaximum are left out
    featuresFrom: PixelImage * PixelImage → TrdFeatures
    featuresFrom(plo, phi) ==
      let m = positionOfMaximum(histogram(plo)) in
        mk_TrdFeatures(m, histogram(plo)(m), histogram(phi)(m))
  instance variables
    destination: @GlobalProcessor;
  methods
    inputImages(plo, phi: PixelImage)
      is not yet specified;
-- obligation: after an invocation of inputImages(pic1, pic2)
--               destination!inputFeatures(featuresFrom(pic1, pic2))
--               should be invoked within a finite time
end TrdMaxVideo;
```

2.2. Selected refinement steps

The first refinement step is to change the image data structure into a sequence of words, where each word is a sequence of bits. This is a simple case of data reification with an invertible retrieve function.

The summation over the bits in one row is now split into a summation over the bits in one word (modelled by the function *bitSum*) and the summation over the words in a row (modelled by the function *blockSum*). To prepare for a distribution of these tasks over different objects the intermediate results are introduced explicitly, and a new method that performs the calculation of the features is specified:

```
methods
  extractFeatures() ==
    if qlo ≠ [] then
    ( [ ext wr qlo, qhi, plo, phi, hlo, hhi, tft;
        post qlo = tl qlo↼ ∧ qhi = tl qhi↼ ∧
             plo = [ bitSum((hd qlo↼)(i)) | i ∈ {1,...,imagesize} ] ∧
             phi = [ bitSum((hd qhi↼)(i)) | i ∈ {1,...,imagesize} ] ∧
             hlo = [ blockSum(plo, r) | r ∈ {1,...,nrows} ] ∧
             hhi = [ blockSum(phi, r) | r ∈ {1,...,nrows} ] ∧
             tft = featuresFrom(hlo, hhi) ];
      destination!inputFeatures(tft) );
thread
  periodic(extractFeatures);
```

The multi-line statement within square brackets is a specification statement: it gives the values of the listed instance variables after its execution in terms of the values before its execution (the values with a harpoon). One observes that the new value of *hlo* depends on the new value of *plo*, and hence these should be calculated sequentially in this scheme. To enable parallel evaluation of these quantities one should calculate *hlo* from the old value of *plo*. If one also follows this idea for the calculation of *tft* from *hlo* and *hhi* one arrives at a pipelined system with three elements: simultaneous evaluation of *featuresFrom* for an image, *blockSum* for the next image, and *bitSum* for the image after that.

At this intermediate level of abstraction we can satisfy the real-time obligation by giving the class a thread and specifying that the method *extractFeatures* should be executed periodically. The time scale for this periodic obligation is specified elsewhere in the class definition. The latency of the system can be expressed in multiples of this time scale.

Next we introduce objects, following the decomposition of the system as sketched above: BitSummer, BlockSummer and FeatureBuilder. The TrdMaxVideo class serves as the interface class to the outside world and determines the connections between the other objects. Those are derived from a generic pipeline element class with methods *shift* and *process* that are to be defined in the subclasses that are the actual pipeline elements. Further it contains a synchronisation statement to limit the amount of concurrency allowed:

```
methods
  shift()
    is subclass responsibility;
  process()
    is subclass responsibility;
sync
  general T = < { (shift; process)* }, { shift, process } >;
```

The trace structure T forces the methods *shift* and *process* to be invoked alternatingly. In the interface class it is specified that this alternating invocation is performed synchronously for all elements of the pipeline in parallel.

The next refinement step is to distribute the data corresponding to one image over the pipeline objects, thus obtaining a finer-grained parallelism: an image is pushed through the pipeline word for word. In stead of an entire image a BitSummer now only holds one word. To be able to construct a retrieve function for an image we have to introduce the concept *state*, and retrieve an image from a number of words in consecutive states. A synchronisation signal is added to the specification of the system in order to determine which sequence of words form an image.

The final step is to transform all data to binary data, from where a hardware specification language could take over. The interfacing of VDM++ to hardware specification languages is the subject of one of Afrodite's other partners.

3. Specification of a global second level trigger

The second objective of the CERN group is the specification of the global second level trigger. The current status is subject to revision in the light of the rapidly evolving LHC detector concepts. In a recent paper[6] existing knowledge is drawn together to obtain an informal requirements specification of the system. We have chosen to specify a class of architectures taking a data driven approach in which the components of the system are designed to keep up with the required operating frequency. The feature extractor specified here is an example of a component fitting this scheme. The alternative would be to make use of asynchronous farms of processors. In the process of requirements specification it was felt that a visual representation of the system might proof helpful. Since the VDM++ tool set does not yet provide such a representation we chose to use the OMT[7] notation for object model and dataflow model views. In figure 1 the object model view of a specification at an intermediate level of abstraction is presented. The feature extractor specified above fits in as one of the "T2 local" objects. At this level of detail the system has been formally specified in VDM++ as far as possible without the ability to specify real-time constraints.

4. Conclusions

VDM++ has the potential to become a useful method for developing complex

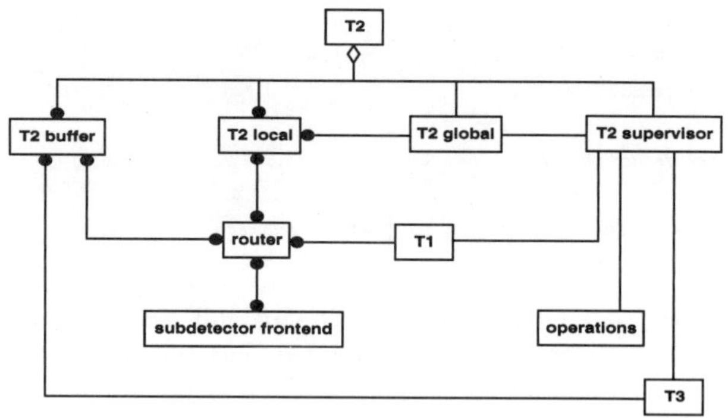

Figure 1: Object model overview of the second level trigger system.

hard- and software systems. Its ability to specify a synchronous system, where one time scale is enough to determine the real-time properties, has been shown in the feature extractor specification. On the other hand the language cannot yet deal with the timing properties of the global trigger, the really challenging case. For that task we will have to wait for the introduction of real-time primitives in VDM^{++}, which is scheduled for 1994.

References

1. C.B. Jones, *Systematic Software Development using VDM* (Prentice Hall International, 1990).
2. E.H. Dürr, A. Duursma, N. Plat, *VDM^{++} Language Reference Manual*, Afrodite report AFRO/CG/ED/LRM/V8.
3. A. Gheorghe, W. Krischner, Z. Natkaniec, *Second level trigger algorithms on pipelined signal processing architectures*, CERN/EAST note 92-17.
4. P. Bialas, J. Chwastowski, P. Malecki, A. Sobala, *Benchmarking with data from the transition radiation detector*, CERN/EAST note 91-11.
5. J. Haveman, *Specification of a Pipelined Feature Extractor in VDM^{++}*, Afrodite report AFRO/UU/JH/TRD++/V1.
6. A.C. Balke, J. Carter, J. Haveman, *Specification of Data Driven Second Level Trigger System* (in preparation), Afrodite report AFRO/CN/JC/DOC/V0.
7. J. Rumbaugh et al., *Object-oriented Modeling and Design* (Prentice Hall, 1991).

OBJECT ORIENTED SOFTWARE DESIGN AND MULTIPLE GRID TECHNIQUES FOR THE SIMULATION OF MICRO STRIP GAS COUNTERS

E.H. Agterhuis
Department of Physics and Astronomy, Utrecht University
P.O. Box 80000, 3508 TA Utrecht, The Netherlands

ABSTRACT

In 1988 Oed et al. published on a concept for a new type of charged particle detector, the MicroStrip Gas Counter (MSGC). Its operation is based on gas ionisation. The charge freed in the ionisation process is collected on microscopically small conducting strips. A good MSGC based detector can be build although it requires a long way of trial and error. Many principles are clear, but their effect in terms of detector response is very hard to predict. Computer simulations of the detector with software that models the physics of the detection principle can be used to solve this problem. The objective is to develop a software tool for the design of MSGCs. Object oriented methods are used to keep the software flexible and easily extendible and much effort is put in the implementation of highly efficient algorithms. In this paper the object oriented design of a solver for the electric field equations based on multiple grid techniques is described. Furthermore a comparison is made with a classical implementation of a solver based on a single grid algorithm.

1. Microstrip Gas Counters

1.1. Geometry and materials

An MSGC basically consists of four components[1]. First there is a substrate layer which is in most cases an insulator. Typical dimensions are a few centimeters along each side. It supports a structure of small parallel conducting strips arranged in anode/cathode pairs with a spacing of a few hundred microns. A conducting plane is mounted at a distance of several millimeters from the substrate layer. The space in between is filled with gas.

As a charged particle passes through the gas it ionises its molecules leaving clusters of free electrons behind. The conducting plane is kept at a strongly negative potential which causes the electrons to drift towards the substrate layer. A strong electric field is maintained between the strips causing an avalanche of ionisation when electrons enter its region of influence. The gain in amount of charge is typically in the order of 10^3. Finally the electrons are captured on the anode strips giving a measurable signal.

1.2. Design parameters

Because it is theoretically possible to detect single freed electrons the inefficiency of an ideal detector can be defined as the probability that no primary

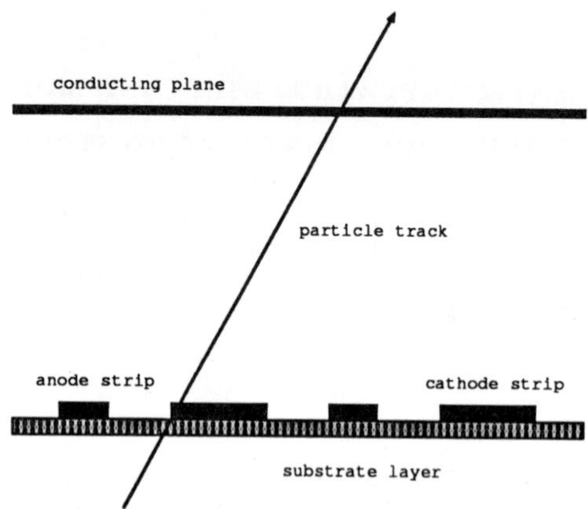

Figure 1: Longitudinal cut through an MSGC

ionisation takes place. According to Poisson statistics this is:

$$P_0 = e^{-<n>} \tag{1}$$

where $< n >$ is the average number of primary ionisations, which is linearly related to the ionisation density of the gas[2]. Because the actual detection limit depends on the sensitivity of the amplifier electronics a high gas gain generally results in high efficiency.

The geometry of the strips plays a very important role in this respect. Reducing the anode/cathode pitch or the anode width results in a stronger electric field and thus in a higher gas gain. Nevertheless the strength of the electric field does not depend in a trivial way on the MSGC geometry. Furthermore it need not be stable in the vicinity of the substrate due to pileup of charge. Therefore the gas gain is hard to predict. Also there is a limit on the strength of the electric field given by the point where the generation of sparks starts.

Resuming it will be clear that a good description of the electric field is very important.

2. The physical and numerical models

2.1. Field equations and their discretization

The electric field is described by Poisson's equation:

$$\nabla^2 \phi = -\frac{1}{\varepsilon} \rho \tag{2}$$

where ϕ is the electric potential, ε is the relative dielectric constant and ρ is the charge density. With a finite volume discretization scheme it can be mapped onto a grid. In two dimensions the equation then takes the form:

$$h_1 \frac{2\varepsilon_j \varepsilon_{j-\vec{e}_2}}{\varepsilon_j + \varepsilon_{j-\vec{e}_2}} \nabla_2 u_j + h_2 \frac{2\varepsilon_j \varepsilon_{j-\vec{e}_1}}{\varepsilon_j + \varepsilon_{j-\vec{e}_1}} \nabla_1 u_j - h_1 \frac{2\varepsilon_j \varepsilon_{j+\vec{e}_2}}{\varepsilon_j + \varepsilon_{j+\vec{e}_2}} \Delta_2 u_j - h_2 \frac{2\varepsilon_j \varepsilon_{j+\vec{e}_1}}{\varepsilon_j + \varepsilon_{j+\vec{e}_1}} \Delta_1 u_j = h_1 h_2 s_j \quad (3)$$

where

$$\begin{cases} \Delta_n u_j \equiv \frac{u_{j+\vec{e}_n} - u_j}{h_n} \\ \nabla_n u_j \equiv \frac{u_j - u_{j-\vec{e}_n}}{h_n} \end{cases} \quad (4)$$

j is an index on the grid, the constants 1 and 2 indicate the dimensions, $\vec{e}$ is a unit vector and h the grid spacing in the indicated dimension and s and ρ are the discrete equivalents of u and ϕ. Taking all coefficients together in a matrix $\mathbf{A}$ this can also be written as:

$$\mathbf{A}\vec{u} = \vec{s} \quad (5)$$

This matrix can be implicitly inverted with the help of a relaxation scheme which lets an initial guess to the solution evolve like:

$$\vec{u}^{t+1} = \mathbf{L}\vec{u}^t + \mathbf{M}^{-1}\vec{s} \quad (6)$$

where

$$\begin{cases} \mathbf{A} = \mathbf{M} - \mathbf{N} \\ \mathbf{L} = \mathbf{M}^{-1}\mathbf{N} \end{cases} \quad (7)$$

A much used decomposition of $\mathbf{A}$ is one where $\mathbf{M}$ is taken to be the upper triangular part of $\mathbf{A}$. The resulting scheme, known as Gauss-Seidel relaxation, is very slow because it involves only local operations. Information travels very slowly across the computational grid.

2.2. Over-relaxation

The traditional Gauss-Seidel method can be given non-local behaviour by modifying equation 6 as follows[3]:

$$\begin{cases} \vec{u}_* = \mathbf{L}\vec{u}^t + \mathbf{M}^{-1}\vec{s} \\ \vec{u}^{t+1} = \omega\vec{u}_* + (1-\omega)\vec{u}^t \end{cases} \quad (8)$$

The intuitive argument for why this will be faster is that with each update one tries to look ahead of the traditional scheme given the tendency of the value at hand to go in a certain direction. This method has $O(N^2)$ complexity if N is the number of vertices[5]. Faster relaxation schemes have a complexity of at best O(NlogN) but involve specific constraints on the geometry of the problem[5].

2.3. Multiple grid methods

For traditional relaxation schemes locality is a disadvantage but multiple grid methods benefit from it. The problem is mapped onto coarser grids where relaxation operators only work on components of the approximated solution with a period comparable to the grid spacing. This results in algorithms of O(N) complexity[3].

The design of a multiple grid solver starts with the definition of a set of grids with decreasing resolution. In most cases coarse grids are produced by reducing the number of vertices by a factor of two in all dimensions. Operators must be defined to transfer information between grids. Transferring information from fine to coarse grids is called restriction. Transfer in the opposite direction is referred to as prolongation. A two grid algorithm will look like[3]:

solve(n);
$\vec{r}_n := \vec{s}_n - \mathbf{A}_n \vec{u}_n;$
choose $\vec{u}^*_{n+1}, \xi_{n+1};$
$\vec{s}_{n+1} := \mathbf{A}_{n+1} \vec{u}^*_{n+1} + \xi_{n+1} \mathbf{R} \vec{r}_n;$
solve(n+1);
$\vec{u}_n := \vec{u}_n + \frac{1}{\xi_{n+1}} \mathbf{P}(\vec{u}_{n+1} - \vec{u}^*_{n+1});$
solve(n);

where $\mathbf{R}$ is a restriction, $\mathbf{P}$ is a prolongation, ξ is a scaling parameter and n indicates a level of coarseness. The differential operator $\mathbf{A}$ must be defined on all levels. If direct discretization is not possible, for instance because of undersampling, Galerkin's approximation[3] can be used to define $\mathbf{A}$ on any grid in terms of its equivalent on a finer grid as:

$$\mathbf{A}^{n+1} = \mathbf{R}\mathbf{A}^n\mathbf{P} \tag{9}$$

Only on the finest grid direct discretization is used.

3. Software design

3.1. Demands

A major demand on the software is flexibility. Without rewriting any part of the program it must be possible to simulate a wide variety of geometries. This implies that any assumptions concerning special features of the physical problem must not be used by the algorithm. Furthermore the software must be easily adaptable. The principle of inheritance, a feature of object oriented design, helps to build new objects and classes with a basic functionality already defined in lower parts of the hierarchical tree. Interface code for manipulation of hidden data helps constructing mutually independent objects. The internal representation of data can be changed provided that the pre- and postconditions of the interface code are preserved[7].

3.2. Class definitions

Two groups of classes were defined. The first implements general concepts which can be re-used in other software. A second group represents the solving algorithm. Figure 2 illustrates a hierarchical structure from each group. The first example shows the definition of a computational grid as a matrix of cells combined with a class Stencil representing the structure of a relaxation operator. The second example shows the definition of transfer operators.

4. Performance

The performance of a C implementation of a solver based on over-relaxation was compared to the performance of the C++ multiple grid solver. Benchmarking was done on a DEC 5000/133 workstation. The test geometry consisted of a small

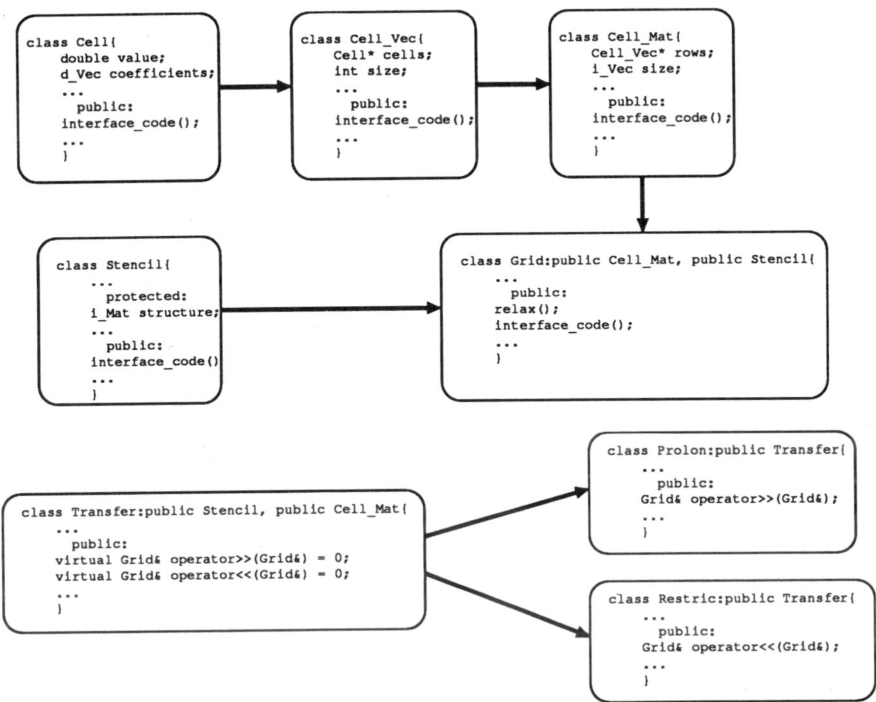

Figure 2: Examples of class definitions in C++ notation. Above the definition of a computational grid. Below the definition of transfer operators.

conducting square inside an area bounded by conducting strips on two opposite sides and periodic boundary conditions on the remaining two sides. Though the complexity of the multiple grid method is less than that of the over-relaxation method the advantage is not dramatic for the tested gridsizes. This is due to the data abstraction used in the C++ implementation. Large addressing overhead results from the scattering of data over computer memory. Furthermore practice shows that Galerkin's approximation gives rise to time consuming solving schemes.

5. Conclusions

A multiple grid algorithm for solving electric field equations for a wide variety of geometries was successfully implemented. Object oriented design techniques have proven to provide a natural way of mapping a problem onto software components. Current research which involves the description of subatomic particles and statistical

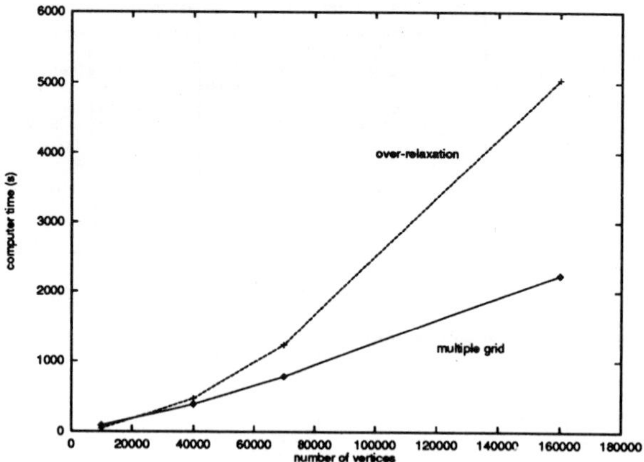

Figure 3: Performance of the multiple grid program compared to the performance of a program based on over-relaxation

processes shows that general classes developed for the Poisson solver can be easily re-used. Although data abstraction lowers the performance the flexibility of the program and the re-usability of separate objects make the software a very good basis for further development.

6. Acknowledgement

This work is financially supported by the Foundation for Fundamental Research on Matter (FOM).

1. A. Oed, *Position-sensitive detector with microstrip anode for electron multiplication with gases*, Nucl. Instr. and Meth. A263 (1988), p. 351.
2. F. Sauli, *Principles of operation of multiwire proportional and drift chambers*, Academic Training Program of CERN, (1977).
3. P. Wesseling, *An introduction to multigrid methods*, John Wiley & Sons, (1991).
4. J. Stoer, R. Bulirsch, *Introduction to numerical analysis*, Springer Verlag, (1980).
5. W.H. Press et al., *Numerical Recipes in C. The Art of Scientific Computing*, Cambridge University Press, (1992).
6. S. Brandt, H. Schneider, *Computer-drawn field lines and potential surfaces for a wide range of field configurations*, Am. J. Phys. Vol 44,No. 12, (1976).
7. B. Stroustrup, *The C++ programming language*, Addison Wesley, (1991).

DEVELOPMENT OF GRAPHIC "USER-TO-COMPUTER" INTERFACE AND SOFTWARE TO OFFER GRAPHIC AND ALPHANUMERIC DATA OUTPUT

ZAITZEV M.A., KISELEV A.E., KISELEV V.P., MERCULOV S.Yu.,MERCULOVA S.R.
Nuclear Safety Institute, B.Tulskay 52, Moscow, 113191, Russia

ABSTRACT

This publication outlines work on processing of graphic and alphameric (text) data in computers employing the VAX-VMS, UNIX, MS-DOS operating systems, describes formats of the GKSM graphic metafiles providing machine-independent processing of the results of calculation of various programs including visualization of graphic preprocessors and postprocessors, generation of complex two- and three-dimensional finite-element calculation grids. Operating the system of different computers calls for standardization of application programs proper, presentation of computed results as well as efficient exchange of information between different computers and data output to various peripherals. A great number of graphic application programs designed to process computed results and oriented to various high-level programming languages (C, FORTRAN, PASCAL) have been already developed. Most widely used in Russia is a package of GRAFOR application programs. As a rule, such program packages support a small number of output peripheral devices and do not allow information to be stored in a machine-independent fashion. So, this job is aimed at agreement of algorithms and low-level programs of graphic application programs with a known standard for processing and storing the graphic information presenting a device-independent package of graphic subprograms based on the International Standards Organization (ISO) and American National Standards Institute (ANSI) Graphical Kernel System.

1. Graphic data display and processing system

The GKS standard was used as a basis for construction of a lower level of the graphical support system. The Graphical Kernel System (GKS) is one of the most widespread standards of machine-oriented graphics. The GKS offers generation, representation, segmenting, conversion and entry of pictures. To perform these functins, the GKS calls for:

- operating stations combining input and output devices;
- coordinate systems and their conversion (an independent coordinate system to construct an object an object on and an equipment coordinate system to describe the picture being displayed);
- output objects (primitives) to construct an output picture;
- segmentation of pictures and conversion of segments;
- a metafile to provide for long-term storage and transfer of pictures;
- a status list of the graphical system;
- levels into which a host of functions are broken down, etc.

The GKS was initially designed as a two-dimensional graphical system. Then a three-dimensional modification, 3D-GKS, compatible with the GKS (from a smaller to a larger system) was developed. Added were three-dimensional primitives and aspects, provision for access and test of algorithms of remoteness of hidden lines and surfaces. To implement the GKS in the VAX computers, provision is made for the VAX GKS standard.

The VAX GKS constitutes a device-independent package of graphical subprograms based on the International Standards Organiza- tion (ISO) and American National Standards Institute (ANSI) Graphical Kernel System. The VAX GKS makes it possible to create - graphical supplements irrespective of the characteristic property of specific peripherals. The VAX GKS standard (level 2) aids all peripherals supported by the VAX-VMS operating system. The peripheral is a graphical logic unit which allows information to be represented as supplements in a device-independent fashion. For connection to a real physical device the latter is identified from the DCL VAX-VMS operating system. The VAX GKS supports various devices, the output metafile of the GKS system being of particular interest. It is an alphameric file of specific structure whose format will be discussed below. The format of such a metafile allows the metafile to be transferred to various computers (series VAX, IBM PC AT and compatible operating graphical stations, such as APOLLO or MAC-II) thus affording machine-independent program processing and allowing the computers and their peripherals to be used to best advantage.

Discussed in this report are programs for executing operations on metafiles in various operating systems. These programs allow operations to be performed efficiently on the data obtained from the GRAFOR system.

2. GKS Standart metafile

The GKS standard employs the GKS metafile as a basic element for storage and subsequent recovery of graphic data. The metafile incorporates complete information on graphic functions used when processing computed results. The information stored in the metafile is machine-independent. It may be processed to be displayed on an optional output device (graphic terminal, plotter, etc.). Besides, the possibility exists of editing pictures, i.e. relocation of separate elements, superposition of information - stored in various metafiles, etc.

The structure of a metafile represents a sequence of logical elements. The GKS metafile comprises a heading followed by data elements. Every element of a metafile, in its turn, contains a logical name of the graphical instruction executed in the main program of a postprocessor or a preprocessor followed by the instruction logical data.

The type of element defined by an integer number (a value lying in the range 0 - 100) corresponds to a specific GKS instruction executed during operation of the postprocessor or preprocessor. The values exceeding 100 are redundant as instructions of the user.

3. Data-processing operations on VAX computer

Processing the GKSM metafile in the VAX-VMS system is performed with the aid of a program developed. Specified as input parameters are the following:

(a) namee of input metafile;

(b) type of output peripheral.

The type of output peripheral defines output of information to a color graphic display or a file in the format of the HP7475 plotter or output of information to a routine driver of the peripheral.

The peripheral routine driver is defined in the following manner. Connection to a real physical device is through its identification from the DCL VAX-VMS system. The VAX GKS supports a number of various devices, such as graphic (VT-340) and alphameric (VT-320) terminals, operating stations of the Tekronix and DEC companies, plotters of the Heulet-Packard company, the LN-03 laser printer, input and output metafiles, a driver in the PostScript language.

To entry information into a specific peripheral before a run of the graphical programs, it is essential that the type of specific output device and the method of connection to it should be defined. Information may be entered directly into a specific device and outputted to an intermediate file. By default the type of peripheral is a terminal provided with direct black-white graphic display. Changing the type of peripheral or direction of addressing is achieved by the respective DCL VAX-VMS instructions:

DEFINE GKS $ WSTYPE TYPE DEFINE GKS $ CONID DRIV

where TYPE is number of the device in the hexadecimal or decimal number system; DRIV is either logical name of the file (in which information is being recorded) or logical name of the peripheral address. Example:

$ ALLOCATE TTB4: $ DEFINE GKS $ CONID TTB4:

Holds (Allocate) a terminal with logical name TTB4 so that the GKS could transfer information to the terminal.

$ DEFINE GKS $ CONID OUTPUT.GRAPHICS

Output of information to a file identified as OUTPUT.GRAPHICS

4. Data-processing operations on personal computers

Since the type IBM PC personnal computers are widely accepted and feature efficient graphic representation capabilities, a system of programs based on the LPIC program has been developed to process data and display the results on the personnal computers in the form of slide films.

The purpose of the LPIC program (LPIC.EXE and LPC.EXE files) is to observe the so-called GKSM metafiles, created by the VAX GKS graphical system, on the EGA monitors of the IBM PC computers. The program is composed of two functional sections, namely a user interface and an interpreter of metafiles. The functions of the user interface are the following:

- output of a list of metafiles in a given directory (subdirectory);
- request for an interpreter to review a metafile;
- directory change;
- exit out of a program.

By default the . pic metafile is thought of as being extended; if this is not the case, specify extension of your metafiles. The interpreter of metafiles offers the following functions:

- reading of information from a metafile and execution of respective graphic functions on the personnal computer display;
- interrupt of operations before execution of the "close a segment" function (accompanied by a sound signal; to resume the operations, depress any key);
- return to the user interface mode of operation when achieving the "end of metafile" function or depresing the Ctrl-z keys.

It should be noted that a number of the VAX GKS functions can not be processed and some are processed but not in strict compliance with the standard. This is most commonly due to inadequate resolution of the EGA monitor. The program is fetched by the LPIC instruction or obtained from the type XTREE, Norton Commander packages observing the rules of their own.

A picture bearing the program name is being displayed for about a second, then the user interface mode is initiated. The upper line represents a name "pattern" of files you are to operate on. When running the program those will be a current disk, directory, subdirectory, an asterisk in place of the file name, .pic - extension. Displayed at the screen center is a "box" where the files corresponding to the "pattern" are indicated. Represented to the right of the box are the keys being active at the moment. Now one may select a file for review with the use of the keys marked "arrow up" and "arrow down". Depressing the Enter key calls for the interpreter of metafiles and the content of the specified metafile is displayed in the form of a picture. The metafile review is over return to file selection for review. The possibility exists of reviewing another metafile, escaping the selection mode (the Esc key), changing the directory and escaping the program. To change the directory, depress the Space key, then enter the name of a new directory by depressing the Enter key. Should the need arise to use the names of files other than *.pic, specify them. To escape the program, depress the Esc key. The keys being active in this case are indicated above the "box". When reviewing a metafile the instructions specified in it are being executed until one of the following two instructions is encountered:

- close segment;
- end of metafile.

In this case a sound signal informs that the operation is interrupted. To restore the operation, depress any key (except for the Shift, Ctrl keys and the like). Reviewing the metafile can be ceased at any instant by depressing the Ctrl and z keys at a time. Errors in the metafile may result in "hang-up" (failure of the program to operate) which fact is indicated by flickering of the disk operation indicator though the picture displayed on the screen does not change. Depress the

Ctrl and z keys at a time to escape this situation.

The LPIC program shared by the partner program used to "catch" the graphic screen (for instance, CAPTURE from the SHOW PARTNER package, FREEZE from the PBRUSH package, etc.) allows computer slide films to be readily created for representation of computed results.

5. Data-processing operations on APOLLO workstations

The APOLLO is a graphical station provided with the UNIX operating system. The station affords multi-box operation (up to 20 boxes initiated at a time, each may be considered as a separate operating unit), a high-quality graphical display (resolution 1280 - 1084, 256 colors), high-speed memory of up to - 32 megabits, several times higher than the memory speed of typical personal computers. A set of programs allowing the GKS metafiles to be reviewed on the APOLLO station has been developed by executing both the user program on the graphical system proper and the programs created on the VAX computers and transmitted through the network.

On the APOLLO station the GKS standard has been embodied in the Domain/GKS system. To review the metafiles created by the Domain/GKS system, provision is made for an interpreter of the GM1 metafiles available in the APOLLO. As distinguished from the VAX, where visible information is being represented on the screen gradually, the APOLLO screen displays the entire picture after a while as soon as the picture is created. If the metafile contains several pictures (i.e. it contains more than one "close segment" GKS instructions) with the GM1 functioning a picture is displayed on the screen while the next picture is being interpreted and only the last one is held on the screen for the time specified in the GMI-time option. A picup program has been developed to review the Domain/GKS metafiles containing several pictures and allowing a picture-to-picture jump upon depression of the Enter key (the cursor should be within the box in the process). Besides, the picup provides for processing of non-standard graphic instructions (with a number over 100) and suppression of a number of instructions. The maximum possible dimensions of the box for a picture to be displayed will be 1024 - 1024 picocells. As distict from the VAX GKS the Domain/GKS APOLLO provides for a continuous stream of data in the metafile, 80 bytes in a line. To this end, programs of the VAX-to apo (VAX GKS in Domain/GKS) and apo-to-VAX (Domain/GKS in VAX GKS) converters had to be developed and realized in the VAX computers.

6. Conclusion

Developed is a system of two-level data processing of application programs allowing graphic representation of computed results and output to various peripherals (including color hard copies), storage and subsequent processing of graphic information in machine-independent format; these latter allow transferring graphic and text information adequately and using various computers and their peripherals most efficiently.

The Experiment Control System of the 3-Spectrometer-Setup at MAMI *

H.Kramer‡, A.Hake‡, V.Kunde‡, C.Martin‡, K.Merle§, S.Steffens§
‡ *Institut für Kernphysik, Becherweg 45, D-55099 Mainz, FRG*
§ *Zentrum für Datenverarbeitung, Becherweg 12, D-55099 Mainz, FRG*

ABSTRACT

The 3-Spectrometer-Setup at the MAMI consists of up to 18 detector systems and 10 magneto-optical elements. To cope with the local separation, the complexity of the apparatus and to provide a high degree of independence to the individual subsystems, our control system has been split into various processes running on different hosts. While MC68030 based VMEbus machines are used for direct hardware access, UNIX Workstation provide the environment for central services and user interaction. The software organization has been standardized up to a high degree and modelled in an object-oriented manner. We have developed a C++ class library for adding new control objects, and a message passing system – called MUPIX – to manage all exchanges of information between them. Even though the architecture of our system is highly decentralized, access to all status and monitor information is possible from any machine in the network. All user interaction may be done on command-level or through our X-Window interface. The whole system is configured via a database.

1. Introduction

The 3-Spectrometer-Setup [1] is a large experimental device to perform electron scattering experiments at the 855 MeV electron accelerator MAMI [2] in Mainz. Each of the three magnetic spectrometers is equipped with a detector system which consists of two pairs of vertical drift chambers, two sets of plastic scintillators and a cerenkov detector. The whole system adds up to 10 magneto-optical elements and about 18 detector systems. The 3-Spectrometer-Setup operates under a control, monitoring and data aquisition system called MECDAS†. As the name implies, there are two aspects of the software: data acquisition and experiment control. Basically the data acquisition part of the software deals with realtime data readout of CAMAC and Fastbus modules, data flow and event-building. This paper focuses on the concepts and implementation of experiment control (ECS) component. It is responsible for monitoring and controlling all technical aspects of the setup, commonly referred to as slow control. In addition the ECS will control and monitor run

*Work supported by Deutsche Forschungsgemeinschaft (SFB 201)
†Mainz Experiment Control and Data Acquisition System

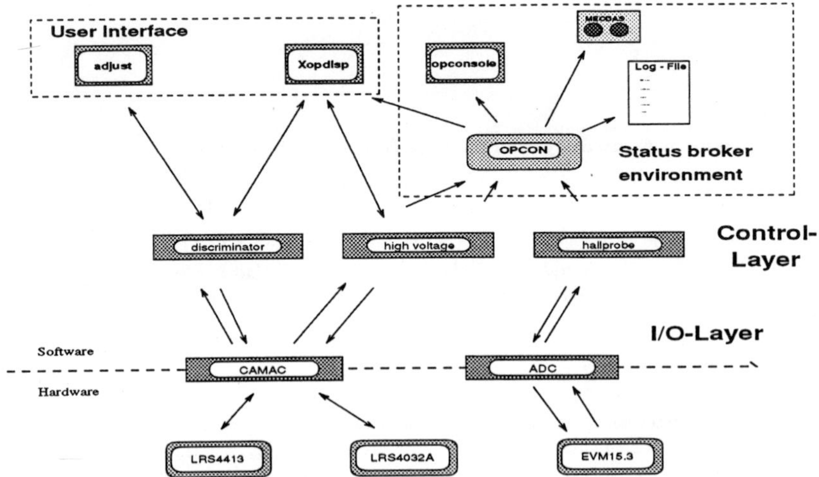

Figure 1: Overview of the 3-Spectrometer-Setup control system

status and data quality in the future.

2. Overview

The 3-Spectrometer-Setup is controlled by a large set of tasks distributed over a heterogeneous computer network. This separation is typical for experiment control systems in high and medium energy physics. It is caused by the complexity of the apparatus and the local distribution of the various devices to be controlled. To keep software development and maintenance costs low the work on one network node is split into pieces which finally form the building blocks of different programs. These programs run in an host environment appropriately chosen for the specific task. While MC68030 based VMEbus machines are used for direct hardware access, Workstations provide the environment for all central services and user interaction. All machines are operating under UNIX now.

An overview of the ECS architecture is given in figure 1. The lowest software layer (I/O-layer) is comprised of objects which are directly related to I/O-hardware modules (VMEbus, RS232C, CAMAC, ...). Also on this layer are objects which have knowledge about the behaviour and characteristics of signal transferring devices (ADC, DAC, ...). The detailed structure of specific experiment devices (e.g. temperature controller, hv-crates) is modelled by objects on the control layer. These objects provide device specific realizations of control and monitoring algorithms. The term "control object" will refer from now on to both – objects on the I/O-layer and on the control layer. Each control object is identified by a unique name. This name is e.g. used to get configuration parameters out of the database. However, note that the configuration database is organized in such a way that a parameter entry is searched first for the specific instance of a class and in case it is not found the search looks for an entry according to its class type. Commands are entered on the top layer, either on command-level, or better suited for non-specialized shift crews, via an interactive X-Window program.

As indicated in figure 1 all control objects send status information to a central status server/broker called `opcon`. Status messages are sent any time a status parameter changes or on a regular basis depending on the object's configuration, which is stored in a database. Note that there might be more than one status broker in the system. The status server always keeps the latest status information in its internal database. Clients may query this database. Of greater importance is the possibility that `opcon`-clients may register interest in status information from specific devices. Such clients are called *monitor clients*. Monitor clients are forwarded all new status messages in which they have expressed interest. Since all control objects are identified by strings, UNIX regular expressions are allowed as a selection criterion. There are three standard monitor clients in the system:

- `opterm` prints selected information on the screen.
- `opconsole` prints debug, warning and alarm messages.
- `optrans` pipes all status information in the eventbuilder in order to save these data together with event data on tape.

`opterm` and `opconsole` normally run on one of the screens in the control room.

3. Classes and Messages

3.1 MUPIX - a message system

In order to solve the interprocess-communication problem we have developed a reliable message passing system called MUPIX (see [3] for a detailed discussion). Within MUPIX senders and receivers are viewed as objects contained in processes distributed over a computer network. These objects address each other by symbolic names. Therefore local transparency is obtained. Address resolution is provided transparently by a name-server process. The name-server maps logical addresses like the symbolic names used in MUPIX onto low-layer transport addresses. This nameserver can also be used for dynamic name binding in other protocol suites like TCP/IP. All names are registered by the objects themselves at runtime. MUPIX sits on top of UDP/IP. A thin protocol layer is added to ensure a reliable message transport. In addition MUPIX allows the mapping of asynchronous events like timer signals onto messages. This avoids the need of asynchronous programming which is difficult for an unexperienced programmer.

3.2 Class Description File

The implementation of control objects proceeds in a standardized way. The first step is writing down the *class description file (cdf)* of a control object. The cdf contains a list of all variables that completely describe the state of the object, in regard to what is of interest for the outside world (e.g. an experimentalist). These variables are called *fields*. Each field is assigned a number: its *field key*. In addition all operations which are possible on an instance of the class are listed. Like fields, operations are identified by an *operation key*. However, the operation key is composed of two numbers: a field key and an action key. For better documentation, the field key is included in the operation key, because most operations access a field's value. Each operation is given together with its input and output parameters. Since all generic programs in the system (e.g. database, user interface, monitor clients) gain knowledge about an object's interface from a cdf, the number of different parameter types is limited to a set of about ten predefined types. For the same reason only one input and one output parameter are allowed. As an example the cdf of a temperature controller is shown in figure 2.

```
#define TEMP_FK 1
#define MSG_SET 1
#define MSG_GET 2

Name:  TempController
Fields:
```

Temperature	float	TEMP_FK

```
Operations:
```

set	FLOAT	INT	TEMP_FK	MSG_SET
get	void	INT	TEMP_FK	MSG_GET

```
class TempContInterface {
public:
    virtual int set (float) = NULL;
    virtual float get (void) = NULL;
};
```

Figure 2: Class Description File and Interface Class of a temperature controller, which is completely characterized by a single parameter temperature. The temperature can be set or read by a computer.

The cdf completely defines a control object for the outside world. It also defines a protocol in terms of field and action keys, which allows the storage of state information in a database or to communicate with an object in a general way. The cdf presented here is an abstract concept. It is actually implemented as a set of database tables and header files.

3.3 Device Classes

Each device is represented by a set of C++-classes as shown in fig. 3. The functionality of a (physical) device is readily modelled by its device class. The public interface of a device class is inherited from a corresponding pure virtual

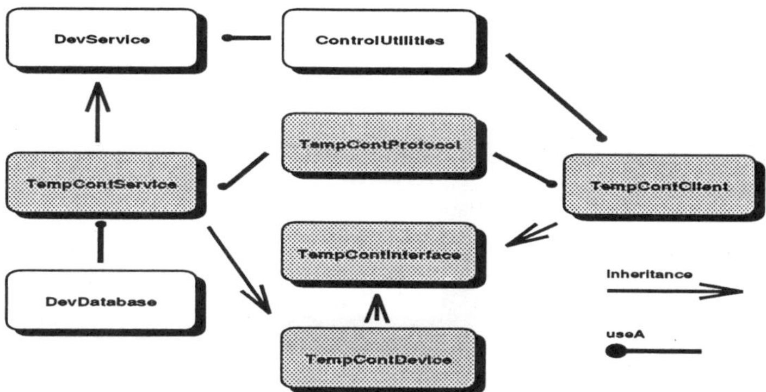

Figure 3: Class relationship of the temperatur controller example. All device specific classes are shaded.

interface class. The mapping from the cdf to the interface class is obvious (see figure 2). The device service class offers the functionality of its base device class network wide. A service registers itself at the MUPIX nameserver and waits for incoming messages, which contain member function calls encoded according to the protocol specified by the cdf together with the corresponding input parameters. The messages itself are encoded in a machine independent format using XDR [4]. Most of a service objects code is independent of a particular control object, because it deals with message-I/O, (de-)coding of messages and the invocation of the functions specified by the delivered operation key. All general code is contained in a class library. The device specific service inherits this code from a class called `DevService` which is part of the library. The particular mapping of an operation key to a member function call is part of a service class implementation which is coded using a set of predefined macros. This step has still to be performed manually. But as the whole procedure has been standardized up to a high degree it could be done automatically. This might be part of future developments. In general all message exchange between objects in different processes is done asynchronously. However, in some circumstances an RPC like behaviour is useful on the client side. Therefore a device specific client class provides the stub-routines which implement the code of the virtual interface functions. A typical control process contains in general a set of services. A class called *DevServiceSet* takes care of the distribution of incoming messages and scheduling of the different services.

4. User Interface

From a user's (experimentalist) point of view the most prominent feature of a system is of course its interface. The 3-Spectrometer-Setup ECS offers two means of user interaction:

- A simple interface program which works on command level and is especially useful in shell scripts (e.g. at system startup time).
- An interactive program for X-Window terminals.

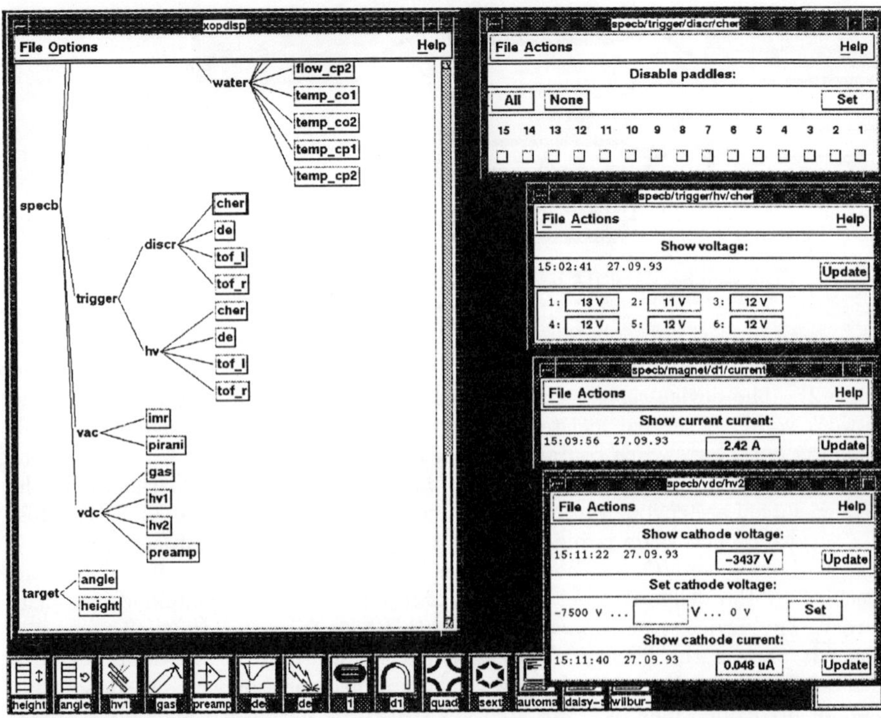

Figure 4: A typical operators display. All devices are presented in a tree. Device specific icons have been created for faster recognition.

The X-Window interface program called `Xopdisp` is written in C++ and based on the OSF/Motif toolkit. OSF/Motif was chosen because of its wide spread usage and the availability of a cheap interface builder (VUIT [5]) when the project started. `Xopdisp` is completely table driven. There is no need to recompile it if a new control object is added to the system. Since the interface of each control object is given by a cdf in a uniform way and the number of data types is limited, it is almost straightforward to write a generic interface programs for the whole system. In case of `Xopdisp` it was in principle only necessary to implement a set of display blocks (widgets) for each of the data types. A few specialized widgets have been added for ergonomic reasons. `Xopdisp` works in two modes: as a direct client of a particular control object and as a monitor client. The latter automatically updates information on the screen without user intervention.

5. Status

The ECS presented here has been operational since the first day of data taking with the 3-Spectrometer-Setup in June 1992. Even though an object-oriented concept was chosen at project start, all early code was written in C. This was due to the lack of good and cheap C++ compilers. Since acceptance of the C++-language has increased and an almost mature C++-compiler is available (FSFs g++) for free, coding is also done in C++. Some old and awkward C code which mimicked C++ features has been replaced. It has been turned out that working with an OOP language like C++ increased productivity. Future developments concentrate on refining our methods. Also of importance will be the replacement of our home grown (unix file system based) database server with some commercial tool.

1. Blomqvist et al. *Experimente mit virtuellen Photonen* in *Jahresbericht 1990-1991, Inst. für Kernphysik, Mainz*
2. H.Herminghaus et al. *The design of a cascaded 800 MeV Normal Conducting CW Race Track Microtron* , *Nucl. Instr. Meth.* **138** (1976).
3. H.Kramer and K.Merle *MUPIX — a Portable Message-System for Distributed Experiment Control Systems* in *Proceedings of the International Conference on Computing in High Energy Physics'92, Annecy, France*
4. Sun Microsystems, Inc. *XDR External Data Representation, RFC1014.*
5. Digital Equipment Corporation *DEC Vuit Users Guide* , Maynard, Massachussetts

LasVegas:

Object oriented Monte Carlo Simulations for the COSY-TOF-Spectrometer

S. Brand

and

H. Brand, H. Freiesleben, J.S. Lange, U. Zielinski
Institut für Experimentalphysik I, Ruhr-Universität Bochum
Universitätsstr. 150, 44780 Bochum, Germany

ABSTRACT

A Monte Carlo Simulation package has been developed in order to meet the special requirements of the COSY-TOF-Spectrometer. Using an object oriented approach it was possible to attain fast progress. The basic class structure of LasVegas is introduced and first applications and results are presented.

1. Motivation

A large Time-of-Flight-Spectrometer is under construction at COSY (KFA Jülich), where we will investigate the pp-Bremsstrahlung. The TOF-Spectrometer will be divided into three components: a so-called quirl (see also fig. 1), a ring and a barrel. Each of them consists of several layers of scintillator paddles. One layer is formed by straight paddles, the other ones by left and right wound scintillators. None of the standard geometry packages put at our disposal the needed geometries such as Archimedian spirals. Hence we decided to develop a Monte Carlo Program Package (LasVegas) according to the special requirements of the COSY-TOF-Spectrometer. In order to attain a clear structured, modular program with the possibility of adding any geometry and medium in a simple manner we chose C++ as programming language. We fully exploit special features of C++ by encapsulating the main characteristics of geometries and media into classes and using inheritance. These proceedings were ideal to achieve fast progress in programming. We began with simple approximations and soon were able to create more complex classes and could replace or complete the old ones in a modular way.

2. Class Structure

The main structure of LasVegas is implemented by three different kinds of base classes (see also fig. 2):

- a *geometry base class*,
- a *medium base class* and

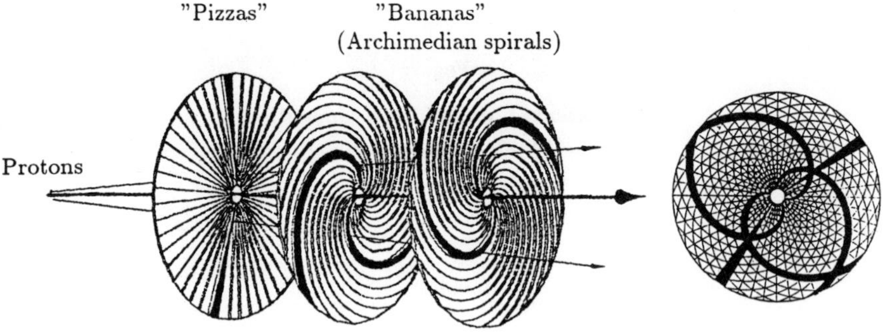

Figure 1: This is a view of the quirl showing all three layers consisting of several straight and wound scintillator paddles. The pattern caused by protons which traverses all three layers is indicated on the right side.

- a *detector base class*, which connects the former two to a complete detector component.

All methods and variables, which are necessary to describe the according group of classes, are virtually predefined in these base classes. Actual geometries or media are created by deriving a child class from the corresponding base class. In the child class the several predefined methods have to be redefined in order to implement classes which describe the real geometry or medium.

This is illustrated by two examples, which are essential for our application:

- The *geometry base class* includes a method *IsPoint()* which returns a true value if a given point is located within the defined geometry. Of course, this method has to be implemented separately for each geometry class.
- The *medium base class* includes the method *Propagate()* which should transport a given particle for an infinitesimal distance. Within the several medium classes these routines may look very differently depending on the kind of medium which is described. It may restrict itself on simply calculating energy loss and thus only varies particle kinematics according to the suffered energy loss. Due to a nuclear interaction this routine also may delete the actual particle or create new particles of a second generation.

After implementation of the adequate geometry and adequate medium this procedure of introducing base classes allows to connect **all** kinds of geometries and **all** kinds of media to one detector component via the *detector base class* (see fig. 2). The *detector base class* contains another method *Propagate()*, which transports the particle through the complete component. To ease the handling of often used geometries such as barrels, child classes were created (*mbarrel*, *mcircle*, see fig. 2), which automatically allocate the according geometry.

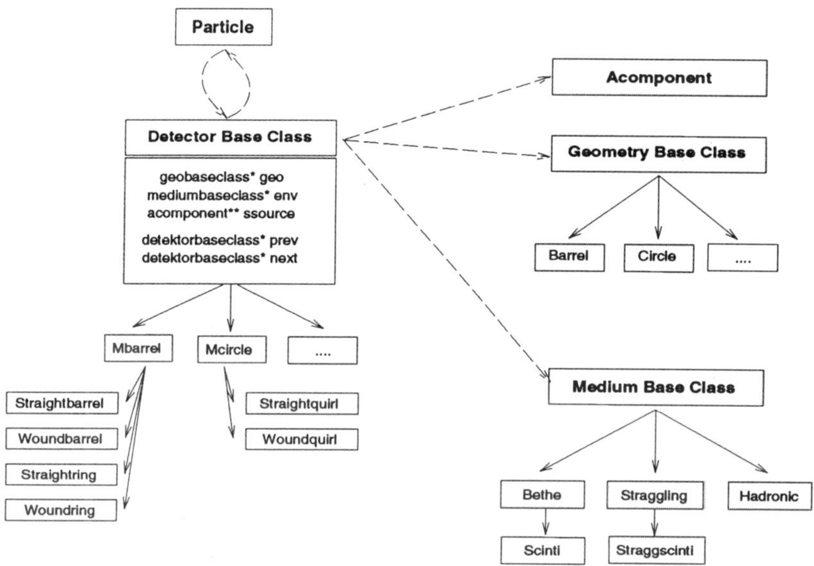

Figure 2: Overview of the LasVegas class structure: all essential classes are shown. Element classes are indicated by dashed, child classes by solid lines.

Furthermore the detector component can be declared as an active component, i.e. a component which gives electronic response, by relating one or more objects of an adequate class *acomponent* to it. In this class all information connected to the detector output is filed. If a detector component is hit, the according object *acomponent* has to be marked, i.e. informations such as energy deposit and time has to be written to it. The declaration of several active components can be used to implement a "static substructure" of one detector component. For using this it is necessary to derive a child class of a *detector base class* and to redefine the method *Propagate()* in order to manage the several active components. This makes it possible to implement an approximation of the components of the COSY-TOF-Spectrometer in a simple manner. Each straight and wound layer of the several components is represented by one of the classes *straightquirl*, *woundquirl*, etc. shown in figure 2. This is already sufficient for most problems. For more complex questions each paddle type needs to be implemented in additional classes.

The *particle* class contains parameters such as mass, charge, energy, etc., needed for particle's characterization. Passive access to this variables is given by a number of public methods; active access is only allowed to the medium classes by declaring them as "friend classes". This is due to the fact that only medium classes may have an influence on particle characteristics. The class *particle* contains two time informations: the time elapsed since the event was started and the time elapsed since the particle was created.

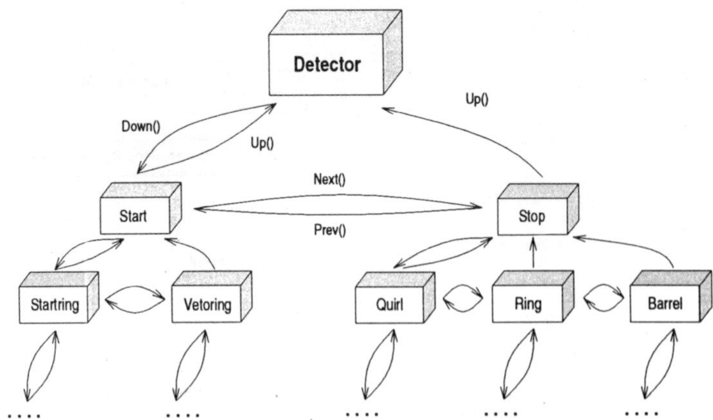

Figure 3: Example for the hierarchical structur of a detector: The Time-of-Flight-Spectrometer can be divided in a Start- and a Stop-Component, which again consists of several subdetectors. The particle list looks very similar, where the different levels represent several particle generations.

3. Particle and Detector Lists

LasVegas makes use of two kinds of lists: a detector and a particle list. Both lists can be structured in a hierarchical way so that a detector can be divided in substructures on several levels (see also fig. 3). Particles can be created dynamically and sorted into the list: the hierachical list represents several generations of particles. A superior class *detector* is able to manage both detector and particle list automatically. The propagation of several particles of one event, which on their way through the detector create new secondary particles, can thus be executed by one function call.

The propagation of a particle takes place at several levels of the LasVegas classes:

- an infinitesimal propagation (→ *medium base class*),
- propagation through a complete component with possibly marking of one or more active components (→ *detector base class or an adequately derived class*) and
- propagation through the complete detector (→ *detector*).

The user can also initialize the detector in a simple manner: a class method *Clear()* for clearing the detector and initializing the according variables exists at all different levels.

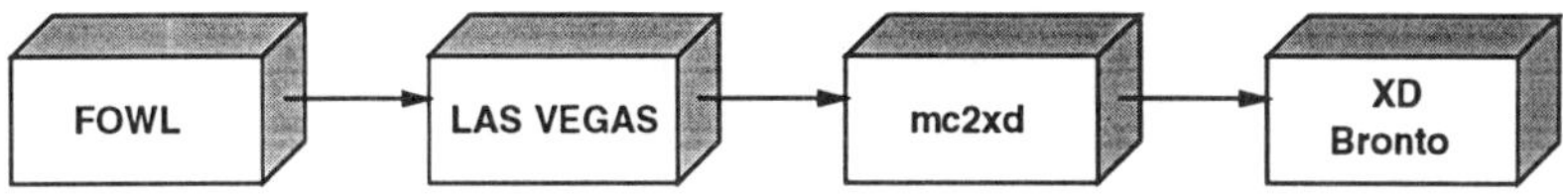

Figure 4: Interfaces for LasVegas: LasVegas gets event values from FOWL via standard input. For Simulations of the COSY-TOF-Spectrometer LasVegas is connected to the analysis software BRONTO by mc2xd via standard output. Any interface can be added within other implementations.

4. User Implementations and Example

It is up to the user to create his own implementation by

- creating media, geometries and active components and relating them to a *detector base class* object,
- linking the detector components together reasonably,
- creating particles and linking them together,
- transporting the particles through the detector, if wanted by means of the according method *Propagate()*,
- creating output by investigating the *acomponent* class members of all detector components.

It has to be mentioned that LasVegas itself doesn't have an event generator. Any event generator can be used by connecting its output to the standard input of the LasVegas implementation and reading in the different parameters of the particles of an event. In the actual implementation mainly the CERN program *FOWL* [1] and parts of the nuclear reaction code HERMES [2] are used (see also fig. 4).

On the other hand LasVegas supplies no tools for event reconstruction either. Any analysis or reconstruction has to be done by the user, perhaps via an interface. In the simulations for the COSY-TOF-Spectrometer the reconstruction is done by BRONTO[1] [3], an extension of the program package *XD* [4] written in C++, too. BRONTO will be the online analysis software of the ppγ-experiment. To connect LasVegas and BRONTO an interface, mc2xd, is necessary.

Figure 5 shows an example of the simulation obtained in this way for the TOF-Spectrometer. For a beam energy of 400 MeV different kinds of events were processed through the detector. From the detector response the missing mass of the unobserved particle was reconstructed. The problem to separate the several reactions may be solved by means of missing mass analysis. A significant improvement however can be achieved by the use of artificial neural networks. This is discussed within another contribution [5] to this conference.

[1] Bochumer Reconstruction Online Tool

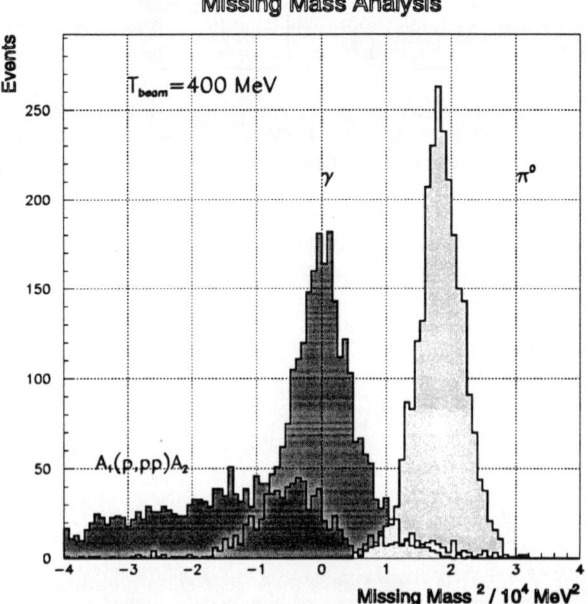

Figure 5: Example of a LasVegas-Simulation: this picture shows various missing mass spectra, reconstructed by BRONTO from events simulated by LasVegas. The reactions $pp \rightarrow pp\gamma$ and $pp \rightarrow pp\pi^0$ were generated by FOWL; the background reactions $pA_1 \rightarrow ppA_2$ were generated using the hadronic module HETC of the nuclear reaction code HERMES.

5. References

1. F. James, *FOWL - a General Monte-Carlo Phase Space Program*, (Program Library W505, CERN 1977)
2. P. Cloth, V. Drüke, D.Filges, G. Sterzenbach, *HERMES - A Monte Carlo Program System for Beam-Materials Interaction Studies*, (Jülich Preprint 2203, May 1988)
3. S. Lange, *BRONTO - reconstruction and analysis software for the ppγ-Experiment at COSY-TOF*, (private communication, Bochum 1993)
4. M. Gasthuber, F. Dohrmann, H.Krause, W. Scobel, *Datenauswertung für das EDDA-Experiment bei COSY*, (DPG-Frühjahrstagung, Mainz 1993)
5. S. Lange, H. Freiesleben, M. Kunze, M. Steinke, *Artificial Neural Networks for Reaction Discrimination in the pp-Bremsstrahlung Experiment at COSY*, (Contribution to the Third International Workshop on Software Engineering, Artificial Intelligence and Expert Systems for High Energy and Nuclear Physics, Oberammergau 1993)

SOFTWARE TOOL FOR TRAINING FIR AND FIR-OS FILTERS

JARKKO NIITTYLAHTI
Research Institute for High Energy Physics in Finland
CERN, CH-1211 Geneva 23, Switzerland

and

SAMI J. INKINEN
CERN, CH-1211 Geneva 23, Switzerland

ABSTRACT

A novel software package for designing FIR (Finite Impulse Response) and FIR-OS (FIR-Order Statistic) hybrid filters has been developed. The design approach is *to train* the filters to give the optimal response to given example signals. This method is adopted from the artificial neural networks techniques. Both analytic and stochastic methods for the training are used. Two novel algorithms were especially developed for this software. The LMA sigmoidal approximation algorithm enables finding the optimal FIR coefficients under the MAE (Mean Absolute Error) criterion using the gradient search, even if the problem is nonlinear. Similarly, the novel soft OS approximation algorithm enables finding the optimum FIR-OS filter using the gradient search.

1. Introduction

In this paper we present novel software tools for designing FIR and FIR-OS filters by training. This design approach has been chosen for two reasons: first, it enables an easy design entry to the most powerful design algorithms and filter structures, and second, it is the only possible way to design filters that are truly optimal for a given arbitrary signal. This tool has been succesfully used to design filters for signal processing in High Energy Physics[1,2].

The design concept is the following. The user prepares a file of training examples. These examples are input-output pairs that describe the desired filter output for a given input in the time domain. These examples are fed to the chosen filter structure, that has initially random values for its variables. Next the filter variable values are selected in such way that the error sum between the actual filter outputs and the desired filter outputs is minimized over all the examples. Thus the design problem is to minimize this multivariable error function.

The minimization of a differentiable function can be done with the gradient search. In the case where there are nonlinearities in the system it is much more complicated to find the optimum configuration. This is because the first derivative of the nonlinear function is not continuous and the gradient search can not be applied. In this case stochastic methods can be used, or the nonlinear function can be approximated with a linear function, and then apply the gradient search. We have implemented both of these methods.

The analytical methods are the Least Mean Square (LMS) algorithm[3], which produces optimal FIR filter under the MSE (Mean Squared Error) criterion, a Least Mean Absolute (LMA) for optimal FIR under the MAE criterion, and a FIR-OS training algorithm with soft OS approximation. The conjugate gradient[4] search is utilized in all of these algorithms. The stochastic optimization methods are the Simulated Annealing method, the Monte Carlo Method, and a Fine Tuning method. This stochastic design is supported with the *XView* user interface. Also a novel graphic curve illustrator is included as a part of the software.

2. FIR Training

The output for a FIR filter is given by

$$y = \sum_{k=1}^{t} w_k x_k \tag{1}$$

where t is the number of filter taps, w_k is the coefficient for the kth tap and x_k is the kth input. For a trainable FIR a training set of n training pair is given, where $\mathbf{X} = \{X_1, X_2, ..., X_n\}$, $X_k = \{x_1, x_2, ..., x_t\}$ is the training input and $\mathbf{O} = \{o_1, o_2, ..., o_n\}$ is the corresponding training output. The error function for the FIR filter in the LMS algorithm in the training set is

$$E = \sum_{i=1}^{n} (\sum_{k=1}^{t} w_k x_{ki} - o_i)^2 \tag{2}$$

where and o_i is the desired output of the ith training pair, and its partial derivatives are

$$\frac{\partial E}{\partial w_l} = \sum_{i=1}^{n} 2x_{li} (\sum_{k=1}^{t} w_k x_{ki} - o_i) \tag{3}$$

where $l = 1, 2, 3, ..., t$.
The minimum point, i.e., the FIR tap coefficients that minimize the error function value, can now be calculated using the conjugate gradient minimization. Because the distance between the actual and desired filter outputs is squared, the resulting filter is optimal under the MSE criterion.

In order to find the optimal filter under the MAE criterion we have to use the absolute distance between the actual and desired output values instead of the squared value in the error function. The abrupt change in the absolute value function at the zero crossing makes it nondifferentiable. In our novel LMA sigmoidal approximation algorithm, we approximate the absolute value function with a suitable differentiable function thus enabling the conjugate gradient search. The approximation is done with high numerical precision, and the error due to the approximation is neglible. The starting point of the gradient search is first obtained by the LMS algorithm, and the resulting LMA filter is the closest local minimum to the starting point. This was found to be a good approximation for the global minimum point, because better solutions could not be found with other starting points. This algorithm is given in details in Ref. 5.

3. FIR-OS Training

The hybrid filters is a class of filters where both linear and nonlinear filters are used. The FIR-OS structure includes several FIR filters and one nonlinear OS filter. The linear FIR filters are arranged to work parallelly in the first of the two filter layers. The Order Statistic filter in the last layer then ranks the outputs of the FIRs in order according to the magnitude and then selects one of them to be the output of the whole network. The output value can be the minimum value, the maximum value, or anything in between, including the median value. To fully yield the hybrid structure it is necessary that the linear filters have different characteristic to each other. The idea is that the FIRs are specialized to give good response at different points of input signal and the nonlinear part selects the best subfilter at each point. For this reason the performance of the hybrid structure is superious in comparison to any individual linear filter.

The training of the FIR-OS filters is not a trivial task because of the nonlinear OS part. However, it is possible to use similar approximation techniques as in the LMA case and use the gradient search to find the optimal filter configuration[6,7]. We have developed a succesful approximative training method for the FIR-OS training. The OS operator selects one of the FIRs for each training input. The FIR coefficients are optimized inside the training subset that is found by this OS selection. The optimization is again the minimization of the error function for the filter output. Now, the error function is formed using soft OS weighting for the outputs of each FIR. The FIR output, that was selected by the OS operator, is multiplied by an acceptance weight, and the others by a rejection weight. The magnitude of these weighting factors is equal (0.5) at the beginning of the procedure, and it is exclusive (1.0 and 0.0) at the end. Thus the OS operator is approximated to be "fuzzy", or soft, at the beginning of the training procedure. During the optimization it is made harder and eventually it is a hard OS operator. The whole optimization is done separately for each OS rank value, and the best configuration is finally chosen. An exhaustive description of the method is given in Ref. 8.

Good results can also be obtained using stochastic optimization methods. The included stochastic algorithms are the Simulated Annealing method, the Monte Carlo method, and a Fine Tuning method. These methods are especially suitable for solving complex nonlinear multivariable optimization problems. The Simulated Annealing[9] method is quaranteed to eventually find the global minimum for the cost function, if the annealing rate is slow enough. However, this powerful optimization method was found to be very slow in most cases. For simple tasks it is faster to use the Monte Carlo method, even if it is not as sophisticated as the Simulated Annealing method. For fastest possible execution time a possibility to use the Monte Carlo method with reduced number of task parameters was added. Finally, in order to have the facility to improve the existing filter parameters, one more optimization method was included to the system. It is a simplified Monte Carlo Method and here it is called Fine Tuning.

The Simulated Annealing method is analog to the annealing process of the solid matter. The main parameters of the SA optimization method are a temperature T and an error function E. The temperature is initially high and it decreases by exponential steps during the optimization process. The optimum values for the error function variables globally minimize the error function. Initially random values are given to the variables and the error function is evaluated. Then a random modification is made to the system, i.e., the value of one variable is changed, and the error function is evaluated again. Now, if the error function is decreased, the modification is accepted. Otherwise the modification is accepted by probability $P = \exp(-\Delta E/kT)$, where ΔE is the difference between the new error function value and the previous value, k is a constant and T is the current temperature. The acceptance probability is large for small error function difference and high temperature. This possibility to accept a modification that leads to decreased performance prevents trapping to a local energy minimum. After a large number of modifications has been made the temperature is decreased. At each temperature a thermal equilibrium has to be reached until the next temperature is given. At this thermal equilibrium the error function value is distributed around value that is high for high T and low for low T, until it reaches the minimum value at the zero temperature. This means that the error function minimum and the optimal values for the error function variables have been found.

The Fine Tuning method is somewhat similar to the SA-method. The main difference is that only good changes in the configuration are accepted and the bad ones are cancelled. For this reason no temperature parameter is needed. A large number of random changes are made to the initial configuration and the best one is chosen. In this case the initial configuration does not need to be a random one. This gives also the possibility to load an existing filter configuration to be improved. The weight values can be quantificated or restricted to a given range. Also the immunity to noise can be improved by adapting the filter to a signal corrupted by noise. However, this method is not able to prevent trapping to a local minimum of the cost function.

In the Monte Carlo method all the variables are changed to new random values when a new trial is made, in contrast to the previous methods, where only one variable is modified at the time. This way there is no way that this method would be trapped in a local cost function minimum.

4. Software

The software was written in C-language for Sun workstations. The user has to prepare a training set as an input file to the optimization programs. For the stochastic design a graphic interface *safir.c* was developed for *XView* environments. The graphic interface provides an easy way of mastering the optimization process (Fig. 1).

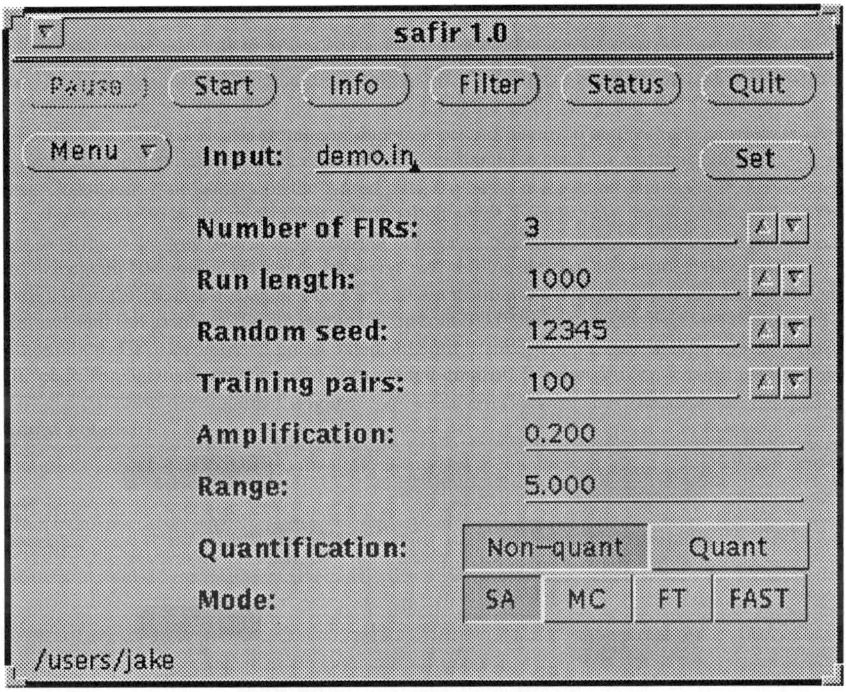

Figure 1: The *safir.c* main window

The first version of the software and the user's guide[10] is available for non-commercial use through anonymous ftp. The connection number is 128.141.8.13 (sunra1) and the directory is ftp/pub/jake. The graphic interface has been compiled using static binding of the graphic libraries. This first edition does not include the analytic FIR-OS design algorithm, that has been implemented as a MATLAB subprogram. The next version will be integrated to comprise all the programs. Also the analytic methods will be supported with graphics. A VHDL compiler 11 is included to provide interface to hardware implementation. This compiler provides outputs in behavioural VHDL, in *Xilinx* compatible hardware VHDL, and in C-language. All comments are welcome to jake@sunra1.cern.ch.

5. Conclusions

In this paper, we have presented a novel software tool for designing FIR and FIR-OS filters by training using analytic and stochastic optimization methods.

The new algorithms that were developed for this software provide new aspect to the adaptive design approach. This software has a graphic user interface for the stochastic design that makes it is easy to use and still allows detailed customizing of the structure. This approach of computer aided design can support an expert designer in his work, but it also gives to the less expertized designers the possibility to use the most advanced filter structures in the area of digital signal processing.

6. References

1. S. J. Inkinen and Y. Neuvo, "A Trainable FIR-OS Hybrid Filter for Precision Pulse Height Measurement of Particle Detector Signals", submitted to *IEEE International Symposium on Circuits and Systems ISCAS -94*, May 30 -June 2, 1994, London, England.
2. S. J. Inkinen and Y. Neuvo, "Trainable Hybrid Filter Structures in Particle Detector Readout Systems", submitted to *VII European Signal Processing Conference EUSIPCO -94*, September 13-16, 1994, Edinburgh, Scotland.
3. B. Widrow and S.D. Stearns, *Adaptive Signal Processing*, Englewood Cliffs, NJ: Prentice-Hall, 1985.
4. W.H. Press, B.P. Flannery, S.A. Teukolsky, W.T. Vetterling, *Numerical Recipes in C - The Art of Scientific Computing*, Cambridge University Press, 1988.
5. J. Niittylahti and S. J. Inkinen, "FIR Filter Design Under the Mean Absolute Error Criterion Using Sigmoidal Derivative Approximation", submitted to the *IEEE Transactions on Circuits and Systems - II: Analog and Digital Signal Processing*, May 1993.
6. F. Palmieri, "A Backpropagation Algorithm for Multilayer Hybrid Order Statistic Filters", *Proc. 1989 IEEE Int. Conference on Acoustics, Speech and Signal Processing*, 1989, Glasgow, Scotland, pp.1179-1182.
7. L. Yin and Y. Neuvo, "Adaptive FIR-WOS Hybrid Filters", *IEEE Int. Symposium on Circuits and Systems*, May 10-13, 1991, San Diego, California, USA, pp. 2637-2640.
8. S. J. Inkinen and J. Niittylahti, "Trainable FIR-Order Statistic Hybrid Filters", submitted to the *IEEE Transactions on Circuits and Systems - II: Analog and Digital Signal Processing*, April 1993.
9. S. Kirkpatrick, C. D. Gelatt and M. P. Vecchi, "Optimization by simulated annealing", *Science*, **220**, pp. 671-680, 1983.
10. J. Niittylahti and S. J. Inkinen, *FIR and Hybrid Filter Design Tools*, CERN/ECP/RD12 Technical Report, 1993.
11. T. Konnunaho, J. Niittylahti, and S. J. Inkinen, "VHDL Code Generator for the SAFIR Filter Training Software" *Proc. of Teollisuusfysiikan päivät*, Tampere, Finland, 7-8 September 1993.

USING CERNLIB, PAW AND GEANT PACKAGES ON PCs UNDER DJGPP

FINE V.
Joint Institute for Nuclear Research
Dubna, 141980, Russia

and

POTREBENIKOVA E.
Joint Institute for Nuclear Research
Dubna, 141980, Russia

ABSTRACT

This paper describes the result of porting the parts of the CERN Program Library (release 93d) on IBM PC AT/386-compatible computers under MS-DOS and GNU-ish tools provided by Free Software Foundation, Inc. This work was supported by CERN, FermiLab and JINR institutes.

1. General Appearance

The first release of CERNLIB libraries and packages for the IBM PC was presented about 2 years ago[1,2,3]. The number of requests for this release and the many positive references underline the usefulness of this work and suggest that the CERN libraries should be maintained and distributed for PCs as for other platforms.

This initial release was made using the NDP compilers from MicroWay Inc. This resulted in a number of problems and restrictions, the most important of which are listed below.

1. **The compilers are expensive (about $1500 per computer)**
2. **Lack of graphics facilities**
3. **Poor and slow exploitation of the virtual memory management features of i386/486 CPUs.**

Since this release, a new software generation has been developed by the Free Software Foundation Inc. In addition, recent "free" kits of GNU tools for MS DOS have included all of the components that are necessary to manage the large HEP Fortran and C sources.

The following "free" tools have been adopted for building CERNLIB on the PC since V93B:

```
DJGPP 1.10
```

```
F2C 12.03.93 + GCC v.2.4.1
Csaba's GRX graphics library 1.02
        by Csaba Biegl, 820 Stirrup Dr,
                  Nashville, TN 37221
```

where:

GCC is the GNU version of a public-domain C compiler and is available on many platforms.

DJGPP is a special DOS Extender and port of the GNU gcc/g++ compiler to the 386/DOS platform. This allows 32-bit protected applications to be run under MS DOS with full virtual memory support.

Various libraries and utilities are provided also.

F2C[4] is a Fortran to C converter

Hardware Requirements:

An 80386-based IBM compatible PC or PS/2, approximately 4-5MB of space on a hard disk and 512K RAM. Extended (or expanded under VCPI) memory and coprocessor are strongly recommended. The PAW package, for example, occupies 2279936 bytes on hard disk. An additional 10-12 MB are required to store the complete CERN libraries. (see Table 1)

Table 1: List of CERNLIB libraries

```
 Directory of D:\DJ\CERN\LIB

LIBPAW   A      1256432 05-28-93  18:57p
LIBGRAF  A      1039352 05-05-93  11:47p
LIBKERN  A       472946 01-06-93  12:08p
LIBPACK  A      2935816 21-05-93  16:36p
LIBMATH  A      1230316 26-05-93  14:00p
LIBGEANT A      2909400 05-05-93   5:48p
LIBGXINT A       382612 06-05-93   4:54p
        7 file(s)   10226874 bytes
```

Supported Environment:

- Up to 128M of extended memory (expanded under VCPI)
- Up to 128M of disk space used for swapping
- SuperVGA 256-color mode up to 1024x768 and VESA VGA
- 80387
- XMS & VDISK memory allocation strategies
- VCPI programs, such as QEMM, DESQview, and 386MAX

- DPMI programs, such as Windows 3.x, OS/2, and QDPMI (without graphics facilities)

Unsupported:

- Multiprocessing (fork())

Table 2 shows the result of using different optimisation modes with the combination of `f2c+gcc` or the NDP compiler using a Whetstone benchmark on an IBM PC AT 486/25 computer.

Table 2: Benchmark done by Whetstone code (KWhet/s) for AT 486/25

	Gcc 2.3.3		NDP 4.0	
Option	single	double	single	double
default	4594,27	4276,76	4880,39	4748,64
Optimize	5216,27	4791,79	5441,97	5293,84

Notes: 1M Whet/s performance is about 0.5 CERN Unit

The folloiwng table (see Table 3) gives timing results for various CERN HBOOK tests on different systems.

Table 3: Time (sec.) for HBOOK4 examples

Example number	PC AT 486 (gcc)	CONVEX 120 (fc)	PC AT 486 (NDP486)	SUN Spark 2 (f77)
1	2.02	1.22	1.54	0.40
2	5.73	4.36	4.17	1.11
3	6.59	3.31	4.23	0.88
4	1.34	0.95	1.81	0.30
5	42.42	46.94	41.69	16.16
6	1.56	1.50	1.65	0.33
7	29.81	4.80	5.77	1.26
Total	89.52	63.12	60.86	20.46

The time to process a single event using GEANT example 1 is given below.

- 47 sec/event for IBM PC AT 386/33 computer
- 10 sec/event for IBM PC AT 486/66 computer
- 11.7 sec/event for SUN SparcII (provided by Rene Brun)

The libraries and modules such as PAW (without network facilities) and interactive GEANT for MSDOS have been distributed with the standard CERNLIB release since V92B. MSDOS and IBM PC features are tagged using the `IF=MSDOS` PATCHY construct.

2. Conclusions

Advantages of "GNU"-version:

- common format of object files for MSDOS as well UNIX

- Independence from the type of graphics hardware and effective exploitation of all current and future video cards.
- Proper virtual memory management.
- Free software tools
- Full support of all HIGZ primitives.
- Unrestricted supported for all types of AT 386 PC computers.
- Possibility of building MSDOS executables and libraries on another platform, e.g. SUN, using cross software. Thus one may use standard Unix tools in the development and installation process without a performance penalty when running on the target system.

All CERN tests and examples have run correctly, including on power PCs and NoteBooks with as little as 1 MB of RAM.

We therefore suggest that GNU-ish tools and f2c are suitable for the HEP community using PCs - both MSDOS and Unix - with the following restrictions:

1. *C* floating point expressions are normally evaluated in double precision. To quote from Kernigan and Ritchie (1978 edition): "All floating arithmetic in C is carried out in double-precision: whenever a float appears in an expression it is lengthened to double by zero-padding its fraction. When a double must be converted to float, for example by an assignment, the double is rounded before truncation to float length".
2. GO32 DOS Extender runs the coprocessor at full precision (64 bits rather than the 53 bits of a double). Therefore many sub-expressions may be evaluated at even higher precision than double.

These points can account for strange results from numerical codes, such as GEANT, run on a PC using the combination of f2c and gcc. The forthcoming g77 - the GNU Fortran compiler - should hopefully solve at least the first of these two problems.

The libraries using gcc can be used on both MSDOS as well as UNIX/PC as gcc produces the same object code and uses a common library format on both systems.

Free software and the F2C converter are currently used to maintain the official release of the CERN libraries for **MS DOS**, **LINUX** (Free UNIX for PC) and **Windows/NT for the DECpc AXP** computer.

The CERN libraries and the PAW and GEANT packages can be obtained via anonymous ftp from the `asis01.cern.ch` ftp server from an appropriately registered site or computer.

3. Downloading the Distribution Files to Your Machine

All CERNLIB files for PCs are distributed by the asis01.cern.ch ftp server. This includes the following files:

```
 Directory of J:\CERN\SYS

CERN     TXT       9022 06-08-93   2:56p
DJGPP110 TXT       7293 06-02-93   2:29p
COPYING  DJ        1625 05-31-93   8:30p
README   DJ       14202 05-31-93   8:44p
COPYING  GPL      12737 03-06-91  10:12p
COPYRGHT TXT       1561 11-16-92   4:51p
PAW      CON       2234 05-28-93   4:26p
PAW      EXE    1050503 05-28-93   4:26p
PAWDEMO  EXE     311469 04-16-93   8:19p
PAWDEMO  CON       6627 04-16-93   8:19p
CERNMLIB CON        407 06-01-93  11:31a
CERNMLIB EXE     495141 06-01-93  11:31a
CERNLIB  CON        507 06-01-93   1:08p
CERNLIB  EXE    1310011 06-01-93   1:07p
CERNGLIB CON        507 06-01-93   1:13p
CERNGLIB EXE     838819 06-01-93   1:12p
CERNBIN  CON        892 06-01-93   2:26p
CERNBIN  EXE     938542 06-01-93   2:26p
GEANT    CON        596 06-02-93   6:10p
GEANT    EXE    1146883 06-02-93   6:10p
       20 file(s)    6149578 bytes
```

These files may be copied over the network as shown below.

```
mkdir \tmp
cd    \tmp

ftp asis01.cern.ch (128.141.201.136)
ftp> cd /cernlib/pc_dos50/93b/exe
ftp> mget README*
ftp> binary
ftp> mget *.exe
ftp> quit
```

The files have been packed using the LHA archiving and compression facility, available in this area (`LHA213.exe`) or from Public Domain area (for instance from `garbo.uwasa.fi`). We have generated self-extracting archives which do not require LHA itself in the unpacking process. We welcome feedback on the distribution methods and procedures used, as well to learn from similar utilities which may be more powerful or user friendly.

4. References

1. V.Fine and S.Kadantsev *Porting PAW on PC's under MS DOS.*
   In Second International Workshop on Software Engineering, Artificial Intelligence and Expert Systems for High Energy And Nuclear Physics, 1992.
2. V.Ivanov, A.Khasanov and V.Palichik *PAW/GKS Implementation on IBM PC under MS DOS.*
   In Second International Workshop on Software Engineering, Artificial Intelligence and Expert Systems for High Energy And Nuclear Physics, 1992.
3. V.Fine, O.Smirnova and T.Strizh. *PAW Using under MS DOS*, Dubna, 1992 - (JINR, P11-92-416)
4. S.I.Feldman, D.M.Gay, M.W. Maimone, N.L.Schryer. *A Fortran-to-C Converter.*
   Compurting Science Technical Report **No. 149**. (AT&T Bell Laboratories, Murray Hill. NJ 07974)

COMPUTING IN THE NINETIES
A SUMMARY OF THE TOPICAL WORKSHOP ON SOFTWARE ENGINEERING AND ROUND TABLE DISCUSSION

MARCEL KUNZE
Institut für Experimentalphysik I, Ruhr-Universität Bochum,
Universitätsstr. 150, 44801 Bochum, Germany,
e-mail:marcel@tau.ep1.ruhr-uni-bochum.de

ABSTRACT

This contribution does not only try to give an overview on software engineering and related topics presented in the talks at the Oberammergau workshop, but also tries to do an abstraction from a set of lively discussions going on at several locations for a long time. Software engineering in the sense of computing management actually seems to span a wide range from the management of super workstation farms and distributed control applications up to questions of software design and the use of software development tools.

1. Introduction

The end of the 80's saw the advent of powerful workstations that offered the solution of complex problems on everybody's desk, that would not even have been attacked with the mainframes of the 70's. Despite the progress in single processor power, one of the major efforts in the 90's will be the construction and the operation of huge compounds of several thousand workstations in a single network. Those workstation farms will be used as the supercomputers to run the software for the LHC (ATLAS) and SSC (GEM, SDC) collaborations. It was proposed to use workstation clusters not only in the domain of offline analysis, but also as a means to run second level online trigger applications[1]. Furthermore, inside the physics experiments, thousands of CPUs are working in parallel in the regime of control, trigger and data acquisition: The Software-engineers of the nineties not only have to worry about the tools to manage complex software projects of million lines of code, but they face the situation that they have to provide and use tools to manage complex pools of CPU power as well.

2. Distributed Computing

One of the basic requirements of project management is delegation. If there is frequent need of delegation of the same task, this very often leads to the installation of specific services which are shared amongst a large community of clients[2]. One of the reasons for the success of the UNIX operating system is, that its network management is entirely based on the idea of having services; there are hundreds of networking services, the most prominent of those being e.g. Telnet or FTP. As a matter of fact, the client/server model is one of the most important principles in distributed computing at all. High level services may aid in the operation and the management of large workstation clusters.

3. High Level Services

In contrast to the basic UNIX network services offered by individual workstations, there are high level services that act on the workstation cluster as a whole. The high level services are bound to one or several workstations of the cluster. The mapping of services to nodes is handled *dynamically* by a management program (e.g. day/night mode of operation). Different services may need different resources and access to special I/O devices. There are two classes: computing services and I/O services.

3.1. Computing Services

The computing services are supporting program development and data analysis
- Interactive services: Management of interactive X-window sessions
- Batch services: Management of job queues
- Parallel services: Management of parallel processing and threads
- Visualisation services: Management of data presentation, e.g. rendering
- Compilation services: Compilation of large program packages
- Debug services: Debugging of large program packages
- Checkpoint services: Management of job migration
- License services: Validation of program usage

3.2. I/O Services

The I/O services handle the input, output, transmission and transformation of data
- File services: Management of directories and disk space
- Tape services: Management of tape storage
- Database services: Management of information
- Print services: Management of text and graphics
- Multimedia services: Management of movies, sound and video conferences
- Trigger services: Management of high level online triggers
- Data acquisition services: Management of online data streams

For the I/O services the evolution of the Fibre Channel Standard (FCS) could play an important integrative role: Using FCS, virtually any bus and backplane technology might be replaced by one single standardized data transport medium and protocol.

A lot of excellent client/server solutions are already commercially available[a], some services are just at the beginning (e.g. tape service). As the usage of a service grows, it is possible to add on further nodes to take part in that particular service. This *scalable* approach allows for continuous growth of a site´s computing power at a moderate cost. The construction of a TeraFlop-cluster until 1996 seems feasible.

---

[a] E.g. LoadLeveler, Workbench, Data Explorer (IBM)

4. Installation of Large Workstation Clusters

The recent advent of DCE (Distributed Computing Environment) eases the setup of large workstation clusters by offering a single system image: Service authorization and I/O services are managed in a transparent way centrally. On the other hand, there are some practical issues that complicate the installation and operation of large workstation clusters ("Superclusters"):

- Bandwidth:
  Current networking technology finds its limit at 1 GBaud. For a supercluster consisting of 5,000 nodes this leads to a worst case average bandwidth of 200 kBaud, a baudrate somewhat lower than AppleTalk. For this reason the transition from a round-robin to a switched network seems unavoidable (packet switched and/or peer-to-peer). The development of giga-switches will be even more important than the offer of fast CPU cycles. In addition, fast switch technology is of major interest in the area of telecommunication as well and shows a rapid development.
- Maintanance:
  Without the availability of special management tools, a supercluster would need about 100 system managers to keep it running. This would mean, that about 10% of the LHC/SSC staff would have to deal with the task of system management full-time.
- Availability of internet numbers:
  The current 32 bit addressing scheme supports 65,536 class B networks only. If everybody in the world starts the construction of superclusters, we would sooner or later run out of internet numbers. The extension to 64 bit addressing seems necessary.

5. Software Development

5.1. Parallel Processing

How is it possible to exploit the capabilities of such a supercluster of workstations? Is it useful to use a supercluster as a massive parallel system ? Most physics applications can be treated the classical way in the sense of farming problems, i.e. parallelism takes place at a coarse grain event level: The same program can be loaded simultaneously on different processors and run in parallel, there is almost no synchronisation overhead.

In contrast to that, however, more ore less all methods imitated from nature (e.g. artificial neural networks, genetic algorithms) and most of the symbolic manipulation problems need to be parallelized on a fine grain level. The calculation of Feynman graphs and QCD lattice calculations would benefit much from scalable parallel computer systems. Compiler technology and programming environments are just coming up; promising results have been shown in this workshop[3].

5.2. Object-Oriented Programming (OOP)

The detectors of particle physics are intrinsically complex devices, and so is the software to deal with them. Complexity very often leads to unforeseen behaviour. Validation and verification, a central issue of software engineering, *must* be nevertheless guaranteed. Due to the importance of the validation process, it has to be the overall strategy of any software development for the proposed LHC/SSC experiments to reduce complexity. How can we manage to write fault-tolerant, robust software that is reusable in an easy way? Very promising in this respect seems to be the paradigm of object-oriented programming[4] (OOP). OOP can be understood in terms of a transition from procedural software design philosophy to a service based software design. The real world deals with objects; the object-oriented paradigm turned out to be especially successful in the conception of any kind of *real-world application*: Encouraging results have been reported on in this workshop in the field of device control[5,6], trigger simulation[7,8] and Monte-Carlo simulation programs[9,10]. The question of which programming language to choose seems less stringend once programmers obey the principles of object-oriented software analysis and design[11] ("object think"). As C++ matures and it is widely available, it might be the premier choice[12].

With all object-oriented languages, however, there still remains the problem of object persistance: it is yet not foreseen to store the objects and their methods to access the data in a standardized way. Different solutions to the problem have been proposed and methods are available, but with SOM[a] (System Object Model), an object oriented toolkit with C and C++ bindings including persistance, a standard way will be available until 1996.

6. Tools

A lot of tools are available to support the management of large software projects[13]. In the field of object oriented software development, object-oriented analysis and design (OOA/OOD) according to Coad and Yourdon[14] and object modelling technique (OMT) according to Rumbaugh are widely accepted methodologies in industry. In this field, tools are available that offer not only the design of software in a comfortable way but also allow for automated code generation[b]. In addition, code management, language sensitive editors and code analyzers, compilation and debugging services are put together jointly in integrated development environments[c]. According to the client/server principle, the distribution of programming services around a workstation cluster and the integration of one´s own specific services into the environment is possible.

For documentation purposes, FrameMaker[d] seems to be a good choice; in addition to perfect WYSIWYG wordprocessing, it offers comprehensive graphics tools and an excellent formula editor. Its hypertext capability eases the generation of hypertext links and thus the posting of information to the World Wide Web (WWW) in HTML format. Furthermore, FrameMaker is available for virtually any platform.

---

[a] SOM is a CORBA compliant technology (Taligent)

[b] E.g. ObjectMaker (Mark V Systems) and Westmount OMT

[c] E.g. Workbench and Integrator (IBM), Softbench (HP)

[d] Frame Technology Corporation, San Jose, CA

7. Database Management Systems

As the amount of information in a physics experiment is increasing, the installation of database programs to manage the flood of information becomes vital. It is not only necessary to store data concerning manufacturing processes and manpower, but also data concerning the detector system itself, its devices and data. The use of databases offers flexibility in design changes of both, the detector hardware and software. There are three main aspects of database applications:

- Detector system database
  A consistent description of the detector geometry, the detector components and materials is needed not only for running the analysis and Monte-Carlo programs offline, but also for the online event display, data acquisition and trigger programs. Furthermore, information is acquired for all kinds of electronic equipment used in an experiment to keep track of e.g. a component´s history.
- Event data database
  Any work related to data summary tapes (DST) for a final physics analysis would benefit from direct access on specific events. Regarding the progress in direct access storage technology, the transition to data summary databases (DSD) seems promising.
- Run history database
  The run history database holds information concerning an experiment´s slow control parameters (e.g. beam momentum, gas mixtures, voltages) for different running periods. This information is of main interest for data analysis and error diagnosis.

A comparative performance measure of relational database management systems (RDBMS) and object oriented database management systems (OODBMS) has been performed[15]. Despite a small degrade in performance, OODBMS seem to fit the physicist´s needs very well. As there are many different databases on the market, exchange of information might however be a problem.

8. Conclusion

The construction of TeraFlop superclusters of workstations seems in reach within a few years. Due to the rapid lifecycle of processor technology, a careful investment in hardware is mandatory. A scalable parallel approach based on the client/server principle in a switched network seems to be a promising way to overcome bandwidth limitations. Tools are needed to manage supercluster services centrally.

Regarding the spectrum of methodologies and software development tools on the market, it seems advisable to jointly form study groups to figure out what is relevant. A lot of work and manpower can be saved by careful investigation of the market. Concerning the cost of commercial software, campus licenses (collaboration licenses ?) should be afforded.

Of major importance in this respect are the FreeHEP software catalogue and the World Wide Web project. Those projects are excellent examples for practice of international collaboration and should become a central forum for the scientific community to offer software products and to discuss experience and trends.

9. References

1. S. O'Neale, *Computing in ATLAS: Status and Plans*, these proceedings
2. J. Burton, *Physics Research Integrated Development Environment*, these proceedings
3. O. Krämer-Fuhrmann, *TRAPPER - A Graphical Programming Environment for Parallel Systems*, these proceedings
4. M. Kunze, *Experience with Mixed Language Programming*, New Computing Techniques in Physics Research II, Ed. D. Perret-Gallix, World Sci. 1992
5. H. Kramer, *The Experiment Control System of the Three-Spectrometer Setup at MAMI*, these proceedings
6. P. Duval, *A Hierachical and Distributed Control Architecture for a HEP DAQ System*, these proceedings
7. J.C. Vermeulen, *Simulation of Data-Acquisition and Trigger Systems in C++*, these proceedings
8. I.C. Legrand, *Real Time, Data Driven Architectures Simulated in Concurrent C++ for the LHC second-level Trigger*, these proceedings
9. S. Brand, *Las Vegas: Obect Oriented Monte-Carlo-Simulation for the COSY-TOF-Spectrometer*, these proceedings
10. E. Agterhuis, *Object Oriented Programming Techniques in Detector Design*, these proceedings
11. P. Coad, J. Nicola, *Object-Oriented Programming*, Prentice Hall, Englewood Cliffs, N.J., 1993
12. B. Stroustrup, *The C++ Programming Language*, 2nd Ed., Addison-Wesley, Reading, Mass., 1991
13. B. Jones, *Software Engineering Techniques and CASE Tools in RD13*, these proceedings
14. P. Coad and E. Yourdon, *Object-Oriented Analysis and Design*, Prentice Hall, Englewood Cliffs, N.J., 1991
15. J. Burton, *Comparative Performance Measures of Relational and Object-Oriented Databases for HEP Data*, these proceedings

ARTIFICIAL INTELLIGENCE

- Symbolic AI
- Evolutionary Algorithms
-Neural Networks

(IV) d^{st} = ODD-INTEGER <-- AUTMATHCAT crossover ------> d^{st} = EVEN-INTEGER

causes

STATISTICS BAYES' - THEOREM (b)

causes

NEURAL-NETWORK SIGMOIDAL-FUNCTION effect a

$1/(\oplus 1+e^{-\Delta/T})$ --AUTMATHCAT Crossover or--> $1/(\ominus 1+e^{-\Delta/T})$

II ≈ II

FERMI-DIRAC (FERMION) QUANTUM-STATISTICS $1/(e^{\hbar w/kT}\oplus 1)$ --AUTMATHCAT Crossover--> BOSE-EINSTEIN (BOSON) QUANTUM-STAT. $1/(e^{\hbar w/kT}\ominus 1) \sim$ "1"/f^1 (a la Hsu!!

III III

(BOUNDARYFUL) ≡ &/III (LOCALITY) ("1"/f^0-WHITE) NOISE≈GENERALIZED-SUSCEPTIBILITY POWER SPECTRUM

(...GLOBALITY...)≡(...BDRY-LESS...) &/III (..."1"/f^1-HYPERBOLICITY...) NOISE≈ GENERALIZED-SUSCEPTIBILITY POWER SPECTRUM

II II

PAULI-EXCLUSION PRINCIPLE HUND'S-RULE TRAPPING IN LOCAL-MINIMA('THE ELEMENTS) NECESSITATING SPECIFICITY-OF-COMPLEXITY TACTICS TO EXTRICATE ONESELF FROM ONE'S OWN ARTIFICIAL...NATURAL-STUPIDITY...MYOPIA OF THE SIGMOIDAL-FUNCTION SWITCHING-"ON" IN THE WRONG DIRECTION...NECESSITATING NON-OPTIMAL SoC-TACTICS:BOLTZMANN-MACHINE & SIMULATED-ANNALING, or GRADED-SLOPE FUNCTION(Treves & Amit)

(c) AUTOMATIC INTERCONNECTION OF ALL LOCAL MINIMA TO ONE ANOTHER AND TO GLOBAL-MINIMUM (IFF IT EXISTS!!!)

& $\delta(w-w_0)=\delta(w-0)$

GENERALIZATION TO:

$\#/(e^{\hbar w/kT}\ominus 1)=\#/(\ominus 1+e^{-\Delta/T})$

WITH

ASYMPTOTIC-LIMIT # --> oo

II

NEURAL-NETWORK "SOFTWARE" BOSE-EINSTEIN C O N D E N S- A T I O N

II

EFFECTIVELY PUSHING "SOFTWARE" "DIGITS" THROUGH LOCAL-MINIMA INTERCONNECTION TUNNELING ie. SUPERFLUIDITY

causes

OPTIMIZING OPTIMIZATION-PROBLEMS O P T I M A L L Y (OOPO)

COMMENTS (?)

(V) Hardware implimentation of "software" BOSE-EINSTEIN C O N D E N S A T I O N requires correct dimensionality (d^{st} EVEN-INTEGER O N L Y!!!) & CORRECT ASYMPTOTIC-LIMIT (# --> oo) to effect HYPERBOLICITY limit as DIRAC DELTA-FUNCTION on Power-Spectrum Amplitude/Magnitude vertical-axis in spectral-density-function & a cheap Bose-Einstein CONDENSABLE working-fluid". Recent convergence of work pn Beckenstein-Hawking-Unruh-effect of Takagi[5] and Garlock and Klimontovich[8] prediction of Bose-Einstein CONDENSATION of L I G H T,along with Chiao et.al.[8] prediction of the di-photon and Chang[9] prediction of condensed-matter fiber-optics Bose-Einstein CONDENSATION of LIGHT mandate OPTICAL-NEUROCOMPUTATION via Bose-Einstein CONDENSATION of LIGHT into a "hardware" superfluidity which just might effect & implimentation predicted NN OOPO via Bose-Einstein CONDENSATION of "software"!!!

1. E.Siegel,Symp.on fractals,MRS Fall Mtg.,Boston(1989)-5 papers!
2. S.MacLaine,Induction (1981);Category-Theory(1981);J.Simpson-pvt.comm.
3. " Takagi-Matsushita-pvt.comm;J.Fuzzy Reasoing (1990)

INTERLOG: CONSTRAINT LOGIC PROGRAMMING ON NUMERIC INTERVALS

Bernard BOTELLA
Dassault Electronique, 55 quai Marcel Dassault
Saint Cloud, 92214 France

and

Patrick TAILLIBERT
Dassault Electronique, 55 quai Marcel Dassault, Saint Cloud, 92214 France

ABSTRACT

INTERLOG is a constraint programming language which enables non-linear constraints and inaccurate data to be handled. This paper describes the characteristics of the language and its underlying principles and gives several examples of applications.

1. Introduction

Physical problems in an industrial environment tend to be solved more and more frequently by artificial intelligence techniques. Among these techniques, constraint logic programming is assuming an ever increasing role, supported by a wide range of research work (1-2). Unfortunately the problem solving methods that are at the core of CLP (linear programming or integer programming) are not suitable for the modeling of certain aspects of physical phenomena, thus:

- data are not only rarely integer values but are also inaccurate. This inaccuracy, which is one of the characteristics of each datum, has to be dealt with rigorously by the problem solver.
- physical phenomena cannot always be expressed in the form of linear problems.

It is for these reasons that we at Dassault Electronique have focused our attention on the facilities offered by Interval Arithmetics (3-4). We began by investigating the use of these techniques for diagnosing analog electronic circuits. This line of research led to the CATS algorithm (5-7), which incorporates the semantics of interval arithmetics into an Assumption-Based Truth Maintenance System (ATMS) with numerous applications ranging from analog circuits to the power supply system of a satellite (8) and now extending to the control of a petrochemical plant (HINT: ♯6447 Esprit project).
The results obtained so far with interval arithmetic encouraged us to extend its application beyond the field of diagnosis; this is the aim of INTERLOG, a Prolog

extension to constraints on numeric intervals, which is the subject of this paper.

2. INTERLOG overview

INTERLOG, whose roots can be traced back to BNR PROLOG (9), is built on a commercially available Prolog system. It allows numeric constraints to be added anywhere in the program and their consequences are drawn automatically as soon as (possibly partial) information is obtained about variables.
For example, if the constraint $Y = X^2$ is added at some time, the program will deduce $Y \geq 0$; if later, as a consequence of some other piece of knowledge, it happens that $Y = X + 2$, then $Y = 4$ and $X = 2$ will automatically be deduced.
Any kind of constraint can be introduced; the ability to handle such non-linear constraints - which is not the case for other well-known systems like CLP (R) (10), CHIP (11), PROLOG III (12) or is limited to polynomial constraints as in CAL (13) - results from the interval propagation procedure which INTERLOG is based on.
Here is an example of a constraint system that can be specified in INTERLOG:

$$\begin{cases} (X,Y) \in [0,1000]^2 & \\ (Z,T) \in [0,\pi]^2 & \\ X \times Y + T - 2 \times Z & = 4, \\ X \times \sin(Z) + Y \times \cos(T) & = 0, \\ X - Y + \cos(Z)^2 & = \sin(T)^2, \\ X \times Y \times Z & = 2 \times T \end{cases}$$

The following results are obtained:

$$\begin{cases} X \in [1.9999, 2.0001] \\ Y \in [1.9999, 2.0001] \\ Z \in [1.5636, 1.5701] \\ T \in [3.1273, 3.1416] \end{cases}$$

which is a rather good approximation of the solution

$$X = Y = 2, \quad Z = \frac{\pi}{2}, \quad T = \pi$$

In order to understand how INTERLOG functions, let us now look at the elementary principles on which it is based. They are three in number:

- interval arithmetic,
- declarativity of constraints,
- interval restriction.

These principles are jointly used in a propagation mechanism.

2.1 Interval arithmetic

As soon as one starts to tackle physical problems the problem of how to represent numeric data arises.

To begin with, it is obvious that when a measuring device gives the value M for a datum with an error E, the only possible conclusion is that the datum belongs to the interval [M - E, M + E]. Being able to represent this datum by this interval is a way of keeping all the available knowledge.
Secondly, when it comes to handling real numbers on computer, the traditional method is to approximate the value using a floating point number. But this choice is not without risks: the slight inaccuracy due to this approximation may, if propagated throughout large computations, distort the results and jeopardize their validity. So here, too, the fact of representing real numbers not by a floating point but by an interval encompassed by two floating points, and of using this interval in all computations, means that the results (which will also be intervals) will be valid.
In general terms an interval allows the degree of knowledge of a numeric datum to be expressed:

- $[a, a]$ when it is known, specific and equal to a,
- $[a, b]$ when it is known to be limited by the two values a and b,
- $]-\infty, +\infty[$ when the datum is totally unknown.

In INTERLOG a numeric datum is therefore represented by an interval. As suggested earlier, in order to take full advantage of this form of representation it is necessary to know how to handle these intervals in computations.
An operation on two intervals is defined as follows:

- $[a, b]$ op $[c, d]$ = the smallest interval that contains (x op y) for every $x \in [a, b]$ and $y \in [c, d]$.
- $f([a, b])$ = the smallest interval that contains $f(x)$ for every $x \in [a, b]$.

In particular:

$$[a, b] + [c, d] = [a + b, c + d]$$

$$[a, b] \times [c, d] = [\min(a \times c, a \times d, b \times c, b \times d), \max(a \times c, a \times d, b \times c, b \times d)]$$

To date, INTERLOG implements multiplication, addition, division, subtraction, power, logarithms, exponential functions, trigonometric functions ($\cos, \sin, tan$).
A first simple yet extremely useful application of INTERLOG is the computation of mathematical expressions on intervals, for instance:

$$\frac{[1.2, 2.35]^2 \times \log[2.15, 3.2]}{cos[13.7, 14.8]} = [1.102, 7.192]$$

2.2 Declarativity of constraints

The second principle on which INTERLOG is based, comes from the notion of constraint. A constraint (in the case of INTERLOG it is an arithmetic equation or inequation) is a relation that links several data and allows a certain number of pieces of elementary knowledge about this data to be defined concisely.

For example, the simple constraint X + logY = 2 contains the following imperative knowledge:

$$\begin{cases} X := 2 - \log Y, \\ Y := \exp(2 - X). \end{cases}$$

In INTERLOG, the user only has to express knowledge declaratively in the form of constraints. It is up to the system to deduce the elementary knowledge involved by these constraints and to use it in the right way and at the right moment.
For example, if the user specifies $X + \log Y = 2$, INTERLOG is able to deduce Y as soon as X is known and vice versa.
This property illustrates the usefulness of constraint programming in knowledge-based systems where it is desirable to represent models as naturally as possible without prejudging how they will be used during reasoning.

2.3 Interval restriction

The third principle underlying INTERLOG is based on the interpretation of equality (or inequality) in numeric equations on intervals.
Let us suppose we have two data X and Y represented respectively by the intervals $[a,b]$ and $[c,d]$. Let us set the constraint $X = Y$. This implies that, in order for there to be a solution, X must also belong to $[c,d]$ and Y must belong to $[a,b]$. The only possible values are therefore in the intersection of these two intervals. If this intersection is empty the system has no solutions. If the intersection is not empty it can now be used to represent X and Y and provides a more accurate piece of knowledge about them.
We can generalize this principle to all constraints and say that each time a constraint is evaluated it either restricts the intervals of the variables it contains or else proves that the system has no solutions.

2.4 Propagation mechanism

We have already seen that once a constraint has been set, it makes possible to specify more accurately the variables it refers to. As long as the constraint is deemed to hold, these variables can be specified by other constraints. This constraint must hence be revaluated to take this new knowledge into account.
Propagation is a technique which stores all the constraints which are deemed to hold and which, each time there is an interval restriction, provokes the revaluation of the constraints that are affected by this modification.
This revaluation can in turn provoke interval restriction which will then trigger the revaluation of other constraints. This iterative process stops once no more intervals can be restricted. Let us take the following example:

$$\begin{cases} (1) \quad Y^2 = 16, \\ (2) \quad Y = \log X, \\ (3) \quad Y = Z^2. \end{cases}$$

From (1) we deduce that $Y \in [-4,4]$. Note that in order to avoid combinatorial explosion during propagation INTERLOG only retains one interval (the smallest

encompassing all solutions) for each variable. It is up to the user to ask explicitly for the list of solutions.
From (2) we deduce that $X \in [exp(-4), exp(4)]$.
From (3) we deduce that $Z \in [0,2]$ and that $Y \in [0,4]$. This new value of Y is propagated and provokes the revaluation of (1).
From (1) we deduce that $Y = 4$ which is propagated in (2) and (3).
From (2) we deduce that $X = exp(4)$. From (3) we deduce that $Z = 2$.

2.5 Example

Let us see how these three principles are used within the propagation mechanism in the case of a simple equation:

$$X + \log X = 0$$

Constraint decomposition gives the following system:

$$\begin{cases} (1) & X := -\log X \\ (2) & X := \exp(-X) \end{cases}$$

Taking X as an unknown - i.e. belonging to $]-\infty, +\infty[$ - equation (2) gives us the fact that $X \in \exp(-]-\infty,+\infty[) = [0,+\infty[$ as the result of the application of interval arithmetic. This new encompassing of X is propagated in (2), giving $X \in [0,+\infty[\cap exp(]-\infty,0]) = [0,1]$. By iterating the same process we get:

$X = [0.36, 1]$, $X = ...$, $X = [0.56706, 0.56719]$ successively.

This iteration stops when no more variables change within a certain degree of accuracy which can be parametrized by the user.

2.6 Synthesis

Using all above-mentioned principles together enables INTERLOG to reason on a system of equations either in order to find the solutions (or at least to get as close to them as possible) or to prove that there are no solutions.
INTERLOG garantees that if the problem has a solution, the solution belongs to the resulting intervals and that if a failure occurs (due to an empty intervals intersection) it is a proof that the system has no solution.
However, when INTERLOG gives a result, it does not garantee that it is the exact (smallest interval) solution nor even that the system has a solution.

3. INTERLOG application examples

In addition to interval computation, INTERLOG may also be used in two types of application, namely system consistency checking and system resolution.

3.1 Consistency checking

How can it be proved that a set of constraints (possibly non-linear) is inconsistent (i.e. has no solutions)?
INTERLOG can solve this problem. Thanks to its propagation and interval restriction mechanisms it can deduce through saturation all the consequences in terms

of intervals on the variables of a system. As soon as it has deduced that a given variable must belong to several intervals without intersections it concludes that the system is inconsistent.
We have used this method to respond to problems arising in the design of devices. A device is an assembly of elementary components and we have the specifications of the device, the characteristics of the individual components and the constraints on the way they are assembled. During a first step all the different ways in which the components are assembled that do not correspond to the device specifications have to be eliminated. Using INTERLOG, this stage can be carried out with the advantage of being able to detect incompatibilities very early on, even before the device is completely assembled, thus avoiding having to consider all possible assemblies.

3.2 System resolution

In this case the system of equations has solutions, the question being to find them or at least to characterize them as accurately as possible.
Let us take an example from pure mathematics, that of the computation of a regular pentagon. This problem of determinating the vertices of a pentagon - knowing one of them P1 and P0 the center of the circle they belong to - can be expressed in the following terms:
Find P2, P3, P4, P5, such that :
distance(P1,P2)=distance(P2,P3)=distance(P3,P4)=
distance(P4,P5)=distance(P5,P1)
and distance(P0,P1)=distance(P0,P2)=distance(P0,P3)=
distance(P0,P4)=distance(P0,P5).

Here is the equivalent INTERLOG program (INTERLOG being an extension of PROLOG, an INTERLOG program is a PROLOG program using some additional predicates):

```
% the "distance" predicate defines the distance between two points
distance(p(Xa,Ya),p(Xb,Yb),D):-
   D**2 :=: (Xa - Xb)**2 + (Ya - Yb)**2, D :>: 0 .
% the "pentagon" predicate defines a pentagon centered on P0
pentagon(p(X0,Y0),p(X1,Y1),p(X2,Y2),p(X3,Y3),p(X4,Y4),p(X5,Y5)):-
   distance(p(X1,Y1),p(X2,Y2),D) ,          % distance(P1,P2) = D
   distance(p(X2,Y2),p(X3,Y3),D) ,          % distance(P2,P3) = D
   distance(p(X3,Y3),p(X4,Y4),D) ,
   distance(p(X4,Y4),p(X5,Y5),D) ,
   distance(p(X5,Y5),p(X1,Y1),D) ,
   distance(p(X0,Y0),p(X1,Y1),R) ,          % distance(P0,P1) = R
   distance(p(X0,Y0),p(X2,Y2),R) ,          % distance(P0,P2) = R
   distance(p(X0,Y0),p(X3,Y3),R) ,
   distance(p(X0,Y0),p(X4,Y4),R) ,
   distance(p(X0,Y0),p(X5,Y5),R).
```

To ask for the pentagon centered on (0,0) and with P1 = (1,0), submit the PROLOG goal:
pentagon(p(0,0),p(1,0),p(X2,Y2),p(X3,Y3),p(X4,Y4),p(X5,Y5)).

The obtained result is:

$$\begin{cases} (X2,Y2) \in [0.306, 0.312] \times [-0.953, -0.950] \\ (X3,Y3) \in [-0.813, -0.804] \times [-0.594, -0.582] \\ (X4,Y4) \in [-0.813, -0.806] \times [0.582, 0.592] \\ (X5,Y5) \in [0.308, 0.310] \times [0.950, 0.952] \end{cases}$$

For this type of problem there are cases where the elementary mechanisms described above do not encompass the solutions sufficiently accurately. However, INTERLOG provides several refinement procedures stemming from our theoretical research on interval propagation (14-15). Similarly, it is possible to code conditional constraints, which are added dynamically when some condition becomes true, or demons, which are activated when a variable is updated.

4. Conclusion

INTERLOG allows reasoning on inaccurate data and the use of non-linear mathematical relations linking the data.
The main kind of application we had in mind when designing INTERLOG was AI programs, in which it helps to:

- draw a link between symbolic and numeric domains,
- improve the declarativity of knowledge

and in which the slightest narrowing of the search space is worthwhile since AI programs may be threatened by combinatorial explosion.
In addition it appeared that INTERLOG is also of value for purely numeric problems where mathematical tools are too clumsy and unwieldy.

5. Acknowledgments

We would like to thank Olivier Lhomme and Serge Varennes for their useful comments and Rosalind Greenstein for the English version of this paper.

References

1. J. Cohen, *Constraint Logic Programming Languages*, Communication of the ACM, Vol. 33, Number 7, 1990.
2. J. Jaffar, J.L. Lassez, *Constraint Logic Programming*, Proc.14th ACM POPL Conf., Munich, 1987.
3. G. Alefeld, J. Herzberger,*Introduction to Intervals Computations*, Academic Press, 1983.

4. R.E. Moore, *Interval Analysis*, Prentice Hall,New Jersey, 1966.
5. P. Dague, P. Deves, O. Jehl, P. Luciani, P. Taillibert,*When oscillators stop oscillating*, IJCAI 91,Sydney, 1991.
6. P. Dague, P. Deves, P. Luciani, P. Taillibert,*Analog system diagnosis*, ECAI 90,Stockholm, 1990.
7. P. Dague, P. Deves, P. Luciani, P. Taillibert, *Model-based Diagnosis of Analog Systems : an Application to Electronic Circuits*, 1st International Workshop on Software Engineering, Artificial Intelligence and Expert Systems in High Energy and Nuclear Physics (IN2P3), Lyon, 1990.
8. P. Deves, C. Fischer, P. Taillibert,*An alternative to Expert Systems for Electrical Diagnosis*, Future Generation Computer Systems, Vol 7 (1991/1992), pp 343-352.
9. W. Older, A. Vellino,*Extending Prolog With Constraint Arithmetic on Real Intervals*, IEEE Canadian conf. on Electrical and Computer Engineering, 1990.
10. J. Jaffar, S. Michaylov,*Methodology and Implementation of a CLP System*, Proceedings of the 4th ICLP, Melbourne, 1987.
11. M. Dincbas, P. Van Hentenryck, H. Simonis, A. Aggoun, T. Graf, F. Berthier, *The Constraint Logic Programming Langage Chip*, in Proceedings of the International, Conference on Fifth Generation Computer Systems, Tokyo, 1988.
12. A. Colmerauer,*An Introduction to Prolog III*, Communication of the ACM, vol. 33 (7), pp. 69-90, 1990.
13. K. Sakai, A. Aiba, CAL: *A Theoretical Background of Constraint Logic Programming and its Applications*, Journal of Symbolic Computation 8, 1989.
14. O. Lhomme, *Consistency Techniques for Numeric CSPs*, Proceedings of the 13th IJCAI, Chambery, 1993.
15. O. Lhomme, *Une consistance partielle pour les problemes de satisfaction de contraintes sur intervalles*, 12eme journee sur les systemes experts et leurs applications,Avignon,1992.

NATURAL LANGUAGE INTERFACE FOR KNOWLEDGE REPRESENTATION SYSTEM FOR MATHEMATICS

Tatjana Yakhno
Institute of Informatics Systems,
Russian Academy of Science, Siberian Division.
Pr.Lavrent'eva, 6, Novosibirsk, 630090, Russia
e-mail: yakhno@isi.itfs.nsk.su

ABSTRACT

In the paper the application of the pragmatic-oriented analysis to the problem of text translation is described. The input text is in the "natural" mathematical language describing the informal proof of the mathematical theorems. The output text is the formal one which is the input for the system ONTIC - the system of automatical verifying of the theorems. The system Lingua.F is used to make such a transformation on the base of pragmatic-oriented analysis.

1. Introduction

In the paper the application of the pragmatic-oriented analysis to the problem of text translation is described. The considered texts are the proofs of the theorems written in the "natural" mathematical language. These texts are transformed into the formal texts which are the input for the system ONTIC -the system of automatical verifying of the theorems. The output text consists in the information of general use, which is kept in the knowledge base of the system.

Such an application demonstrates the fruitfulness of the pragmatic-oriented analysis and the necessity to use linguistic information (vocabulary, grammar) in the knowledge bases of intellectual systems. This information makes the background for synthesis of natural language explanation and interfaces in the programming environment for mathematicians.

2. Brief ONTIC Descriprion

ONTIC is an interactive system for developing and verifying mathematics [1]. This system was designed with both motivations in mind - an attempt was made to make the system pragmatically effective verification system and the same time to make the system a rough model of human mathematical cognition. The ONTIC system should be judged on two independent grounds relative to these two goals. First, one can evaluate the system as an engineered device for verifying proofs by attempting to use the system for that purpose. Second, one can attempt to evaluate the system as a cognitive model by judging the similarity between natural language arguments acceptable to people and formal arguments acceptable to the system.

By a "natural mathematical argument" it means a proof written in a natural language, such as English, that would be acceptable as a fully worked out proof in a text book or journal article. A natural mathematical argument consists of a sequence of natural language statements. The human reader is expected to use his knowledge and intelligence to see that each step clearly and necessarily follows from the previous steps [1].

The ONTIC system cannot read natural language - before an argument can be verified it must be translated into machine readable form. The ONTIC system manipulates formulas in the formal language called Ontic. The language is designed to be as similar as possible to natural language while still being simple and mathematically precise. Most atomic formulas in the Ontic language consists of a subject "noun phrase" and a predicate "verb phrase".

Every expression in the Ontic language belongs to exactly one of five syntactic categories; an expression is either a term, a formula, a function expression, a function expression, a type expression, or a type generator expression. Terms are expressions that denote objects. A formula is an expression which denotes one of the of the Boolean truth values true or false. A function expression denotes a mapping from objects to objects. Each function expression takes a fixed number of arguments and returns an object. Type expressions are predicates of one argument. A type generator expression denotes a mapping from objects to types. Each type generator expression takes a fixed number of arguments and returns a type. Let us consider the fragment of the proof:

```
Let P be a partially ordered set in which least upper
bound of S exists for all S in P. For H in P , let K be
the set of all lower bounds of H ...
```

In ONTIC language this fragment takes the form:

```
(IN-CONTEXT ((LET-BE P POSET)
             (SUPPOSE (FORALL ((S (SUBSET-OF (U-SET P))))
                        (EXISTS (LEAST-UPPER-BOUND-OF S P))))
              (LET-BE H (SUBSET-OF (U-SET P)))
              (PUSH-GOAL
               (EXISTS (GREATEST-LOWER-BOUND-OF H P))));
 (IN-CONTEXT
     ((LET-BE K (THE-SET-OF-ALL (LOWER-BOUND-OF H P)))
              .......
```

The main idea of our research is to make a programming environment for mathematician which allows him to work in traditional mathematical language. As an instrument for such an interface we use the system Lingua.F - the environment for creating natural language interfaces.

3. Brief Lingua.F Description

The Lingua.F programming environment gives the user the possibility to create the natural language interface (NLI) in different subject domains using the semantic of this domain. The main complication of NLI constructing consists in the absense of formal demands for such an interface. It is possible only to describe the subset of natural language adequate to the given problem [2].

NLI constructing has the iterative character and as a result we'll have a number of NL prosessors every next is more adequate to the problem we solve. So the strong question is to have the technology for such processors constructing.

Lingua.F consists of the tool kit allowing the user to create the NLI using the linguistic information and semantics of the subject domain. This NLI can be used as independent module in any programming system.

NLI consists of the number of production systems modules every of which corresponds to the next stages:

- lexical analysis (using the vocabulary);
- word combination collecting;
- main analysis (pragmatic-oriented analysis);
- natural language phrase generation.

Every stage uses its own production system language. Let us consider this stages in more details on the example of Ontic language.

Lexical analysis is the first stage of transformation when the lexems of the input natural language text are transformed into the sequence of components. The components are described as the main data types of the subject domain. In this subject domain more than twenty types of the components are picked out. For example, the components for any binary predicate and function has the structure:

```
PRED2     / name: string, arg1: string, arg2: string /

FORMULA   / type: string, body: string /
```

The next stage is the word combination collecting. For example, such kind of word combination as "partially ordered set" will get the next information from the dictionary:

```
/ partially ordered set /
   ( TYPE )
  [ NAME = 'POSET'; NARG = 0; ].
```

The main analysis is the principal stage when the sequence of the components is transformed into the tree of parsing (the tree of immediate constituents). Here is the special production rules language by means of which we describe the dependences in given subject domain. Let us consider the example:

```
/* this rule  is from the group which  defines the symbols
   type which doesn't mention in the text. It follows from
   the  formulas. Ex.  Let P  be poset,  S subset  of P.
   N belongs to S. It defines the types of S and N */

(SYMBOL, PRED2, SYMBOL) : (not (defined (type.1))) and
                          ( type.3='POSET')
              -> (1) [new = 0; type = '(SUBSET-OF (U-SET '+
                     name.3 + ')']].
```

This example illustrates the type of the production rules used in the system. For given subject domain we have made more than 100 rules.

The generation stage. After the parsing tree is constructed the system generates the output formal representation of the given text. The rules of this stage have the next semantics. In the left hand side of the rule we indicate the root of the subtree and its descendants. In the right hand side we indicate the fragment of the string which can be obtained as a concatenation of the objects attributes.

```
(FORMULA/ FF,FF) :->
        (0) [body =  '('+ type.1 +' ' +  value.1 +' ' +
                     '('+ type.2 + ' ' + value.2 + '))'].
```

4. Conclusions

The approach used in the Lingua.F was successfully used in many different application such as NLI for the data bases, input language for expert systems and so on [3]. The advanced technology of the Lingua.f system gives the user the ability to construct the interface in very short period of time.

References

1. D.McAllester, *ONTIC: A Knowledge Representation System for Mathematics.* (Doct. Dis. of the Department of Electrical Engineering and Computer Science, 1987).
2. S.Trapeznikov, *The system for Development of Linguistic Processors.* (Artificial Intelligence II, Methodology, Systems, Applications. North-Holland, 1987), p.329-335.
3. S. Trapeznikov and T. Yakhno, *From Linguistic Processors to Expert Systems Designer.* (Russian Conference in Artificial Intelligence, 1988), p.527-530.

EFFICIENT IMPLEMENTATION OF DECLARATIVE KNOWLEDGE REPRESENTATION IN PROBLEM-ORIENTED ES SHELL

Tatjana Yakhno
Institute of Informatics Systems,
Russian Academy of Science, Siberian Division.
Pr.Lavrent'eva, 6, Novosibirsk, 630090, Russia
e-mail: yakhno@isi.itfs.nsk.su

and

Sergej Grinberg
Institute of Informatics Systems,
Russian Academy of Science, Siberian Division.
Pr.Lavrent'eva, 6, Novosibirsk, 630090, Russia
e-mail: green@isi.itfs.nsk.su

ABSTRACT

In the paper the structure of the DIGEN shell is described. This shell is intended for building diagnostic expert systems as everyday well organized activity.

The main idea of the DIGEN is that at the knowledge acquisition stage all problem knowledge is described by expert in the object-oriented style with help of special tools. Then the rule-based representation which is used on the consultation stage is generated automatically and is hidden from expert. We suggested an approach of complete compilation of knowledge base into procedural code as the alternative to interpretive approach for reasoning.

Key-words: diagnostic systems, object-oriented knowledge acquisitions, compilation of the knowledge base.

1. Introduction

When creating new expert systems building tools (ESBT) one of the main problem is to choose the adequate formalism for knowledge representation.

Now a wide spectrum of these formalisms is available - from procedure-oriented to declarative ones and their various combinations [1]. And the fact is that declarative methods of knowledge representation are enjoying wide popularity today.

One of the reasons to abandon from procedure-oriented approach is that it provides a rather low level of knowledge representation. For example, production rules were a widely used formalism for knowledge representation some years ago. It allows a complex process controlled by unsteady environment to be described in a natural manner [2].

But besides the well-known advantages, this approach has serious shortcomings:

- a rule-based description of the problem domain does not embrace all structural components of the system under observation;
- a large amount of rules complicates a clear view of the process flow;
- besides, often it is rather easy and natural for an expert to express his knowledge in the traditional form "If.. then..", but he is faced with serious problems trying to write down this knowledge with the help of concrete production language.

However it should be noted that in practice the rule-based representation is supported with the most effective implementations.

Declarative knowledge representation has a number of advantages which force ESBT developers to show preference to it. Declarative description gives to developers an opportunity to describe the denotative semantic of problem domain in natural manner, declarative knowledge bases are well-structured and easy to understand.

But main shortcoming of declarative representation is that its implementation has rather low run-time efficiency because of interpretative nature of this formalism. In this paper we propose a frame-oriented approach as example of declarative representation in which we make an attempt to use the advantages of production rules. As a small remark it should be noted that we make no differences between object and frame and below they will mean the same [3],[4].

Object-oriented approach gives us an opportunity to describe easy and naturally the structural organization of a domain under investigation in the form of different kinds of hierarchy (in particular, in "a kind of" hierarchy).

The hierarchical frame organization made it easy to distribute "properties" so that those being shared were stored in the hierarchy at the place covering the maximal subset of frames sharing them.

Due to this organization, the frame hierarchy is an efficient storage scheme, because the shared properties are not replicated in each place they hold true. But they are inherited by all frames which are below in the hierarchy. The inheritance mechanism allows a kind of interaction among different parts of this system to be defined.

2. The Main Idea

During last year we have carried out into practice a number of experiments and got a firm conviction that object-oriented approach was the most convenient one for knowledge description in knowledge bases. As a result we decided to build the ES shell, in which to maintain the object-oriented style of knowledge representation, but at the same time to obtain the effective implementation of ES constructed in the shell.

To solve this problem we had to use the following restrictions. We decided to make the problem-oriented shell instead of universal one. The rejection from universal shell lets us to pick out such slots in frames, which had common semantics for the given problem orientation and did not depend on the concrete subject domain.

After that the operational semantics of such slots can be built into the shell in the form of production rules, for which there are quite effective implementations.

The main idea is that at the knowledge acquisition stage all problem knowledge is described by an expert in the object-oriented style by means of the special tools. Then the rule-based representation used at the consultation stage is automatically generated and is hidden from the expert.

The effective use of knowledge-based system technology to solve the complex real-time problems requires maximal performance of these systems for processors with the conventional architecture.

The use of such systems in some applications has been limited by the quality of existing implementations. This is particularly true for applications which must deal with large amounts of data in very short period of time.

That is why we suggested the complete compilation of knowledge base into procedural code as an alternative to interpretive approach to reasoning [6].

3. Brief DIGEN Description

This approach has been realized in the shell DIGEN (DIagnostics and GENesis) intended for building ES in diagnostic problems. The study of wide spectrum of diagnostic ES gave us the conclusion that generally in diagnostic process we were interesting in indications of anomalous states, their causes and effects and recommendations which could eliminate these states.

For the illustration let us consider the simplified example. The anomalous state can be described by means of the next frame:

$$\begin{array}{cc} \multicolumn{2}{c}{STATE:} \\ \textit{name of state} & : \\ \textit{indications} & : \\ \textit{causes} & : \\ \textit{recommendations} & : \end{array}$$

where some other states or parameters can be the slot's values.

Let some state be described:

$$\begin{array}{cl} STATE: & \\ \textit{name of state} & : ST_1 \\ \textit{indications} & : P_1, P_2, \ldots, P_n \\ \textit{causes} & : R_1, R_2, \ldots, R_k \\ \textit{recommendations} & : REC_1, REC_2, \ldots, REC_m \end{array}$$

This frame can be transformed into the structural set of production rules:

$$
\begin{array}{ccc}
INDICATIONS & CAUSES & RECOMMENDATIONS \\
P_1 \rightarrow ST_1 & R_1 \rightarrow ST_1 & ST_1 \rightarrow REC_1 \\
P_2 \rightarrow ST_1 & R_2 \rightarrow ST_1 & ST_1 \rightarrow REC_2 \\
\ldots & \ldots & \ldots \\
P_n \rightarrow ST_1 & R_k \rightarrow ST_1 & ST_1 \rightarrow REC_m
\end{array}
$$

By means of separate groups of production rules we can solve the different subtasks in the subject domain (for example, diagnostics, prognosis etc.). This production rules representation of the knowledge base can be transformed into the discriminative net, which is used for effective pattern match (for example like in RETE-algorithm) [5].

Roughly speaking our DIGEN shell consists of two main parts: a programming environment to allow the user to create and analyze the knowledge base in declarative formalism; and a compiler to convert an object-oriented form of the knowledge base into C code.

The suggested scheme was realized in the shell DIGEN, which was successfully tested on the tasks of medical and technical diagnostics. Now we continue investigations in the following directions. First, just the same approach can be used in the problem-oriented shells, which deal with the problems quite different from diagnostics ones. Our plans are to work out the language for description of the slot operational semantics. Using such a language we can turn to the problem-independent shell keeping the efficiency of ES generated in the shell.

4. References

1. Waterman D.A. *A Guide to Expert Systems* (N.Y.: Addison - Wesley Publishing Company, 1986).
2. Davis R. et al. *Production Rules as a Representation for Knowledge-based Consultation Program* (Artificial Intelligence. 1977. V.8), pp. 15-45.
3. Ten Dyke R.P., Kunz C. *Object-oriented Programming* (IBM System Journal. 1989. V.28, N.3), pp. 465-478.
4. Fikes R., Kehler T. *The Role of Frame-based Representation in Reasoning* (Comm. ACM. 1985. V.28, N.9), pp. 904-920.
5. Forgy C.L. *RETE: A Fast Algorithm for the Many Pattern/object Pattern Match Problem* (Artificial Intelligence. 1982. V.19, N.1), pp. 17-37.
6. Highland F.D., Iwaskiw C.T. *Knowledge Base Compilation* (Proc. IJCAI'89, 1989) pp. 227-231.

ZEX — AN EXPERT SYSTEM FOR ZEUS

Ulf Behrens, Mariusz Flasiński, Lars Hagge, Kars Ohrenberg
DESY Deutsches Elektronen-Synchrotron
22603 Hamburg, Germany

ABSTRACT

ZEX is an expert system which is designed to support operation of the ZEUS experiment. A prototype version of the expert system, ZEX-P, has been implemented successfully and is operating since the beginning of 1993. Experience gained in the design and implementation of ZEX-P are reported and the resulting constraints for implementing the final system are discussed.

1. Introduction

ZEUS is a high-energy physics experiment located at HERA, the electron-proton colliding facility of DESY. It consists of a general purpose magnetic detector, which is controlled and read out by a highly parallel distributed data acquisition (DAQ) system.

The nature of the ZEUS experiment required the components of the DAQ system to be constructed independently at different places. Thus, expert knowledge about the DAQ system is spread over many human experts at many institutes. Because of this situation, becoming an expert of the entire DAQ system requires a long training period, and the complexity of the system still provokes misunderstandings and operator errors.

During the first data taking period, experts from most institutes needed to be on location to integrate and debug their systems. Now, the principally operating experiment bears the danger of experts leaving to their home institutes and being unavailable in case they are needed.

Therefore it was decided within the ZEUS collaboration to construct an expert system which supports experiment operation. The goal is to store the knowledge of the different experts in the experiment control system and make it available to the operators on shift to reduce their required expertness. Task of the expert system is to detect anomalous behaviour in any part of the experiment, to trace errors to their origin and to recover the system as quick as possible. These tasks are reflected in the three parts of the expert system:

- a monitoring process surveying the data acquisition system for symptoms of anomalous behaviour,
- a diagnostic process which tries to find the origin of anomalous behaviour by detailed evaluation of the data acquisition status,

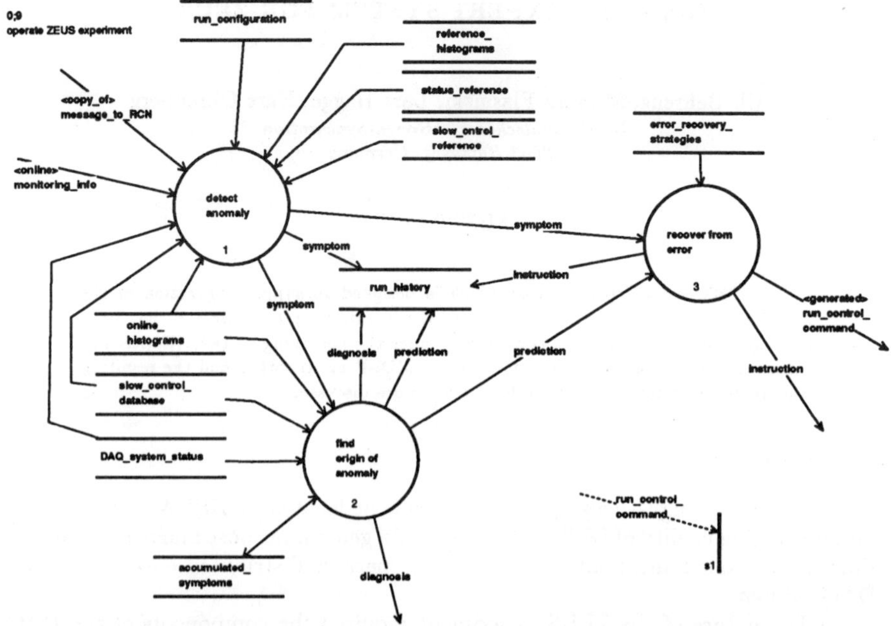

Figure 1: Data Flow Diagram showing the architecture of the ZEUS expert system. Bubbles represent processes, arrows define dataflows and bars define stores.

- a control process to recover the data acquisition system from errors

Our aim has to be to drastically shorten the training period for operators and to increase both efficiency and reliability of experiment operation.

2. Expert System Prototype ZEX-P

Constructing an expert system for a high energy physics environment raises several questions: Are the analysis methods and algorithms powerful and fast enough to fulfill their tasks? Will the system be maintainable, or does it require a set of experts to operate an expert system? Are people willing to use an expert system, i. e. believe its diagnosis and allow it to issue commands to the DAQ system?

It was decided to implement a prototype version of ZEX to study these questions. The prototype is connected to the DAQ system and running online in the same way as ZEX will be, but operates only on a small subset of the input available to ZEX.

ZEX-P is embedded in the online DAQ system of the ZEUS experiment,

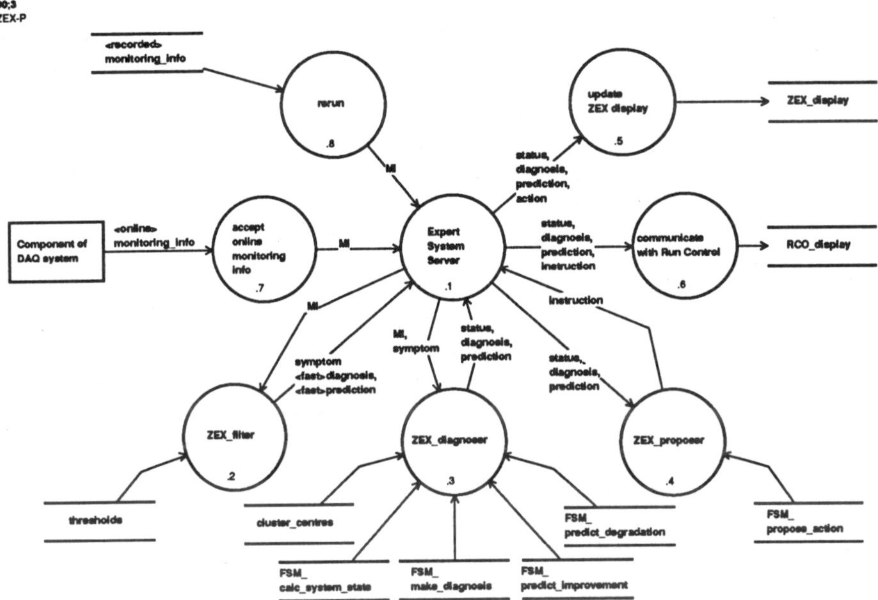

Figure 2: Data Flow Diagram defining the prototype implementation.

using monitoring information collected by the Eventbuilder. The Eventbuilder is a central part of the DAQ system and has the task to combine the dataflows of the individual components to a complete event record for each event accepted by the global second level trigger. For further filtering, these event records are distributed by the Eventbuilder over six branches of third level trigger processor nodes [1]. Because of its connections to nearly all components of the DAQ system and because of its powerful monitoring capabilities, ZEX-P was based on the monitoring information of the Eventbuilder.

The input available to ZEX-P includes the accept rate of the second level trigger, response time and output buffer loads of each component subsystem, the data throughput and buffer loads of the Eventbuilder and the interface loads at the different branches of third level trigger processors. Errors detectable by ZEX-P include rate and bandwidth problems, synchronization problems, component subsystem failures and blockage of the main data stream.

ZEX-P was implemented using a server based architecture (fig. 2). A re-run process, input and output tasks and the different analysis processes are connected as clients to a central Expert System Server.

Three analysis processes implement the symptom finding, diagnostics and control tasks defined in fig. 1 [2]. A filter process extracts symptoms from the monitoring information of a component subsystem. A diagnoser combines the symptoms from several components into the system state, derives a diagnosis and sometimes risks a prediction on the evolution of a run. A proposer evaluates status, diagnosis and prediction to provide the shift crew with instructions on how to proceed.

The diagnoser is based on the nearest cluster center method. Monitoring information fed into ZEX-P is transformed into features and the diagnoser combines the different features into the system state using the nearest cluster center method. The underlying idea of this approach is that objects of similar type tend to cluster in a feature space. By inserting an unknown object into the feature space, it can be classified by associating it with its nearest cluster. In case of ZEX-P, different system states correspond to clusters in the feature space of the DAQ system. This way, the actual system state can be determined by calculating the status vector in the feature space and finding its nearest cluster center.

Syntactic pattern recognition is applied to the system state to evaluate the system trajectory in the feature space. Syntactic pattern recognition is based on formal grammars and automata. The set of possible system states is treated as an input alphabet with the known errors forming a language over this alphabet. The diagnoser simply parses a given series of system states, i. e. checks whether it belongs to the language. An overview of the pattern recognition methods used for the ZEX-P implementation can be found in [3].

The language describing the DAQ system errors can be defined in a way that every error corresponds to a unique string of system states. Thus, by connecting appropriate actions to the identification of an error, the parser can also be used for case specific treatments. Fig. 3 illustrates this similarity between finite state machine systems and rule based systems (It results from the fact that both types of systems are conceptually based on formal grammars and automata). Groups of states correspond to contexts, transition are described by rules.

Expert systems are generally developed and tested with the help of simulators. This often leads to problems as simulators usually differ from reality. In case of ZEX extensive logfiles of monitoring information from the first run period of the experiment were available. A "rerun"-process was written which used the logfiles to replay the behaviour of DAQ system during data taking. Thus ZEX-P could be implemented and tuned in quasi-online mode.

3. ZEUS Expert System ZEX

All analysis processes of ZEX-P are implemented as finite state machines. This implementation has been chosen in order to test the capabilities of finite state machines in real-time expert systems. They turn out to be reliable and fast, but as the addition of error strings to the language usually results in a multiplication of system states, they are difficult to maintain and can be modified by trained experts only.

The knowledge about a data acquisition system in a high energy physics

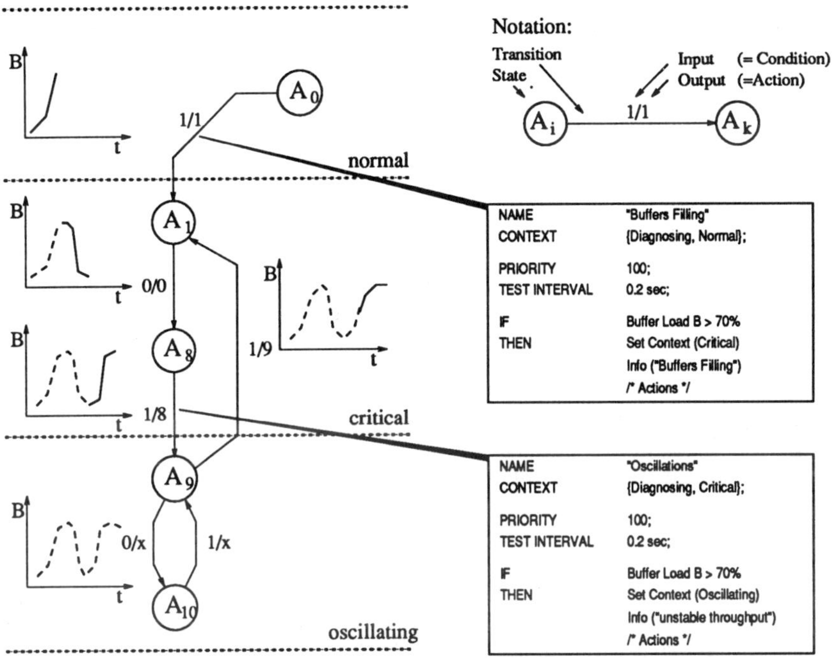

Figure 3: Part of the FSM calculating the system status. An arbitrary monitoring quantity B is tested for critical values and oscillating behaviour. The right hand side shows, how the same functionality could be obtained with a rule based system.

experiment is changing frequently. This requires to code this knowledge in the form easily readable and expandable. Unfortunately, finite state machine systems do not fulfill this requirement. Therefore, it was decided to switch to a rule based system for further development of ZEX.

There is a long list of tools [G2, NexpertObject, OPS5, CLIPS, ART-IM, ...] which allow for rule based knowledge storage, but the boundary conditions in the ZEUS environment drastically reduced the number of alternatives. Finally only RTworks [4] was seen as a tool which fulfills our requirements to a high level:

- real-time performance: monitoring a complex real-time system requires to be able to detect correlations of symptoms in time. This is only possible with fast quasi real-time reaction of the tool.
- powerful reasoning mechanisms: forward and backward chaining, firing rules according to a context, a priority or a test interval, a library for reasoning over time is mandatory.

- a distributed, modular and flexible architecture: if the expert system has to be combined with an existing distributed system, the expert system has to be fit to the system to be surveyed and not the other way round. This means that the tool has to be able to reflect the distributed nature of the system and that existing interfaces, existing preprocessing methods have to be easily linkable to the tool.
- user-friendliness: the tool has to be easy to use without having to learn any specific language.
- powerful interfacing and visualization mechanism to allow the developer to concentrate on knowledge acquisition and implementation. An integrated tool is preferred which offers an easy-to-use high level graphical interface.
- maintainability: the knowledge has to be stored in a way that it is readable and understandable by untrained human experts without major efforts. Because of the large number of human experts in our environment and the not yet final system this is the only possible way of keeping the knowledge up-to-date without a large number of expert system experts.

RTworks has proven within a test installation that it provides a user-friendly, easy-to-handle development environment which allows for speeding up the process of designing and implementing an expert system. An experimental version of ZEX has been prepared with RTworks in four weeks, whereas the group needed about nine months for preparing the prototype version ZEX-P of a similar functionality with conventional tools (C language, TCP/IP, OSF Motif). (There are more detailed diagnostic conclusions, but no graphical visualization of the DAQ system performance in ZEX-P.)

Fig. 4 shows the layout of the system which will be set up for the data taking period in 1994. Split up in sub-expert-systems for the different main parts of the ZEUS online system, ZEX will be put on top of the existing systems [Fig. 5]. ZEX will provide a common environment for surveying data taking, slow control and data quality monitoring. The top level part of ZEX with an embedded run control will combine the knowledge about the subsystems to an overall system understanding which allows to efficiently operate the detector under well-known conditions.

4. Conclusion

A prototype of the ZEUS expert system, ZEX-P, has been built using conventional programming techniques. The implementation has proven the benefits of a "rerun" facility to simulate the DAQ system by using real monitoring information of previous runs. ZEX-P has also shown that the applied paradigm (formal grammars and automata) can be used successfully, but a rule based system is the appropriate technique for implementing such systems.

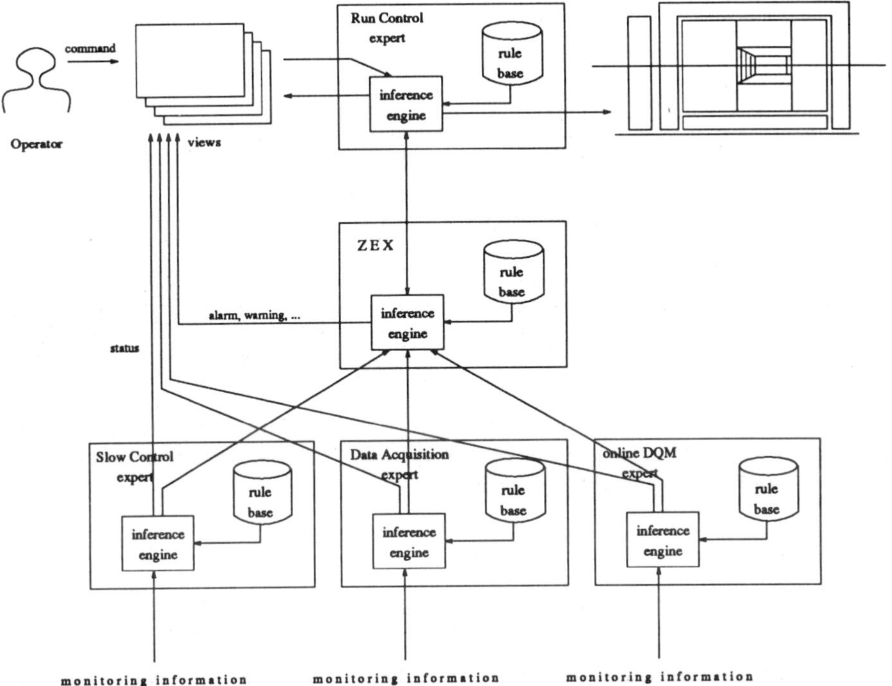

Figure 4: Layout of ZEX for the next data taking period. Sub-expert-systems with individual knowledge bases are surveying encapsulated parts of the experiment. Their output drive the central inference engine which is responsible for the overall diagnosis and control of the experiment.

First steps towards the final system have been based on the expert system tool RTworks and it was obvious that the use of a commercial tool drastically reduces development efforts by supporting knowledge acquisition and maintainance and providing powerful interface environments.

Finally, our experience with ZEUS and ZEX-P has shown that control systems for complex setups should not be planned without inference mechanisms.

References

[1] Ulf Behrens, Lars Hagge, and Wolfgang O. Vogel. The Eventbuilder of the ZEUS Experiment. *Nucl. Instr. & Meth.*, 1993. DESY preprint DESY 93-008.

[2] Ulf Behrens, Mariusz Flasiński, Lars Hagge, and Kars Ohrenberg. ZEX – An Expert System for ZEUS. In *Proc. Eighth Conference on Real-Time Applications in Nuclear,*

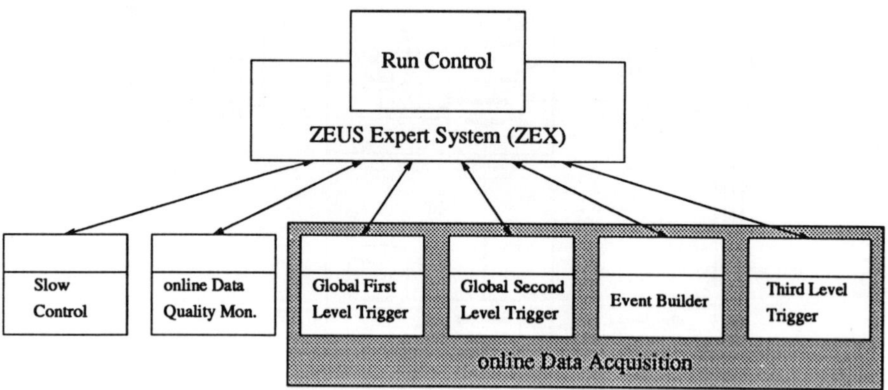

Figure 5: Interfaces of ZEX. ZEX will be put on top of the existing central parts of the system. Embedding a new run control system, ZEX will provide a homogenous operator interface to control the experiment.

Particle and Plasma Physics, 1993.

[3] Ulf Behrens, Mariusz Flasinski, and Lars Hagge. ZEXP - The ZEUS Expert System. DESY–Report 92-141, DESY, Notkestr. 85, D-2000 Hamburg 52, 1992.

[4] Talarian Corporation, Mountain View, USA. *RTworks V2.1 Users's Manual*, December 1992.

DELOS
an object oriented expert system for error diagnosis in particle detectors

JÖRG P. NOWAK
Department of Physics, University of Wuppertal, Germany
nowak@wrcd1.urz.uni-wuppertal.de

ABSTRACT

Todays high energy physics research requires complex particle detectors for achieving the desired precision or to find new physics. So detectors have reached a new level in size as in complexity. Unfortunately, those detectors can fail so an automated error detection and analysis is essential because the operating crew of the detector isn't often build up of experts.
DELOS (**Delphi Operation Support**) is an expert - system for a certain part of the DELPHI detector at CERN the so-called forward drift chambers FCB. DELOS follows a case oriented approach and is realized by applying the principles of object-oriented programming.
It is divided into two major parts: DELOS the expert system itself and the graphical user interface. This paper describes the ideas behind DELOS and parts of its realization. The user interface will not be discussed here. [1]

1. the detector and its hierarchy

As mentioned above the FCB chamber is part of the detector DELPHI. It consists of approximately 7500 sense wires, organized in 2 times 12 planes. The wires are grouped in blocks of 8 wires and are the input of a preamplifier. 16 preamplifiers are connected to one of the 144 multiplexer units - these units are connected via sc. LTDs to an I/O system called FASTBUS as shown in figure 1.
We can see that this detector has two unique points:

1 there is a treelike structure of the components and
2 the detector consists of a large number of identical parts.

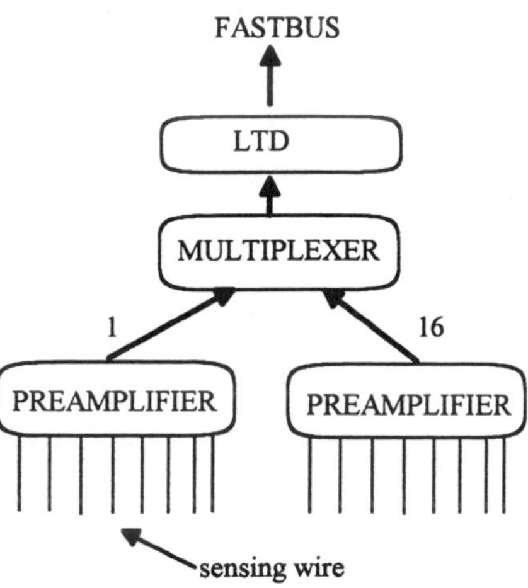

Figure 1. the detector hierarchy

[1] A last remark on the graphical user interface: due to this division DELOS can easily be ported to different machines - at this moment it is running on DEC ULTRIX ,and MS WINDOWS. Text based UIs are also possible.By adding a Client/Server frontend you can also get a network friendly system.

Therefore the number of possible errors is limited - but they can occur anywhere in the detector. This "boundary condition" reduces the number of well suited expert systems.The object oriented approach is a well suited concept for this problem, firstly it maps the hierarchy of the real world to the programs world - and secondly there is a relativ small number of different objects. Each object - or in the case of DELOS - error is unique but more on that later.

2. why C++

2.1. C++ versus pure AI languages

Now I'd like to discuss the question why I use C++.[2]A full size AI language like e.g. OPS5 is designed for a wide and general use, but in some situations where one only needs a small set of abilities when constructing a highly specialized system or construct a hybrid *in the meaning:AI features and features of older procedural languages* one has to carry such a huge overhead so that the benefits of the language can vanish .
So the idea was to rewrite the needed AI parts in C++ or better, to choose an expert system model that also suites C++ well. As a result we have an increase in performance, in maintainability and a decrease in size .
The factor maintainability is very important in the universitarian environment where the programming staff changes very often and documentation and clarity of the programs is needed.

2.2. C++ and multiple language programming

C++ allows multiple language programming. Which means, that different errors - which are classes in DELOS - can call functions written in different languages. So it is possible to add a logic - tool written in PROLOG or LISP, call a neural network for pattern recognition or add fuzzy sets where it seems appropriate as shown in figure 2. So even old FORTRAN code can be reused.
This is not so easy to do with pure AI languages. We can see: you can give each error its perfect tool for diagnosis.
I belief that multiple language programming will boost the quality of programming dramatically.

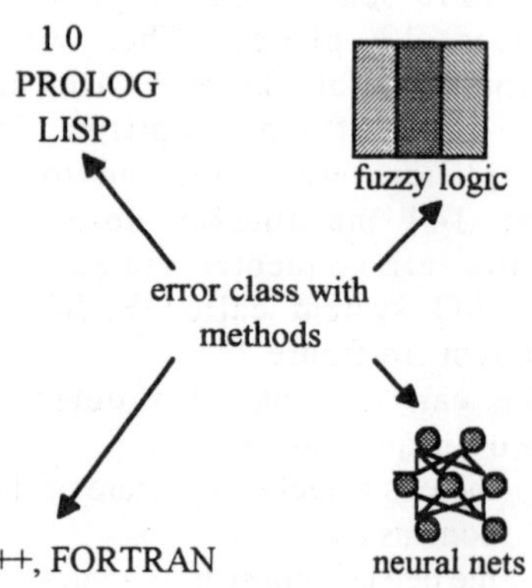

Figure 2: multiple language programming

[2] C++ according to AT&T level 2.1. ANSI C++ or level 3 is still not that common .

As a last point I'd like to mention the fact that programs written in C++ can easily be ported to other platforms .
The idea is: place all machine dependent code into classes or objects and access those code only by class methods.
This results in a consistent user interface of those classes (the interface of the class looks the same on all systems). So major parts of the program need no change. The machine dependent code is strictly localised and encapsulated.

3. how DELOS works

As mentioned earlier DELOS consists of three modules PRECHECK, C.O.I.M. and the LOCALIZER (see figure 3). The input of DELOS is the output of a program called FCBANA which gives a large set of histograms to the user. Histograms, which contain information like *wiremap with dead wires, efficiency vs. wire* and so on.

3.1. PRECHECK

In a first step the first of the three modules of DELOS- called PRECHECK - performs a test on a user defined subset of the histograms by comparing them with a reference. Its aim is to reduce the input to the next module by passing only those histograms further, which show some form of derivation. PRECHECK mostly works with the histogram class, so I want to show you the class hierarchy of one of the main classes - the histogram class - now.
The histogram class is derived from the class for basic database access and handling and a class to hold the data plus methods for data access. The database class is responsible for database[3] access , for finding and retrieving the histograms. To do that it is derived from another class witch holds an arbitrary amount of index positions.

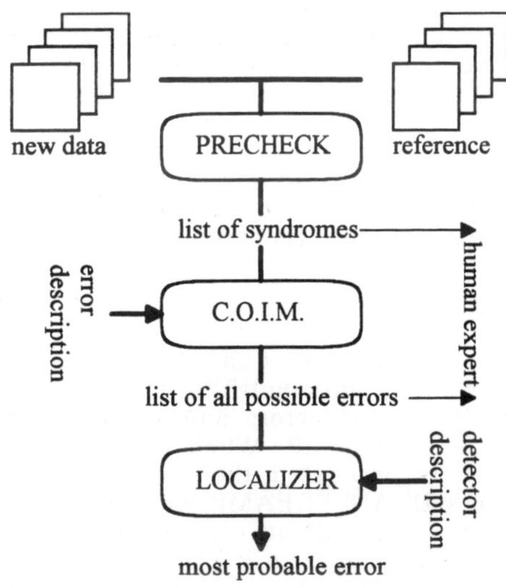

Figure 3: how DELOS works

[3] the database for the histograms is accessed by an index file

When the data is retrieved it is stored in a special array which is derived from the VALUE class. The VALUE class holds one element and its error e.g. 5 plus minus 2.
Allowing operator overloading the question if two histograms are statistical equal can be written by the familiar statement

H_IN == H_REF.

Any of those - non equal - outgoing histograms is called a *symptom* - it is known that something is wrong but not what. A symptom which can be explained is called an *error.*
If a histogram is found that has no reference at all the human expert is informed.

3.2. C.O.I.M.

This list of symptoms is the input of the **case oriented inference machine** which tries to build up a conflict set of all possible errors.
To do that it processes the list of known cases or errors. Each case holds a list of histograms to be checked and the methods to check them. These methods try to verify or falsify an error.
There is also access to more data, like the history of the error (e.g. how often it has occurred) maintenance intructions (e.g. whether the faulty part can be repaired now) and so on.
The location (the ID) of the faulty part in the detector is determined by the methods. In contradiction to pure case based systems - which are selflearning systems - the error classes of DELOS are written by the human expert - therefore the term case oriented.

3.3. example

Let us now look at an example: imagine one of the preamplifiers is broken. We have several errors and the errorclass DEAD_PREAMP. When the list of errors is processed the methods of the class DEAD_PREAMP will look for an eight wire hole in certain private data or histograms by using its own methods. If the eight wire hole - which is represented by the grey erea in figure 4 - is confirmed , the location or number of the preamplifier is determined. But, there is a problem: the data for a single wire fault will show some derivation , too. To find the real error these informations are passed to the last stage of DELOS - the LOCALIZER.

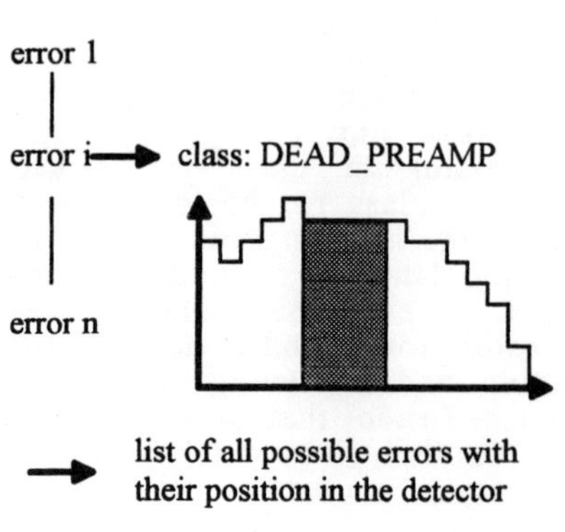

Figure 4: an example

3.4. LOCALIZER

The LOCALIZER module tries to map these potential errors to the detector-tree using a detector description. It tries to find the most probable error by using the heuristic:

find a part that dominates all other

This is reasonable, because e.g. the possibility of a broken preamplifier is much higher than the possibility of a failure in eight adjacent wires at the same moment.

4. conclusion

To end this paper we come to the conclusion, that applying the ideas of object oriented programming and case oriented system design leads to an expert system which is scaleable from a small basic "skeleton" to one with sophisticated features. It is fast , flexible easy to expand , maintenance friendly and runs on multiple platforms and is well suited for any system with large amounts of identical parts in a treelike structure. Due to its open architecture is can also be used for testing new methods. I finish with an outlook to further development: it is planned to shift the error classes to an object oriented database in one of the next versions and to construct these object in a graphical development environment.

AN EXAMPLE OF DEEP KNOWLEDGE REPRESENTATION FOR THE BEAM–LINE OPERATION

V. UKHOV
Computer Science Department, Institute for High Energy Physics
142284, Protvino, Moscow region, Russia

ABSTRACT

Model–based technique has been taken for deep knowledge representation. The beam–line tuning time was used as one guideline in deciding, which model to employ. Model verification is based on tolerances and parameter fits. Hypothesis seizing and adjustment forms diagnostic procedure.

1. Introduction

Traditional ES technique was in difficulties when we have tried to apply it to the beam–line control and diagnosis. Non serial character of machines forms the cost problem. The time problem has its origin in relatively frequent beam–line modification. One way to get out of difficulties was to use the deep knowledge. And one way of doing this was resorting to the model–based technique (figure 1).

Influence net [1], MAD/TRANSPORT [2,3] and other models were under consideration. The beam–line tuning time was used as a guideline in deciding which model to employ. This value depends on the amount of information about machine under the control. So, the more detail model we have, the less time it's need to adjust the beam-line. But overly detail model is a wrong model, because there is no way to know precisely all parameters of the real machine. On the other hand our PC orientation imposed some limitations on the model complexity.

Linear in beam dynamics model has been chosen to describe the beam–line in the model-based expert system DALS [4]. This model is more powerful than Influence Net and less complex than ones in MAD or TRANSPORT. Near the working point it is reasonably accurate, so, precision tuning is fast. And this model is qualitatively good far away, it follows that rough adjustment isn't slow. The DALS model (figure 2) consists of linear beam dynamics model, linear or nonlinear magnet and detector models.

2. DALS operation

Existence of the traditional control system [5] is a fundamental condition for DALS installation. Functional scheme (data flow diagram) of the ES DALS is shown on figure 3. The beam–line data sets are placing into the archive after preliminary filtering. The verification module tests model and archive data mutual correspondence. In the event of faulty correspondence diagnostic procedure activates. The

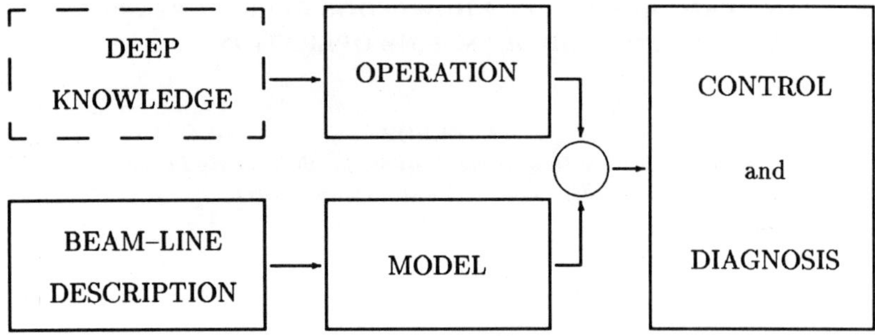

Figure 1: Model–based technique application.

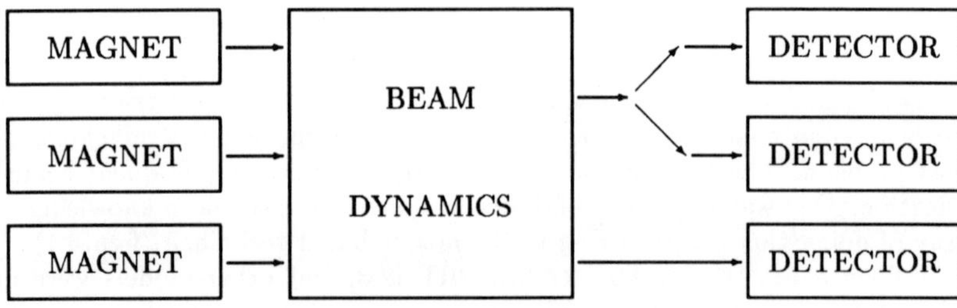

Figure 2: An example of DALS model.

diagnostic module adjusts model and recover model–data correspondence. The tuning module conduct beam–line correction. It uses information about the machine from model and from archive.

The precision agreement between archive data and model couldn't be archival because of:

- The limit of model accuracy.
- Noise in the beam–line parameters.
- Excursion of the real parameters from the model ones.

Beside filtering there are two pure strategies to solve this problem. The first one is to place tolerances on model parameters. Forcing them to fit the model is another item. In DALS the combined strategy has been used. Due to the linearity of beam propagation model all additional disturbances are accumulated in special vector $\vec{u}$. Tolerances are placed on some parameters and force are applied to vector $\vec{u}$ and to parameters of detector models to fit them.

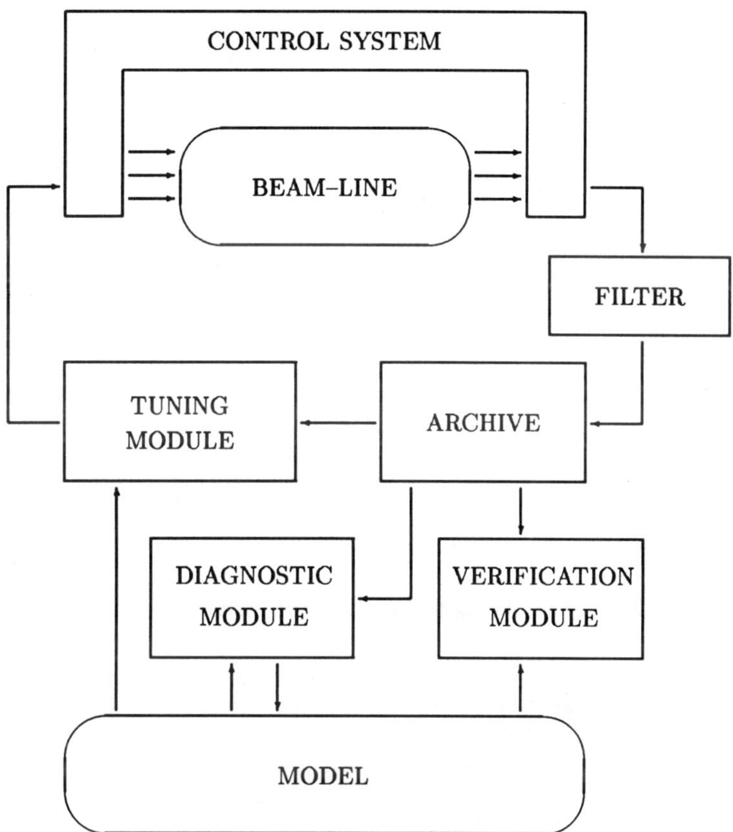

Figure 3: DALS data flow diagram.

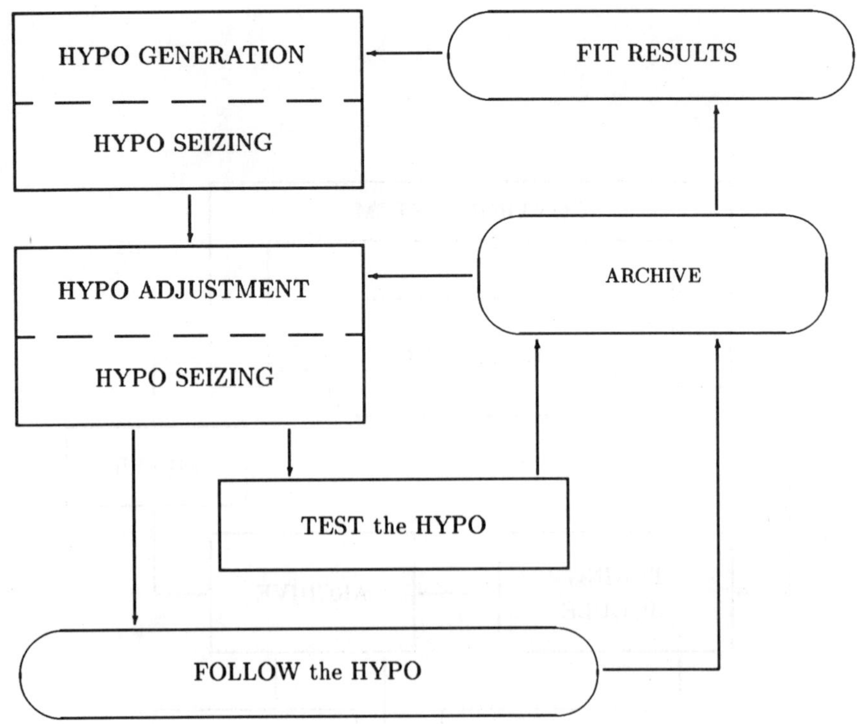

Figure 4: Hypothesis propagation.

The outline solution makes effective algorithms for linear optimization compatible with non–linear detector models. It allows to integrate beam loss monitor into the DALS in particular.

3. Diagnostic procedure

The diagnostic process is providing through hypotheses propagation (figure 4). In the beginning, hypothesis is a beam–line model with some unknown parameters, which obtain definite values after hypothesis adjustment. Hypothesis indicates what devices are wrong. The variable η accounts for preliminary opinion of the hypothesis:

$$\eta = \left(\prod_i P_i\right)\left(\prod_j Q_j\right)\left(\prod_k R_k\right)\left(\prod_l U_l\right) \tag{1}$$

P_i is a probability degree for element i breakage. Q_j, R_k and U_l estimate

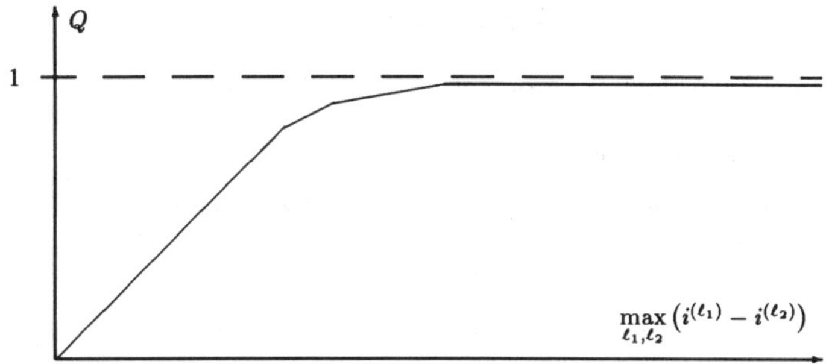

Figure 5: The dependence for Q.

probability for breakage to manifest itself in the archive data. For example (figure 5):

$$Q_j = \frac{2}{\pi}\arctan\left(\frac{\max_{\ell_1,\ell_2}\left(i_j^{(\ell_1)} - i_j^{(\ell_2)}\right)}{F^{\lambda}\Delta i_j}\right) \tag{2}$$

Where $i_j^{(\ell)}$ is a current value for magnet j in archive data set ℓ, Δi_j is a tolerance on current for magnet j, F^{λ} is a scale parameter.

Preliminary hypothesis seizing allows to adjust the more probable ones first. The variable ζ is kept to a minimum by means of hypothesis adjustment:

$$\zeta^{\nu} = \sqrt[\nu]{\frac{1}{N_k}\sum_{k=1}^{N_k}(\varrho^k)^{\nu}} \quad , \quad \nu > 1 \tag{3}$$

Where ϱ^k is a mean squared distance in model–data fit for detector k. It was found experimentally that 3 is a credible value for ν. For finish hypothesis evaluation the variable ζ^{∞} is used:

$$\zeta^{\infty} = \lim_{\nu\to\infty}\sqrt[\nu]{\frac{1}{N_k}\sum_{k=1}^{N_k}(\varrho^k)^{\nu}} = \max_k \varrho^k \tag{4}$$

There was no chance to employ ζ^{∞} for hypothesis adjustment because of its derivative gaps. Hypothesis adjustment takes a lot of computing time. Archive compression decreases this time significantly. Two data sets ℓ_1 and ℓ_2 are treated as equal and first of them temporary excluded from archive if:

$$\forall\, j \quad |i_j^{(\ell_1)} - i_j^{(\ell_2)}| < F^{\delta}\Delta i_j$$

and if:

$$\forall\, k \quad |d_k^{(\ell_1)} - d_k^{(\ell_2)}| < F^{\delta}\Delta^0 d_k$$

Where $d_k^{(\ell)}$ is a signal value for detector k in archive data set ℓ, $\Delta^0 d_k$ is a tolerance on signal for detector k, F^{δ} is a scale parameter.

There are two criteria for finish seizing of hypothesis:

- $\left(\prod_i P_i\right)$ — probability of group element breakage.
- ζ^∞ — result of hypothesis adjustment.

They were combined into next one. On the first step hypotheses are divided on two groups by condition:

$$\zeta^\infty < \varrho^\mathsf{T} \tag{5}$$

Where ϱ^T is a fit threshold. The next step is hypothesis sorting by μ according to the equation:

$$\mu = \frac{\varrho^\mathsf{T}}{\zeta^\infty}\left(\prod_i P_i\right) \tag{6}$$

4. Conclusion

A fascinating solution for integration of linear optimizing algorithms and nonlinear detector models was found. A conceptually simple and fast way to adjust beam–line model was developed.

5. Acknowledgements

The author would like to thank Stanislav Klimenko and Boris Volkov for supporting and encouraging this work.

6. References

1. E. Malandain, *Application of a diagnostic Expert System in the Accelerator domain* (Prep., CERN/PS 89–52, 1989).
2. M. Lee, S. Clearwater, *GOLD: Integration of Model-Based Control Systems with Artificial Intelligence and Workstations* (Prep., SLAC–PUB–4396, 1987).
3. M. Lee, *GLAD: a Generic Lattice Debugger*, Proceedings of the International Conference on Accelerator and Large Experimental Physics Control Systems (KEK Proceedings 92–15, 1992).
4. V. Ukhov, B. Volkov, *Code DALS*, Computer Codes in Accelerator Domain (DESY M–92–07, 1992), p. 14.
5. F. Perriollat, *Accelerator Controls*, 1988 CERN School of Computing (Prep., CERN 89–06, 1989).

STATUS OF THE MODEL–BASED EXPERT SYSTEM FOR BEAM–LINE CONTROL AND DIAGNOSIS, RADIATION MONITOR INTEGRATION

A. AFONIN
U–70 Accelerator Department

V. UKHOV and B. VOLKOV
Computer Science Department

Institute for High Energy Physics,
142284, Protvino, Moscow Region, Russia

ABSTRACT

Article reports the model–based ES DALS evolution. To build in ES the radiation monitor model is a next problem under consideration. The experimental radiation monitor data for U–70 fast extraction are represented. The partially qualitative and partially quantitative model is describing.

1. Introduction

The BEAM project guide–line is to elaborate algorithms and programs for assistance in beam–line operation. The general task is to attain and to keep the efficiency of the beam extraction at the maximum value. We need to reach minimum of the beam losses another words. This task is provided through some steps: selection of the regime; tuning of the extraction elements; consulting with model programs; trajectory optimization based on position detectors or beam loss monitors. The ES DALS is a part of the project BEAM.

2. DALS features

Model–based ES DALS is designed for rough and precision beam trajectory correction. Initially DALS includes position monitor, magnet and beam dynamics models [1]. With time embedded algorithms and heuristics have been developed. Graphical interface for manual control has been introduced.

DALS can be adopted to a new beam–line by model modification. DALS is used as adviser or operate automatically. Graphical interface maps operational information (figure 1) and allows manual control of the beam–line (figure 2). To build beam loss monitor model into DALS is a next problem under consideration. These monitors are very effective, because they don't intersect the primary beam and so they are lossless themselves. We began our work from the model for U–70 extraction system.

3. Beam loss experiments

At present time the system for beam loss measurements is ordinary for high intensity accelerators. On the U–70 proton synchrotron of IHEP [2] this system consists of scintillators with photomultyphiers placed along a circumference of accelerator and along extraction trajectory up to external target. This system is intended for registration of charged radiation component with

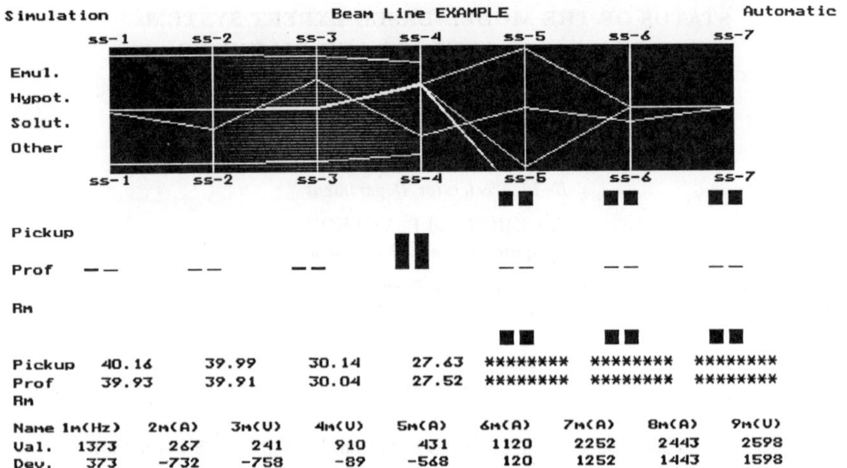

Figure 1: DALS automatic regime.

relatively high energy. Radiation induced by interactions of primary beam with the material of extraction elements, beam detectors, vacuum foils, and beam–gas scattering.

The system of the beam loss monitor displays the beam loss along trajectory of extracted beam. The picture of the beam loss during normal, tuned regime of operation is considerably different from the picture during breakage in some elements of accelerator complex or during process of tuning. One of the most important parameter of the high intensity accelerator operation is efficiency i.e., small loss.

So, if we finished the procedure of optimization of extraction regime and have complete picture of beam loss distribution along the beam line we so have the possibility to use this picture as a basis for optimization. The background for such optimization is dependence of beam loss from displacement of the beam around ideal trajectory. Characteristic property of this function is plateau, which corresponds to the minimum of beam loss. The width of this plateau depends on the beam parameters and on aperture of extraction elements. It will change from one place along the beam way to another, from one mode of operation to another, and from one run to another (figure 3,a). In some places the curve can be non symmetric, for example beam loss on the first septum magnet (figure 3,b). Left side of this curve corresponds to the loss of extracted beam and right side – to circulating beam.

It should be pointed out that when displacement of beam trajectory increases the beam loss signal increases too, but after some moment the beam loss signal will decrease. This is explained by different factors. The most important of them is displacement of the maximum of scattering diagram in reference to beam loss monitor location. This property of function demands some special attention before the function employment to beam-line optimization.

Attention is given to the fact that if the beam loss occur in one place, several monitors can "see" it. One of the preliminary goal, before we try to make a model for complete set of monitor, is to understand real behavior of all beam loss monitors during normal operation. The experimental data were collected during some period of extraction system running. The regime of operation was tuned initially and after that real changes in regimes of accelerator and extraction elements give the reaction of all beam loss monitors. We found that the behaviour of monitors was different. For some of them there were correlations, and for some of them there were no correlation, (figure 3,c–f). Our

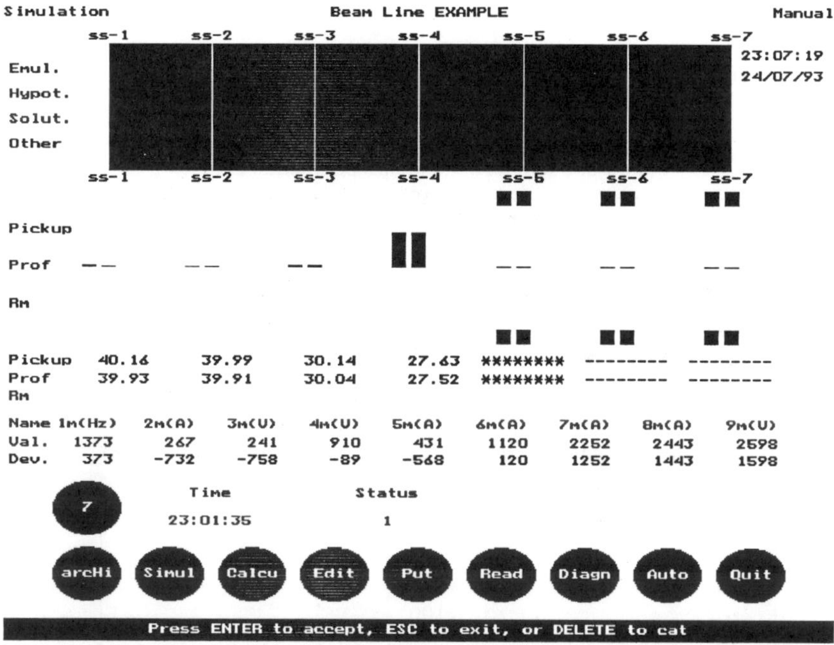

Figure 2: Archive examination in manual regime.

attempts to find simple (linear) model for collective behavior of U–70 beam loss monitor have not met with success. We'll try to test rather sophisticated method in nearest runs.

4. Beam loss approximation

The precise description of beam loss monitor functions is often troublesome. These functions vary with time in form and magnitude. One way to get out of the difficulty is to use more qualitative description. For the case of great trajectory deviation it is:

- To sketch the cover with sequence of characteristic points (typically extremums).
- To set limits on they relative positions.

This description is using to examine the cover and to find the main minimum. When it is done another description is employed:

- The function is approximated by plateau and two parabolas (figure 4).
- The limitations on parameters of approximation cover are imposed.

The last description allows to perform consequent (section–after–section) tune of the beam trajectory, by optimizing beam position on loss function.

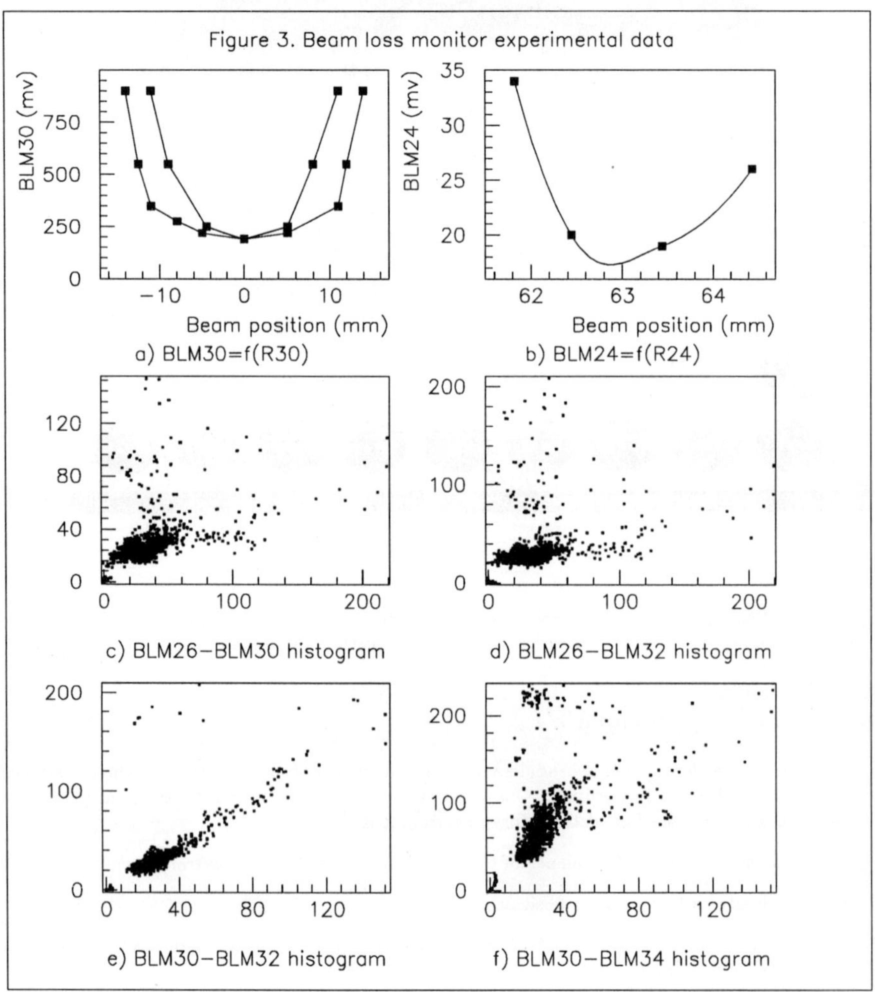
Figure 3. Beam loss monitor experimental data
BLM30 (mv)
750
500
250
0
-10
0
10
Beam position (mm)
a) BLM30=f(R30)
BLM24 (mv)
35
30
25
20
62
63
64
Beam position (mm)
b) BLM24=f(R24)
120
80
40
0
0
100
200
c) BLM26-BLM30 histogram
200
100
0
0
100
200
d) BLM26-BLM32 histogram
200
100
0
0
40
80
120
e) BLM30-BLM32 histogram
200
100
0
0
40
80
120
f) BLM30-BLM34 histogram

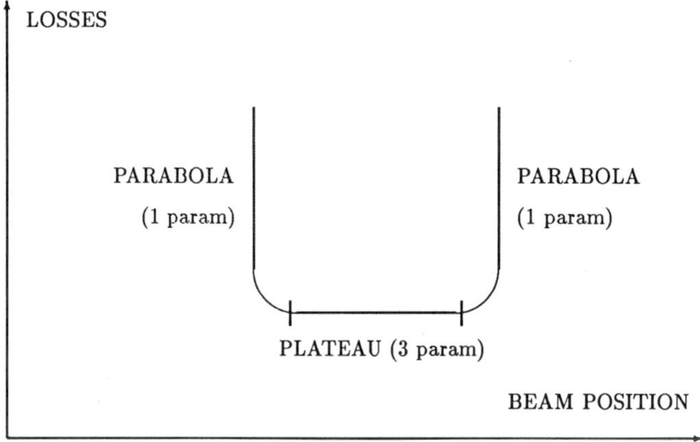

Figure 4: Beam loss approximation near working regime.

The collective beam–loss monitor model, providing it is known, allows to accelerate tuning process significantly [3]. Unfortunately, now we hadn't found yet simple collective model to use it for U–70 fast extraction and have the plans to accumulate more experimental data during accelerator operation.

5. Conclusion

DALS has been tested on the set of beam–line models and on the series of runs. The adequate efficiency has been achieved in all cases. The model for single beam loss monitor has been implemented.

6. References

1. V. Ukhov, B. Volkov, *BEAM: an expert system for control and diagnosis of particle channel equipment*, New Computing Techniques in Physics Research II (World Scientific, 1992).
2. Yu. Ado, et al., *Proceedings of the 14th International Conference on High Energy Accelerators* (Tsukuba, Japan, 1989), p. 235.
3. S. Lapitsky et al., *Possible Approach to Creation of Systems for High Energy Proton Beam Loss Measurement*, Proceedings of the 3-rd Particle Accelerator Conference EPAC-92 (Berlin, 24-28 March, 1992), p. 1052-1057.

USE OF A LEARNING PROGRAM FOR TRIGGER SENSITIVITY STUDIES

SCOTT H. CLEARWATER
Dynamics of Computation Group, Xerox Palo Alto Research Center, 3333 Coyote Hill Road Palo Alto, California 94304, USA

and

YONGWON LEE
Computer Science Department, University of Pittsburgh, Alumni Hall, Pittsburgh, Pennsylvania 15260, USA

ABSTRACT

In a novel application of a rule learning program, we are studying ways to assist and automate the design of triggers in high energy physics experiments. The advantage of using a rule learning program over manual design is that the program can automatically test many more design possibilities. However, unlike many learning programs that rarely give an indication of the robustness of their answers, ours provides the designer with critical sensitivity information needed for designing complicated experiments. Our initial study, which is still in progress, is to understand how to design an event trigger which is robust with respect to various kinds of random and systematic errors in the models used to design the experiment. We call this technique "Performance Sensitive Learning".

1. The Trigger Design Problem

The trigger system of a High Energy Physics (HEP) experiment ultimately decides which events will be saved for later analysis and which will be forever lost. Thus, it is essential for the success of any HEP experiment that the trigger be well-understood. By well-understood we mean that not only are the trigger efficiencies known, but that the uncertainties and therefore the sensitivities are known. These uncertainties arise from both the unavoidable statistical nature of the underlying physical processes and the difficult to understand, systematic errors resulting from imperfect models and detector biases. As these systematic errors are often unavoidable, the best that can be done is to insure that the trigger rules will be able to maintain a high efficiency over a wide range of different conditions. In this paper we describe a technique called Performance Sensitive Learning which automatically learns trigger rules in terms of trigger attributes that are assured of being insensitive, in terms of overall trigger efficiency, over a wide range of trigger attribute errors. This means that the event detection efficiency will be relatively stable over a range of possible conditions. Just as important, the event rejection efficiency will also be stable. This is crucial since many backgrounds at the trigger level rise exponentially fast and thus a small bias in the trigger could let in many more background events and overload the online data acquisition system.

2. Performance Sensitive Learning

Performance Sensitive Learning (PSL) overcomes the weakness of many design approaches, including many learning systems, that rely purely on maximizing efficiency without regard to the sensitivity of the efficiency with respect to the assumptions made in the design model[1-4]. Performance sensitive rule learning takes into account the expected tolerances of the experimental measuring devices to produce a more robust rule, and, thus, an experiment with a greater chance of success. Also, unlike some other sensitivity analysis procedures, performance sensitive rule learning provides an explicit trigger design that can be directly incorporated into hardware or software trigger logic. Further, the attribute ranges over which the performance of the trigger exceeds some minimum are explicitly provided.

We used the learning program Rule Learner version 4 (RL4) to test PSL, which given a set of examples, produces IF-THEN type rules expressed in terms of the event attributes that classify a certain class[5]. The condition part of the rule (the "If" part) is a conjunction of attribute-marker pairs. For example, a rule with two conjunctions has the form IF $(x > 5)$ AND $(y < 3)$ THEN event is in class z. $(x > 5)$ is an attribute-marker pair with event attribute x which must be greater than the marker value of 5 for this conjunction to be true.

The examples are divided into two classes, one class being the class to be learned (called positive examples, i.e. bottom quark events) and the other class being examples from another class (called negative examples, i.e. background events). RL4 explores a tree of possible rules driven by a set of performance constraints pre-defined by the user. These performance criteria define the minimum fraction of positive examples a rule must cover (the number of true positives divided by the number of positive examples, TP/P) and the maximum fraction of negative examples a rule may cover (the number of false positives divided by the number of negative examples, FP/N). In the HEP domain the preformance thresholds correspond to trigger rates. There is also a constraint on the rule complexity, that is, the number of conjunctions of attributes that may appear in a rule. The definition of a class by RL4 is a set of disjunctive (ORed) rules of the above type.

The performance of RL4 has been augmented by adding Performance Sensitive Learning which provides additional information about the class being learned. This additional information expresses the sensitivity of the rule performance with respect to variations in the attributes of the rule. Unlike other learning systems, and this is the power of PSL, the data used to learn the rules need not come from the same distribution as the data in the real experiment in order for the rules learned with PSL to have a high probability of satisfying the performance constraints. A rule returned with the PSL procedure includes the attribute ranges for which the rule is likely to satisfy the performance thresholds and are valuable in determining the sensitivity of a rule. In other words, the valid range for a marker m_i in a rule is given by $[m_i - \Delta m_i^-, m_i + \Delta m_i^+]$ where the Δm_i's are such that the performance of the rule is not degraded below the performance constraints. Before a rule is judged to be satisfactory the range must be large enough to satisfy the constraint $\Delta m_i^+ + \Delta m_i^- > \delta_i$, where Δm_i is as defined earlier, δ_i is related to the resolution of attribute i, e.g. $\delta_i = 2\sigma_i$ where σ_i is the resolution of attribute i.

Alternatively, we can choose the attribute ranges and compute what effect such a change will have on the performance of a rule, δP, which is basically the partial derivative of the performance with respect to the attribute value $\frac{\partial P}{\partial a_i}\delta a_i$, where P is some measure of the performance of the rule such as classification accuracy.

3. Implementation

We used a sample of Monte Carlo events from the H1 detector at HERA[6]. The event sample included several different processes, bottom quark events, neutral current events, and background events consisting of beam-wall interactions. Each event was described in terms of 6 Boolean and 17 numeric attributes. These attributes are the usual type of trigger quantities relating to energy deposition location, quantity, and asymmetry. In the studies reported here we looked at various kinds of errors occurring in the numeric attributes only.

4. Results

The initial results show that incorporating sensitivity into rule learning for HEP trigger rules gives quite different results from what would be expected without incorporating any sensitivity. Although the overall trigger efficiencies are comparable in both cases assuming no additional bias in the test data, the particular rules learned in each case are quite different. For example, using performance sensitive learning, components of the apparatus that are less reliable than others have a smaller chance of appearing in a rule. This results in a more robust trigger. To illustrate the sensitivity problem we have also compared PSL to a well-known decision tree program C4.5[7].

Several different experiments were performed. First, gaussian noise was added to all the numeric attributes using the particular attribute value for the event as the mean and a fraction of that value for the standard deviation. Second, noise from a uniform distribution was added, again with the mean of the distribution being the attribute value in each example. Third, a fraction of the average attribute value (using all training examples) was subtracted from each attribute. Lastly, a fraction of the average attribute value was added to each attribute. The first two experiments tested the effect of random errors using two different distributions and the latter two experiments tested the effect of systematic errors. Adding random noise simulates the effect of an apparatus that has poorer resolution than the model, not an uncommon occurrence. Adding systematic noise simulated the effect of erroneous offsets or biases in the apparatus. There were about 2500 total examples used in the study, about half of which were bottom quark or neutral current events (the two classes of positive events) and the rest beam-wall events (the negative events). The positive and negative examples were divided into three sets to perform a 3-fold cross validation test. The procedure was as follows:

0) Choose the type of noise to study, i.e. Gaussian, uniform, or systematic,

1) Two of the sets were used for learning and the third for testing,

2) Add noise to each attribute of the test (third) data set by the same amount according to the type of noise in step 0), e.g. 10% systematic addition,

3) Test the accuracy of the rules found in step 1) on the noisy examples generated in step 2),

4) Repeat steps 2) and 3) 100 times and compute the overall accuracy,

5) Repeat steps 2)-4) for each noise level 0.1 to 1.0 of the attribute value,

6) Repeat steps 0)-5) for each type of noise.

The results are shown in the figure below.

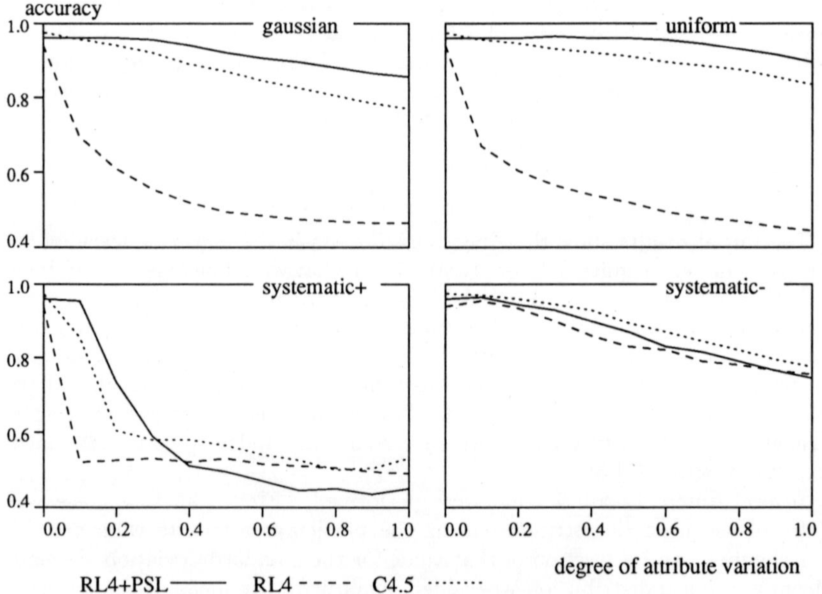

Figure 1: Comparison of Learning Results: Learning Bottom Event Trigger Rules (upper left) Effect of additional Gaussian noise. (upper right) Effect of additional uniform noise. (lower left) Effect of adding systematic noise. (lower right) Effect of subtracting systematic noise.

Fig. 1 shows the results of learning the class "bottom quark event" using the beam-wall events as background. For each plot the vertical axis is the classification accuracy defined as $(TP + TN)/(P + N)$, where TN is the number of "true negatives" (negative examples correctly classified). The horizontal axis is the fraction of the attribute value that the attribute is smeared to simulate the effect of an error. As can be seen from the plots above the rules learned with PSL maintain a higher level of performance than the other techniques over wide ranges and classes of added noise. In the case where a systematic amount of noise was subtracted there was little difference between any of the learning methods. This is due to the fact that interesting HEP events tend to have higher attribute values than their background counterparts. Similar results were found when learning the class "neutral current event" again using the beam-pipe events as background.

An example of a rule found by RL4 for H1 is:

IF (E_BARREL > 0.95) AND (E_BACKW $\leq$ 18.8) AND (E_PLUG $\leq$ 137.0) THEN Event is Bottom-quark event.

This particular rule correctly identified 54.5% of the positive events and incorrectly classified 1.8% of the background events. An example of a rule in terms of Performance Sensitive Learning is:

IF (DC_MULT_TOT > [7-2, 7+2]) AND (E_IF $\leq$ [50.7-28.6, 50.7+39.8]) THEN Event is Bottom-quark event.

This rule correctly identified 66.2% of the bottom quark events and incorrectly classified 0.8% of the background events. Note the relatively wide values that each attribute can attain and still maintain the overall performance thresholds. An example of an overly sensitive rule flagged by PSL is:

IF (DC_HIGHPOS > [0-0, 0+0]) AND (E_BEMC $\leq$ [18.1-6.6, 18.1+5.0]) THEN Event is Bottom-quark event.

While this rule correctly identified 79.3% of the bottom quark events and misclassified only 0.8% of the backgrounds, note the sensitivity of the rule with respect to the attribute DC_HIGHPOS, there is no room for error!

While it is unlikely that an experiment will have all of these errors occurring simultaneously with the same degree of noise, our results unequivocally show that unless sensitivity is incorporated into the learning process the output of the learning program will very likely be more sensitive to errors in the modelling, a dangerous proposition for any experiment. Further, even without the added noise the PSL rules performed as well as the rules without PSL. In other words, there is no trade-off in performance by using PSL.

There are other interesting differences between the rules learned by RL4 and by RL4 with PSL.

1) There were no rules that were the same in the two sets.

2) There were some attributes that were used by RL4 rules, but which were not used in the PSL rules. This is evidence that these attributes were too sensitive to be reliable at the trigger level.

3) The length of rules found with PSL typically involved more attributes

than the RL4 rules.

5. Discussion

We have demonstrated a method for generating trigger rules that takes into account the sensitivity of the apparatus and co-dependencies of the trigger attributes to thereby create rules whose sensitivity is well-understood. This provides the experimenters with more confidence in their results and takes a step towards greater automation of scientific experimentation.

The basic lesson of Performance Sensitive Learning is this: Sensitivity should be incorporated into the learning process rather than ignored or done as a completely separate phase. By doing this the output of a learning program will be much more useful to the end users.

PSL can also be applied to other learning systems such as neural networks. Neural network PSL involves testing whether the classification accuracy is insensitive to the learned set of weights, that is, how broad is the optimum? If it is too narrow then the network will be overly sensitive to changes in the inputs and accuracy will suffer.

6. Acknowledgements

We thank Albert DeRoche of the H1 collaboration for his assistance in providing us with data and the trigger thresholds.

7. References

1. S. H. Clearwater and E. G. Stern, *Computer Physics Communications* **167** (1991) 159.
2. S. H. Clearwater, W. E. Cleland, and E. G. Stern, in *Proceedings of the Symposium on Detector Research and Development for the Superconducting Super Collider*, ed. T. Dombeck, V. Kelly, and G. Yost **Fort Worth** (1990) 485.
3. Y. Lee and S. H. Clearwater, in *Proceedings of the Fourth International Conference on Tools for Artificial Intelligence*, (IEEE, 1992).
4. S. H. Clearwater and Y. Lee, in *Proceedings of the Second Workshop on Computational Learning Theory and Natural Learning Systems*, (1992) (in press).
5. S. H. Clearwater and F. J. Provost, in *Proceedings of the Second International Conference on Tools for Artificial Intelligence*, (IEEE, 1990) pp. 24-30.
6. B. Schwarzschild, *Physics Today* **March** (1992) 21.
7. J. R. Quinlan *C4.5: Programs for Machine Learning*, (Morgan Kaufmann, San Mateo, 1993)

OPTIMIZATION ALGORITHMS IMITATING NATURE

FRANK KURSAWE
Department of Computer Science, University of Dortmund
44221 Dortmund, Germany

ABSTRACT

This survey introduces optimization algorithms imitating certain principles of nature in order to find the global optimum of a given problem. Especially those domains are worth copying where nature has found 'stable islands' in the 'turbulent ocean' of possibilities, like annealing processes, central nerve systems and biological evolution. These approaches start to make their way today, although the underlying ideas already occurred relatively early, measured in scales of computer science. In a rather general and informal summary, this paper tries to convey similarities, differences and applicability of the algorithms derived from those ideas. An introduction to optimization has intentionally been left out. Instead, the reader's intuitive knowledge is appealed to.

1. Introduction

During the last years, people have started to pay more attention to algorithms imitating certain aspects of nature for problem solving purposes. On the one hand, this phenomenon goes back to the increasing computing power which allows the simulation of models becoming more and more complex. On the other hand, why not make use of those languages and strategies for modeling and solving real–world problems nature has been providing and successfully using for the last 4.5 billion years on our planet?

The correspondence between three–dimensional projections of the Mandelbrot set and rugged glaciers does not occur just by chance, but both express the universal law of 'nonlinearity' which rules on the border between order and chaos. This bears incisive consequences on the computability and predictability of systems in which nonlinear relations occur: Despite their strong determinism, the smallest change in starting conditions or the smallest interference with the ongoing process may be reinforced dramatically resulting in a seemingly open or 'chaotic' outcome. The metaphor of a butterfly's beat of wings — at least theoretically — causing a change of weather often illustrates this reaction. But additional feedback and selection processes may 'tame' incalculable systems. Thus, they get the opportunity of *self–organization*: The 'final' state of a system becomes the starting–point for new formations, potentially threatening disturbances may not only be weakened and dominated, but may even trigger further development.

By means of a more general descriptive language the different scientific communities grow together more closely: One common language may describe the behaviour of physical, chemical, biological, medical and even economical and sociological processes. The outbreak of a panic depends on the number headless people in the beginning just as well as the victory of the VHS video system over the Beta video system (which was said to be superior technically) depended on a small original majority of people preferring VHS [4]. Maybe, we are witnessing a change of paradigm: The Arts' and Sciences' traditional approach of classifying everything into drawers could be overcome by a 'systems science' trying to reveal the conditions being responsible for the formation of structures in the material world. The ideas of quantum physics have already shaken the mechanistic view of life, the knowledge gained by nonlinear dynamics may finally detach the old theories. For those, however, who have looked at the Sciences as a replacement for religion, this changeover may be painful. For we do *not* know what we are doing when interfering with natural systems comprising nonlinearities. Even the best intentions can lead to unbounded consequences.

2. Simulated Annealing

The freezing in and crystallization of a fluid or, in general, anything melted serves as the natural metaphor for the *simulated annealing* optimization algorithm. Given enough time, a crystal structure on a minimum energy level is reached. But if one anneals too quickly, this state cannot be reached and one obtains an inhomogeneous structure of a higher energy. For example, it takes several months to cool down the mirror of a classical, large telescope, the optical quality of which mainly depends on the homogeneity of the glass body. The slightest tension would deteriorate the observations.

The simulation of natural annealing processes on a computer is based on the so-called *Boltzmann probability distribution* $Prob(E) \sim \exp(-E/(k * T))$. In words: The system's energetic state in thermal equilibrium (at temperature T) is subject to a probability distribution over all possible energetic states E (k denotes the Boltzmann constant which relates energy to temperature). A system at low temperature can contain quite a lot of energy, although the probability of such an event decreases exponentially. Transferred to the artificial world of optimization, this implies that one has a chance to leave a local optimum again, find a better one and finally the global one [27, 1]. That is why the energy cannot decrease monotonically. Only by allowing intermediate deterioration, one gets the chance to find a better result than just the local minimum located next to the starting-point (see Figure 1).

The underlying idea of this method goes back as far as 1953. Metropolis et al. applied it to thermodynamic models: A system's energy changes from level E_1 to level E_2 with probability $p = \exp(-(E_2 - E_1)/(k * T))$. If $E_2 < E_1$ holds, the probability is 'cut off' at $p = 1$, meaning that any modification of the system resulting in a lower energy level will be accepted. In order to apply this algorithm without a thermodynamic context, one needs the following elements [12]:

- an appropriate description of the system's states,
- a mechanism generating random variations of these states,
- an objective function E to minimize and

- a control parameter T — the temperature's analogy — and a schedule decreasing T by a certain amount (usually a factor < 1) after a fixed number of successful changes.

As a more detailed example, we will now apply the simulated annealing approach to the *travelling salesman problem* (TSP). This problem consists in visiting n cities exactly once and returning to the starting–point on a tour (path) as short as possible. Although being of low practical relevance in the original form, it serves as the most intuitive representative for all sorts of scheduling problems which are highly relevant today in production planning. Furthermore, the TSP has become a standard measure relating the performance of optimization algorithms. It also represents the class of *NP–complete* problems, the members of which have one characteristic in common: The time necessary to solve such a problem *exactly* grows exponentially in n (the size of the problem). Trying out all possible tours returning to the starting city for only $n = 30$ cities on a computer being able to sequentially generate and evaluate $1\,000\,000$ tours per second requires $\approx 1.4 * 10^{17}$ years. The age of our universe 'only' comes to $\approx 20 * 10^9$ years. Therefore, even much faster computers will not help solving this problem. So, we will always have to rely on robust heuristic strategies when dealing with NP–complete problems. Furthermore, the TSP implies a highly *multimodal* objective function, i. e. a function with more than one local optimum (see also Figure 5, right). But what does one have to do in order to apply the simulated annealing heuristic?

- The cities are numbered from 1 to n, and each of them is located at position $(x_i,\ y_i)$, $i = 1, \ldots, n$. The system's states comprise all possible permutations of the order in which a certain tour visits the cities.
- A tour can be modified in two ways:
  - A piece of the tour is taken out and inserted again at the same position in reverse order.
  - Again, a piece of the tour is cut out, but this time, it is inserted at a randomly chosen position.
- Naturally, the length of the tour serves as the objective function.
- Finding a good cooling schedule often is the hardest part and may require some experiments. One should choose the starting temperature with T_{start} being significantly larger than the largest ΔE one has observed during the experiments. If — as in Figure 1 — the cities' locations are mapped to a unity square, e. g. 0.5 would represent a good initial value for T. The temperature remains unchanged on one level as long either $100 * n$ variations have been proposed regardless of their effect or $10 * n$ better configurations, i. e. shorter tours have been found. By then multiplying T with a constant factor < 1.0, the system cools down more or less slowly. One might also think of some rule varying this factor over time depending on the schedule's success so far. If no more improvements have occurred on one temperature level, the search terminates.

The following figure displays the final tour and the cooling process over time, respectively, for $n = 100$ cities.

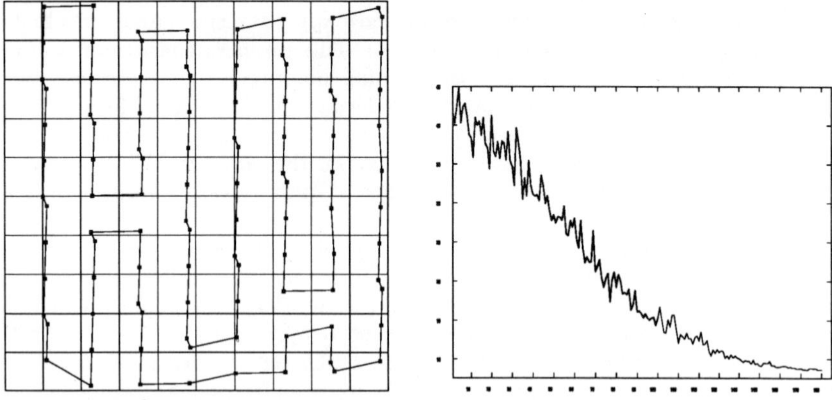

Figure 1: Simulated Annealing (length of tour: 8.5)

3. (Artificial) Neural Networks

This notion comprises all kinds of information processing which imitates the way natural nerve systems operate: Many *neurons* work together in parallel being able to communicate with each other via synapses and axons. In a human brain, each neuron can receive the signals of up to 10 000 partners. And even neurons located far away may be reached via only 10 'relays'. Their working principle can be thought of as follows: A neuron registers the incoming signals, sums them up in a weighted sum and communicates this 'value' to those neurons connected to its outlet:

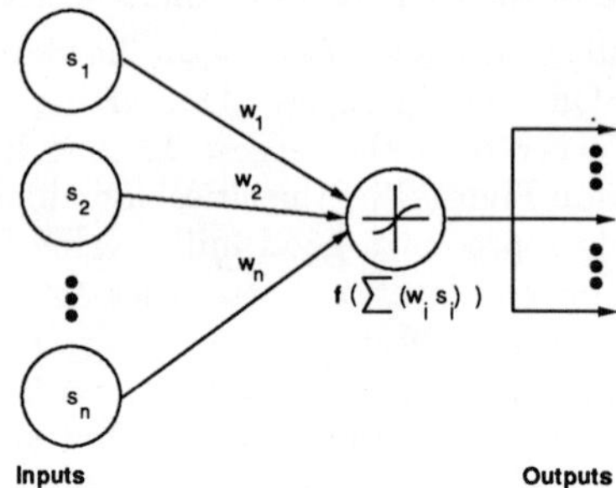

Figure 2: Working scheme of an artificial neuron

For a child, learning consists in a decay of many synapses and a reinforcement of those links activated by some pattern. This simple 'model' — proposed by Hebb [8] already in 1949 — may be sufficient for computer scientists to play around with, but it should be mentioned here that the processes really going on in a brain are still very much subject to speculation. We do know that the communication takes place with the help of at least 30 messenger substances and that not only one type of neuron exists, but many. By the way: If the human brain was simple enough for us to fully understand it, we would be too simple to do this job (E. Pugh).

Neural networks can build a framework for a learning controlled by successes and failures. Although sharing many tasks with the artificial intelligence (AI) community (pattern recognition or completion, associative memories, feature detection, fault–tolerant systems, language processing), they differ significantly from the traditional way of knowledge processing and representation: The knowledge is no longer located in e. g. the single rules of an expert system, but it is rather spread over the whole structure like in a hologram.

Already in 1958, Rosenblatt has developed the 'perceptron' [20]. Some of its properties were quite unique at that time: This simple model was capable of learning and tolerating faults. While classifying patterns, it displayed a certain ability to abstract by responding correctly to unknown inputs, i. e. patterns it had not been trained with. Furthermore, it could be proved that this learning strategy is capable of learning all patterns it can possibly classify in a finite number of steps. But in 1969, there was a serious set–back, when Minsky and Papert discovered and proved the main weakness of a perceptron: It is necessary to adapt the synaptic weights with an accuracy growing exponentially with the problem's complexity. By mistake, they assumed this result (criticism) would hold for this type of network with more than two layers, too [15]. Until the beginning of the 80ies, this field of research could not recover. But then, the works of Hopfield [11] and Rumelhart, Hinton and Williams [22] have called it back to life and have given it an impetus it still thrives upon today.

The main parts of a neural network are

- the neurons (working elements) defining
  - a set of activating states,
  - an activation function,
  - an output function depending on the present state,
- a topology linking the neurons (the network) and
- a (learning) rule how to adapt the synaptic weights.

A certain choice of those parts mentioned above results in a certain type of neural network. Today, one can distinguish the following most important models [16, 18]:

- A *perceptron* consists of only one layer of a fixed number of neurons. Its task is to classify into categories the input patterns, each containing m characteristics. Whether a given network is capable of solving a certain problem, is not guaranteed a priori: The patterns have to be separable by a $m-1$ dimensional hyperplane. Spirals being intertwined cannot be separated by a straight line; therefore, the perceptron has no chance to succeed. By simply adding another feature to the data set (here: a third dimension), one may often overcome this situation. Even such a simple problem like calculating the 'exclusive or' of two binary inputs lies beyond the capabilities of this network type. The four points $(0,0)$, $(1,1)$ — these constituting (output) class 0 — and $(0,1)$, $(1,0)$ — the members of class 1 — cannot be separated appropriately by a straight line. But neural networks with more than one layer go beyond the perceptron's limited scope:

- The multi–layered *perceptron with back-propagation* ('backprop') probably represents the best known network model today. Training a net of this type has become possible since Rumelhart, Hinton and Williams [22] have managed to generalize the delta–rule, applicable only for nets with one input and one output layer, resulting in the recursive back–propagation learning algorithm. Now, neural networks can perform nearly arbitrary mappings between inputs and outputs, because the additional layers somehow encode and compress the information coming in. Figure 3 illustrates the topology (structure) of such a net:

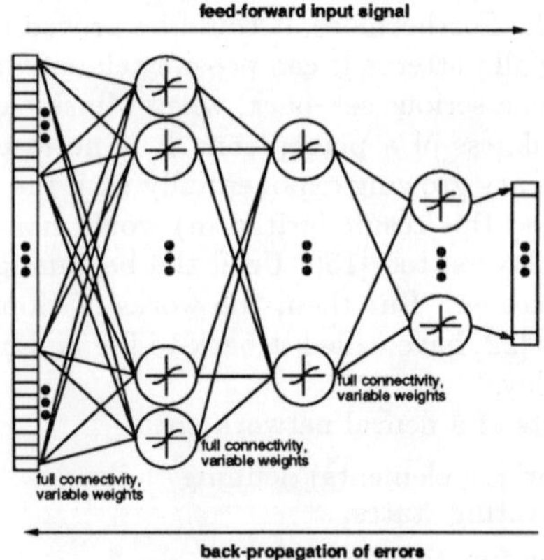

Figure 3: Back–propagation network with three layers

The following 'sigmoid' function often defines the output of one single neuron depending on all its synaptic inputs $\sigma(x) = 1/(1+e^{-x})$. Starting with randomly chosen initial values of the synaptic weights, the algorithm tries to adjust those weights in such a way that the deviation between input and expected output becomes smaller and smaller. If one chooses the stepsize (learning rate) appropriately, the algorithm follows the path of steepest descent (gradient) of the error function. For a more detailed discussion, see e. g. [6].
The back–propagation algorithm is applicable to neural networks with an arbitrary number of hidden layers. But the time needed to train such a network

grows enormously. Furthermore, since using a gradient algorithm, the training phase can get stuck in a local optimum differing from the global one. If one uses perfect, i. e. undisturbed patterns for training, the resulting net will have great difficulties in deciding about an input being slightly different. Moreover, it is still unknown in general, how many layers with how many neurons one needs in order to solve a given problem, and what will happen if the network is oversized.

- *Hopfield-nets* bear much resemblance to the idea of simulated annealing introduced in chapter . In 1982, Hopfield [11] could prove that the behaviour of a certain network type, which has later been named after him, corresponds to a physical system containing a large number of elementary magnets ('spin glasses'), thus leading to a common energy function. Hence, analytical methods from physics could be used to describe this sort of network. Accordingly, the maximum information capacity of a Hopfield net comprising n neurons calculates to $2 * n$. But this number decreases, if one wants the net to recognize patterns subjected to noise as well.
  All network models presented so far rely on the same basic building blocks. A feedback loop constitutes the new characteristic of Hopfield–nets: The outputs of the one–layer net are fed into the inputs again. In this way, the net can restore the missing parts or parts disturbed by noise step by step. A program working in a traditional way could only compare a certain number of stored patterns to the present input and then respond with the least deviating pattern. A well trained *associative memory*, on the other hand, determines its answer according to an abstract and holistic similarity criterion being unknown beforehand.
- *Kohonen's (self-organizing) feature maps* have been designed as two dimensional layers of neurons [13]. In contrast to back–propagation networks, any incoming signal only causes an amount of activity limited spatially in the following way: As in natural brains, similar signals only stimulate neighbouring (adjacent) neurons. Again using randomly chosen initial synaptic weights, an input vector – also selected randomly from the training set — represents the stimulus. The neuron feeling most 'responsible' for this input moves a little bit towards the vector and drags some of its neighbours along. The number of those neighbours decreases over time. Of course, the speed of this reduction strongly influences the quality of the results. On the one hand, if the neighbourhoods are made smaller too quickly, the network runs into 'stress' and 'freezes' in a state not being the global optimum. On the other hand, one gives away computing time. The analogy to chapter is rather evident.
  As an example, we use the travelling salesman problem again: A network of neurons is supposed to learn a path as short as possible visiting all cities. In each iteration, a town $C_{x,y}$ chosen at random drags the nearest neuron $N_{x,y}^{closest}$ a little towards it, thus deforming the net which has started as a circle. A certain number of neighbours also moves towards the city. The further away

these neurons are from the 'best match' neuron, the smaller their co–operation. During the iterations, this feature vanishes more and more, and one can observe the net first shaping into a rough structure, but the neurons then get closer and closer to the cities until there is correspondence. Naturally, having started with more neurons than cities, several neurons will lie on one city in the end. Figure 4 (left) tries to convey an impression of what has gone on. This example only contains 20 cities for clarity reasons. One can see the starting circle, then the approximate shape of the tour and finally the global best tour in this case. This is not always guaranteed, though. But with one run only lasting for about 10 seconds, one can afford several runs.

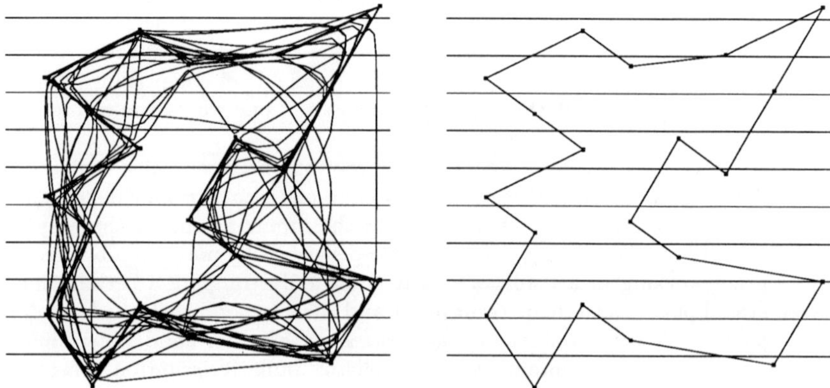

Figure 4: Development of a tour, best tour

In the field of robot control, this algorithm yields good results, too, without having to undergo any changes: The cities now represent positions where a robot is supposed to drill holes. So, visiting those positions on a short tour is desirable in order to minimize abrasion.

- At the moment, *Carpenter / Grossberg–classifiers* are probably the models being closest to what is really going on in natural brains. They include feedback loops, a short–term and long–term memory and a function controlling the alertness, i. e. lowering it if the outside world does no longer provide any stimuli. For the first time, a model — like a brain — is capable of learning continuously without destroying (forgetting) the previous knowledge. So, in one area, you can observe islands of stability, whereas other regions produce seemingly chaotic activity. Grossberg's *adaptive resonance theory* (ART) laid the foundations of this model in the 70ies. But presenting further details would go beyond the scope of this paper.

4. Evolutionary Algorithms

Because both algorithms presented in this chapter — *evolutionary strategies* (EA's) and genetic algorithms (GA's) — imitate certain principles of organic evolution, a common introduction seems appropriate.

Despite some of nature's astonishing results — from an engineering point of view — the idea to imitate certain principles of nature in order to create a new optimization algorithm has been ridiculed and rejected rigorously for the following reasons:

- Evolution proceeds very slowly compared to the lifetime of human beings, and therefore, it can only be observed very rarely.
- There is no global 'objective function'.
- A mutation–selection scheme with a stochastic description (modelling) of disturbances appears to waste resources compared to algorithms which specialize in a certain type of problem.
- When first trying to imitate evolution on a computer, the slogan 'Survival of the fittest' (singular!) was taken too literally. By the way, this slogan attributed nearly automatically with Charles Darwin was meant to be a provocation by Herbert Spencer.
- The element of chance inherent in evolution often leads to a misunderstanding: Simply 'rolling the dice' could indeed not have produced the diversity of complex living systems observable today. Accordingly, taking random samples from the high dimensional parameter space of an objective function in order to find the global optimum *must* fail (*Monte–Carlo search*). One can even prove that an algorithm enumerating the search space completely always performs better, because it avoids visiting a point more than once.

But by looking at evolution as a cumulative, highly parallel sieving process, the results of which pass on slightly modified into the next sieve, the amazing diversity and efficiency on earth no longer appears miraculous. The point is to isolate the main mechanisms which have led to today's world and which have been subjected to evolution themselves. Inevitably, nature has 'invented' a mechanism allowing individuals of one species to exchange parts of their genetic information in order to meet changing environmental conditions in a better way (recombination or crossing–over). With this principle and others, nature managed to 'learn' even complex models of the environment: For instance, at branching points, the ratio of the diameters in a system of blood–vessels is close to the theoretical optimum given by the laws of fluid dynamics ($2^{-1/3}$). For other examples, see [19].

Genetic algorithms and evolution strategies originated independently from each other on both sides of the Atlantic Ocean in the middle of the 60ies competing with other (global) optimization algorithms. Later, their emergent properties became a topic of interest, too. In both methods, a population tries to 'learn' (find) the global optimum of the system to be optimized collectively, imitating the following principles:

- population
- inheritance (recombination or crossing–over)
- mutation
- selection

Mutation and recombination may be regarded as the 'engine' of the vehicle evolution, whereas selection would provide the 'steering wheel' in this metaphor.

4.1. Evolution Strategies

Rechenberg and Schwefel [17, 23] developed the *evolution strategies* when they wanted to optimize technical objects like a nozzle. No closed form analytical objective function was available, and hence, no applicable optimization method existed at that time. So, these strategies resemble a person experimenting, in the worst case not knowing anything about the object (black box situation). All there is to play around with is a finite number of regulators which have to be tuned in such a way that the output meets some kind of optimality criterion.

The direction and the stepsize of the variations are crucial for all optimization methods. Figure 5 illustrates that these values have to be adapted all the time until the convergence has been reached. Starting far away from the optimum, large steps are advantageous in the beginning, but later, they have to become smaller as the algorithm approaches an extreme point. Figure 5 (right) illustrates the situation of an optimization method using another metaphor: It resembles hillclimbing in foggy conditions, higher peaks might exist across the valley.

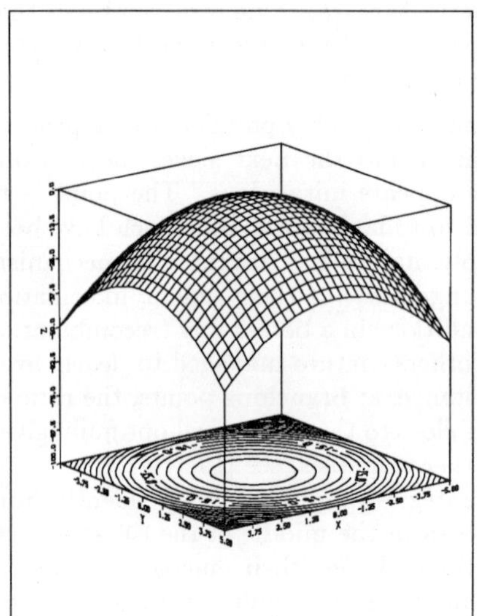

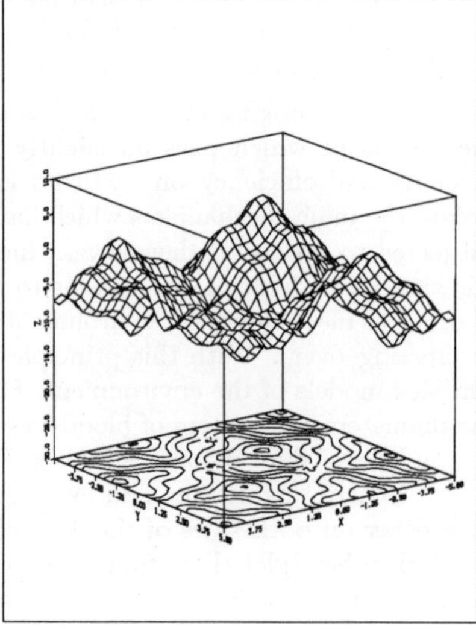

Figure 5: Topologies of objective functions

Schwefel [23, 24] generalized a two–membered or $(1+1)$ strategy to become the multimembered or $(\mu \dagger \lambda)$ evolution strategy which will now be explained in more detail. An individual consists of the following 'genes':

- Real–valued *object variables* x_i have to be tuned by recombination and mutation in such a way that an objective function reaches its global optimum. Referring to a metaphor mentioned previously, these are the regulators of a black box.
- Real–valued *strategy variables* or stepsizes σ_i determine the mutability of the x_i. They represent the standard deviation of a $(0, \sigma_i)$ Gaussian distribution being added to each x_i as an undirected mutation. With an expectancy value of 0, the parents will produce offsprings similar to them on average. In order to make a doubling and a halving of a stepsize equally probable, the σ_i themselves mutate log–normally distributed from generation to generation. Inside those stepsizes, the population hides the internal 'model' it has made of its environment so far. Hence, learning takes place on two levels.
- Depending on an individual's x_i, the resulting objective function value $f(\underline{x})$ serves as the 'phenotype' (fitness) in the selection step.

The μ parents produce λ offsprings per generation in three steps:

1) Object and strategy variables of an offspring are put together according to the type of recombination chosen. Each parent has an equal opportunity to reproduce.
2) First, the strategy variables mutate, and their new values are used for mutating the object variables.
3) According to the strategy chosen, the selection then takes place: In a plus strategy, the μ best of all $\mu + \lambda$ individuals survive to become parents in the next generation. Using the comma variant, the selection takes place only among the λ offsprings. The second scheme is more realistic and therefore, more successful, because no individual may survive forever which could at least theoretically occur using the first variant. Again (see chapter), a strategy allowing intermediate deterioration performs better. Only by 'forgetting' individuals with a good phenotype, which may have been achieved with an internal model that is no longer appropriate for further progress, a permanent adaptation of the stepsizes can take place. This becomes particularly evident if the location of the optimum changes over time.
   By choosing a certain ratio μ/λ, one can determine the convergence property of the evolution strategy: If one wants a fast local convergence, one should choose a small ratio, but looking for the global optimum, one should decide in favour of a 'softer' selection.

Evolution strategies are adaptable to nearly all sorts of problems in optimization, because they need very little information about the problem — especially no derivatives of the objective function. The objective can 'hide' the result of a simulation run, it does not have to be given in a closed form. This also holds for the constraints which may represent the outcome of a finite elements method (FEM)

[5]. Evolution strategies have been adapted to vector optimization problems [14], and they can also solve NP–complete problems like the TSP [2, 21].

Using $f(\underline{x}) := \sum_{i=1}^{30}(i * x_i^2)$ as the objective function, $n = 30$ stepsizes have to be tuned properly with respect to each other in order to achieve maximum progress. Figure 6 displays the performance of different strategies:

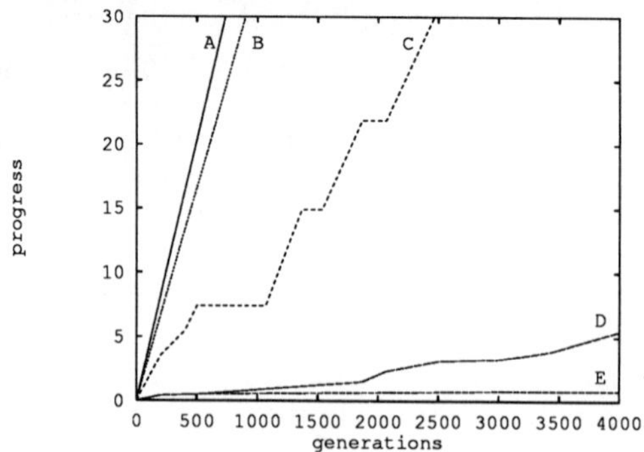

Figure 6: Learning of a scaling

Curve E represents a $(1, 10)$ strategy which fails because the convergence velocity decreases reciprocally to n. If only one offspring survives, an individual operating in a subspace of R^n will prevail over others. Increasing λ does not help, but the simultaneous increase of μ and λ leads the way. Curves D $((3, 6))$ and C $((6, 30))$ display the impact of recombination on convergence velocity. But only the comparison with the best variant can show its true capabilities. Curve A already starts with optimum stepsize relations ($\sigma_i = c/\sqrt{i}$) and can therefore abandon recombination. A $(15, 100)$ evolution strategy starting 'dull' and having to gain this knowledge on the fly nearly proceeds at the same speed (B). But how many offsprings should survive to become the parents of the next generation? Varying μ with a constant $\lambda = 100$, Figure 7 answers this question.

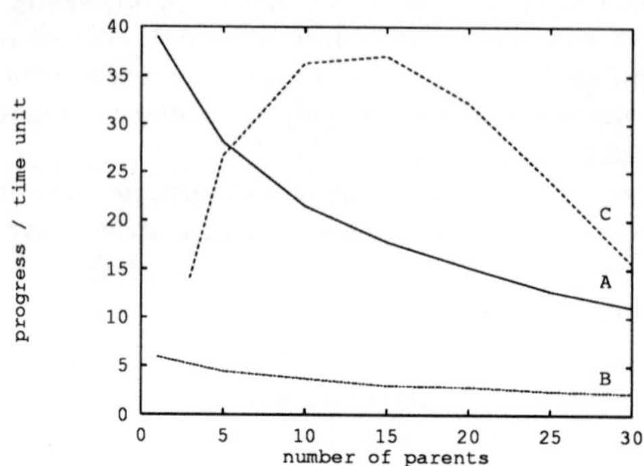

Figure 7: Optimum selection pressure

Curve A again represents a strategy with perfect knowledge calculated analytically, whereas in B, randomly chosen relations of the σ_i were fixed. In curve C, the stepsizes could self–adapt. Not surprisingly, $\mu = 1$ proved to be the best choice for A and B where only one stepsize had to adapt, because the relations had been fixed. On the other hand, one should choose μ between 12 and 20 if one wants learning to take place (C). Two observations from Figure 7 are remarkable: The $(15, 100)$ strategy converges nearly as fast as variant A, and it performs better than a $(15, 100)$ strategy with perfect knowledge. One can regard this phenomenon as a *synergetic* effect: 15 'fools' perform better *collectively* than the same number of 'specialists'.

To sum up, the self–adaptation capability of evolution strategies depends on the following factors [25]:

- The population has to be sufficiently large. Not only the temporary best should be allowed to reproduce, but a set of better individuals. Biologists have coined the term 'requisite variety' being necessary to prevent a species from becoming poorer and poorer genetically and eventually dying out.
- Within a population, the individuals should recombine their knowledge with that of others (co–operate).
- In order to allow better internal models (stepsizes) to provide better progress in the future, one should accept deterioration from one generation to the next. A limited life–span in nature is not a sign of failure, but an important means of preventing a species from 'freezing' genetically.

4.2. Genetic Algorithms

In principal, the scope of *genetic algorithms*, which have been developed by John Holland [10] taking up Bremermann's earlier idea of a binary coding [3], is identical to that of evolution strategies. Despite their common roots, differences exist and will be mentioned in the following [9]. An overview over the range of applications can be found in [7].

- The coding (usually binary) of the object variables represents the biggest difference. The knobs of the black box have turned to switches [7]. In order to calculate an individual's fitness the bitstring has to be decoded first, thus also determining the interval where one suspects the optimum.
- Mutation and recombination or crossing–over, respectively, form the genetic operators, but with a different importance than in ES: Crossing–over which typically occurs with probability 0.6 represents the main search operator, whereas mutation occurring only with probability 0.0001 prevents the total loss of an allele (0 or 1) in the population at a certain position. Both probabilities are determined exogenously, and they do not adapt during the search, although a decrease of the mutation rate seems appropriate if many bits have already been set correctly. Furthermore, a self–adaptation does not work using the standard GA selection schemes.
- Genetic algorithms do not generate surplus offsprings like ES (and nature) do.

- Reproduction probabilities proportional to the individual's contribution to the overall fitness of the population can lead to the dominance of a good individual in the following generation. Such a behaviour may already occur in simple, unimodal topologies and slow down the convergence velocity. Of course, other selection schemes have been developed, e. g. ranking where a certain number of offsprings is assigned to an individual depending on its relative fitness (rank) in the population. But for self–adaptation on a second level to work, one has to introduce an extinctive selection like the one in ES.

The following table sums up the main differences between GA's and ES's:

genetic algorithms	*evolution strategies*
binary coding	real–valued variables
limited search space because of decoding	(in principle) unlimited search space
recombination most important operator	mutation most important operator, recombination for the self–adaptation of the stepsizes
no collective learning of strategy parameters	collective learning of strategy parameters
learning on one level	simultaneous learning on two levels

5. Summary / Outlook

Especially for problems which cannot be solved (optimized) analytically, the new methods presented in this paper can provide results not obtainable otherwise. Convergence proofs exist for simulated annealing, genetic algorithms and the $(1+1)$ evolution strategy, but they are not of much practical relevance due to the unlimited time they assume. But in engineering applications, there often is no need to know for sure about the global optimum, because doing better than before is sufficient.

Finally, the algorithms presented here should be looked upon from a more abstract level in order to stress the aspects they have in common:

- All these methods can be written down mathematically in very few lines. One has to provide only a few (strategy) parameters as the framework in which learning can take place.
- All algorithms strongly rely on chance modifying states and 'proposing' them to the selection component. The emergent order is achieved in the long run by selectively stabilizing 'good' states.
- In all cases, the knowledge is distributed over the respective structure. Intelligence occurs as a collective phenomenon.

It seems as if the notions 'self–organization' and 'evolution' only describe the same phenomena either from a physicist's or a biologist's point of view. They both describe those reactions in which a system with many attractors steers towards a stable one, successfully defending it against disturbances — within certain limits of course.

The aim of this overview was the introduction of algorithms still being capable of optimizing those models which do no longer fulfill the traditional mathematical prerequisites like smoothness and differentiability. The freedom thus gained in the modelling phase has to be paid for by a loss of convergence certainty and velocity. Of course, the simplex algorithm or one of its variants is the most efficient method if the objective function and the constraints are linear. But do linear models reflect the complex relations in the real world correctly?

6. References

1. E. H. L. Aarts and J. H. M. Korst, *Simulated Annealing and Boltzmann Machines*, Wiley & Sons, Chichester, 1989.
2. P. Ablay, *Optimieren mit Evolutionsstrategien*, Spektrum der Wissenschaft, 1987, pages 104–115.
3. H. J. Bremermann, *Optimization through evolution and recombination*, in: Self–Organizing Systems, M. C. Yovits, G. T. Jacobi and D. G. Goldstein (Eds.), Spartan Books, Washington, D.C., 1962, pages 93–106.
4. J. Briggs and F. D. Peat, *Turbulent Mirror — An Illustrated Guide to Chaos Theory and the Science of Wholeness*, Harper & Row Publishers, New York, 1989.
5. I. Campos Pinto, *Wissensbasierte Unterstützung bei der Lösung von Optimierungsaufgaben*, PhD thesis, Universität Dortmund, Fachbereich Informatik, 1989.
6. J. E. Dayhoff, *Neural Network Architectures: An Introduction*, Van Nostrand Reinhold, New York, 1990.
7. D. E. Goldberg, *Genetic Algorithms in Search, Optimization and Machine Learning*, Addison–Wesley, Reading, Massachusetts, 1989.
8. D. O. Hebb, *The Organization of Behaviour*, Wiley & Sons, New York, 1949.
9. F. Hoffmeister and T. Bäck, *Genetic algorithms and evolution strategies: Similarities and differences*, in: Schwefel and Männer [26], pages 445–469.
10. J. H. Holland, *Adaptation in Natural and Artificial Systems*, University of Michigan Press, Ann Arbor, Michigan, 1975.
11. J. J. Hopfield, *Neural networks and physical systems with emergent collective computational abilities*, in: Proc. Nat. Acad. Sci. USA, 1982, pages 2554–2558.
12. S. Kirkpatrick, C. D. Gelatt and M. P. Vecchi, *Optimization by simulated annealing*, Science, **220** 1983, pages 671–680.
13. T. Kohonen, *Self-Organization and Associative Memory*, Springer, Berlin, 1984.
14. F. Kursawe, *A variant of evolution strategies for vector optimization*, in: Schwefel and Männer [26], pages 193–197.
15. M. Minsky and S. Papert, *Perceptrons*, MIT Press, Cambridge, Massachusetts, 1969.

16. Y. H. Pao, *Adaptive Pattern Recognition and Neural Networks*, Addison–Wesley, Reading, Massachusetts, 1989.

17. I. Rechenberg, *Evolutionsstrategie: Optimierung technischer Systeme nach Prinzipien der biologischen Evolution*, Frommann–Holzboog, Stuttgart, 1973.

18. H. Ritter, T. Martinetz and K. Schulten, *Neuronale Netze*, Addison–Wesley, Bonn, 1990.

19. R. Rosen, *Optimality Principles in Biology*, Butterworths, London, 1967.

20. F. Rosenblatt, *Principle of Neurodynamics*, Spartan, New York, 1962.

21. G. Rudolph, *Global optimization by means of distributed evolution strategies*, in: Schwefel and Männer [26], pages 209–213.

22. D. E. Rumelhart, G. E. Hinton and R. J. Williams, *Learning internal representations by error propagation*, in: Parallel Distributed Processing I: Foundations, D. E. Rumelhart and J. L. McClelland (Eds.), MIT Press, Cambridge, Massachusetts, 1986, Chapter 8.

23. H.-P. Schwefel, *Numerische Optimierung von Computer–Modellen mittels der Evolutionsstrategie*, Birkhäuser, Basel, 1977.

24. ——, *Numerical Optimization of Computer Models*, Wiley & Sons, Chichester, 1981.

25. ——, *Collective phenomena in evolutionary systems*, in: 31st Annual Meeting of the International Society for General System Research, Budapest, 1987, pages 1025–1033.

26. H.-P. Schwefel and R. Männer (Eds.), *Parallel Problem Solving from Nature, Proceedings of the 1st PPSN–Workshop, Dortmund, 1990*, **Volume 496 of Lecture Notes in Computer Science**, Berlin, 1991, Springer.

27. P. J. M. van Laarhoven and E. H. L. Aarts, *Simulated Annealing: Theory and Applications*, D. Reidel, Dordrecht, 1987.

EVOLUTIONARY CONTROL OF A POPULATION OF INFERENCE ENGINES

ANDREAS HEMKER
Physics Department, University of Wuppertal
42097 Wuppertal, Gauss-Str. 20, Germany

ABSTRACT

After a short introduction in the problem of reconstructing particle collisions and an overview about artificial intelligence problem solving methods, we build a naive knowledge based system for the reconstruction of particle collision events. A more detailed investigation of the problem domain shows that such a conventional knowledge-based system would not work, because the inference engine is not able to handle the complexity of the problem space. Evolutionary algorithms show us a way out of the dilemma. We use an evolutionary algorithm as a meta system to control a population of inference engines. The results show a clear advantage of the hybrid architecture.

1. The Reconstruction of Particle Collisions

Our initial motivation was to build a system, which is able to reconstruct the physical processes after the annihilation of particles in collider rings like LEP[1]. After the collision of electron and positron several physical processes are running up and a great number of mostly short living particles is produced. Only the final products of the physical processes can be identified with the help of large detector systems.

An obvious approach to analyse such a collision event is to try to run the invisible physical processes in an event backwards. Starting from the detected event state, an event is reconstructed step by step using only particles and interactions predicted by the standard model. Physical laws are used to restore the data (momentum, energy) of the short–living particles. If the reconstruction process succeeds, then in principle any question about an event can be answered. If no satisfactory reconstruction of an event can be found, then this could be a strong hint that unknown physical phenomena have appeared. In this case the point can be located, where an event, which probably contains new physics, does not match the model.

In the bubble chamber experiments of the fifties particle physicists were able to interpret an event only by visual inspection. They reconstructed the processes in an event bearing in mind the model of particle physics. This "manual" proceeding is of course in our days not feasible anymore. The higher collision energies lead to very complex event topologies and the progress in theoretical understanding requires from such a potential event analyser an enormous spectrum of detailed knowledge.

The standard model knows hundreds of particles and each one can emerge from dozens or hundreds of interactions.

The usual way out is to replace the physicists by a piece of software which does the work for you. But to my knowledge nobody has gone this way. The question arises, why nobody tries a full reconstruction of an event, although no other method could give results of a higher quality. A first reason is that the frequently used programming language FORTRAN neither has the formalisms to represent the knowledge of the theory, nor has the methods to apply the knowledge in a reconstruction process.

2. Artificial Intelligence

This is now the point where Artificial Intelligence (AI) techniques come into the play. In principle we can distinguish two different paradigms in the AI field[2]. The conventional symbol processing approach of AI operates with high level entities which are locally represented, this means one symbol stands for one concept. In general, the problem solving process is based on a chain of logical inferences driven by a centralized control. Furthermore it is typical for the symbol processing approach that knowledge base and inference engine are separated and that the knowledge processing happens in a strongly sequential mode.

With the rise of artificial neural networks the dynamical paradigm of AI comes to the foreground[3]. In dynamical systems there are in general numerical values which evolve with time according to a finite set of non–linear equations. Variables like connection weights or activation thresholds in neural networks do not have a pre–defined meaning, but instead of specialize during a learning phase. This kind of knowledge processing is often called distributed or sub–symbolic, because high level concepts are represented simultaneously by a set of variables, for example activation pattern in neural networks. It is less the way knowledge is represented what differs dynamical systems from the symbol processing approach, as it is the way knowledge is processed. The problem solving process of dynamical systems is based on their self–organisation ability. Instead of a centralized control, all units of a dynamical system operate simultaneously under their own control. Each unit obeys only a few simple rules and these rules determine the behaviour of the whole system. The inherent parallelism of dynamical systems makes an implementation on a multi–processor architecture obvious.

3. A Knowledge–based System for Event Reconstruction

The decision which of the two paradigm is better to use for the reconstruction task is at first sight simple. The components of the standard model, these are a database of particle properties and particle interactions, physical laws and constraints, are the typical ingredients of a knowledge–based system. One of the hardest tasks in designing a knowledge–based system, the knowledge acquisition process, can be omitted in our case, as the physical knowledge is available and collected in Monte Carlo (MC) generators. The database of the MC generator only has to be converted into the knowledge representation schemes known from the symbol processing approach. For example, a production rule is a natural formalism to represent a physical interaction.

The application of the knowledge base on the measured data of an event is done by the inference engine. The problem is now, that the inference engine has to cope with considerable combinatorial problems resulting from the backward application of the model. In each cycle the inference engine has to match the set of physical interactions against the set of particles with unknown physical origin. This results in a huge set of conflicting reconstruction steps, where most of them are wrong. The conflict resolution module of the inference engine has to pick up one of the few correct ones. But there are no strong hints which could be the right reconstruction steps. In principle the inference engine has to guess the physical processes, which are running up spontaneously in nature. If the inference engine implements an irreversible control strategy like hillclimbing then only one mistake can lead the reconstruction process into a dead end. One simple way to handle the indeterminism of nature is to let the inference engine check systematically all the different processes which may have happened in an event. But a formal proof shows that the reconstruction task is NP–complete[4]. This means that the problem space increases exponentially with the input data. For a typical event of the LEP experiment the number of possible reconstruction steps is greater than 10^{36}. This explains why a brute force search algorithm does not make sense.

If you want to use a knowledge–based system in a NP–complete problem space then you have to reduce the complexity by incorporating additional knowledge, e.g. meta knowledge to focus the global search into promising areas of the problem space and domain specific heuristics to perform a fine–tuned local search. Otherwise the inference engine either makes a random walk through the problem space, or run immediately into the next sub optimum. But meta knowledge for the reconstruction task is not available and the existing heuristics are misleading in most cases. The consequence is that the deterministic and sequential control strategies of the symbol processing approach would have no chance to find the global solution in an acceptable time scale.

4. Evolutionary Algorithms

It is mother nature which shows us a way out of the dilemma. The evolution principle of nature tells us that each species tries to adapt to a complex and changing environment in the best possible way. Only the fittest among a population of species survive and are allowed to reproduce. The features of each species are encoded in the "make–up" of their chromosomes. The chromosomal make–up can be altered when parents reproduce.

From a technical viewpoint the adaptation process of nature can be seen as a random–directed optimization method. Evolutionary algorithms simulate the interplay of selection and recombination steps to generate better and better solutions. The best known elvolutionary algorithms are the evolution strategy and the genetic algorithm[5]. Like neural networks, evolutionary algorithms belong to the class of dynamical systems. The self–organisation property of evolutionary algorithms originates from the sequence of selection and recombination steps applied to a finite population of individuals. The pressure of selection leads to a quality improvement in the next generation of solutions, while the application of recombination operators introduces permanently new kinds of individuals. This innovation process is not completely randomized, because each offspring contains well proved properties of the previous generation.

5. The Hybrid Approach

The parallelisation potential and the ability to handle the complexity of huge problem spaces made evolutionary algorithms interesting for us. But before we can apply evolutionary mechanisms, we have to make some preparation. The first step is to have a population of inference engines instead of only one. When activated, each inference engine acts independently and selects randomly or probabilistically reconstruction steps. In general each inference engine follows its own reconstruction route. Although you can have a large population of inference engines, it is extremely unlikely, that one of them finds at the first go a complete reconstruction of an event. In fact, an inference engine reconstructs some of the original particles, as well as it makes a number of wrong inference steps, which will guide the inference process to a dead end. This is what we expect from a hillclimbing procedure.

The population of inference engines is now controlled by the evolutionary meta system. I termed the evolutionary meta system "evolution engine" because the problem solving process is not based on logical inferences but on trial and error. The basic mechanisms of the evolution engine are the selection step and the recombination operator. The selection step prefers above averaged solution candidates for the application of the recombination operator and additionally performs something like a filtering for the reiterated application of the inference engines. The task of the recombination or crossover operator is to exchange informations between different solution candidates on a syntactic level. The crossover operator inherit chains of reconstruction steps from different parents to a common offspring. This is the constructive effect of the crossover operator. On the other side the crossover operator randomly breaks up inference chains at the crossover points. Nevertheless, this destructive effect is desirable because inference steps might be wrong.

6. Results

The reconstruction system is implemented on the Connection Machine CM–2 which belongs to the class of SIMD† computers. The following results are based on a set of hundred simulated e^+e^- events produced by the MC generator LUND 7.3 at the Z^0 mass. The reconstruction system has been applied to the problem of quark flavour tagging to show the physical relevance of the approach. For B–quark events the reconstruction system reaches an optimal 100%:100% efficiency purity ratio. In comparision with it the results of a neural network and a Fisher discriminant analysis never exceed a 80%:70% efficiency purity ratio. Beyond the better ratio the reconstruction system gives results with a higher quality. Because of the model based reasoning each classification can be explained in contrast to the other methods.

The next experiment shows that the design decision to have an evolution engine controlling a population of inference engines really makes sense and that furthermore a synergistic effect can be observed. By switching off the evolutionary meta system, one can estimate its contribution to the final result. If each inference engine acts independently then 54.2% of all physical processes are correctly reconstructed on average. Under the control of the evolution engine one gets a much better result. About 96% of all processes are reconstructed by a comparable amount of CPU time. This shows that the power of the system is based on the interaction of the inferential and the evolutionary module

7. Conclusion

The results have shown that the incorporation of an evolutionary module can be a way to overcome one of the weaknesses of knowledge–based systems, which is attributed to the restricted control regimes of their inference engines. To cope with huge problem spaces, a conventional inference engine needs meta knowledge to focus the search quickly in promising areas of the problem space. The evolution engine replaces the missing meta knowledge about the reconstruction task by the parallel proceeding and the meta heuristics selection and recombination. The evolution engine prefers above averaged solution paths and inherits chains of inference steps from different parents to a common offspring. It is the self–organisation property of the evolutionary meta system, which guides the search, instead of goal–directed heuristics. In the long term, the evolution engine automatically prefers the correct reconstructions and makes use of them in innovative new combinations.

8. References

1. A. Hemker, in *New Computing Techniques in Physics Research II*, ed. D. Perret-Gallix (World Scientific, 1992).
2. K.-H Becks and A. Hemker, in *Proceedings of the 1991 CERN School of Computing*, **CERN 92-02**, (1991).

† Single Instruction Multiple Data

3. R. Serra and G. Zanarini, *Complex Systems and Cognitive Processes* (Springer Verlag, 1990).
4. A. Hemker, *A knowledge-based genetic algorithm for the reconstruction of physical events*, Preprint **WUB-DIS 92-11** Physics Department, University of Wuppertal (1992), in german.
5. see the contribution of F. Kursawe in this book.

ADJUSTING PARAMETERS OF GENETIC ALGORITHMS BY FUZZY CONTROL RULES

Andreas Bergmann and Wolfram Burgard
Department of Computer Science III, University of Bonn, Römerstraße 164
D-53117 Bonn

and

Andreas Hemker
Physics Department, University of Wuppertal, Gauß-Straße 20
D-42097 Wuppertal

ABSTRACT

The practical application of systems based on evolutionary concepts is often complicated by the problem of tuning the system parameters. Depending on the parameter setting the evolutionary algorithm converges prematurely or makes a long term random walk through the solution space. A successful problem solving process is only guaranteed on the small edge between these extremes. In this paper we present a new method applying fuzzy control rules for the dynamic guidance of a genetic algorithm. This approach allows the incorporation of available knowledge about reasonable parameter settings. The fuzzy control rules adjust the inversion, crossover and mutation rate to achieve a desired population diversity. Experimental results of a comparison with an (1+1)-evolution strategy demonstrate that fuzzy control allows better approximation of the target entropy and results in better perfomance of the underlying genetic algorithm.

1. Introduction

In [3, 4] we describe a system for the reconstruction and analysis of e^+e^- collision events at LEP/CERN. The power of this system steems from the cooperation of a hill climbing component, which is based on the standard model of elementary particle physics, and a genetic algorithm (GA). The hill climbing module generates solution candidates for the reconstruction task and the GA provides for selforganization of the suggested solutions.

In general a GA [8] is an efficient and widely applicable optimization technique. The real performance of a GA in a concrete application is to a high degree determined by the setting of its system parameters such as population size, mutation rate and crossover rate. Since an optimal parameter setting depends on the properties of the unknown solution space [5] and because of nonlinear dependencies between different parameters, finding a reasonable parameter setting is an immense task even for experienced users. An unsuitable parameter setting may quickly lead to loosing balance between goal directed improvement of the solution and consideration of new information. In this case the optimization process may either converge prematurely to a suboptimal solution or become an inefficient random walk through the solution space. Particularly this problem arises in complex applications as the reconstruction system, since only small solution populations can be employed.

Besides these problems, a fixed parameter setting usually is not optimal over the whole run, because the GA initially has to find the high performance regions of the solution space, which must be explored more and more deeply in the sequel. Therefore several different techniques have been developed for dynamic control of parameters of GA.

Heuristic methods like the M-Heuristic [7] and Scheduled Sharing [9] utilize available knowledge about the evolution of single parameters during the search process and about dependencies between parameters. The M-Heuristic chooses in each generation a mutation rate for every single gene by relating mutation rate, population size, length of the representation of a solution and a further time dependent parameter. Scheduled Sharing extends the sharing model adjusting the niche size according to a convergence schedule borrowed from simulated annealing. Different experiments demonstrated that these methodes indeed provide better results than a GA using a fixed parameter setting. One disadvantage is that the parameters are tuned deterministically over time whereby the properties of evaluation functions of different search problems are scarcely regarded. Furthermore, it is not clear how different heuristics can be combined.

Evolutionary parameter control techniques go another way utilizing selforganization capabilities to attain optimal parameter settings. For evolution strategies (ES) [10] it is a natural approach to represent parameters locally in every individual thereby making them object to selection and recombination. Bäck [1] tried to carry this mechanism forward to GAs. The analysis of a special GA using this method showed, that at every time the optimal mutation rate is represented in the population. But this procedure is only applicable to parameters with meaning to single individuals such as the mutation rate. Other parameters like population size or niche size, however, are only meaningful on population level. As the evaluation only depends on the quality of the solution represented in the corresponding individual, the success of a good parameter setting cannot be ensured. Another argument to favour global parameter control methods is that a new parameter setting effects all individuals at once, thereby rendering an immediate reaction on changed situations in the search process.

A global evolutionary control technique uses a population of individuals, all representing a parameter setting. The fitness of such individuals is determined by the performance the underlying GA could achieve. To evaluate a parameter setting at least one generation of the underlying GA has to be computed. For efficiency reasons an evolutionary control method must be able to find a new, suitable parameter setting by testing very few parameter combinations. In contrast to an ES a GA itself is not suitable for this task, because it needs several individuals in the population to gain progress. Being conceptually identical to GA, ES show differences in detail allowing an optimization process using a population consisting of only one individual ((1+1)-ES) [2].

Because of the small populations in the reconstruction system involving the risk of premature convergence we applied an (1+1)-ES to adjust some of the parameters of the GA in order to approximate an idealized diversity function [4]. As a result the number of correctly reconstructed physical processes was increased significantly. The fact that incorporating existing knowledge is very difficult, has turned out as a disadvantage of this control method.

This shows the need for an efficient, global and dynamic control method oriented at performance measures reflecting the optimization process, and utilizing available, often qualitative knowledge about parameter settings. In this paper we present a control system based on fuzzy logic with all demanded properties and compare it with an (1+1)-ES.

2. Fuzzy Control of a Genetic Algorithm

The term fuzzy control stands for a regulation mechanism based on fuzzy logic and fuzzy set theory [11]. Whereas the value of the characteristic function of a traditional set can be either 0 or 1, it can take any value of the interval [0,1] for a fuzzy set. The basic idea of fuzzy control is to incorporate human expertise in the form of heuristic rules of thumb to build control systems. The vagueness of the linguistic and generally qualitative statements in these rules is represented by fuzzy sets. The fuzzy regulation process is divided into three phases. At first the generally crisp input values have to be fuzzified, i.e. must be turned into a fuzzy representation by computing their membership values. The fuzzy inference is done by determining for every rule to which extend the input values match the rule conditions. Conjunctions and disjunctions are computed using intersection and union operators. An inference operator is applied to evaluate the conclusion part of every rule. Finally, the defuzzification component turns the generated fuzzy output values into crisp ones. Because of their smooth interpolative behaviour fuzzy regulation mechanisms provide a stable control, even if the available knowledge is inexact, incomplete or contradictory. To evaluate the capabilities of this technique

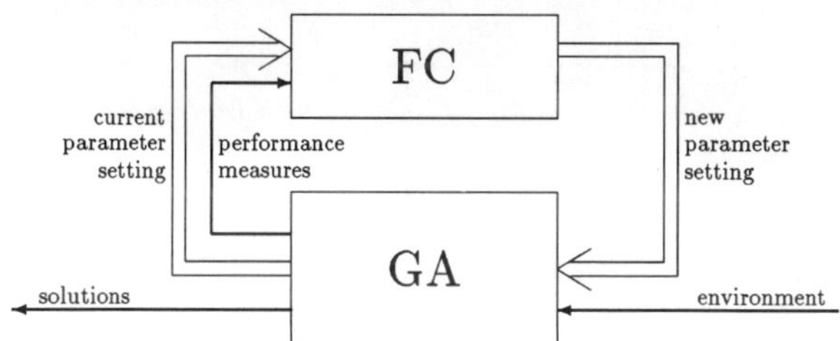

Figure 1: A fuzzy controlled GA

for the control of GAs we designed a system (fig. 1) to control the crossover, inversion and mutation rate of a GA solving travelling salesman problems [6]. The aim of the control procedure is to achieve better performance by approximating an idealized diversity function at any time. For this purpose it uses knowledge about the coherence of entropy evolution and change in single parameter settings as well as about favourable parameter constellations with respect to GA performance.

3. Experimental Results

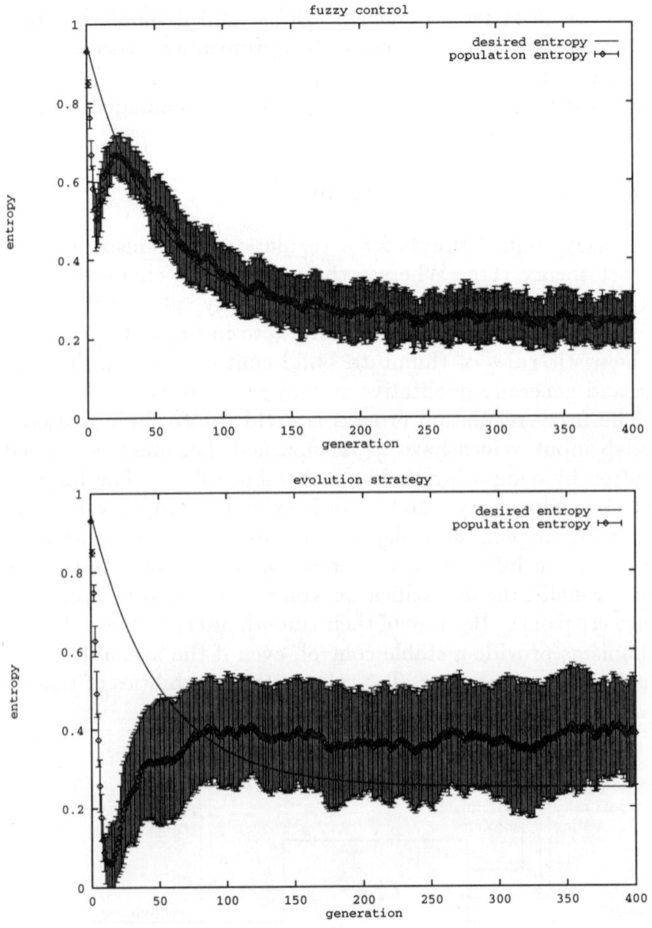

Figure 2: Approximation of a diversity function

To evaluate the fuzzy controlled GA (GA^{FC}) we compared it with an GA controlled by an (1+1)-ES (GA^{ES}). As an example, figure 2 shows the average results of both systems applied to 10 TSPs each using 5 different starting populations. Initially the three controlled parameters were all set to 0. In contrast to GA^{ES} the GA^{FC} is able to adjust the parameters quickly thereby reaching the desired entropy soon and approximating it well during the search process. On all considered diversity functions the control error of the GA^{FC} is significantly smaller at the 0.1% level according to the t-test than the control error of GA^{ES}. The reason is that the evolutionary control method has not got enough parameter information to react immediately to changed situations. Several generations of the underlying GA have to be computed to permit an appropriate evolution of parameters. The GA^{FC}, however, is able to decide on a given entropy error, in which direction and with which amount a parameter should be changed. Figure 2 also shows an ideal-

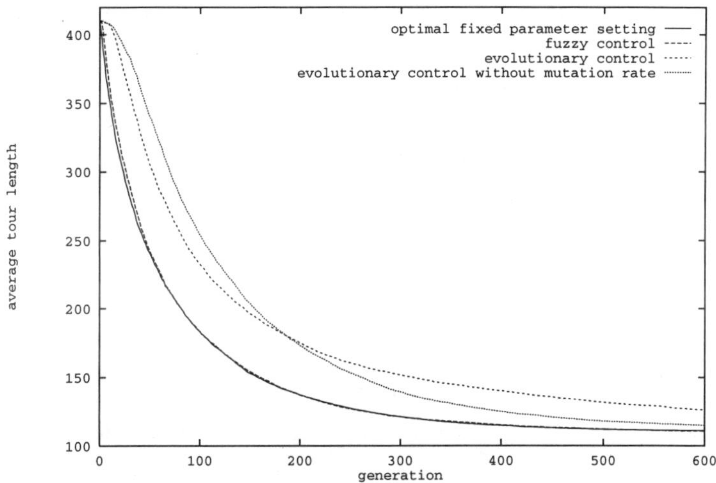

Figure 3: Evolution of tour length on different GAs

ized diversity function, which starts with a high value and decreases exponentially to a value that guarantees an entropy allowing further progress in the optimization process. The performance values shown in figure 3 steem from GA runs using this diversity function. The GA^{FC} clearly exceeds the GA^{ES} and is as good as a GA with a nearly optimal fixed parameter setting (GA^{opt}). In every run GA^{FC} and GA^{opt} both found tours very close to the shortest. The weakness of the GA^{ES} comes from the fact that it chooses parameter values solely oriented at the entropy and negatively influencing the performance. As mentioned above this is avoided in GA^{FC} due to appropriate rules.

4. Conclusions and Future Work

In this paper we demonstrated that control mechanisms for GAs should permit a dynamic and global tuning of the system parameters. Performance measures that show the progress of the optimization process as well as available strategy knowledge should be considered. Fuzzy control turned out to be a good tool for this task. Performing the control of a GA our fuzzy controller showed clear advantages over an (1+1)-ES. A substantially better approximation of diversity functions could be obtained, because the knowledge based technique is able to react flexible and goal directedly to changes in environment. Furthermore, knowledge about appropriate parameter constellations ensures achieving the entropy without unnecessary performance loss of the GA.

Actually we are working on the integration of fuzzy control in the reconstruction system. The first step lies in the acquisition of knowledge about parameter settings considering additional performance measures and analysing different diversity functions. Developing the knowledge base of a fuzzy controller is a complex and time consuming task. Due to this we will consider methods to automatically generate optimal rule sets.

5. References

1. T. Bäck, *Self-Adaptation in Genetic Algorithms*, In Proc. of the 1st European Conf. on Artificial Life (MIT Press, 1992).
2. T. Bäck and F. Hoffmeister, *Genetic Algorithms and Evolution Strategies: Similarities and Differences*, In First Int. Workshop on Parallel Problem Solving from Nature, University of Dortmund (1990).
3. K.-H. Becks, W. Burgard, A. B. Cremers and A. Hemker, *A Genetic Algorithm for the Reconstruction of Physical Events*, In Cognitiva, Proc. of the Int. Conf., Madrid (Elsevier Science Publishers B.V., 1990).
4. K.-H. Becks and A. Hemker, *Evolution Engines and Artificial Intelligence*, Int. Journal of Modern Physics C, Physics and Computers (World Scientific Publ. Comp., to appear).
5. R. K. Belew and W. E. Hart, *Optimizing an Arbitrary Function is Hard for the Genetic Algorithm*, In Proc. of the 4th Int. Conf. on Genetic Algorithms and their Applications (Morgan Kaufmann Publishers, 1991), pp. 190-195.
6. A. Bergmann, *Adaptive Control of Strategy Parameters of Genetic Algorithms*, Diploma Thesis, University of Bonn (1993), in German.
7. J. Hesser and H. Männer, *Investigation of the M-Heuristic for Optimal Mutation Probabilities*, In Parallel Problem Solving from Nature 2 (Elsevier Science Publishers B.V., 1992), pp. 115-124.
8. J. H. Holland, *Adaptation in Natural and Artificial Systems* (The University of Michigan Press, Ann Arbor, Michigan, 1975).
9. M. E. Palmer and S. J. Smith, *Improved Evolutionary Optimization trough Scheduled Sharing* (Technical Report, Thinking Machines Corporation, 1991).
10. I. Rechenberg, *Evolution Strategy* (Frommann Verlag, 1973), in German.
11. M. Sugeno, *An Introductory Survey of Fuzzy Control*, Information Sciences 36 (1985), pp. 59-83.

SOLVING OPTIMIZATION PROBLEMS WITH EVOLUTIONARY ALGORITHMS

SIEGFRIED HAHN*
Physics Department, University of Wuppertal
42097 Wuppertal, Germany

and

K. H. BECKS, A. HEMKER
Physics Department, University of Wuppertal
42097 Wuppertal, Germany

ABSTRACT

Evolutionary Algorithms (EAs), which are based on the model of natural evolution, have been successfully used for difficult optimization tasks[1]. The main representatives of this class of algorithms are the Genetic Algorithms (GAs), developed by J. Holland, and the Evolution Strategies (ES), developed by I. Rechenberg[2] and H.P. Schwefel[3]. We report on the use of GAs for tuning Monte-Carlo-Generator-Parameters to measured data. The global character of the search strategy leads to a variety of different solutions and reveals the multimodal character of this optimization problem. The analysis of the parameter sets and correlations between the parameters yields additional information about the search space and allows the identification of the global optimum.

1. Genetic Algorithms (GAs)

Genetic Algorithms[4], which were developed by J. Holland[5] in 1975, refer to a family of computational models inspired by evolution. They are often used for difficult optimization problems, since their robustness in respect to complicated multimodal, non-differentiable, discontinuous, and noisy functions with high dimensionality.

GAs are based on a collective learning process within a population of individuals, each of which represents a search point in the space of potential solutions to a given problem. GAs don't work on the continuous parameters of the problem itself, but on a discrete coding (binary or gray coding) of the parameters. Each individual corresponds to a bitstring, often referred to as chromosome, coding one point of the search space. This coding allows a problem independent optimizing strategy, which operates strictly on a

* e-mail: hahns@vxcern.cern.ch

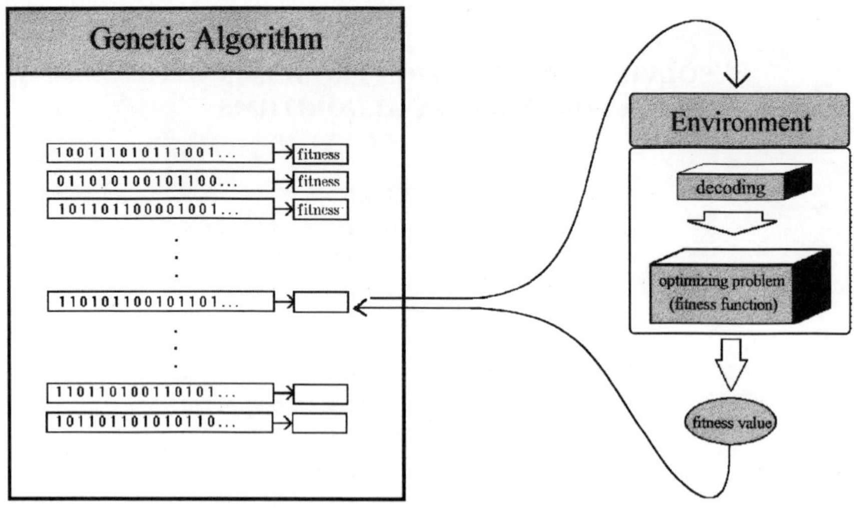

Figure 1: GAs process a population of bitstrings, each coding a search point of the parameter space. They operate on a problem independent, syntactic level through manipulating chromosomes or single bits. The function to be optimized and the (problem dependent) decoding function is supplied by the environment, which can (from GAs view) be regarded as a black box, providing a feedback of the quality of the individuals (fitness value).

syntactic level, not requiring any information about the search space nor any expert knowledge. The optimizing problem itself and the problem dependent decoding mechanism are supplied by the environment, which can, from GAs view, be regarded as a black box, providing the fitness value as a feedback of the quality of the individuals (see figure 1).

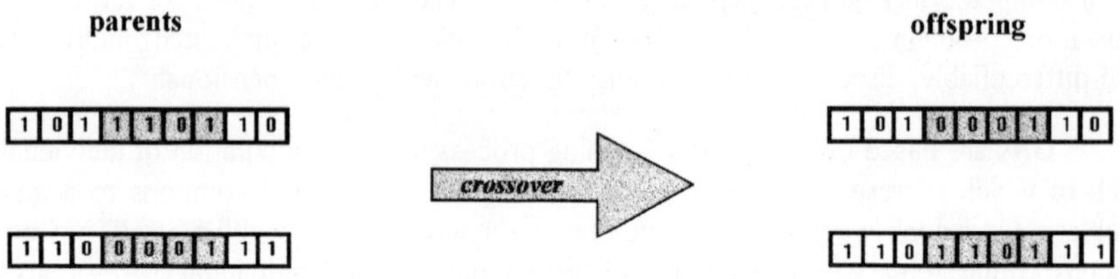

Figure 2: Illustration of the crossover operator. A randomly selected substring gets exchanged between two parent chromosomes. Two descendants are created, each carrying partial information of both parents.

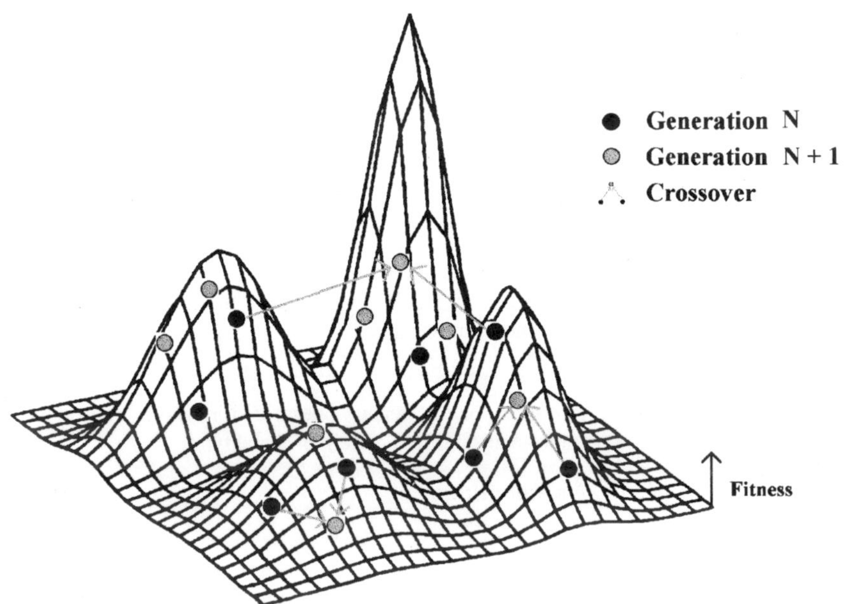

Figure 3: Example of a 2-dimensional fitness landscape to illustrate the operation principle of the crossover operator. Crossover enables the exchange of information between individuals in different regions of the search space, in particular partial solutions of the optimizing problem can be recombined to build a global solution. Since the descendants will have a worse fitness as well as a better fitness, progress in the populations average fitness will be achieved through the selection procedure.

A GA typically starts with a randomly initialized population. The individuals get evaluated and their function value determine their fitness, i.e. their opportunity to create offspring and to reproduce their genetic information for coming generations. This is done by the means of the genetic operators crossover, reproduction and mutation. The crossover mechanism (see figure 2) allows the exchange of information between the individuals, resulting in a globally working optimizing strategy (i.e. not a local hillclimbing mechanism) as illustrated in figure 3. In comparison to crossover, the mutation operator is of minor importance. By flipping randomly the value of a single bit of the population with very low probability, information, which has been lost through sampling errors, gets reintroduced into the population. Since the selection process favors those individuals of higher fitness to reproduce more often than worse individuals, the population evolves towards better and better regions of the search space. The algorithm can be summarized as follows:

Initialization

All individuals of the initial population are initialized at random.

Evaluation of the initial population

All chromosomes get decoded, function values and fitness are calculated from the resulting parameter values.

WHILE (termination criteria not satisfied) DO

Select

To create a new population, individuals get selected from the current one with a probability according to their fitness values.

Create Offspring

Offspring creation through applying genetic operators (crossover, reproduction, mutation) to the parents according to a certain probability.

Evaluation of the new population

ENDDO

2. Advantages of Genetic Algorithms

In comparison to conventional optimizing strategies, Evolutionary Algorithms, in particular the Genetic Algorithms, have several advantages:

- There are nearly no restrictions on the search space. GAs work on non-differentiable and discontinuous functions.
- GAs can handle problems with high dimensionality and highly nonlinear, multimodal functions.
- GAs are robust against statistical fluctuations.
- GAs are a global search strategy, not a local hillclimber.
- GAs can easily be implemented on parallel computer systems.

3. Tuning of Monte Carlo Generator Parameters

The analysis of data in HEP-experiments implies a good correspondence between Monte Carlo generated events and measured data. The growing precision of the measured data in LEP-experiments require a constant improvement of this correspondence, since the theoretical errors of the physical parameters to be determined, directly depend on the quality in describing the data through Monte Carlo simulation programs. To achieve the desired precision, a fine-tuning of several model-dependent, free parameters of the Monte Carlo generators (e.g. QCD-parameters, fragmentation parameters) is required.

Comparison of the Monte Carlo Generator output with real data is done by comparing various distributions of event shape variables and single particle spectra derived from the generated Monte Carlo with real data distributions which have to be corrected for detector effects. To measure the similarity between Monte Carlo and data distributions, one usually calculates the χ^2 value between them, which is given by

$$\frac{\chi^2}{D.F.} = \frac{1}{N_D} \sum_{j=1}^{N_D} \left[\frac{1}{N_j} \sum_{i=1}^{N_j} \frac{\left(N_{DATA}^{(i)} - N_{MC}^{(i)}\right)^2}{\left(\Delta N_{DATA}^{(i)}\right)^2 + \left(\Delta N_{MC}^{(i)}\right)^2 + \left(\Delta N_{SYST}^{(i)}\right)^2} \right]$$

where N_D is the number of considered distributions and N_j is the number of individual bins of the jth histogram. $N^{(i)}$ is the content of the ith histogram bin either on Monte Carlo or data distributions, and $\Delta N^{(i)}$ the statistical respectively the systematic error of the ith histogram bin. The systematic error comes dominantly from the systematics of the hadronic event selection and the correction procedure. The statistical error of the generated Monte Carlo has to be included due to the limited number of events which can be generated. The dependency of the χ^2 value and its statistical error on the number of generated events is shown in figure 4.

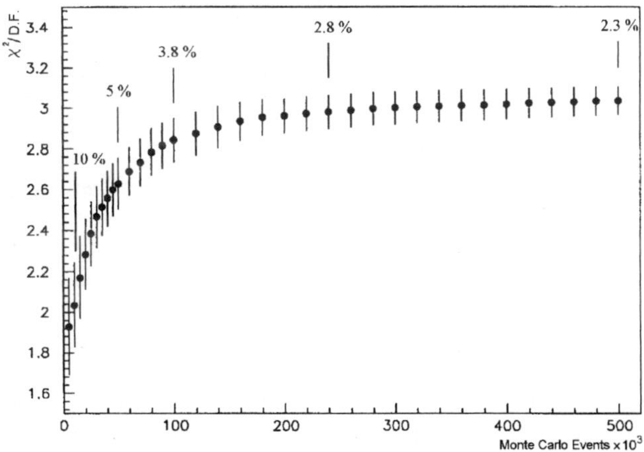

Figure 4: χ^2 value and statistical error as a function of the number of generated events.

Here, data distributions of 8 different observables have been compared to those generated by the JETSET 7.3 Partonshower generator. The error is only very slowly decreasing with the number of generated events. For a sample of 50000 events, whose generation takes approximately 1 CPU hour on a DEC 3000 workstation, the relative statistical error in χ^2 is about 5 %, for 500000 events, the error is still more than 2 %. One has to consider, that the optimization procedure requires the evaluation of χ^2 values for several hundred different sets of parameters, or even more, depending on the number of parameters to be tuned. Therefore a lot of computer power is required. Furthermore, the used fit algorithm has to be robust against statistical fluctuations.

The optimization task is to find the set of model parameters $\vec{x}$ for which

$$\vec{x} \quad \rightarrow \quad \chi^2(\vec{x}) \overset{!}{=} \min$$

In general, the number of parameters N_P lies between 4 and 10, depending on the model to be tuned. Most of these parameters are highly correlated and show multimodal behavior so that deterministic algorithms are likely to be mislead. Therefore, conventional algorithms like the so called matrix method, where at each point of a grid in the parameter space a χ^2 value is calculated, require to restrict the search space to a small local area, where linearity can be assumed. This implies a human expert, who "knows" good starting values for the parameters in advance. Another disadvantage of the matrix method is the strong dependency on the number of function calls on the number of parameters to be tuned. Since the dimensionality of the grid grows with N_p, the number of function evaluations grows with the power of N_p. Furthermore, lacking robustness against statistical fluctuations doesn't allow tuning many parameters in parallel, instead of this often a "single parameter variation based on visual inspection of systematic trends"[6] is performed.

4. Tuning the JETSET 7.3 Partonshower Generator using Genetic Algorithms

Since the lack of suitable conventional algorithms for tuning Monte Carlo Generators, GAs are a very interesting alternative. GAs maintain a whole population of search points by exploiting correlations between the individual members, therefore they are very efficient in particular on problems with high dimensionality. Through the global character of the search strategy, GAs maintain complicated multimodal problems with no restriction on the parameters fit range. GAs do not require any problem specific expert knowledge and can be seen as a nearly objective optimization method. In contrary to conventional deterministic algorithms, the "weakness" of the probabilistic sampling method of GAs causes tolerance against sampling errors and therefore robustness against statistical fluctuations. This allows a partial, approximate evaluation of worse performing individuals with reduced statistics. Only good performing parameter sets get an expensive evaluation with high statistics in order to discriminate slight differences in χ^2, a feature which significantly reduces computer time.

In a first approach, we tuned the following 5 QCD and fragmentation parameters of the JETSET 7.3 Partonshower generator (PS), which is the most frequently used Monte Carlo Generator at LEP experiments:

Parameter	Fit Range
LUND a	0.1 ... 0.8
LUND b	0.1 ... 0.8
σ_q	0.1 ... 0.5
Q_0	0.5 ... 3.5
Λ_{LLA}	0.1 ... 0.5

Table 1: Tuned parameters of the JETSET 7.3 PS generator.

For the fragmentation of the heavy b- and c-quarks the Peterson fragmentation function was used, but the parameters ε_c and ε_b have not been optimized. Instead of this, the parameter values from the DELPHI tuning were used.. We performed a fit to the DELPHI data from 1992, using distributions of the event shape variables thrust, oblateness, sphericity and aplanarity and of the single particle variables $p_\perp^{in}, p_\perp^{out}, x_p$ and rapidity. Due to the global search strategy of the GA we found many possible solutions with different parameter values, even multiple solutions within one optimization run. Here, a parameter set was considered to be a solution, if the evaluated χ^2 value was less than the χ^2 of the parameter set from the DELPHI tuning from 1992.

The fits were performed with parallel evaluation on a cluster of 9 DEC 3000 workstations. Altogether we performed 12 optimization runs yielding 92 different solutions[7]. The parameter values for the DELPHI tuning of 1992 and for the parameter set with the best χ^2 value are given in table 2.

Parameter	JETSET 7.3 Default	DELPHI-Tuning 92	Optimized Values
LUND a	0.500	0.180	0.276
LUND b	0.900	0.340	0.585
Λ_{LLA}	0.400	0.255	0.341
σ_q	0.350	0.395	0.368
Q_0	1.000	1.300	1.504

Table 2: Optimized JETSET 7.3 PS parameters. Given are the default values of the simulation program, the values from the DELPHI tuning 1992, and the best parameter set found by GA.

The resulting χ^2 values for the individual distributions are listed in table 3.

Distribution	**DELPHI-Tuning 92**	**Optimized Values**
Thrust	3.36	0.95
Oblateness	1.59	0.93
Sphericity	2.72	0.53
Aplanarity	2.63	1.28
$p_{\perp}^{in}$	2.81	2.65
$p_{\perp}^{out}$	11.5	5.73
x_p	2.40	1.55
Rapidity	3.36	1.67
Sum	**3.79**	**1.90**

Table 3: $\chi^2/D.F.$ values of the considered distributions for the parameters of the DELPHI-tuning 1992 and for the optimized parameter set.

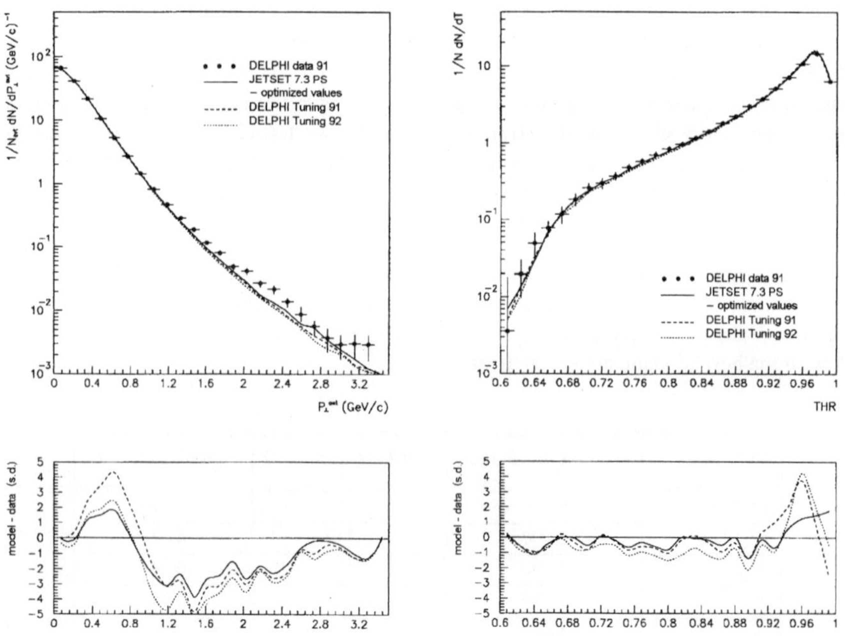

Figure 5: $p_{\perp}^{out}$ and thrust distributions for the JETSET 7.3 PS generator in comparison to DELPHI data 1991. The lower plots show the difference of data and Monte Carlo distributions in units of the standard deviation.

For most of the listed distributions the χ^2 value for the optimized parameter set is significantly reduced. In particular the χ^2 for the thrust distribution in the range $THR > 0.9$ and the $p_{\perp}^{out}$ distribution over the whole range of values show an extraordinary improvement (see figure 5).

5. Analyzing the Structure of the Parameter Space

The fact, that we have so many different solutions, is really surprising and suggests further investigations. Figure 6 shows the distributions of the parameter for the transversal fragmentation σ_q and for the difference of the LUND parameters |a-b|. Shown are the parameter values of the individuals which were considered to be a solution against the relative fitness value associated to the individual.

Both plots reveal the multimodal structure of the parameter space. While one can see solutions with the difference |a-b| over a wide range of values, we have two distinct ranges of values for σ_q. For both parameters, the individuals with the highest fitness values are in the region near the default values of the Monte Carlo program, while the individuals with parameter values close to the values of the DELPHI tuning 92 have smaller fitness values, indicating that this region is not the one of the global optimum. But this could also originate from randomly missing the parameter sets with high fitness, since we have evaluated only a very small fraction of all search points.

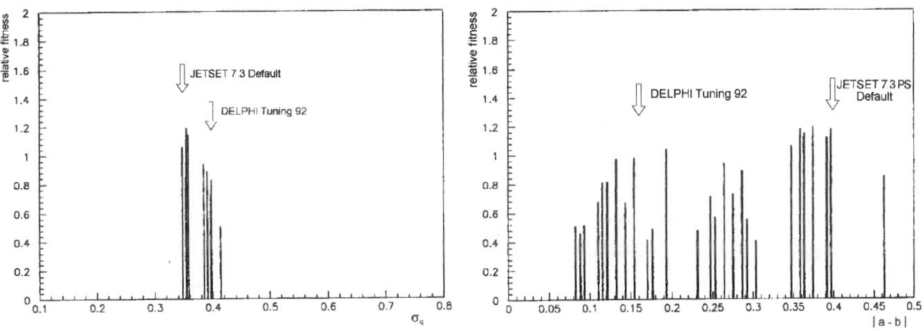

Figure 6: Distribution of the parameter σ_q (left picture) and the absolute difference of the LUND-parameters |a-b| (right picture) in the whole tuned interval. Shown are the parameter values of the individuals which were considered to be a solution. In this and in the following figures the scale on the ordinate gives the relative fitness value associated to the specific parameter set. Additionally, the parameter values of the DELPHI Tuning 92 and the default values of the Monte Carlo generator are indicated.

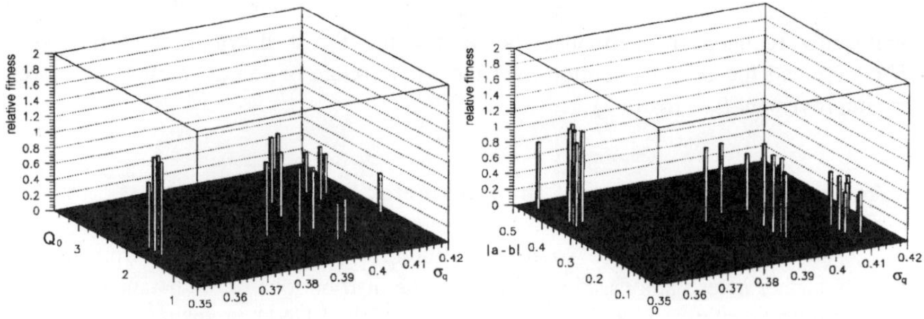

Figure 7: Correlations between the parameters σ_q and Q_0 (left picture) and between σ_q and |a-b| (right picture).

For a further investigation, we are mainly interested in the parameter σ_q. We have two clearly distinct regions of values for this parameter. The default value of the JETSET 7.3 generator, which originates from a fit to lower energy data, belongs to the region with the smaller σ_q values, while the value from the DELPHI tuning 92 belongs to the region of higher σ_q values. The parameter σ_q, which describes the width of the distribution of the transverse momenta of fragmenting hadrons with respect to the initial quarks, should be in first order independent of the center of mass energy. In order to decide, which of these regions is the right one, we will have a look at correlations between σ_q and other model parameters. Figure 7 shows the correlations between σ_q and Q_0 and between σ_q and |a-b|. For both correlations we see a broad region of solutions for high σ_q values and a narrow one for the small σ_q values. The same effect can also be seen for correlations between σ_q and other parameters.

In figure 8 correlations between other parameters are shown for the two distinct σ_q ranges. The left column shows the correlation between the Partonshower cut off parameter Q_0 and the difference of the LUND parameters |a-b|. In the whole σ_q range we have a broad region of different solutions. For the range of big σ_q values it is nearly the same picture, but if we restrict the range to small σ_q values, the number of solutions is greatly reduced to a small and narrow region of parameter values. Most interesting is the correlation between Q_0 and the QCD parameter Λ_{LLA} (shown in the right column), two parameters which are highly correlated. In the whole range of σ_q values we have two distinct regions of combinations between these parameters. The same is the case for $\sigma_q > 0.375$. But for the range with $\sigma_q < 0.375$ we get a unique region of solutions. This shows that the region of small σ_q values belongs to the global optimum while the solutions with a big σ_q value, e.g. the parameters of the DELPHI tuning 92, belong only to a suboptimal region of the parameter space.

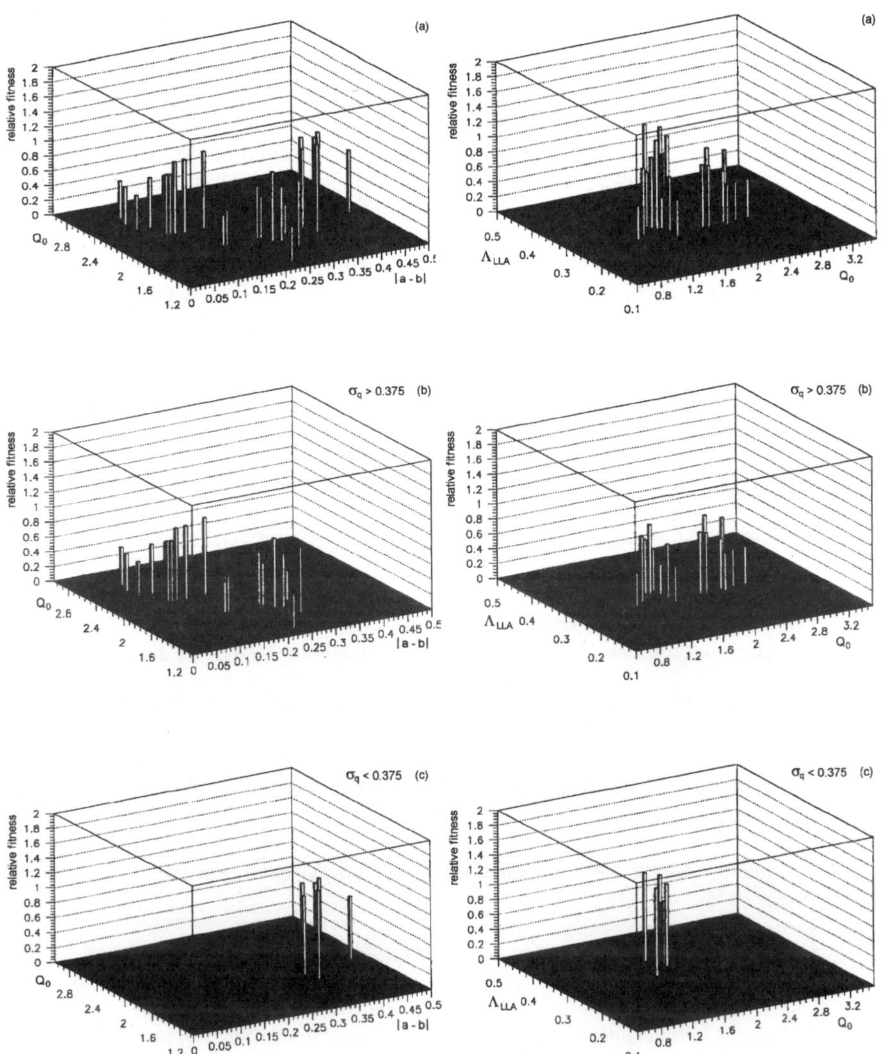

Figure 8: Correlation between the parameters Q_0 and the difference |a-b| (left column), and between Λ_{LLA} and Q_0 (right column). In (a) the correlations are shown for the whole σ_q range, in (b) the correlations are shown for $\sigma_q > 0.375$ and in (c) for $\sigma_q < 0.375$.

6. Conclusions

We have used Genetic Algorithms for tuning Monte Carlo Generators. GAs offer an automated procedure for this optimization task which doesn't require any problem specific expert knowledge. The inherent parallelism of the optimizing strategy allows the parallel optimization of a big number of parameters. The evaluation of the parameter sets can be performed in parallel on a workstation cluster. Partial, approximate evaluation of worse performing individuals can significantly reduce the amount of CPU time.

Due to the global search strategy, the algorithm found many solutions with different values for the parameters and reveals the multimodal structure of the search space. Inspection of the distribution of the parameters and their correlations yield additional information on the search space and allow the identification of the global optimum. It can be shown, that the parameter values of the DELPHI tuning 92 belong to a suboptimal region of the search space.

7. References

1. T. Bäck, F. Hoffmeister, H.P. Schwefel, *Applications of evolutionary algorithms.* University of Dortmund, 1992.

2. I. Rechenberg, *Evolutionsstrategie: Optimierung technischer Systeme nach Prinzipien der biologischen Evolution.* Frommann-Holzboog Verlag, Stuttgart, 1973.

3. H. P. Schwefel, *Numerical Optimization of Computer Models.* Whiley, Chichester 1981.

4. D. Goldberg, *Genetic Algorithms in Search, Optimization and Machine Learning.* Addison-Wesley, 1989.

5. J. Holland, *Adaptation in Natural and Artificial Systems.* University of Michigan Press, 1975.

6. OPAL Collaboration, *A measurement of Global Event Shape Distributions in the Hadronic Decays of the Z^0.* CERN-EP 90-48.

7. S. Hahn, *Optimierung von Monte Carlo Generator Parametern mit Genetischen Algorithmen. University of Wuppertal.* WU D 93-24.

Trends in Neural Network Research and an Application to Computer Vision

ENNO LITTMANN
Department of Information Science, Bielefeld University
P.O.B. 100 131, D-33615 Bielefeld, FR Germany

ABSTRACT

In recent years, artificial neural networks (ANNs) begin to turn from a 'black box' technique into a scientifically well-founded research field. While originally inspired by models of biological neurons, current ANNs are part of non-linear statistics, too. The paper gives a summary of current trends in neural network research. Typical representatives of the main neural network types are presented and their advantages and problems are discussed. Major fields for applications of neural networks are robotics and computer vision. A hybrid system is presented combining classical and neural techniques to extract hand orientation from real images.

1 Introduction

In recent years, neural networks have become established as a major tool for tackling problems from such diverse fields as pattern recognition, medical diagnosis, composition of music, task scheduling, speech recognition and generation, process control, game playing and even forecasting of bond rates. But although it sometimes seems that neural networks are advertised as an "universal algorithmic cure" that is able to solve almost any problem at hand, serious efforts are made to improve the understanding of the basic strengths and weaknesses of the neural network approach. A summary of these current trends in neural network research will be given in section 2.

Most of the successful current applications use neural networks to implement some kind of signal transformation that cannot be computed satisfactorily by any other means, be it for lack of knowledge about the underlying process, or because a more traditional approach would turn out more complicated or expensive. For such types of tasks, neural networks offer a highly attractive approach, and since this type of task may occur in a wide variety of fields, neural networks have proven useful in very diverse domains. Typical representatives of the main neural network types are presented in section 3 and their advantages and problems are discussed.

Major fields for applications of neural networks are robotics and computer vision. If issues of perception or movement control are concerned we have nowhere come even remotely close to the information processing capabilities of even simple animals, such as bees or birds. Therefore, to understand the working principles of neural networks, the fields of robotics and computer vision seem to be particularly adequate and promising areas, both for studying the kind of information processing tasks that are well solved by biological networks, and for exploring ways to replicate these capabilities with artificial neural networks. One of the key problems is the robust recognition, grasping and manipulation of objects in an arbitrary environment. Part of this task is the visual control of hand movements. In section 4 a hybrid system is presented that combines classical methods and neural networks to extract the hand orientation from real images.

2 Current Trends

The recent years were mainly dominated by contributions showing what problems could be tackled by neural networks *in principle*. The claim of biological plausibility was gradually abandoned in favour of networks that solved actual problems. Thus, the close relationship of most of the NN algorithms with classical methods known from statistics is more and more rediscovered and inspires the analysis of the properties of ANNs as well as improvements in their performance. Currently, we can distinguish five main streams of research:

- Faithful modeling of biological neural systems on various levels
- Technical NNs based on and improved by statistical and probabilistic methods
- Optimization of the NN design
- New learning methods for real-world (real-time) applications
- Solution-oriented hybrid systems

2.1 Biological neural systems

In former years, the biological plausibility of a NN model was a widely used criterion to judge the "neural" character of an approach. Today, the mentioned argument has lost some of its importance, since the opinion is widely spread that if nature needed some architecture or algorithm it will have found ways to realize it in neural structures. The new challenge is the faithful modeling of biological data. Such models could be able to prove hypotheses about the biological system by synthesis and serve as tools to optimally design and guide further biological experiments. Examples are models to explain the synchronization phenomena, networks consisting of spiking neurons, or even modeling the behavior of a synapse itself. All these models are limited to only one or few levels within the whole scope of possible description levels reaching from membrane processes to high-level vision or language understanding.

2.2 Technical neural networks

Even nowadays, NN research still has a touch of alchemy. We do not know what type (and structure, size ...) of network should be used for what type of data, we do not even know if there are rules for this at all. Well-known methods are applied in the context of neural networks to analyze the properties of given data with interpretable results, thus providing a firm basis for further research. Investigations on the use of mutual information to analyze the importance of the input data for the output [16] will help to select optimized feature vectors. Also, probabilistic methods are used to optimize learning procedures wherever possible. The implementation of an on-line version of the EM algorithm for the training of expert networks by Jordan and Jacobs [6] is an example for this. Well-known numerical techniques can be enhanced with adaptive methods to yield a network architecture with a great transparency.

2.3 Optimization of the NN design

A high-tech neural network based on the heuristics of its designer - this is no longer a satisfying situation. Furthermore, it should be noted that in most current applications the neural component mostly consists of a single or a small number of "neural modules", usually of moderate size and complexity, and it is amazing that such small systems can already perform the tasks that have been demonstrated so far. This may give us some notion of what performance may be expected if we manage to move artificial neural systems closer to the level of complexity that we observe even in simple biological neural systems.

Three basic approaches deal with this problem. *Genetic and evolutionary algorithms* can be used not only to train networks but also to select and evolve network structures from network populations (for an introduction, cf. [3]). *Incremental approaches* like the cascade correlation approach by Fahlman [1], the cascade architecture by Littmann and Ritter [10], or the GCS by Fritzke [2], start with very small networks that grow until a certain criterion is met. The opposite approach is represented by *pruning methods* like the optimal brain surgeon [4], where "unnecessary" weights or nodes are culled from an initially oversized network.

2.4 New learning methods

The overwhelming part of publications is concerned with supervised learning so far. Real-world applications, however, require on-line learning from few examples, processing of time-varying signals or predictive control. A special issue of IEEE Transcations on NN on recurrent networks is about to appear. Autonomous systems need unsupervised learning approaches like reinforcement learning (Q learning, Dyna Q). A further key issue are strategies to learn actively, to explore the environment, instead of passively transforming examples into weights. The vast amount of information we receive if we exploit visual information leads to the demand for methods to guide some focus of attention in order to reduce the information to a reasonable amount.

2.5 Solution-oriented hybrid systems

Although many projects aim at applications, the number of neural networks being used in the field is rather small. Often, the NN approach does not lead to a substantial performance improvement so that there is no need to abandon the established, well-understood techniques. So far, we cannot predict reliably what performance can be expected and what effort must be made to solve a given task with a NN approach.

Thus, if we aim at applications for the practical use, we must accept that ANNs are employed for only part of a complex task. Since there is a lack of methods to train very large systems, problems must be decomposed into parts as it can be done e.g. by semantic networks. Wherever explicit knowledge or exact methods are available they definitely should be used to reduce the complexity of the remaining problems. Whenever the decomposition is too complex or no exact knowledge about the underlying processes available, ANNs should be taken into account. This requires a close cooperation of experts from various fields.

3 Backpropagation vs. Self-Organizing Maps

The two most important network types in supervised learning are multi-layer perceptrons (MLP) trained by error-backpropagation (or variations) and self-organizing maps with the extension to learning vector quantization (LVQ) or local linear maps (LLM).

3.1 Multi-layer perceptrons

Simple perceptrons have been proved to be restricted to the solution of linearly separable problems. These limitations do not apply to feed-forward networks with "hidden" layers between the input and output layer and nonlinear units at least in the hidden layer. With the reinvention of a suitable learning rule by Rumelhart et al. [15] the training of such networks became tractable. The majority of current research projects involves MLPs or related approaches.

The layers consist of units that calculate a weighted sum of their inputs followed by some nonlinear operation in the case of nonlinear units. For the adaptation of the weights a cost function is defined. Using gradient descent, the resulting error can be backpropagated through the network and the weights be updated according to the delta rule (for a full account, cf. [5]).

3.2 Local linear Maps

LLM networks have been introduced by Ritter ([12, 14]). They are related to self-organizing maps [7] and the GRBF-approach. The architecture is computationally efficient and learns fast. Basically, it approximates a non-linear transformation by a set of locally valid, linear mappings (Fig.2), where mapping r ($r = 1 \dots N$, N being the number of "units" of the network) is given by

$$\mathbf{y}_r = \mathbf{w}_r^{(out)} + \mathbf{A}_r(\mathbf{x} - \mathbf{w}_r^{(in)}). \tag{1}$$

Here, $\mathbf{x}$ denotes the input vector, $\mathbf{w}_r^{(in)} \in \mathrm{I\!R}^L$ and $\mathbf{w}_r^{(out)} \in \mathrm{I\!R}^M$ ("input" and "output weight vector") are elements of the input and output space (of dimensionality L and M, resp.), and $\mathbf{A}_r$ is a $M \times L$-matrix. The output of the network is a weighted superposition of the outputs $\mathbf{y}_r$ of the individual maps. In the simplest case, the contribution of the map s with the minimal distance in the input space is used ("winner-take-all"-network). This leads to a tesselation of the input space.

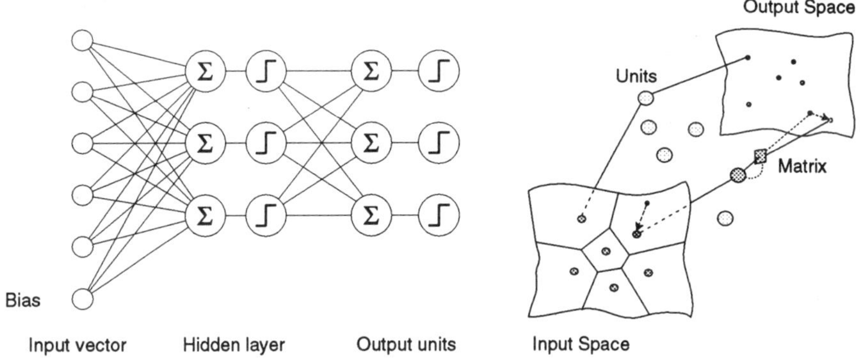

Figure 1: Multi-layer perceptron

Figure 2: LLM Network

To perform the required transformation between input and output space the necessary values of the network parameters are learned during a training phase. For this purpose correct input-output pairs $(\mathbf{x}^{(\alpha)}, \mathbf{y}^{(\alpha)})$, $\alpha = 1, 2, \ldots T$, of a set of T training samples are presented repeatedly and in a random sequence. The input and output weight vectors as well as the matrix coefficients are adapted according to the following simple error-correction rules:

$$\begin{aligned}
\Delta \mathbf{w}_s^{(in)} &= \epsilon_1(\mathbf{x}^{(\alpha)} - \mathbf{w}_s^{(in)}), \\
\Delta \mathbf{w}_s^{(out)} &= \epsilon_2(\mathbf{y}^{(\alpha)} - \mathbf{y}^{(net)}) - \mathbf{A}_s \Delta \mathbf{w}_s^{(in)}, \\
\Delta \mathbf{A}_s &= \epsilon_3(d_s^2)^{-1}(\mathbf{y}^{(\alpha)} - \mathbf{y}^{(net)})(\mathbf{x}^{(\alpha)} - \mathbf{w}_s^{(in)})^T.
\end{aligned}$$

During this adaptation process each training step parameter ϵ_i decays exponentially from a large initial value to a small final value.

3.3 Comparison

Main features of the MLP are the fast recall due to the inherent parallelity and the forming of internal representations in the hidden layer(s). The problems of local minima and slow learning progress have given rise to a variety of partially useful enhancements like the inclusion of momentum terms or higher order learning rules such as quickprop. The main problem, however, the optimal structure and learning rate for a given task, is far from being solved.

LLMs are very fast to train and are quite robust against getting stuck in local minima. This is due to the unsupervised forming of the input quantization and the guaranteed convergence of the linear mapping performed by the matrices connected to the output vectors. The configuration of the input reference vectors reflects the topology of the input space. The main disadvantage is the long search time needed to find the "winner node" or, if more nodes contribute to the output, the calculation of many outputs instead of one.

4 A Hybrid System for Hand Posture Recognition

One of the main fields for applications of ANNs is signal processing and interpretation. Recognition and description of objects is one of the main problems in computer vision. Furthermore, this task is one of the key capabilities required for the successful operation of both biological organisms and artificial robots. A key problem in our robot lab is the visual recognition of a hand and the characterization of its orientation. This poses a challenging vision problem the solution of which is of great practical interest, for instance to facilitate the control of multifingered anthropomorphic manipulators.

In the following, a hybrid system for hand posture recognition will be presented including not only neural networks but also classical tools from image processing, both approaches together organized in a semantic network. The sequence demonstrates how the method, originally limited to computer-rendered or at least presegmented images, is integrated into a system that works with real images.

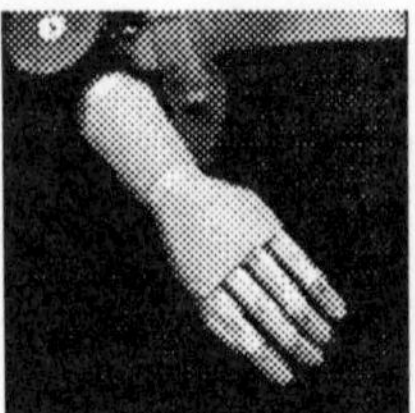
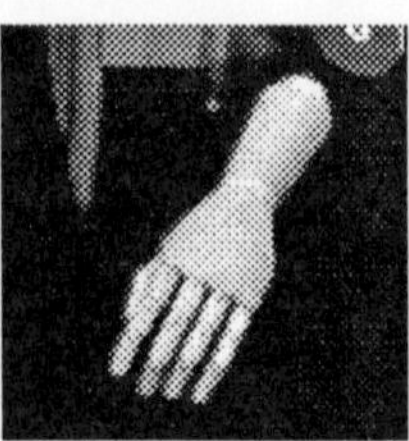

Figure 3: Possible hand orientations with extreme values for roll and yaw whereas pitch was fixed at 0°. (roll,yaw) have the values (a) (35°, −55°), (b)(−35°, −55°), (c)(−35°, 0°), (d)(35°, 0°).

4.1 Hand posture recognition

In order to extract the hand orientation from images only containing a hand with a black background, the required mapping of a suitable feature space into the orientation space is learned by a LLM network. The images show hands within a certain range of permitted orientations (Fig. 3). First, the image is preprocessed. After thresholding, a laplace filter operates on the image and negative filter values are clipped. The resulting image mainly contains the edge information of the original image and forms the basis for a subsequent operation with gabor filters. These filters are orientation sensitive and well suited to exploit the edge information. We use a 3×3 "filter grid" centered at the area of interest with four different orientations at each grid location. The resulting 36-dimensional feature vector forms the input for the LLM network (see also [11]). At present, this network calculates two of the three relevant angles—roll and yaw—, whereas the third (pitch) is a constant through all images.

While this method yields good results for images with "no" background, it does not work with arbitrary backgrounds.

Figure 4: Complex scenes with a roboter hand

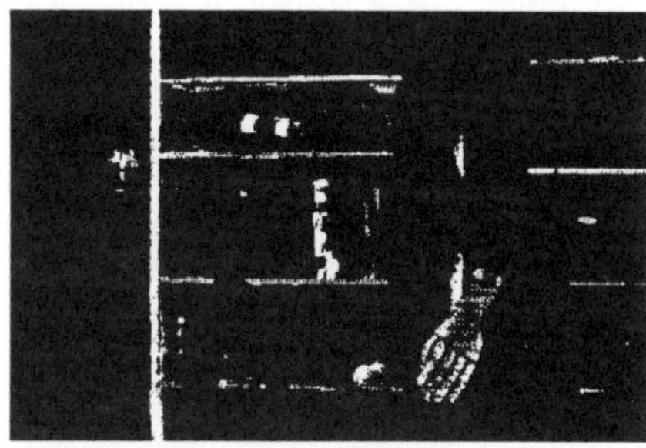

Figure 5: One-step skin-color activity map for the image of Fig. 4

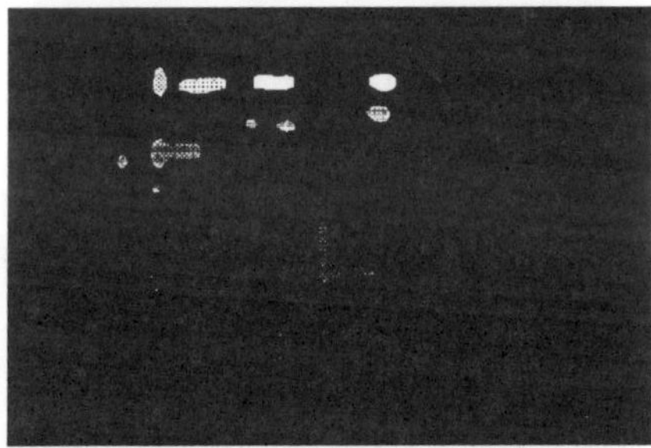

Figure 6: Segmented image with instances of hand-arm-complex (every instance is marked by one grey value)

Figure 7: Optimal hand instance

4.2 Hybrid image segmentation

Thus, a segmentation of the hand must precede the orientation recognition. The first part is realized by a LLM network, too. A network is trained to map the local color information in the surround of each pixel onto a real-valued "skin-color activity" value (see Fig. 5 for an example). The training is based on samples from subregions of two of the training images. The 9-dimensional input for each pixel contains the RGB values of the pixel itself as well as the average RGB values within a 3×3 and a 5×5 neighborhood, thus introducing a moderate amount of spatial context information into the network.

In the hybrid approach, the network provides a "skin-color activity map" that is the basis for the detection of the hand-arm-complex with classical and knowledge-based methods. These include first a median filtering to eliminate small regions of activity. Based on the histogram of the filtered skin-color activity map a threshold is calculated. For the resulting binary image a region labeling is performed (see Fig. 6). Each labeled region is transformed into a normalized orientation by a Karhunen-Loeve transformation. Finally, the region with the smallest distance to a subspace spanned by "eigenhands" derived from a training set of hands is selected. This method is adopted from the eigenface approach (cf. [17]). Based on general knowledge about the shape and proportions of hands and arms the hand is extracted from the complex. The resulting region is remapped and represents the area of interest for evaluating the orientation of the hand (see Fig. 7). From this image the hand orientation can be extracted according to the approach mentioned above. With this hybrid system, we can extract the hand region from all 200 images of our database.

Figure 8: Complex scene in low resolution

Figure 9: Segmented low resolution image

Figure 10: Candidate regions in low resolution

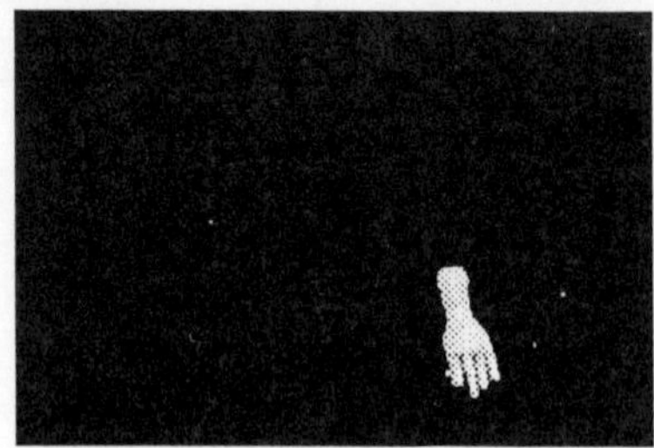

Figure 11: Segmented high resolution image

4.3 Neural image segmentation

In real images, there are lots of single pixels with RGB values identical to those of hand pixels. This leads to a noisy segmentation result. Therefore, we chose a two-step process to accelerate and improve the neural segmentation. First, the resolution is decreased by a factor of 10 (Fig. 8). The segmentation of this blurred image with a special network is very fast (Fig. 9). Now only regions with at least four connected pixels are selected as candidate regions. Using a threshold with low sensitivity, we achieve high specifity and get almost exclusively hand pixels (albeit not all of them)(Fig. 10). In the second step, we use the candidate regions and their surrounding as masks for the segmentation of the original image (Fig. 11). This yields a significant reduction in the number of pixels to be filtered with the full resolution. After this step, the approach yields only one (and the correct) hand region for 190 of 200 images. The further steps of the hybrid approach have not been replaced yet, but investigations are under way.

4.4 Lighting calibration

Additionally, the result of the segmentation by the LLM-network as shown in Fig. 5 suffers from the variations in the images introduced by changing lighting conditions, camera quality and calibration.

One possibility to achieve more robustness against such effects is a *soft-calibration* of the image. The idea is similar to the white-calibration of standard cameras. We need a reference table including white and maybe colored spots to be seen in at least one of the images of each series with constant conditions. Then, this calibration information could be used to transform the actual RGB values of the image pixels into the virtual RGB values they would have had, if the image had been taken under the conditions the neural networks had been trained for.

This task is not particularly difficult if there are many reference points available. The aim, however, is to find ways to calibrate the images without pre-known reference points. Rather, it should be based on extreme values of color, hue, saturation, or intensity in the images and use these points to parametrize the mapping for the soft-calibration. An architecture that seems to be particularly suited for this task is the parametrized self-organizing map (PSOM) proposed by Ritter [13].

5 Conclusion

The sacrifice of the biological plausibility claim has opened the neural network research for a variety of new methods. In particular, well-understood statistical tools have found entrance to the design and investigation of neural algorithms and help to improve our understanding of the working principles. Furthermore, the still existing skepticism towards the capabilities of ANNs could be reduced, while the knowledge of neural network experts about similarities to classical approaches has increased.

In our application we do not claim any biological plausibility. We have decomposed the problem of the detection of complex objects in real images into subtasks like the recogniton of hand orientation, image segmentation, or color calibration. Each subtask employs classical techniques as well as neural networks wherever it appears to be convenient. With this system, we actually can solve the task under some constraints but have lost the ease with which anyone of us can find and track hands in an almost arbitrary environment. The implementation of systems including neural networks to solve actual problems is necessary to justify the research in this field and to investigate the computational power of the algorithms.

Nevertheless, the biological model should never be forgotten. Lighting invariance or segmentation are symbolic names for properties that probably will not exist in nature in this distinctness. We will have to learn how to compose and how to train large systems solving a variety of problems simultaneously. Thus, we might learn to exploit those synergetic effects that our own capabilities are based on.

6 Acknowledgements

This work was supported by the German Ministry of Research and Technology (BMFT), Grant No. ITN9104AO. Any responsibility for the contents of this publication is with the author.

References

1. Fahlman, S.E., Lebiere, C. (1989), "The Cascade-Correlation Learning Architecture", in *Advances in Neural Information Processing Systems 2*, ed. D.S. Touretzky, pp. 524–532, Morgan Kaufman Publishers, San Mateo, CA.
2. Fritzke, B. (1993), "Growing Cell Structures — A Self-Organizing Network for Unsupervised and Supervised Learning", TR-93-026, ICSI, Berkeley, CA.
3. Goldberg, D.E. (1989). *Genetic Algorithms*, Addison-Wesley, Redwood City, CA.
4. Hassibi, B., Stork, D.G. (1993). "Second Order Derivatives for Network Pruning: Optimal Brain Surgeon", in *Advances in Neural Information Processing Systems 5*, eds. C.L. Giles, S.J. Hanson, J.D. Cowan, pp. 188–195, Morgan Kaufman Publishers, San Mateo, CA.
5. Hertz, J.A., Krogh, A.S., Palmer, R.G. (1991), *Introduction to the Theory of Neural Computation*, Addison-Wesley, Redwood City, CA.
6. Jordan, M.I., Jacobs, R.A.(1993), "Hierarchical mixtures of experts and the EM algorithm", TR-9301, MIT Computational Cognitive Science, Cambridge, MA.
7. Kohonen, T. (1984), *Self-Organization and Associative Memory*, Springer Series in Information Sciences **8**, Springer, Heidelberg.
8. Kohonen, T. (1990), "The Self-Organizing Map", in *Proc. IEEE* **78**, pp. 1464–1480.
9. Kummert, F., Littmann, E., Meyering, A., Posch, S., Ritter, H., Sagerer, G. (1993). "A Hybrid Approach to Signal Interpretation Using Neural and Semantic Networks", *Proc. 15. DAGM-Symposium 1993, Lübeck*, eds. S.J. Pöppl, H. Handels, pp. 245-252, Springer, Heidelberg.
10. Littmann, E., Ritter, H. (1993). "Generalization Abilities of Cascade Network Architectures", in *Advances in Neural Information Processing Systems 5*, eds. C.L. Giles, S.J. Hanson, J.D. Cowan, pp. 188–195, Morgan Kaufman Publishers, San Mateo, CA.
11. Meyering, A., Ritter, H. (1992), "Learning 3D-Hand Postures from Perspective Pixel Images", in *Artificial Neural Networks 2*, eds. I. Aleksander, J. Taylor, pp. 821-824, Elsevier Science Publishers (North Holland).
12. Ritter, H. (1991), "Learning with the Self-organizing Map", in *Artificial Neural Networks 1*, eds. T. Kohonen, K. Mäkisara, O. Simula, J. Kangas, pp. 357-364, Elsevier Science Publishers (North-Holland).
13. Ritter, H. (1993), "Parametrized Self-Organizing Maps", in *Artificial Neural Networks 3*, eds. S. Gielen, B. Kappen, pp. 568-575, Springer, Heidelberg.
14. Ritter, H., Martinetz, T., Schulten, K. (1992). *Neural Computation and Self-organizing Maps*, Addison-Wesley, Reading, MA.
15. Rumelhart, D.E., Hinton, G.E., Williams, R.J. (1986), "Learning Representations by Back-Propagating Error", in *Nature*, pp. 533–536.
16. Steil, J. (1993). "Markoffsche Felder und wechselseitige Information in der Bildverarbeitung", Diploma thesis, Bielefeld University, FR Germany.
17. Turk, M. Pentland, A. (1991). "Eigenfaces for Recognition", in *J. Cog. Neuroscience* **3**(1), pp. 71–86.

CLASSIFICATION WITH MULTI-MODULAR NEURAL NETWORKS: COMPARISON OF TWO EXAMPLES

J.PRORIOL

Laboratoire de Physique Corpusculaire

CLERMONT-FERRAND

ABSTRACT:We present some multi-modular neural networks methods of classification of hadronic events produced at LEP .
The results of a cooperative MLP and RBF(radial basis function) neural network and of a multi-MLP neural network are discussed and compared to the MLP results.

1-INTRODUCTION

Various researchers have recently advocated the use of multi-modular neural networks [1,2,3,4,5,7];various networks were used:Multi-Layer Perceptrons(MLP),Time Delay Neural Networks(TDNN),Learning Vector Quantization(LVQ) or Radial Basis Functions(RBF).The most common applications have been character recognition[7] or speech [1,2,5].The idea underlying these efforts is to get improved performances from the cooperation of networks:for example,a MLP is first used on the raw noisy data,then a LVQ or a RBF network is trained on the MLP-extracted features and will be able to perform a better classification than the last layer of the MLP.

Most often these architectures are trained sequentially,one module after the other.However,such a training procedure is sub-optimal:global optimality can be obtained by cooperatively training all modules together [4].

Our aim is to classify very noisy data coming from High Energy Physics (HEP) problems [6]:events produced by the collisions of particles (e^+e^-) at LEP and detected by the detector ALEPH.In a work reported previously [9],the comparison of the performances of conventional techniques (Discriminant Analysis),and various Neural networks:MLP and LVQ was done.

In this paper,we compare the performances of a MLP and of a cooperative MLP-RBF neural network :the first part of the processing is done by the first layers of a MLP,and the final classification by a RBF.Instead of using the usual initialisation by a k-means method [7],we have initialized by small random numbers.Our model is also different from [8]:we first use a MLP and then a RBF structure.Our last multi-modular neural network is a multi-MLP network.

2-MLP-RBF (cooperative)

With this method we use a cooperative learning between the 2 modules MLP and RBF.

2-1 Theory

The architecture is built up from two modules.First a MLP,with L layers of neurons;the MLP used for classification

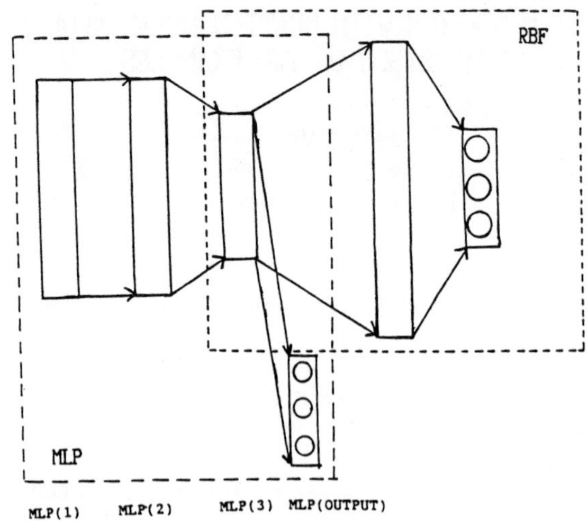

RBF(1) RBF(2) RBF(OUTPUT)

Figure 1:MLP-RBF archtecture

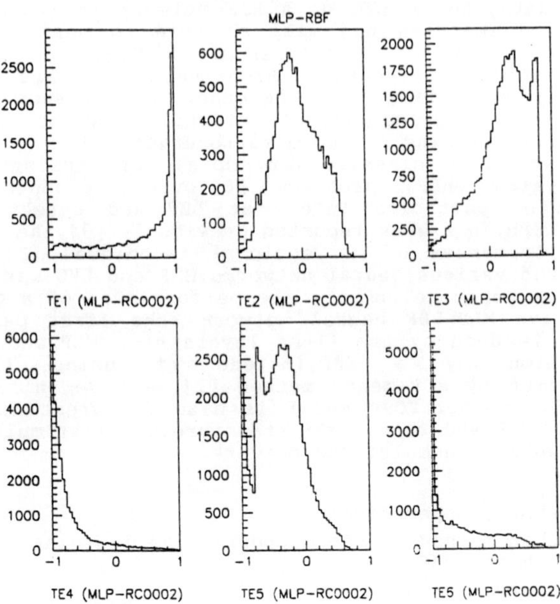

Figure 2:Histograms of the outputs of the MLP-RBF

TE1:b events TE4:not b(udsc) events
TE2:C events TE5:not c(udsb) events
TE3:uds events TE6:not uds(bc) events

would have L+1 layers,but here we have stripped the last layer,and only retained the features computed by the MLP which are used as inputs to the RBF.The RBF has two layers,the first one is a set of gaussian kernels,and the other is the standard MLP output layer.The architecture of the neural network is presented on figure 1.MLP(1) is the MLP input layer.MLP(2) and MLP(3) are the MLP hidden layers.The MLP(output) layer is not considered.The MLP(3) layer is the RBF(1) input layer which feeds the RBF(2) layer built with kernels.The RBF(output) layer is similar to MLP one.

In our application,the MLP-RBF is built from a 20-20-8-3 MLP network which is not trained.With the 8 variables of the last MLP hidden layer,21 kernels are computed to the 3 output neurons.

The feed-forward calculations is done according to the standard MLP [10].The last hidden layer of MLP is connected to the gaussian kernels according to the relation:

$$a_i = \sum_j R_{ij} \exp\left(-\frac{\|\mathbf{x}-\mathbf{x}^j\|^2}{\sigma_j^2}\right)$$

where i is the index of output units,$\mathbf{x}^j$ is the center of the unit ,σ_j its width and R_{ij} the weights.The output is computed by the relation $y_k=F(a_k)$ where F is a sigmoid.

The error computed on the output layer is propagated backwards through the RBF and the MLP layers[12].

The initialization of the different weights of the MLP and RBF is done with small random values;The σ values are initialized with the value 0.75.

2-2 Results

We intend to classify hadronic events into 3 classes:b quark events,c quark events and light quark (uds) events.We have generated 150 variables for each event,and we have chosen 20 discriminating variables [9].The training was done with a set of 30000 events(10000 for each class).The test was done with a different set of 73376 events.

The best results obtained with MLP-RBF give $(73.5\mp0.2)\%$ for the percentage of correct classification for 3 classes.The curves of purity/efficiency are similar to MLP ones [9],they are given in figures 2-4.

3- THREE MLP

We have studied another method to use a multi-network system:the 3 parallel MLP network.

3-1 Theory

The 3 classes of HEP events strongly overlap.To get a more efficient classification,it is interesting to follow a method used for the handwritten digits recognition [11]:the training is done for all sets of 2 classes: b/c, b/uds,c/uds.The inputs are different for the 3 sets,and the outputs of the 3 MLP are different.During the test,we combine the 3 outputs to get the final classification.On figure 5,we present the architecture of the system for the learning and the test.Each MLP is a 20-20-8-2 one.

The outputs of the 3 MLP are:y_b^1,y_c^1 for the first set;y_b^2,y_{uds}^2 for the second set and y_c^3,y_{uds}^3 for the third

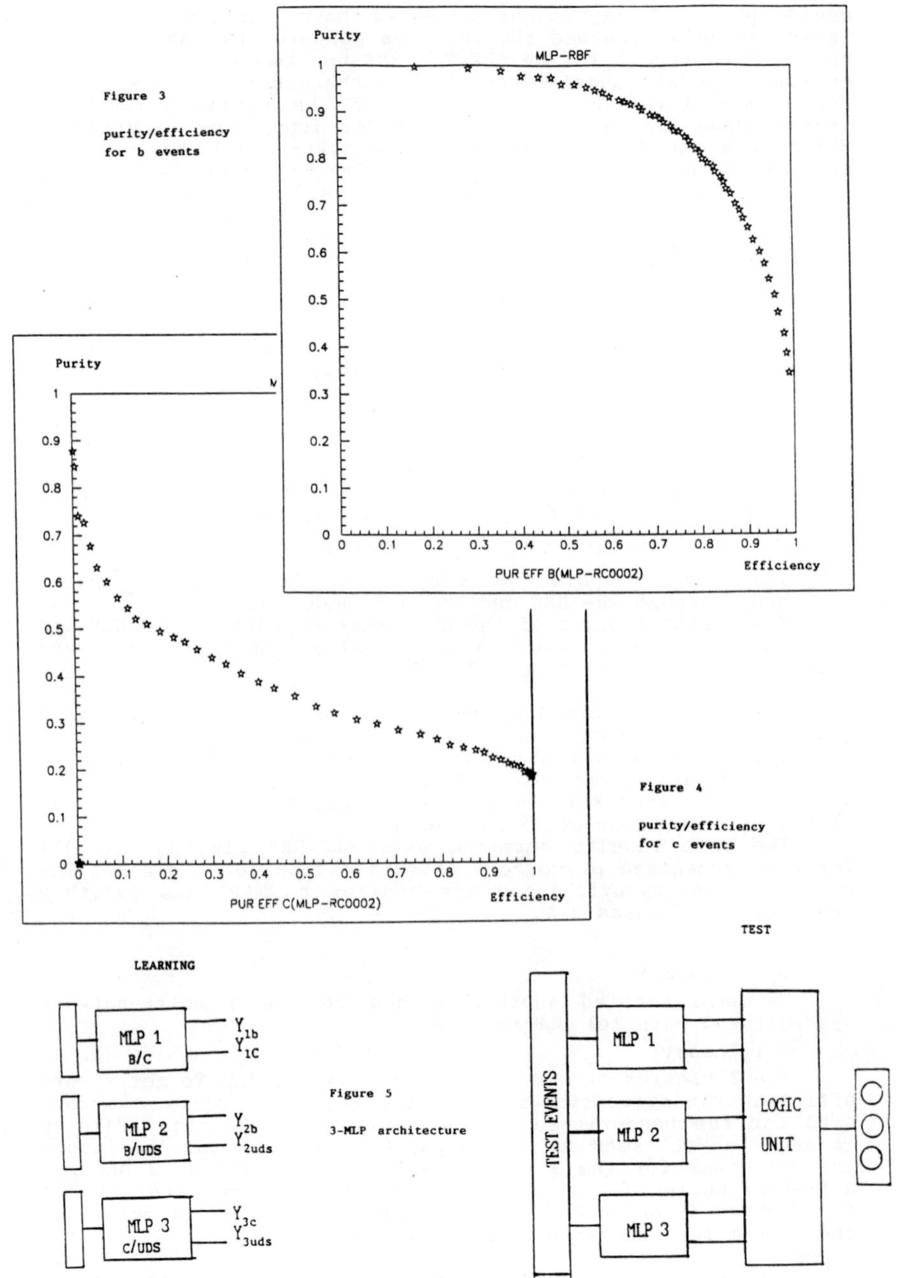

Figure 3

purity/efficiency for b events

Figure 4

purity/efficiency for c events

Figure 5

3-MLP architecture

set.The outputs combinations in the test phase are given by:

$Y_b=y_b^1*y_b^2$, $Y_c=y_c^1*y_c^3$,$Y_{uds}=y_{uds}^2*y_{uds}^3$.

3-2 Results

The training is done with 3 sets of 20000 events(10000 per class).The test set is similar to MLP-RBF one.

The shape of the curves purity/efficiency is similar to previous ones.

The percentage of good classification is (73.7∓0.2)% for 3 classes.

4- CONCLUSION

To perform learning in all the cases,we have used a learning set composed of the same number of events for the different classes.

We have compared different types of multi-modules neural networks.The most efficient are the 3-MLP neural network and the cooperative of MLP-RBF neywork;but the last one is very easy to use,easier to handle than the 3-MLP network.

Neural network indeed improve upon conventional methods.And multi-modular architectures are more efficient than simple ones.

The different results are gathered in table 1.

NAME	percentage 3 classes	TOP Coefficients[13] b	c
DISC ANALYSIS	67.2∓0.3	0.98	0.33
MLP	71.2∓0.2	0.98	0.44
MLP-RBF(coop)	73.5∓0.2	0.98	0.45
3-MLP (prod)	73.7∓0.2	0.98	0.44

Table 1:Results for different methods:

-Percentage of well classified events:3 classes

-TOP coefficients [13],for b and c classes

5-ACKNOWLEDGMENTS

This work was initiated by a suggestion of Professor F.FOGELMAN-SOULIE to apply to high energy phycics problems the multi-modular neural network architectures;several discussions on neural networks proved very stimulating.

This paper is a part of a work going on in the ALEPH collaboration:I would like to thank the physicists of the collaboration for their support.

6-REFERENCES

[1]Y.BENGIO,R.DE MORI:Glogal optimization of a neural-hidden Markov model hybrid.*Trans.on neural networks* 3*(1992)252*

[2]Y.BENNANI,P.GALLINARI:A modular connectionist architecture for text-independant talker identification.*IJCNN'91 vol 2,p 857*

[3]M.de BOLLIVIER,P.GALLINARI,S.THIRIA:Cooperation of neural networks and task decomposition.*IJCNN'91,Seattle,vol 2 p573*

[4]L.BOTTOU,P.GALLINARI:A framework for the cooperation of learning algorithms.In *Neural Information processing systems,NIPS'90,R.P.LIPPMANN,J.E.MOODY,D.S.TOURETZKY eds,Morgan Kaufmann,vol 3 (1991) p 781*

[5]H.BOULARD,N.MORGAN:A continuous speech recognition embedding MLP into HMM.*NIPS'89,D.Touretzky ed,Morgan Kaufman,vol 2 (1990) p 186*

[6]B.DENBY:The use of neural networks in high energy physics.*Neural Computation,(to appear)*

[7]F.FOGELMAN-SOULIE,B.LAMY,E.VIENNET:Multi-modular neural network architecture for pattern recognition:applications in optical character recognition and humanface recognition.*Int. Jl Pattern Recognition and Artificial Intelligence(to appear)*

[8]S.LEE: In Neural networks for signal processing,*B.Kosko ed,Prentice Hall,Englewood Cliffs (1992)*

[9]J.PRORIOL,J.JOUSSET,C.GUICHENEY,A.FALVARD,P.HENRARD,D.PALLIN ,P.PERRET:Tagging b quark events in ALEPH with neural networks.*In Proceedings of the Elba workshop:Neural networks:from biology to high energy physics. O.Benhar, C.Bosio, P.del Giudice,E.Tabet,eds. (1991)p 419*

B.BRANDL,A.FALVARD,C.GUICHENEY,P.HENRARD,J.JOUSSET,J.PRORIOL: Multivariate analysis methods to tag b-quark events at LEP/SLC *Nuclear Instruments and methods A324(1993)307*

J.PRORIOL Classification of hadronic events in e^+e^- collisions *Nuclear Instruments and methods (To appear)*

[10]D.E.RUMELHART,J.L.Mc CLELLAND:Parallel distributed processing:explorations in the microstructures of cognition, *MIT Press,(1986)*

[11]S.KNERR,L.PERSONNAZ,G.DREYFUS:From theory to silicon:an efficient procedure for the design of neural classifiers and its application to the automatic recognition of handwritten digits.*NEURO-NIMES'91,EC2,PARIS(1991)*

[12]J.PRORIOL *Multi-modolar neural networks for the classification of e^+e^- hadronic events Preprint*

[13]K.H.BECKS,F.BLOCK,J.BREES,P.LANGEFELD,F.SEIDEL:B- quark tagging using neural networks and multivariate statistical methods.A comparison of both techniques.
Nucl Inst Meth A329(1993)501

A NEWTON-TYPE ALGORITHM FOR MULTILAYER PERCEPTRON TRAINING

V.V.IVANOV, B.PUREVDORJ, I.V.PUZYNIN
Laboratory of Computing Techniques and Automation, Joint Institute for Nuclear Research, Moscow, 101000/Head Post Office, P.O. Box 79, Russia

ABSTRACT

First- and second-order methods for feed-forward multilayer networks training are considered. A Newton-type algorithm is proposed and compared with the common back-propagation algorithm. It is shown that this algorithm provides better learning quality. Some recommendations for the usage are given.

1. Introduction

Today artificial neural networks are widely used in high energy physics [1]. They are applied to both constructing high-level triggers providing effective discrimination of background events in the real time of experiment and processing collected data, e.g. track recognition, mass reconstruction, primary and/or secondary vertex identification.

In particular, feed-forward multilayer [1,2] networks are widespread. Usually network learning is implemented with the back-propagation algorithm [3] based on the gradient descent. However, this algorithm does not ensure high recognition accuracy (see discussion below). It is important for problems in experimental particle physics, where rare events are separated from suppressing background.

2. Problem statement

2.1. Network architecture

Multilayer feed-forward network (multilayer perceptron) involves an input layer corresponding to analyzed data, an output layer corresponding to results and hidden ones.

The output signal $\vec{y}$ is given by

$$y_i = f(a_i), \qquad a_i = \sum_j \omega_{ij} h_j + \theta_i,$$

$$h_j = \begin{cases} f(a_j) & \text{for hidden layers,} \\ x_j & \text{for input layer,} \end{cases}$$

$$f(x) = g(x/T),$$

where x_j – input data, ω_{ij} – neuron connection weights, θ_i – thresholds, T – temperature, g – transfer function. We choose $g(x) = [1 + tanh(x)]/2$.

The learning procedure consists in finding weights minimizing the energy functional

$$E = \frac{1}{2}\sum_p \left(\vec{y}^{(p)} - \vec{t}^{(p)}\right)^2,$$

where $p = 1, \ldots, N_{train}$, N_{train} – number of training patterns, $\vec{t}$ – target value of output signal.

The network is simulated on IBM PC AT386/486 in the NDP [4] environment using JETNET 2.0 package developed in the LUND University [5].

2.2. Classification problem

We use neural networks for classifying samples from two overlapping multidimensional Gaussians with densities f_1 and f_2. A network with the number of input neurons equal to the dimension of the events and one output neuron is taken. The learning set contains an equal quantity of events of both kinds. The target value is set to 0 for the 1st distribution and 1 for the 2nd one.

Denoting $y = F(\omega, x)$ one has in continuous limit

$$E = \int \left((F(\omega,x))^2 f_1(x) + (1 - F(\omega,x))^2 f_2(x)\right) dx.$$

Let us take

$$opt(x) = \frac{f_2(x)}{f_1(x) + f_2(x)}, \qquad D = \{x | f(x) = f_1(x) + f_2(x) \neq 0\}.$$

Then

$$E(opt + \epsilon) = E(opt) + \int \epsilon(x)^2 f(x) dx.$$

Introducing in D the norm $y = \left(\int y^2(x) f(x) dx\right)^{\frac{1}{2}}$, we see that minimization of E is equivalent to finding the function y nearest to opt. We consider accuracy of minimization as a learning quality.

3. First-order methods

In the back-propagation (BP) method [3] weights are calculated according to the algorithm:

1. Initial values are chosen randomly in the range $[-\omega_0, \omega_0]$,
2. Weight updates are given by

$$\Delta\vec{\omega}_n = -\eta \nabla E_n + \alpha \Delta\vec{\omega}_{n-1},$$

$$\frac{\partial E}{\partial \omega_{ij}} = d_i h_j, \qquad d_i = \begin{cases} (y_i - t_i) f'(a_i) & \text{for output layer,} \\ \sum_k d_k \omega_{ki} f'(a_i) & \text{for hidden layers.} \end{cases}$$

BP differs from the pure gradient descent in ∇E_n being calculated on a subset of the training set containing N patterns and a momentum α, important for ravine functions [3]. Efficiency of BP depends on the network parameters. However, it is difficult to formulate accurate criteria for choosing them.

BP comes to the gradient descent when $N = N_{train}$ and $\alpha = 0$. In this case learning goes very slow. The value of η must be small and attempt to find it minimizing $E(\eta)$ at $0 < \theta \leq \eta \leq 1$ does not affect the result strongly. The gradient descent can be expedient near the solution due to less oscillations.

Though BP is rather efficient, it has the following shortcomings:

- the method does not provide accurate approximation of the minimum;
- the learning parameters are often to be chosen empirically;
- it is difficult to detect reaching the neighborhood of the minimum;
- the gradient norm reduces slowly and is substantially non-zero leading to large oscillations.

4. Second-order methods

4.1. Newton method

In the Newton method weights are updated according to

$$\Delta\omega = -\eta(\nabla^2 E)^{-1}\nabla E,$$

where the Hessian matrix $\nabla^2 E$ is given by

$$\frac{\partial^2 E}{\partial\omega_{ij}\partial\omega_{mn}} = d_i\frac{\partial h_j}{\partial\omega_{mn}} + \frac{\partial d_i}{\partial\omega_{mn}}h_j.$$

The method is not applicable for this problem since $\nabla^2 E$ is signvariable and nearly singular in the neighborhood of the minimum. Strictly speaking, the minimum is not completely degenerate, but the Hessian matrix is ill-conditioned near the minimum. Typical conditionality values $|\lambda|_{max}/|\lambda|_{min}$ (λ – eigenvalues) are 10^3 at the initial point and $10^5 - 10^6$ (even 10^{12}) in the neighborhood of the minimum.

The use of regularization /7/

$$\Delta\omega = -\eta H^{-1}A^*\nabla E,$$

$$H = A^*A + \alpha\|\nabla E\|^2, \qquad A = \nabla^2 E.$$

ensures convergence only with a large α ($\alpha \sim 1$) and does not provide better quality than a gradient descent. The matrix H is too ill-conditioned when the regularizing addition reduces to zero. Therefore, H cannot be inverted with acceptable accuracy.

4.2. Newton method with Hessian correction

When the Hessian matrix has poor qualities the following method is used

$$\Delta\omega = -\eta H^{-1}\nabla E,$$

where H is positive-definite and well-conditioned. When $\|(\nabla^2 E)^{-1}\| > A$ in the small neighborhood of the minimum, where A – a maximal acceptable value, $H \not\to \nabla^2 E$ with approaching the minimum.

The matrix H is taken as [6]

$$H = \nabla^2 E + \mu I,$$

where μ ensures that eigenvalues of $H \geq \delta > 0$. The choice $\delta = 1$ gives good results. In this condition H^{-1} is a squeezing operator.

The steplength η is calculated by means of one-dimension-al optimization of $E(\eta)$. It must be noted that the CPU time is spent mostly on the calculation of the Hessian matrix. Stabilization of η near 1 indicates, as a rule, that the solution is close. It is also convenient to calculate η as [8]

$$\eta_{n+1} = \eta_n \frac{||\nabla E_{n-1}||}{||\nabla E_n||}, \quad \theta \leq \eta_{n+1} \leq 1,$$

where η_0 must be sufficiently small. However, it slows down the performance.

The method has the following merits:

- it provides the learning quality unachievable with BP. It is demonstrated by the value of E and the gradient norm, which is much closer to zero;
- as a rule, the convergence speed does not depend strongly on the network size, the number of training patterns or complexity of the recognition problem;
- the solution of more complex problems, e.g. separation of strongly overlapped distributions or reliable recognition with a small learning set, is possible;
- the learning parameters are chosen automatically.

Large computational expenses are a demerit of the method. BP requires memory proportional to n and the CPU time per epoch – to $n \times N_{train}$, where n – the number of weights. The Newton method requires memory proportional to n^2 and the CPU time – to $n^2 \times N_{train}$ (when $n \ll N_{train}$). Let us note that the learning speed is not a crucial factor if the network is used many times.

5. Comparison of the learning algorithms

As mentioned above, the feed-forward network is used for classifying two multidimensional Gaussians. The network configuration is 2–6–1. The following parameters are chosen: $\omega_0 = 0.5$, $T = 1$; and for BP: $\eta_0 = 0.1$, $\alpha = 0.5$, $N = 10$.

We present the performance of different methods in the table 1. The denotations are: BP – back-propagation, GRAD – pure gradient descent, NW – Newton method with Hessian correction, REG – Newton method with regularization (used after 10 epochs of BP), *Example* – standard deviations of classified 2-dimension-al Gaussians (mean values are set to zero), *Limit* – Bayesian limit of recognition, *Epoch* – the number of epochs needed to reach the Bayesian limit and become stable, E_{min} – the least value of E reached during 100 (* – 1000) epochs and divided by N_{train}, $||\nabla E||$ – the gradient norm at the end of learning.

Table 1: Performance of different algorithms of the feed-forward network learning

Method	*Example*	N_{train}	*Limit*	*Epoch*	E_{min}	$\|\nabla E\|$
BP	1/0.3	1000	76,1	7	0.160*	1.0 – 20
NW				7	0.155*	0.047
GRAD				107	0.161*	0.48
REG				–	0.159*	0.0014
BP	1/0.7	1000	58,5	52	0.235	2.0 – 40
NW				4	0.229	0.13
BP	1/0.8	1000	55,3	–	0.242	4.0 – 40
NW				6	0.238	0.092
BP	1/0.3	100	76,1	19	0.144	0.4 – 4.0
NW				17	0.138	0.065
BP	1/0.5	100	66,1	–	0.182	0.2 – 5.0
NW				25	0.142	0.26
BP	1/0.3	5000	76,1	4	0.164	3.0 – 60
NW				7	0.163	0.10

The figure 1 shows learning curves for BP and the Newton-type method.

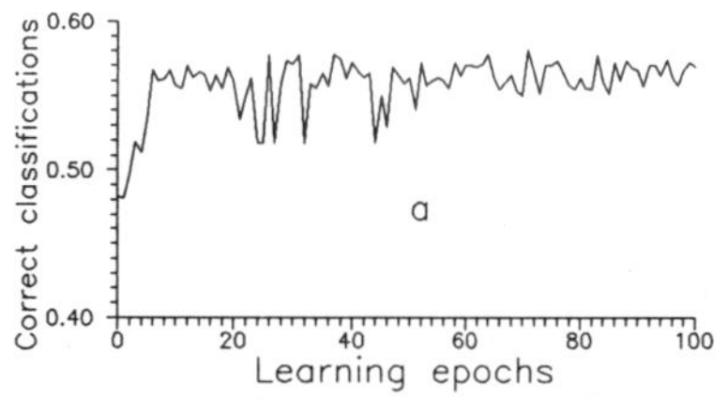

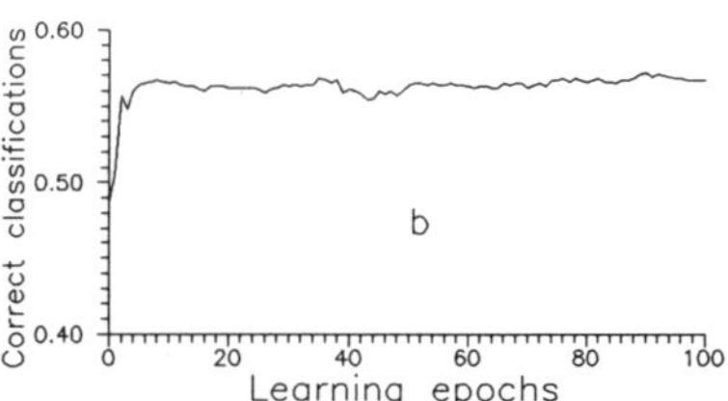

Figure 1: Fraction of correct classifications as a function of learning epochs for $Example = 1/0.7$ and $N_{train} = 1000$: a) BP, b) Newton-type method

6. Conclusion

The results of comparison of different learning algorithms are as follows.

1. When the learning parameters are optimal, BP is rather fast and requires a small memory size.
2. The pure gradient descent is expedient only near the solution.
3. The pure Newton method is not applicable.
4. The Newton method with the Hessian correction requires more computer resources than BP, but provides better learning quality and is simpler in use.

BP is better for big networks used for recognition of complex patterns, but the Newton method with the Hessian correction is good for relatively small networks when maximal recognition accuracy is needed.

References

1. B.Denby: *"Tutorial on Neural Networks Applications in High Energy Physics: 1982 Perspective"*. In Proc. of the Second International Workshop on *"Software Engineering, Artificial Intelligence and Expert Systems in High Energy Physics"*. January 13-18, 1992 L'Agelaude France-Telecom La Londe-les-Maures (France). New Computing Techniques in Physics Research II, edited by D.Perret-Gallix, World Scientific, 1992, p.287.
2. B.Humpert: *"A Comparative Study of Neural Network Architectures"*, Comp. Phys. Comm. 58, 1990, p.223.
3. D.E.Rumelhart, G.E.Hinton, R.J.Williams: *"Learning Internal Representations by Error Propagation"* in D.E.Rumelhart & J.L.McClelland (Eds.), *Parallel Distributed Processing: Explorations in the Microstructure of Cognition.* Vol. 1: *Foundations.* MIT Press, 1986.
4. NDP – FORTRAN Reference Manual. Microway Inc., Kingston, Massachusetts.
5. L.Lönnblad, C.Petterson, T.Rögnvaldsson: *"Pattern Recognition in High Energy Physics with Artificial Neural Networks - JETNET 2.0"*, Comp. Phys. Comm. 70, 1992, p.167.
6. M.Minoux: *Programmation Mathématique.* Bordas et C.N.E.T. – E.N.S.T., Paris, 1989.
7. V.V.Ermakov, N.N.Kalitkin: *"An Optimal Step and Regularization of the Newton Method"* (in Russian), Zhur. Vych. Mat. i Mat. Fiz., Vol.21, 2, 1981, p.491.
8. T.Janlav, I.V.Puzynin: *"On the Convergence of Iterations Based on the Continuous Analog of the Newton Method"* (in Russian), Zhur. Vych. Mat. i Mat. Fiz., Vol.32, 6, 1992, p.846.

USE OF A MULTILAYER PERCEPTRON IN THE CDF TOP SEARCH IN THE MULTIJET DECAY MODE

BRUCE DENBY
INFN Sezione di Pisa, Via Livornese 582/a, 56010 S. Piero a Grado, (Pisa) Italy

and

JEFF WYSS, ANDREA CASTRO
University of Padova, Dipt. di Fisica, Via Marzolo 8, 35100 Padova, Italy

ABSTRACT

This paper reports on the application of a Multilayer Perceptron (MLP) used as a tool in the CDF top search for top decays to multijets. Although the efficacy of the network on the final data sample was reduced due to a different initial set of cuts from those used in the training data, the network output was found to be correlated with certain physics and 'common sense' variables. This gives confidence that an MLP is indeed a useful tool for High Energy Physics data analysis, since it is sensitive to rational physics variables that are easily understandable. When the network is retrained on the new data, it will provide an optimal way to place cuts upon these variables.

1. Introduction

Top antitop pairs should be produced at hadron colliders by gluon fusion and by quark antiquark annihilation. The multijet decay mode is that in which the W's from both members of the t-tbar pair decay into quark jets. The final state will then consist of 4 quark jets from the W's, two more jets from the b's, plus possible additional jets from gluon radiation, for a total of at least 6 jets. This channel accounts for 44% of t-tbar decays, however the background from multijet QCD processes is very high; for example, for events with at least 120 GeV of summed transverse energy, the S/B ratio is about $1/10^4$. The utility of this channel hinges upon our ability to reduce this overwhelming background.

2. Definition of Data Set and Loose Cuts

The results presented are based upon 21.4 pb^{-1} of data from the 1992/1993 run of the CDF Experiment. The data is first subjected to a set of 'loose' cuts which cuts simply select clean data with a high $\sum E_T$, multijet character.

- $N_{jets}^{0.7cone} \geq 6 \quad (|\eta^{jet}| < 2.4,\ E_T^{jet} \geq 10GeV)$

- $\sum_i E_T^i \geq 150\ GeV$ (sum over towers)

The S/B is still quite low after the loose cuts, of the order of 1/800. In order to further reduce the background, a number of approaches were tried.

3. Tight Cuts

First were the 'tight' kinematical cuts, a set of 'ad-hoc' topological cuts designed to reject most of the background while retaining a high efficiency for top:

- $\sum_i E_T^i \geq 60\ GeV + 15\ GeV * N_{jets}^{0.7cone}$ (sum over cone 0.7 jets)
- mean $|\eta|$ of the jets ≤ 1
- mean $|p_T/\eta|$ of the jets ≤ 1
- $N_{jet}^{0.4cone} \geq 6$

After the tight cuts, the S/B is of the order of 1/100. This S/B is of course still not good enough to do a physics analysis. In the top search, events passing the tight cuts (plus perhaps additional cuts) are examined for the presence of a possible secondary vertex, as detected in the silicon microvertex detector (SVX), which could be evidence for the presence of a b-decay from the top. For the purpose of the present paper, however, we shall restrict our discussion to kinematical cuts before the application of the SVX information.

4. Multilayer Perceptron

Next, a Multilayer Perceptron (neural network) was tried as a kinematical discriminant. The network uses the E_T, η, and ϕ of the six leading jets in each event as input and has 5 hidden units. It was trained to distinguish 1989 multijet data (assumed to be pure background) from Isajet or Herwig Monte Carlo top data. Figure 1 shows the output of the network on its test set consisting of data from the 1988-89 run of CDF (dashed) and the Isajet Monte Carlo (solid); top–like events peak to the right while background events peak to the left. In an an earlier work it was shown the the 5 node network was more effective than a Fisher linear discriminant [1]. For the 1988-89 data, the neural network alone achieved better S/B for a given efficiency than was achieved using the tight cuts, as shown in figure 2. The figure shows the reciprocal of the S/B versus the total global top efficiency after loose cuts, for the MLP alone (black dots) and for tight cuts plus MLP (open squares). The vertical line shows the efficiency for loose cuts alone. This result was to be expected since the MLP is a way of achieving the optimal separation between classes.

For the 1992-93 CDF run, the tight cuts were changed to:

- $\sum_i E_T^i \geq 70\ GeV + 20\ GeV * N_{jets}^{0.7cone}$ (sum over cone 0.7 jets)

- mean $|\eta|$ of the jets ≤ 1
- $N_{jet}^{0.4cone} \geq 6$

In addition, the multijet trigger requirements were changed, and the requirement of a single event primary vertex was removed. The neural network was not retrained after these cuts were changed. As a result of these differences, when the MLP was applied to the 1992-93 data, its performance was substantially reduced as compared to the tight cuts. The main effect can be seen in the first two frames of figure 3 which show signal (right) and background (left) MLP output distributions, for loose cut data, which are somewhat shifted to the right as compared to those figure 1.

The remaining frames of figure 3 show the effect of increasing the minimum E_T requirement for the jets. Jets from top decays should be rather stiff as compared to jets from ordinary QCD processes; therefore, cutting harder on jet E_T should select events which are more top–like. This qualitative understanding is confirmed by the MLP output which shows, in the left column, a significant change in the output for the background as the E_T cut is raised. At 30 GeV E_T, the background distribution looks almost like the initial top distribution. As shown in the right column, the top distribution is also skewed to the right, but considerably diminished in size, as E_T is raised. Thus, in accord with an intuitive understand of what top events should look like, we see that raising the E_T cut on the jets does selects events which look more top–like, as classified by the MLP. This cut alone, however, is not very efficient for top.

5. Comparison of Kinematic Cut Methods

This effect can be be seen further in figure 4. This plot shows the number of 1993 CDF events remaining versus the expected number of 160 Isajet top events remaining after 4 types of cuts. The solid curve (black squares) is generated by starting with loose cut data (off the page at 17,000 events) and cutting harder and harder on E_T to generate the curve shown. Clearly, at no point on this curve is the S/B as good as taking the tight cuts along (black dot marked 'tight alone').

Although the 1989 trained MLP alone is not competitive (due to the different cuts) on the 1993 data, it can still be applied to advantage *after* the tight cuts. doing so generates the dashed curve which joins the black dots. This curve shows that the S/B can be improved about a factor of 2 while retaining about 80% efficiency for top.

The additional discrimination of the MLP was a motivation to try to find additional 'physics' cuts which could improve S/B after tight cuts. One such cut is the so called 'extra tight' cut:

- $\sqrt{\hat{s}}/\sum E_T < 1.4$

which essentially requires consistency with the decay of a heavy object. Applying this cut produces the diamond point in figure 4 labelled, 'x–tight alone'. It is interesting to note that this point lies almost exactly on the curve produced by cutting on the MLP output after applying the tight cuts. This suggests that the residual discrimination of the MLP is in fact highly correlated with this $\hat{s}$ cut. A comparison of MLP output of tight events before and after the tight cuts confirms this correlation, i.e., the extra tight cut shifts the mean of the MLP output distribution from .16 to .36, i.e., towards higher top probability. Once again, the neural network output probability is seen to be highly correlated with rational physical variables. The MLP is simply a way of cutting in an optimal way upon a combination of such variables.

Also shown in the figure for comparison is the result (asterisk) after applying more conventional optimised cuts on sphericity and aplanarity after the tight cuts.

6. Conclusion

An MLP based on the E_T, η, and ϕ of the six leading jets has been trained on 1988-89 CDF data and Monte Carlo top data. On the 1988-89 test set, the MLP outperformed the S/B obtained with the standard 'tight' cuts. On the 1993 data, due to a different set of cuts, the MLP was less effective, but can still be used as a cutting variable on top of the tight cuts. Furthermore, the MLP output cut was shown to be correlated with a 'common sense' cut on the minimum E_T of the jets in the selected data and with a 'physics' cut based on the $\hat{s}$ in the event. The MLP is thus shown to be a valuable tool which, when retrained on the 1993 data, will provide an optimal way of selecting a top enriched sample of events.

Acknowledgements

The authors acknowledge the rest of the CDF collaboration for all the steps involved in making the multijet data available, and in particular the University of Padova hadronic top group.

Reference

1. W. Badgett et al., Proceedings of the conference, Computing in High Energy Physics, Annecy, France, September, 1992, to appear.

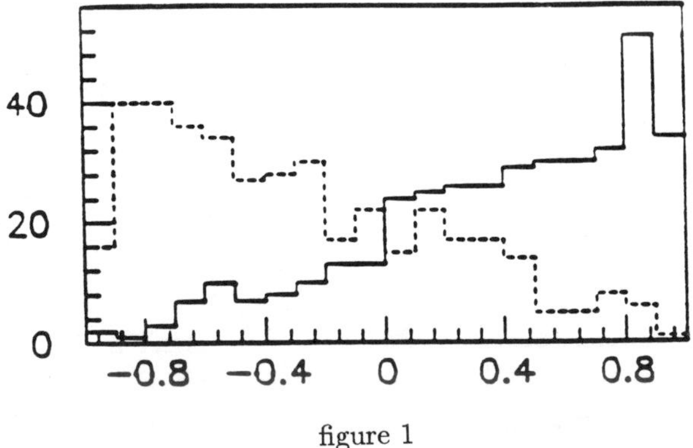

figure 1

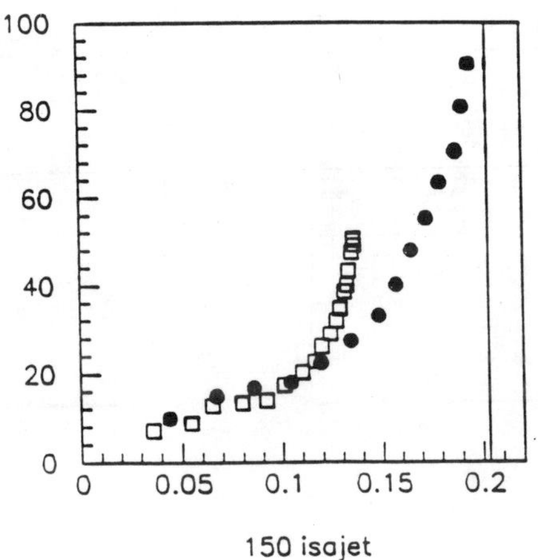

figure 2

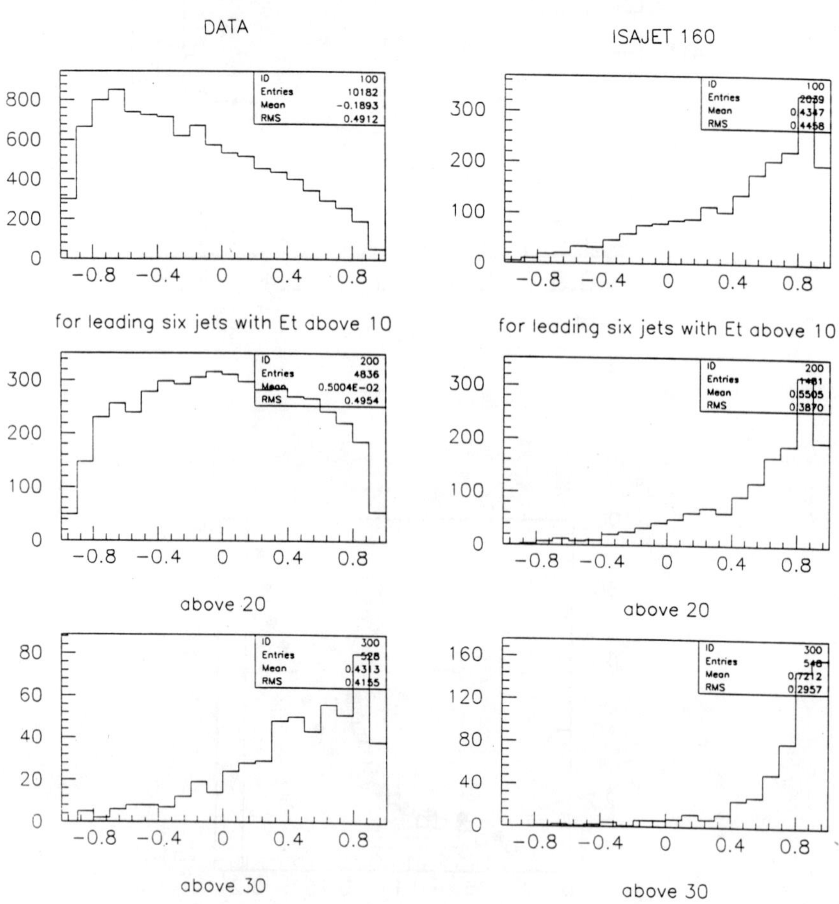
DATA
ISAJET 160
ID 100
Entries 10182
Mean -0.1893
RMS 0.4912
for leading six jets with Et above 10
ID 200
Entries 4836
Mean 0.5004E-02
RMS 0.4954
above 20
ID 300
Entries 528
Mean 0.4313
RMS 0.4155
above 30
ID 100
Entries 2039
Mean 0.4347
RMS 0.4458
for leading six jets with Et above 10
ID 200
Mean 0.5505
RMS 0.3870
above 20
ID 300
Mean 0.7212
RMS 0.2957
above 30

figure 3

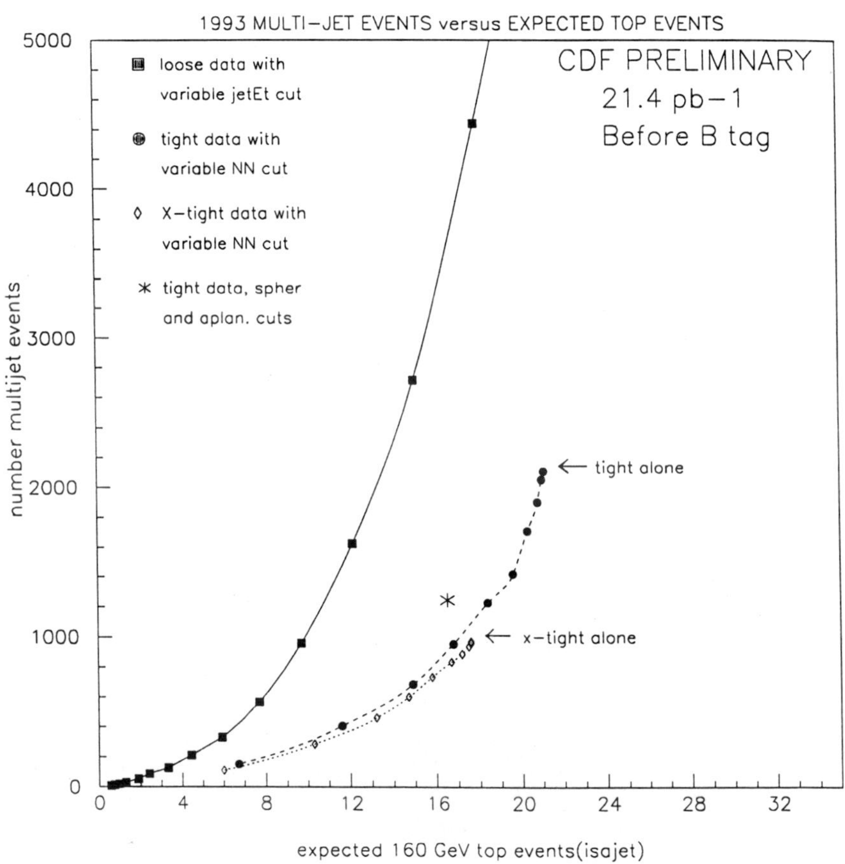
1993 MULTI-JET EVENTS versus EXPECTED TOP EVENTS
loose data with variable jetEt cut
tight data with variable NN cut
X-tight data with variable NN cut
tight data, spher and aplan. cuts
CDF PRELIMINARY
21.4 pb-1
Before B tag
tight alone
x-tight alone
number multijet events
expected 160 GeV top events(isajet)
0
1000
2000
3000
4000
5000
0
4
8
12
16
20
24
28
32

figure 4

b-Tagging Using Shape Variables in the Hadronic Decays of the Z^0

DELPHI Collaboration
P.Branchini[1], *M.Ciuchini*[1], *G.Cosmo*[2] *U.Daurù*[2],
A.DeAngelis[2], *P.DelGiudice*[1], *L.Lyons*[3], *A. Saccavini*[2], *F. Stringhetta*[2]

Abstract

The performance of b-taggers, based on Multidimensional Analysis applied to variables related to the shape of hadronic events, is analyzed on a sample of about 166,000 hadronic decays of the Z^0 collected by the DELPHI detector at LEP during 1991. The classification efficiency is computed from the data, by using tagging on hemispheres.

(Preliminary analysis on 166,000 events)

[1] Istituto Superiore di Sanità and INFN Sanità, Roma
[2] Dipartimento di Fisica dell'Universita' di Udine and INFN Trieste
[3] Department of Physics, Oxford University

1 Introduction

The selection of as pure as possible subsamples of hadronic decays of the Z^0 in which the primary branching is into $b\bar{b}$ pairs is of primary interest in LEP physics. Methods so far explored to achieve this are based on:

1. the shape of the multihadronic events [1];
2. the presence of leptons with large transverse momentum with respect to the event axis [2];
3. the presence of tracks with large impact parameter with respect to the primary vertex [3].

Method 2. is limited by low efficiency, since the branching ratio of b into muons is of the order of 10%. Method 1. is generally limited by low efficiency, and by a large systematic error in estimating the performance. The systematic error has usually been estimated using simulations, and the results are sensitive to the models assumed for the description of the hadronization process.

In this paper, we discuss the possibility to extract a discriminant variable from many variables related to the shape of a multihadronic event, and to evaluate its performance in an almost simulation independent way, using a feed-forward Neural Network as a multidimensional classifier.

In the past three years many attempts have been done to use Neural Networks as a tool for classification of jet events according to the flavour of the parent quarks (for a review of the results recently obtained see Ref. [6] [7] [8]). These methods achieved performances which are competitive with standard analysis in terms of efficiency and purity; they share, however, a weakness in the dependence on Monte Carlo, that makes the estimation of systematic errors particularly tricky.

This motivated us to explore a strategy (see for example [5] for a previous application of this method) based on the independent tagging of suitably defined emispheres in which events are divided. In this way, an almost Monte Carlo independent tagging can be obtained, paying the price of a worsening of the performances.

We report here on preliminary results obtained in this direction.

2 Experimental Procedure and Event Selection

The sample of events used in the analysis was collected during 1991 by the DELPHI detector at the LEP e^+e^- collider, operating at centre of mass energies around the Z^0 peak.

A description of the apparatus can be found in Ref. [9]. Features of the apparatus relevant for the analysis of multi-hadronic final states (with emphasis on the detection of charged particles) are outlined in Ref. [10]. The present analysis relied on the information provided by the central tracking detectors.

The central tracking system of DELPHI covers the region between 25° and 155° in polar angle, θ, with reconstruction efficiency near 100%. The average momentum resolution for the charged particles in hadronic final states is in the range $\Delta p/p \simeq 0.001p$ to $0.01p$ (p in GeV/c), depending on which detectors are included in the track fit.

Charged particles were used in the analysis if they had:

(a) momentum larger than 0.1 GeV/c;

(b) measured track length in the TPC greater than 30 cm;

(c) θ between 25° and 155°;

(d) relative error on the measured momentum smaller than 100%.

Hadronic events were then selected by requiring that:

(α) the total energy of the charged particles in each hemisphere (θ above and below 90°) exceeded 3 GeV;

(β) the total energy of the charged particles exceeded 15 GeV;

(γ) there were at least 5 charged particles with momenta above 0.2 GeV/c.

In the calculation of the energies, all charged particles have been assumed to have the pion mass.

A total of 166K events satisfied these cuts. Events due to beam-gas scattering and to $\gamma\gamma$ interactions have been estimated to be less than 0.1% of the sample; background from $\tau^+\tau^-$ events was calculated to be less than 0.2%.

The influence of the detector on the analysis was studied with the simulation program DELSIM [11]. Events were generated with the JETSET 7.3 Parton Shower Monte Carlo program [12] (JETSET PS in the following) with parameters tuned as in [13]. The particles were followed through the detailed geometry of DELPHI giving simulated digitizations in each detector. These data were processed with the same reconstruction and analysis programs as the real data.

3 The Double Tagging Method

The major drawback of multidimensional methods is that the dependence on the simulation introduces on the physical observable a systematic error which is harder to estimate compared to the single variable case.

In order to overcome this problem, a double tagging method was used in the present analysis.

Hadronic events were split into two hemispheres according to the plane perpendicular to the thrust axis, then only hemisphere-defined variables were considered. Let ϵ_b be the probability of tagging a hemisphere in a b event, and ϵ_l the same probability in u, d, s, c events. In the hypothesis in which the two hemispheres are statistically uncorrelated the following equations hold:

$$f_1 = \epsilon_b R_b + \epsilon_l (1 - R_b) \tag{1}$$
$$f_2 = \epsilon_b^2 R_b + \epsilon_l^2 (1 - R_b) \tag{2}$$

where f_1 is the fraction of tagged hemispheres and f_2 is the fraction of events in which both hemispheres are tagged. The quantities ϵ_b and ϵ_l can be determined from the equations above, by assuming $R_b = 0.217$ as predicted by the Standard Model.

It would be possibile to estimate R_b using a third equation coming from a lepton tagged sample but this possibility will not be considered here.

The correlations between hemispheres, and possible differences in efficiency between the u, d, c, s quarks, modify equation (2) in such a way that

$$f_2 = \epsilon_b^2 (1 + c_b) R_b + \epsilon_l^2 (1 + c_l)(1 - R_b) \tag{2'}$$

The coefficients c_b and c_l have to be determined by simulation:

$$c_j = \frac{f_{j,2}}{f_{j,1}^2} - 1 \tag{3}$$

where $j = b$ or l.

The purity π of a selected sample of b-tagged events (i.e., of events in which both hemispheres are tagged as b) can be obtained from ϵ_b and ϵ_l through

$$\pi = \frac{\epsilon_b^2(1+c_b)R_b}{\epsilon_b^2(1+c_b)R_b + \epsilon_l^2(1+c_l)(1-R_b)} . \tag{4}$$

4 Variables Used for the Classification

Among many variables which have been considered for feeding the network, we have chosen ten variables on the basis of their discriminating power according to the F test. Impact parameter has been excluded because our intention is to combine the forthcoming results with the one obtained Ref [4] using impact parameter variables.

This procedure resulted in the following list of input variables:

1. The sphericity S^β calculated after a boost $\beta = 0.96$ along the thrust axis.
2. The aplanarity A^β calculated after a boost $\beta = 0.96$ along the thrust axis.
3. The sum of the momenta parallel to the event axis, after the boost.
4. The sum of the products of the momentum components parallel to the event axis, times the perpendicular components.
5. The sum of the p_T^2.
6. The longitudinal momentum of the most energetic particle.
7. The invariant mass M_{1234} of the four most energetic particles.
8. The directed sphericity S_{1234} of the four most energetic particles. For a set Q of tracks in a jet, this variable is defined as

   $$S_Q = \frac{\sum_Q p_t^2}{\sum_Q p^2}$$

   where the p's are the momenta in the rest frame of the set Q and the p_t's are their components perpendicular to the thrust direction in the laboratory frame.
9. The directed sphericity S_{123}.
10. The directed sphericity S_{124}.

The values of their F-test from the simulation, where

$$F = \frac{|<x_b> - <x_{udsc}>|}{\sqrt{\sigma_b^2 + \sigma_{udsc}^2}} \tag{5}$$

are listed in Table 1.

5 The Neural Network

We adopted a feed-forward Neural Network architecture trained with Backpropagation. Both the architecture and the learning procedure have become by now quite well known, so we

Variable	F
2	0.259
1	0.236
8	0.215
6	0.209
10	0.207
9	0.128
5	0.071
4	0.061
7	0.051
3	0.013

Table 1: ***Value of the F-test for the variables used as an input to the multidimensional classifiers.***

refer to the literature for details (see [14]), just giving here a brief sketch of them and a list of the choices made for the parameters.

The Neural Network we used has 10 input variables, strictly coming from the shape of the event.

Each node in the input layer was associated with one of the input variables listed above. There were 2 hidden layers of 10 and 8 nodes and one output node. Each unit gives as output a sigmoid function of the weighted sum of its inputs.

The training procedure performs a gradient descent in the space of the weights with respect to a quadratic cost function quantifying the discrepancy between the value obtained at the output for each training event and the one conventionally fixed as the target value for the corresponding class (in our case 1 for b and -1 for udsc). The process is controlled by the "learning strength parameter" η and the "momentum" α [14]. Each updating step in the space of weights, computed by gradient descent, is multiplied by η and added to the previous step, multiplied by α.

The weights were updated every 10 events, chosen at random from the two classes $u\bar{u}+d\bar{d}+s\bar{s}$ $+c\bar{c}$ and $b\bar{b}$, in such a way that, on average, there was an equal number of events from each of the two classes.

Changing the parameters η and α during the training is convenient in order to allow for a fast movement in the space of weights in the early stage of training, and to obtain a controlled approach to the minimum at the later stage. For this reason, the learning and momentum parameters were decreased and increased respectively after every 3,000 updates (an "epoch") according to the rule:

$$\eta_t = \eta_{t-1} \times (\eta_{min}/\eta_{t-1})^{k_\eta}$$

$$\alpha_t = \alpha_{t-1} \times (\alpha_{max}/\alpha_{t-1})^{k_\alpha},$$

where η_{min} and α_{max} are the minimum (maximum) allowed values for the parameters, and subscript t $(t-1)$ refers to the epoch number. Exponents k_η and k_α were set to 0.05 and 0.14, respectively. Given the finite value of the weight change, the gradient descent might lead to an occasional increase of the error value. In this case, the parameters were reset to their initial values.

The architecture of the network is summarized in Table 2, together with the parameters used in the training phase.

Nodes in the hidden layers	10,8
α (training)	0.4 - 0.9
η (training)	0.05 - 0.0001

Table 2: *Characteristics of the NN.*

After training, the network was exposed to a new 'test' set of events; the behaviour of the error function with respect to this new set shows the error the network makes in generalizing to new data.

It is relevant to monitor the generalization error for trainings of increasing lenghts; when the generalization error starts to increase while the training error is still decreasing, the NN has 'overlearned' the training sample and its ability to generalize is degraded.

The system was trained on a set of 20,000 simulated events, generated with the JETSET Parton Shower Monte Carlo model (JETSET PS), where each event is described using variables related to a single hemisphere. By monitoring the behaviour of the generalization error, the training was stopped after 400,000 updates. After the network has been trained, its performance can be assessed in terms of signal efficiency ε_s (number of events correctly classified as belonging to a given class over the total number of events in that class), and purity p (number of correctly classified events in a given class over all the events classified as belonging to that class). The test sample consisted of about 50,000 simulated events, generated by using JETSET PS and 50,000 real data taken during 1991 using the DELPHI detector.

The distribution of the output of the network is plotted in Figure 1(a) for b and $udsc$ events. In Figure 1(b), the distribution of the network output on single hemispheres is shown for simulated and real data. In Figure 1(c), the fraction of events for which the network output is larger than a given cut in both hemispheres is plotted for real and simulated data.

The calculation of the coefficients defined in Eq. (3) gave the results plotted in Figure 2. Each event is classified testing the Network separately on both the hemispheres and considering an event as $b\bar{b}$ only if it is classified as b on both the hemispheres. The number of events generated in each class corresponded to the Z^0 hadronic branching fractions in the Standard Model.

The curve giving purity of the selected sample versus efficiency for the selection of b quark pairs is given in Figure 3. The shape plotted as black markers refers to real data, while the solid line is for the simulation. Purity and Efficiency are computed using the solutions of the equations (1) and (2') described in Section 3. The error bars are computed allowing a 100% variation of the c_j.

6 Summary and Conclusions

Using the independent information coming from suitably defined emispheres in jet events proves to be a viable strategy to obtain a multidimensional tagging method based on Neural Networks, which is less sensitive to Monte Carlo assumptions than previously achievable. Further analysis is required to better quantify the influence of the correlation coefficients on the classification in terms of efficiency and purity; this work is presently in progress. As one of the developments planned for the near future, we also expect to obtain better performances by the combined use of this method together with an independent, single particle tagging based on impact parameter ([4]).

Acknowledgements

We are greatly indebted to our technical collaborators and to the funding agencies for their support in building and operating the DELPHI detector, and to the members of the CERN-SL Division for the excellent performance of the LEP collider. Part of this work has been supported by INFN through the ANNETTHE initiative.

References

[1] W. Braunschweig et al., *Zeit. Phys.* **C44** (1989) 1, P. Abreau et al., *Phys. Lett.* **B281** (1992) 383

[2] D. Decamp et al., *Phys. Lett.* **B259** (1990) 236, P. Abreau et al., *Zeit. Phys.* **C56** (1992) 47, B. Adeva et al., *Phys. Lett.* **B261** (1991) 177, P.D. Acton et al., *Zeit. Phys.* **C55** (1992) 191

[3] The ALEPH Collaboration, CERN-PPE/93-108, June 1993, T. Baroncelli et al, DELPHI Note 93-93

[4] T. Baroncelli et al, DELPHI Note 93-93

[5] The ALEPH Collaboration, CERN-PPE/93-113, July 1993

[6] "New computing techniques in physics research II", D. Perret-Gallix ed., World Scientific 1992

[7] "Neural networks: from biology to high energy physics", supplementary issue of the *International Journal of Neural Systems*, World Scientific 1993

[8] "Computing in high energy physics", CERN 92-07, Geneva 1992

[9] P. Aarnio et al. (DELPHI), *Nucl. Instr. and Meth.* **A303** (1991) 233.

[10] P. Aarnio et al. (DELPHI), *Phys. Lett.* **B240** (1990) 271.

[11] *DELSIM User's Guide*, DELPHI 89-15 PROG 130, CERN, February 1989. *DELSIM Reference Manual*, DELPHI 89-68 PROG 143, CERN, September 1989.

[12] T. Sjöstrand, Comp. Phys. Comm. **27** (1982) 243, ibid. **28** (1983) 229; T. Sjöstrand, *PYTHIA 5.6 and JETSET 7.3*, CERN-TH.6488/92, September 1992.

[13] W. de Boer et al., IEKP-KA/91-07, Karlsruhe, June 1991.

[14] J. Hertz, A. Krogh, R.G. Palmer, "Introduction to the theory of neural computation" Addison-Wesley 1991

[15] Fisher.

APPLICATION OF NEURAL NETWORKS IN THE ARGUS EXPERIMENT FOR THE ANALYSIS OF B- AND TAU-PHYSICS*

M. JOSWIG, H. KOLANOSKI, S. KUBLUN, H. THURN, D. WEGENER and S. WESTERHOFF
Institut für Physik, Universität Dortmund,
D-44221 Dortmund, Germany

(Presented by H. KOLANOSKI)

ABSTRACT

We present analyses of e^+e^- annihilation data taken with the ARGUS detector in the energy range of the Υ resonances. The selection of $\Upsilon(4s)$ decays and leptonic τ decays was studied with feed-forward multilayer networks using the backpropagation algorithm for the training. The correlated decay of $\tau^+\tau^-$ pairs was analysed with a feed-forward network to determine a weak coupling constant.

1. Introduction

Until the beginning of this year the ARGUS experiment[1] took data at the e^+e^- storage ring DORIS II in the energy range of the Υ resonances ($E_{e^+e^-} \approx 10$ GeV). The most important physics contributions were made to the understanding of the 3rd fermion generation studying b quark and τ lepton decays.

In this paper we present the application of neural network algorithms for the selection of B decays[2] and τ events[3] and for the determination of an electroweak parameter in τ decays[4]. The studies reported here used feed-forward nets with at most 2 hidden layers and the backpropagation algorithm as implemented in the JETNET program[5]. The training was done mainly with simulated events and rather standard settings for the learning parameters (step size, inertia term, updating frequency).

2. Selection of $\Upsilon(4s)$ Events

The properties of the b-quarks have been extensively studied by analysing the decays of B mesons[6]. At e^+e^- colliders associated $B\overline{B}$ production is observed in decays of the $\Upsilon(4s)$ resonance:

$$e^+e^- \to \Upsilon(4s) \to B\overline{B}.$$

In competition to the resonance production is the production of a continuum of light $q\overline{q}$ pairs. At the $\Upsilon(4s)$ the signal of $B\overline{B}$ events to the background of continuum events is about 1:4. We have studied the separation of the resonance decays from

*This work was supported by the German Bundesministerium für Forschung und Technologie under contract number 056DO57P.

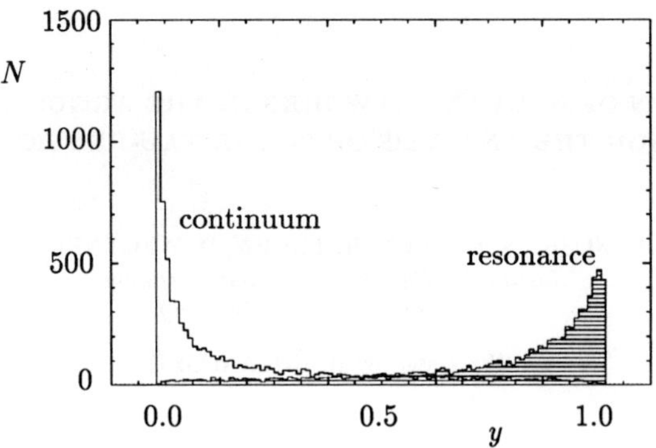

Figure 1: The two output distributions for simulated continuum and resonance events.

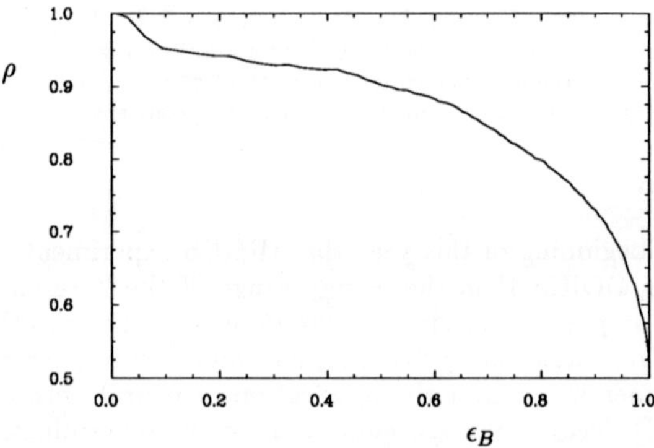

Figure 2: Purity versus selection efficiency for $B\overline{B}$ events. The 'purity' is defined as the fraction of real $B\overline{B}$ events in the sample.

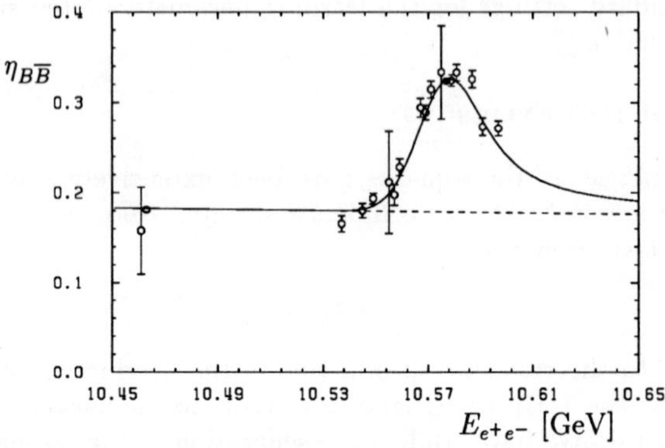

Figure 3: The fraction of events classified as $B\overline{B}$ events for an energy scan in the region of the $\Upsilon(4s)$ resonance. The full curve is a fit to the resonance plus the continuum (dashed line: continuum only).

the continuum using a feed-forward neural network together with backpropagation training.

The most characteristic difference between both event classes is found in the event topology: Since the $\Upsilon(4s)$ resonance is just above the $B\overline{B}$ threshold the B mesons are produced nearly at rest and therefore decay isotropically. The continuum events, on the other hand, have dominantly back-to-back two-jet topologies. Therefore topological quantities, like sphericity or Fox-Wolfram moments, are important input variables for the net. Other used variables exploit the somewhat different multiplicities (higher in resonance decays), the occurrence of leptons and kaons in the weak decays of B mesons and kinematical differences (e.g. the maximal momentum of a particle in resonance decays is restricted to half the B mass. Altogether 20 characteristic variables were chosen as network inputs, normalized such that each variable approximately had values in an unit interval.

The evaluation of the results for various network configurations led finally to a 20-20-1 configuration. This net was trained with simulated events to the target output values $y = 1$ for the resonance reaction and $y = 0$ for the continuum. After about 300 learning epoches with about 30000 events typical output distributions as in Fig.1 were obtained. Varying the y cut the efficiency for $B\overline{B}$ selection could be tuned as a function of the remaining continuum background (Fig.2). A cut at $y = 0.5$ yielded a selection efficiency of $(84.6 \pm 0.5)\%$ and a background rejection of $(80.0 \pm 0.6)\%$. This has been applied to experimental data from a scan over the resonance. A signal-to-background ratio of about 1:1 at the resonance peak could be reached as shown in Fig.3.

The peak cross section obtained with the neural network is $\sigma = (0.87 \pm 0.01)\ nb$. With the standard procedure of fitting the resonance on top of the full (unsuppressed) background $\sigma = (0.86 \pm 0.03)\ nb$ was obtained which has a substantially larger error than with the network selection. However, the background suppression introduces a larger dependence on the simulation leading to an additional systematic error of about 3%.

3. Selection of Leptonic τ Decays

This analysis aimed at the determination of the product of the leptonic branching ratios of τ leptons, $B_e \cdot B_\mu$, studying the τ production and decay process

$$e^+e^- \rightarrow \tau^+\tau^- \rightarrow \mu^+\nu_\mu\overline{\nu}_\tau\, e^-\overline{\nu}_e\nu_\tau.$$

The network had to recognize

- 2 leptons in the event,
- missing energy and momentum due to the four undetected neutrinos

and to distinguish the wanted reaction from background with similar signatures. The most difficult background came from other τ decays and two-photon reactions.

The final analysis was done using a 15-15-1 feed-forward net. The 15 input variables were the momentum components of the two tracks, associated particle

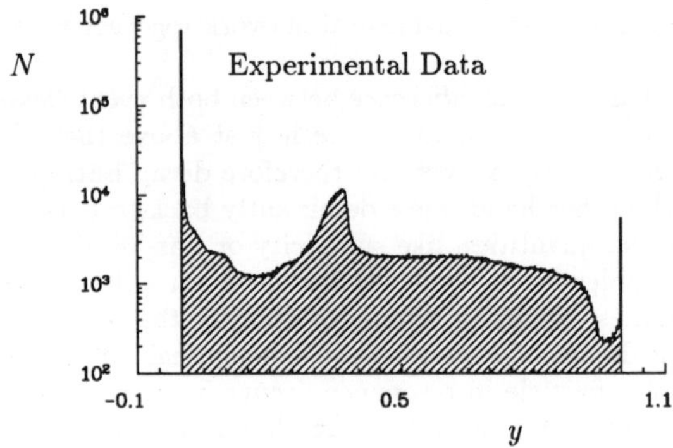

Figure 4: Network output for experimental data (leptonic τ decay candidates).

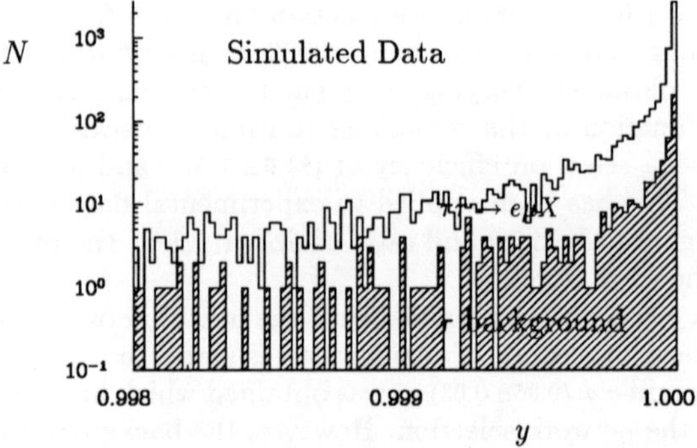

Figure 5: Network output near $y = 1$ for simulated leptonic τ decays (open histogram) and simulated background from other τ decays (shaded histogram).

identification quantities derived from shower energies, muon chamber hits, specific energy loss and time-of-flight measurements, and in addition kinematical quantities for the whole event, like missing mass and the direction of the event momentum. The necessary number and the optimal choice of input variables was determined from studies of different network configurations. These studies also showed a major advantage if the variables of each lepton species were assigned to specific input nodes.

The net was trained using 5000 simulated events composed of the studied reaction and the background in approximately the expected composition. After about 500 learning epoches a selection efficiency above 90% for the leptonic τ decays was reached. However, the network output distribution for the experimental data in Fig.4 shows no clear separation between signal and background. The bumpy structure of the distribution was found to be due to the different background contributions, e.g. the bump near $y = 0.4$ is due to the two-photon reaction.

It turns out that signal and background cannot be completely separated, which was also found in conventional analyses of the same reaction[7]. Figure 5 shows that even near $y = 1$ the background from other τ decays persists. Varying the cut on the output variable y a reasonable signal-to-background ratio was found for $y_{cut} = 0.985$. With this cut the selection efficiency was

$$\epsilon_{sel} = 0.247 \pm 0.003$$

for a remaining background of 9%. The final result was

$$B_e \cdot B_\mu = 0.0295 \pm 0.0007 \pm 0.0015.$$

The two errors are statistical and systematical, respectively. The systematic error is estimated from the variation of $B_e \cdot B_\mu$ when changing the y-cut and using different network input variables.

In a conventional analysis[7] using one- and two-dimensional selection cuts an efficiency $\epsilon_{sel} = 0.147 \pm 0.003$ was reached for a remaining background of only 2%. Since the background is in this case much smaller the two efficiency numbers are not quite comparable. However, it seems as if both methods could yield similar signal-to-background ratios. Nevertheless, one may see advantages in using a neural network:

- The cuts in the conventional analysis have been fine-tuned in a tedious and time-consuming way. Network training offers a more automatic optimisation of cuts (and takes into account multidimensional correlations).
- The evaluation of inconsistencies between Monte Carlo simulations and the data can be made more systematically by varying the y-cut.

The systematic uncertainties due to incorrect simulations affect both the conventional and the network approach. It should be emphasized that a net training with an incorrectly simulated data set does not necessarily lead to a wrong data analysis as long as the efficiency corrections are done for this specific (may be not optimal) network correctly.

4. Determination of Electro-Weak Coupling Parameters from $\tau^+\tau^-$ Decay Correlations

In this work we investigated possibilities of extracting parameters of statistical distributions by the neural network approach. In contrast to the above discussed examples this is not a classification (yes-no) problem but the aim is to calculate a continuous number. In case of a linear dependence of the distribution $p(\vec{x})$ on the parameters λ_i,

$$p(\vec{x}|\vec{\lambda}) = \sum_i \lambda_i p_i(\vec{x}),$$

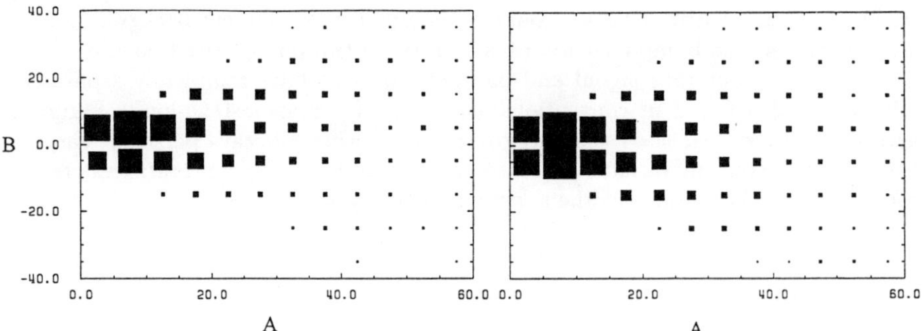

Figure 6: Distribution of events as a function of the A and B terms simulated for γ_{AV}^2=1 (left) and γ_{AV}^2=0 (right).

the distribution of the net output can be linearly composed of the output distributions for the different $p_i(\vec{x})$. Thus the λ_i can be determined from a fit of simulated output distributions to the output distribution of the data. Using neural networks compared to, e.g., a likelihood analysis has the advantage that the $p_i(\vec{x})$ have not to be explicitely known but the net can be trained with simulated data.

This method has been applied to the determination of the vector-axialvector interference in correlated hadronic decays of τ pairs. The angular distribution of the final state in

$$e^+e^- \rightarrow \tau^+\tau^- \rightarrow \pi^+\pi^0\overline{\nu}_\tau\pi^-\pi^0\nu_\tau$$

can be described by[8]

$$p(\vec{\alpha}|\gamma_{AV}^2) = A(\vec{\alpha}) + \gamma_{AV}^2 B(\vec{\alpha})$$

where $\vec{\alpha}$ is a set of 11 kinematical variables and γ_{AV} is the normalized product of the vector and axialvector charged current couplings. The B term can be positive and negative which generates according to the equation above a higher event density for positive B and a lower for negative B (Fig.6).

The distribution for any γ_{AV}^2 can be obtained as a linear combination of the two distributions with $\gamma_{AV}^2 = 0$ and 1:

$$p(\vec{\alpha}|\gamma_{AV}^2) = \gamma_{AV}^2\, p(\vec{\alpha}|1) + (1-\gamma_{AV}^2)\, p(\vec{\alpha}|0) = \gamma_{AV}^2 (A+B) + (1-\gamma_{AV}^2)A$$

The idea is now to train a feed-forward net to distinguish the distributions $p(\vec{\alpha}|1)$ and $p(\vec{\alpha}|0)$. This can be achieved by requiring the target output (1 output node) to be $y = 1$ and $y = 0$ for $p(\vec{\alpha}|1)$ and $p(\vec{\alpha}|0)$, respectively. After such a net training output distributions for $p(\vec{\alpha}|1)$ and $p(\vec{\alpha}|0)$ are obtained as shown in Fig.7a. The two distributions are not very distinguished due to the relatively small asymmetry observed in Fig.6.

There are two additional experimental complications to be taken into account: Because of the escaping neutrinos the τ direction can only be reconstructed

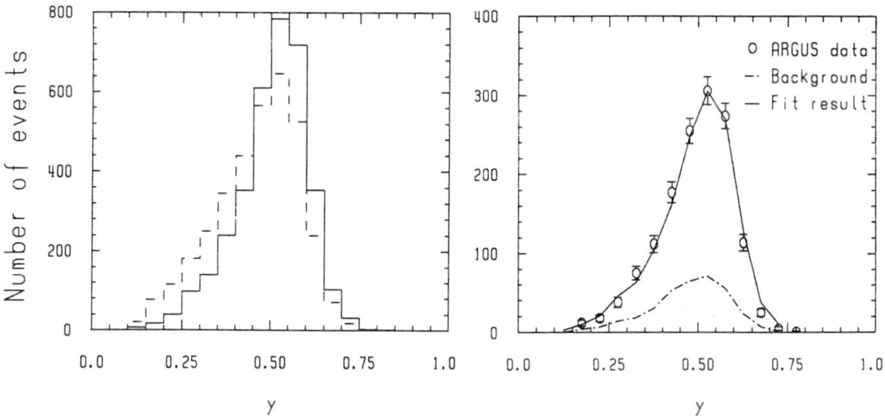

Figure 7: a) Network output distribution for simulated events with γ_{AV}^2=1 (full line) and γ_{AV}^2=0 (dashed line).
b) Network output distribution for the 1412 ARGUS events (points with error bars) and the background. The curve is the result of the fit described in the text.

up to a two-fold ambiguity (and even that is only true if one neglects radiative corrections). This yields for each event two sets of kinematical variables and hence also two (A, B) sets. We have finally used a 4-10-4-1 net with A, B, A', B' as the four inputs (Fig.7a was obtained with this net). The second complication is a quite substantial background of about 25%, mainly from other τ channels. This background could be simulated and was also fed through the net to generate its output distribution.

Figure 7b shows the network output distribution for the selected 1412 events together with the simulated, properly normalized background. The data distribution was now fit by the two reference distributions in Fig.7a and the background distribution. The latter was held fixed in the fit. The only free parameter, $|\gamma_{AV}|$, was determined to be:

$$|\gamma_{AV}| \ = \ 0.96 \pm 0.05 \pm 0.04 \ \ \text{(prelim.)}$$

The errors are statistical and systematical, respectively. The systematical error is mainly due to the uncertainty in the background determination. No systematic effects specific to the use of neural networks have been identified.

The network result can be compared to a likelihood analysis using the complete (but uncorrected) matrix element. The preliminary result from this analysis is $|\gamma_{AV}| = 1.00 \pm 0.03 \pm 0.03$. With the likelihood method the detector and radiative effects cannot easily be included and can only be corrected for by detailed simulation studies. In this respect the network approach is advantageous since all simulated effects are naturally included by the training. It is under investigation if the statistical error in the network result can be decreased further, e.g. by improving the statistics of the distributions used in the fit and/or by using the kinematical variables explicitely rather than the A and B terms.

5. Summary and Conclusion

We have successfully applied feed-forward networks to select $\Upsilon(4s)$ events and leptonic τ decays. The advantage of using neural networks is most obvious for the selection of $\Upsilon(4s)$ events. These events differ in each of the chosen variables (multiplicities, topology, ...) often only by a small amount and therefore a symbiotic view of all the variables is particularly important. On the other side, in the selection of leptonic τ decays the particle identification plays the dominating role and the decision seems to be to less an extend multi-dimensional. In this case one may consider to put more basic information on particle identification onto the net rather than preprocessed classifiers.

The third example shows that neural networks can be used to determine linear coefficients of distributions. In this method the necessity of fitting the output distributions has little 'neural appeal'. More studies are necessary to search for networks which could provide more directly the solutions and maybe also the generalisation to non-linear parameters.

In the studied applications the neural networks have provided satisfactory problem solutions, sometimes better than what could be reached by other methods. The identified advantages of the neural analysis approach are:

- quasi-automatic learning in a multi-dimensional space;
- natural use of simulated or measured data: the distributions have not to be explicitely known;
- the signal-to-background ratio can be smoothly adjusted by varying the cut on the network output;
- this cut on the output allows also to check systematically differences between simulation and experimental data;
- by the intrinsic parallelism of neural algorithms neural networks are optimal for use on the trigger level.

A final remark: from our experience neural network classification is not less transparent than any other multi-dimensional analysis technique. There are methods to analyse the network behaviour and to visualize what a network is doing. One can for example find out what the relevance of the different input variables is. In most cases one finds no surprises: a good physicist would choose the same variables.

References

1. ARGUS Coll., H. Albrecht et al., Nucl. Inst. Meth. A275 (1989) 1.
2. M. Joswig, diploma thesis, University of Dortmund, 1993.

3. S. Westerhoff, diploma thesis, University of Dortmund, 1993.
4. S. Kublun, diploma thesis, University of Dortmund, 1993.
5. L. Lönnblad, C. Peterson and T. Rögnvaldsson, *'Pattern recognition in high-energy physics with artificial neural networks: JETNET 2.0'*, Comput. Phys. Commun. 70 (1992) 167.
6. H. Schröder, *'Physics of the B Mesons'*, Rep. Prog. Phys. 52 (1989) 765.
7. ARGUS Coll., H. Albrecht et al., Z. Phys. C53 (1992) 367.
8. H. Thurn and H. Kolanoski, submitted to Z. Phys. C, (DESY 93-056), 1993.

TAGGING HEAVY MAJORANA NEUTRINOS AT LEP200 USING A FEED-FORWARD NETWORK WITH CONSTRAINT FIT INPUTS

J. DAHM, C. VOIGT
Physics Department, University of Wuppertal
D-42097 Wuppertal, Germany

and
K.-H. BECKS, J. DREES
Physics Department, University of Wuppertal, Wuppertal, Germany

ABSTRACT

The neural network (NN) technique was used to find heavy neutrino events in a large WW-background at LEP200. Because of a small production cross section in the signal and the fact that the ratio background-signal is at least 26, high efficiencies *and* high purities are needed. We used a feed-forward network with the common back-propagation (BP) algorithm. The set of input variables contains besides some simple quantities the results of constraint fits for signal and background. The neural network approach compared with conventional methods of cutting, using the same set of variables, has shown a much better tagging performance for the heavy neutrino class.

1. Introduction

This work deals with the problem of extraction of heavy neutrino (HN) events at LEP200. The existence of such kind of HNs would imply nonvanishing rest masses of light neutrinos as well. Constraints from low energy measurements imply HN masses heavier than the Z^0 mass. This forbids the detection at LEP 1 energies [1]. HNs can only be identified using their decay products whereas the simultanously produced light neutrino escapes undetected. The same final state configuration occurs in semileptonic WW-events, and in the so called single W-production respectively. The processes are:

$$\begin{array}{lccl} e^+e^- \to & \nu HN \to \nu l W & \to \nu l \; jets \\ e^+e^- \to & WW & \to \nu l \; jets \\ e^+e^- \to & \nu l W & \to \nu l \; jets \end{array}$$

Due to detector specifications, the clearest signal should arise from the decay into a high energetic muon and a W-boson. Because of this one can drop the third channel as there are no isolated high energetic muons in the final state. Thus the extraction turns out to be a typical two-class problem. Within the classification there are two

major difficulties:
First, small production cross sections demand high efficiencies and secondly high purities are needed to answer the question of HN existence on a sufficiently high confidence level.
The HN tagging problem turned out to be a hard one since no strategy just based on cuts on some variables characterizing the event seems viable. It seems necessary to exploit a full multidimensional data structure and multivariable-correlations are likely to play a major role. NNs have often proven capable to successfully cope with this kind of problem.

2. The Network Input

2.1 Data Generation

We took PYTHIA 5.6 out of the LUND Monte-Carlo package [2] for the production of the WW-events and our own Monte-Carlo generator for the HNs [3]. We assumed the largest possible mixing of 0.1 [1] and a center of mass energy of $200GeV$. The detector was simulated with the FASTSIM simulation package [4].

2.2 The Variables

It is nessecary to reconstruct both the WW and the HN using constraint fits [5]. Besides energy and momentum conservation the HN mass and the W mass were given as constraints. Because of this, the set of input variables consists of directly measured and fitted quantities. The invariant rest mass of the HN was rescaled according to standard rescaling procedures [6]. Notice the fact that signal and background were fitted under the assumption to be a HN and a WW event respectively.
The input variables are:

- χ^2_{HN} : Resulting χ^2 value for the HN hypothesis out of the constraint fit
- χ^2_{WW} : Resulting χ^2 value for the WW hypothesis out of the constraint fit
- E_μ : '*Measured*' muon energy
- θ^f_μ, ϕ^f_μ : *Fittet* μ angles with respect to the positive z-axis
- $\theta^f_{HN}, \phi^f_{HN}$: *Fittet* HN angles
- θ^f_W, ϕ^f_W : *Fittet* W angles
- θ_{HN}, ϕ_{HN} : Reconstructed HN angles by four-momentum conservation
- θ_W, ϕ_W : Reconstructed W angles (LUCLUS [2])
- E_{j_1} : Energy of the most energetic jet (jet-finding by LUCLUS)
- E_{j_2} : Energy of the second jet
- $\alpha_{\mu W}$: Angle between μ and reconstructed W

- m_{HN} : Rescaled mass of reconstructed HN

Because of the unknown HN mass it was necessary to produce a train-/test set for six different mass hypotheses. Table 1 shows the development of the signal-background ratio as the HN production cross section drops with increasing HN mass. The validity of the sketched procedure is illustrated in figure 1. It shows that

Table 1: Background/signal ratio for the six investigated mass hypothesis

HN mass [GeV]	ratio
100	25.7
110	31.3
120	37.6
130	43.0
140	48.9
150	63.4

the constraint fit provides minimal χ^2 values in case of the correct mass assumption whereby it is possible to fix the HN mass in advance.

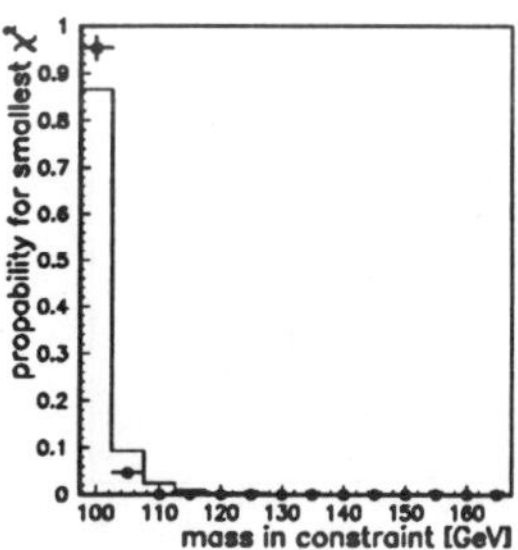

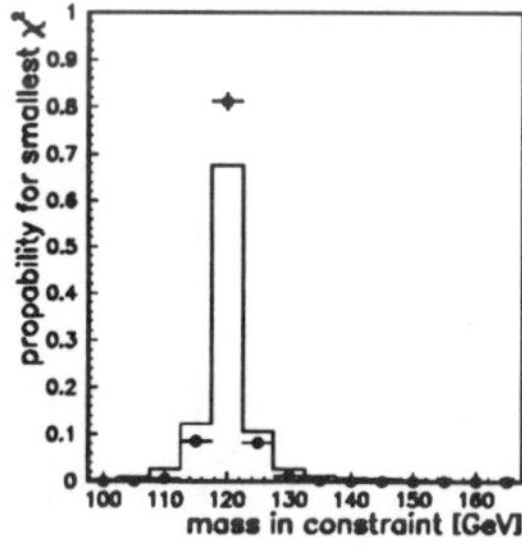

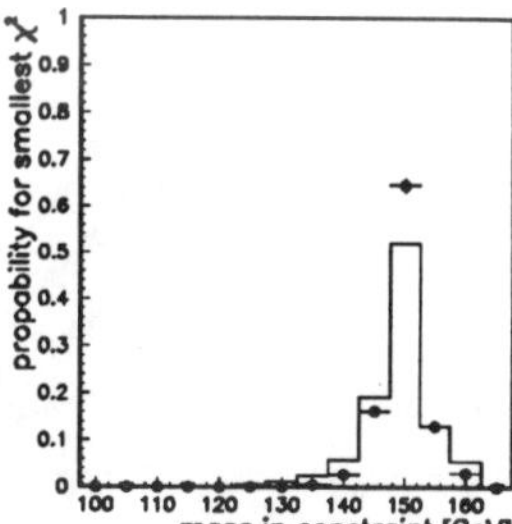

Figure 1: Mass plot, solid line: no cuts, dots: $\chi^2_{HN} \leq 3$. L.h.s.: generated $m_{HN} = 100GeV$, middle: generated $m_{HN} = 120GeV$, r.h.s.: generated $m_{HN} = 150GeV$.

3. The Network

We used a three layered standard feed-forward NN with 17 input nodes, 10 hidden nodes and 1 Output node to feed the net with the whole set of inputvariables. The output was -1 for WW-events and +1 for HN-events.

4. Training Procedure

For each mass hypothesis we prepared a training set of 3500 HN-events and

3500 WW-events. The training was done using the standard back-propagation algorithm [7] with a learning rate $\eta = 0.003$, an additional momentum term $\alpha = 0.7$ and a temperature $T = 1.0$ in the sigmoid activation function. The weights were initialized in the range of $[-0.1, 0.1]$. We trained 300 sweeps and updating was done after presentation the net the whole trainset once. Training was interrupted after 30 sweeps and the NN was testet with an independent test set consisting of 3000 HN-events and 3000 WW-events.

5. Results

First of all we tried a simple strategy based on cuts on some variable and non linear cuts in two dimensional subspaces (Simple-Cut-Method, SCM) for the six assumpted masses. For a given set of cuts we calculated the efficiency(ϵ) and

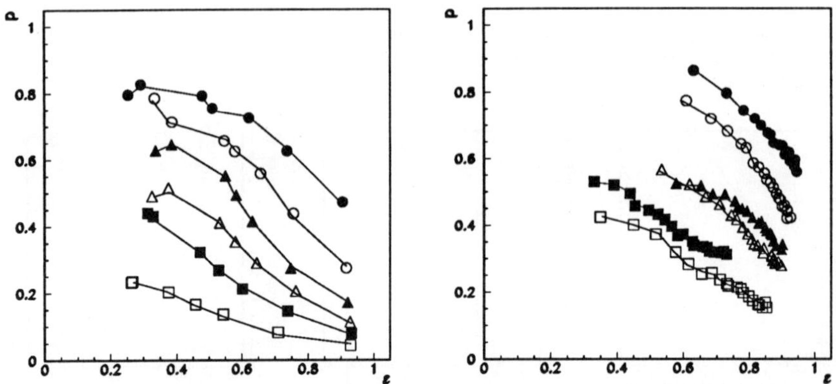

Figure 2: Purity versus efficiency for six mass hypothesis; l.h.s.: simple-cut-method; r.h.s.: neural network. ●: $100GeV$, ○: $110GeV$, ▲: $120GeV$, △: $130GeV$, ■: $140GeV$, □: $150GeV$.

purity(p). They are defined as follows:

$$\epsilon = \frac{\text{number of correct signal events}}{\text{number of all signal events}}$$

$$p = \frac{\text{number of correct signal events}}{\text{number of all patterns classified as signal}}$$

In general and especially for higher HN masses the performance of this method was rather poor (see figure 2 left side). Both, this fact and the problem's multidimensional character persuaded us to attempt the neural network approach. After training the network we calculated the performance quantities mentioned above for

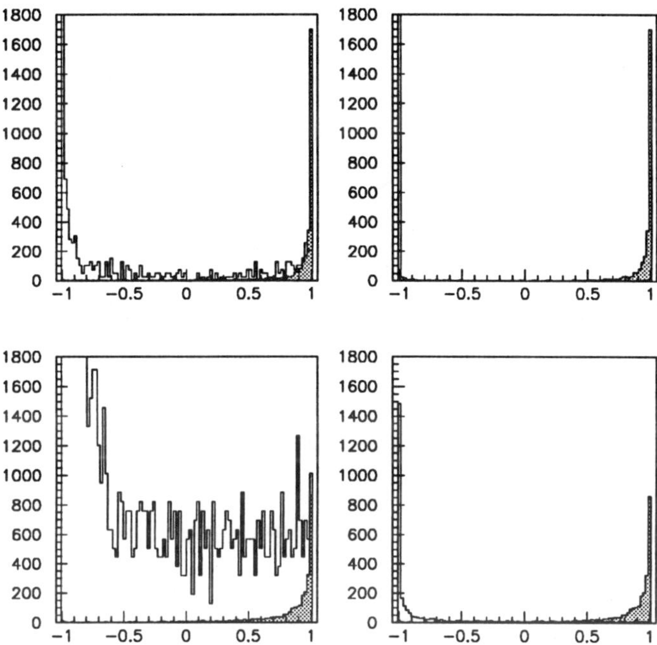

Figure 3: Networkoutput distribution; l.h.s.: real background-signal ratio; r.h.s.: 1:1 ratio; top: $m_{HN} = 100GeV$; bottom: $m_{HN} = 150GeV$.

different cuts in the output space of the test sample. Figure 2 shows our results for the six different mass hypothesis. It is an important fact as one can see in figure 3 that the convincing network performance is not responsible for the relatively low tagging ability but it is the background-signal ratio itself. Compared with our first applied method the average gain in tagging ability is approximately 15% (see figure 4, table 2). To calculate the cross sections with a given purity it is necessary to take into account the efficiency of the constraint fit as well. Thus the expected cross section becomes,

$$\sigma_{exp} = \epsilon_{meth} \cdot \epsilon_{fit} \cdot \sigma_{theo}$$

where ϵ_{meth} denotes the efficiency of the method (NN or SCM) and ϵ_{fit} the fit efficiency.

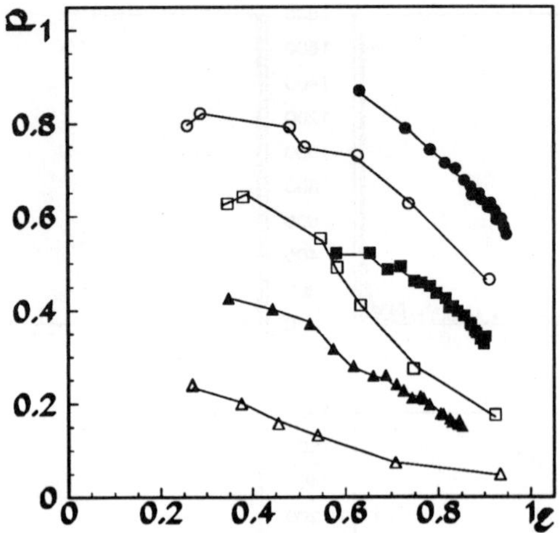

Figure 4: Purity versus efficiency. Simple-cut-method: ○: 100GeV, □: 120GeV, △: 150GeV. NN-method: ●: 100GeV, ■: 120GeV, ▲: 150GeV.

Table 2: Efficiencies and purities for different mass hypothesis

mass[GeV]	p	ϵ^{SCM}	ϵ^{NN}
100	70%	65%	84%
120	50%	57%	73%
150	24%	26%	71%

6. Summary and Conclusion

We have shown that it is useful to apply the NN approach in this case of new physics where no strategy based on cuts on some variables is available. We reached a good seperation of the two classes in a relatively short time. The next problem is the extraction of HNs out of the overwhelming large Z^0/γ-background. For this purpose a FULLSIM Monte-Carlo simulation is necessary as secondary vertices (not implemented in FASTSIM) play an important role. Such kind of tagging problems investigated here may appear quite frequently at LEP200 e.g. in the Higgs search. In our opinion it should be an interesting idea to use Self-Organizing Networks to extract kinematical features and to study correlations of the investigated variables.

7. References

1. W. Buchmüller, C. Greub, *Phys. Lett.* **B256** (1991) 465.
2. T. Sjöstrand, *Comp. Phys. Comm.* **39** (1986) 347.
   T. Sjöstrand, M. Bengtsson, *Comp. Phys. Comm.* **43** (1987) 367.
   H.-U. Bengtsson, T. Sjöstrand, *Comp. Phys. Comm.* **46** (1987) 43.
3. C. Voigt, to be published as Diploma thesis (Wuppertal 1993)
4. J. Cuevas et. al., *Fast Simulation for DELPHI* (DELPHI 87-26 Prog-71).
5. N. J. Kjaer, R. Moeller, *Reconstruction of invariant masses in multi-jet events* (DELPHI 91-17 PHYS 88).
6. ECFA Workshop on LEP 200, *Proceedings Vol.1* (CERN 87-08 49).
7. D. E. Rumelhart, G. E. Hinton, R. J. Williams, *Learning Internal Representation By Error Back Propagation* (Parallel Distributed Processing 1, MA: MIT Press, Cambridge 1986).

FEATURE EXTRACTION USING SELF-ORGANIZING NETWORKS

J. DAHM, C. VOIGT
Physics Department, University of Wuppertal
D-42097 Wuppertal, Germany

and

K.-H. BECKS, J. DREES
Physics Department, University of Wuppertal, Wuppertal, Germany

ABSTRACT

Self-Organizing networks are used to extract kinematical features and to study correlations of high dimensional variable spaces in new physics areas. This method is applicated in the search for heavy neutrinos at LEP200. We will show that the extracted knowledge improves the distinction between heavy neutrino candidates and background.

1. Introduction

Every new generation of particle accelerators or even any upgrade in energy of existing ones implies the possibility of many new physics processes. Thus it is necessary to develop new criteria typical for each one in particular. For that purpose one is forced to transform a high dimensional data structure down into a reduced representation where a strategy based on cuts on some variables is viable. Here Self-Organizing networks provide by means of a non-linear mapping the ability to simplify the occuring structure . We illustrate this method in the case of heavy neutrino (HN) search at LEP200 [1]. The classification turned out to be hard due to the fact that the neutrino is unstable and decays into a charged lepton and a W-boson which further decays hadronically into jets. The final state configuration occurs in semileptonic WW-events and in the so called single W-production, too. The small signal/background ratios demand high efficiencies and high purities. This requires very discriminant variables.

2. Description of the Method

2.1 Self-Organizing Networks

This neural network type is an unsupervised learning scheme, i. e. there is no feedback from the environment to control what the network output should be. The network has to discover patterns, features, regularities, or correlations in the input data for itself and codes them in the output.
The architecture is shown in figure 1. The input nodes x_k are fully connected with each node h_j in the outputlayer. For each output unit there is a weight vector $\vec{w}_j$

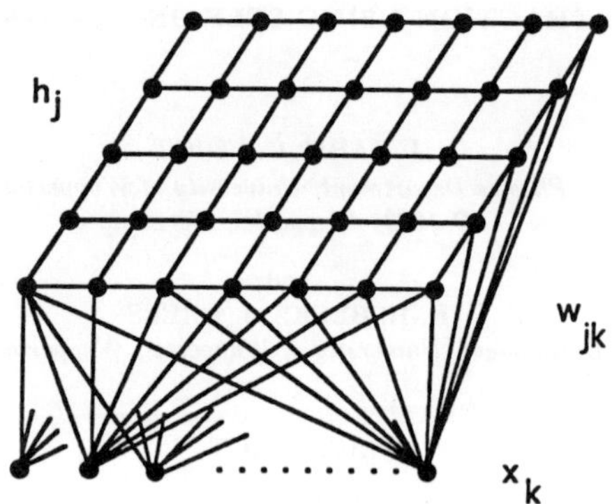

Figure 1: Architecture of a Self-Organizing network

which has the same dimension as the input vector. Thus every weight has a physical meaning.
The network operates by competitive learning. Here for each pattern presented to the network the responses of the feature nodes h_j are computed and a winner h_m is determined. Finding the winner amounts to finding the $\vec{w}_j$ which is closest in profile to the input vector $\vec{x}$. This weight vector is updated in such a way that the h_m will become more sensitive to the presented pattern. As a result each feature node represents a region in the input space.
If the output units are arranged in a geometrical way (collective self-organization) such as a two-dimensional array, similar input patterns give maximum response at nearby output nodes. Thus the network performs a mapping that preserves neighborhood relations. A more detailed description can be found in [2],[3].

2.2 Feature Extraction

As the dimension of the weight vectors is equal to that of the input vectors the result is extremely easy to analyse. Inspecting the weights of each output node gives the specific features of the assigned input patterns.
Here we want to extract features of two event-classes which have to be distinguished. At first the distributions of the classes in the feature map have to ensure the possibility of separation with respect to the input quantities. For the distinguishing purpose one has afterwards to inspect and compare weight vectors of typical nodes of the two classes. Typical in this sense means sensitive for one class but not for the other.
Due to the relationship of the feature nodes the network keeps some kind of "stiffness". Therefore in the common two-dimensional case the map is approximately a plane aligned in the directions of the two variables characterized by the greatest variances. In this case all further information of the remaining input variables is nearly suppressed. To overcome that problem these variables have to be removed

from the input space and can be used in general for the distinction. This procedure has to be repeated until the variances of the remaining input quantities become similar. At this stage one can start the real analysis.

3. The Network Input

3.1 Data Generation

We took the PYTHIA 5.6 package out of the LUND Monte-Carlo program [4] for the production of the WW-events and implemented our own Monte-Carlo generator for the HNs [5] as an external process for PYTHIA. Both processes were simulated including the effects of initial state radiation. We assumed the largest possible mixing of $\xi = 0.1$ [6], a center of mass energy of $200\,GeV$, and an invariant mass for the HN of $100\,GeV$. The detector was simulated with the FASTSIM simulation package [7].

3.2 The Variables

Part of our set of input variables was the result of a constraint fit [8] and a rescaling procedure [9] whereas the rest was taken directly from the FASTSIM output. In the following we denote them as fittet (superscript f) and 'measured' quantities.
The initial input variables are:

- χ^2_{HN} : Resulting χ^2 value for the HN hypothesis from the constraint fit
- χ^2_{WW} : Resulting χ^2 value for the WW hypothesis from the constraint fit
- E_l, E_l^f : Lepton energies
- E_W, E_W^f : Energies of the W-boson
- E_{HN} : Heavy neutrino energy
- $E_{j_1}, E_{j_1}^f$: Energies of the most energetic jet
- $E_{j_2}, E_{j_2}^f$: Energies of the second jet
- m_{HN} : Rescaled mass of reconstructed HN
- m_W : Rescaled mass of reconstructed W
- $M_{j_1}, M_{j_1}^f$: Invariant masses of the first jet
- $M_{j_2}, M_{j_2}^f$: Invariant masses of the second jet
- $\theta_{HN}, \theta_{HN}^f$: Polar angles of the HN
- θ_W, θ_W^f : Polar angles of the W
- θ_l, θ_l^f : Polar angles of the lepton
- α_{lW} : Angle between lepton and reconstructed W

4. Application to Heavy Neutrino Tagging

First of all we used a simple strategy based on cuts on the variables mentioned above and non linear cuts in two dimensional subspaces. For each cut configuration purities and efficiencies according to the real signal/background ratios [1] are given in figure 2.

A further exploitation of the kinematical information was achieved by using the

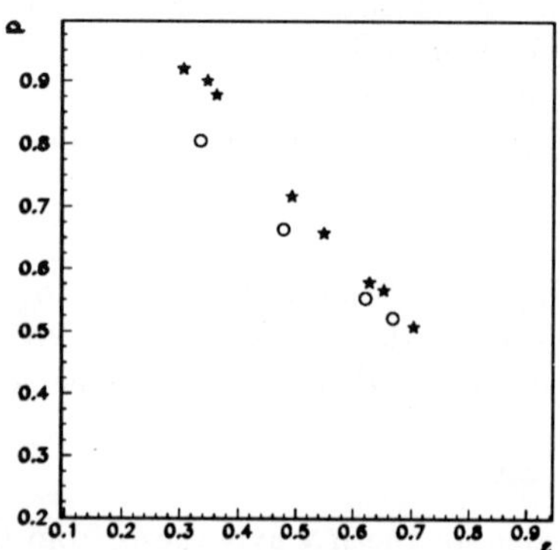

Figure 2: Purity p versus efficiency ϵ. Open circles: original cuts; stars: new developed cuts.

Self-Organizing network with a two-dimensional feature map. Besides the two χ^2-values from the constraint fit all other variables formed the inputspace. After the typical train and test procedure we inspected weight vectors of typical signal and background nodes. At first, all variables being very similar for both classes were removed. So the input space neither contains very discriminant nor useless quantities for the distinction.

Then we trained the network with this reduced input. Afterwards, pair correlations between measured and fittet variables were studied. Examplary features are shown in table 1. Furthermore we inspected the behavior of the weights along the direction perpendicular to the main separation line (see figure 3). Table 2 shows some weight vector components. In the last row the decrease and increase of the quantities is denoted as *behavior*. Due to that behavior the following discriminant variable was constructed:

$$dv = \frac{E_l + M_{HN} + E_W^f}{M_W + E_W} \tag{1}$$

Linear cuts only on dv, the new variables listed in table 1 and the χ^2-values were

Table 1: Extracted features studying single variables and pair correlations

Signal	Background	Features
$E_l \simeq E_l^f$	$E_l \geq 2\,E_l^f$	$\hat{E}_l = \frac{E_l - E_l^f}{E_l^f}$
$E_W^f \simeq E_W$	$E_W^f \geq 1.8\,E_W$	$\hat{E}_W = \frac{E_W^f - E_W}{0.8\,E_W}$
$\Theta_W - \Theta_W^f \leq 7$	$\Theta_W - \Theta_W^f \leq 7$	useless
$\Theta_{HN}^f \simeq \Theta_{HN}$	$\Theta_{HN}^f \simeq 2\,\Theta_{HN}$	$\hat{\Theta}_{HN} = \frac{\Theta_{HN}^f - \Theta_{HN}}{\Theta_{HN}}$
$\Theta_{HN} \notin [40, 130]$	$\Theta_{HN} \in [40, 130]$	$\tilde{\Theta}_{HN} = \frac{\Theta_{HN} - 85}{5}$
$\Theta_W \notin [40, 130]$	$\Theta_W \in [40, 130]$	$\tilde{\Theta}_W = \frac{\Theta_W - 85}{5}$

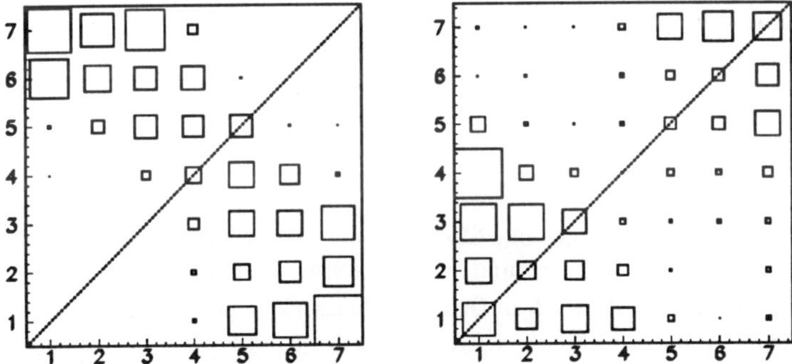

Figure 3: Distribution of signal (left) and background (right) events in the feature map.

Table 2: Behavior of some variables

Node	E_l $[GeV]$	M_{HN} $[GeV]$	M_W $[GeV]$	E_W^f $[GeV]$	E_W $[GeV]$	$E_{j_1}^f$ $[GeV]$	$E_{j_2}^f$ $[GeV]$
(1,1)	71.0	118.0	43.0	109.0	54.0	25.0	84.0
(2,2)	65.0	117.0	48.0	105.0	60.0	25.0	80.0
(3,3)	46.0	113.0	64.0	99.0	78.0	30.0	69.0
(4,4)	28.0	102.0	76.0	100.0	96.0	52.0	47.0
(5,5)	27.0	102.0	76.0	102.0	97.0	72.0	31.0
(6,6)	50.0	111.0	60.0	108.0	75.0	84.0	24.0
(7,7)	62.0	114.0	49.0	111.0	63.0	87.0	24.0
behavior	⤴	⤴	⤵	⤴	⤵	↑	↓

used to distinguish between the HN-signal and the WW-background. The result is given in figure 2.

5. Summary and Conclusion

We used successfully a Self-Organizing network to increase the tagging performance in a two-class separation problem by about 7%. It enabled us to construct new discriminant variables in two different ways. First we concentrated on pair correlations extracted from weight vectors of typical signal and background nodes, secondly we studied the node dependent behavior of some quantities by crossing the feature map perpendicular to the seperation line.

6. References

1. J. Dahm, C. Voigt, K. H. Becks, J. Drees, *Tagging Heavy Majorana Neutrinos at LEP200 using a Feed-Forward Network with Constraint Fit Inputs*, in this Volume.
2. J. Hertz, A. Krogh, R. G. Palmer, *Introduction to the Theory of Neural Computation* (Addison-Wesley 1991).
3. C. Peterson, T. Rögnvaldsson, *An Introduction to Artificial Neural Networks* (LU TP 91-23).
4. T. Sjöstrand, *Comp. Phys. Comm.* **39** (1986) 347.
   T. Sjöstrand, M. Bengtsson, *Comp. Phys. Comm.* **43** (1987) 367.
   H.-U. Bengtsson, T. Sjöstrand, *Comp. Phys. Comm.* **46** (1987) 43.
5. C. Voigt, to be published as Diploma thesis (Wuppertal 1993)
6. W. Buchmüller, C. Greub, *Phys. Lett.* **B256** (1991) 465.
7. J. Cuevas et. al., *Fast Simulation for DELPHI* (DELPHI 87-26 Prog-71).
8. N. J. Kjaer, R. Moeller, *Reconstruction of invariant masses in multi-jet events* (DELPHI 91-17 PHYS 88).
9. ECFA Workshop on LEP 200, *Proceedings Vol.1* (CERN 87-08 49).

FEEDFORWARD NEURAL NETWORKS AS AN ON-LINE PATTERN RECOGNITION TOOL

H.M.A. Andree, W. Lourens, A. Taal
Department of Physics and Astronomy, Utrecht University
P.O. Box 80000, 3508 TA Utrecht, The Netherlands

and

J.C. Vermeulen
NIKHEF-H, P.O Box 41882, 1009 DB Amsterdam, The Netherlands

ABSTRACT

Strictly layered feedforward neural networks are explored as recognition tools for energy deposition patterns in a calorimeter. Networks consisting of linear threshold units are generated by a constructive learning algorithm, the Patch algorithm. The back-propagation algorithm, using analogue neurons, is applied as a non-constructive counterpart. Generalization capabilities of the neural networks are compared to a Parzen-window classifier and a 1-nearest-neighbour classifier.

1. Introduction

The study of feedforward neural networks presented in this paper is carried out within the context of the CERN-EAST collaboration (Embedded Architectures for Second-level Triggering,DRDC/RD-11)[4]. A widely applied learning algorithm is the back-propagation algorithm[19], which requires a fixed network architecture during training. A learning algorithm that is also able to add units to the network is called a constructive learning algorithm[5, 6, 7, 8, 9, 11, 12, 13, 14, 17, 21]. For each training problem, an input set of patterns with their desired network outputs (or output labels) is given. This set is divided into a training set and a test set. A learning algorithm tries to modify or construct a neural network that classifies all patterns in the training set according to their output label. The generalization capability of a network is measured by the percentage of correctly classified patterns in the test set. Results obtained by applying Bayes' decision rule are the best one can hope to achieve. Since this rule is optimal, it is important to determine for any method its closeness to Bayes' decision rule. From the field of non-parametric statistics we used Parzen-windows[15, 3] and a 1-nearest-neighbour classifier[3, 2]. Furthermore, we used an improved version of the Patch algorithm[1], which builds feedforward neural networks consisting of linear threshold units. As a non-constructive counterpart the back-propagation algorithm[19], extended with a momentum term, is applied. For the above methods the percentage of correctly classified patterns in the test set is compared as a function of the training set size and the method of pre-processing. This comparison is carried out for the recognition of energy deposition patterns in a calorimeter.

2. Statistical decision rules

2.1. Bayes strategy

An input set $\{(\mathbf{p},\theta)|\mathbf{p}\in P \text{ and } \theta \in I\}$ can be constructed by assigning an output label θ to each pattern $\mathbf{p}$ in a set P of observations, where $I=\{1,2,...,M\}$ is the index set of the set of different categories $\{C_i: i\in I\}$. The optimal Bayes' decision rule assigns to a measurement $\mathbf{p}$ the label θ=i with the largest conditional probability, $Pr(\theta=i|\mathbf{p})$. This results from Bayes' law

$$Pr(\theta = i \mid \mathbf{p}) = \frac{Pr(\mathbf{p} \mid \theta = i) \cdot Pr(\theta = i)}{\sum_{j=1}^{M} Pr(\mathbf{p} \mid \theta = j) \cdot Pr(\theta = j),}$$

where $Pr(\theta=i)$ the a priori probability of category C_i, $Pr(\mathbf{p}|\theta=i)$ the probability of getting a measurement $\mathbf{p}$ from category C_i, and $Pr(\theta=i|\mathbf{p})$ the a posteriori probability of a measurement $\mathbf{p}$ belonging to category C_i. In practice all $Pr(\mathbf{p}|\theta=i)$, and $Pr(\theta=i)$ are unknown: a non-parametric classification problem.

2.2. Parzen-windows

Parzen[15, 3] showed how to construct a family of estimators which asymptotically approache the unknown underlying probability density function $f_\theta(\mathbf{p})$ of patterns $\mathbf{p}$ of category C_θ. The estimate $f_{\theta;n}(\mathbf{p})$ of $f_\theta(\mathbf{p})$ from a training set of n independent samples $(\mathbf{p},\theta)$ is for instance the gaussian:

$$f_{\theta;n}(\mathbf{p}) = \frac{1}{(2\pi)^{d/2} \cdot \sigma^d} \cdot \frac{1}{n} \cdot \sum_{i=1}^{n} \exp\left[-\frac{(\mathbf{p} - \mathbf{p}_{\theta;i})^2}{2\sigma^2}\right]$$

with n the number of training patterns from category C_θ, d the dimensionality of the pattern space, and σ the window width. For the problem discussed, a proper value of σ for a set of n patterns from category C_θ is easily found by a binary search. As proven by Parzen the expected error decreases as the estimate is based on a larger training set. Taken the probabilities of the different categories equal, Bayes' decision rule reads: If for all $j\neq i$ $f_{i;n}(\mathbf{p})>f_{j;n}(\mathbf{p})$ decide $\mathbf{p}\in C_i$. To use Parzen-windows, probabilistic neural networks are applied as discussed by e.g. Specht[20].

2.3. The 1-nearest-neighbour rule

An easy to implement classifier, although computer time consuming, is the 1-nearest-neighbour classifier. If one defines E^* as the expected error of Bayes' decision rule, the expected error E of the 1-nearest-neighbour classifier is bounded by $E^*\leq E\leq 2\cdot E^*(1-(M/(M-1))\cdot E^*)$ where M is the total number of different catagories[2, 3]

3. Description of the constructive learning algorithm

Section 3.1 contains a concise description of the Patch algorithm[1]. In section 3.2 the improvements with respect to this algorithm are discussed.

3.1. The Patch algorithm

A network generated by the Patch algorithm is considered to be completed if for the last layer a bijective mapping exists between its internal representations and the output labels. The construction of a layer k is governed by a quality function $Q(k)=-Q_1(k)+\alpha\cdot Q_2(k)$. Herein $Q_1(k)$ is the number of pairs of internal representations on the previous layer (k-1) with different output labels and an identical internal representation on layer k. If $Q_1(k)$ is equal to zero, layer k is faithful and no more neurons have to be added to this layer. $Q_2(k)$ is the number of pairs of internal representations on the layer k-1 with an identical output label and an identical internal representation on layer k. The Patch agorithm tries to maximize the quality function Q(k) with α chosen such that $Q_1(k)$ dominates the quality function. This maximization is done in weight space. Any linear threshold unit is defined by the values of its weights and its bias and thus corresponds to a point in d+1 dimensional space. Normalized neurons $\mathbf{w}$ (i.e. $|\mathbf{w}|=1$) correspond to points on a unit hypersphere in weight space. To each pattern $\mathbf{p}^{(i)}$ of the training set a hyperplane $h^{(i)}=w_0+w_1\cdot p_1^{(i)}+w_2\cdot p_2^{(i)}+...+w_d\cdot p_d^{(i)}=0$, corresponds. These hyperplanes partition the surface of the unit hypersphere into patches (polyhedral regions). For any pair of points (neurons) on the same patch the value of the quality function Q(k) equals. It therefore suffices to chose an arbitrary representative from a patch to evaluate Q(k). The algorithm now proceeds as follows:

1. Choose an initial neuron $\mathbf{w}$ and calculate the quality function Q(k).
2. Choose a random direction $\mathbf{w}'$, and calculate Q(k) for all combinations $\mathbf{w}+\lambda\cdot\mathbf{w}'$ with distinct quality. The new neuron $\mathbf{w}$ is chosen from the patch with the largest value of Q(k) along this trajectory.
3. If N_{opt} trajectories are evaluated without an improvement of Q(k) the neuron $\mathbf{w}$ is added to the layer under construction. Otherwise continue with step 2.
4. If the layer k is not faithful, $Q_1(k)\neq 0$, proceed with step 1.
5. If the number of different internal representations equals the number of output labels, the network is completed. Otherwise, start the construction of a new layer and proceed with step 1.

3.2. Improvements of the Patch algorithm

By ommiting the demand of a faithful first layer, smaller networks with the same generalization can be generated. The quality function Q(k) is maximized with respect to the first half of the training set. The first layer is considered to be finished if no neuron can be found that improves the generalization on the second half of the training set by a certain factor. The search for the optimal patch can be enhanched by Powell's method[16]. At the best patch found so far, the next trajectory $\mathbf{w}+\lambda\cdot\mathbf{w}'$ on the hypersphere is determined by calculating $\mathbf{w}'$ from a set of conjugate directions. The random choice of a neuron in a patch does not affect the size of the networks. But the best generalization can be expected from the neuron with maximal distance to any neighbouring patch. We used the QuadProg algorithm by Ruján[18] to calculate this neuron of maximal stability.

4. The calorimeter data recognition problem

The data set described in[10] consists of calorimeter events using the detector description known in simulation as 'Eagle B'. Electro-magnetic and hadronic cells are of the same size, $0.02\times2\pi/300$ in (η,ϕ)-space, and are precisely aligned. Only events, electrons and QCD jets, are selected which passsed a first-level trigger condition. A trigger window of 4×4 cells slides in steps of 0.1 in both variables over the entire calorimeter range. Two adjacent electro-magnetic cells, both not touching the edge of the window, are defined as a cluster region. The remaining electro-magnetic cells (12) and all corresponding 16 hadronic cells form a veto region. The criterium used is a minimum of 25 GeV for clusters and a maximum of 5 GeV for the sum of all veto cells (after thresholding individual level-1 cells at 1 GeV). Pileup was added after the first-level selection and only slightly distorted the energy spectrum in equal ways for both event types.

5. Results

Pre-processing of the patterns may improve the performance of the learning algorithm with respect to the network size, generalization, and CPU-time requirement of the learning algorithm:

1. For all cells we either fed the energy E or its logarithm, log(1+E)
2. The patterns were translated such that the centre pixel of the electro-magnetic data field of each pattern coincides with the centre of gravity of the electro-magnetic part of the corresponding event.

We tried training sets of different sizes (N). Each training set consists of a number of electrons (N_{el}) and a number of QCD jets (N_{QCD}). These numers were in the range N=50, i.e. $(N_{el},N_{QCD})=(25,25)$, to N=1000, i.e. $(N_{el},N_{QCD})=(500,500)$.

Every classification method is trained on 20 randomly drawn training sets of size N. The mean generalization $\langle G\rangle$ of a method is the average of the generalizations on the corresponding test sets, i.e. the remaining patterns present in the input set but not in the training set.

Results for electron-hadron recognition are given in figure 1. The results shown are obtained for input sets (training set and test set) of events without pileup contamination. Using a training set of clean data, no pileup contamination, and a test set of data with pileup, decreases the generalization for all methods applied. This deterioration in performance, about 3.0%, could not be compensated by training on data with pileup only. In the case of logarithmic pre-processing all methods perform better. The Patch algorithm, in contrast to the other methods, is almost insensitive to whether or not logarithmic pre-processing is applied. (This is due to the search for the optimal patch on the hypersphere in weight space, logarithmic pre-processing does only affect the shape of the patches.) By applying

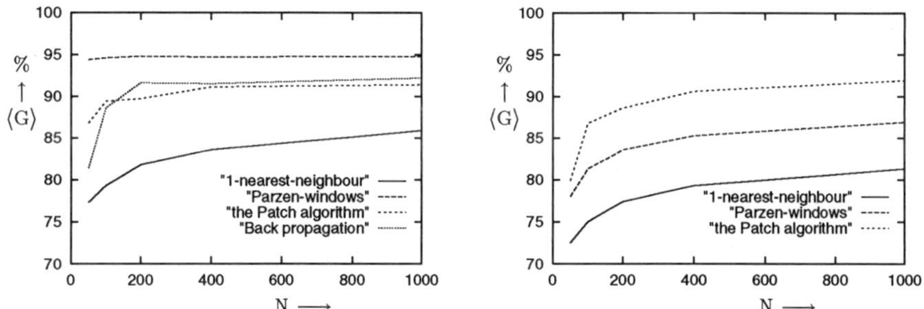

Figure 1: The mean generalization, ⟨G⟩, of electron-hadron recognition as a function of the training set size N. The left figure shows results with logarithmic pre-prossesing, the right figure shows results with the original gray values.

logarithmic pre-processing the Patch algorithm and the back-propagation algorithm perform about equally well, both with respect to the generalization as well as to convergence to the absolute minimum. Almost all patterns in the training set are correctly classified. The network sizes differ. The Patch algorithm yields networks of a single neuron (i.e. larger networks did not contribute to a better generalization). For the back-propagation algorithm networks of 4 neurons were chosen. The implementation of the other two methods, probabilistic neural networks implementing Parzen-windows and the 1-nearest-neighbour approach, grows linearly with the number of training patterns used. The performance of the back-propagation algorithm was measured only for logarithmic preprocessing. Using the original gray values for the training patterns, the back-propagation algorithm always converged to a local minimum far away from the optimal solution. No measures were undertaken to escape from a local minimum, since the probability to enter a better minimum is very small for the given training sets, and increasing CPU-time is required for larger training sets.

6. Conclusions

Pattern recognition by means of feedforward neural networks generated by a constructive learning algorithm, and networks trained by the back-propagation algorithm is studied. Both methods perform about equally well with respect to the network size, but only if proper pre-processing is applied. The Patch algorithm has the advantage to converge to an optimal solution, if N_{opt} is large enough, with small networks and, it is rather insensitive to pre-processing. For the back-propagation algorithm the search for the optimal network size and the algorithm parameter values is difficult and time consuming. For the statistical methods, Parzen-windows and the 1-nearest-neighbour classifier, the complexity of a hardware implementation grows linearly with the training set size. For the problem considered, the complexity of neural networks generated by the Patch algorithm and the back-propagation algorithm does not depend on the training set size. This favours to hardwire the networks from these algorithms above those from the other methods.

7. Acknowledgement

The authors thank R.K. Bock spokesman of the EAST/RD11 collaboration of CERN for the oppertunity and support given to fulfil this study. This work is made possible by financial support from the Foundation for Fundamental Research on Matter (FOM).

8. References

1. G. T. Barkema et al., *Network* **1** (1993), p. 393.
2. T. M. Cover and P. E. Hart, *IEEE Trans. on Inf. Theory* **13** (1967), p. 21.
3. R. O. Duda and P. E. Hart, *Pattern Classification and Scene Analysis,* (John Wiley & Sons, New York, 1973).
4. EAST (RD-11), Proposal and Status Report, (CERN/DRDC 90-56, 1990) and (CERN/DRDC 92-11, 1992).
5. S. E. Fahlman and C. Lebiere, *Advances in Neural Information Processing Systems* **2** (Morgan Kaufmann Publishers, 1990) p. 524.
6. M. Frean, *Neural Computation* **2** (1990), p. 198.
7. S. I. Gallant, *IEEE Trans. on Neural Networks* Vol. **1** No. **2** (1990), p. 179.
8. M. Golea and M. Marchand, *Europhys. Lett.* **12** (1990), p. 205.
9. S. A. J. Keibek et al., *Europhys. Lett.* **18** (1992), p. 555.
10. G. Klyuchnikov et al., (EAST-note 92-23, 1992).
11. M. Marchand et al., *Europhys. Lett.* **11** (1990), p. 487.
12. D. Martinez and D. Estve, *Europhys. Lett.* **18** (1992), p. 95.
13. M. Marchand and M. Golea, *Network* **4** (1993), p. 67.
14. M. Mézard and J-P. Nadal, *J. Phys. A. Math. Gen.* **22** (1989), p. 2191.
15. E. Parzen, *Annals of Mathematical Statistics* **33** (1962) p. 1065.
16. W. H. Press et al., *Numerical Recipes in C. The Art of Scientific Computing* (Cambridge University Press, 1988).
17. P. Ruján and M. Marchand, *Complex Systems* **3** (1989), p. 229.
18. P. Ruján, *J. Phys. I France* **3** (1993), p. 277.
19. D. E. Rumelhart and J. L. McClelland, *Parallel Distributed Processing Vol. 1 and 2* (MIT Press, 1986).
20. D. F. Specht, *Neural Networks* **3** (1990), p. 109.
21. J. A. Sirat and J-P. Nadal, *Network* **1** (1990), p. 423.

AN ALGORITHM FOR IDENTIFYING SECONDARY VERTICES

V.V.IVANOV
Laboratory of Computing Techniques and Automation, Joint Institute for Nuclear Research, Moscow, 101000/Head Post Office, P.O. Box 79, Russia

and

G.B.PONTECORVO
Laboratory of Nuclear Problems, Joint Institute for Nuclear Research, Moscow, 101000/Head Post Office, P.O. Box 79, Russia

ABSTRACT

Variables permitting effective selection of events containing a secondary vertex are proposed. Reconstruction of an event and searching for the primary and secondary vertices are not required for this purpose. If a neural network is utilized for classification of events in the space of these variables, then the identification of signal and background events will practically be totally unambiguous.

1. Introduction

The technique for identifying the secondary vertex in studies of production and decay mechanisms of short-lived particles comprising s, c, b-quarks is the keystone in formulating and performing experiments in the field of intermediate and high energy physics. A number of interesting algorithms has been proposed with the aim of solving the problem [1,2].

The authors of ref. [1] reduced the information on the tracks from each individual event detected in the region of the vertex detector to the variables D (the impact parameter) and Φ (the inclination of the track with respect to the incident beam). They also proposed passing from D and Φ to more effective quantities – to moment functions of these variables, which are subsequently input to the 3-layered perceptron utilized for discrimination of background events. In another work [2] the rejection of background events is based on a statistical test of the hypothesis that the event has a sole vertex. The test is performed in two steps: 1) estimation of the location of the vertex; 2) application of the Kolmogorov test for checking that the considered event is consistent with the zeroth hypothesis.

At present an experiment is being prepared by the DISTO [3] collaboration for studying spin effects in the reaction

$$\uparrow pp \longrightarrow pK^+Y$$

with the polarized proton beam of Saturne (Saclay, France). At the beginning a detailed study of the reactions $pp \longrightarrow pK^+\Lambda^0$, $pp \longrightarrow pK^+\Sigma^0$ will be performed. For identification of the indicated reactions and for separating them from background processes with large cross sections, for instance, from the $pp \longrightarrow pp\pi^+\pi^-$ channel, advantage is taken of the fact that the Λ^0 (from both reactions) cover a distance of about several centimeters before decaying.

2. The experimental setup and event simulation

The layout of the DISTO experiment is presented in Fig. 1. The DISTO

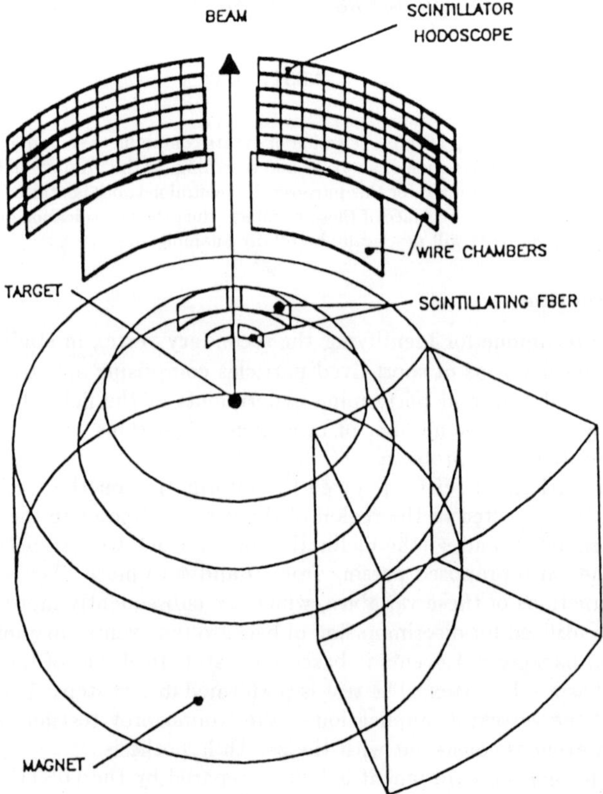

Figure 1: Layout of experimental apparatus

apparatus consists of 10 cylindrical detectors in two arms situated symmetrically about the beam direction[1]. They cover a scattering angle of 45^0 in the horizontal plane and a dip angle of $\pm 20^0$. For the selection of events involving the production

[1] A detailed description of the experimental setup is presented in proposal /3/.

of lambda and sigma particles and the rejection of background events a two-level trigger will be used.

The first-level trigger is for selection of events by their multiplicity: only four-prong events are selected. It was proposed in the project /3/ to construct the second-level trigger on the basis of four RISC processors, which will be used for geometrical reconstruction of each event with the aim of finding a lambda or a sigma decay vertex. For its realization one must perform real-time reconstruction of the event to search for both vertices. This procedure requires much time and cannot be completed with good accuracy during acceptable time. Clearly, any simplification accelerating event analysis will inevitably lead to some loss of good events. For these reasons we have undertaken attempts to find alternative approaches to the solution of the considered problem.

For solving the problem at issue a data file was generated with the aid of the LACYL code (simulating processes in the DISTO experimental setup) for the reaction

$$pp \longrightarrow pK^+\Lambda^0,$$

containing the simulated detector information (particle positions, energy losses, and flight times) for each event, in which at least two positively charged particles crossed the detectors of one of the arms and one negatively charged particle was detected in the other arm. Moreover, each such event contained data relevant to the energy and momentum components of the beam proton, the coordinates of the interaction vertex, the components of the scattered proton momentum p_1, of the K^+ and of the Λ^0 momenta, the coordinates of the Λ^0 decay vertex and the components of the momenta of the Λ^0 decay products – the decay proton momentum p_2 and the π^- momentum.

It must be noted that in the coordinate system chosen for the apparatus the OZ axis coincides with the primary beam direction, and the OX axis lies in the horizontal plane, while the OY axis is in the vertical plane. The following dimensional units are used below: centimeter, radian.

3. Algorithm for identifying the secondary vertex

The production kinematics of an unstable particle gives rise to a certain peculiarity in the event geometry. It consists in the momenta of the unstable particle decay products being mainly not directed along the primary beam. This results in the momenta of the secondary particles produced in the decay of the unstable particle having directions differing essentially from those of the particles emitted from the primary vertex. Thus, the idea arose to utilize the coordinate of the intersection point of a straight line (coinciding with the particle track) with the direction of the primary beam as a parameter permitting revelation of the existence of a secondary vertex in the event being considered. To a certain extent this quantity is similar to the impact parameter D, but here, unlike ref. /1/, the dispersion of the track about the YOZ-plane (or the XOZ-plane) is to be used as the measure.

The distributions of the Z-coordinates of intersection points of particle trajectories in the YOZ-plane[2] are presented in Fig. 2a for tracks corresponding to the proton p_1 and the K^+ (background events), while the respective distributions for proton p_2 and the π^- from the Λ^0 decay (signal events) are presented in Fig. 2b. They are seen to differ significantly.

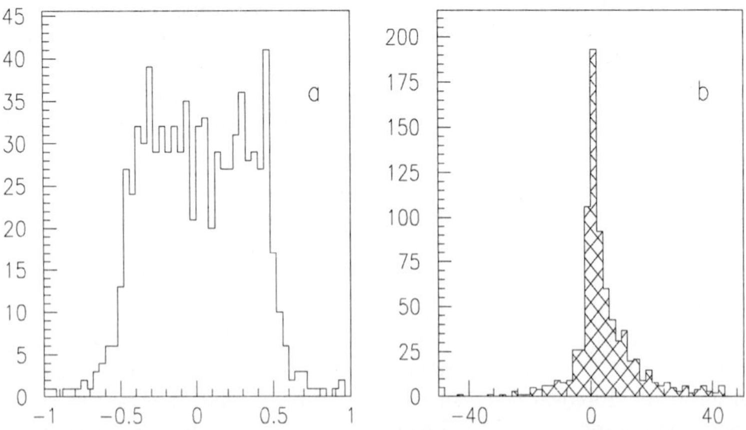

Figure 2: Distributions of the Z-coordinates of intersection points of particle trajectories with the direction of the primary beam in the $Y0Z$-plane for background (a) and signal (b) events

By analogy with ref. /1/ we shall now pass to the moment variables

$$\xi_{02} = \frac{1}{N}\sum_{i=1}^{N}(D_z - \bar{D_z})^2 \quad and \quad \xi_{11} = \frac{1}{N}\sum_{i=1}^{N}(\Phi - \bar{\Phi})(D_z - \bar{D_z}),$$

where $\bar{D_z}$ and $\bar{\Phi}$ are the mean values of D_z and Φ (set equal to zero), and N is the number of tracks of the considered sort. As the variable D_z we took the distance from the intersection point of the track with the axis of the beam to the origin of the coordinate system. In Figs. 3a and 3b the joint distributions of the random variables ξ_{02} and ξ_{11} are presented for practically all the background events and for those signal events (shaded in the Figures) that happened to be in the same region as the background events.

For the chosen critical value of $\xi_{02} = 0.5$ the number of background events interpreted as signal events (error of the second kind) amounted to 37, while the amount of signal events that happened to be among the background events (error of the first kind) equaled 57. The corresponding quantities for the variable ξ_{11} in the range of $|\xi_{11}| \leq 0.1$ were 5 and 126. Now, if a neural network is utilized for

[2]The influence of the magnetic field on a charged particle is very small in the YOZ-plane, so the trajectories of particles in this projection are almost linear; it makes more simple their searching and parameter determination in the real-time of the experiment.

classification of events in the space of the variables ξ_{02} and ξ_{11}, then the identification

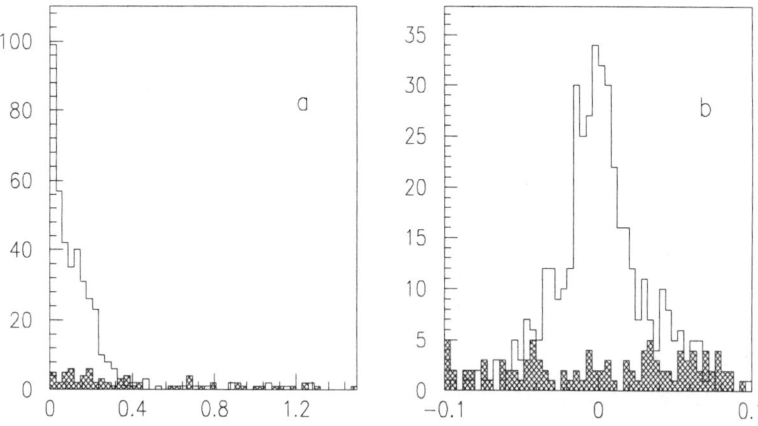

Figure 3: Distributions of the random variables ξ_{02} (a) and ξ_{11} (b) for background events and for those signal events (shaded in the Figures) that happened to be in the same region as the background events

of signal and background events will practically be totally unambiguous (see Fig. 4).

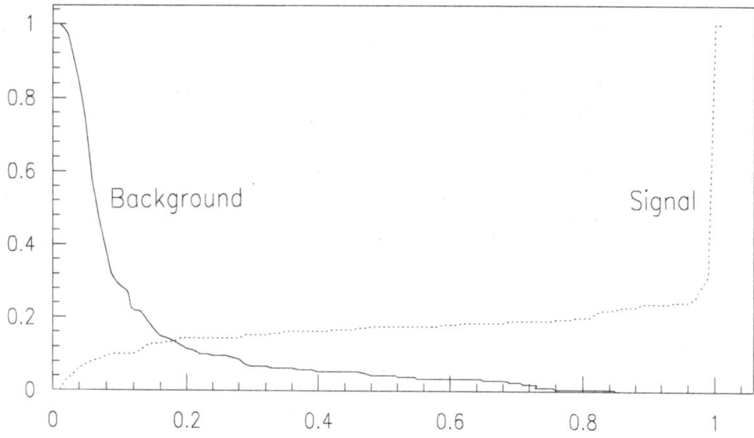

Figure 4: Plots of accumulated probability $F(y) = \Pr\{y < y_g\}$ for signal events and of dependence $1 - F(y)$ for background events

4. Conclusion

A comparative analysis of event classification in the space of variables mentioned above and of variables proposed in ref. /1/ has shown the advantage of the former, what was displayed mostly in the XOZ-projection /4/.

Thus, the new variables, and especially the quantity ξ_{02}, permit effective selection of events containing a secondary vertex. Reconstruction of an event and searching for the primary and secondary vertices are not required for this purpose, just as there is no real need of utilizing a neural network for event selection. If, however, one additionally takes advantage of a neural network, then practically total rejection of background events may be achieved. Quite clearly, the proposed algorithm does not exclude its application together (in parallel) with the method suggested in ref. /3/.

5. Acknowledgements

We are grateful the DISTO collaboration for fruitful discussions and support. One of the authors, V.V.Ivanov would like to thank the International Science Foundation in Washington for financial support of his participation in the Third International Workshop on Software Ingineering, Artificial Intelligence and Expert Systems for High Energy and Nuclear Physics.

6. References

1. L.Gupta et al.: *"Neural Network Trigger Algorithms for Heavy Quark Event Selection in a Fixed Target High Energy Physics Experiment"*. FERMILAB-Pub-91/17, FNAL, Batavia, 1991.
2. G.A.Ososkov, N.A.Russakovich, N.I.Chernov: *"Algorithm for fast analysis of multi-track events involving the production of short-lived particles"*. JINR Preprint P10-90-367, Dubna, 1990 (in Russian).
3. J.Arvieux et al.: DISTO collaboration, 1991, Proposal 213 at Saturne.
4. V.V.Ivanov and G.B.Pontecorvo: *"Simple and Efficient Method for Secondary Vertex Searching"*. JINR Preprint P10-93-271, Dubna, 1993 (in Russian); Submitted to the International Conference on "Programming and Mathematical Methods for Solving Physical Problems", Dubna, 14 – 19 June, 1993.

AN APPLICATION OF CELLULAR AUTOMATA AND NEURAL NETWORKS FOR TRACK RECONSTRUCTION IN HIGH ENERGY PHYSICS

I. KISEL and G. OSOSKOV
Joint Institute for Nuclear Research, Head Post Office, P.O. Box 79, 101000 Moscow, Russia

The study of functioning of such a detector as a multiwire proportional chamber has led to necessity of taking into account its discrete structure. As a result of using additional discrete information one can construct a cellular automaton for track filtering. Next natural step in data processing is global extension of cellular automaton to modify rotor model of neural network. The energy function of a network in this modification contains only one "cost" term because of taking into account parametric form of tracks. Use of cellular automaton to preliminary selection of data allows us to create an initial network configuration with the energy close to its global minimum. We study also robustness of the neural network to heavy contaminated raw data and stability to the growing event multiplicity. The algorithms were tested on real and simulated events obtained from the ARES spectrometer with satisfactory results.

1. Discrete features of MWPC

As was mentioned in Charpak's pioneer works [1], the discrete structure of such a detector as a multiwire proportional chamber (MWPC) bears sufficiently rich information about track points and its slope at each of them.

Using the ARES spectrometer (JINR, Dubna) as an example [2], let us consider two discrete features of a multiwire proportional chamber. First, the shape of external equipotential lines in a proportional chamber allows us to draw between wires imaginary equidistant planes perpendicular to a plane of wires. Second, due to electronegative additives in the chamber gas the distribution function of the electron collection is box-shaped, i. e. on both sides of the wire plane there exist two planes, so that all electrons that appeare inside them are collected by a wire. But any of electrons outside them are absorbed by the gas. Both these features allow us to consider the chamber to be a chain of imaginary rectangles, each of them surrounding a signal wire. Two parallel sides of each rectangle are determined by the "equipotential" planes, two other sides are formed by the planes of maximal electron collection.

Whenever a charged particle track hits a rectangle, a wire inside it produces a signal. If a track crosses a few adjacent rectangles in the chamber, they all work, forming a cluster. In the latter case the track crosses the left side of the lower rectangle and the right side of the upper one or vice versa. Otherwise it would either not have crossed all the hit rectangles or have touched one of missing ones. The most important feature of our model of the discrete detector is the possibility of solving approximately the inverse problem of the track reconstruction. Knowing a cluster structure, one can make conclusion about a range of possible slopes of track passing through the cluster. Besides, when the slope is fixed, the cluster structure determines a possible range of points, where the track crosses the plane of signal wires.

This simple model is a satisfactorily good approximation for description of cluster forming effects in proportional chamber [3,4,5,6]. The difference between the real physical processes and our simplified model is easily taken into account in a software implementation.

2. A cellular automaton filter

At the same time the analysis of real experimental data obtained with the ARES spectrometer required the necessity of prefiltering these data from noise hits. The problem of prefiltering arose because of high rate, specific features of the beam duty cycle, chamber and electronics malfunctions, presence of delta-electrons, returned tracks and other difficult-to-check reasons. Common requirements for a prefiltering procedure are: high degree of event cleaning from noise hits; restoring of hits missing because of chamber malfunctions; grouping of filtered hits according to their possible belonging to different tracks; simplicity and high speed of the filtering algorithm needed for on-line implementation; suitability for parallel processing.

The accurate consideration of the detector discrete structure and necessity to hold these requirements brought us to the idea of using well developed methods of cellular automata. Let us briefly formulate the specific features of our filtering automaton:

- Every cell is either cluster or phantom (missing cluster) and therefore has four states: real cluster or phantom and living cell or dead one.
- Cells in adjacent chambers are neighbors only if an admissible track could pass through them.
- A cell birth occurs if this cell has at least two neighbors in different adjacent chambers. A cell death occurs in two cases: if this cell has less than two neighbors (small noise) or if it has more than four neighbors (hard noise).
- Evolution is clocked, i. e. all cells change their states simultaneously. Each step of the automaton evolution consists of two distinctive stages: birth at the first stage and death at the second one.

Final stage of the automaton evolution is attainment of stable or cyclic state. In order to sense the automaton arrival in stable or cyclic state we suggest calculation of the checksum CRC (cyclic redundancy check) for every generation and stop iteration in the case of the coincidence of checksums related to different generations.

The other features or functioning details of our automaton are just the same as for conventional cellular automaton.

The cellular automaton has been tested with real three-track events obtained in experiments during the search for the forbidden decay $\mu^+ \to e^+e^+e^-$ [7] and the study of the rare decay $\pi^+ \to e^+\nu_e e^+e^-$ [8].

The cellular automaton rejects noise quite well. Prefiltering procedure eliminates on the average 65 – 70% of the total noise. So high cleaning allows us to determine a number of tracks according to the number of clusters. The analysis of the rest of noise hits shows that they are mainly clusters from δ-electrons lying too close to tracks and having cluster structure similar to track clusters. Besides, the automaton can decrease the volume of data needed for further processing almost by a factor of 3: it rejects about 20% of events being non-three-tracks and decreases almost 50% of the number of experimental points from proportional chambers due to noise rejection.

The advantages of the automaton are its simplicity and high speed of work. The processing of one event by our automaton takes about 1/25 sec for our 3 MIPS computer. Use of special computer board with parallel hardware implementation can increase speed up to 1500 events per second. Due to its parallelism, simplisity and high speed of work the described cellular automaton for track filtering is suitable for parallel computing and could be also used as trigger in the on-line mode.

The result of the cellular automaton working, which is a by-product but very important for further processing, consists in grouping experimental points according to their possible belonging to a track. It allows us to create easy an initial network configuration with the energy close to its global minimum. Another global method of initializing neural network is suggested in [9].

3. Modified rotor model of neural network

Let us put a unit vector $\vec{s}_i$ (rotor) at each measured point and choose a two-rotor interaction energy in such a way that the neural network has to place vectors $\vec{v}_i = \langle \vec{s}_i \rangle_T$ tangentially to track and effectively diminish $\vec{v}_j$ for all j that lie apart from true trajectories.

Let two points i and j lie on the circle. Then, tangential vectors are connected by the expression $\beta_i = -\beta_j$, where β_i, β_j are the angles between these vectors and the chord r_{ij}. The partial energy function for the pair (i,j) can be expressed as:

$$E_{ij} = -\vec{s}_i \vec{s'}_j, \tag{1}$$

where $\vec{s'}_j$ is the mirror reflection by w_{ij} of $\vec{s}_j$ relative to r_{ij}. The energy E_{ij} has a minimum for all pairs of points placed at one track. The globality of calculating the total energy function is due to the possibility of summarizing all pair energies E_{ij}

$$E = -\frac{1}{2}\sum_{ij} \vec{s}_i w_{ij} \vec{s}_j = -\frac{1}{2}\sum_i \vec{s}_i \vec{H}_i \tag{2}$$

without any additional constraint terms. MFT-dynamics appears here naturally because of the long range interaction. The equation of dynamics [10]

$$\vec{v}_i = \frac{\vec{H}_i}{|\vec{H}_i|} \frac{I_1(|\vec{H}_i|/T)}{I_0(|\vec{H}_i|/T)}, \tag{3}$$

minimizes energy function (2).

For better convergency of the MFT-dynamics it is necessary to invent special robust-type restrictions that are not present explicitly in the conventional MFT-algorithm. During the calculations of the mean field $\vec{H}_i$ as vector sum, neurons lying "too far" must be excluded. To realize this we use the modified robust approach. We amplify the repulsive force explicitly by multiplying w_{ij} by a special robust multiplier responsible for the "soft" rejecting of all outlying neurons. As such a multiplier we use the famous Tukey's bi-weights [11]

$$w_{opt}(\Delta\varphi_{ij}) = \begin{cases} \left(1 - \left(\frac{\Delta\varphi_{ij}}{\Delta\varphi_{cut}}\right)^2\right)^2 & , |\Delta\varphi_{ij}| \leq \Delta\varphi_{cut}; \\ 0 & , |\Delta\varphi_{ij}| > \Delta\varphi_{cut} \end{cases} \tag{4}$$

where $\Delta\varphi_{ij}$ is the angle $(\widehat{\vec{s}_i, \vec{s'}_j})$.

The average curvature κ is also controlled to satisfy the cut-off criteria.

Besides the variation of the angle β we have one more degree of freedom, namely the coordinate of a neuron inside the central cell of the cluster. To control this parameter the simplest dynamics was chosen: MFT-dynamics with zero temperature[1]. The coordinate of a rotor was assigned to the point where the local field $\vec{H}_i$ has maximum (if this point is outside of a central cell, it is moved to its nearest outer border).

As it was mentioned above, our initial rotor values are designed to set neural network in a vicinity of the global minimum. Therefore, it is reasonable to substitute the simulated annealing recommendation to use a decreasing temperature sequence by applying the only temperature T_{opt} optimally chosen for all iterations. In this case we can use this temperature as a cut-off parameter to keep tracks having enough experimental points (neurons).

Besides that, we try to prevent loosing of some crossing tracks by inventing the possibility to support at any point supposed to be a cross more than one neuron.

The modified rotor model of neural network was tested on the set of 10^4 real three-prong events from the ARES spectrometer with satisfactory results [12].

The modified rotor model of neural network was also tested on simulated events obtained on the basis of the ARES spectrometer parameters. All the events contain only straight tracks and have 100% signal-to-noise ratio. An example of 30-track event is presented in Fig. 1. In this event there are about 300 points corresponding to tracks and 300 noise points. The result of the application of the modified rotor model one can see in Fig. 2. Here rejected points are denoted by dots, but active neurons are shown as vectors with lengths proportional to their activation levels. Outer lines indicate simulated straight tracks. In this case 251 noise points are rejected by our network and the rest 49 are attached by tracks (on the average 1.6 per track).

[1] It is possible due to the small range of the parameter, therefore only one minimum can exist.

Figure 1: An example of 30-track event with 100% signal-to-noise ratio.

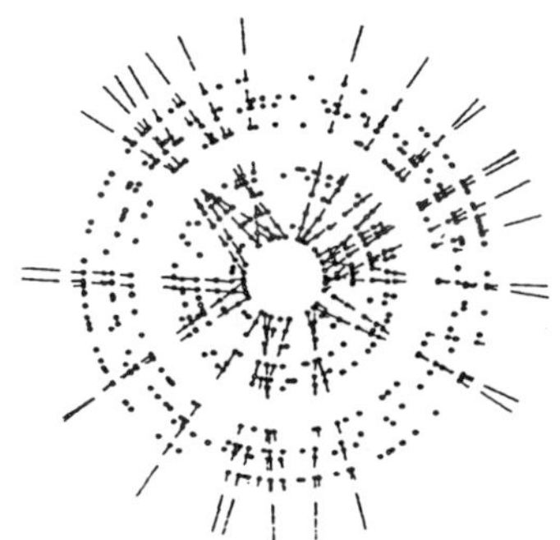

Figure 2: The result of the application of the modified rotor model of neural network to data presented on the previous figure.

4. Conclusion

The method of track filtering described above is the first application of cellular automata to handling of data from multiwire proportional chambers. The results obtained with the real data are quite successful: the high degree of cleaning from noise and restoring of missing clusters lead to a threefold reduction of input information with data grouping according to their belonging to different tracks. The percentage of useful events is increased and their further processing is considerably simplified and accelerated.

According to their nature the cellular automata are ideal objects for making evolution algorithms parallel. Even the minimal parallelism (at least one processor per chamber) increases the speed of calculation by an order of magnitude. So the described cellular automaton for track filtering can be successfully used in parallel computers and also as trigger in the on-line mode.

Note another interesting application of cellular automaton as random number generator: the logic of a cellular automaton placed on a 64x64 thor must guarantee that each cell changes its state with the probability 0.5 (one can reach that, for instance, changing the state every time when the number of "living" neighbors is odd).

In conclusion, we can say that the global character of the modified rotor model of neural network together with certain robust restrictions made it possible to achieve fast convergency and good track resolution. Moreover, this algorithm can work in the case of many-prong events without any special complications. We are convinced that our algorithm can also be applied to the case of an arbitrary magnetic field configuration.

References

1. G. Charpak, *Ann. Rev. Nucl. Sci.* **20** (1970) 195.
2. V. Baranov et al., *Nucl. Instr. and Meth.* **B17** (1986) 438
3. I. Kisel and G. Ososkov, *Proc. Int. Conf. on Computing in High Energy Phisics CHEP-92*, Annecy, France, **CERN-92-07** (1992) 646
4. N. Chernov, A. Glazov, I. Kisel, E. Konotopskaya, S. Korenchenko and G. Ososkov, *Comp. Phys. Commun.* **74** (1993) 217.
5. A. Glazov, I. Kisel, E. Konotopskaya and G. Ososkov, *Nucl. Instr. and Meth.* **A329** (1993) 262.
6. A. Glazov, I. Kisel and G. Ososkov, *Commun. of JINR* **E10-93-123**, Dubna, 1993.
7. V. Baranov et al., *J. Phys. G: Nucl. Part. Phys.* **17** (1991) S57.
8. V. Baranov et al., *Sov. J. of Nucl. Phys.* **v. 55, N 11** (1992) 1644.
9. S. Baginyan, I. Kisel, E. Konotopskaya and G. Ososkov, *Commun. of JINR* **E10-93-86**, Dubna, 1993 (subm. to *Comp. Phys. Commun.*).
10. C. Peterson, *Preprint LU TP* **90-6**, Lund, 1990.
11. G. Ososkov, *Proc. 2d Intern. Tampere Conf. in Statistics*, Tampere Univ. Press (1987) 615.
12. A. Glazov, I. Kisel, E. Konotopskaya, V. Neskoromnyi and G. Ososkov, *Commun. of JINR* **E10-92-352**, Dubna, 1992 (subm. to *Comp. Phys. Commun.*).

A FUZZY RADON TRANSFORM FOR TRACK RECOGNITION

J. BLOM and C. TH. A. M. DE LAAT
Department of Computer Topics in Physics, University of Utrecht,
Buys Ballot Laboratory, Postbus 80000, 3508 TA Utrecht, The Netherlands

and

P. G. KUIJER
Subatomic Physics Department, University of Utrecht,
Buys Ballot Laboratory, Postbus 80000, 3508 TA Utrecht, The Netherlands

ABSTRACT

In this contribution a fuzzy Radon transform is shown for application in ALICE and ATLAS (typical track density: 8000 tracks in 1 unit of rapidity). Resolution is introduced by the "broadening" of the matching tracks in the Radon transform, which is obtained by making a convolution of the matching tracks with a Gaussian kernel. In a good approximation, an analytical expression for the fuzzy Radon transform is given. An example of two track separation with noisy input is added.

1. Introduction

The use of the *Radon* transform, or *discrete*, the *Hough* transform, in particle physics is not new [1, 2, 3]. Sometimes the concept of *resolution* is introduced to gain more stability. A Radon transform is defined by the family of line integrals over a class of trajectories where the integrand is mostly a (measured) density. The output of the Radon transform is given as a function of the trajectory parameters; it defines a transform from lab space to track parameter space.

Due to noise, drift and track errors (no perfect track, discretisation errors) the original Radon transform is intrinsic *instable*. Therefore the concept of *resolution* is introduced in the generalised *fuzzy* Radon transform by making a convolution of the original Radon transform with a *blur kernel*. Contrary to the common practise we use a *Gaussian kernel* which has mathematical [4, 5] advantages.

The fuzzy Radon transform is especially useful if detectors with anode strips in stereo pairs are used, such as wire chambers and microstrip gas chambers: the influence of ghost hits in these detectors upon the fuzzy Radon transform is low.

2. Radon transform

2.1. Definition

As stated above, the *Radon* transformation is defined as

$$R_{\mathcal{P}}(\mathcal{P}_T) \equiv \int_T \rho_{\mathcal{P}}(\vec{r}(\tau, \mathcal{P}_T))\, d\tau \tag{1}$$

where $\rho_{\mathcal{P}}(\vec{r})$ is the *density function* which serves as input for the Radon transform; the set $\mathcal{P} \equiv \{p_1, \ldots, p_r\}$ denotes the set of (*measured*) *parameters* of the density function; the set $\mathcal{P}_T \equiv \{p_1^T, \ldots, p_s^T\}$ contains the parametrisation parameters of a formal *trajectory* T with parametrisation $\vec{r}(\tau, \mathcal{P}_T)$.

It appears to be convenient to introduce the so called *trajectory density* ρ_T

$$\rho_T(\vec{r}, \mathcal{P}_T) \equiv \int_T \delta(\vec{r} - \vec{r}(\tau, \mathcal{P}_T))\, d\tau \tag{2}$$

With Eq. (2) the Radon transform Eq. (1) can be written as a space integral

$$R_{\mathcal{P}}(\mathcal{P}_T) = \int_{\mathbb{R}^n} \rho_{\mathcal{P}}(\vec{r})\, \rho_T(\vec{r}, \mathcal{P}_T)\, d\vec{r} \tag{3}$$

2.2. Charged track trajectories

We will assume a constant magnetic field in the z-direction, $\vec{B} = B\hat{z}$. The vertex is positioned at the coordinate system origin. Energy losses are neglected. So in three dimensions a track is a spiral emerging from the vertex with emission angle ϕ and *signed* curvature κ (both in the xy-plane) and with constant velocity in the z-direction ($\tau > 0$)

$$\begin{aligned} x &= \kappa^{-1}\{\sin(\phi + sgn(\kappa)\,\tau) - \sin\phi\} \\ y &= \kappa^{-1}\{-\cos(\phi + sgn(\kappa)\,\tau) + \cos\phi\} \\ z &= \gamma\tau \end{aligned} \tag{4}$$

2.3. Detector density function

We will also assume that the density function ρ, which serves as input for the Radon transform, is determined by detectors which deliver track output positions. It can be approximated by

$$\rho(\vec{r}) = \sum_{i=1}^{N} \lambda_i\, \delta(\vec{r} - \vec{r}_i) \tag{5}$$

where i is the index within the detector array which contains N detectors; $\lambda_i \in \{0, 1\}$.

The Radon transform for the spirals becomes with Eq. (5) substituted in Eq. (1)

$$R(\kappa, \phi, \gamma) = \sum_{i=1}^{N} \lambda_i \int_0^{\tau_m} \delta(\vec{r}(\tau, \kappa, \phi, \gamma) - \vec{r}_i)\, d\tau \tag{6}$$

where τ_m is the maximum value of the parametrisation value. Only particles which are exactly on the track $\vec{r}(\tau, \kappa, \phi, \gamma)$ contribute to the Radon transform.

3. Embedding in a multi-resolution structure

3.1. Motivation

The presented Radon transform is sensitive to errors which may be introduced by: 1. no perfect track form; 2. discretisation errors; and 3. detector noise.

Gyulassy et al. [2] generalised the Radon transform Eq. (3) to a so called *fuzzy Radon transform* with the introduction of a *resolution-dependent* kernel $K(\vec{r};\sigma)$

$$R_{\mathcal{P}}(\mathcal{P}_T;\sigma) = \int_{\mathbb{R}^n}\int_{\mathbb{R}^n} \rho_{\mathcal{P}}(\vec{r})\, K(\vec{r}-\vec{r}\,';\sigma)\, \rho_T(\vec{r}\,',\mathcal{P}_T)\, d\vec{r}\,'\, d\vec{r} \tag{7}$$

In Eq. (7) σ is called the *resolution parameter*: a larger value of σ means a lower resolution level.

We can introduce a so called *fuzzy track density* $\rho_T(\vec{r},\mathcal{P}_T;\sigma)$ such that the form of the Radon transform equation is again identical with Eq. (3)

$$\rho_T(\vec{r},\mathcal{P}_T;\sigma) \equiv \int_{\mathbb{R}^n} K(\vec{r}-\vec{r}\,';\sigma)\, \rho_T(\vec{r}\,',\mathcal{P}_T)\, d\vec{r}\,' \tag{8}$$

Gyulassy et al. [2] used a Lorentzian form for the kernel function $K(\vec{r};\sigma)$, however we prefer a Gaussian kernel. For n-dimensions it is given by

$$K(\vec{r};\sigma) = G_n(\vec{r};\sigma) \equiv \frac{1}{(2\pi\,\sigma^2)^{\frac{n}{2}}} e^{-\frac{\|\vec{r}\|^2}{2\sigma^2}} \tag{9}$$

There are many mathematical arguments for the use of the Gaussian kernel with respect to the filtering defined in Eq. (8). The most important is the conservation of *causality* [4] when the resolution parameter is enlarged. An important practical argument is the strong noise reduction [5].

3.2. Fuzzy Radon transform for a spiral

If the definition of the track density Eq. (2) is substituted in the expression of the fuzzy track density Eq. (8), the fuzzy Radon transform becomes for a spiral

$$R\,(\kappa,\phi,\gamma;\sigma) = \sum_{i=1}^{N} \lambda_i \int_0^{\tau_m} G_3\left(\vec{r}\,(\tau,\kappa,\phi,\gamma) - \vec{r}_i;\sigma\right) d\tau \tag{10}$$

From Eq. (10) it follows that also particles in the *neighbourhood* of the track (with respect to τ) contribute to the *fuzzy* Radon transform.

3.3. Reduction by using local gauge coordinates

Eq. (10) is (for each particle) a rather complicated expression in the track parameters $\{\kappa,\phi,\gamma\}$ and the ith particle coordinates (x_i,y_i,z_i). Reduction in Eq. (10) can be achieved by expressing Eq. (10) in the *particle dependent* coordinate system $\{x',y',z'\}$ which is obtained with the following coordinate transformation: a translation $\vec{r}\mapsto\vec{r}-\vec{r}_i$ followed by a rotation in the xy-plane over an angle of $\phi + sgn(\kappa)\,\tau_i$, where τ_i is the value of the track parameter which gives the track position closests to the particle coordinates $\vec{r}_i$. With this local *gauge* coordinate transformation the ϕ and x_i dependencies are removed: the (invariant) distance d_i between track and ith particle in the xy-plane is the remaining spatial parameter in this plane.

Minimalisation of d_i gives for τ_i

$$\tau_i = sgn\,(\kappa)\,\arctan\left(\frac{y_i\sin\phi + x_i\cos\phi}{\kappa^{-1} + x_i\sin\phi - y_i\cos\phi}\right) + n\pi \tag{11}$$

The redundancy in Eq. (11) can be resolved for $0 \leq \tau_i < 2\pi$ by comparing the distances between track and ith particle for the positions at τ_i and $\tau_i + \pi$.

The track parametrisation becomes in gauge coordinates

$$\begin{aligned} x' &= |\kappa|^{-1} \sin \tau' \\ y' &= \kappa^{-1} (\eta_g - \cos \tau') \\ z' &= \gamma \tau' - z_g \end{aligned} \tag{12}$$

with

$$\begin{aligned} \tau' &\equiv \tau - \tau_i \\ \eta_g &\equiv \sqrt{(\kappa\, x_i + \sin \phi)^2 + (\kappa\, y_i - \cos \phi)^2} \\ z_g &\equiv \gamma\, \tau_i - z_i \end{aligned} \tag{13}$$

Please note that the minimum distance in the xy-plane between track and the ith particle is found for $\tau' = 0$; it is given by $d_i \equiv |y'| = |\kappa^{-1}(\eta_g - 1)|$.

Performing the gauge coordinate transformation in Eq. (10) and substitution of Eq. (9) and Eq. (12), together with the approximation $\cos \tau' \simeq 1 - \tau'^2/2$ gives an *analytical expression* for the *gauge* fuzzy track density ρ_{T_g}

$$\rho_{T_g}(\eta_g, z_g, \kappa, \gamma; \sigma) \simeq \frac{1}{2\pi\, \sigma^2\, \tau_m} \cdot \frac{|\kappa|}{\sqrt{\eta_g + \kappa^2 \gamma^2}}\, e^{-\frac{1}{2\sigma^2}\left[\frac{(\eta_g - 1)^2}{\kappa^2} + \frac{\eta_g\, z_g{}^2}{\eta_g + \kappa^2\, \gamma^2}\right]} \tag{14}$$

The fuzzy Radon transform is again the sum over the gauge fuzzy track densities of the separate particles as in Eq. (6) and Eq. (10).

The $\cos \tau'$ approximation in Eq. (14) is valid if $|\kappa|\sigma \ll 1$ which is always the case in a detector with sufficient resolution to be able to reconstruct the tracks.

4. Example

In figure 1 an example is given of two artificial tracks which differ in curvature value. The parameter values are $r \equiv \kappa^{-1} = 2 \pm 0.03\,\text{m}$, $\phi = 60$ deg and $\gamma = 1.5\,\text{m}$. To the coordinates of these ideal tracks random drift values are added. The resulting tracks are displayed in figure 1.

In the images of figure 2 each row contains the fuzzy Radon transform with the following standard deviations from the top to the bottom row $\sigma = 0.5\,\text{mm}$, $\sigma = 1.8\,\text{mm}$ and $\sigma = 5\,\text{mm}$. In each image row γ is varied, in the columns of each image r is varied and in its rows ϕ is varied. The original track parameters are in the center image of the center row. The images of each image row are separately scaled. The discretisation steps are $\Delta r = 0.017\,\text{m}$, $\Delta\phi = 0.5$ deg and $\Delta\gamma = 0.05\,\text{m}$.

We see clearly in the middle row of figure 2 that there are two local maxima in the Radon transform: they are found for the correct track parameters. The used

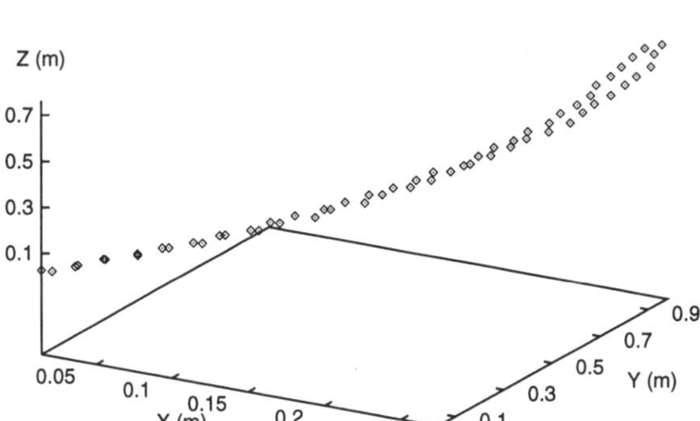

Figure 1: Two artificial tracks with added random drift noise are shown. Only the curvature values differ in the artificial tracks: $r \equiv \kappa^{-1} = 2 \pm 0.03\,\mathrm{m}$.

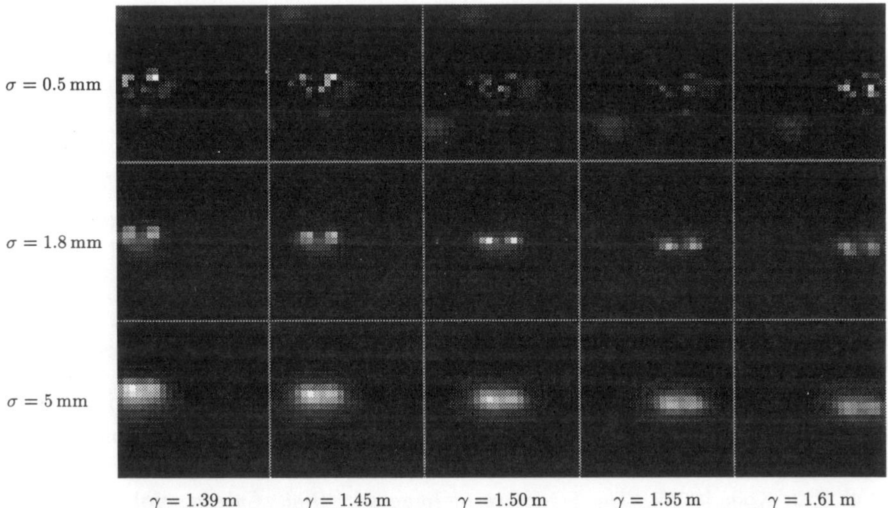

Figure 2: The image rows display the fuzzy Radon for three standard deviations which are increasing from the top to the bottom row. In each image row γ is varied, in the columns of each image r is varied and in its rows ϕ is varied.

value of σ has about the same size as the mean drift distance.

5. Conclusions

The fuzzy Radon transform with resolution embedding introduced by the Gaussian kernel proves to be a stable way to obtain the track parameters from noisy detector input as is shown in the example. The standard deviation of the used Gaussian kernel should have about the same size as the size of the mean drift distance (detector resolution)

The differences between the algorithm of the fuzzy Radon transform we use and the algorithms commonly used, depends from the way resolution is introduced in general. Often the fuzzy Radon transform is regarded to be an *energy function* (the blur function is a potential [2]), however we feel that an interpretation as an multi-resolution embedding (section 3) is more naturally, because detectors are measuring physical observables. Anyhow with our interpretation there exists a firm mathematical description which leads to the most fundamental blur kernel to use: the Gaussian kernel.

Another difference with the commonly used algorithms and our method is that often the formal integral in the fuzzy track density over the trajectory has been replaced by a contribution which is only determined by the *intersection* of the formal track with the detectors. This method has two disadvantages: 1. an implicit minimum resolution is introduced by the physical meaning of the feature "intersection" and 2. a reduction by using gauge coordinates as in subsection 3.3 is not possible anymore. Please note that our analytical reduction may also lead to a better numerical stability.

6. References

1. A. L. Yuille, K. Honda, and C. Peterson, *Particle Tracking by Deformable Templates* in *Proc. IEEE INNS International Joint Conference on Neural Networks* **vol. 1** (Seattle, WA, July 1991) pp. I 7–12.
2. M. Gyulassy and M. Harlander, *Elastic Tracking and Neural Network Algorithms for Complex Pattern Recognition* in *Computer Physics Communications* **no. 66** (1991) pp. 31–46.
3. M. Ohlsson, C. Peterson, and A. L. Yuille *Track Finding with Deformable Templates – The Elastic Arms Approach* in *Computer Physics Communications* **no. 71** (1992) pp. 77–98.
4. J. J. Koenderink, *The Structure of Images* in *Biol. Cybern.* **vol. 50** (1984) pp. 363–370.
5. J. Blom, B. M. ter Haar Romeny, A. Bel, and J. J. Koenderink, *Spatial Derivatives and the Propagation of Noise in Gaussian Scale-Space* in *J. of Vis. Comm. and Im. Repr.* **vol. 4** (March 1993) pp. 1–13. Also: 3DCV Technical Report 91-03.

DISTORTION INVARIANT OBJECT RECOGNITION UNDER DRASTICALLY VARYING LIGHTING CONDITIONS

MARTIN LADES
Institut für Neuroinformatik, Ruhr-Universität Bochum
44780 Bochum, Germany

JOACHIM BUHMANN
Institut für Informatik II, Universität Bonn, Römerstraße 164
53117 Bonn, Germany

FRANK EECKMAN
Institute for Scientific Computing Research
Lawrence Livermore National Laboratory
Livermore, California 94550, U.S.A.

ABSTRACT

Distortion invariant object recognition under drastically varying lighting conditions poses one of the great challenges in computer vision. This paper describes a face recognition system which is equipped with a silicon retina for image preprocessing. It matches stored prototypes to new images in an elastic fashion. Each face prototype is represented as an attributed graph with feature vectors attached to vertices and distance vectors labeling edges. The link structure is a minimum spanning tree. Vertices are placed at image locations with high saliency. Under drastically changing lighting conditions the recognition system with a silicon retina shows a 30% increase in recognition performance compared to the control setup with a gain controlled CCD camera.

1. General Algorithmic Approach

Face recognition is recognized as one of the difficult object recognition tasks in computer vision. Local distortions due to changing facial expressions or perspective and changes in illumination have to be discarded to find the correct prototype for a face. The nature of the distortions encountered in this object identification problem favors an elastic matching procedure to adapt different face prototypes to image features. The sensitivity to lighting conditions is most efficiently reduced by a retinal preprocessing as performed in the first layers of the retina. In this paper, we describe a comparative face recognition experiment with two different sensors — a silicon retina and a gain-controlled CCD camera. The experiment is designed to measure the increase in recognition performance due to retinal preprocessing. For both camera types the face recognizer employs an elastic matching strategy to identify the unknown face.

Our elastic matching system for object recognition is inspired by a novel concept for neural computing, the dynamic link architecture model of brain function[7]. Related work on elastic template matching was reported by Burr[3] for handwritten character recognition and by Yuille[8] for identifying parts of faces. The system was reformulated in an algorithmic fashion to improve its performance. This section gives an overview of the basic elastic matching procedure for object recognition. The structure of the recognition system consists of (i) preprocessing with a silicon retina, (ii) model extraction and (iii) the elastic matching process which results in the classification of unknown faces. The following sections will describe in sequence the functionality provided by the chip prototype, the experimental setup for the retina chip experiments, the experimental results and the structure of the program package. Extensions to the method e.g. for hierarchical, size invariant recognition are reported elsewhere [2].

1.1. Preprocessing

Image preprocessing in the face recognition system is motivated by neural receptive fields found in visual cortex of the primate visual system. A filter bank of selfsimilar filter functions which are generated from the mother wavelet by dilation and rotation is employed for image preprocessing. A mathematical description of the filter bank are wavelets based on Gabor functions

$$\psi_{\vec{k}}(\vec{x}) = \frac{\vec{k}^2}{\sigma^2} \exp\left(-\frac{\vec{k}^2\vec{x}^2}{2\sigma^2}\right) \left[\exp\left(i\vec{k}\vec{x}\right) - \exp\left(-\frac{\sigma^2}{2}\right)\right] \tag{1}$$

The advantage of the Gabor filters (apart from a small correction to remove the DC component) is minimal variance in the frequency and spatial domain. Equation (1) defines a filter positioned at $\vec{x}$ with wave vector $\vec{k}$ for the carrier frequency and σ as the width of the envelope function. Our filter bank covers 5 frequency channels with 8 different orientations each. It is designed to compensate for the $1/f$ decrease in the power spectrum of natural images. Altogether the filter set covers the frequency band relevant for our face objects (see Fig. (1)). A feature vector is formed at the position of every pixel composed of the modulus of all filter responses at that coordinate. A feature vector, also called jet in the following, carries information of the local power spectrum. The final result of the preprocessing stage is a map of feature vectors, a feature image.

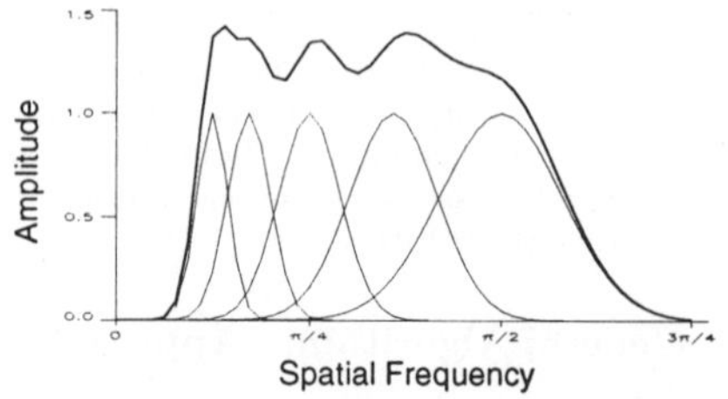

Figure 1: Frequency space coverage with Gabor filters. Five filter functions for one direction with halve octave spacing and section through the 2D sum of all filters.

1.2. Model Extraction

The map of feature vectors is the basis to extract an object model. Due to redundancy in the feature map and for computational reasons, it is desirable to transfer only a subset of salient feature vectors to the object model. We use the jets at the vertex positions of a sparse graph and attach them as node labels to the graph. The image compression property and smoothness of Gabor wavelets warrant an efficient object representation. The links of the object graph are labeled with the distance vectors between its vertices.

A first version of the algorithm[5] with regular grids for the graph topology (see Fig. (2a) achieves good recognition results despite discounting some of the sampling aspects. The model extraction in that work positioned a rigid master graph model such that a similarity measure between the master model and the new input is maximized. To stress the physics aspect of the algorithm, the similarity measure can be viewed as a potential which is minimized by a steepest descent procedure. Local minima do not pose a problem since the feature maps are sufficiently smooth. Almost any face model can serve as the master model because of the structural similarity of different faces.

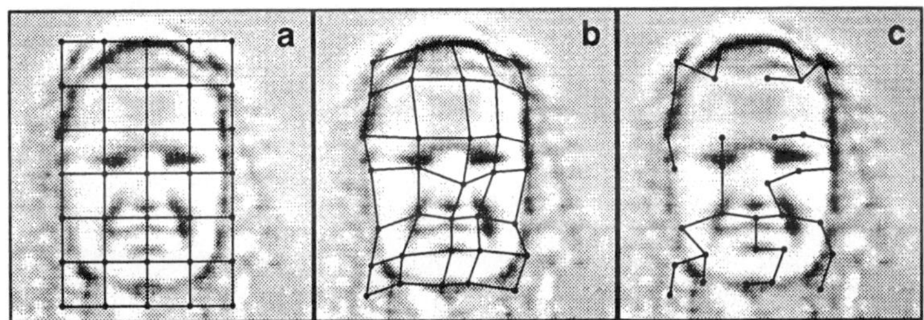

Figure 2: Successive stages of adaptive template extraction on a retina chip image: crude segmentation by the global move of a rigid template graph (a), elastic repositioning of graph vertices to salient points in the image (b), reduction of graph structure to a minimal spanning tree (c).

Nodes were not explicitly positioned to capture highly salient image points. The features of a graph model therefore did not represent a face optimally. Another stage in the model extraction process is required which maximizes jet length by elastic deformation of the template prior to extraction. This node placement step improves the information content of each node in the model since large moduli of the Gabor wavelet reflect structural image information, i.e., edges. The deformation process is constrained by the requirement that nodes and their attached feature vectors should not encode redundant information. Therefore, nodes should be spaced further than a minimal distance apart, as shown in Fig. (2b) for a typical case. The nodes are placed approximately on local maxima of jet magnitude in the feature image, the position of multiscale edges in the original image.

To boost performance the link structure of an object model is reduced to its essentials with Kruskal's minimum spanning tree algorithm[4]. It selects the structure with the shortest path length of all links which connects all nodes (the result is shown in Fig. (2c)).

1.3. Elastic Deformation and Classification

The elastic matching dynamics optimizes the cost function

$$\mathcal{H}^M\left(\vec{x}_i^I\right) = \frac{\lambda}{2}\sum_{\langle i,j\rangle}\mathcal{D}_l(\Delta\vec{x}_{ij}^I, \Delta\vec{x}_{ij}^M) - \sum_i \mathcal{S}_v\left(\vec{J}(\vec{x}_i^I), \vec{J}_i^M\right) \quad (2)$$

$$\mathcal{S}_v(\vec{J}^I(\vec{x}_i), \vec{J}_i^M) = \frac{\vec{J}^I(\vec{x}_i)\cdot\vec{J}_i^M}{\|\vec{J}^I(\vec{x}_i)\|\,\|\vec{J}_i^M\|} \quad (3)$$

$$\mathcal{D}_l(\Delta\vec{x}_{ij}^I, \Delta\vec{x}_{ij}^M) = \left(\Delta\vec{x}_{ij}^I - \Delta\vec{x}_{ij}^M\right)^2 \quad (4)$$

which is composed of the feature similarity term (3) and the topology constraint term (4). $\mathcal{S}_v$ measures the similarity between the feature vector $\vec{J}_i^M$ of vertex i in an object model M and its corresponding jet $\vec{J}^I(\vec{x}_i)$ at position $\vec{x}_i$ in the input feature image I. $\mathcal{S}_v$ is independent of linear contrast changes by construction. $\mathcal{D}_l$ characterizes the dissimilarity of a distorted link $\Delta\vec{x}_{ij}^I$ connecting vertex i and j in the image to its original, undistorted version $\Delta\vec{x}_{ij}^M$ stored in the object model. The cost function $\mathcal{H}$ combines the topological and feature terms into a measure of distance between the graph M and the input I. λ offers the possibility to adjust the relation between the two cost terms.

We now minimize that distance for all object models stored in our database by elastically deforming them until they optimally fit a given input image. All object graphs are at first positioned on the coordinates yielding the optimal average feature costs for a rigid master template. This is equivalent to the minimum of $\lim_{\lambda\to\infty}\mathcal{H}$ for the master model and performs a crude image segmentation. Single vertices of each model are then moved in a Monte Carlo fashion to coordinates showing improved costs. The result is a rubber-sheet like adaptation of each object model to the input data. The optimization scheme for the whole object gallery is hierarchical. All graph models are distorted only a few Monte Carlo steps before the half of the object gallery showing the best costs is selected to continue the deformation procedure while the rest is discarded. Distortion and selection are iterated until a fixed small number of models is left. Only those are adapted to the input until the equilibrium between topological and feature costs is reached.

A classification stage collects the distance measures calculated during the hierarchical distortion process and ranks the object models accordingly. The best model, the one closest to the input, is proclaimed to be the person found if a significance threshold is passed. We calculated several significance measures based on the cost distribution[5].

2. Retina Chip

The analog VLSI retina chip from Caltech is designed with CMOS transistors in a novel subthreshold mode as silicon equivalent to ion channels. The chip models the outer plexiform layer of the vertebrate retina to implement a synaptic network. The excitatory elements in the receptor layer are coupled by diffusion channels to implement low pass filtering. A second layer of neural elements (the horizontal cells) with a larger diffusion constant feeds back the local background intensity as shunting inhibition and, thereby, normalizes local contrast. The two coupled layers regulate local contrast over more than five orders of magnitude. In contrast to a biological retina the chip's receptive fields widen with intensity, i.e. the low pass filtering increases. A blooming effect became visible when the circuits were saturated. Both side effects have been solved in the following chip generation.

The intensity profile of Fig. (3c) shows that not much image information in CCD video imagery is left on the dark side of the face. The chip prototype clearly preserves information which is irretrievable from CCD images by a software solution alone.

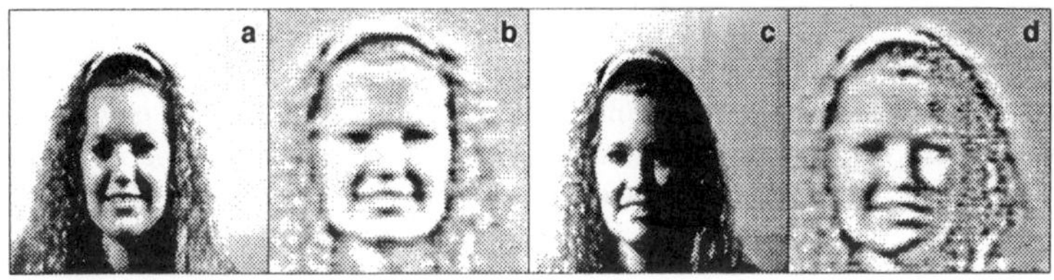

Figure 3: Example pictures taken with a gain controlled CCD camera and the retina chip prototype. (a) and (b) are taken with spotlight illumination from both sides, (c) and (d) with only one spotlight.

3. Experimental Setup

To compare the face recognition system equipped either with a silicon retina or a gain-controlled CCD camera, a laboratory experiment was set up with controlled illumination conditions. As seen in Fig. (4) two spotlights are directed on a face of a test person from the left and right side. A datacube video board provided the means for quickly taking pictures from the nonstandard chip video interface as well as a normal gain-controlled CCD camera. In the case of the conventional camera, the pictures were reduced from NTSC format to 128×128 pixels. The images from the 90×92 pixel retina chip were further processed to reduce artifacts introduced by the on-chip readout circuitry.

4. Recognition Results

The recognition results are summarized in Table (1). Table (1a) gives recognition results under even lighting conditions, but with distortion caused by changing the viewing direction of the test person approximately 30 degrees to the side. Clearly

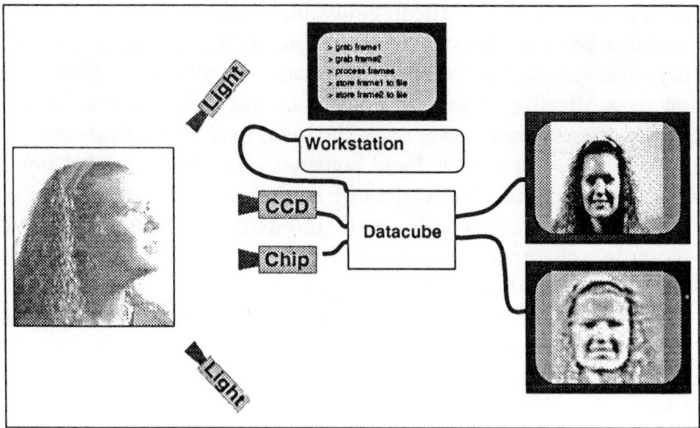

Figure 4: Schematic setup of the retina chip experiment.

the tested retina chip has a disadvantage in an evenly illuminated environment due to its lower resolution by almost half the number of pixels and the band pass filtering occurring for the contrast normalization. The situation changes drastically when the

Table 1: Face recognition results: (a) in a well illuminated environment, viewing direction of test subjects deviated up to 30 degrees from the model gallery. (b) in an environment with drastic illumination changes, test subjects looked straight into the camera. Recognition rates in parenthesis refer to the more stringent significance criterion that all matches should be discarded as insignificant if the correct prototype is not in the database.

	false positive rate	silicon retina	conv. CCD
a	100%	83.5%	89.0%)
	< 10%	80.7(65.1%)	88% (74.3)%
	< 5%	74.3(58.7%)	84.4 (66.0)%
	< 1%	59.6(48.6%)	82.5 (60.5)%
b	100%	94.5%	82.5%
	< 10%	93.6% (82.5%)	78.9% (56.0%)
	< 5%	93.6% (76.1%)	75.0% (44.9%)
	< 1%	90.8% (65.1%)	59.6% (39.4%)

light is switched off on one side so that the normal camera can not adjust to the large local differences. The retina chip provides improved recognition rates by up to 30% if significance is checked. The results are shown in Table (1b. Recognition rates in parenthesis refer to the more stringent significance criterion that all matches should be discarded as insignificant if the correct prototype is not in the database.).

The complete recognition process for a data base of 100 subjects runs on a NeXT workstation (68040 processor) in slightly less than one minute: 40 seconds for the wavelet based preprocessing and 17 seconds for the elastic matching stage with 35 vertices per graph model. The complexity scales logarithmically with the number of face models.

5. Program Structure

Our program modules are object oriented and provide further possibilities for internal restructuring and organization of the object model data base. The C++ language classes are based on a generic template to alleviate readability. The basic graph structures use the LEDA class libraries[6]. The design followed the natural data hierarchy of our object recognition problem. The preprocessing of an input image is done in regular data structures until the array of feature vectors extracted from the wavelet filter responses forms a feature image. The graph classes are as independent as possible from those block structures and start at the feature level. Our C++ node class offers different functions to compare its instances to features in an input feature image or other node instances. It includes the transformations of the feature vector in a node instance e.g. to compensate for affine transformations of the comparison target. Additional book-keeping of coordinate traces in a node during repositioning operations in the course of the elastic matching procedure is intended to aid dynamic tracking of a feature over image sequences. The hierarchy of graphs continues with a second graph order at the CollAGE level that will allow more complicated hierarchical compound object models in the future. The Gallery class supports the building of a structured database from these complex object models.

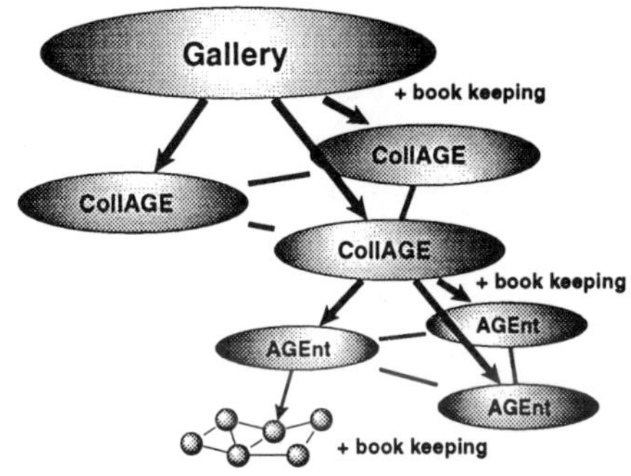

Figure 5: Aggregated Graph Entities (AGEnts), Collections of AGEnts (CollAGEs) and Gallery constitute a fully dynamic graph hierarchy of several orders. Each level of the hierarchy adds book keeping of cached information to minimize tree traversal needed for elastic operations.

Acknowledgements

It is a great pleasure to thank K. A. Boahen for providing us with the retina chips. We acknowledge stimulating discussions with C. von der Malsburg and C. Mead. This work was supported by the German Ministry of Science and Technology (ITR-8800-H1) and by the U.S. Department of Energy, Lawrence Livermore National Laboratory (W-7405-Eng-48).

1. K. A. Boahen and A. G. Andreou, *A Contrast Sensitive Silicon Retina with Reciprocal Synapses* in *NIPS 91 Proceedings* (IEEE, 1992).

2. J. Buhmann, M. Lades and C. von der Malsburg in *IJCNN San Diego Proceedings* (IEEE, 1990), II 411-416.
3. D.J. Burr, *IEEE Tr. PAMI* **3** (1981) 708–713.
4. J.B. Kruskal, *Proc. American Math. Society* **7**, (1956) 58–50.
5. M. Lades, J. C. Vorbrüggen, J. Buhmann, W. Konen, C. v.d. Malsburg and R. P. Würtz, *IEEE Tr. on Computers* **42** (1993) 300–311.
6. K. Mehlhorn, S. Näher, *LEDA, a Library of Efficient Data Types and Algorithms*, TR **A 04/89**, FB10, Universit"at des Saarlandes, (1989).
7. C. von der Malsburg, *Internal Report, MPI phys. Chem.*, (1981).
8. A. Yuille, *Neural Computation* **2** (1990) 1–24.

MODEL NEURAL NETWORK AND THRESHOLD CALORIMETER TRIGGERS FOR b-QUARK EVENTS SELECTION IN FIXED TARGET EXPERIMENTS

V. V. PALICHIK, V. M. SEVERYANOV, V. N. SHIGAEV
Laboratory of Computing Techniques and Automation
Joint Institute for Nuclear Research, 141980 Dubna, Russia

Ju. A. BUDAGOV
Laboratory of Nuclear Problems
Joint Institute for Nuclear Research, 141980 Dubna, Russia

and

N. O. POROSHIN
Department of High Energy Physics
Moscow Engineering and Physics Institute, 115409 Moscow, Russia

ABSTRACT

Classification power of a set of calorimeter triggers for b-quark events selection is investigated on the basis of Monte-Carlo simulated data for fixed target experiments on 3 TeV proton beam. Characteristics of a neural network trigger and threshold triggers with modified decision functions are shown to remarkably surpass characteristics of commonly used simple threshold triggers. Some variations of the neural network architecture were examined in order to improve the trigger.

1. Introduction

Fixed target experiments on Accelerating Accumulating Complex (UNK, Serpukhov) possess a number of unique possibilities[1], but there exists the problem of complex background conditions, which is characterized by relation $\sigma_b/\sigma_{in} \approx$ $\approx 3 \cdot 10^{-5}$ for the "Multiparticle Spectrometer" (MPS) experimental setup[2]. One of the ways to solve such a problem is creation of triggers with effective background rejection property. Due to big similarity in data between physically interesting events and background events in TeV-range, experiments require more sophisticated triggers than commonly used. Using simple cuts is no longer sufficient.

This paper contains comparative investigation of characteristics of commonly used threshold triggers, their modifications, and neural network trigger for selection of b-quark events related to B-mesons production (termed B-events for short).

Electromagnetic and hadron calorimeters are assembled in pairs and named vertex and forward calorimeters correspondingly. In this work simulated data were used only for the model forward calorimeter of $200cm \times 200cm$ size located at 30 m from the target.

The main characteristics of triggers are efficiency of B-events recognition ε_B, efficiency of non b-quark events (background) registration ε_{bg} (usually background rejection factor $R = 1/\varepsilon_{bg}$ is used), enrichment coefficient Q which by definition indicates the change in number of B-events to number of background events ratio after triggering. The curve Q versus ε_B is the most easy-to-interpret and informative one. In a sense, $Q(\varepsilon_B)$ function is universal operational characteristic and allows one to compare different triggers easily.

It should be noted that the results we have obtained are based on the simulated data and relate to certain energy of the incident beam (3000 GeV) and to certain geometry and parameters of the MPS setup. The simulation data files contain the forward calorimeter information for 500 B-events with $p+p \to B+X$ processes including B-mesons in the final state and for 1000 background events $p+p \to X$ without b-quarks. These processes are generated by PYTHIA 5.3 program. The passage of particles through the setup regions is simulated by GEANT 3.14 and GHEISHA packages taking into account all physical effects due to the nature of the particles and their interactions with the matter and the magnetic field. Cut was at 100 MeV for e^+, e^-, γ and 50 MeV for other particles. Magnetic field is oriented along Y-axis in reference system.

In searching for optimal neural network trigger we used the back-propagation multilayer feed-forward artificial neural network from JETNET 2.0 package[3].

2. Threshold triggers

Our main goal was mathematical construction of a model neural network trigger that would satisfy certain requirements: it should effectively suppress background events, its architecture should be rather simple, its overall performance

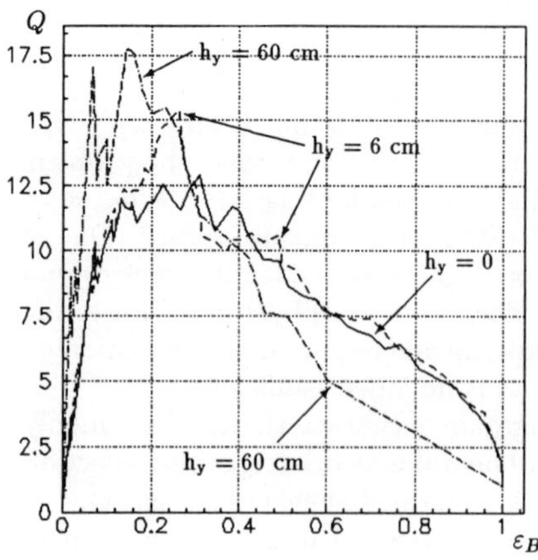

Fig. 1: Q vs ε_B for triggers on E_{ty} ($|Y| > h_y$)

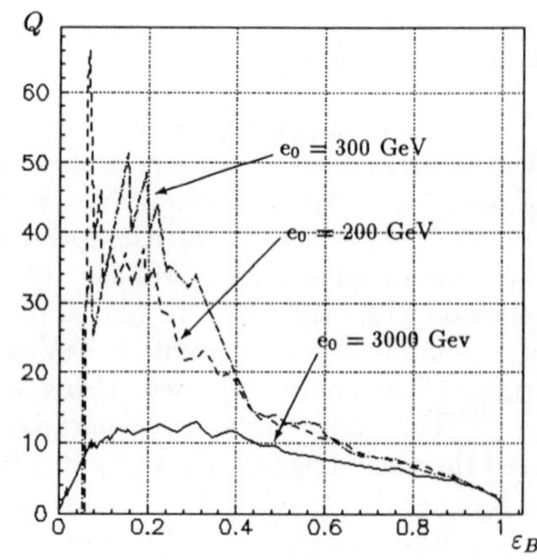

Fig. 2: Q vs ε_B for triggers on E_{ty} ($e < e_0$)

should be potentially high after implementation in hardware. Consequently the information from 10000 cells of our model calorimeter should be preprocessed by a simple and fast procedure and thus reduced to a small set of feature values used as the neural network input data.

Each feature has its own specific frequency distribution in both classes of background and B-events. Feature evaluation functions may naturally be used as

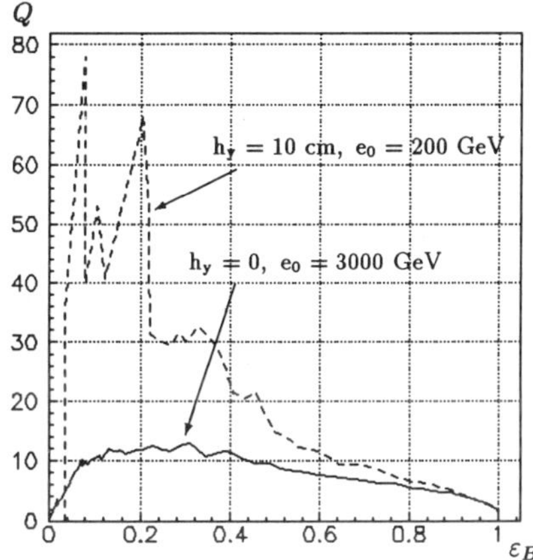

Fig. 3: Q vs ε_B for triggers on E_{ty} ($|Y| > h_y, e < e_0$)

Fig. 4: Q vs ε_B for triggers on E_{c44} (solid line) and E_{tot} (dashed line)

decision functions in simple threshold triggers. Function candidates were chosen upon physical reasons, the choice being in certain aspects limited by the data registration model. In particular the search for active cell clusters with view to identify particles for subsequent use in decision functions has not been included into data preprocessing because only sums of energies deposited in corresponding pairs of electromagnetic and hadron calorimeters are accessible. Therefore in the course of our investigation only functions of integral type which reflect macroscopic attributes of deposited energy distribution within the calorimeter were considered.

Let us consider some of the threshold triggers.

Radial transverse energy trigger is $E_{tr} = \sum_k e_k r_k$, where e_k - energy deposition in k-th cell, r_k - distance between the incident beam and the center of k-th cell. Operational characteristic of the trigger is presented in fig. 5 (line 1). It follows that in efficiency range $0.8 - 0.9$ the background rejection factor is two times lower compared to requirements of MPS project[2]. Three modified decision functions have been tested in order to improve background rejection property of the trigger on E_{tr}: $E_{tr}(e < e_0) = \sum_k e_k r_k$ for $e_k < e_0$, $E_{tr}(r > d) = \sum_k e_k r_k$ for $r_k > d$, and $E_{tr}(r > d,\ e < e_0) = \sum_k e_k r_k$ for $r_k > d$ and $e_k < e_0$. Operational characteristics for triggers with modified decision functions of E_{tr}-family are presented in works[4,5]. Figures 1, 2, 3 present operational characteristics of another similar family of threshold triggers based on Y-component of transverse energies. Triggers of E_{ty}-family allow one to reach higher background rejection rates for low values of efficiency ε_B. Explanation for the fact as well as optimal values of e_0 and d-parameters of modified decision functions are given in works[4,5].

Total energy trigger is $E_{tot} = \sum_k e_k$. At efficiency $\varepsilon_B = 0.96$ the background rejection factor is 3. Trigger on partial energy sum $E_{c44} = \sum_k e_k$ for $r_k < r_0$ was taken as a modified version of E_{tot} trigger. Optimal value of r_0 corresponds to summing up energies deposited in the central 4×4 calorimeter cells. For efficiency $\varepsilon_B = 0.99$ the E_{c44} trigger suppresses background events by a factor of 2. Figure 4 presents operational characteristics of threshold triggers based on decision functions E_{tot} and E_{c44}.

3. Neural network trigger

To select the optimal neural network trigger we have examined different combinations of input parameters and such neural network characteristics as the number of neurons in hidden layers, "temperature" T, weight updating mode (with and without Langevin noise in learning), learning rate η.

We found no substantial improvement in trigger performance when network characteristics mentioned above varied. In specifying other parameters of neural network we followed recommendations of JETNET authors.

For testing generalization ability of the neural network we have split our Monte-Carlo data into two equal parts: the first part used for neural network training (learning sample) and the second part - for testing neural network as a trigger (test sample). Twenty parameters were estimated for each event. Various combinations of these parameters were used as input data for neural network.

In our research we have most thoroughly examined a one hidden layer neural network architecture with $10 + 2 * i$ hidden neurons in it, where i is the number of input parameters. The output layer had one neuron, the activation function of each neuron was a sigmoidal one.

Neural network training procedure was subdivided into epochs and cycles: an epoch contains a given number of training cycles and one cycle for testing neural network as a classifier. The learning rate η was gradually decreased : from 0.01 for the first epoch to 0.005 – 0.001 for the last epoch. Training sessions including up to 20 epochs with 10 – 20 cycles in an epoch were carried out. It turned out that the combination of 4 epochs with 10 cycles in an epoch was sufficient for getting low enough classification error.

Into the reduced set of input parameters for neural network trigger we included total energy E_{tot} deposited in the calorimeter by all particles of an event, energy E_{c44} deposited in central 4×4 cells of the calorimeter, radial transverse energy E_{tr} (total and modified), Y-component of transverse energy E_{ty} (total and modified), and maximum of absolute magnitudes of active cells Y-coordinates. Parameters were normalized in [0,1] interval on learning sample before passing them to the network. Experiments with the model neural network have shown that E_{tot}, E_{c44}, E_{tr} and E_{ty} ($|Y| > 10\ cm,\ e < 200\ GeV$) could be used as optimal set of input parameters for the neural network calorimeter trigger.

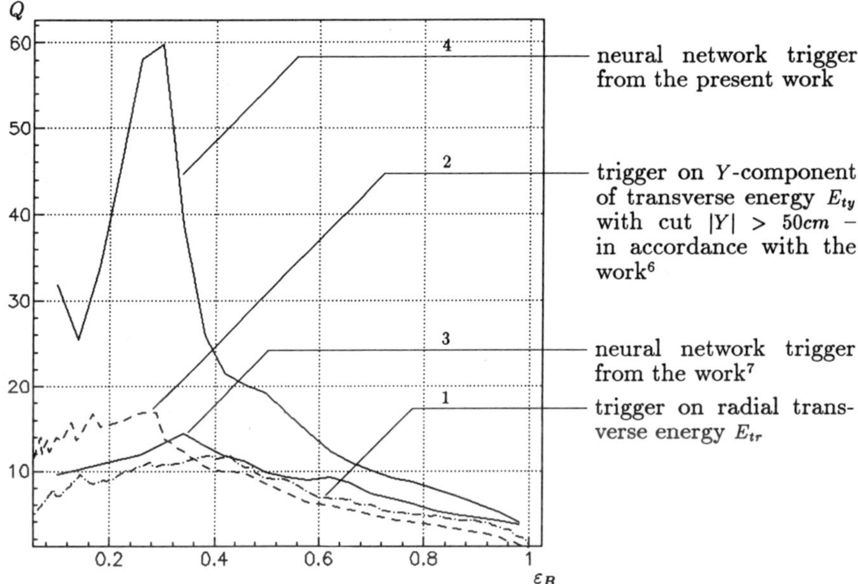

Fig. 5: Q vs ε_B for several triggers

Operational characteristic of the neural network trigger with optimal parameters on input is presented in fig. 5 (line 4). We arrive to a conclusion that enrichment coefficient $Q = 8 - 10$ for B-events at efficiency $0.7 - 0.8$ can be achieved. Background rejection factor is $10 - 15$ with only small loss of B-events ($20 - 30$ %).

In fig. 5 the operational characteristics of E_{tr} trigger (line 1) and the neural network trigger from the work[7] (line 3) are presented too. In work[7] the total energy and 3 values averaged on active cells (mean energy, mean transverse energy and mean distance from the center of the calorimeter) were used as input to the neural network. Line 2 corresponds to a trigger in accordance with the work[6]. It should be noted that in the range $\varepsilon_B = 0.7 - 0.8$ all these three triggers have enrichment coefficient $Q = 4 - 6$ that compares unfavorably (approximately two times lower) with the results of the present work.

4. Conclusion

It is shown that modified threshold and neural network triggers provide higher background rejection rates in the whole trigger efficiency range for b-quark events selection. For a particular fixed target experimental setup (MPS) background rejection factor in the range $10 - 200$ at trigger efficiency $0.2 - 0.8$ can be achieved with the proposed version of neural network trigger operating on the forward calorimeter data without using data from other detectors. This background rejection factor is 1.5 – 5 times higher compared to results at UNK energies published elsewhere.

The lateral and longitudinal jet dimensions and data from the vertex calorimeter were not taken into account in this work. These factors and use of data from other detectors of the MPS setup are object of future investigations.

5. Acknowledgements

We would like to express our gratitude to Stanislav Tokar for fruitful discussions and useful suggestions.

6. References

1. *The Accelerating Accumulating Complex on 3000 GeV*, *Preprint IHEP 93-27*, (IHEP, Serpukhov 1993) (in Russian).
2. A. M. Zaitsev, *Workshop "Physics on UNK"*, (IHEP, Serpukhov 1989), p. 27.
3. L. Lönnblad, C. Peterson, T. Rögnvaldsson, *Computer Physics Communications* **70** (1992) 167.
4. Ju. A. Budagov, V. V. Palichik, N. O. Poroshin, V. M. Severyanov, V. N. Shigaev, *JINR, B1-10-93-116*, (JINR, Dubna, 1993) (in Russian).
5. Ju. A. Budagov, V. V. Palichik, N. O. Poroshin, V. M. Severyanov, V. N. Shigaev, *JINR, R10-93-140*, (JINR, Dubna, 1993) (in Russian).
6. G. A. Kolomensky, G. Z. Obrant, A. I. Smirnov, *Preprint 1699 LINP*, (LINP, Leningrad, 1991).
7. V. V. Ivanov, N. O. Poroshin, V. M. Severyanov, *Matematicheskoe modelirovanie,* **4** *(1992) No. 8, 94–106* (in Russian).

Neural Networks Applied to a Second-Level Trigger Based on Calorimeters

J.M. Seixas , L.P. Caloba, M.N. Souza,
A.L. Braga, A.P. Rodrigues and H. Gottschalk
COPPE/EE/UFRJ, C.P. 68504, Rio de Janeiro 21945, Brazil

Presented by J.M. Seixas

ABSTRACT

A second-level trigger system for LHC based only on calorimetry is analyzed using neural networks. Calorimeter data is obtained from Monte Carlo simulations and all events used have passed a simulated first-level trigger operation.

It is shown that sending to the network the cell contents of the calorimeter in a region of interest, one can achieve 97% electron efficiency, while 10% of jets are misclassified as electrons. For higher performance a feature closely related to the effective radius of the shower development inside the calorimeter is derived. When this feature is fed into network, less than 6% of the jets are misclassified.

1. Introduction

The event selectivity is known to be quite critical for the LHC experiment. The event rate of 100 MHz coming from the collider machine has to be dramatically reduced by the triggering system. Detailed physics simulations show that a second-level trigger system is desired to achieve further event rate reduction after the first-level operation.

The second-level system takes information from different detectors and its operation has been conceived in two steps [1] . The first comprises the extraction of features from each detector that participates in the trigger system and the second one combines all features extracted into a global decision unit. Here, we consider the feature extraction problem for calorimeters. The features to be extracted can be found by searching for the best calorimeter variables one could send to the input nodes of a neural network discriminator capable to perform electron/ jets separation based only on calorimeter information.

Monte Carlo simulations have been used to generate events for the second-level operation [2] . A large sample of QCD jets were generated using PYTHIA. Single electrons were generated with PYTHIA in the same η range as the QCD jets, simulating an electron energy profile quite close to the one of jets. The calorimeter has cells of size $0.02 \times 2\pi/300$ in η-Φ. The detector has electromagnetic (e.m.) and hadronic sections with cells of the same size and precisely aligned. Only barrel cells were considered.

On the sample data a first-level trigger algorithm that has been proposed for the ATLAS experiment [3] was applied. After the first-level filtering, pileup was added. Minimum bias events were generated with PYTHIA and an average of 20

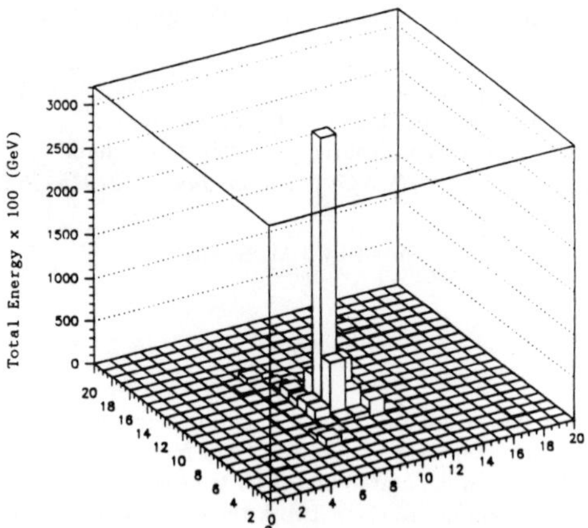

Figure 1: Typical jet event without pileup.

such events were combined to simulate high-luminosity operation. The first-level algorithm was then applied again. In this manner, two data sets were available for analysis, covering low (no pileup) and high-luminosity scenarios, with a total of 1057 (995 when pileup was added) jets and 1634 (1625) electrons. Each event is described by means of two matrices of 20x20 elements, one corresponding to the energy deposited in each cell of the e.m. part of the calorimeter and another to the hadronic one. This information comes from a region of interest (ROI) identified in the detector by the first-level trigger system.

The way we defined the training set for the neural network design was the following. The total number of events in each data file was split in order to fill 10 boxes; each one with (barely) the same number of events. Training was performed on five randomly selected boxes of each particle class, with testing being applied to the remaining boxes that did not participate of the training phase. We then performed 10 groups of independent tests, by selecting randomly the boxes participating of the training phase in each test. The mean value of these 10 results is the figure we quoted for the performance of the discriminator. In all tests a network with 10 hidden layers was used.

2. Matrix Analysis

The first feature investigated was the energy profile in the ROI. Figure 1 shows a typical jet event. As most of the cells located at the outermost part of the ROI sample very low (or even null) energy, we searched for a subregion of interest. In order to define it, we performed a simple pre-processing on the input data. The cell of maximum energy deposition was identified in an event by event basis and the subregion was built around it. Following this way, we formed an 11x11 matrix of total energy, adding the corresponding e.m. and hadronic matrices. Therefore,

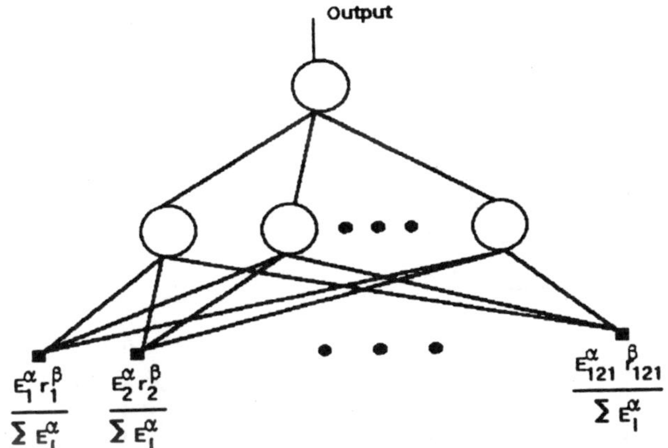

Figure 2: Neural network system using the R_p n-dimensional function as an extracted feature

each event was described by a vector having 121 components which was sent to the input nodes of the network.

The performance achieved with such an arrangement was an electron efficiency of 97% with 10 (12) % of jets being misclassified, using events without (with) pileup. The input data normalization used takes the total energy in the subregion under analysis as the normalizing factor for all cells. Other features having similar discriminating power were also investigated [4].

3. Effective Radius

A quantity R_p

$$R_p = \frac{\sum_{i=1}^{121} E_i^\alpha r_i^\beta}{\sum_{i=1}^{121} E_i^\alpha} \tag{1}$$

is known to be a good measurement of the effective shower width [5]. Here E_i represents the energy deposited in the i-th cell, r_i is the Euclidean distance between the center of the i-th cell and the cell of maximum deposited energy and α, β are parameters which can be adjusted. Again, we use the 11x11 ROI description. This quantity has also been used as part of an algorithm [2] for electron/jet discrimination. The values for α and β used there were 1 and 2, respectively.

In this section we develop a feature which is an extension of this basic idea. The R_p function, which maps the input vector space onto the real line, can be generalized in order to define the R_p n-dimensional function:

$$\begin{aligned} R_p : X &\rightarrow Y \\ (E_i, r_i) &\rightarrow \frac{E_i^\alpha . r_i^\beta}{\sum_{i=1}^{121} E_i^\alpha} \end{aligned} \tag{2}$$

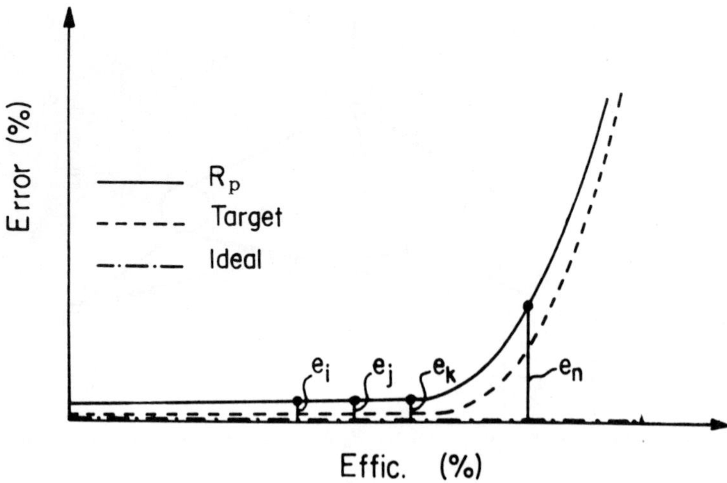

Figure 3: Norm of the error function

Thus the R_p n-dimensional function maps each cell contents in an energy weighted profile, using the radius as a weighting parameter. The image of such a mapping is sent to the neural network input nodes. Figure 2 shows the arrangement.

We have searched for the optimum value of the adjustable parameters in order to have the best performance in terms of electron/jet separation. To achieve this goal, we minimized the distance between the error function for jets misclassified as electrons obtained by the application of R_p mapping and a target one (see Figure 3). The target curve was defined by scaling down by 50% the one obtained in [2]. The idea in setting this target curve was to use a high-performance discriminator, which was not based on neural networks and has been applied on the same data we used, as a reference that we would like to improve. We can also mention a special case of the target curve when we take an ideal curve of null error. Then, the measured distance is just the norm of the error vector of the R_p discriminator.

Mathematically, we have

$$\| f \| = \sqrt{\sum_{i=1}^{N} [er(eff_i) - ert(eff_i)]^2} \tag{3}$$

where er (.) and ert (.) are the error functions for jets of the R_p and the target discriminators, respectively, and the parameter eff is the efficiency function for electrons.

Figure 4 shows how the norm (with respect to the target curve) varies according to different values of α and β for both scalar and n-dimensional R_p functions. Optimum value selection of $\alpha = 0.64$ (0.7) and $\beta = 1.2$ (1.0) for scalar (n-dimensional) R_p can be extracted from the figures. Similar behavior is found

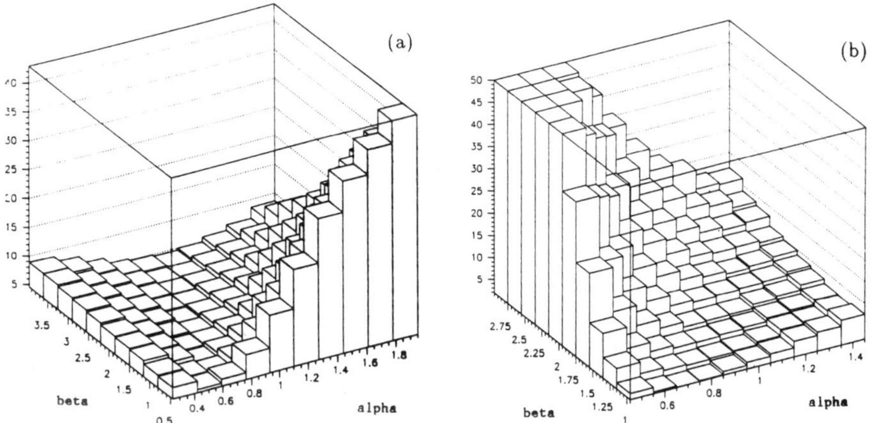

Figure 4: Lego plots for the norm of the error using R_p functions: (a) scalar and (b) 121-dimensional.

using the ideal curve as a target.

In which results are concerned, when we use the optimum scalar R_p, less than 10% of jets were misclassified when an efficiency for electrons of 97% is kept and pileup is not included. This shows that taking the norm as a figure of merit for the determination of the optimum values for α and β works as a fine tuning, as the discrimination obtained with $\alpha = 1.0$ and $\beta = 2.0$ misclassifies 12% of jets under the same conditions.

The best performance is achieved using the R_p 121-dimensional function, with which less than 6 (9) % of jets are misclassified for the same electron efficiency and using events without (with) pileup. For even higher electron efficiency, say 99%, less than 7 (12) % of the jets are misclassified. Figure 5 shows the output of the neural network for testing events in the case of feeding its input nodes with the R_p 121-dimensional mapping and without pileup.

4. Conclusions

The analysis of a second-level triggering system based on simulated calorimetry data was carried out. The system was realized by neural networks and different calorimeter features were tested. A subregion of interest was defined comprising 121 cells of the calorimeter. Sending to the neural network input nodes the total energy deposited in each one of such cells, an electron efficiency of 97% was achieved, while 10 (12) % of jets were misclassified in case of pileup of 0 (20) minimum bias events. The use of this feature results in good performance and requires few pre-processing (to find the cell of maximum energy deposition).

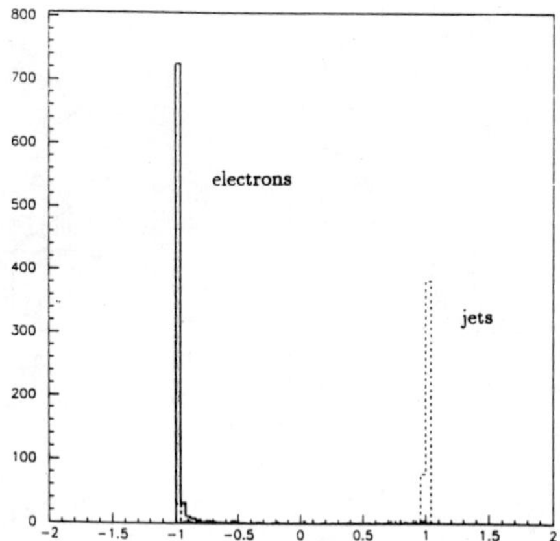

Figure 5: Output of the neural network using 121-dimensional R_p function

The best performance can be obtained applying extra processing. The R_p multi-dimensional function is able to reach 97% of efficiency on triggering on electrons with an error of less than 6 (9) % for jets identification. The optimum values of the R_p parameters can be determined using a method that uses the norm of the error function for jets as a figure of merit.

Acknowledgements

We are thankful for the support that has been provided to this work by Conselho Nacional de Desenvolvimento Científico e Tecnológico (Brasília, Brazil) and CERN. We would like to thank the RD11 Collaboration for providing the simulated data sets. In particular, we express our best thanks to R.K. Bock, F. Block, W. Krischer, G. Klyuchnikov, I.C. Legrand and J. Carter, all from CERN, for fruitful discussions concerning this work.

5. References

1. RD-11 Collaboration. **CERN/DRDC 93-12** (1993).
2. G. Klyuchnikov et al. **CERN-ATLAS/DAQ-007** (1992).
3. RD-27 Collaboration **CERN-ATLAS/DAQ-005** (1992).
4. J.M. Seixas et al. *Proc. of the Fourth Int. Conf. on Calorimetry in High-Energy Physics, Elba, Italy* (1993) (to be published).
5. D. Acosta et al. *NIM* **A305** (1991) 55.

A FEED FORWARD NEURAL NETWORK FOR RECOGNITION OF FLUCTUATIONS IN ELECTROMAGNETIC SHOWERS

THOMAS DEGENER

Fakultät für Physik und Astronomie, Lehrstuhl I für Experimentalphysik, Ruhr-Universität Bochum, Universitätsstr. 150, 44801 Bochum, Germany

e-mail: thomas@tau.ep1.ruhr-uni-bochum.de

ABSTRACT

In order to correlate energy deposits in the electromagnetic calorimeter of the Crystal Barrel detector with particles, neighbouring crystals with energy deposits are grouped to clusters. These clusters are searched for local maxima. Then particles from the primary reaction are assigned to these local maxima. A few percent of these maxima are just due to fluctuations in the electromagnetic cascade. A feed forward neural network is used to separate these fluctuations from energy deposits of photons.

1. Introduction

The Crystal Barrel detector has been in operation since October 1989 at the Low Energy Antiproton Ring (LEAR) at CERN. The detector system consists of an electromagnetic calorimeter of 1380 CsI(Tl) crystals surrounding a cylindrical Jet-driftchamber. A more detailed description is given in reference[1]. The experiment is designed to study exclusive charged and neutral final states in antiproton ($\bar{p}$) proton (p) annihilation at $\bar{p}$-beam momenta up to 1940 MeV/c. This contribution concentrates on neutral final states.

The first step in the analysis of photon data of the Crystal Barrel calorimeter is to group neighbouring crystals with an energy deposit above 1 MeV to form clusters. These clusters are searched for local maxima to resolve photons of which the electromagnetic showers are overlapping. It is assumed that every local maximum corresponds to a photon, if the energy deposit in the whole cluster exceeds 10 MeV. If more than one local maximum exists in the cluster, the central crystal of each local maximum must have more than 10 MeV, to be considered a photon. This assumption is correct in many cases, but some local maxima are just due to fluctuations in the electromagnetic cascade. These are often referred to as electromagnetic split-offs. Using the above definition of a photon candidate, Monte Carlo studies show that about 5% of all candidates are just due to fluctuations. Most split-offs appear in clusters containing more than one local maximum, called fluctuations of the 1st kind in the following. A fluctuation of the 2nd kind is a split-off that forms its own well separated cluster. Especially fluctuations of the 2nd kind of low energy are hard to distinguish from low energy photons, because a kinematic fit will almost ever fit an extra low energy photon. For a detailed analysis of antiproton proton annihilation it is vital to identify every photon correctly and to reduce the background due to fluctuations. The simplest method to eliminate many fluctuations is to increase the threshold values in the definition of a photon. But in this case low energy photons get lost. Two conventional methods, SMART[2] and DOLBY C[3] based on conventional cuts were developed. Using these methods, however, a model is introduced. A feed forward neural network as a model-free estimator is a completely different approach to solve the split-off problem. The input parameters of the network should be chosen as simple as possible to avoid the calculation of „difficult", time consuming functions.

2. Implementation of the networks

The networks described in the following paragraphs are generated with the Aspirin/MIGRAINES[4] software tools. Aspirin is a parser that offers a declarative language to describe feed forward neural networks. It generates C-code that can be easily integrated into any application. The main advantage of the parser is that many different network structures can be generated and tested in a short time: The code of the network is treated as a black box. Migraines is a tool to move through such a network interactively, i.e. one can have a look at the weights at any iteration and dump them to a file.

The networks used for this contribution are trained with the standard error back-propagation algorithm. The change of a weight is calculated by:

$$\Delta w_{ji} = -\alpha \frac{\partial E}{\partial w_{ji}} \tag{1}$$

The error E is the absolute difference between the output of the network and the target value:

$$E = |t - \text{out}| \tag{2}$$

The input to node j is a sum of weighted outputs from previous nodes connected to node j, plus a bias weight:

$$\text{in}_j = \sum_i w_{ij} o_i + w_{0j} \tag{3}$$

The node functions of all hidden and output nodes are sigmoidal functions.

After convergence, the networks are recompiled with static weights and then linked to the Crystal Barrel off-line code. As the off-line software is written in FORTRAN and Aspirin generates C-Code, a mixed language programming[5] approach is used.

3. Description of the network

The network is structured hierarchically: it consists of two sub networks (Sum9 and 9Xtal) with one single output node, connected to the input layer of a top network as shown in Fig. 1. The output of the top network is used to separate photons and fluctuations. This hierarchic structure will be described in detail in the following.

3.1. Input and output of the sub network Sum9

First a 5x5 matrix is filled with the energy deposits of the central crystal of a local maximum and its surrounding neighbours (Fig. 2). Then nine times a sum of nine is calculated as indicated by the squares in Fig. 2, in each of the four corners, in between them, and in the central region. Each sum of nine is normalized by the total energy of the matrix. These normalized sums of nine are the input to the network „Sum9". For example the

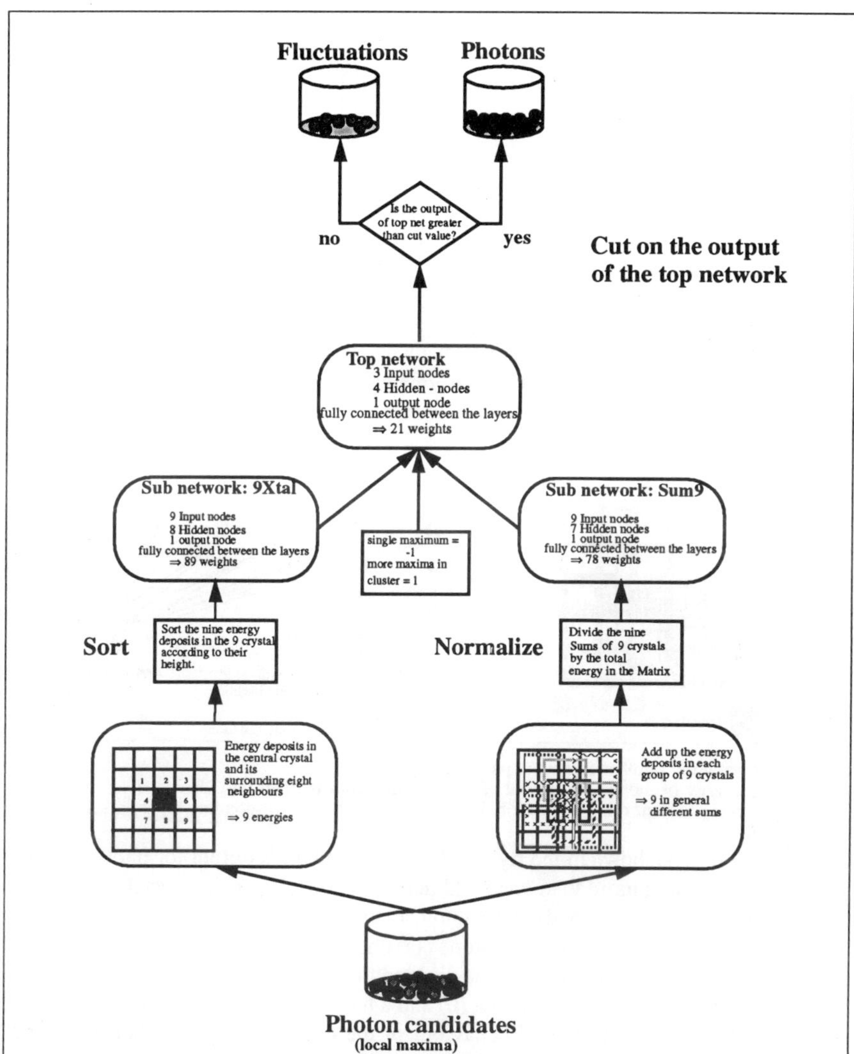

Fig. 1 Flow chart of the whole network

input of the first input node is calculated from the nine crystals in the upper left corner of the matrix by

$$I_1 = \frac{\sum\limits_{\phi = \phi_c - 2,\, \theta = \theta_c - 2}^{\phi_c,\, \theta_c} E_{xtal}(\phi, \theta)}{\sum\limits_{\phi = \phi_c - 2,\, \theta = \theta_c - 2}^{\phi_c + 2,\, \theta_c + 2} E_{xtal}(\phi, \theta)} = \frac{\sum\limits_{\phi = \phi_c - 2,\, \theta = \theta_c - 2}^{\phi_c,\, \theta_c} E_{xtal}(\phi, \theta)}{E_{25\ Xtals}} \tag{4}$$

The network has nine input, seven hidden and one output node. It is fully connected between the layers and has one „bias" weight for each hidden node and for the output node, this results in 9•7+7+7+1=78 weights. The training set consists of 2000 photons and 2000 fluctuations of the 1st kind of a $p\bar{p} \rightarrow 3\pi^0 \rightarrow 6\gamma$ Monte Carlo at a $\bar{p}$-beam momentum of 1940 MeV/c. After convergence, the performance is tested with eight „unknown" reference Monte Carlo data sets. The performance does not depend significantly on a specific γ-multiplicity in the final state, but it works a lot worse for low energy fluctuations of

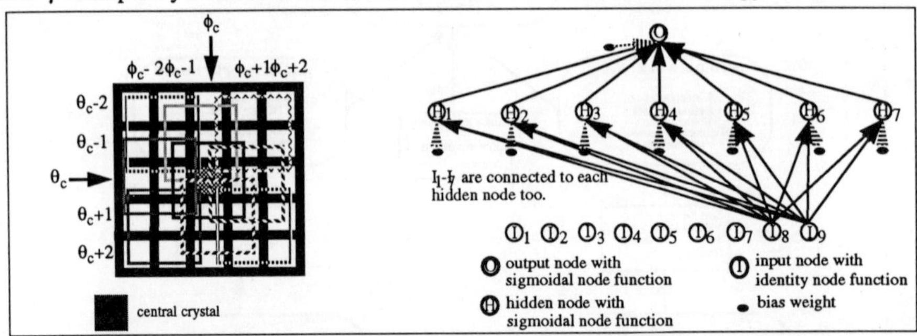

Fig. 2 Sub network Sum9 and its input parameter space.
The energy of the central crystal of a local maximum is filled into element E_{33}, the rest according to their relative position. Each sum of 9 has its corresponding input node on the right.

the 2nd kind. Fig. 3. shows in the right column the network's output for photon candidates in clusters containing more than one local maximum on a logarithmic scale. A higher output value, nearby 1, corresponds to a photon, a lower one, nearby 0, to a fluctuation. There is an order of magnitude between photons classified correctly and incorrectly (upper right corner in Fig. 3). For fluctuations of the 1st kind this is reversed, there is an order of magnitude between fluctuations with a low (0) and a high (1) network output (lower right corner in Fig. 3). This is the desired behaviour of the network. On the other hand, the left column in Fig. 3 shows the network's output for clusters with only one single photon candidate. Obviously there is no class separation at all, because nearly all fluctuations have an output value close to 1, just as the photons. The main reason for this is that the network is trained with fluctuations of the 1st kind only. But even when it is trained with both kinds of fluctuations, the performance, especially for cluster with low energy (E_{Cl} < 40 MeV) and a

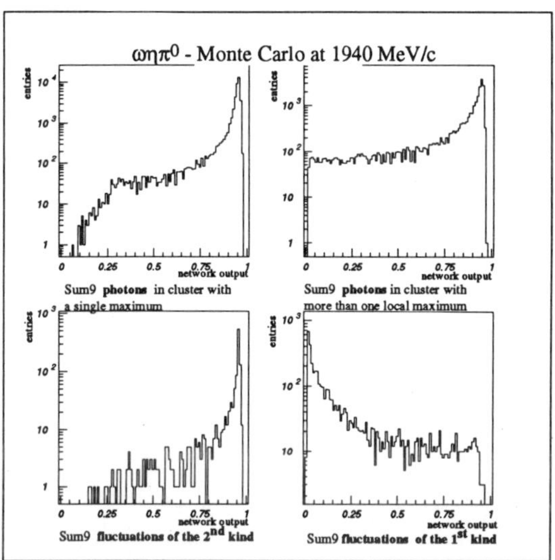

Fig. 3 Network output of sub network Sum9 for an „unknown“ $\omega\eta\pi_0$ - Monte Carlo data set. The first row shows the network's output for photons, the second for fluctuations. The right column shows photon candidates in clusters with more than one local maximum, the left column clusters with a single maximum.

single maximum, does not improve. These cluster have only a few crystals (mostly less than 10) with an energy deposit: „Averaging“ over nine crystals and dividing by the total energy causes that all input parameters apart from the central sum are more or less the same. Therefore different input parameters for clusters with a single maximum are needed.

3.2. Input and output of sub network 9Xtal

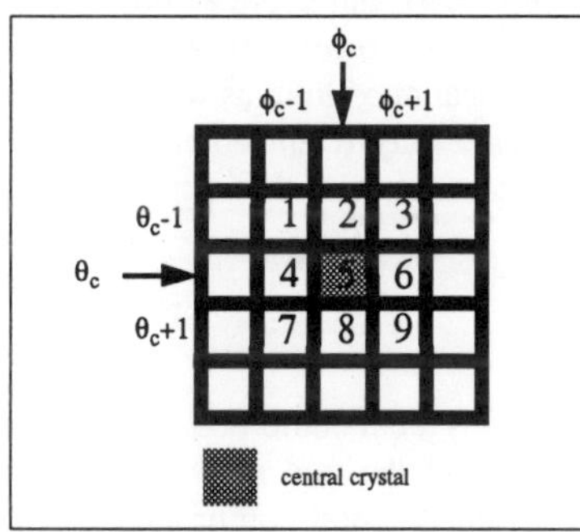

Fig. 4 Input parameter space of sub network 9Xtal.
The same matrix as in Fig. 2 is used, but this time only the central crystal and its 8 nearest neighbours are fed into a network.

The motivation for the second sub network is to obtain a classifier for clusters with a single local maximum, low energy and only a few crystals with an energy deposit. A reasonable input parameter space of the network would be the energy of the central crystal and its eight nearest neighbours (Fig. 4). It turned out that this network only converged after sorting the energies. Sorting the energies causes in many cases a sort of „rotational invariance“ of the input data; it preserves the information of the energy distribution. The first input node gets the highest and the second input node always gets the second highest excitation and so on. The convergence of the training shows that

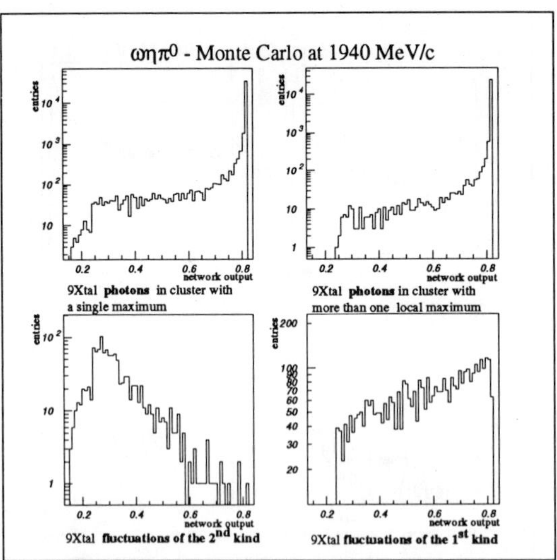

Fig. 5 Network output of sub network 9Xtal for an „unknown“ $\omega\eta\pi_0$ - Monte Carlo data set. The first row shows the network's output for photons, the second for fluctuations. The right column shows photon candidates in clusters with more than one local maximum, the left column clusters with a single maximum.

there is a difference in the ratios of energy between fluctuations and photons.

The network has nine input, eight hidden and one output node, therefore 8•9+8+8+1=89 weights are trained. The left column in Fig. 5 shows the network's output for photons and fluctuations of the first kind. In contrast to the left column in Fig. 3 it shows the desired behaviour, nearly two orders of magnitude more photons with high output values (1) than low output values (0) (upper left in Fig. 5), and about an order of magnitude more low than high output values for fluctuations (lower left in Fig. 5). This network is tested for clusters containing more than one local maximum as well, although it is trained with clusters containing one photon candidate only (right column in Fig. 5). There is some class separation, but far worse than the network with the sum of nine input (compare lower right pictures in Fig. 3 and Fig. 5).

3.3. Input and output of the top network

Having two networks for both kinds of fluctuations a cut could be found by visual inspection of the two-dimensional space of the outputs of the two networks described above (see upper row in Fig. 6). That would mean to find a good compromise between rejecting as many fluctuations and to keep as many photons as possible. For this purpose, the feed forward top network was trained with the outputs of the two sub networks. The sub networks with fixed weights are just used as modules that exhibit structures in the input data. Although the whole structure performed well already, a third input node to the top network is added. A bias of -1 corresponds to a photon candidate in its own well separated cluster and 1 for a local maximum in a cluster with more maxima. The top network has three input, four hidden and one output node, such that 3•4+4+4+1 = 21 weights are trained. The output of the network is shown in the lower row of Fig. 6. Now there are more than two orders of magnitude more photon candidates identified correctly (fluctuations and photons) than wrongly independent of the number of local maxima in the parent cluster.

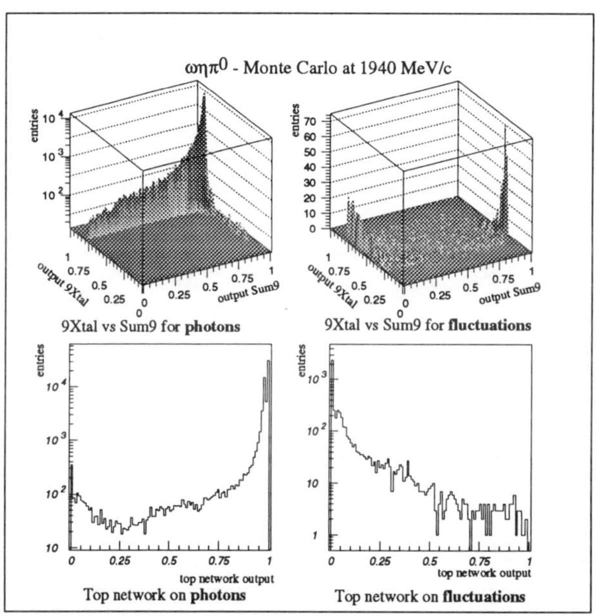

Fig. 6 Input and output of the top network
The upper row shows a two-dimensional projection of the three-dimensional input parameter space of the top network for photons and fluctuations respectively. The lower row shows the network output for photons and fluctuations.

4. Performance tests with Monte Carlo data

A better representation of the network's performance than the network's output is given by the following ratios:

Efficiency for photons:
$$P_G = \frac{\text{Number of remaining photons}}{\text{Total number of photons}} \tag{5}$$

Efficiency for fluctuations:
$$P_S = \frac{\text{Number of remaining fluctuations}}{\text{Total number of fluctuations}} \tag{6}$$

A measure for the quality of the data after a cut are:

Background:
$$B = \frac{\text{Number of remaining fluctuations}}{\text{Number of remaining local maxima}} \tag{7}$$

Purity: $$P = \frac{Number\ of\ remaining\ photons}{Number\ of\ remaining\ local\ maxima} = 1 - B \qquad (8)$$

The network's performance is tested with eight different decay channels of antiproton-proton annihilation at 1940 MeV/c beam momentum: $\bar{p}p \rightarrow \omega\eta\pi^0 \rightarrow 7\gamma$, $\omega\omega \rightarrow 6\gamma$, $\pi^0\pi^0\gamma \rightarrow 5\gamma$, $\pi^0\pi^0\pi^0\pi^0 \rightarrow 8\gamma$, $\pi^0\omega \rightarrow 5\gamma$, $\pi^0\pi^0\pi^0 \rightarrow 6\gamma$, $\eta\eta\pi^0 \rightarrow 6\gamma$, $\pi^0\pi^0 \rightarrow 4\gamma$. There is no significant dependence on the decay channel and the γ-multiplicity in the final state as shown in Fig. 7. The efficiency P_G does not vary more than 1 % and P_S, P and B are varying even less, although the initial background and purity, when no cut is applied (cut value = 0) varies up to 2.5 % in the data sets. Even for a very low cut value on the top networks output node (about 0.05) about 50 % fluctuations are rejected and less than 1 % photons are lost. A cut at 0.25 is sufficient to reduce the background in the remaining data below 1 % (more than 90 % of all fluctuations are rejected) without losing more than 2.5 % of the photons. Fig. 8 shows P_G and P_S for the decay channel $\bar{p}p \rightarrow \omega\eta\pi^0 \rightarrow 7\gamma$ in the most relevant regions

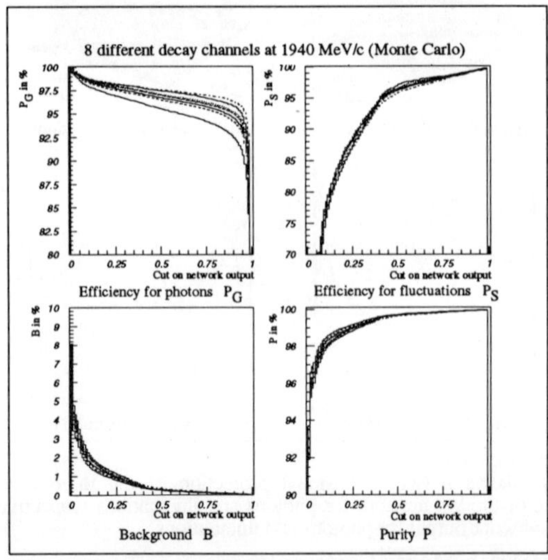

Fig. 7 Efficiency for photons P_G and fluctuations P_S, Background and Purity for Monte Carlo data at 1940 MeV/c on an enlarged scale. 8 different decay channels are shown: $\bar{p}p \rightarrow \omega\eta\pi^0 \rightarrow 7\gamma$, $\omega\omega \rightarrow 6\gamma$, $\pi^0\pi^0\gamma \rightarrow 5\gamma$, $\pi^0\pi^0\pi^0\pi^0 \rightarrow 8\gamma$, $\pi^0\omega \rightarrow 5\gamma$, $\pi^0\pi^0\pi^0 \rightarrow 6\gamma$, $\eta\eta\pi^0 \rightarrow 6\gamma$, $\pi^0\pi^0 \rightarrow 4\gamma$.

between 95 % and 100 % for P_G and 88 % to 100 % for P_S in comparison to two conventional methods. For the same efficiency for photons P_G about 4% (6%) more fluctuations are identified correctly by the net. The amount of events lost due to incorrect classification of photons goes roughly with (1 - P_G) to the power of the multiplicity selected. Therefore a too „hard“ cut (cut value above 0.8) for high purity would result in a small data set.

5. Performance test with real data

As the network was trained with Monte Carlo data, it is very important to check if it behaves well on real data without creating artifacts. Fig. 9 shows an invariant mass spec-

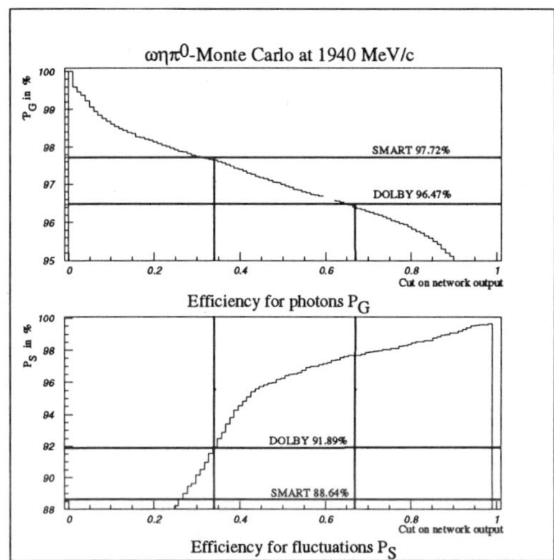

Fig. 8 $\bar{p}p \to \omega\eta\pi^0 \to 7\gamma$ Monte Carlo data at 1940 MeV/c.P_G and P_S vs. cut value on the network's output. The horizontal lines indicate the performance of SMART and DOLBY C (both methods were developed and optimized with data at 200 MeV/c $\bar{p}$-beam momentum) on the same data set. The vertical lines show the cut values for which the network has the same efficiency as the conventional methods.

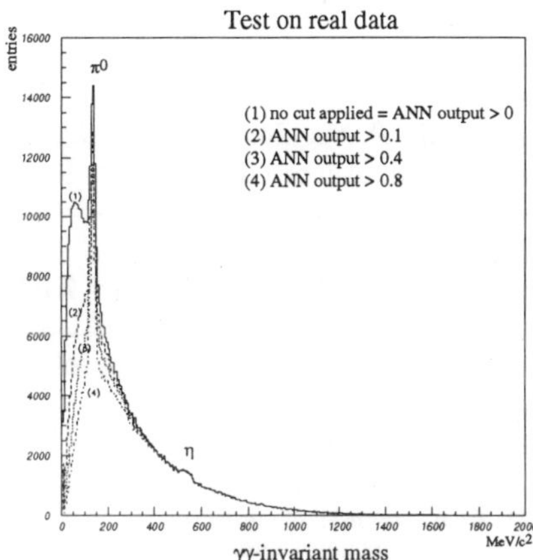

Fig. 9 $\gamma\gamma$ - invariant mass of $\bar{p}p$-annihilation for data taken at 1940 MeV/c beam momentum.
The original spectrum (0) and for three different cuts (2)-(4) on the network's output is shown.

trum of $\gamma\gamma$-pairs for data of $\bar{p}p$-annihilation at 1940MeV/c $\bar{p}$-beam momentum for neutral final states. The huge bump at low mass is due to fluctuations: there is no known, lighter particle, decaying into two photons, than a pion. The bump vanishes already for a very „soft" cut (0.1) on the network's output, whereas the amount of $\gamma\gamma$-pairs inside the π^0-region remains the same even when the „hardest" cut (at 0.8, for this cut the efficiency for photons is about 95 %) is applied. The small η signal is not effected as well. Fig. 10 shows the π^0-region for the same data but with a much finer binning. The network acts like a smooth regulator that is adjustable for special needs. If the background due to fluctuations causes no problems in a specific analysis of the data, a low cut value can be chosen. On the other hand, if very high purity is required, a higher cut value

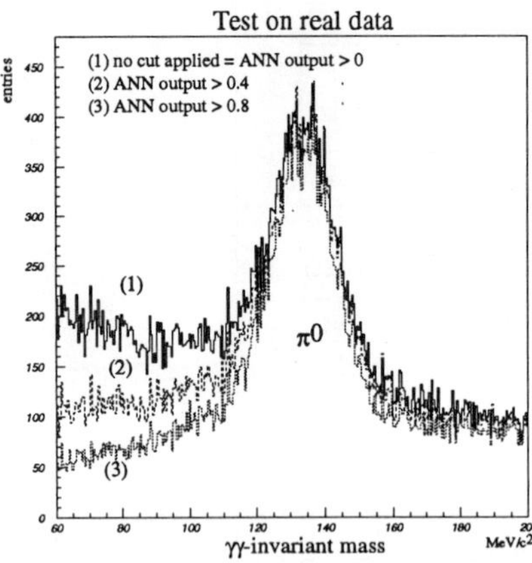

Fig. 10 γγ - invariant mass for the same data as in Fig. 9 in the π^0-region and very fine binning.

is the better choice. Furthermore, the correct classification of an event, before application of a kinematic fit, saves a lot of computing time.

6. Conclusion

A feed forward neural network is a powerful tool for recognition of fluctuations in an electromagnetic shower. The background due to split-offs is reduced from 6%-8% below 1%. There is no significant dependence of the network's performance on a specific intermediate state of antiproton-proton annihilation or on the γ-multiplicity in the final state. The method is model independent and can be used for events at the whole range of antiproton beam momenta that are analysed with the Crystal Barrel detector (200 MeV/c - 1940 MeV/c) so far. The performance of the network is as good as the performance of the conventional methods based on conventional parameter cuts for data of antiproton-proton annihilation at 200 MeV/c $\bar{p}$-beam momentum. For higher beam momenta the network has a superior performance.

References

1 E. Aker et al., *The Crystal Barrel Spectrometer at LEAR*, Nucl.Inst., **A321**(1992)6.

2 J. Salk, *SMART - Split-off recognition in pure neutral events*, CB-note 182, 1991, unpublished.

3 N.Hessey, *DOLBY C*, CB-note, 1992, unpublished.

4 R.R.Leighten et al., The MITRE Corporation, *The Aspirin/Migraines Software Tools* Rel.V5 - User's Manual, 1991.

Aspirin is available free of charge via anonymous ftp:

ftp.cognet.ucla.edu
in the directory:
alexis
in the file:
am6.tar.Z.

Russell Leighton
The MITRE Corporation
7525 Colshire Dr.
McLean, Va. 22102-3481
INTERNET: russ@dash.mitre.org
leighton@mitre.org

5 M.Kunze, *Experience with mixed Language Programming*, Proceedings of the 2nd international workshop on software engineering, artificial intelligence and expert systems for HEP (World scientific, Singapore, 1992).

ARTIFICIAL NEURAL NETWORKS FOR REACTION DISCRIMINATION IN THE PP-BREMSSTRAHLUNG EXPERIMENT AT COSY

J.S. Lange

and

H. Freiesleben, M. Kunze, M. Steinke and M. Strunk
Institut für Experimentalphysik I, Ruhr-Universität Bochum
44780 Bochum, Germany

1. Detector and Experiment

One of the major facilities at the new COSY accelerator at the Forschungszentrum Jülich is the wide-angle time-of-flight-spectrometer TOF. This spectrometer allows to determine position and energy (by time-of-flight) of charged particles via scintillator techniques. Properties of neutral particles have to be deduced via reconstruction algorithms. One of the first experiments at TOF will be the investigation of pp-Bremsstrahlung ($pp \rightarrow pp\gamma$), which provides essential information about off-shell effects in a relativistic description of nucleon-nucleon-potentials. In this experiment two protons will be detected by the TOF-spectrometer and the photon can be reconstructed via 4-momentum conservation.

2. The Problem of Reaction Discrimination

It is of major importance to discriminate as accurate as possible the pp-Bremsstrahlung reaction from any background reactions. In the energy range 280 MeV $\leq T_{beam} \leq$ 600 MeV two background reactions have to be considered: At first the production of neutral pions $pp \rightarrow pp\pi^0$ in the LH_2-target and and at second nuclear reactions $pA_1 \rightarrow ppA_2$ in target foils, producing two outgoing protons passing the pp-Bremsstrahlungstrigger. Both reactions have rates (number of reactions per incoming beam particle) which are larger than that of the Bremsstrahlung by a factor of up to 10^2 at T_{beam} = 600 MeV.

3. Standard Analysis

The standard means for discrimination of reactions is a missing mass reconstruction. The missing mass is calculated by $m_3{}^2 = E_3{}^2 - p_3{}^2$, where the unobserved particle has the mass m_3, the total energy E_3 and the momentum p_3. E_3 and p_3 result from 4-momentum conservation. Because of finite detector resolution the missing mass distributions observed do usually not allow to separate the three reactions completely (c.f. fig. 4, solid line histogram).

4. Analysis using an Artificial Neural Network

A feedforward network of the following architecture was trained to discriminate

$pp \to pp\gamma$ from the two background reactions mentioned: 5 input nodes, 1 hidden layer with 8 hidden nodes, 3 output nodes.

Input: The aim of the study is to improve the missing mass analysis, hence E_3 and p_3 must be used as input values for the neural network. Further input values are deduced parameters which turn out to be most suited for discrimination, since their physical information is lost in the course of a missing mass analysis:

- ϑ_3, the polar angle of the third outgoing particle in the lab frame, which can be larger for a photon than for a π^0,

- $\vartheta_1 + \vartheta_2$, the sum of the polar angles of the two observed protons in the lab frame, which is smaller in case of a π^0 production than in case of the pp-Bremsstrahlung, because production of mass means loss of kinetic energy,

- $\Delta\phi_{12}$, the so called acomplanarity, which can be larger in case of pion production (following the same argument as above).

Thus, we have a 5-dimensional input vector space. This turns out to be sufficient for a first attempt and will, perhaps, be increased to more dimensions in future. Each reaction has a special signature in the 5-dimensional input vector space (e.g. $pA_1 \to ppA_2$ events turn out to be mainly quasielastic).

Output: There are three output nodes - one node for each reaction. The network is trained orthogonally and the output target vector is a triple, i.e. (1,0,0) for the pp-Bremsstrahlung, (0,1,0) for $pp \to pp\pi^0$ and (0,0,1) for $pA_1 \to ppA_2$. After completing the learning phase, each output node reacts on a particular input vector with a value $0 \leq v_i \leq 1$, i=1,2,3.

Hidden Layer: It turned out that one hidden layer is sufficient for a discrimination on a level of 1%, i.e. this accuracy is achieved by a cut by a 4-dimensional hypersurface in the 5-dimensional input vector space. The number of hidden nodes was varied in test runs and was finally chosen to 8 as a compromise of runtime and accuracy of reaction discrimination.

Training was typically performed with 10000 events per reaction type via the backpropagation algorithm. Events for training were generated by phase space simulations for $pp \to pp\gamma$ and $pp \to pp\pi^0$ and simulations of $pA_1 \to ppA_2$ reactions using the hadronic module HETC of the nuclear reaction code HERMES [1]. All events had to pass the detector structure, implemented in detail in the C++ based geometry code LasVegas [2]. Simulations were tested by utilizing different event generators, e.g. those of GEANT 3.15 [3][4] and MC4 [5].

The response of the trained network can be examined in several ways:

- With regard to the output node values while confronting the neural network with reference data of all three reaction types: ability of the neural network to correctly recognize events of a special reaction type (c.f. fig. 2),

- with regard to missing mass distributions with certain cuts: definite improvements of the analysis including the neural network compared to the standard missing mass analysis (c.f. fig. 3, fig. 4).

5. Technical Details

The feedforward network was set up using the Aspirin/MIGRAINES code [6]. We exploit the advantages of UNIX i.e. the offline analysis program (reconstruction of E_3, p_3, ϑ_3, $\vartheta_1+\vartheta_2$ and $\Delta\phi_{12}$) and Aspirin/MIGRAINES are - during both the neural network training phase and the final analysis - running as parallel processes, which communicate via named pipes (fifo-streams[1]). This means that the network decision is not called as a subroutine, but is separated from the analysis package, which is a feature with several advantages: At first, during analysis one still has full read and write access to all network parameters (complete MIGRAINES shell running). At second, if necessary, during analysis training can still be performed. At third, the network part can easily be exchanged for another module of, perhaps, another network topology. First attempts are on their way using a Hopfield type network using the code NSL [7] and a Kohonen map type net.

Convergence was reached very fast (5-6 epochs within a CPU-time of less than 60 s on an IBM RS6000/320H Workstation (AIX 3.2.2) (c.f. fig. 1).

6. Future Plans

Our application of a neural network differs from several other approaches (i.e. standard pattern recognition problems) in so far as complex input parameters are used which were deduced by means of a detailed reconstruction process. It is, to our knowledge, one of the first neural network applications in medium energy physics and one of the first neural network applications in offline analysis. The complete analysis including the network part is fast enough (even in the training phase $\simeq$60 ms / event) not only to analyze offline, but also online during beamtime for control purposes. Online - in this context - does not mean trigger applications, but visualization and control of the experimental status and progress. The good results lead to some new projects for the near future: Firstly, search for alternative or additional input parameters which may increase the quality of reaction discrimination (e.g. the fraction E_1/E_2), secondly a combination of a kinematic fit (presently examined by the Jülich part of the TOF-Collaboration) and the neural network and thirdly perhaps usage of other network topologies like networks of Hopfield or Kohonen type.

References:

1. P. Cloth, D. Filges, R.D. Neef, G. Sterzenbach, Ch. Reul, T.W. Armstrong, B.L. Colburn, B. Anders, H. Brückmann, *HERMES - A Monte Carlo Program System for Beam-Materials Interaction Studies*, Jülich Preprint 2203, May 1988.

[1]The stream of data is handled according to the "first in - first out" principle.

2. S. Brand, H. Brand, H. Freiesleben, J.S. Lange, U. Zielinski, *Las Vegas: Object-Oriented MC Simulation for the COSY-TOF-Spectrometer*, see this Workshop, Group: Software Engineering.
3. R. Brun, F. Bruyant, M. Maire, A.C. McPherson, P. Zanarini, *GEANT 3*, CERN **DD/EE/84-1**, Sept. 1987.
4. B. Naumann (FZ Rossendorf), *GEANT Setup for the COSY-TOF-Spectrometer*, private communication.
5. P. Cloth, V. Drüke, D. Filges, K. Kilian and A. Röser, *Monte Carlo Simulation of Medium Energy Detectors at COSY Jülich*, **MC93**, Int. Conf. on Monte Carlo Simulation in High Energy and Nuclear Physics, Feb.22nd-26th, Tallahassee, Florida, USA.
6. J.S. Lange, A. Röser *MC4EXT Interface Subroutine Package for the MC4 Monto Carlo Code*, unpublished.
7. R.R. Leighton, *The Aspirin/MIGRAINES Software Tools / User's Manual*, MITRE Corporation publ. MP-91W00050, Dec. 1991.
8. A. Weitzenfeld, *Network Simulation Language NSL Vers. 2.1*, Technical Report 91-05, Brain Simulation Laboratory, Center for Neural Engineering, University of Southern California.

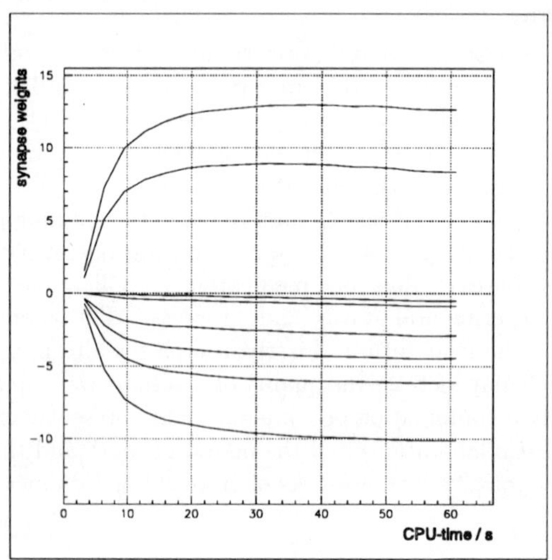

Fig. 1: Convergence of the weights of the synapses connecting the hidden nodes with one output node in the final network design (5 input, 8 hidden, 3 output nodes)

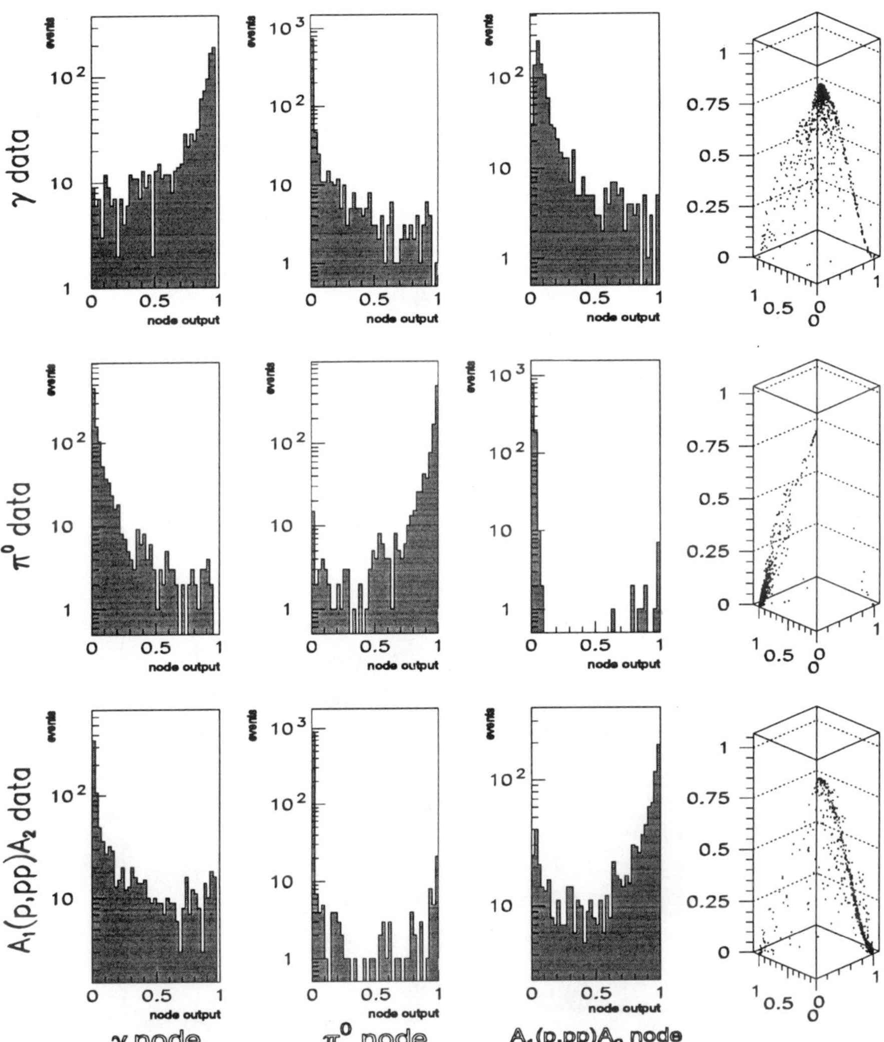

Fig. 2: Response of the trained network when confronted with reference data (top row $pp \rightarrow pp\gamma$, middle row $pp \rightarrow pp\pi^0$, bottom row $pA_1 \rightarrow ppA_2$): values (left columns) and correlations (right column) of the three output nodes

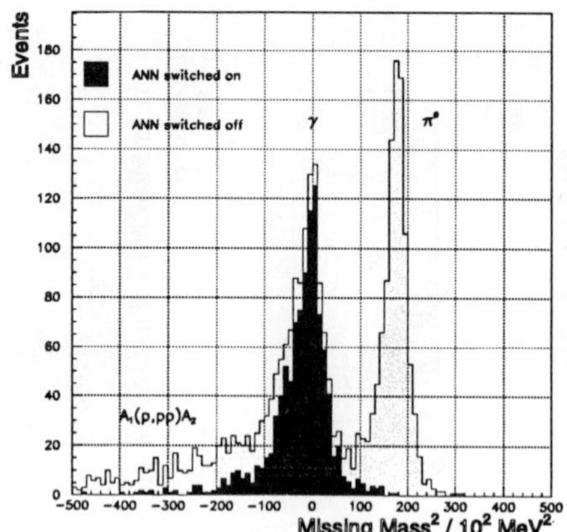

Fig. 3.: Missing mass plot: data of all three reactions with and without filtering by the neural network

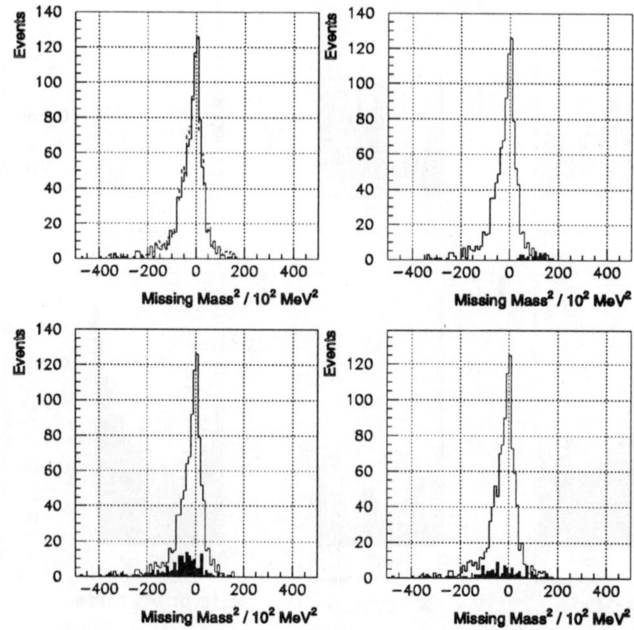

Fig. 4.: Missing mass plots: comparison between pure pp-Bremsstrahlung data (solid line histogram in each plot) and data of all three reactions filtered by the neural network (top left, dashed line histogram, there is hardly any difference), $pp \rightarrow pp\pi^0$, falsely recognized by the network as pp-Bremsstrahlung (top right, black histogram), $pA_1 \rightarrow ppA_2$, falsely recognized by the network as pp-Bremsstrahlung (bottom left, black histogram) and pp-Bremsstrahlung, interpreted by the network as one of the other reactions (bottom right, black histogram)

The Application of Artificial Neural Networks Techniques in SATURNE Experiment LNS284

Paul Fuchs
Laboratoire National SATURNE
91191 Gif-sur-Yvette Cedex, France

ABSTRACT

A large acceptance detector is proposed in order to study rare decays of the η meson. It will be installed at the η tagging facility of Saturne, where the η are produced with the $pd \rightarrow {}^3He\ \eta$ reaction. Neural networks are well suited for the analysis of high-dimensional data spaces. We discuss the possible applications of neural networks for this experiment and focus on one specific application : the reconstruction of photon kinematics from calorimeter data.

1. Introduction

1.1. Motivation for the experiment

The branching ratios of the rare decays of the η meson are not well known. With the source of tagged η's installed at the Saturne synchrotron of Saclay[1], considerable improvement can be achieved on most of the decay modes. The η are produced with high intensity by using the $pd \rightarrow {}^3He\ \eta$ reaction near threshold. The ${}^3He$ are detected in the magnetic spectrometer SPES2 to tag the η.

In order to study decay modes where three or more particles are produced, we propose to build a large acceptance detector (3π) consisting essentially of a barrel of pure crystals of CsI, to detect photons and charged particles, and a barrel of thin plastic scintillators to identify charged particles. By triggering only on ${}^3He$, all decay modes will be recorded at the same time, the selection being made off line. We are specifically interested in $\eta \rightarrow \pi^0\gamma\gamma$ (test of higher orders of Chiral Perturbation Theory), $\eta \rightarrow \mu^+\mu^-\gamma$, $\eta \rightarrow e^+e^-\gamma$ and $\eta \rightarrow \mu^+\mu^-$ (electromagnetic structure of the η meson and test of the Vector Meson Dominance model) and $\eta \rightarrow e^+e^-\pi^0$ (test of C conservation in the electromagnetic interaction).

The CsI calorimeter is inspired on the "Crystal Barrel" used at LEAR[2]. It consists of 1020 crystals of pure CsI, with the shape of truncated pyramides pointing to the target. Each crystal covers 6° both in polar and azimutal angles. The crystal lenght is 30 cm (16.1 L_R), which provides good energy resolution.

The tracker consists of 120 plastic scintillators parallel to the barrel axis. They will be read by photomultipliers at both ends to get the longitudinal coordinate. The plastic scintillators, 0.5 cm thick, will signal charged particles and help with their identification through dE/dX.

1.2. Possibilities for neural network applications

Neural network algorithms are often used for event classification and non-parametric function approximation. The method works particularly well if there are many input parameters on which the analysis is to be based. We can make the following list of problems for which a neural network approach is beneficial for this experiment :

- The reconstruction of the position of impact of the particle at the calorimeter based on the distribution of deposited energy.
- Particle identification ($\gamma/n/\pi/\mu$) based on the energy deposition in both the tracker and the CsI crystals.
- The suppression of background through a neural network based event classification.
- The suppression of accidental hits through an analysis of hit patterns in the calorimeter.

In order to gain experience with neural networks, we have studied in detail the problem of position reconstruction for photons hitting an array of hexagonal BGO calorimeter modules which were used in experiment LNS258 for the measurement of the branching ratio BR($\eta \to \gamma\gamma$).

2. The application of neural networks to the reconstruction of the impact position of a photon at the calorimeter

2.1. *The problem*

The calorimeter consists of hexagonal BGO crystals with a width of 5.5 cm and a length of 20 cm. Photons from the decay of the ηmeson originate at the target with a typical energy of 300 MeV. They get absorbed in the calorimeter. The problem is to determine energy and momentum of the photon from the pattern of deposited energy.

Due to the size of individual modules and the high mass of BGO, the electromagnetic shower generated by a photon of 300 MeV energy is completely contained within a cluster of seven elements - the module with the highest energy deposition and its six nearest neighbours.

One can immediately define the total energy and a first approximation of the position of impact.

$$E_{tot} = \sum_{i=0}^{6} E_i \qquad \vec{x}_{ews} = \sum_{i=0}^{6} \vec{x}_i \frac{E_i}{E_{tot}} \tag{1}$$

where E_0 is the maximal energy deposition with $E_1 \ldots E_6$ denoting the energy deposition in the neighbouring modules (counted anti-clockwise). The vector $\vec{x}_{ews}$ is simply the energy-weighted sum of the positions of the modules within the cluster evaluated at a certain depth within the crystal where the longitudinal shower profile peaks.

The fully reconstructed position $\vec{x}_s$ is the result of a general fit to the lateral shape of the electromagnetic shower and will be the standard with which we compare the neural network results. Without going into details, we simply write :

$$\vec{x}_s = \text{empirical fit to the data} \tag{2}$$

2.2. The input data to the neural network

We have tried a number of possible input data vectors depending on how much a priori knowledge about the symmetry and orientation of the shower we want to offer to the neural network. The four classes of input vectors which have been investigated are :

- **I** is defined as the set of normalized energies $\frac{E_i}{E_{tot}}, i = 0 \ldots 6$ plus a bias unit which is always set to a value of -1.
- **S** is defined as the set of energy asymmetries with respect to the six symmetry axes of the BGO element, e.g.

$$s_1 = \frac{(E_6 + E_1 + E_2) - (E_3 + E_4 + E_5)}{(E_6 + E_1 + E_2) + (E_3 + E_4 + E_5)} \tag{3}$$

- **A** is defined as the set of estimated angles for the given event.

$$a_x = \frac{x_{ews}}{z_{tgt}} \qquad a_y = \frac{y_{ews}}{z_{tgt}} \tag{4}$$

  where z_{tgt} is the distance between target and BGO detector. This set of inputs can be added to the other data sets to generate the input vectors **AI** and **AS**.

2.3. The desired output from the neural network

We can train the neural network to learn the position of impact relative to the central element of the cluster (denoted by **CX**)

$$\vec{x}_{nn} = \vec{x}_{real} - \vec{x}_0 \tag{5}$$

or we can teach the network to learn the correction to the first approximation of the position $\vec{x}_{ews}$ (denoted by **EX**)

$$\vec{x}_{nn} = \vec{x}_{real} - \vec{x}_{ews}. \tag{6}$$

2.4. The structure of the neural network

There are two structures which have been studied. The first structure is a fully connected multi-layer-perceptron with two hidden layers. The transfer function chosen for the hidden layer is given by $f(x) = tanh(x)$, where x is the sum of weighted inputs. The output layer is linear since we like to describe a linear function with in principal arbitrary range. The sizes of the hidden layers have been fixed to 6 neurons in the first hidden layer and 3 neurons in the second. We denote this structure by **Input NNL**.

The second structure is a hybrid structure combining a two-layer neural network with a least-squares fitting stage. The neural network has an input layer and a nonlinear output layer with 3 neurons (transfer function as before). The output of this neural network is trained to compute the projection of the true impact position onto three symmetry axes of the BGO element. We then make a least-squares fit to the hypothesis that the output of the network is indeed the projection of a single point $\vec{x}_{impact}$. The reason for choosing such a complicated structure is both historical as well as motivated by the hope that the information learned by the network is directly interpretable in terms of the lateral shape of the electromagnetic shower. We denote this structure by **Input NLF**.

2.5. Training and testing

The simulation package GEANT has been used to generate three data samples for the training and testing of the neural network algorithm.

- **X** Photons originating from a point source iluminate the horizontal axis of the detector only, i.e. $y_{real} = 0.0$ by definition.
- **Y** Photons originating from a point source iluminate the vertical axis of the detector only, i.e. $x_{real} = 0.0$ by definition.
- **D** Photons originating from a point source iluminate a diagonal axis of the detector only, i.e. $x_{real} = y_{real}$.

Training was performed on the data sample **D**, whereas the performance of the algorithm was tested on all three data samples. The size of the training sample was kept at 500 events. For larger training samples, the neural network not only learned to represent the impact position as function of energy deposition but also the distribution of data itself, i.e. the diagonal line across the detector. The training of the neural network on data distributed over all of the detector is still in progress and results are not yet available. The test sample size was typically 2000 events.

The neural networks were trained with the traditional backpropagation algorithm with a learning term and a momentum term. The weights are updated on-line, i.e. after the presentation of each single event. Furthermore, we update the learning parameters after each epoch (i.e. after 500 events have been presented for training). If the network was able to improve its description of the desired output, the learning parameters were increased by 5 %, otherwise they were decreased by 25 %. This resulted in very stable learning behaviour and the optimal solution could be reached after typically less than 50 epochs.

2.6. The results

The performance of the studied algorithms were evaluated on the basis of the following variables :

- $\sqrt{<\chi^2>_x}$, which is the sum of squares of deviations from the desired output averaged over the test sample **X**. Similar variables were also computed for the other two data sets, **Y** and **D**.

Table 1: Results for the neural network types which have been studied, together with the performance of the standard algorithms. The units are cm.

Type of network	$\sqrt{<\chi^2>_x}$	σ_x	σ_y	$\sqrt{<\chi^2>_y}$	$\sqrt{<\chi^2>_d}$
$\vec{x}_{ews}$	0.72	1.04	0.15	0.84	0.91
$\vec{x}_s$	0.77	0.78	0.67	0.80	0.79
INNL CX	0.60	0.65	0.44	0.64	0.64
INNL EX	0.57	0.62	0.36	0.67	0.65
ASNNL EX	0.63	0.65	0.57	0.65	0.63
ASNLF CX	0.72	0.70	0.69	0.78	0.63
ASNLF EX	0.63	0.69	0.51	0.65	0.63

- σ_x,σ_y are the values of the standard deviation read off from a Gaussian fit to the distributions for $(x_{nn} - x_{real})$ and $(y_{nn} - y_{real})$ for the test sample **X**. The variable $\sqrt{<\chi^2>_x}$ can be interpreted as the mean of σ_x and σ_y.

The results are listed in table 1.

3. Discussion of the results

3.1. The traditional versus the neural network approach

The problem of reconstructing the position of the impact of the photon on the calorimeter is one to which both traditional as well as neural network algorithms can be applied. A priori there is no reason to expect the neural network algorithms to performe better than an optimal fit to the data since the same information is used in all cases. The essential advantage of using neural networks in this application is due to the non-parametric nature of the algorithm. Because of the simplicity of the algorithm it is possible to explore a large region of the function space which is explored by the neural network (by changing input data or the size of the hidden layers etc.). In contrast, the traditional approach requires a specific choice of function representation before the fit to the data can be attempted and there is no garantee that this choice allows for an optimal fit. The neural network is more likely to find the really optimal representation of the desired output function in this application than the traditional approach.

3.2. The choice of output reference

In most cases, the choice of **EX** leads to better results than the choice of **CX**. Even in cases, where the average chi-squared variables like $\sqrt{<\chi^2>_x}$ look good, the response for **EX** turns out to be less biased towards the center of the detector element than if one chooses the axis of the cluster as reference (**CX**).

3.3. The choice of inputs

It is not clear from the results which have been presented so far, whether the inclusion of angular information (**AI** or **AS**) leads to improved results. In order to objectively evaluate this question, a more complete survey has to be done. The same conclusion holds true for the benefit of preprocessing the inputs **I** to generate the inputs **S**.

In summary, one may conclude on the basis of the present results that the choice of inputs is not critical for this application (this is another advantage coming from the non-parametric nature of the neural network approach).

3.4. The choice of network structure

We have been able to compare a traditional fully-connected multi-layer perceptron with a hybrid system combining a neural network with a least-squares procedure. The results favour neither one of these approaches over the other. This is interesting from the point of view that the second hidden layer of the **Input NNL** network and the first layer of the **Input NLF** network allow for different internal realisations even though the output is the same. One might think then that these internal realisations represent different characteristic features of the problem and that one might employ these features in later applications, which perform particle identification for instance.

4. Future directions

The study of neural network algorithms has until now been limited in application to simulated data. It remains to be seen how these algorithms perform on real experimental data under the presence of gain variations and pile-up. It might also prove useful to study other network structures such as receptive fields and shared weights.

5. Acknowledgements

I am very greatful to my collaborators Alex Efendiev (JINR) and David White (UCLA) for valueable discussions and the development of standard position reconstruction algorithms. I am also greatful to the whole Eta collaboration for their continued support and encouragement.

6. References

1. B. Mayer, in *Rare decays of light mesons*, ed. B. Mayer (Frontières, Gif-sur-Yvette, 1990) p 161.
2. E.Aker et al., *NIM in Phys. Res.* **A321** (1992) 69

Neural Network Architectures for Small Angle Muon Triggers in DZero

Michael R. Fortner
Northern Illinois University
Dept. of Physics, DeKalb, IL 60115, USA

ABSTRACT

The use of neural networks as part of a trigger for the DZero small angle muon detector is considered. The problems with the existing muon trigger at small angles is discussed, and a variety of network architectures that might be implemented are presented.

1. The DZero Small Angle Muon System (SAMUS)

The DZero detector is a large general purpose detector for proton-antiproton collisions at Fermilab's Tevatron [1]. Muons are identified by measurements in proportional drift tubes on either side of a magnetized iron toroid. There are a total of five toroids: a central toroid covering $|\eta| < 1.1$ and two toroids in each end covering $1.1 < |\eta| < 2.4$ and $2.4 < |\eta| < 3.4$ (3 - 10 degrees). The small angle muon system (SAMUS) is designed to measure the muons that pass through this last toroid.

The small angle muon system is comprised of about 6000 individual tubes of 3 cm diameter [2]. The tubes are grouped into six stations, three on each end of the detector. Each station consists of two layers of tubes oriented in each of x-, y-, and stereo u-directions. The tubes in the two layers are offset to improve coverage and to help resolve drift ambiguities. On each end, one station is situated on the interaction side of the toroid and two stations are on the other side as in Fig. 1. Measurements from the three stations together are used to reconstruct the muon momentum.

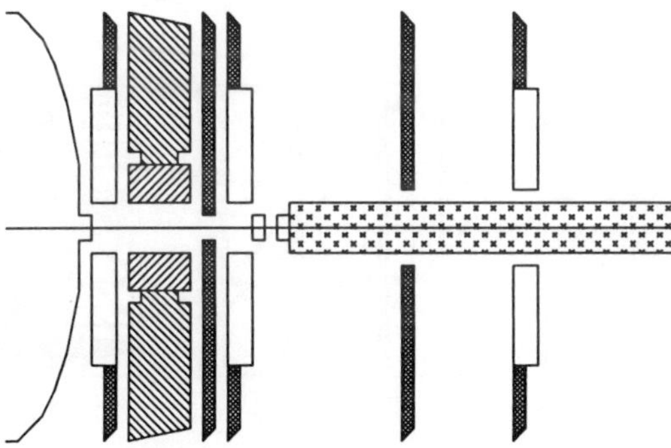

Figure 1. Side view of the DZero detector in the forward direction. The SAMUS detectors are the unshaded rectangles. Other featuresinclude the calorimeter (left), the toroids (diagonal hatches), and the quadrupole (crosses).

Tracking in the small angle region is complicated by the background to the signal muons. The major source of background is due to sprays of soft particles as a result of very forward jets interacting with objects at small angles such as the beam pipe, quadrupole magnets, calorimeter cryostat walls and edges of the muon toroid. These particles show up in the SAMUS stations as hits in time with the collision but generally uncorrelated from station to station. The average multiplicity per event in each station varies from 40 hits in the station directly behind the toroid to 70 hits in the station most downstream, adjacent to the quadrupole magnet. A typical event in each of three stations is shown in Fig. 2.

2. Triggers in the SAMUS detector

The present trigger for SAMUS consists of two stages of hardware trigger designated level 1 and level 1.5 [3], and a software trigger designated level 2. The level 1 trigger operates in less than 2 μsec after beam crossing and uses groups of tubes 12 cm wide as hodoscopes. The level 1.5 trigger operates in 3 to 30 μsec after beam crossing and uses both layers of tubes in a station to form hodoscopic elements 1.5 cm wide. At level 2 the trigger has the full drift information available to reconstruct the muon.

The efficiency of a trigger to select muons of a particular momentum is largely dominated by the multiple scattering of muons in the iron toroid. An ideal detector of perfect resolution cannot turn on any sharper than the existing level 1.5 trigger as shown in Fig. 3. Improvements to efficiency well above threshold are limited by the need to keep the very low momentum hit combinations from passing.

The existing trigger gets a rejection factor of 30 at level 1, an additional factor of 15 at level 1.5 and a further rejection of 500-1000 at level 2. This rejection is almost completely due to the trigger's ability to reject the background of spray in the chambers. One finds that the acceptance of this combinatoric background follows closely the resolution of the elements used in the trigger. As the pixels shrink the background is rejected.

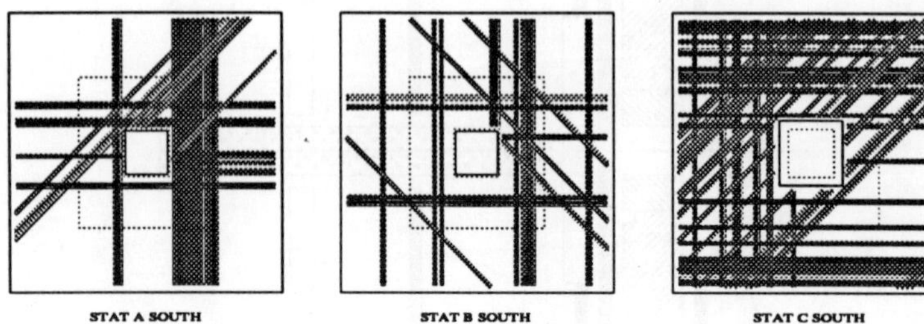

Figure 2. Hits for a single event in each of three SAMUS stations.

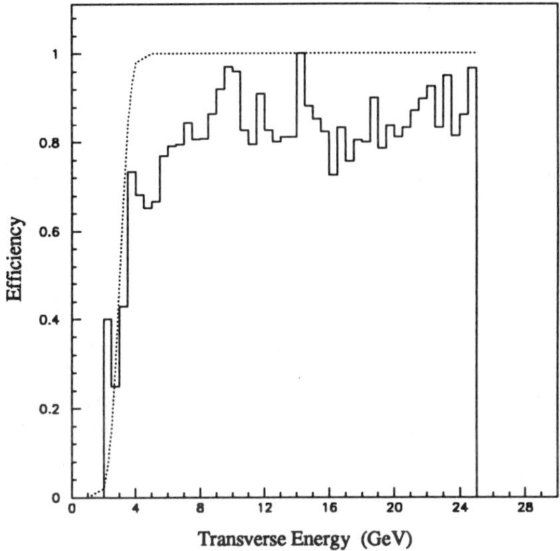

Figure 3. Efficiency of the muon level 1.5 trigger as a function of transverse momentum for a threshold set at 3 GeV/*c* (solid line). The dotted line is for an ideal detector with multiple scattering due to the iron toroid.

For collider operation in 1996 and beyond, DZero is preparing to see an increase in luminosity from 5×10^{30} to 5×10^{31} cm^{-2} sec^{-1} and a decrease in bunch crossing time from 3500 nsec to 132 ns. The muon elctronics including SAMUS will be rebuilt and additional rejection will be needed at earlier stages of the trigger to handle the higher luminosity. A pixel detector is planned for installation in the ends of the detector and simulations indicate that a rejection factor of 300 at level 1 may be achieved. The level 1.5 trigger will need to be significantly better to provide a factor of 10-100 rejection over this new level 1 trigger. Using the actual drift information is as opposed to a latch set on the tube would effectively transfer some of the rejection now in level 2 to level 1.5.

3. Neural Network Architectures for SAMUS Triggers

One technique to process the full drift information at level 1.5 would be to use neural networks in some arrangement. Some studies of neural network triggers for wide angle triggers [4] but not at small angles. Naively one would take all the inputs into a single network and have it output the momentum of the muon to the overall level 1.5 trigger. This is the equivalent of the level 2 software. Unfortunately, with 3000 wires one one end of the detector, there are clearly too many inputs for a single networks to handle.

The natural step to take is to subdivide the network into separate regions and report a momentum (if any) from each. Splitting the SAMUS detector into four quarters cuts the number of inputs into each network in half. For an arbitrary n x n partition of the detector of N total inputs, there are n^2 networks each with N/n inputs. To restrict SAMUS to networks of no more than 64 inputs each would require more than 2000 networks in this scheme. This number is too many for a practical application.

A better partition is to use the natural symmetry of the detector as three separate 2-dimensional projections in x, y and u. In this case the divisions of the projections are each 1-dimensional, so a division of one projection into n sections requires n networks with (1/3)N/n inputs. The entire detector can then be divided into 3n networks with N/3n inputs. Restricting SAMUS to networks of no more than 64 inputs would result in 48 networks per end, a much more reasonable number.

A software simulation of one of these individual networks was made using a 400 event training sample made of mixed minimum bias events and minimum bias events with a simulated muon added to the data. With 100 good tracks in the data to compare, the efficiency is consistent with the existing level 1.5 trigger. A much larger data set will be generated to make continued studies of the rejection and efficiency. Implementation may be in networks or in direct hardware based on the simulated network.

One other architecture to consider is to use the new level 1 pixel information to select data from just those drift tubes that transsect the pixel. Since typical events rarely have more than a few level 1 roads set, the number of networks to process the data may be very few, equal to the maximum number of level 1 roads set that one wants to process in parallel. This may be the cheapest and easiest to implement if the input information can be extracted in a short time scale, with implementation in either a neural network device or in a DSP using a network derived algorithm. This architecture will also be the subject of further study.

4. Acknowledgements

I would like to thank Dmitri Denisov of SSCL, Oleg Eroshin of IHEP, Russia, Herman Haggerty of Fermilab and Ken Johns of the University of Arizona for their assistance. This work is supported by a grant from the Department of Energy.

5. References

1. S. Abachi et al., "The DZero Detector", submitted to *Nucl. Inst. & Meth.*, Fermilab PUB 93/179.

2. Yu. Antipov et al., "Drift tubes for the SAMUS muon spectrometer of the D0 detector at FNAL", *Nucl. Inst. & Meth.*, A297 (1990) 121.

3. M. Fortner et al., "The VME-Based D0 Muon Trigger Electronics", *IEEE Tran. Nucl. Science*, 38:2 (1991) 480.

4. M. Fortner, "Analog Neural Networks in an Upgraded Muon Trigger for the DZero Detector", Proceedings of the Second International Workshop on Software Engineering, Artificial Intelligence, and Expert Systems for High Energy and Nuclear Physics, La Londe des Maures, France, January 1992.

TWO APPLICATIONS OF A NEURAL NETWORK CHIP TO PARTICLE DETECTORS

C.S. LINDSEY, P. HÖRNBLAD, Th. LINDBLAD
Royal Institute of Technology, Dept. of Physics-Frescati, Stockholm, Sweden
B. DENBY
Fermi National Accelerator Laboratory, Batavia, Illinois
G. SZEKELY, J. MOLNAR
Institute of Nuclear Research, Debrecen, Hungary
Å. EIDE
Ostfold College, Halden, Norway
S.R. AMENDOLIA, A. SPAZIANI
University of Sassari and INFN Pisa, Italy

ABSTRACT

We review two recent studies of applications of an analog neural network chip to particle physics detectors. In the first study the chip filters ADC data from a drift chamber to find track hits. In the second study simulated fixed target interactions produce tracks through a silicon strip detector. The hit information goes to an associative memory which finds the track parameters. These parameters go to the chip which indicates whether the event contains a secondary decay or not for use in B triggering. The two studies show the usefulness of such a chip in both low and high stages of the event processing stream.

1. Introduction

Several recent reports discuss the use of the Intel VLSI neural network chip for various real-time processing tasks in particle experiments such as track finding and triggering[1,2]. Currently at least two experiments use the chip in their trigger systems[3,4]. The advantages of such a chip include fast parallel processing and implementation of trained networks that carry out complicated pattern recognition tasks. In this report we discuss studies of two particular applications: signal filtering of pulse heights from a drift chamber and recognition of secondary decay events in a fixed target experiment using track information from a silicon strip detector.

The Intel Electrically Trained Analog Neural Network (ETANN) chip[5] provides 64 inputs, 16 internal biases, and 64 neurons (a threshold amplifier with a sigmoidal transfer function). Each input connects to 64 synapses. Each synapse contains two floating gates. A system similar to an eeprom programmer sets the non-volatile gate charge to obtain a given synapse weight. With a circuit similar to the analog Gilbert multiplier, a synapse outputs a differential current proportional to the product of the input voltage and the difference in threshold voltages of the two gates. Each neuron receives the current sum of a column of 64 input*synapse and 16 bias*synapse products. The chip contains two 80x64 arrays of synapses. With different timing and control setups, plus use of the second synapse array, a single chip can implement three feed-forward networks: (1) 64 inputs and 64 output neurons; (2) 64 inputs, 64 hidden and 64 output neurons; and (3) 128 inputs and

64 output neurons.

We use the Intel development system with the iDynamind emulator to first simulate the network on a pc[6]. When the software network reaches a satisfactory performance level on test pattern files, we download the weights to the chip. Further training with *chip-in-the-loop* (CIL) achieves performance improvements because the software can never perfectly emulate the analog processing of the chip. Previous papers discuss in more detail the steps required to train the chip[1,2].

2. Track Filtering

The first study examined filtering of pulse heights from a drift chamber[7]. The ICARUS solar neutrino experiment uses a liquid argon drift chamber tracking detector[8]. Track ionization in the chamber volume produces free electrons which drift, under a HV electric field, to a grid of sense wires at one end of the chamber volume. The drift time to reach the sense wires gives the position of the track in the z coordinate. The position of the sense wire gives the y coordinate of a hit. A second grid of induction wires lies at right angles to the sense wires. The nearest induction wires receive an induced pulse when the electrons reach the sense wire. The position of the induction wire provides the x coordinate of the hit.

We received data from a test setup with 192 wires of a chamber, with each wire connected to a fast ADC[9]. For a given cosmic ray event, an FADC gives a pulse height reading every 200ns, producing 2048 numbers for each wire. The middle curve in figure 1 shows the raw output of single wire FADC. The sharp transitions indicate track hits at positions proportional to the time bins. The data processing must locate these transitions to obtain the drift coordinate for the tracking algorithms. Figure 2 shows an event in one sector (projected onto a 2-D plane). The grey scale darkens in proportion to the pulse height. The lower plot shows the same event but after processing the event with a filter that only gives large output at sharp transitions. The mathematical filter uses the standard deviation formula:

$$\sigma = \sqrt{\sum_N (x/N)^2 - (\sum_N (x/N)^2}$$

With a window of N=9 channels, the formula moves through the 2048 values and produces an output as shown in the lower curve of figure 1. With a microprocessor the filtering of 192 wires takes 1.6s.

The ETANN network to do this filtering consisted of 64 inputs, 64 hidden units, and 64 outputs. After conversion to analog, 64 FADC values enter the network inputs. The 64 outputs correspond to the same 64 input channels, However, in the training patterns, only the outputs at the transitions contain non-zero activation in the targets. For 2048 FADC values, 32 such 64 channel windows must enter the ETANN.

The emulator trained on 2000 patterns. After reaching satisfactory error levels, we downloaded the weights to the ETANN and continued with CIL training. The upper plot in figure 1 shows the output of the ETANN for a test pattern. The mathematical filtering shows more noise but we deliberately made ETANN target patterns flat except at positions of the transitions.

The performance of the ETANN filtering compares well with the mathematical filter. Furthermore, by training on a flat background, it learns to ignored the noise which the mathematical filter would pass. A hardware system of ETANN chips could result in a significant increase in the processing rate. For example, if the 2048 values were compress by factor of 4, the resulting 512 channels could feed into 8 ETANNs in parallel. The filtering time would then take about $5\mu s$ (+ DAC time). Processing all 192 wires would then take about 1ms, a factor of 1600 improvement over the current cpu system. Furthermore, a second stage of ETANNs trained as described in reference 2 to find the position of Gaussians, could produce the hit coordinates as well.

3. Secondary Vertex Finding

In the second study we examine recognition of secondary decay events using tracking information from a silicon strip detector[10]. Figure 3 shows a setup for a proposed experiment. Interactions in the target produce primary particles, some of which decay before traversing the silicon strip detector. The high background rate, large hit multiplicity, short decay lengths, etc., make triggering on secondary decays of heavy flavour particles quite difficult. Previous work (e.g. ref. 1) discussed track finding with neural networks. Here, however, an associative memory (basically a fast lookup table, see ref. 11) finds the angle Φ and impact parameter D (figure 3) for each track. The neural network receives the parameter pairs of tracks found in the ZX and ZY planes. (The associative memory does not correlate the tracks found in the two planes.) The network output indicates whether the event contains a secondary vertex. The signal to background ratios vary from 1:1000 (hadron colliders) to 1:50000 (fixed target expt.)

Figure 4 shows the D vs Φ distributions for 2 events: a background and a signal event. The points for the primary tracks all lie along a flat line. The points from a secondary decay lie along a line at an oblique angle with respect to the line of primary points[12].

The network contains 64 inputs, 64 hidden and two outputs. The first output fires for background events and the second output fires for signal. The first 16 inputs take the D values of ZX tracks and inputs 17-32 take the Φ values for those tracks. Similarly, the inputs 33-48, 49-64 receive, respectively, the D and Φ values for the ZY tracks. The number of tracks average from 6-7 for background and 7-8 for signal with rms of around 3. We scaled the inputs to the range -1.0 to 1.0 and inputs with no tracks receive 0.0 input value. Note that the net must deal with varying numbers of activated inputs from event to event, with the first few inputs of each group of 16 almost always activated and the last few entries almost never activated.

We trained the network on 4000 background and 800 signal events and tested with 1000 background and 200 signal. As before, we first train the software emulator and then download to the chip for chip-in-the-loop training. We measured the performance with both *strict* and *loose* criteria. The strict criteria required consistent background and signal outputs (i.e. background event: BG output > 0.0 and SIG output < 0.0, signal event: BG output < 0.0 and SIG > 0.0). The loose criteria simply took the value of the signal output (i.e. BG event : SIG < *cut*, SIGNAL event: SIG output > *cut*, where we choose a particular cut value).

With strict cuts the ETANN gave a BG efficiency of about 92% correctly categorized and 48% signal efficiency. Figure 5 shows the loose cut results for the software and the chip as a function of the cut on the signal output. The tightest cut in figure 5 shows that one can reach 99% efficiency on BG with about 30% acceptance of signal. These results compare well even to offline analysis. If implemented in an experiment this system could provide high enrichment of the data with heavy flavor events.

4. Acknowledgements

We thank S. Ventura for providing us with the raw data for the filtering study.

5. References

1. C. S. Lindsey et al.,*Nucl. Instr. and Meth.* **A317**(1992) 346-356.
2. T. Akkila et al.,*Nucl. Instr. and Meth.* **A327**(1992)573-579.
3. W. Badgett et al.,*A neural network calorimeter trigger used in CDF*, Conference Record of the IEEE 1992 Nuc. Sci. Sym., Orlando Fla. Oct.27-31, 1992.
4. R. Odorico,*Heavy Flavor Discrimination by Neural Networks*, IEEE 1992 Nuc. Sci. Sym., Orlando, Fla. Oct.27-31, 1992. In print, *IEEE Trans. Nucl. Sci.* See also Odorico these proceedings.
5. 80170NX Electrically Trainable Analog Neural Network, Data Booklet. Intel Corp. 2250 Mission College Boulevard, Santa Clara, Ca. 95052-8125.
6. NeuroDynamX, Inc. 319 B Raymondale, South Pasadena, ca. 91030.
7. P. Hörnblad et al., *Filtering Data from a drift-chamber detector using neural networks*, in print, *Nucl. Instr. and Meth. Sec. A.*
8. Icarus collaboration,"ICARUS 1: An optimized, real time detector of solar neutrinos“, *Servizio Documentazione dei Laboratoi'ri Nazionali di Frascati.*
9. Sandor Ventura, CERN and Padua, *priv. comm.*.
10. Denby et al., *Investigation of a VLSI Neural Network Chip as Part of a Secondary Vertex Trigger*, in print, *Nucl. Instr. and Meth. Sec. A.*
11. S.R. Amendolia et al., *Nucl. Inst. and Meth.* **A315** 446-448.
12. L. Gupta et al., *Pattern Recognition* **25**(1992) 385-400.

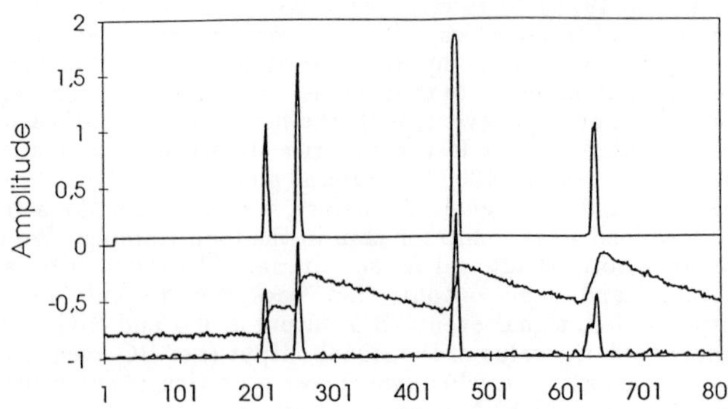

Figure 1. The middle curve shows the unfiltered FADC values versus time (200ns/bin). The bottom curve shows the output of the mathematical filter. The top curve shows the output of the ETANN.

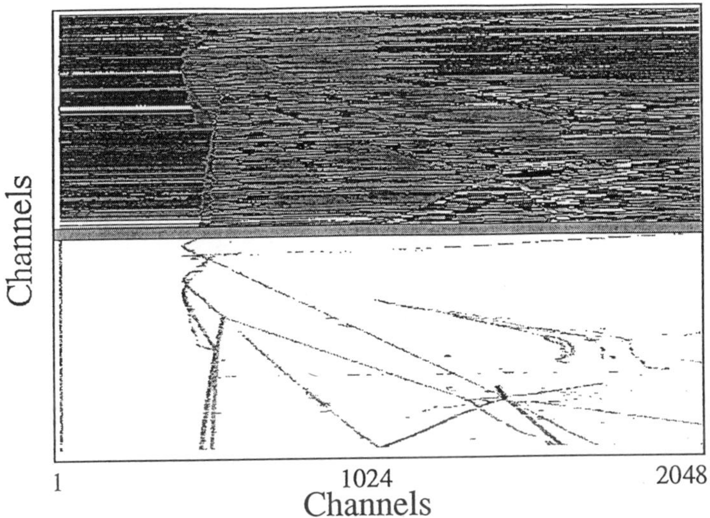

Figure 2. An event showing all 2048 FADC values for all 192 wires in the top plot. The plot indicates FADC value in proportion to grey scale. The bottom plot shows same event after filtering.

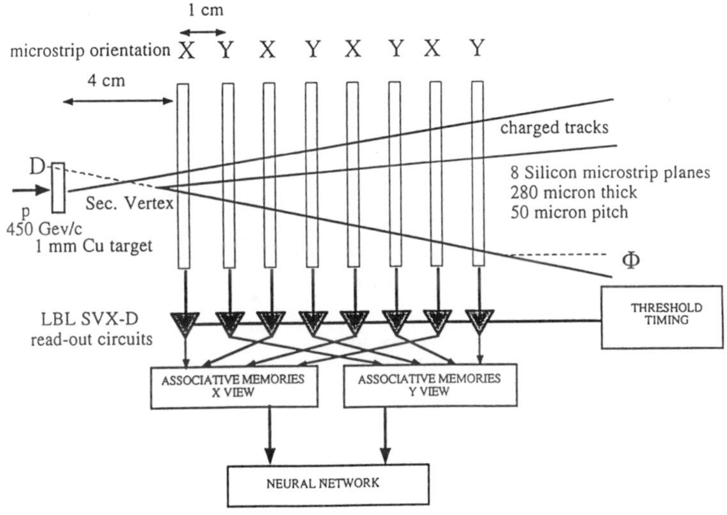

Figure 3. A proposed layout for a fixed target heavy flavour experiment.

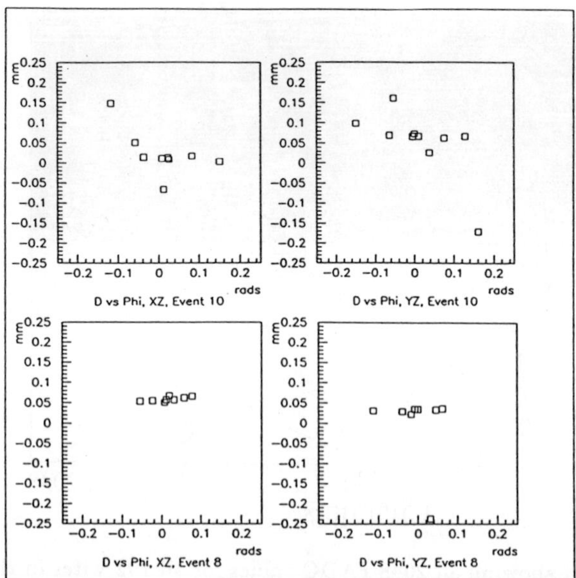

Figure 4. Signal (top) and Background (bot.) events in D-Phi space for each plane.

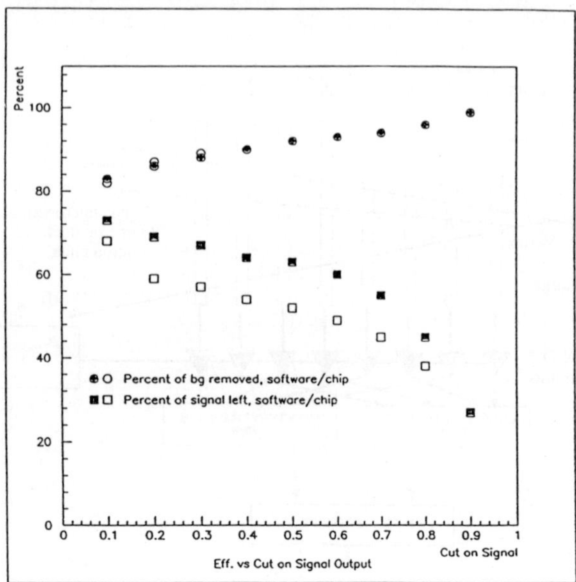

Figure 5. Identification efficiency versus output cut (fraction of -1 to 1 range).

RESULTS FROM AN ON-LINE NEURAL TRIGGER WITHIN A FIXED TARGET EXPERIMENT FOR THE PRODUCTION OF BEAUTY PARTICLES

C. BALDANZA, F. BISI, A. COTTA-RAMUSINO, I. D'ANTONE,
L. MALFERRARI, P. MAZZANTI, F. ODORICI, R. ODORICO, M. ZUFFA

INFN ANNETTHE
Via Irnerio 46, 40126 Bologna, Italy
E-mail: odorico@bo.infn.it

and

WA92 COLLABORATION

BOLOGNA/INFN, CERN, DUBNA JINR, GENOA UNIV./INFN, LONDON ICL, MOSCOW LEBEDEV PHYS. INST., PISA UNIV./INFN, ROME UNIV. I/INFN, ROME UNIV. II/INFN, SOUTHAMPTON UNIV.

ABSTRACT

Results from a neural trigger hosted by experiment WA92, beauty particle production from 350 GeV negative pions on a fixed Cu target, are presented. The neural trigger has been used to send on a special data stream events to be analyzed with high priority, adding to events with electrons and muons also events selected by a non-leptonic signature exploiting microvertex detector information. The neural trigger module consists of a VME crate hosting two MA16 digital neural chips from Siemens and two ETANN analog neural chips from Intel. The crate also includes boards for the preprocessing of input variables, for interfacing to the WA92 apparatus and for monitoring by a PC. The four neural chips run independently, for cross-checks and to show relative performance and stability. During the experimental run, only the ETANN chips were operational, while the MA16 were still waiting for completion of the debugging of the chip microcode. Input patterns of up to 32 variables can be processed, although 16 variables were found to be enough for the problem at hand. They are preprocessed in the neural trigger module from angles and impact parameters of tracks determined by the WA92 Beauty Contiguity Processor. Training of the feed-forward nets was done on a workstation. A final tuning for ETANN with chip-in-a-loop training was unnecessary, since a single layer net (i.e. a Fisher discriminant) turned out to be adequate and not outperformed by a two layer net. Event samples used for training derive from the WA92 1993 production run and consists of events already accepted by the standard non-leptonic trigger. Using the event reconstruction program, they have been divided in two classes: one containing events with a secondary C3 vertex (three tracks with sum of charges equal to ± 1), and the other containing the remaining events. Certification of the neural trigger selections was similarly done on independent events from the 1993 run. The neural trigger operated for two continuous weeks in the 1993 run. For an acceptance of 18% for C3 events the neural trigger yields an enrichment of 6.5, to be multiplied by that already provided by the standard non-leptonic trigger (somewhat above an order of magnitude). The response time of the neural trigger module is 5.8 μs.

1. Introduction

The idea of developing signatures for the dominant non-leptonic decays of heavy flavor particles by exploiting correlations within the associated events was put forward some time ago[1,2]. The technique proposed at that time was Fisher's linear discrimination[3], which corresponds to a simple feed-forward neural network with one input and one output layer and a single output neuron. The recent appearance of neural microprocessors, able to implement multi-layered neural networks with response times of a few microseconds, has stimulated efforts for the realization of on-line neural triggers capable of exploiting non-leptonic signatures of heavy flavors in high luminosity particle accelerators. Projects of this type were announced at the IEEE 1992 Nuclear Science Symposium[4-6]. So far, the only hardware development presented consists of a commercially available VME card, ELTEC SAC-711[7], hosting the ETANN chip[8] which has been used off-line with simulated events[9].

The neural trigger module, whose realization and on-line running results are presented here, consists of a VME crate utilizing the neural chips ETANN (analog) from Intel[8] and MA16 (digital) from Siemens[10] which has operated within the WA92 experiment at CERN[11] during the 1993 run. WA92 is an experiment looking for the production of beauty particles by a π^- beam at 350 GeV/c impinging on a Cu target. Task of the neural trigger is to send events to be analyzed with high priority on a special data stream, the Fast Stream, which previously collected events with appropriate semi-leptonic signatures only, exploiting a purely non-leptonic signature. Target of the neural trigger are events with secondary C3 vertices, i.e. branching into three tracks with sum of the electric charges equal to ± 1. At the time of running only the ETANN part of the crate was operational, while the MA16 part was still waiting for completion of the debugging of the MA16 microcode. The training of the ETANN chip was done using 1993 WA92 data, with C3 events certified by the Trident event reconstruction program. Input to the neural crate was provided by the Beauty Contiguity Processor (BCP) of Darbo and Rossi[12], which determined tracks and their impact parameters on-line, using data from the silicon microstrip Vertex Detector. The BCP data, arranged in 5x64-bit hit-maps, were preprocessed within the neural crate to yield 16 input variables for the neural chips. The neural trigger has been in operation for two weeks, during which its stability was verified.

2. Input from the WA92 Apparatus

The part of the WA92 trigger detector relevant for the input to the neural chips consists of 6 silicon-microstrip planes of the Vertex Detector (VD), measuring the z coordinates of tracks, where the z axis has the direction of the magnetic field (bottom-up direction). The last plane is positioned at about 41 cm downstream of the target and has a square shape with a 5 cm side, centered around the beam. The hit information from the 6 planes is processed by the Beauty Continguity Processor (BCP) developed by Darbo and Rossi[12], which reconstructs tracks and their impact parameters (IP), with a global response time of about 40 μs[13]. IP is defined as the signed difference between the z value of the track extrapolated back to the target-center plane and the beam z position. The BCP supplies to the Neural Trigger (NT) 5 words of 64-bit each, plus a standard go-ahead termination word. Each one of the 5 words directly represents the hit-map of tracks on the last (6th) VD layer, divided in 64 bins, for tracks falling within a given IP window. The 5 IP windows used are, in μm: $-200 < IP < 200$, $200 < IP < 400$, $-400 < IP < -200$, $400 < IP < 900$, $-900 < IP < -400$. The BCP could supply up to 8 hit-maps, with a resolution which can reach 256 bits. Data transfer times would be correspondingly increased. Simulation has shown that the arrangement we eventually employed is adequate for the task and does not lead to any significant loss of performance.

The BCP also makes a first level decision about accepting the event. Defining as secondary tracks those with $|IP| > 100$ μm and as primary the other tracks, the Standard Non-Leptonic Trigger (SNLT) requires the presence of 3 primary tracks, 2 secondary tracks and 1 track in the Butterfly Hodoscope with high p_T (defined in a detector dependent mode, roughly corresponding to > 0.6 GeV/c along the z axis). The Neural Trigger considers only events which have already been accepted by the SNLT.

3. Neural Trigger Hardware

3.1. Generalities

The Neural Trigger module consists of an Extended VME crate (VME9U boards). Its overall structure is shown in Fig. 1. There are 9 boards in the crate:

- one VIC 8250 module from CES, interfacing the VME Bus to an IBM-compatible personal computer;
- one Input Interface Board, interfacing to the WA92 apparatus;
- 4 Preprocessing Boards, with 2 Preprocessing Units each, each unit calculating variables for a given IP window;
- 2 MA16 Boards, with one MA16 chip per board;
- one ETANN Board, hosting two independent ETANN chips.

Boards communicate by means of 3 buses:

- Interface Bus (J3), connecting the Input Interface Board to the Preprocessing Units, working at a clock-frequency of 4 MHz;
- Processor Bus (J2), to access the ETANN and MA16 Boards, working at a clock-frequency of 8 MHz;
- VME Bus (J1&J2), to access diagnostic registers on the boards and to control all boards from the PC.

All inputs from the WA92 apparatus enter through connectors in the front panel of the Input Interface Board. Outputs to the WA92 trigger logic and data acquisition exit through connectors in the front panels of the MA16 and ETANN Boards.

3.2. VIC Board

The VIC 8250 board from CES connects the VME Bus to a VBAT 8218 card, also from CES, installed in the PC. By means of C++ instructions it is thus possible to access for read and write operations VME addresses from the PC. Specific C++ programs have been developed for configuring, testing and monitoring the crate from the PC. Also, programs have been developed for simulating on-line running conditions, and having recorded experimental inputs or simulated inputs passed through the crate with collection of the corresponding outputs.

3.3. Input Interface Board

This wire-wrap board receives inputs from the WA92 apparatus and controls their transfer to the rest of the crate, Fig. 2.By 2x64-pin TTL differential connectors it receives the 5 64-bit words from the BCP, containing the 64-bit hit maps of the last layer of the VD for each one of the 5 IP windows, plus a standard 64-bit go-ahead termination word. The latter is necessary because the BCP starts the data transfer for each map as soon as it has calculated it, and may subsequently decide not to accept the event. In this case elaboration of the event information by the Neural Trigger is aborted. The TTL differential signals are converted to TTL unipolar. Through LEMO connectors it also receives 5 NIM signals (inhibit, start, burst, clear, strobe) controlling the operating mode of the crate. Left unused in the 1993 run are 8 NIM signals carrying information from the Butterfly Hodoscope, signalling the presence of high-p_T tracks.

A strobe signal signals the presence of a valid data word in input. A start signal controls the start of the data transfer, whose termination is indicated by a clear signal. The burst signal remains active for the full duration of the beam burst. When it is not active it is possible to initiate from the Interface Board the test procedures for the entire crate. More in general, the Interface Board plays the role of master for the whole crate by controlling the data flow on the Interface and Processor Buses.

The board may also receive inputs from the PC via the VME Bus, mimicking on-line running conditions.

A block scheme of the Interface Board is shown in Fig. 2.

3.4. Preprocessing Boards

There are 4 printed-circuit boards of this type, each one hosting 2 independent Preprocessing Units. Each Preprocessing Units outputs the input variables to the neural chips for a given IP window. With 5 IP windows, only 5 such units are actually used. Determination of the input variables is done by segmenting the 64-bit hit-maps in 4 adjacent groups of 16 bits each. Each 16-bit group is sent in input to a 16-bit addressable with 8-bit output EPROM (64K), on which a Look Up Table (LUT) is loaded, giving in output the number of non-zero bits and the OR of the bits. All the OR bits from the 4 EPROM's are sent to a fifth EPROM, loaded with a LUT which can be designed to give in output: i) either the number of the four groups which have been hit; ii) or the number of the two groups in the bottom half of the hit-map which have been hit; iii) or the number of the two groups in the upper half of the hit-map which have been hit. Thus, for each IP window, 5 variables are calculated from which one can pick up the variables to be actually sent to the neural chips. The EPROM I/O is arranged so as to also calculate similar variables for 8-bit groups. But they turned out to be not significant and were left unused. More in general, one can program the EPROM's with a different choice for the LUT's. The configuration which was finally adopted turned out to be the optimal one among several attempted in the preliminary studies.

A block scheme of the Preprocessing Boards is shown, in Fig. 3.

3.5. MA16 Boards

There are two identical boards each one hosting a MA16 neural chip[10]. The MA16 chip has been supplied by Siemens as a prototype, it is not yet a commercial product. The MA16 Board has been developed in collaboration with J. Beichter, Siemens AG, Munich. At the time of the experimental run, debugging of the MA16 microcode was still in progress.

The MA16 is a digital microprocessor of systolic type. It accepts input variables with 16-bit precision and has an internal precision of 48 bits. It has been designed to run up to a clock frequency of 70 MHz. Because of the lack of commercially available external components working at that speed, it has been actually certified to run up to 50MHz. It needs: i) external storage for the neural weights, ii) an external LUT for the neural transfer function, iii) an external controller for synchronization with I/O devices. We have used 16-bit, 50 MHz EPROM's (CY7C259 from Cypress) to store the weights and the controller sequences. RAM's would have been more practical, but we were not able to find anyone working at that speed. Exploiting the fact that the MA16 is designed to handle 4 independent inputs simultaneously, whereas we can use only one at the time in our trigger application, exploiting the idle time in the sequential output we accommodated the transfer function LUT on two 16-bit addressable, 20 MHz EPROM's (CY7C286 from Cypress) with 8-bit output.

The MA16 Board has been designed to accommodate 16 input variables, a 2-layer feed-forward neural net with a maximum of 15 hidden nodes, a Fisher discriminant exploiting the spare node left by the 2-layer net (the MA16 can accommodate a maximum of 16 nodes). The MA16 receives input from a multiplexor, which directs to the MA16 input from the Processor Bus at the first cycle of the 2-layer net, and output from the LUT at the second cycle. A block scheme of the board is shown in Fig. 4.

Output from the board consists of: i) 2x32-bit words sent in TTL differential through 2x64-pin connectors to the WA92 data acquisition system, containing 16-bit output from the 2-layer net, 16-bit output from the Fisher discriminant, copies of the trigger bits, crate status information; ii) one 2-layer net and one Fisher discriminant trigger bits, sent as NIM signals to the WA92 trigger logic, which are obtained by comparing the corresponding output words with threshold values set in the appropriate registers of the two comparators. The response time of the board at 50 MHz is 5.2 μs. Adding to it the time spent in the Interface and Preprocessing Boards, which is 3.2 μs, one has a total response time for the Neural Trigger module of 8.4 μs.

One can also have input to the board sent and the corresponding output collected via the VME Bus, connected to the PC.

3.6. ETANN Board

The board hosts two independent ETANN chips. Intel's 80170NX Electrically Trainable Analog Neural Network device is an analog chip which is commercially available together with a development system, the iNNTS[14]. Neural weights are stored in the chip in EEPROM-like fashion. iNNTS includes a hardware box (the trainer) where to insert the chip, which is interfaced to an IBM-compatible personal computer, and software to be run on the PC so as to allow read and write operations for the neural weights and processing of the input patterns through the chip. The chip can operate in 1-layer mode, accepting a maximum of 128 input variables, and in 2-layer mode with a clocked-back feedback operation, accepting a maximum of 64 input variables. In both cases, it has 64 independent outputs (thus, a maximum of 64 hidden layer nodes in the 2-layer mode). We have measured response times of 1.5 μs for the 1-layer mode and of 4.5 μs for the 2-layer mode. We have then chosen to allow safer processing times of 2.6 μs for the 1-layer mode and of 6.1 μs for the 2-layer mode. Precision of the chip is quoted[8] as being equivalent to 6-7 bits. Operatively, with our set-up, after conversion of the output to a 8-bit digital number (range 0-255), we have found during the experimental run a maximum deviation of 4 units for a high response, around or above the threshold value of about 146 for triggering, and a maximum deviation of 6 units for a lower response (likely due to the larger degree of cancellations), when routinely sending a prefixed set of test patterns chosen so as to uniformly cover the output range.

A critical feature of ETANN is its dependence on the values of the power supply voltage (about 5 V) and of the temperature. In particular, one must care that these values remain stable when moving the chip from the iNNTS trainer to the board. In order to eliminate these problems, we have developed an ETANN Controlled Environment (ECE) Unit consisting of a daughter board detachable from the main board and fit to be inserted in the iNNTS trainer. In the ECE Unit, ETANN is serviced by a voltage regulator, stabilizing its power voltage to within $\Delta V \leq 5$ mV, and a temperature controller stabilizing its temperature at 18 °C within $\Delta T \leq 1$ °C. The temperature controller is based on Peltier's cells from MELCOR, and absorbs up to 15 W versus the 5 W dissipation of ETANN. Because of its high power consumption, the temperature controller has a separate power supply.

Input variables from the Preprocessing Boards are converted to analog signals for ETANN by 16 two-fold 8-bit Digital to Analog Converters (DAC's). Since they are relatively large in size (8 cm by 4.5 cm), they are arranged in separate daughter boards.

The range of the input variables is 0-16, therefore 8-bit input precision is sufficient. Each two-fold DAC can convert two input variables simultaneously, with a loading time of 0.125 μs imposed by the Processor Bus data transfer rate of 8 MHz. Thus 32 variables can be accepted by the board. With the present crate set-up, though, only 16 variables are sent via the bus, and each one is converted by one half of a DAC. That was done to have the bus structure uniform with what required by the MA16 Boards. The 8-bit input is converted to an analog signal from 1.6 to 3.2 V, with zero corresponding to $V_{REFI} = 1.6$ V, as required by ETANN. The output from ETANN, ranging from 0 to 3.2 V, is converted by Analog to Digital Converters (ADC's) to an 8-bit digital number in the range 0 to 255. The converted output is compared with a preset threshold value for triggering in a comparator, which leads to the generation of the corresponding trigger bit as a NIM signal. The board is set-up to handle two independent outputs from ETANN, associated to two output nodes. That is meant to accommodate two independent neural nets in the chip, e.g. a 2-layer net and a Fisher discriminant (passed twice through the transfer function). We have found, though, that such an arrangement is to be discouraged because of the cross-talk between the nets, due to the non-exact zero setting of the weights connecting the two nets (iNNTS uses a recursive procedure to get weights approaching zero with some precision). That has a marginal effect when a single net is loaded, because of the inactivity of the unused nodes. Eventually, as discussed in the next Section, we have loaded on the chip one single-layer net corresponding to a Fisher discriminant (passed once through the transfer function). The parameter controlling the transfer function slope has been set at $V_{GAIN} = 4.0$ V. In the iNNTS trainer, we have set the zero reference value for the output at $V_{REFO} = 1.6$ V.

Output from the board consists of: i) 2x32-bit words sent in TTL differential through 2x64-pin connectors to the WA92 data acquisition system, containing the two 8-bit outputs from the Fisher discriminants loaded on the two ETANN's, the two 8-bit outputs from the two other nets which can be accommodated on the ETANN's (left unused), copies of the trigger bits, crate status information; ii) the Fisher discriminant trigger bits from the two ETANN's, sent as NIM signals to the WA92 trigger logic. Adding to the response time from the chips the time spent in the Interface and Preprocessing Boards and data transfer (i.e. including the loading time of DAC's), which is 3.2 μs, one has a total response time for the Neural Trigger module of 5.8 μs for the single-layer net we have actually used. That would grow to 9.3 μs for a 2-layer net.

One can also have input to the board sent and the corresponding output collected via the VME Bus, connected to the PC. That is done by means of input registers which can be periodically loaded with diagnostic test patterns and output latches holding the corresponding responses of the board for VME readout. Prominent analog operating parameters and various status bits can also be similarly monitored via VME.

An artist's view of the board, showing the arrangements for the daughter boards, is given in Fig. 5. A block scheme of the board is shown in Fig. 6.

4. Neural Network Architecture and Training

In principle, we could have set as target for the Neural Trigger the direct recognition of beauty particle decays, defined by simulated events obtained by an event generator with added GEANT simulation of the WA92 apparatus. However, we refrained from doing that mainly for two reasons: i) although such simulated events were available, they were suitable for off-line studies, after some event selection, and were not appropriate for quantitative tuning of a neural trigger, also because of the general dependence of event generators on model assumptions which are especially critical for the relatively low mass beauty particles and for a complex nuclear target like copper; ii) the necessity of depending on graphical scanning, with the long waiting times involved, to have indications on the actual performance of the neural trigger.

Taking into account the effective necessities of the experiment, which required the analysis of many tens of millions of events to extract the few thousands affordable by graphic scanning, we rather focused on the enrichment in events which are selected for that sake. Especially interesting, from this point of view, are events with C3 secondary vertices, branching into three tracks with sum of electric charges equal to ± 1. A C3 vertex is a signature for $D \to K\pi\pi$ decay, and thus a non-leptonic signature for beauty particles. Certification of these events comes from the Trident event reconstruction program. By applying the event reconstruction program to real data obtained with a given experimental set-up, one gets the necessary event training samples to tune the Neural Trigger for enrichment in C3 events with the same experimental set-up. In this way one avoids altogether dependence on theoretical modeling and only relies on the event reconstruction program. This is the target we have chosen for the Neural Trigger. In doing that the Neural Trigger does not act alone, but in conjunction with the first level Standard Non-Leptonic Trigger (SNLT, see Section 2): only events already accepted by the SNLT are worked out by the Neural Trigger. By concentrating on SNLT events only, the Neural Trigger can better focus on correlations present within the events which discriminate signal from background. An example of them is offered by angle-IP correlations for tracks. As shown in Fig. 7 the angular distribution of tracks is strongly correlated to the sign of IP for C3 events, while this correlation is marginal for background events. The SNLT&NT non-leptonic signature has been used to send events to be analyzed with high priority on a special data stream, the Fast Stream, where they lay together with events selected by semi-leptonic signatures.

In order to train the neural net, one not only needs to have samples of signal and background events certified by the event reconstruction program, but also to have the corresponding input patterns to the neural chips calculated from them. As specified in Section 2, input variables to the neural chips are calculated from the BCP output, but the latter is not recorded on the event tapes. To recover it we used the BCP simulator, which exactly reproduces the BCP calculations[15], applied to the raw data on tape.

In order to design the neural net architecture and to chose the input variables, we could not wait for the 1993 data being made available, but we had to rely on the 1992

WA92 data obtained with a somewhat different SNLT set-up. Since the 1993 SNLT included in part the angle-IP correlations exploited in the preliminary neural net studies on the 1992 data, it turned out that some of the input variables which had been chosen had become irrelevant. Performance of the Neural Trigger has degraded to some degree because of that. Only input variables simple to calculate with the available hardware were taken into consideration. The basic strategy in devising them was to have them convey multi-scale information on the hit maps provided by the BCP. We first considered the maximum resolution made available by the BCP: 256 bits. By directly studying the corresponding hit maps for a set of events, however, we saw that 64-bit hit maps were enough. Using such a resolution had the advantage of reducing transfer times and simplified the collection of the hit maps from the BCP by employing an already existing but unused 64-bit output connector. We then considered variables counting the number of hits with 1-bit, 4-bit, 8-bit, 16-bit, 32-bit resolutions. We submitted them to significance tests based on Fisher discrimination[16,17]. It turned out that the significant variables were those associated with 1-bit and 16-bit resolutions. The study was made by varying the IP window segmentation. Eventually, we found that an adequate IP window segmentation was the following one (the BCP allows for a maximum of 8 IP windows):

#IP	Range (μm)
1	-200 < IP < 200
2	200 < IP < 400
3	-400 < IP < -200
4	400 < IP < 900
5	-900 < IP < -400

As to the choice of variables, we found that the following 16 variables were adequate according to significance tests based on Fisher discrimination. Each 64-bit map (#IP = 1-5) is divided in 4x16-bit groups, ordered bottom-up in z, and on the basis of that the following variables are defined:

K16MAP(i)	= #hits in group i = 1-4	(range: 0-16)
K16	= #groups hit	(range: 0-4)
KDW16	= #groups in bottom half hit	(range: 0-2)
KUP16	= #groups in top half hit	(range: 0-2)

The 16 variables which have been chosen are:

#	VARIABLE	#IP
1	K16	1
2	K16MAP(1)	1
3	K16MAP(4)	1
4	KDW16	2
5	K16MAP(1)	2
6	KUP16	3
7	K16MAP(3)	3

8	K16MAP(4)	3
9	KDW16	4
10	K16MAP(1)	4
11	K16MAP(2)	4
12	K16MAP(3)	4
13	KUP16	5
14	K16MAP(2)	5
15	K16MAP(3)	5
16	K16MAP(4)	5

The slight asymmetry between windows with opposite IP is a reflection of the similar asymmetry observable in angle-IP correlations, Fig. 7. That is due to the fact that the BCP looks for the beam particle starting from bottom up and that there is a substantial fraction of events with two beam particles, one of which is just passing through. If the latter is picked up by the BCP as the actual beam particle in the event, one gets a spurious determination of the IP of tracks leading on average to the above asymmetry.

One can argue against reliance on significance tests based on Fisher discrimination for choosing the input variables, regarding them as appropriate when discussing the separation of the bulk of the event class distributions but too insensitive to the tail structure of these distributions, which is in principle relevant in high enrichment applications. Being aware of that, in our preliminary neural net studies we made several tries with larger samples of input variables, taken from the wide original pool. But we got no significant improvement of the results.

For the neural net architecture we considered the two options offered by ETANN: a 2-layer or a 1-layer net, the latter being equivalent to a Fisher discriminant. As discussed at some length in[17], the optimal type of neural net totally depends on the problem at hand. If the two class distributions one has to discriminate have a gaussian shape with similar correlation matrices, no classifier can outperform the Fisher discriminant. We made preliminary software studies simulating some of the ETANN restrictions, as represented by the software handling the iNNTS development system. Specifically, weights and thresholds were requested to stay between -2.5 and 2.5 with input variables limited to ± 1. No useful way was found to simulate the limited weight precision. After training by standard back-propagation, the 2-layer net gave results slightly better (of the order of 20%) than a 1-layer net. But, after loading the nets on ETANN and having made some chip-in-loop training of the 2-layer net (that is not necessary for a 1-layer net), we found that the small margin in favor of the 2-layer net evaporated. Looking in detail at how event patterns were processed in the 2-layer net, we realized that the net had organized itself so that discrimination between the two event classes was largely due to small differences between the responses of the first layer which entered cancellations performed by the second layer node. Thus, weight precision had a large incidence on the results. We found no way of forcing back-propagation to direct the organization of the net to a structure less sensitive to weight precision. (Procedures to compensate for the limited weight precision by separately changing the transfer function slopes of nodes[18] are not applicable to ETANN, which has a single parameter, V_{GAIN}, to control all the, nominally, equal slopes.) We had other reasons pressing us toward a 1-layer architecture: i) simplicity of training, since a 1-layer net, i.e. a Fisher discriminant, requires only inversion of the pooled-over-classes correlation matrix for that sake; ii) avoidance of chip-in-loop training, which may burn out nodes in ETANN, the latter being required essentially because of the variance of the transfer functions of nodes, which becomes irrelevant for a 1-layer net with one output node; iii) increased robustness towards changes in the operating conditions of the trigger detector, e.g. due to temperature excursions which can change alignments and the level of electronic noise, since the net is less fine-tuned to details of the shapes of the event class distributions; iv) shortening of the response time of the Neural Trigger.

In conclusion, we adopted a 1-layer net architecture with one output node. Its weights (i.e. the Fisher vector components) were determined on a conventional workstation by a simple matrix inversion operation and then loaded on ETANN by means of iNNTS. The procedure was fast and straightforward, in compliance with the strict time restrictions we had to meet during the experimental run. The training event samples we used consisted of 3000 C3 events and 10000 non-C3 events, accepted by the SNLT. The collection of the C3 event sample required the running of the Trident event reconstruction program for about 4 days on 3 alphavaxes 3000/400. C3 vertices are defined by the Trident event reconstruction program, release 11, requiring in addition: i) IP in the y-z plane (orthogonal to the beam direction) > 60 μm, ii) distance in x from the primary vertex > 6 σ, iii) x of the vertex not before the target and not beyond the microvertex detector, iv) at least two tracks with |IP| along the z direction > 20 μm.

5. Results from the Experimental Run

The Neural Trigger, after its training was completed, operated for two continuous weeks within the apparatus of WA92. The results we present are based on a sample of events collected about two weeks after the acquisition of events used for training. It consists of 6403 C3 events and 246746 non-C3 events accepted by the SNLT and certified by the Trident reconstruction program. Fig. 8 shows the values of C3 event acceptance and enrichment obtained by the Neural Trigger when varying the lower cutoff on the neural net output, ranging from 0 to 255 (8-bit output). Fig. 9 shows the same results but with the C3 event enrichment directly plotted toward the C3 acceptance. The results refer to the output from one of the ETANN chips, the other ETANN chip giving consistent results.

Above a certain value the C3 enrichment flattens out when increasing the cutoff on the neural net output. Setting the cutoff to a value corresponding to the beginning of the plateau, one obtains for a C3 event acceptance of 18% a C3 event enrichment of about 6.5, corresponding to a rejection factor for the background of about 36. The enrichment obtained by the Neural Trigger should be multiplied by that provided by the SNLT, which is somewhat larger than an order of magnitude, to get the total enrichment for C3 events given by the non-leptonic trigger.

6. Conclusions

A Neural Trigger device based on the ETANN chip has been demonstrated by two weeks of continuous running in the experiment WA92.

A non-leptonic trigger for heavy flavor particles has been shown to be viable.

7. Acknowledgements

This research has been supported in part by a computer grant by CINECA, Bologna.

A. Benvenuti has suggested the idea of demonstrating neural network techniques within experiment WA92, and has contributed, together with D. Bollini and F.L. Navarria, at the early stages of this project. The MA16 project is being made possible by the continuous support of J. Beichter, N. Brüls and U. Ramacher, of Siemens AG, Munich. C. Lindsey has made us familiar with a number of features of ETANN. F. Degli Esposti, M. Lolli, P. Palchetti and G. Sola, of the INFN Electronics Laboratory of Bologna, have given relevant contributions to the hardware development. M.L. Luvisetto, F. Ortolani and E. Ugolini have helped us in sorting out a number of problems with C++ programming. One of us, R.O., acknowledges useful conversations with R. Battiti about neural net training procedures.

8. References

1. R. Odorico, *Proc. of the 1982 DPF Summer Study on Elementary Particle Physics and Future Facilities*, Snowmass, Colorado, 1982, p. 478; R. Odorico, *Phys. Lett.* **120B** (1983) 219; G. Ballocchi and R. Odorico, *Nucl. Phys.* **B229** (1983) 1.
2. A. Cherubini and R. Odorico, *Z. Physik* **C 47** (1990) 547, **C53** (1992) 139.
3. R.A. Fisher, *Annals Eugenics* **7** (1936) 179.
4. R. Odorico, Invited Talk at the *1992 Nuclear Science Symposium,* Orlando, Florida, 1992, Conf. Record p. 822; *IEEE Transactions on Nuclear Science* **40** (1993) 705.
5. P. Ribarics, Invited Talk at the *1992 Nuclear Science Symposium,* Orlando, Florida, 1992, Conf. Record p. 825.
6. W. Badgett et al., Invited Talk at the *1992 Nuclear Science Symposium,* Orlando, Florida, 1992, Conf. Record p. 841.
7. *SAC-700/800 User's Manual*, ELTEC Elektronik, GMbH, Mainz, Germany.

8. *80170NX Electrically Trainable Analog Neural Network Data Booklet*, Intel Corp. 2250 Mission College Boulevard, Santa Clara, CA 95052-8125, USA.
9. B. Denby et al., *Nucl. Instr. and Meth.* in press.
10. U. Ramacher et al., *Proc. Int. Conf. on Systolic Arrays*, Killarney, Ireland (Prentice Hall, 1989), p. 277.
11. M. Adamovich et al., WA92 Collaboration, CERN/SPSC 90-10 (1990), *Nucl. Phys.* (Proc. Suppl.) **B 27** (1992) 251.
12. G. Darbo and L. Rossi, *Nucl. Instr. and Meth.* **A 289** (1990) 584.
13. C. Bruschini et al., *1992 Nuclear Science Symposium, Orlando,* Florida, 1992, Conf. Record p. 320.
14. *iNNTS Neural Network Training System User's Guide*, Intel Corp. (1992).
15. C. Bruschini, *University of Genoa Thesis* (1992).
16. M. Kendall, A. Stuart and J.K. Ord, *The Advanced Theory of Statistics*, Vol. 3, 4th ed. (C. Griffin &Co. Ltd., London).
17. P. Mazzanti and R. Odorico, Univ. of Bologna, Dept. of Physics Report DFUB 92/15, *Z. Physik* C in press.
18. M. Hoehfeld and Scott E. Fahlman, *IEEE Trans. Neural Networks* **3** (1992) 602.

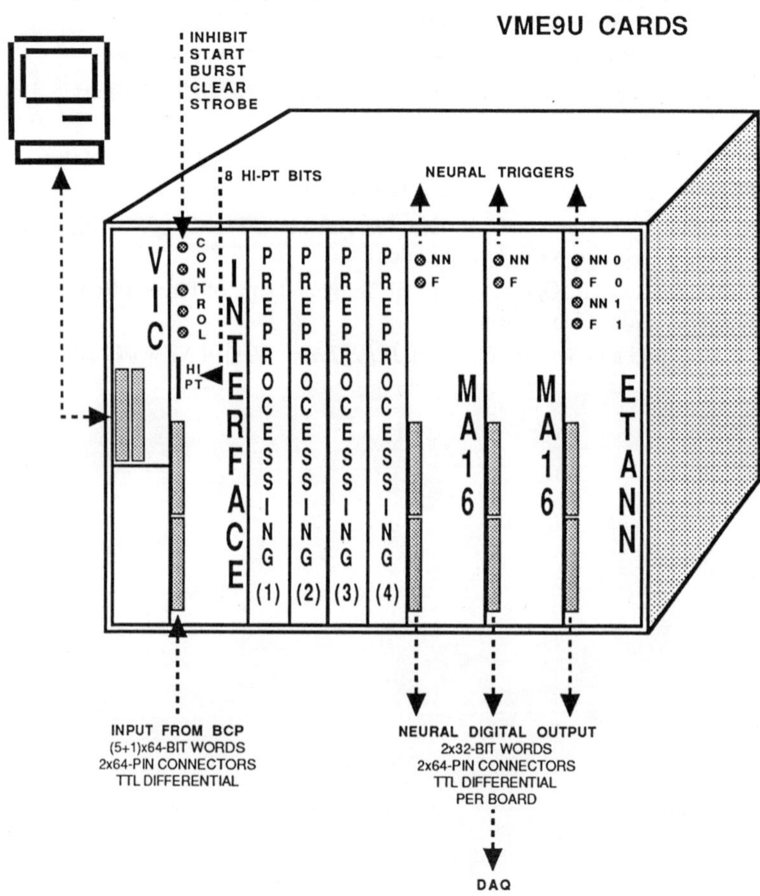

Fig. 1. Schematic view of the Neural Trigger VME crate.

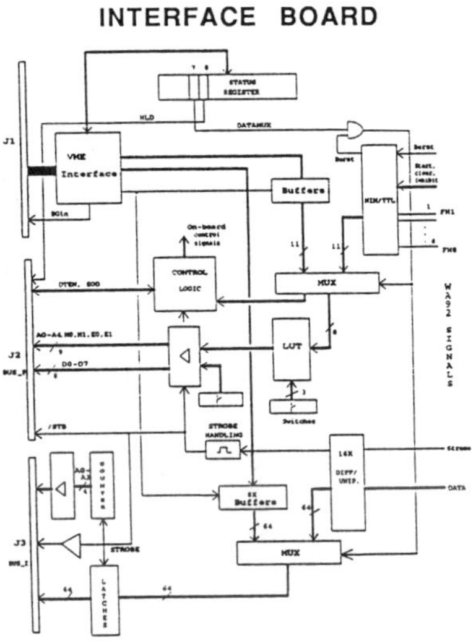

INTERFACE BOARD

Fig. 2. Block scheme of the Interface Board.

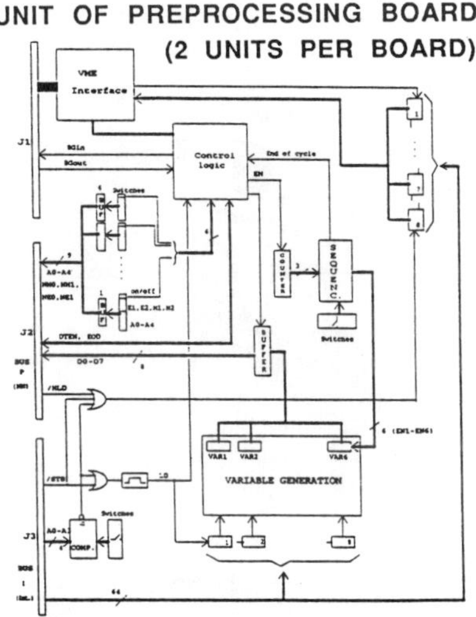

Fig. 3. Block scheme of a Preprocessing Board.

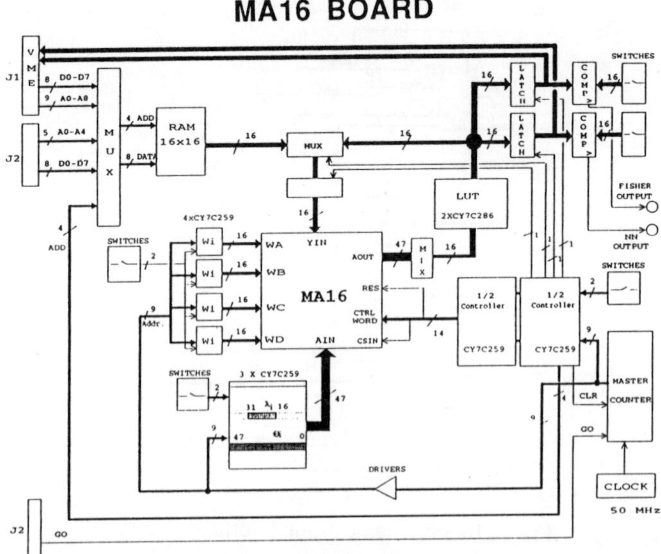

Fig. 4. Block scheme of the MA16 Board.

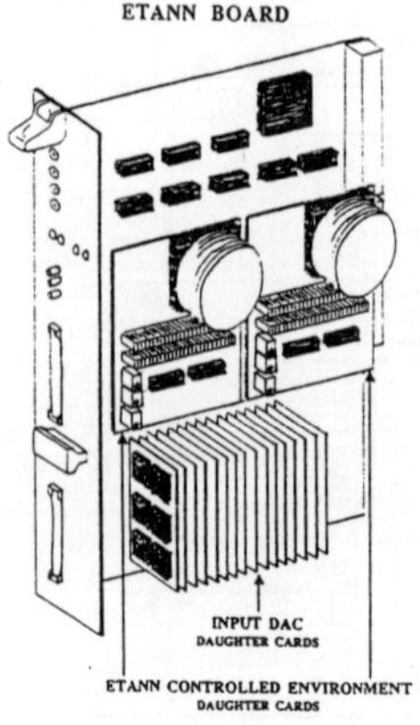

Fig. 5. Artist's viev of the ETANN Board.

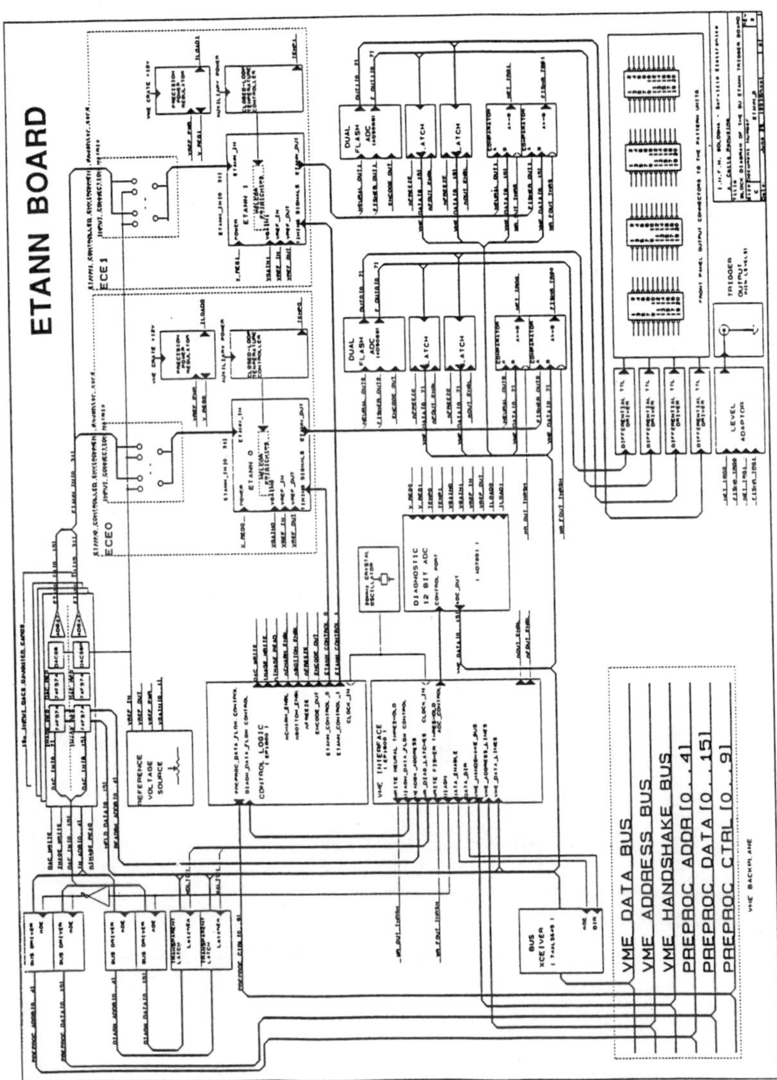

Fig. 6. Block scheme of the ETANN Board. The VME and Preprocessing bus entries refer to the VME Bus and Processor Bus, respectively.

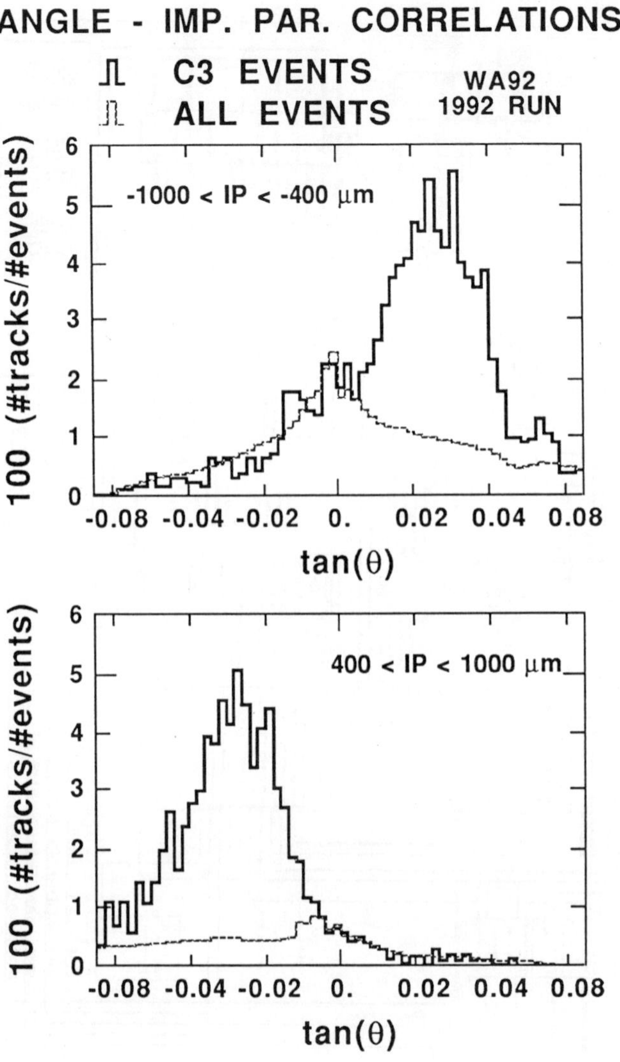

Fig. 7. Angular distribution of tracks as seen on the last plane of the micro-vertex detector for two IP windows. The angle θ is the angle of the track in the plane of the magnetic field and the beam with respect to the beam direction.

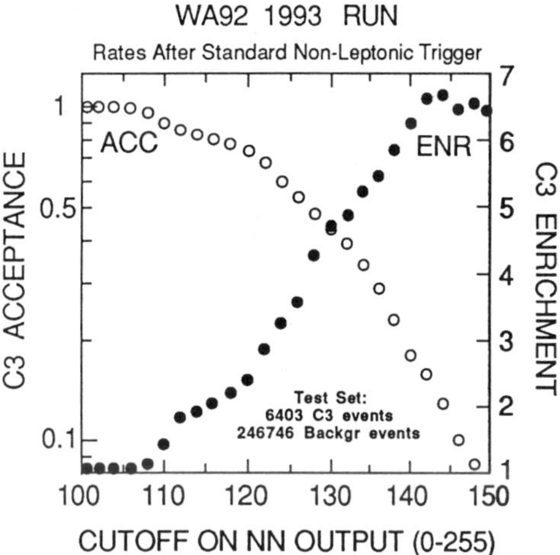

Fig. 8. C3 event enrichment and acceptance obtained when varying the lower cutoff on the Neural Trigger output (range 0-255), for the WA92 1993 data.

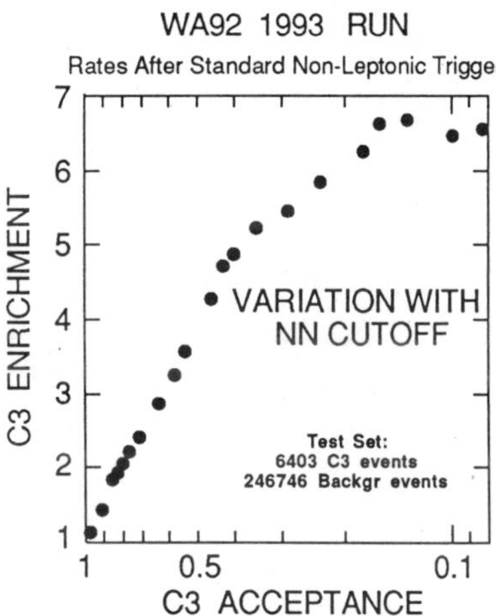

Fig. 9. Same as in Fig. 8, with C3 event enrichment plotted directly versus the corresponding C3 acceptance.

CONCEPT FOR A FAST SECOND LEVEL TRIGGER USING A NEURAL NETWORK ARCHITECTURE FOR THE H1-EXPERIMENT AT HERA

C. Kiesling[2], J. Fent[2], W. Fröchtenicht[2], D. Goldner[1], A. Gruber[2], T. Kobler[2], H. Kolanoski[1], T. Krämerkämper[1], J. Möck[2], P. Ribarics[2]

[1]*Institut für Physik, Universität Dortmund, 44221 Dortmund, Germany*

[2]*Max-Planck Institut für Physik, Föhringer Ring 6, 80805 München, Germany*

ABSTRACT

At the HERA *ep* collider the expected machine background rates exceed by several orders of magnitude the typical collision rates from *ep* physics reactions. One of the great challenges for the *ep* experiments at HERA is therefore to find methods suppressing the high rate machine background efficiently without cutting heavily into the physics. We present here the concept, the design, and the status for a second level hardware trigger based on the neural network architecture. The design is centered around commercial VLSI hardware. In this concept we not only concentrate on the hardware realisation but also address the problems of efficiently selecting "known" physics as well as preparing such a trigger for "new" physics never presented to the network.

1. Introduction

At full luminosity, the HERA Electron-Proton-Collider at DESY will challenge experiments with a few 100 kHz of interaction rate, mostly from collisions of protons with the rest gas in the beam pipe (beam-gas) or with components of the beam transport system (beam-wall). Most of the beam-gas reactions are characterized by very low transverse momenta of the secondary particles produced and can therefore be suppressed efficiently at the first trigger level, e.g. by requiring a minimum transverse energy seen in a calorimeter. Beam-wall events are largely reduced by observing that their origin is mostly far off the beam orbit and upstream of the experimental zone. Fast scintillator triggers (time of flight counters and up-stream veto walls) are therefore a good protection against this kind of background. Both beam-gas and beam-wall backgrounds can also be recognized by reconstructing, using tracker information, the event origin. For genuine *ep* collisions, the origin is confined to the interaction region given by the length of the proton beam, about $\pm$ 50 cm around the nominal point of collision.

Despite all these precautions a large remaining background rate of several kHz is escaping fast detection by the first level trigger system. Since a typical logging rate of events to tape for a modern high energy physics detector system is confined to a few Hz (usually below 10 Hz due to the large amount of data from these detectors) the trigger system has to reduce this primary rate of several kHz down to the required few Hz, without losing substantially in the physics acceptance. The H1 experiment has tried to meet this task by building a multilevel trigger system with two hardware and two consecutive software stages.

While the software trigger levels can accept up to 100 Hz due to computing time limitations, the two preceeding hardware levels have to bring down the rate to this level. This means for the second level hardware trigger that a rate reduction of order 10 to 50 with respect to the output rate of the level 1 trigger needs to be achieved. Due to time limitations, the level 2 trigger only has access to the trigger information from level 1. The level 2 trigger therefore has to use this information in an "intelligent" way. Here is where the neural network idea enters: The level 1 trigger information is viewed as a pattern or feature vector which is analysed by a standard feed-forward neural network. Offline, the network has to learn to discern the patterns of physics reactions from those of the background by the usual training methods (e. g. backpropagation). Since the network's decision is taken at a hardware level one is forced to very high speed computations and neural networks with their inherent massive parallelism are ideally suited for this task. Detailed investigations within the H1 experiment using Monte-Carlo data [1] have shown that feed-forward networks trained on the first level trigger information are indeed able to obtain the necessary reduction in background rate while keeping high efficiency for the physics reactions.

We present here the concept, the design, and the status for a second level hardware trigger based on the neural network architecture. The design is centered around digital neural network chips which have become available quite recently. The integration of the network trigger into the overall trigger scheme of the H1 experiment is also addressed. After a short description of the trigger information available in the H1 experiment at level 2 we present the reasoning for the network architecture we have chosen to tackle the level 2 trigger problem. Next we show how these basic networks are used as building blocks of a complex and powerful rejection machine. Finally the hardware realisations, well in progress, are presented.

2. Triggering the H1-Experiment at HERA

A modern, large particle detector system at a new accelerator complex such as H1 at HERA is built with the intention to serve as a general purpose facility and be prepared to be sensitive for the expected as well as potential new phenomena. In this spirit, a large variety of detection principles are employed which cover efficient detection and measurement of hadronic particles as well as of photons and leptons (electrons and muons).

For particle tracking, H1 uses an arrangement of drift and proportional chambers which surround the interaction point, complemented by transition radiation detectors in the forward (proton) direction. The tracker system is surrounded by a central liquid argon calorimeter and two warm calorimeters to cover the extreme forward and backward parts of the solid angle. The calorimeter is enclosed by a superconducting solenoid with diameter of about 6m, followed by an instrumented

Table 1: Summary of H1 Multilevel Trigger System

Level	Information	Use	Timing	Max. Rate
L1	8 bit energy $\sum$'s tracking flags	discriminators hardwired logic	deadtime free 2.2 μs pipeline	5000 Hz
L2	8 bit energy $\sum$'s tracking flags	pattern recogn. with NN	20 μs	100 Hz
L3	full trigger data	μ based pattern recogn.	2 ms (max)	50 Hz
L4	full data	RISC farm	30 ms	10 Hz

magnetic flux return yoke for muon measurement. In the forward direction, a warm toroidal magnet, sandwiched between sets of drift chambers, allows additional measurement of high energy muons. For photoproduction reactions and for luminosity measurements an electron tagger and a photon detector are installed about 30 m and 100 m away (upstream of the proton beam) from the interaction region. Further details on the H1 detector are given elsewhere [3].

Each of the detector components (subdetectors) provides a set of first level triggers to a central trigger box. Since the HERA bunch crossing frequency is 10 MHz (corresponding to a time interval of 96 ns between possible interactions), the level 1 triggers must be able to derive a triggger decision each 96 ns. This is achieved by storing the information for each bunch cross (BC) in a digital pipeline, clocked at 10 MHz, for a maximum of 30 BC's. After about 2 μs (22 BC's) a trigger decision is formed and the detector readout is started, knowing the exact bunch cross where the event happened. At this point the primary deadtime starts, no further triggers can be accepted until the event buffers are fully read out or a fast clear from a higher trigger level has been issued rejecting the event.

For the subsequent level 2 hardware trigger a decision time of 20 μs is foreseen. This time is chosen by the request to digest a maximum of 5 KHz from level 1 while keeping the deadtime below 10 %. For the level 2 trigger no further subdetector information is available. Therefore a more "intelligent" use of the trigger information is appropriate, exploiting, most importantly, the correlations among the various trigger quantities. The output of the level 2 trigger must not exceed 100 Hz which is the maximum rate for the third level software trigger running on dedicated microprocessors. Finally, level 4 has at its disposal the complete event information and can apply more complex algorithms based on raw detector data, running on a farm of RISC processors. The output rate of level 4 is the one dumped to permanent tape and cannot exceed, given the amount of data per event, more than about 10 Hz. A summary of the four stages of triggering is shown in Tab. 1.

During the first two running periods with HERA only the levels 1 and 4 were realized. This was sufficient since the beam currents were low and the background rates moderate. Acceptable physics efficiency could be obtained with trigger thresholds set as to provide about 30-40 Hz into level 4. With the expectation of reaching design luminosity for HERA in the near future, which means an increase of proton beam currents by almost a factor of 100, one is forced to study more complex and fast trigger decision algorithms. Since fast digital neural VLSI hardware has become available recently, the realisation of networks using the digital trigger information from the H1 experiment can now be attempted.

3. Neural Network Architectures

If one considers neural networks for a trigger applications, a number of criteria need to be fulfilled in order to accept this technique at such a sensitive point of the event selection chain. The basic task of a trigger is to classify: An event is either interesting ("physics") and will be kept, or uninteresting ("background") and will be rejected. At any level, the functioning of an experimental trigger should be understandable and must be controllable and reproducible. For a trigger application such as the one sketched above a three-layer feed-forward neural net has the required properties. Among many others, one of the most important arguments for the choice of this network architecture is its functional transparence, i.e. its ease to be "understood".

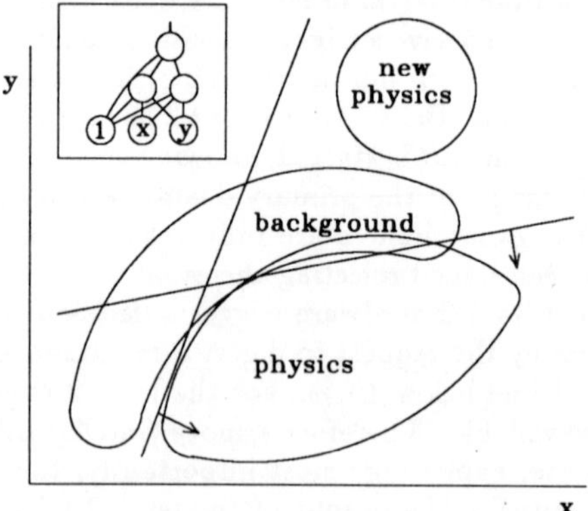

Figure 1: Geometrical visualisation of a feed-forward neural network with one hidden layer.

Concerning classification, feed-forward neural networks can be shown to approach the Bayesian limit, thus providing an optimal classification estimator. Furthermore, a three-layer feed-forward net with one input layer (the various trigger quantities), one hidden layer and one output node (the final decision) can be easily interpreted: Each of the hidden nodes acts mathematically as a separating plane in the multi-dimensional space spanned by the trigger variables. This is most evident by the mathematical prescription to generate the activity a_i of node i, fed by inputs v_j

$$a_i = \sum w_{ij} v_j + \theta \quad ,$$

which is equivalent to the definition of a separating plane in the input space with a normal vector along $\vec{w}_i$. Such a separating plane establishes a classification in "physics" or "background", depending on which side ("above" or "below") of the plane the event is located. The output o_i of the hidden node i which is fed to the final output node is given by applying a transfer function

$$o_i = f(a_i)$$

where $f(x)$ usually is taken as the sigmoid function. This type of transfer function which rises smoothly, but in a nonlinear way, from zero to one defines a "transition region" where the definition of "above" and "below" is subject to choice: The exact mathematical definiton of "above" and "below" depends on the cut in the value for the output o_i.

Using many such hidden nodes or separating planes, a volume can be constructed (not necessarily closed !) containing the wanted physics events. The decision of an event being located in this volume or outside is taken at the final layer. A cut on the output o_{fin} of the final node uniquely defines the separating volume and therefore makes the network completely deterministic and reproducible, once the computations are done in digital form. For simplicity we assume here that the supervised training attributes the value "1" as the desired output o_{fin} for physics reactions while $o_{\text{fin}} = 0$ is desired for background reactions. It should be mentionned that *one* single hidden layer is general enough to cope with any topological arrangement of events in the input space. Although several hidden layers sometimes lead to faster convergence during the learning phase, such more complex and geometrically less transparent net topologies can always be modelled by a net with a single hidden layer containing a sufficently large number of nodes. This simple geometrical interpretation of single-hidden-layer feed-forward neural networks has been rigorously verified in a number of interesting toy problems [2].

Given two sets of events belonging to either physics or background, well-known learning strategies exist (see e. g. [4]) which supply the weights w_{ij} once a network topology has been chosen. As a simple example, the network operation is depicted in Fig. 1 for a two-dimensional trigger space spanned by the variables x and y. The third input node is the bias. Each separating line corresponds to one hidden node (see upper left insert in the figure for the corresponding network topology), the physical region selected (output yields essentially "1") is the lower right part below the two separating planes (see arrows). It is clear that the "new physics" not presented to the network will be rejected as background. This potential problem will be addressed and solved below.

Apart from their physical appeal, feed-forward network algorithms have another important property relevant to realtime applications such as triggering: Their execution time is entirely fixed by the number of inputs and nodes in the hidden layer and is therefore independent of the values of the input data. It is a single-pass execution with a predetermined amount of computation time. This is very much in contrast to other popular feed-back nets (e. g. Hopfield nets).

Feed-forward nets have been used extensively within the H1 Collaboration to study their usefullness in separating background from physics. These studies were done with Monte-Carlo and with real data. Details on these investigations can be found elsewhere [4, 5]. In particular, the nets were able to reduce the background rate by at least a factor of 10, thus meeting the requirements for a level 2 hardware trigger. The trigger information used in these nets comes from all major subdetectors of the experiment, such as the calorimeters, the tracking detectors and the muon chambers. These studies have shown that the networks can find the trivial correlations as well as some highly non-trivial ones which enable cutting into the background and rescue the wanted physics.

4. Principles for an ANN-Trigger

The basic arguments for using artificial neural networks at the hardware trigger level are

- to exploit the multidimensional correlations among the trigger quantites and thus to recognize background from physics reactions with a higher degree of accuracy compared to simple low-dimensional analyses
- to take advantage of the high degree of flexibility offered by neural network architecture, especially concerning the choice of trigger variables and
- to take advantage of the high speed of network calculations which are intrinsically parallel and therefore much faster than standard scalar computations.

Neural nets in a trigger application will always work in the so-called recall phase: The network topology as well as the weights have been pre-determined by offline analyses and are down-loaded to the prospective hardware for execution. The delicate question is, *how* to arrive at the topology and at the weights. In principle, fully connected feed-forward networks being fed with *all* the available trigger information can be successfully trained provided the correct training sets are supplied and sufficiently exhaustive training periods have been granted. This "brute force" approach, however, seems not very attractive in view of the prohibitive computing times both for training and recall. At the same time it is not expected that the various physics reactions one wants to trigger on need all the same trigger quantitites to be distinguished from the background.

It therefore seem reasonable to study the different physics reactions by themselves and develop nets to select these reactions with trigger information relevant to them. This results in smaller and easier to handle networks. Our investigations

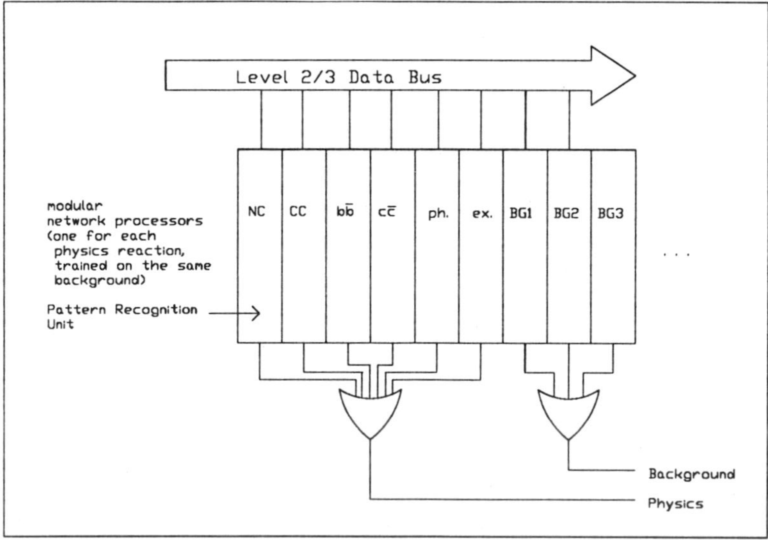

Figure 2: **Principles of the Neural Network Trigger: Physics Recognition and Background Rejection**

[1] have shown that *small* nets trained for *specific* physics reactions, working all in parallel, are more efficient compared to a single larger net trained on all physics reactions simultaneously. Most importantly, putting these nets to a real trigger application, the degree of modularity is very helpful when a new trigger for a new kind of physics reaction is to be implemented: There is no need to retrain the other nets, the new physics net is simply added to the group of the others. In the same way an improved Monte Carlo for a certain reaction will only lead to re-optimize that specific selection net and will have no effect on the other nets in the system.

A major drawback using supervised learning strategies seems to arise from the preceeding discussion: The various networks will only efficiently tag those physics reactions which have been offered for training to the networks. In this spirit, "new physics" will generally not or only inefficiently be classified as physics by any of the physics selection nets. This difficulty can be overcome by recalling the geometrical interpretation of the networks considered here. The separating planes in the trigger space can be arranged in a way as to completely *encapsulate* the background. This procedure does not rely on any physics input, it only rests on the topological properties of the background in trigger space. A straightforward algorithm to enclose the volume of background has been proposed by us and has been intensively investigated [6]. The algorithm proceeds with the intention to minimize the encapsulated space using a principal component analysis, since events within this volume will be considered as background. Depending on the point density of the background events in trigger space, several such volumes could be necessary (e. g. when the background space is fragmented). In order to save nodes in the hidden layer, a special transfer function has been employed to provide *two* parallel separating planes per hidden node [6]. Although this encapsulation might now seem to be the only necessary type of net to use, one should consider that physics "close" to the background will not efficiently be triggered with such a scheme. Since specific physics channels have not been considered in the encapsulation algorithm, the corresponding boundaries are not optimized for physics. The background encapsulation is more thought of a safety channel in order to not throw away physics one has not thought of before hand.

Finally, a word on the training samples should be spent. In order to carry out a meaningful training special care needs to be taken in the selection of data samples for physics and background. In modern, complex high energy physics experiments the trigger efficiencies for the reactions under study usually need to be determined from Monte Carlo caculations. The same is true for the neural nets: Learning will almost exclusively be done on physics samples generated by Monte Carlo methods, partly because the full phase space is accessible, partly because, for rare and interesting processes, the statistics is missing. If the detector response is well understood, this procedure is viable.

A different story is the beam-gas and beam-wall background: Here mostly soft hadron production processes on nuclei are dominating which are neither known from experiment nor from theory. The Monte Carlo road is blocked here. On the other hand, real data can be easily obtained from the machine itself, either by so-called pilot bunches for each kind of beam, which do not have a partner bunch from the other beam in the interaction region, or by dedicated separated beam runs. Pilot bunches are routinely introduced at HERA and separated beam runs could be launched for a crosscheck. In any case, pure background training samples, particularly useful for the background encapsulation algorithm, can be readily obtained.

The final algorithm for arriving at a trigger decision using the outlined neural network architecture is the following (see also Fig. 2):

1. build an OR from all physics selectors: OR(phys)
2. build an OR from all background rejectors: OR (back)
3. reject the event, unless
   - OR(phys) is true, or
   - OR(phys) is false, but OR(back) is also false

   (the latter condition is a signal for potential new physics)

Table 2: Comparison of the two digital VLSI neural network chips available.

	CNAPS (Adaptive Solutions)	MA16 (Siemens)
PN's	64	16
clock	20 MHz	40 MHz
memory	4 kB	(external)
inputs	8 bits	16 bits
weights	16 bits	16 bits
MAC/s	960×10^6	800×10^6
VME board	∃	-
Software	∃	under development

5. Hardware Realisation

In high energy physics only analog neural network realtime applications exist so far [7]. Since the H1 trigger information is entirely digital it seems most natural to consider an all-digital network solution. Apart from the missing burdon to reconvert digital signals back to analog signals, the digital technique is preferred on grounds of stability, controllability and reproduceability. Happily enough, two digital VLSI neural network chips have become available recently (both fast enough to meet the 20 μs time constraint for level 2): The MA16 chip [8] from the Siemens semiconductor laboratory (Germany), and the CNAPS chip [9] from Adaptive Solutions (USA). The relevant data of the two chips are shown in Tab. 2. While both chips have approximately the same computing power (Multiply-Accumulates per second), the Siemens chip has advantages when learning is considered on the chip. On the other hand, Adaptive Solutions have in their program a VME board built around the chip, together with the necessary software. The limitation of 8 bits for the input into the network in the CNAPS chip is not a problem since most of the trigger information (e. g. the calorimeter energies) is in 8 bit accuracy. Other information not available in 8 bit form could be transformed appropriately in a "data box" (see below).

A typical hardware realisation of the neural network level 2 trigger for H1, independent of the type of chip finally used, is depicted in Fig. 3: A sufficient number of chips (between 1 and 5 for the two candidates) is arranged on a VME board ("NN-board") which can model a neural net with a maximum of 64 inputs, 64 hidden nodes, and 1 output node. A VME crate houses the NN-boards, one

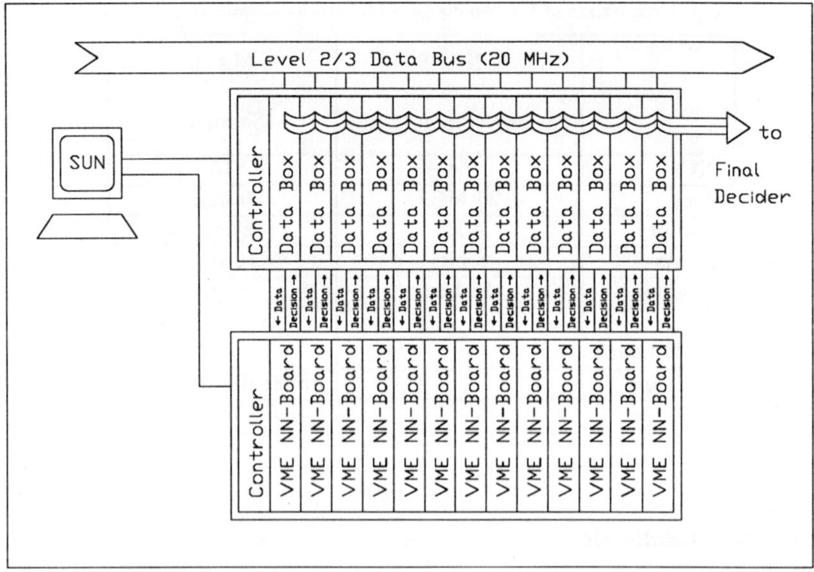

Figure 3: **Schematic overview of VME crates for the Neural Network Trigger. Note that the outputs of the VME NN-boards are fed back to the data boxes and the daisy-chained to the Final Decider.**

board for each physics reaction or for a background region. Each NN-board has a companion data server board ("data box") which supplies exactly the foressen trigger information for the neural net modelled onto the board. Both the data boxes and the NN-boards are loaded from a host computer via micro-processor controller boards.

A 20 MHz data bus, divided in 8 parallel groups, each 16 bits wide, provides the trigger information from the various subdetectors to the levels 2 and 3. The transfer is devided chronologically into two parts: During the first 3.2 μs all data with fixed format (trackers, muon system etc.) are transferred, corresponding to a maximum of 32 words (16-bit) for each group. Then the zero-suppressed data (e. g. energies from the calorimeter) are sent. After 5 μs all data are available. During the time of data transfer the data boxes select precisely the information from the bus which is needed for their companion NN-boards.

After 5 μs the data are loaded completely onto the NN-boards and the data boxes give start signals to the NN-boards which, after another 10 μs present their outputs back to the data boxes. These in turn supply their signal to a Final Decider which proceeds as sketched at the end of the previous section.

The trigger rate from such a system can be easily controlled by choosing the cuts of acceptance in the final output values o_{fin}: A selection bit from a specific NN-board/data box pair is raised only if $o_{\text{fin}} \geq o_{\text{cut}}$. This cut is made in the data box with the value supplied by the companion NN-board. Other rate limiting measures can be taken in the Final Decider, such as prescaling or bypassing of trigger decisions.

6. Conclusions

We have presented the concept for a second level neural network trigger to be realized in the H1 experiment at HERA. The network technique is very well suited for a realtime trigger application. It has been demonstrated that pattern recognition on the basis of the first level trigger information largely enhances the rejection power against background in comparison to classical (low-dimensional) methods. Furthermore, with the availability of digital VLSI neural network chips a realisation fast enough for trigger purposes (about 10 μs decision time) can be attempted now. The trigger system will consist of many networks running in parallel, each specialized for a certain physics reaction ("physics selectors"). In addition, special nets will be trained on pure background only to open the window for new physics, unseen by the physics selectors. These nets serve as "safety channels" to guard against losing "new" and therefore unseen physics.

The project is well under way and works towards completion by the end of 1994.

7. References

1. P. Ribarics, Proceedings of the Workshop on Neural Networks: From Biology to High Energy Physics, Elba, 1991, p. 383 Workshop on Artificial Intelligence, Elba, 1991
2. A. Gruber, MPI/PhE/93-07, Max-Planck-Institut für Physik, December 1992
3. F. Brasse , Proceedings of the 26th International Conference on High Energy Physics, Dallas, 1992, p.1849
4. T. Krämerkämper et al., these proceedings
5. A. Gruber, J. Möck et al., these proceedings
6. P. Ribarics et al., these proceedings
7. C. Lindsay, these proceedings
8. U. Ramacher et al., Design of a 1st Generation Neurocomputer, in: VLSI Design of Neural Networks, Kluwer Academic Publishers, Nov. 1990
9. CNAPS Server by Adaptive Solutions Inc., data sheet supplied by Cromemco GmbH, 61350 Bad Homburg

CONSTRUCTED NETWORK FOR BACKGROUND ENCAPSULATION IN THE SECOND-LEVEL TRIGGER OF THE H1-EXPERIMENT

P. Ribarics[2], J. Möck[2], J. Fent[2], W. Fröchtenicht[2], A. Gruber[2], D. Goldner[1]
C. Kiesling[2], T. Kobler[2], H. Kolanoski[1], T. Krämerkämper[1]

[1]*Institut für Physik, Universität Dortmund, 44221 Dortmund, Germany*

[2]*Max-Planck-Institut für Physik, Föhringer Ring 6, 80805 München , Germany*

ABSTRACT

An algorithm is described which uses only background data to define an encapsulating neural network for the L2 trigger processor of H1. The algorithm is based on a clustering procedure. The trigger pattern vectors - consisting of the L1 trigger elements - are partitioned into predefined number of clusters which are enclosed into hyperboxes with principal component axes. This approach provides closed decision boundaries which gives protection against the loss of unknown physics. The method yields high effeciencies on the low luminosity physics currently under study and a typical background rejection factor of ≈ 10. Having a representative background sample, the encapsulation network gives a complementary or safety channel to the optimally trained neural networks which use MC physics in addition to fit their parameters. The algorithm is fast which makes it possible to implementation an adaptive system, following the changing background conditions.

1. Trained Networks

It is assumed that the trigger information for the events of interest is described by a set of variables $\vec{v} = \{v_j\}$. The aim of the classification task is to construct a function $F = F(\vec{v})$ such that

$$F(\vec{v}) = \begin{cases} 1: & \text{if event is physics} \\ 0: & \text{if event is background} \end{cases}$$

Feed forward neural networks trained by backpropagation provide a straightforward scheme for the approximate construction of such functions $F(\vec{v})$, given event samples for which the classification values are known.

The standard backpropagation algorithm however has two drawbacks: it does not provide necessarily closed decision boundaries and needs a physics sample for the training. The lack of closed and tight decision boundaries around the background sample increases the probability of rejecting physics which was not in the training sample. On the other hand a physics sample is difficult to obtain. Simulated physics assumes a stable and reliable MC program, which is normaly not available at the beginning of the experiment. The physics sample is difficult to collect from off-line analysed data because of the low rates.

In the following chapters an alternative neural network generation method is described which does not have the mentioned drawbacks.

2. Background Encapsulation

2.1 Constructed Networks

The feed-forward network architecture can implement arbitrary cut systems. A hidden node defines a cut plane in the multidimensional space and the output node combines these planes into a volume. If explicit knowledge about the geometry of the multidimensional point distribution is available one can use it instead of the backpropagation to define the parameters of the network.

Our underlying assumption is that the background point distribution in the multidimensional trigger space is well represented by the sum of compact clusters. The background sample is then partitioned into clusters - without any assumption about the physics - and closed volumes are defined, each of them fully containing one of the clusters.

This will not guarantee an optimized cut for a known physics class - here a backpropagation trained net is assumed to give the best selection efficiency. The background encapsulating network will serve as a safety channel for unknown physics, because due to the closed decision surfaces only those parts of the multidimensional trigger space are rejected which were sufficiently sampled by background data. Unpopulated parts are not discarded.

Our proposed way to define a closed volume around the background points is to define a hypercube whose axis are the principal components of the background point distribution within a cluster. The motivation for this is to find the directions along which the variance of the distribution is small and the encapsulation is most efficient. An essential technical requirement for the encapsulating net is that it should contain a small amount of internal parameters compared to the available data to achieve good generalization ability.

2.2 Cluster Analysis

Cluster analysis is a way to find natural groupings in a set of data. Samples in one cluster are more like one another than like samples in other clusters. This involves two separate issues - how should one measure the similarity between samples, and how should one evaluate a partitioning of a set of samples into clusters.

The most obvious measure of the similarity between two samples is the Euclidean distance between them. The result of clustering depends on this choice of Euclidean distance as a measure of similarity because it implies that the feature space is isotropic. Consequently, clusters defined by Euclidean distance will be invariant to translations or rotations - rigid body motions of the data points - they will not be invariant however to linear transformations in general. A simple scaling of the coordinate axes can result in a different grouping of the data into clusters. This is of no concern for problems in which rescaling is a meaningless transformation. As the trigger data of H1 is given on a somehow arbitrary scale and decision bits are mixed up with FADC 8-bit informations, if clusters are to mean anything they must be invariant to a general linear transformation.

Suppose that we have N patterns that we want to partition into exactly C clusters. One way to make this into a well defined problem is to define a criterion function that measures the clustering quality of any partition of the data. Then the problem is one of finding the partition that extremizes the criterion function.

Our criterion function for clustering is a scale invariant sum-of-squared-error criterion. If N_i is the the number of patterns in the i-th cluster, than

$$F = \sum_{i=1}^{C} \sum_{p=1}^{N_i} (\vec{v}^{\,p} - < \vec{v}_i >)^T S^{-1} (\vec{v}^{\,p} - < \vec{v}_i >)$$

where S^{-1} is the inverse of the total scatter matrix S

$$S = \sum_{p=1}^{N} (\vec{v}^{\,p} - < \vec{v} >)(\vec{v}^{\,p} - < \vec{v} >)^T$$

and $< \vec{v}_i >$ is the mean of the trigger pattern vector in a given cluster i

$$< \vec{v}_i > = \frac{1}{N_i} \sum_{p=1}^{N_i} \vec{v}^{\,p}$$

F has a simple interpretation. For a given cluster the mean vector $< \vec{v}_i >$ is the best representative of the samples in the cluster in the sense that it minimizes the sum of the squared lengths of the error vector $\vec{v} - < \vec{v}_i >$. (The inverse scatter matrix has a normalizing effect and transforms the point distribution into a sphere-like one.) Thus F measures the total squared error incurred in representing the N samples by the C cluster centers. The optimal partitioning is defined as one that minimizes F.

F is invariant to the nonsingular linear transformations L of the data, which can be easily shown by substituting the individual transformation rules: $\vec{v} \to L\vec{v}$, $S \to LSL^T$, $S^{-1} \to L^{-1^T} S^{-1} L^{-1}$.

This invariance means that if different groupings can be obtained by scaling the axes or by applying any other linear transformation, then all of these groupings will be exposed by the invariant procedures. This has certain drawbacks: The criterion function is more likely to possess multiple local extrema, and is correspondingly more difficult to extremize.

With the criterion function selected, clustering becomes a well defined problem: find those partitions of the set of samples that extremize it. Since the sample set is finite, there are only a finite number of possible partitions, theoretically the clustering can be solved by exhaustive enumeration.

However, in practice such an approach is unthinkable because of the huge number of possible clusters ($\approx C^N/C!$). The most frequently used method is the iterative optimization [1]. The basic idea is to find some reasonable initial partition and to move samples from one group to another if such a move will improve the value of the criterion function.

The steps of the clustering procedure are then the following:

1. Select an initial partition of N samples into C clusters randomly. Compute F and the cluster means $< \vec{v}_i >$.
2. Select the next candidate pattern $\vec{v}$. Suppose $\vec{v}$ belongs to cluster i.
3. If $N_i = 1$ go to 6; otherwise compute

$$F_j = \begin{cases} \frac{N_j}{N_j+1}(\vec{v}- < \vec{v}_j >)^T S^{-1}(\vec{v}- < \vec{v}_j >) & j \neq i \\ \\ \frac{N_i}{N_i-1}(\vec{v}- < \vec{v}_i >)^T S^{-1}(\vec{v}- < \vec{v}_i >) & j = i \end{cases}$$

4. Transfer $\vec{v}$ to cluster k if $F_k < F_j$ for all j.
5. Update $< \vec{v}_i >$ and $< \vec{v}_k >$.
6. If there was no transfer in N attempts, stop; otherwise go to 2.

One question that plagues all hill-climbing procedure is the choice of the starting point. Repetition with different random initial selections can give some indication of the sensitivity of the solution to the starting point.

2.3 Principal Components

Principal Components Analysis is a statistical technique for calculating the major directions of variation of a set of data points in some high dimension space. It is used when no hypothesis has been formulated as to which dimensions constitute the most relevant information.

The principal components can be derived the following way. We calculate the scatter matrix C for each cluster.

$$S_i = \sum_{p=1}^{N_i} (\vec{v}^{\,p} - < \vec{v}_i >)^T (\vec{v}^{\,p} - < \vec{v}_i >)$$

This is a symmetric matrix. After diagonalization we get an orthogonal system of eigenvectors with the corresponding eigenvalues. The eigenvectors are called the principal components of the point distribution. The principal components corresponding to the largest eigenvalues are the directions of greatest variance. In pattern recognition applications these principal components are the important ones and one normally ignores the principal components belonging to small eigenvalues because they do not contain interesting information.

In our case it will be just the opposite. The directions belonging to small eigenvalues are the important ones because along these directions can we impose the strongest constraints.

We proceed the following way. The direction belonging to the smallest eigenvalue is taken. Let calculate the scalar products between this normal and the background pattern vectors and determine the range they occupy on this projection. Continuing the procedure with the principal components which belong to eigenvalues of increasing size the ranges finaly will define a closed volume fully containing the points of the cluster.

Patterns which lie outside of this volume are triggered as physics events.

The mapping of this hypercube onto a neural network is straigthforward. For an N-dimensional input vector N hidden neurons are needed. Instead of the sigmoid a combination of two Heavyside functions (falling and rising edges) is used as transfer function. To each direction a hidden neuron is assigned whose weight vector is one of the eigenvectors and the threshold function cuts out the range which is occupied by the background on this projection. The weights of the output neuron are identically 1 and its threshold is zero. In this configuration if any of the hidden units is active, the point is outside of the hypercube, the network will accept it

The internal parameter space of the method is small. Assuming an N dimensional input N-1 parameters define the rotation into the system of the principal axes and 2N parameters are needed to fix the positions of the decision planes along the axes. The total number of parameters is (3N-1) per cluster, which is small compared to our usual sample sizes. This again supports the good generalization ability of the encapsulation.

One neural network is assigned to each cluster, they run parallel on different hardware boards and the trigger decision for physics is the AND of the outputs of the individual networks.

3. Results on Real Data

The encapsulation method is sensitive to the background definition, it is based on the accurate identification of the event class to be suppressed. In order to get a representative sample of background the L4 rejected data sample was cleaned up from physics contamination by using the L5 classification. The trigger information used to the encapsulation and the preselected data samples were described elsewhere in this proceedings. [2], [3]

To check the generalization ability of the method we divided the background sample into 2 parts. The first or training sample - containing 80 % of the events - was used to set up the encapsulating hyperboxes on the clusters and the rest served as an independent test set. The training set was partitioned into 1,3 and 5 clusters using the iterative cluster algorithm, which always converged within 50 presentations (23 in average) of the full training sample.

By increasing the size of the background sample the encapsulating volumes will expand because they are defined by the extrem projection values on the principal component axes. This extrem values on the other hand are determined by a few points which are far from the cluster center. In order to get a statistically stable solution a 3σ cut is applied on the distance to cluster center distribution, the tails are removed and the encapsulating hyperboxes are defined on the rest of the cluster.

The algorithm is fast, it runs in less than 1 minute on a IBM 6000 workstation. This makes the implementation of an adaptive system, following the changing background conditions , possible.

Tab. 1 is a summary of the efficiencies obtained. With a relatively small number of clusters a background suppression of ≈ 10 can be achieved with physics

Table 1: Efficiencies on real data.

	ENCAPSULATION			BACKPROPAGATION			
Events	1 cluster	3 clusters	5 clusters				
background	0.0085	0.0152	0.0114	0.12	0.004	0.008	0.13
L4 kept events	0.39	0.48	0.50	0.65			
γ-p	0.99	0.99	1.00		0.99		
NC	0.99	0.99	0.99			1.00	
CC	0.88	0.92	0.93				0.97
enclosed volume	2.5×10^{28}	5.2×10^{25}	9.6×10^{22}				
l2 rejection factor	11.8	9.25	9.17				

efficiencies above 0.9. This is a factor 2 better then the present L4 rejection factor (≈ 5) and will certainly increase as the L4 cuts will become more selective.

The results are comparable with those from backpropagation trained networks, these later are shown on the right side of the table.

4. Conclusions

We have shown that an algorithm can be defined which uses only background data to generate an encapsulating neural network for the L2 trigger processor of H1. This approach provides closed decision boundaries which gives protection against the loss of unknown physics. The method yields high effeciencies on the low luminosity physics currently under study and a typical background rejection factor of $\approx$ 10 given a background sample from L4 with rejection factor ≈ 5.

5. References

1. R.O. Duda and P.E. Hart, *Pattern Classification and Scene Analysis* (John Wiley and Sons, New-York, 1973)
2. C. Kiesling et al., *Concept for a fast Second Level Trigger Using a Neural Network Architecture for the H1-Experiment at HERA*, these proceedings.
3. A. Gruber et al., *Performance of Backpropagation Networks in the Second Level Trigger of the H1 Experiment*, these proseedings

PERFORMANCE OF BACKPROPAGATION NETWORKS IN THE SECOND-LEVEL TRIGGER OF THE H1-EXPERIMENT

A. Gruber[2], J. Möck[2], J. Fent[2], W. Fröchtenicht[2], D. Goldner[1], C. Kiesling[2]
T. Kobler[2], H. Kolanoski[1], T. Krämerkämper[1], P. Ribarics[2]

[1] *Institut für Physik, Universität Dortmund, 44221 Dortmund, Germany*
[2] *Max-Planck-Institut für Physik, Föhringer Ring 6, 80805 München , Germany*

ABSTRACT

The second level trigger of the H1–experiment has to reduce the background rate by an order of magnitude and keep as much physics as possible. It is shown on Monte-Carlo data that we can cope with this challenge. The investigations were performed with a systematic variation of several network parameters. We also used real data from 1993 runs and were able to show the required performance. A new measure for the importance of input variables is presented which facilitates the interpretation of the network cuts.

1. Monte-Carlo investigations

1.1. The parameters

The task of the second level trigger (L2) is to reduce the background events by an order of magnitude. The background in the H1–experiment is mainly due to interactions of the protons with residual gas in the beam pipe (beam-gas) and collisions of off-momentum protons with elements of the beam line (beam-wall). The first day physics reactions of the HERA accelerator consist of γ-proton (γ-p), neutral current (NC) and charged current (CC) events. Details on the simulation of the events can be found in these proceedings[1]. The size of the training and test samples are shown in Tab.1.

Table 1: Physics and Background Monte-Carlo Files

	Training Sample	Test Sample
Charged Current	3000 events	2500 events
Neutral Current	2000 events	1400 events
γ–proton	3000 events	3000 events
Beam-Wall	3000 events	2500 events
Beam-Gas	3000 events	3400 events

The training of the feedforward networks, consisting of 1 hidden layer and one output node was performed with the Aspirin/MIGRAINES software package on IBM/RS6000. For the weights we used an initial width of 2.0 and did an online-learning, i.e. updated the weights after each pattern presentation. We carried out one million iterations. Our input data was normalized in a sense that for each quantity the interval $[-3\sigma, +3\sigma]$ of the distribution was mapped to $[-1, +1]$.

We trained the networks with combinations of different learning parameters (0.1, 0.01, 0.001, 0.0001), different momentum factors (0.0, 0.5, 0.9, 0.99) and different numbers of hidden nodes (1, 10, 30, 64). First we trained each physics class separately against the combined background (beam-wall and beam-gas), then we mixed all physics classes into one sample and trained it against the background. We used two sets of input variables, one using only tracker information (TRACK) and the other making use of calorimeter energies and a coarse z-vertex histogram (HISTCAL), see Tab. 2. A description of the detector can be found elsewhere in these proceedings[2].

Table 2: Quantities used for training

TRACK (35 quantities)	
Z–Histograms :	**Maximum, position and sum of ...**
5 × 3 quantities	Drift chambers (histograms with fine binning)
1 × 3 quantities	Proportional chambers (histogram with coarse binning)
Drift Chamber r–Φ–Trigger :	
4 quantities	Number of tracks with high/low momentum, positive/negative charge
1 quantity	Total number of tracks
Backward Proportional Chamber :	
4 quantities	Number of strips hit in BPC layer a,b,c,d
1 quantity	Total number of hit strips
Muon Chambers :	
2 quantities	Number of hits in forward muon detector behind/in front of toroid
4 quantities	Number of hit muon modules in forward/backward endcap/barrel
HISTCAL (25 quantities)	
cpvtx(1–16)	16 bins of Proportional Chamber Z–Vertex Histogram
Calorimeter :	
eif	Energy in inner forward part of LAr–Calorimeter
efb	Energy in forward barrel part of LAr–Calorimeter
ecb	Energy in central barrel part of LAr–Calorimeter
ex	x – component of energy in the LAr–Calorimeter
ey	y – component of energy in the LAr–Calorimeter
ebemc	Energy in Backward Electromagnetic Calorimeter (BEMC)
eclu	Energy of hottest cluster in the BEMC
etag	Energy in the electron tagger
gtag	Energy in the photon tagger

1.2. Results for Monte-Carlo data

The results of the best networks are shown in Tab. 3. In this context "best" is defined as the maximum of the ratio between the kept physics and the kept background assuming a cut at 0.5 on the output node. In case of equal performance, the simpler network architecture was selected.

The background rejection is always at the 99% level. The physics efficiency is satisfactory for γ-p with TRACK information and for neutral current with HIST-

CAL information. It is reasonable that neutral current events rely on calorimeter information and thus reach only about 70% efficiency without it. An analogous argument holds for γ-p events which are difficult to trigger without enough track information. The low charged current efficiencies can be increased up to 85% by using more information (track and calorimeter).

Table 3: Results from best networks. The rows give the values for the learning parameter (lp), momentum (mom), number of hidden nodes (nod). The rejected background (backgr) and kept physics for the test samples are given in %.

trained on	lp	mom	nod	backgr	physics	CC	NC	γ–p
TRACK								
Charged Current	0.01	0.99	64	98.2	69.7	—	27.4	87.7
Neutral Current	0.1	0.5	64	98.7	72.8	42.6	—	94.9
γ–p	0.1	0.5	64	99.4	97.6	34.9	39.5	—
CC, NC, γ–p	0.1	0.99	64	93.2	85.3	74.3	78.0	97.8
HISTCAL								
Charged Current	0.1	0.9	10	99.8	47.6	—	92.8	9.9
Neutral Current	0.1	0.99	10	99.9	93.8	11.9	—	6.7
γ–p	0.001	0.5	30	99.3	9.8	62.8	11.3	—
CC, NC, γ–p	0.1	0.99	10	99.5	40.8	48.8	93.8	10.0

2. Real data

2.1. Event selection

In the last level (L4) of the H1 trigger system it is decided to keep or reject an event. For realistic second level studies it is necessary to use also the L4-rejected events which are recorded as monitor triggers. It is a problem in real data applications to define background and physics. In the H1 experiment a first classification is done after the reconstruction of the data. These classes still have an overlap, i.e. an event can be a member of different classes. Thus it was necessary to clean up the background classes by requiring *"L1-keep & L4-reject & not in any physics class"*. The data sets that were used are presented in Tab. 4. They were split into training and test sets of equal size.

Table 4: Data sets for physics and background

Data Class	Time	Events
Physics		
Neutral Current (low Q^2)	summer 93	7145
Neutral Current (high Q^2)	summer 93	4959
Charged Current	summer 93	6072
γ–proton	summer 93	4380
Background		
L4-reject (cleaned up)	18.–22. Aug. 93	5728
L4-reject	18.–22. Aug. 93	7145

The training was done with two sets of input variables, a small one with 58 quantities, using the Liquid Argon calorimeter (26 quantities), track triggers (14 quantities), backward electromagnetic calorimeter (4 quantities), muon wire chambers (10 quantities) and the time-of-flight veto system (4 quantities). The large set with 70 variables used information from the electron and photon tagger and some more quantities in addition which are not available at the second level so far.

2.2. Results for real data

The results of the best 3-layer feedforward networks trained against the cleaned up L4-reject sample are shown in Tab. 5. The numbers describe the efficiencies on the physics files in percent calculated with a cut at 0.5 on the output node. The table also shows the percentage of rejected background events. A realistic L2-rejection factor is calculated from the L4-reject sample. The number of hidden nodes is always half the number of input variables.

Table 5: Networks trained with physics classes against *"L4-reject (cleaned up)"*. Numbers for physics and background are given in percent, "reject" denotes the rejection factor.

	58 quantities			**70 quantities**		
trained on	background	physics	reject	background	physics	reject
NC (low Q^2)	99.2	99.4	7.5	99.2	99.8	11.2
NC (high Q^2)	94.0	92.1	5.8	95.3	93.2	6.9
Charged Current	89.7	89.7	4.6	86.5	90.1	4.5
γ–proton	91.2	89.6	6.2	99.7	99.4	40

In the case of neutral current events with low Q^2 we reach a high efficiency for physics and rejection rates of the desired order of magnitude. The increased performance for the γ–p network with 70 variables can be traced back to the utilisation of the electron tagger which was crucial for the selection of these events.

The network for charged current physics has low efficiencies and rejection factors. Reasons for this are the soft cuts for the selection and the changing trigger settings at the first level leading to a charged current sample with a lot of background events. The efficiencies increase if we use events from only a small period with good detector conditions and stable trigger settings. The same arguments hold for the neutral current class with high Q^2.

2.3. Special purpose networks

Two more networks were trained to have an estimate for the power of an L2 network trigger. The first question was, whether L2 can take over the work of the L4 trigger. The network was simply trained by using L4-reject events as background and L4-kept events as physics. The same sets of input variables were used. The results are displayed in Tab. 6. The numbers give the percentage of events with an output value greater than 0.5.

Table 6: Efficiencies of network trained to separate L4 classified events.

Event Class	58 quantities	70 quantities
L4-kept	75.9	81.3
L4-reject	30.4	22.3

Due to the fact that L4 reaches a rejection factor of four we can only achieve a factor of two with these results (weighting the reject efficiencies by this factor). Thus the work of L4 which is based heavily on reconstructed tracker information can not be done on L2.

The second task was to classify beam-gas events correctly. Background was defined by the off-line classification. All the events not belonging to the beam-gas class were assumed to be physics. This corresponds to an emulation of the off-line classification. The numbers in Tab. 7 give the percentage of events with an output value greater than 0.5.

Table 7: Efficiencies of network trained to separate beam-gas from non–beam-gas.

Event Class	58 quantities	70 quantities
non–beam-gas	84.4	91.5
beam-gas	8.1	10.2
Neutral Current	98.3	100.0
γ–proton	52.6	78.9

If we consider also the efficiency on neutral current and γ–proton events we can conclude that a raw beam-gas identification can be performed by the second level trigger.

3. Relevance of variables

In order to see what the network is focusing on and what the important variables are we introduced a measure for the relevance of an input quantity. The basic idea is to calculate the change of the output value if one input variable is fixed to its mean value. In order to get a meaningful value, the difference is summed up over the set of training patterns and divided by the total number of patterns to get a normalization. The relevance R_i of the i-th input variable is defined as

$$R_i = \frac{1}{n_{pattern}} \sum_{j}^{n_{pattern}} \left(out(\vec{x}_j) - out(\vec{x}_j\big|_{x_{j,i} = \bar{x}_i})\right)^2 \tag{1}$$

where $\vec{x}_j$ is the j-th input vector, $\bar{x}_i$ is the mean value of the i-th component and $n_{pattern}$ denotes the number of patterns.

The larger the relevance value the more important is the quantity. This can be seen as an example in Fig. 1 where the 28 biggest values are plotted for the network trained on γ–proton physics. On the left we can see the network with 58 variables and it is clearly demonstrated that the energies in the inner forward part of the Liquid Argon calorimeter play the major role (first 6 quantities).

Figure 1: Relevance of the input variables in the γ–p network.

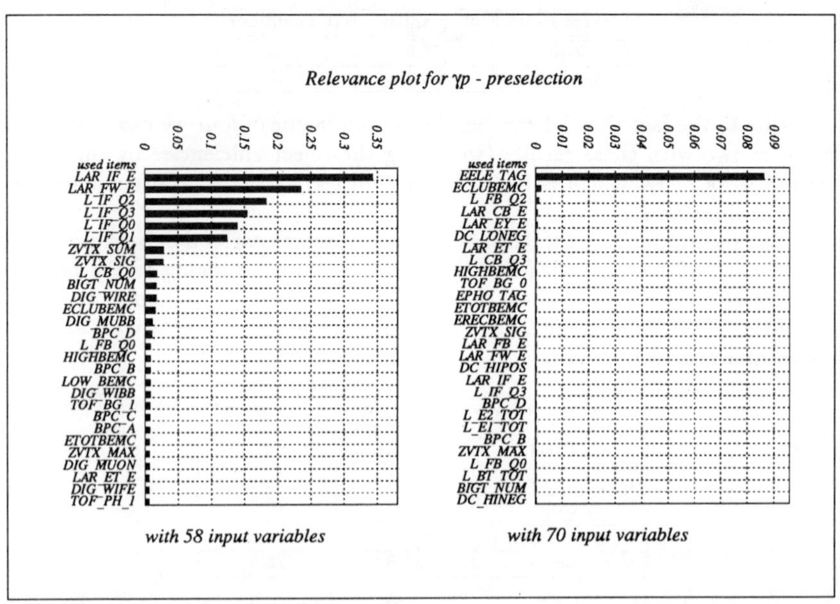

On the seventh and eighth position are quantities from the Z–vertex trigger. These quantities are absolutely reasonable for the triggering of γ–proton events. On the right side we can see the most important variables of the 70 input set. The decision of the network is clearly dominated by the electron tagger which has the greatest relevance value. It was already mentioned before that the electron tagger is a clear sign for γ–p events and must obviously contribute in such a way if included in the input set.

4. Conclusions

We have shown both on Monte-Carlo data and on real data that a feed forward network with one hidden layer and one output node can reduce the background rate by an order of magnitude. The rejected physics events are in an acceptable range. The performance of the network can be increased by using more important information and cleaner physics samples in the real data case.

5. References

1. T. Krämerkämper et al., *Artificial Neural Networks as a Level 2 Trigger for the H1 Experiment – Learning Strategies and Visualization*, these proceedings.
2. C. Kiesling et al., *Concept for a fast Second Level Trigger Using a Neural Network Architecture for the H1–Experiment at HERA*, these proceedings.

ANN AS A LEVEL 2 TRIGGER FOR THE H1 EXPERIMENT - LEARNING STRATEGIES AND VISUALIZATION -

T. KRÄMERKÄMPER[1], J. FENT[2], W. FRÖCHTENICHT[2], D. GOLDNER[1], A. GRUBER[2] C. KIESLING[2], T. KOBLER[2], H. KOLANOSKI[1], J. MÖCK[2], P. RIBARICS[2]

[1]*Institut für Physik, Universität Dortmund, 44221 Dortmund, Germany*

[2]*Max-Planck-Institut für Physik, Föhringer Ring 6, 80805 München , Germany*

ABSTRACT

An Artificial Neural Network (ANN) as a second level trigger is currently under development for the H1 experiment. The hardware is based on a feed-forward multilayer network architecture of the size 64-64-1. In this contribution the training software, learning strategies and the verification and visualization of the net decisions will be discussed.

1. Concepts

In the H1 experiment the trigger decisions are made on four levels, L1 to L4, stepwise decreasing the event rate to gain time for more sophisticated trigger decisions on the next level. The first two levels have to be hardware solutions due to the time limitations of 2.3μs and 20μs. The L2 trigger is not yet implemented into the experiment. We have proposed to use on this level a hardware realisation of an ANN using a highly parallel architecture based on the Siemens Neural Signal Processor MA16 or the Adaptive Solutions CNAPS processor[1].

The training of the nets is done offline with ANN simulation software. Each net is trained to identify a specific physics process (e.g. neutral or charged current events, heavy-flavour events etc.) and to reject as much as possible from the background reactions, mainly beam-gas and beam-wall scattering. At least one of the nets will follow a complementary approach by identifying the background and leaving as much space as possible for unknown physics[2].

2. Training

2.1. Physics at HERA

The most important classes of physics events that HERA provides are given in table 1 with their rates at design luminosity. The trigger system has to select these events out of the background reactions which are produced with much higher rates. The background is dominated by proton induced events, mainly scatterings at the beampipe and collisions with residual gas atoms. Electron induced events (beam wall, beam gas, synchrotron radiation) and cosmic ray events play a subordinate role.

Table 1: Event classes and rates

Process	Rate (at Design Luminosity ($10^{31}cm^2s^{-1}$))
Neutral Currents ($e^- + p \rightarrow e^- + X$)	≈ 3 Hz
Charge Currents ($e^- + p \rightarrow \nu_e + X$)	$\approx 10^{-3}$ Hz
Photoproduction ($\gamma + p \rightarrow X$)	≈ 120 Hz
Beam Wall	$\approx 10^5$ Hz
Beam Gas	$\approx 10^5$ Hz

Table 2: Generated events, each consumes 70 to 140 cpu seconds

Class	Events	L1 triggered events	L1 Efficiency
Neutral Currents	5779	5309	91.9 %
Charge Currents	5867	5268	89.8 %
Photoproduction	6564	6239	95.1 %
Beam Wall	10973	5657	48.5 %
Beam Gas	37872	6515	82.8 %

The L2 trigger should reach a suppression of at least 1:20 with an efficiency for the detection of physics events close to 100%, in particular for rare but interesting classes like high-Q^2 events (events with a high momentum transfer between the e^- and the proton).

2.2. Monte Carlo Simulation

The performance of the proposed L2 trigger system was studied by extensive Monte Carlo simulations. Several thousands of events in different physics and background classes were generated (tab. 2). The training data set for the L2 simulation was made up of events passing the L1 trigger step. The L1 trigger simulation used the setup for the '93 summer run period. As can be seen from table 2 the L1 rejection was not sufficient to reach the required L1 trigger rate of about 1000 Hz at the design luminosity. For future studies we have to define a reasonable L1 trigger setup complying with the high-luminosity requirements.

2.3. ANN Simulation

Our first studies with ANNs were done with the JetNet 2.0 package[3], which contains a set of FORTRAN subroutines for the simulation of multi-layer perceptron and learning vector quantization networks. We later turned to the Aspirin/MIGRAINES 6.0 package[4] which includes a compiler for the high-level language Aspirin for defining networks. The compiler produces from a given Aspirin file C source code and an executable program which is equipped with a text-oriented user interface. These programs are nearly 25 times faster then JetNet 2.0 and still 2 to 5 times faster than JetNet after an optimization[5].

With Aspirin/MIGRAINES 100 epochs training of a typical 64-47-1 ANN with 9000 training patterns and testing every epoch with 6000 test patterns took some 12000 cpu seconds on a RS-6000/550 cluster. The use of neural processors in the near future will reduce the required real time by several orders of magnitude (at least three).

2.4. Input Variables

The H1 detector sends a large amount of trigger quantities to the L2 system, briefly summarized in the following: The tracking system delivers several histograms about tracks linked to positions along the beam axis (z-vertex histograms), two-dimensional histograms for intercepts of particles in planes in the forward and backward regions as well as derived vertex significances. The calorimeters give mostly FADC information for the so-called bigtowers, which are groups of calorimeter cells pointing to the interaction zone. We also obtain E_{trans}, E_{tmiss}, the number of e^- candidates in the main calorimeter and information from the backward electromagnetic calorimeter (BEMC) single electron trigger (BSET), which strongly indicates neutral current events. From the outer part of the detector, the muon system, the trigger will get bit patterns of hit segments. The trigger quantities from two scintillator walls in front of the detector (from the proton side) allow to discriminate between particles coming from the interaction point or the proton beam halo. This is the main device to reduce proton induced background. The electron tagger system, localized roughly 100m before the interaction zone detects electrons from photoproduction events scattered at very small angles.

The limited number of net inputs makes a substantial reduction of the data to characteristic variables necessary. The L2 trigger hardware selects and preprocess the data to be fed into the networks.

2.5. Learning

Learning with a simple backpropagation algorithm (update per pattern, linear decrease of the learning rate from 0.1 to 0.001) yields good results. Smarter algorithms might become necessary for the very difficult future tasks of distinguishing heavy flavour events (e+p $\rightarrow f\bar{f}$ + X) from background. Furthermore, we studied the structure and discriminating power of the data with self-organizing-network algorithms with acceptable results[6].

3. Results and Verification

Each network was trained with 3000 events of one physics class, 3000 beam wall and 3000 beam gas events in random order. The trigger reached good efficiencies without any fine-tuning (table 3). The overall background suppression after a logical OR of the three networks is 1:32.28, but still with respect to the not appropriate L1 setup. The efficiencies in table 3 apply to events with a "natural" distribution (e.g. for neutral current events $\sim Q^{-4}$) and Q^2_{min} = 4.5 GeV^2 as implemented in the generator programs. It is, however, important to have high efficiencies for rare but interesting events like high-Q^2 neutral current events. Since

Table 3: Results

Class	Physics efficiency	Beam wall eff.	Beam gas eff.
Neutral Currents	98.47 %	99.20 %	99.61 %
Charged Currents	84.57 %	98.48 %	98.07 %
Photoproduction	98.68 %	99.60 %	99.41 %
Self-organizing-network:			
Neutral Currents	89.24 %	99.71 %	

Table 4: Efficiency for different neutral current classes

Neutral current class	Efficiency for training with natural Q^2 distribution...	...and after enriching with high-Q^2 events
$Q^2_{min} = 4.5$ GeV^2	98.47 %	97.04 %
$Q^2_{min} = 300$ GeV^2	87.58 %	97.86 %
$Q^2_{min} = 10.000...60.000$ GeV^2	95.52 %	96.54 %

the neutral current training set had only one event with Q^2 above 300 GeV^2 we simulated neutral current events (in their "natural" distribution) with $Q^2_{min} = 300$ GeV^2 and $Q^2_{min} = 10\,000$, $20\,000...60\,000$ GeV^2 (the maximum Q^2 for HERA is 98 600 GeV^2).

Checking with these data the networks trained with $Q^2_{min} = 4.5$ GeV^2 data shows that the network has an efficiency gap for Q^2 between 300 and 1000 GeV^2, but it generalizes well at higher Q^2 (table 4). To ensure high efficiencies for all Q^2 values we enriched the training files with high-Q^2 events and obtained then a uniform efficiency for the whole Q^2 interval (table 4). In the future we will investigate the possibility to map a characteristic physics quantity (like Q^2) into a linear network output by taking this value as training target. Choosing a linear output node would then allow to discriminate high-rate trigger with respect to the physics relevance of the particular events.

Further checks were done to ensure, that the trigger decision is not made by a single detector component. Therefore, the efficiencies of the networks which had been trained with the complete input pattern, were checked by switching of (set to 0) detector parts, one at a time (table 5). The networks are surprisingly stable if a subdetector fails. In most cases a failure of the subdetectors will have only a small effect on the trigger decision. Very surprising is the small influence of "BSET" (= electron in backward region). Most of the neutral current events and only a few of the background events have a BSET signal, but obviously the network learned more than only discriminating with respect to this quantity. The reason for the minor relevance of the calorimeter systems is the low rate for events with hits in the main calorimeter. This should change if we use other than natural distributions

Table 5: Switch-off check

Switch off group:	Loss of efficiency in %								
	Neutral Currents			γp			Charged Currents		
Tracker:									
CIZ/COZ	-0.10	0.15	0.15	-8.20	0.40	0.35	-11.05	-1.11	0.49
CIP/COP	0.51	-5.72	-0.79	0.76	-5.57	-3.11	4.33	-3.20	-4.25
CJC1/2	-0.05	0.35	0.0	-0.71	-0.05	0.15	-16.50	0.99	0.84
BPC	-0.51	0.20	0.05	-8.40	0.25	0.35	-1.83	0.12	0.25
Calorimeter:									
Elec1,Et, Etmiss	-1.43	0.45	0.0	0.0	0.0	0.0	-9.78	0.33	0.10
BSET	-9.17	0.0	0.0	0.0	0.0	0.0	-0.15	0.0	0.0
PLUG	0.0	0.0	0.0	0.0	0.0	0.0	-0.26	0.0	0.0
LAr	0.0	0.0	0.0	0.0	0.0	0.0	-0.20	0.0	0.0
BEMC	0.0	0.0	0.0	0.0	0.0	0.0	-0.15	0.0	0.0
Time-of-flight:									
ToF	-25.61	0.0	0.05	-6.31	0.0	0.0	-0.10	0.0	0.0
Muon:									
Forward	0.0	0.10	0.0	0.0	-0.05	-0.05	-0.10	0.0	0.0
IRON	0.10	-0.20	0.0	0.05	-0.05	-0.10	-2.40	0.42	0.49
Electron tagger:									
LUMI	-0.15	0.0	0.0	0.0	0.0	0.0	-0.41	0.0	0.0

with their rare high-Q^2 events.

4. Visualization

In order to judge the reliability of a network it is important to understand how the networks used the various inputs. Also tuning becomes more efficient with a better understanding of the particular networks. Since the decision criteria of the networks are coded by the weights our aim was to visualize how the weights act together with the different input variables. Pure weight plots have the disadvantage to be normalization dependent. Due to the different inputs with arbitrary scales the weights for each input contain an unknown normalization factor. Hence, we have defined a "scaled weight"

$$w'_{ij} := \langle x_i \rangle * w_{ij}$$

where $\langle x_i \rangle$ is the mean of the input i over all input patterns. The "scaled weight" is the average contribution of input i to the node j in the next layer. To show the effect of the output layer weights the hidden nodes were ordered with respect to their mean contribution to the output node input. Therefore the node with the

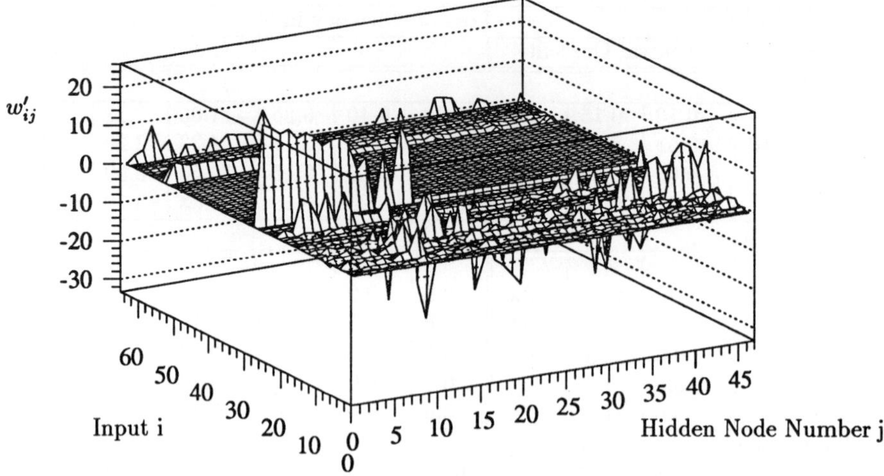

Figure 1: All neutral current events. The inputs are tracking informations (1 to 21), calorimeters (22 to 49), time-of-flight (50 to 54), muon (55 to 61), tagger (62 to 64) and node bias (65).

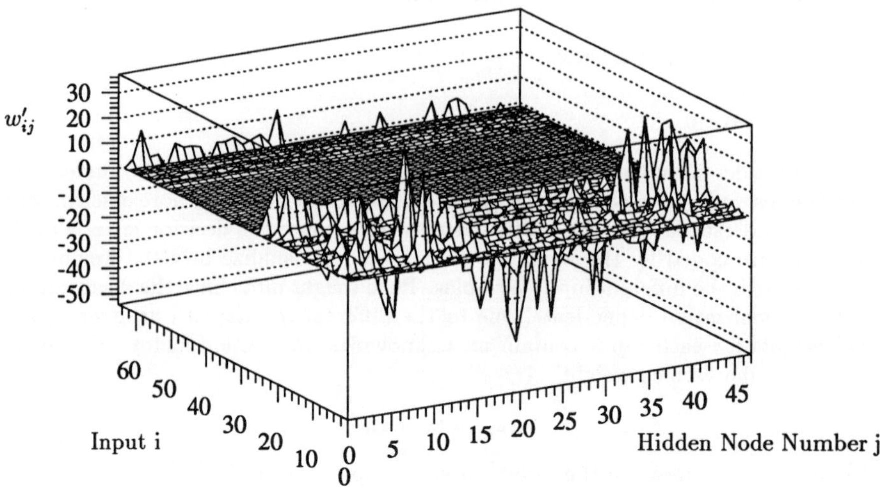

Figure 2: Only those neutral current events without a BSET signal.

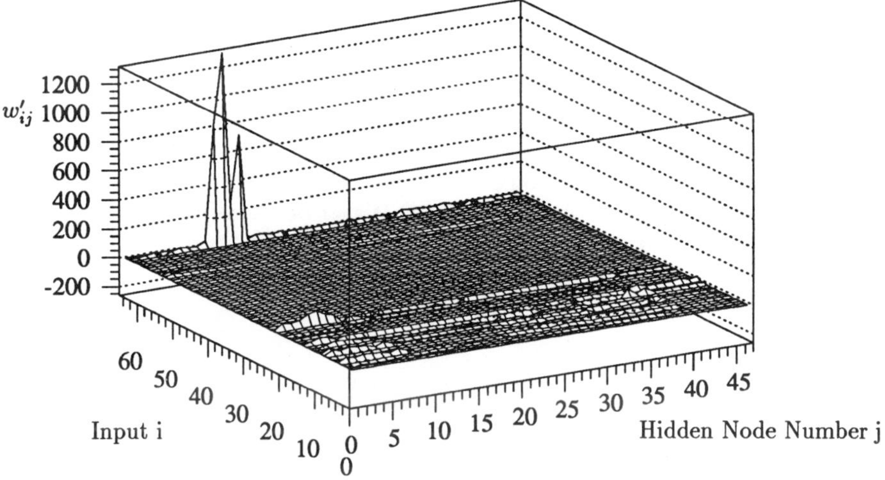

Figure 3: All photoproduction events. The inputs are tracking informations (1 to 21), calorimeters (22 to 49), time-of-flight (50 to 54), muon (55 to 61), tagger (62 to 64) and node bias (65).

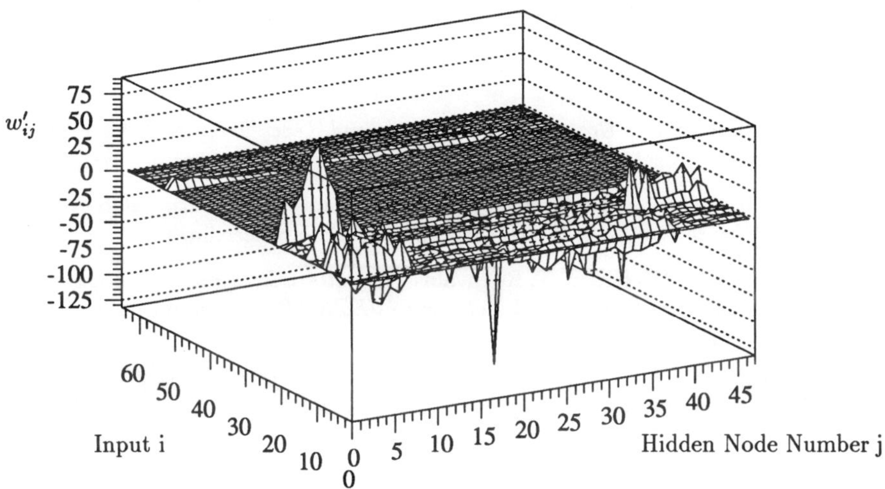

Figure 4: Only those photoproduction events with no electron tagger signal. The trigger efficiency is still the same.

lowest number has the most positive, the node with the highest number the most negative contribution. We plot the w'_{ij} as a function of the input number i (i = 1..64) and the hidden node number j (j = 1..47).
In fig. 1 and 2 this was done for all neutral current events and for those events with no BSET signal (mainly high Q^2 events). Usually the BSET (input 28) dominates the trigger decision (fig. 1). But for those events with no BSET the same network finds its decision also from the tracking information alone (fig. 2).
The situation for photoproduction events is very similar. In this case a signal from the electron tagger (input 64) would fix the trigger decision (fig. 3). However, if there is no electron tagger signal the decision is made up by the other informations, in this case mainly from the track detectors (fig. 4). Note that the network is the same in both cases.

5. Conclusions

We have investigated the training and performance of an ANN to be build in hardware as a level-2 trigger in the H1 experiment. High trigger efficiencies for different physics classes and a good background rejection have been achieved. The networks are found to be stable against failure of detector components. The investigation of efficiencies for high-Q^2 events showed that in general rare but interesting reactions have to be specially treated, e.g. by enriching the training sample with these events. Our studies have shown a way to visualize what a network is doing. Looking this way into a network we found quite interesting structures - in contrast to the common notion of a network as a black box.

6. References

1. C. Kiesling et al., *Concept for a fast Second Level Trigger Using a Neural Network Architecture for the H1–Experiment at HERA*, these proceedings.

2. P. Ribarics et al., *Constructed Network for Background Encapsulation in the Second Level Trigger of the H1-Experiment*, these proceedings.

3. L. Lönnblad, C. Peterson, T. Rögnvaldsson, *Pattern Recognition in High Energy Physics with Artificial Neural Networks - JETNET 2.0*, Comput. Phys. Commun. 70 (1992) 167.

4. A. Wieland, R. Leighton, W. Morgart, *Aspirin for MIGRAINES*, Proceedings of the 1988 International Neural Network Society Conference.

5. J. Möck, private communications.

6. K. Flaton, R. Peffer (Institut für Neuroinformatik, Bochum), private communications.

SUMMARY OF WORKSHOP B: AI

CLARK S. LINDSEY
Royal Institute of Technology, Dept. of Physics-Frescati, Stockholm, Sweden
lindsey@msivax.physto.se

ABSTRACT

I briefly review some of the highlights of the AI workshop. The sub-categories of this workshop included symbolic AI (4 talks), evolutionary algorithms (3 talks) and neural networks (23 talks).

1. Symbolic AI

The sessions under the heading of symbolic AI contained four talks. Two talks concerned expert systems for on-line monitoring of experimental detectors. Behrens discussed an expert system called ZEX for the Zeus detector at HERA. It aims to provide monitoring, diagnois and recovery for problems in the on-line data taking. Previously the system used a finite state machine method. An updated version will instead replace the FSM with more easily understandable rules placed in a commercial expert system shell. A prototype system currently exists and a full system becomes available in 1994.

Novak presented the DELOS expert system designed for error diagnois in particle detectors. It uses an object oriented design and allows for multi-language programming amd multi-platform support. Comparison of acquired data to reference values generates symptoms if it finds anomalies. Then it produces a list of possible errors and instructions for repair. It currently runs under DEC Ultrix and MS-Windows.

Botella presented work on constraint programming with *intervals*. An upper and lower limit defines each variable. One can build and solve nonlinear equations by propagating the intervals, as opposed to the usual case of propagating errors. Such a program could find use in industrial design where materials and parts come in fixed tolerance intervals. Blom discussed a modified Radon transform for finding tracks which takes into account the finite resolution of the measured spacepoints.

2. Evolutionary Algorithms

Evolutionary or genetic algorithms use methods similar to biological systems to sample a functional space with a population of states (initially chosen at random) described by a *genome*. The genome contains a string of numbers which encode the attributes of the state. According to some fitness criteria (e.g. value of a function that one wishes to maximize), those states with the highest fitness become parents to the next generation through combinations of the parental genomes (i.e. crossover of sections of the genomes.) If all goes well, after a finite set of generations one or more states with very high levels of fitness emerge. The evolutionary and genetic algorithms differ in the details of such things as how the states are encoded (e.g. string of binary values for the GA vs continuous values for the EA) and in the selection tests.

The session contained three talks in this area. Bergmann discussed the use of fuzzy logic to control the GA parameters during the running of the algorithm. These controls should improve the learning speed and the likelihood of finding the global maximum. Henker discussed evolutionary control of a system of inference engines. Each inference engine lay on a Connection Machine node. One application he studied involved reconstructing an event where each engine represented one possible set of initial reaction products and their decays.

Hahn described a GA used to tune a MC generator to Delphi e+e- data. The genomes encoded five parameters of the JETSET MC via strings of binary digits (gray coding.) The summed value of the chi-squares of the comparisons of eight histograms of observables for MC and data became the fitness function. A population consisted of 30 genomes, each representing a run (using the corresponding JETSET parameters) containing 50000 events. The GA selection, crossover, and mutation steps created a new generation (i.e. new sets of the 5 MC parameters.) Each new genome then required a new run. A solution represented any case where the summed chi-square lay below the 1992 Delphi conventionally tuned value. Solutions occurred in as few as 8 generations.

3. Neural Networks

A large majority of the talks workshop B involved neural networks. I divide these talks into two broad categories: hardware and software NNW's.

3.1 Hardware Neural Networks

I divide the hardware category further into three sub-categories according to the distance from use in experiments: active experimental hardware, advanced design stage, and early design or prototype stage.

In the active experimental sub-category, Denby discussed the NNW's now in use at the CDF experiment at the Tevatron proton-antiproton collider, and Odorico discussed the recent demonstration test at the WA-92 fixed target experiment at CERN. Both of these experiments use the Intel analog neural network chip (*ETANN*), although WA-92 will test a digital chip in the furture.

The CDF NNW system extracts signals from the existing analog calorimeter electronics which finds energy clusters centered on a single large tower energy deposit (*seed* tower). The NNW's receives 25 electromagnetic calorimeter and 25 hadronic calorimeter signals corresponding to 5x5 arrays centered around the seed tower. These 50 signals feed into the NNW inputs of the chips. The system currently contains 3 chips (on separate fastbus boards), each performing a different network.

One network finds photons in the central calorimeter and another network finds isolated EM showers in the plug calorimeter. These two networks embody simple linear weighting and sum processing and do not arise from trained networks. Results presented show that these two networks performed satisfactorily in the 92-93 run. The third chip contains a trained network to identify jets from B decays but further analysis of the results remains. In the next run, CDF will also install a trained network to identify tau decays.

CERN approved a demonstration experiment to take place at the WA-92 detector (an existing experiment to search for beauty in fixed target interactions) to investigate the usefulness of hardware neural networks. Odorico showed results from the first data taken with the NNW during the most recent run. In the experiment, a negative pion beam (350GeV) impinges on a Cu target. A silicon strip detector lies just beyond the target and records track points. A previously developed track processor does fast on-line track finding in the silicon strip detector and produces hit-maps in 5 separate impact parameter ranges. For the NNW processing, the hitmaps feed into lookup tables which produce a somewhat higher level of data such as the number of groups of hits in each impact parameter bin, number of hits in each group, etc. Sixteen such values go to the NNW chip (after D/A conversion for the analog chip). The output of the chip indicates whether the event contains a secondary vertex from a heavy particle decay. The chip carries out a one layer linear Fisher discriminate classifier in about $3\mu s$. A two layer trained network produced similar results but took twice as long to process. The results show that the NNW chip provides an enrichment of 6.5 at an efficiency of 18%.

Four talks given by Kiesling, Krämperkämper, Gruber and Ribarics discussed an ambitious NNW project for the H1 experiment at the HERA electon-proton collider in Hamburg. There the true ep collisions look similar to the background collisions of the protons with residual beamgas and beampipe materials. Such collisions produce a spray of particles in the forward proton direction. In the early running of the accelerator, it ran far below the design luminosity and a simple first level trigger sufficed for data taking. However, a trigger with greater discriminating power grows in importance as HERA moves towards its design luminosity. The plan discussed here uses NNW's to do this separation in the second level trigger stage. The inputs include various quantities found in the first layer such as values of energy deposits in sections of the calorimeters. Several NNW's, each trained to identify a particular signal (e.g. neutral current, ccbar, etc.) or background, run in parallel. The system OR's all of the NNW outputs to make the trigger decision. Currently they plan to use a digital network and to install the system by fall of 1994.

To insure that the NNW's do not eliminate possible new physics not included in the background training sets, the H1 training restricts the networks to eliminate only known clusters in the input space. New physics outside of these clusters can then still fire the trigger.

Several talks discussed early design or prototypes of hardware NNW's. Fortner discussed designs for a small angle muon trigger in the upgraded D0 experiment at the Tevatron. Seixas talked about a neural network calorimeter trigger for LHC experiments. Lindsey discussed a network to identify events in a fixed target experiment that contain a secondary vertex. He also showed results of a hardware neural network to filter pulse height data from a drift chamber. Palichik discussed a b-quark trigger in a fixed target experiment. Call showed designs of a satellite based NNW to identify lightning strikes in real time from rf signals using DSP's to implement the network.

3.2 Software Neural Networks

The other broad category of neural network talks involved those primarily concerned with implementation or study of networks in software. The sub-categories here also fall into three areas: off-line event selection, detector analysis and general NNW studies.

Event selection with neural networks in off-line analysis involves the use of a NNW to find those cuts on the event variables that give a high signal/background ratio. Usually such analysis involves a search in multi-dimensional space and can take considerable time and effort when done by conventional methods (e.g. cuts on distributions in 1 and 2-d histograms.) One can rely on a NNW to find the best cuts or simply use the NNW as a guide to better conventional cuts. Depending on the type of network used (e.g. a multi-layer network trained by back-propagation or a one layer feature map), one can intrepret the NNW selection with varying degrees of difficulty. Also, simply knowing that a network found superior cuts can spur one to look for better conventional cuts. The NNW doesn't necessarily provide an quick and easy answer. One usually must experiment with various combinations and preprocessing of the event variables to find the best S/BG rato.

Kolanoski discussed NNW's for b meson and tau lepton finding in data from experiments at the e+e- Argus experiment. Denby showed the analysis techniques in use at CDF to search for top quarks in multi-jet events. Lange discussed networks to separate pp-bremstrahlung events from background at the COSY accelerator. Branchini discussed off-line selection of b quark events. Dahn talked about ways of searching for heavy neutrinos at LEP 200.

Another area of off-line NNW analysis involves application of NNW's to pattern recognition in specific detectors. For example, Degener discussed a NNW to identify those clusters of energy in a calorimeter, although appearing as clusters from real photons, actually originated, or split-off, from nearby clusters. Distributions of the energy signals fed into the network which learned to identify the false gammas from the real ones. Similarly, Fuchs used a NNW to find the center position of a gamma shower using the energy distribution in the calorimeter cells as inputs. Ivanov discussed improved secondary vertex finding with NNW's and Kisel discussed track reconstruction with NNW's.

Finally, in the general NNW area, Taal and Ivanov discussed a couple of new learning algorithms. Lades showed results of studies into invariant object recognition where the illumination levels vary. This latter work also involved use of a hardware neural network called the silicon retina.

4. Discussion

In the expert systems area it appears that interest has dampen somewhat from the early enthusiasm, at least in the area of on-line experimental data monitoring. However, the two talks here indicate that substantial investments of time and manpower continue. A sucessful demonstration at a large experiment like Zeus could certainly go a long way to eliminating doubts in the practicality of these systems.

The evolutionary and genetic algorithms show great promise as tools for those optimization problems that resist conventional techniques. For function spaces with many local minima, these algorithms show superiority over conventional gradient descent methods. For large number of variables they can take a long time to find solutions. Also, one normally needs to run many times since they may not find the optimum solution in every case. However, for some problems, one may not find any other method to work at all.

The quantity and quality of the NNW talks indicate that this area continues to grow and mature. There now exist two experiments with working NNW hardware, producing not only useful physics results but also providing experience on the practical problems of implementing these new systems. So far these hardware networks provide ancillary triggers to the many other conventional triggers running simultaneously on the experiments. However, the H1 NNW triggers will provide the core selection triggers for the whole experiment and will become crucial to the success of the experiment as the accelerator luminosities reach design values.

The NNW hardware discussed at the workshop included the Intel analog chip, digital chips from Adaptive Solutions and Siemens, and implementation of networks with DSP's. Each of these options provide various tradeoffs in computational precision, expense and complexity. The ideal NNW chip for high energy physics applications has yet to appear. Such a chip would posess two layers of about 100 neurons each, precision of 8 bits or more in the weights and inputs, processing speed through both layers of $< 1\mu s$, and cost in the hundred $US range.

The use of NNW's as analysis tools continues to expand. Rather than just a black box, NNW's can provide insight into multi-dimensional selection problems. One can use the NNW's as the central selection tool or as a guide and check on conventional analysis methods. One can select from various NNW architectures and learning algorithms to find those best suited to a given analysis.

Acknowledgements

I thank Elena Wildner and Frank Kursawe for help with this summary.

SYMBOLIC MANIPULATION

– Automatic/Semi-Automatic Computations

– Traces and Gamma5

– Diagrams and Loops

– Miscellaneous

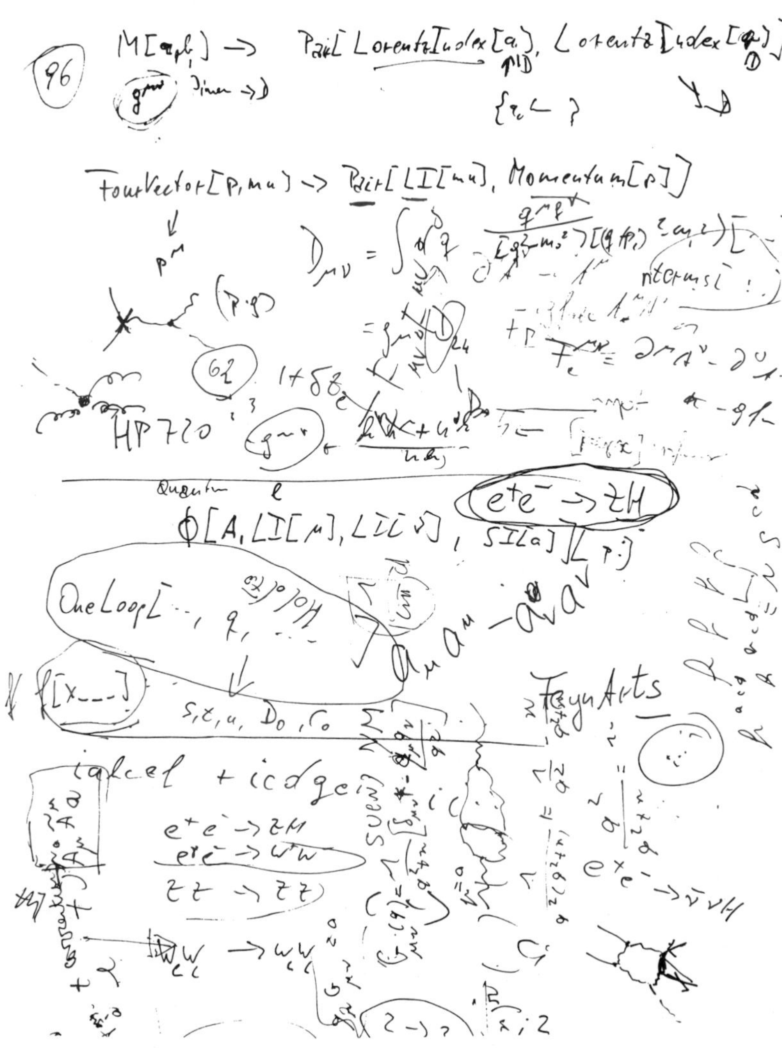

SYMBOL MANIPULATION: EARLY HISTORY AND PRESENT PROSPECTS †

M. VELTMAN
Department of Physics, University of Michigan
Ann Arbor, Michigan 48109, USA

ABSTRACT

An overview of the early development of symbol manipulation is presented. Some presently perceived needs are discussed.

1. Theoretical Need and Technical Opportunity

The history of symbol manipulation by computers (I prefer the shorter name "algebraic programs") is of course tightly bound to the technical evolution of computers and system software. There is the not so unusual coincidence of need and opportunity, which happened in the early sixties. I will describe some of that as I experienced it.

In a real sense computers made their entrance in physics with the IBM 709 computer. Of course, mathematicians had been busy with machines for some time, and they developed many concepts and ideas that are part of what we have today. But their systems lacked a hard to define quality; the nearest way I can think of to describe it is that they were not user oriented. The IBM machine was meant as a commercial enterprise, to be used for business and science. It was there to serve us, not the other way around.

The IBM 709 came with a batch processing system, a line printer, magnetic tapes and a substantial amount of memory (32 k words of 36 bits, one character was 6 bits). A Fortran compiler and an assembler were available. Input was with punched cards, not punched paper tape. No one could possibly have made a substantial program using paper tape and teleprinters. The operating system was really quite advanced; apart from the possibility of parallel processing it was not that different from what we lived with till the advent of personal computers. In other words, since 1961 the necessary equipment and system software was available.

The need for algebraic programs arose very specifically in particle physics. In 1948 Feynman introduced his way of calculating radiative effects in quantum electrodynamics, and gradually these techniques were expanded into the domains of weak and strong interactions. These techniques however gave rise to more and more complicated expressions that soon taxed even the most arduous individuals. We (the particle physicists) simply ran into a wall.

† Invited talk at the Third International Workshop on software engineering, artificial intelligence and expert systems for high energy and nuclear physics, Oberammergau, Germany, Oct 4-8 1993.

2. Early Beginnings

In the beginning of 1963 a number of young theorists, mostly CERN fellows, were debating things at a coffee table on the CERN terrace. Among them Michael Levine, now at Carnegie Tech and myself; I do not remember who else. I had just finished a rather complicated calculation on vector boson production by neutrino's, and complained about the endless number of times that I had to check the algebra before finally getting a correct result. We also discussed the situation of the calculation of the gyromagnetic ratio $(g-2)$ of the muon and the electron. That subject was in real difficulty. Quite naturally the idea came up to do this type of work by computer, and in fact I believe Mike Levine decided then and there to start working on $g-2$. I was very involved with other, more numerical, things around the CERN neutrino experiment and did not start till September 1963, at SLAC, Stanford, California. I have not followed in any detail what Levine did; certainly in the beginning he did not attempt to write a more general purpose program, but appeared to focus on $g-2$. At the end of the sixties he and Kinoshita completed the calculation of the sixth order correction, but I have very little detailed knowledge about that. Let me just say that algebraic processing played a big role as a tool in subsequent developments of the g-2 calculations, notably through the very ingenious work of Kinoshita, Remiddi and Barbieri.

Already quite some time before 1963 I became very interested in the field theory of vector bosons. Lee and Yang, and Lee published seminal work on this subject. In particular Lee had analysed the one loop correction to photon–vector boson coupling. It was very complicated. The challenging problem was to investigate further, to consider more complicated cases such as vector boson scattering. That was clearly beyond human reach. So I set out to attack that problem by computer.

Let me pause here a moment to recall a few important circumstances. For many years to come many, if not all, physicists thought that the issue was to take traces of strings of γ-matrices. In fact, as late as 1978 Stephen Wolfram entered the field writing a trace calculating program. The problem that I considered did not contain a single γ-matrix. Facilities for trace-taking were constructed much later. The $g-2$ problem that Levine considered contained many γ's, and it may well be that he was for some time sidetracked by that issue. It was my luck that I did not need to worry about traces and therefore concentrated on what were, in hindsight, the real issues.

One might well ask why computer scientists were not more active in the field. The first answer is that they were not problem oriented. The second is that they actually did something, and it perfectly illustrated the situation. Just about at that time McCarthy invented LISP, and you could call that the mathematician's view of symbol manipulation.

3. Program Construction

I set out to do things in Fortran. The aim was to develop a tool for analysing the theory, of which the particular case that I considered was only an example. Therefore I tried, from the start, to develop a general purpose system. Consequently I wanted the input to resemble as closely as possible the expressions one would write on paper. Thus the first step was to develop code that would read an expression and translate it into suitable binary patterns. That was the first phase.

The expression that I started from was this:

$$\Big[\delta_{\alpha\beta'}[2P_{\mu'} - Q_{\mu'}] - \delta_{\alpha\mu'}[\zeta Q_{\beta'} + P_{\beta'} - \xi(P_{\beta'} - Q_{\beta'})] - \delta_{\beta'\mu'}[-\zeta Q_\alpha + P_\alpha - Q_\alpha - \xi P_\alpha]\Big]$$

$$\Big[\delta_{\beta'\alpha''} + A(P_{\beta'} - Q_{\beta'})(P_{\alpha''} - Q_{\alpha''})\Big]$$

$$\Big[\delta_{\alpha''\beta''}[P_\mu + P'_\mu - 2Q_\mu] - \delta_{\alpha''\mu}[\zeta(P_{\beta''} - P'_{\beta''}) + P_{\beta''} - Q_{\beta''} - \xi(P'_{\beta''} - Q_{\beta''})]$$

$$- \delta_{\beta''\mu}[\zeta(P'_{\alpha''} - P_{\alpha''}) + P'_{\alpha''} - Q_{\alpha''} - \xi(P_{\alpha''} - Q_{\alpha''})]\Big]$$

$$\Big[\delta_{\beta''\alpha'} + B(P'_{\beta''} - Q_{\beta''})(P'_{\alpha'} - Q_{\alpha'})\Big]$$

$$\Big[\delta_{\alpha'\beta}(2P'_{\mu'} - Q_{\mu'}) - \delta_{\alpha'\mu'}[-\zeta Q_\beta + P'_\beta - Q_\beta - \xi P'_\beta] - \delta_{\beta\mu'}[\zeta Q_{\alpha'} + P'_{\alpha'} - \xi(P'_{\alpha'} - Q_{\alpha'})]\Big]$$

The actual input in SCHOONSCHIP form was:

```
(D(AL,BEP)*{2*P(MUP) - Q(MUP) }
- D(AL,MUP)*{Z*Q(BEP) + P(BEP) - X*(P(BEP) - Q(BEP)) }
- D(BEP,MUP)*{- Z*Q(AL) + P(AL) - Q(AL) - X*P(AL)} )
* (D(BEP,ALPP) + A*( P(BEP) - Q(BEP))*(P(ALPP) - Q(ALPP) ) )
* (D(ALPP,BEPP)*{P(MU) + PP(MU) - 2*Q(MU) }
- D(ALPP,MU)*{Z*(P(BEPP) - PP(BEPP))
+ P(BEPP) - Q(BEPP) - X*(PP(BEPP) - Q(BEPP)) }
- D(BEPP,MU)*{Z*(PP(ALPP) - P(ALPP)) + PP(ALPP) - Q(ALPP)
- X*(P(ALPP) - Q(ALPP))} )
* (D(BEPP,ALP) + B*(PP(BEPP) - Q(BEPP))*(PP(ALP) - Q(ALP)) )
* (D(ALP,BE)*(2*PP(MUP) - Q(MUP))
- D(ALP,MUP)*{ - Z*Q(BE) + PP(BE) - Q(BE) - X*PP(BE)}
- D(BE,MUP)*{Z*Q(ALP) + PP(ALP) - X*(PP(ALP) - Q(ALP))} )
```

At that time only upper case characters could be used.

The expression is a product of polynomials:

$$(a_1 + a_2 + \ldots)(b_1 + b_2 + \ldots)\ldots(c_1 + c_2 + \ldots).$$

The individual terms, a_1, a_2, ..., b_1, $b_2\ldots$ etc. contained algebraic factors such as coupling constants, along with vectors and delta's. Given such an expression in memory two problems had to be solved:

- working out a distribution routine that would systematically generate all products: first $a_1 b_1 \ldots c_1$, then $a_2 b_1 \ldots c_1$ and so on;

- working out a multiplication routine that would implement the algebraic rules such as $\xi \cdot \xi = \xi^2$ and $p_\mu q_\mu = (pq)$.

Thus I developed a routine that systematically emitted all the products, one by one. Another routine analysed the products and implemented the algebraic rules. In the process I kept on inventing conventions to denote the various quantities involved. That then was phase two of the project. Thus now, one by one, the resulting products were generated and of course I took care that identical terms would come out identical, so factors would be reshuffled as necessary (for example $a \cdot b$ remained $a \cdot b$ while $b \cdot a$ would become $a \cdot b$ as well).

The next step was to collect the terms in an output store. There every new term was to be compared with whatever was already there so that for example $a \cdot b + a \cdot b$ became $2a \cdot b$. This required some organization as it turned out that this operation tended to be time consuming. I there introduced the idea of factoring out certain combinations of factors, which was also very useful for other reasons.

Finally the last step was to translate the output to readable form.

After I had all of that going I was confronted with a useless result. Momentum conservation implied that many terms could be expressed as linear combinations of others, and there was clearly great redundancy. I had to find a way to implement the various identities by machine as well, simply because the problem was already by far too large to handle by hand. I learned a very important lesson: if you start doing things by computer you must then do everything by computer to the very end, because the intermediate expressions are usually too large to handle by hand.

It is in this context that I invented the idea of a substitution. Whenever a term would come out of the multiplication routine it was inspected for factors for which some algebraic expression had to be substituted. Like for instance

$$(pq) = (pk) - (pr) .$$

Thus a term of the form

$$ab(pq)cd$$

became

$$ab((pk) - (pr))cd .$$

But that looked precisely like the way the input was defined: a product of polynomials. Thus at this point the whole thing became a matter of carefully implementing a loop.

For completeness I should mention that in the way that I presented the expression above momentum conservation is already worked out. But I do not want to get into too many details here.

To this day, these are basically the essentials of my algebraic program SCHOONSCHIP. It is perhaps useful to point out an important feature of the above approach that has escaped the attention of quite a few programmers, and that is the efficient use of computer memory. I will explain this by way of an example.

Consider the expression

$$(a+b)(c+d)\,.$$

Let there moreover be an identity, say

$$c = e + f\,.$$

Now the distribution routine produces ac, bc, ad and bd one by one. The idea now is not to generate all these terms and then start working on them, but to process them one by one till the very end. Thus first ac is produced. Then $c = e + f$ is substituted. This produces ae as the first term which is stored. Next the second term af is produced and stored, and only then, going back to the beginning, is the term bc generated. In memory one therefore finds at the moment the first term ae is about to be stored:

$$(a+b)(c+d)$$

$$ac$$

$$a(e+f)$$

$$ae$$

This is to be contrasted with the case were everything is worked out at every step:

$$(a+b)(c+d)$$

$$ac + ad + bc + bd$$

$$ae + af + ad + be + bf + bd$$

which is then stored.

Thus for a given calculation SCHOONSCHIP needs very little space to contain the intermediate results. At any level only one term is considered. I see this as one of the most important features of SCHOONSCHIP. It introduces automatically the concept of a "collect" command (called *yep in SCHOONSCHIP language), namely defining the stage where everything is to be collected before further processing is done.

Another important aspect was the magnitude of the output. By that time, autumn of 1963, I was working on an IBM 7090, a transistorized version of the 709. The machine had as mentioned before a memory of 190 k 'bytes' (one byte being 6 bits). The output from my first problem was by far too large to be contained in that memory, and it was necessary to devise some system using magnetic tapes. So, from the first version of SCHOONSCHIP on mass storage was used.

I need not elaborate here on the continued expansion of the identity - substitution idea. It's importance is obvious. There is one limitation: one can only work on one term at a time. Thus things requiring the study of some combination of terms cannot easily be accomodated. For example, it is hard to implement identities such as $a^2 + 2ab + b^2 = (a+b)^2$. That would need additional routines working on the output as a whole. This is very difficult in all cases, and in general forbiddingly time consuming.

As said earlier, in the beginning I tried to do things in Fortran. This became more and more difficult, not in the least because Fortran had no character manipulation facilities. First I made little assembler routines for those tasks, but gradually the whole program became nothing but a succession of calls to those subroutines, and in the end it was much more expedient to write everything in assembly code.

4. Developments

It is interesting to note a parallel development, namely the writing of REDUCE by Anthony Hearn. At the time Tony was also at Stanford, and we talked about things on an almost daily basis. Tony argued that LISP would be a good language to do this work, and he started to construct REDUCE along those lines. He was not considering any specific physics problem, and for all I know he mainly followed what I did, trying to show that LISP was the best tool. There is no doubt however that I also profited from these discussions. In the end his REDUCE program was much like SCHOONSCHIP, with a few important differences:

+ REDUCE is not tied to any specific machine;
- Speed is much less (typically two orders of magnitude);
- Size is limited by the handling capabilities of LISP. No mass storage was used by LISP for a long time to come;
- REDUCE did not use the term-by-term technique and therefore needed much more memory.

These last two points made REDUCE, in my opinion, a not so useful program. It has taken at least ten years for LISP to catch up, and for all I know, up to now, memory use by REDUCE is very uneconomical and slow. Its applications have therefore been limited to relatively small problems.

There can however be no doubt that further development of algebraic programs owes much to REDUCE. I must add however that I have never studied systems such as MACSYMA, and I do not know how and when they came into existence. I believe that ASHMEDAI, a program developed later by Levine, was done much along the lines of SCHOONSCHIP. ASHMEDAI was completely written in Fortran, and could therefore be ported relatively easily. Again, I have not studied these systems, and I truly cannot judge to what extent they have built on SCHOONSCHIP's designs and ideas.

In 1965 I rewrote SCHOONSCHIP for the CDC 6000 series computers. Occasionally other physicists than myself also started using it. In fact, the first manual that I made dates from 1965, applying to the IBM 7090 version.

From 1966 onwards more and more people turned to algebraic programs for help in solving their problems. It was not as easy as today, and nobody saw any commercial interest. In those days no one paid for software. That came much later, in the second half of the seventies.

The last rewrite of SCHOONSCHIP was in 1983, for a Motorola 68000 based system. To give an idea of the work involved: once a decent assembler was available (actually, I had to write that too) it took about three months to translate from CDC to Motorola machine code. Right now SCHOONSCHIP takes about 100 kB.

Of course, there have been many other developments. Stephen Wolfram, with SMP and MATHEMATICA, has correctly understood and implemented many concepts leading to a very succesful commercialization. There the accent is much less on size and speed capabilities, and I think it is a mistake to use such programs for large scale problems such as encountered in present day high energy physics. This I believe also applies to MAPLE, of which I know however too little to make any serious statement. The program FORM by Jos Vermaseren is a machine independent program, written in C. It is very much along the lines of SCHOONSCHIP, including large scale capabilities.

5. Future Prospects

As for future developments I see a real need for specialization. At this time everyone is making very general purpose programs in order to reach a large public. But as theoretical progress is made different techniques are needed. Here there is a real difficulty. The people that develop programs are, with few exceptions, not the same as those that do specific research. In my view there is no way that program developers can anticipate the needs of the researcher. I will try to give an example of such a need that so far nobody has addressed in an adequate way. I have in a modest way tried to do something in SCHOONSCHIP, but I have no pretension that it is in any way a complete solution.

The problem that I am referring to arises in calculations on weak interactions. There one must deal with a proliferation of particles (of the order of 30 at this time) and interactions. For a given process there may be a few types of diagrams, and as an example I take the triangle diagram.

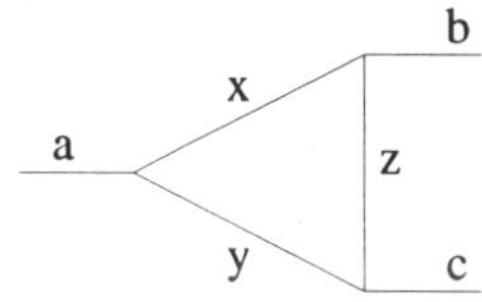

The problem is this. The external particles a, b and c are given. Find all combinations x, y and z such that the vertices axy, bxz and cyz exist. In other words, there is a summation:

$$\sum_{x=1}^{30} \sum_{y=1}^{30} \sum_{z=1}^{30} E(x, y, z),$$

where E is a function that is zero if the vertices involved do not exist, and otherwise is the appropriate expression for the diagram involving those vertices. As you can see this may easily become a large problem. I call this sum a 'character sum', as

I use characters to designate particles. Somewhere there is a list of all existing vertices (all existing 'words' formed from three characters in alphabetic order), and in a systematic way SCHOONSCHIP uses that list to generate all possible terms in the sum.

For now the list of vertices contains about one hundred entries, including weak and electromagnetic but not strong interactions. The system developed for this works satisfactorily. The next step that one would like to take is to extend it to grand unified theories which I believe should be studied more closely. Then the number of entries grows to something on the order of one thousand, and so far I have not been able to do it. It should be mentioned that there are many additional problems in generating the terms E and processing them, none of which has been addressed adequately. I am sure that most people are not even aware of the nature of the difficulties. I leave this as a challenge to my successors: compute two-, three- and four-point one-loop diagrams for the SU(5) grand unified theory. And when you have done that I have another question: what about gravitation?

Toward an Automatic Computation of Cross Sections Including One-loop Diagrams

MINAMI-TATEYA Collaboration
Junpei FUJIMOTO, Tadashi ISHIKAWA, Setsuya KAWABATA,
Yoshimasa KURIHARA, Yoshimitsu SHIMIZU, Jian-Xiong WANG[1]
National Laboratory for High Energy Physics (KEK)
Oho 1-1, Tsukuba, Ibaraki 305, JAPAN
Kiyoshi KATO, Nobuya NAKAZAWA
Kogakuin University
Nishi-Shinjuku 1-24, Shinjuku, Tokyo 160, JAPAN
Toshiaki KANEKO
Meiji-Gakuin University
Kamikurata 1518, Totsuka, Yokohama 244, JAPAN
Hidekazu TANAKA
Rikkyo University
Nishi-Ikebukuro 3-34-1, Toshima, Tokyo 171, JAPAN

1 Introduction

The progress of high-energy physics enables us to study experimental data with high statistical accuracy. This, in turn, requests theorists to calculate the corresponding cross sections for a number of multi-body reactions at least in the lowest order, and sometimes beyond in the perturbative series. This urges us to develop an expert system for the perturbative computation in high-energy physics.

We have already developed the GRACE system for the computation of the matrix element in the tree level. We briefly sketch its feature since the details of the system, GRACE-T, has been reported elsewhere[1]. After the generation of the Feynman diagrams, it generates the FORTRAN code which calculates the relevant amplitude numerically by calling the CHANEL library functions. These CHANEL[2] functions calculate each of elements which compose of the Feynman diagrams, i.e., wave functions, vertices, and propagators. In the generated FORTRAN code, the value of matrix element is calculated by summing the values of all amplitudes and taking the square of the sum.

GRACE-T is applied to various processes and it has proved to work well for the calculation of matrix elements in the tree level. Since it has interface to the adaptive Monte-Carlo integration program BASES and the event-generation manager SPRING,

[1] Address after October 1993, IHEP, Beijing, China.

cross sections and other distributions for physical quantities up to event generation can be obtained under a realistic cut in kinematics[3].

The natural extension of the study is to include the loop diagrams, which is essential to understand the high-statistical data in the existing and future colliders. This paper reports the status of GRACE-L1 whose target is the automatic computation of one-loop radiative corrections. The diagram generator and the REDUCE source code generator are written in the standard C. Since the use of BASES/SPRING is common to GRACE-T, the explanation for this part is skipped in this paper.

2 Graph generation

The graph generator in GRACE can produce Feynman diagrams in the standard model, QED, electroweak and QCD. It generates diagrams including up to 5-loops[4]. Generation of tree and loop diagrams is done at the same time. In order to avoid the duplicate count, we use the algorithm which is the modification of that used for the decision of graph isomorphism[5]. After generating diagrams, it can eliminate one-particle reducible diagrams and tadpole diagrams. In case of loop diagrams, the diagrams for counter terms in the same order of coupling constants are also generated.

Drawing utility program is also provided to display graphs on X-window system interactively. Following operations are possible.

Forward	**B**ackward	**Q**uit
Greater size	**S**maller size	**L**abel
Next process	**P**revious process	**T**ext

3 Matrix elements

For a concrete example for which GRACE-L1 is to be applied, we consider the standard electroweak theory. The present system handles the infrared singularity by introduction of fictitious small mass, which generally violates the gauge invariance in the case of QCD. Though GRACE-L1 generates graphs including quarks and gluons, its applicability is restricted to those without gluon self couplings in the loop. At present, only vertices which exist in the standard model are supplied. Extension to include non-standard vertices will be possible. Here we follow the on-mass-shell scheme for the renormalization of electroweak theory [6].

Two methods can be considered for the calculation of loop diagrams. One is to process the amplitude and calculate its numerical value. Another possibility is to calculate the product of a loop diagram and a tree diagram algebraically and after that the expression is evaluated numerically. The first one is the way used in GRACE–T and it has the advantage that the large scale computation is possible because the number of manipulation is proportional to the number of diagrams while it increases as the square of the number in the latter case.

As a pilot version of the system for automatic computation of loop diagrams, GRACE-L1 is designed to rely on the latter method. In order to avoid the complicated loop integrals, the system only works in the 'tHooft-Feynman gauge. The calculation is done in dimension $n(= 4-2\epsilon)$ for the regularization of ultraviolet divergence. The interference term between a loop diagram and the conjugate of a tree diagram has the following form

$$M_j = \int \frac{d^n l}{(2\pi)^n} \frac{N(l)}{(l^2-m_1^2)((l+p_1)^2-m_2^2))\cdots((l+p_1+\cdots+p_{j-1})^2-m_j^2))} \quad (1)$$

where N is the numerator and j is the number of propagators in the loop. After the introduction of Feynman parameters and an appropriate shift of loop momentum, it can be transformed into the following forms:

$$M_2 = \int dx \int \frac{d^n l}{(2\pi)^n} \frac{l^2(g_0+\epsilon g_1)+(f_0+\epsilon f_1)}{(l^2-D_2+i\varepsilon)^2} \quad (2)$$

$$M_3 = \int dxdy \int \frac{d^n l}{(2\pi)^n} \frac{l^2(g_0+\epsilon g_1)+f_0}{(l^2-D_3+i\varepsilon)^3} \quad (3)$$

$$M_4 = \int dxdydz \int \frac{d^n l}{(2\pi)^n} \frac{(l^2)^2(h_0+\epsilon h_1)+l^2(g_0+\epsilon g_1)+f_0}{(l^2-D_4+i\varepsilon)^4} \quad (4)$$

Here we only confine the calculation up to the four-point functions. In the above, functions f, g, h are polynomials of Feynman parameters.

Some FORTRAN modules are required for the computation. For the two-point function, we have prepared the functions which give the value of renormalized two-point functions. Another one is the module to calculate the coefficients of counter terms for vertices in the electroweak theory. Further, the library is needed to calculate the numerical value of loop integrals in the above equations from the masses of internal lines, the momenta of external lines and the values of coefficients in the numerator polynomials.

The procedure for the automatic computation of the matrix element in the one-loop order is as follows:

1. The external particles are specified in the file `in.prc`. Diagrams in the tree level and those in the one-loop level are generated. We refer the n-th tree diagram and n-th one-loop diagram as `an` and `bn`, respectively. A diagram including two-point function is generated as a diagram with the insertion of a renormalized two-point function to an internal propagator. Diagrams with vertex counter term are also generated.

2. The source programs for REDUCE are generated automatically. For the product of `an` and the conjugate of `am`, we have `anxm.red`, and for `bn` and `am`, we have `bnxm.red`.

3. In the REDUCE program, the calculation is done as follows. If there is a fermion line, the trace is taken first assuming that γ_5 anticommutes with every other gamma matrix. After the shift of loop momentum l, we replace $l^\mu l^\nu$ by $g^{\mu\nu} l^2/n$ and contract vector indices inside the loop in n-dimension. Remaining vector indices are contracted in 4-dimension. This procedure allows to define the vertex counter terms independent of the process.

4. Invoking REDUCE, the FORTRAN subroutines `anxm.f` and `bnxm.f` are generated. The latter, `bnxm.f` calculates the coefficient functions f, g, h in the eqs.(2)–(4) from the kinematical variables and couplings and it gives value of the corresponding matrix element by calling the functions which compute the one-loop integrals with numerator.

5. Other FORTRAN subroutines are also generated automatically. They include the manager to sum the terms, the functions to prepare physical parameters, and modules which make interface to the BASES integration package.

4 Example: $e^-e^+ \to HZ$

As an example to check the performance of the system, we have computed the differential cross section for $e^-e^+ \to HZ$[7]. Here we neglect the coupling of scalar particles to a few light fermions. generated are 97 diagrams including three vertex counter terms and two self-energy diagrams for vector bosons. For the numerical calculation, we take $W = 500\text{GeV}$, $M_Z = 91.17\text{GeV}$, $M_W = 80\text{GeV}$, $M_H = 150\text{GeV}$, and $m_t = 130\text{GeV}$. We calculate the differential cross section by the generated FORTRAN program as is shown in Fig.1. By use of GRACE–T, we also generate the matrix element of $e^-e^+ \to HZ\gamma$ and integrate it above $k_0 > k_{cut}$ where k_0 is the energy of photon. The soft photon factor is given as follows.[2]

$$
\begin{aligned}
d\sigma_{soft} &= R_{soft} d\sigma_{tree} &(5)\\
R_{soft} &= \frac{\alpha}{\pi}\left[2(L-1)\log\frac{m_e}{\lambda} + \frac{1}{2}L^2 - \frac{\pi^2}{3} - 2\ell(L-1)\right], &(6)\\
& L = \log(W^2/m_e^2), \quad \ell = \log(W/(2k_{cut})).
\end{aligned}
$$

Summing these contributions, the one-loop correction, the hard-photon contribution and the soft photon term, we obtain the radiative correction to $e^-e^+ \to HZ$.

In order to check the system, we have made the following examination.

- The amplitudes for the all diagrams are prepared by hand and the similar procedure is performed by use of FORM[8] independently. We have compared the contributions diagram by diagram with those obtained automatically.

- For the numerical value of the loop integral, we have a few packages, those developed by our group[9] and that based on the method described in Ref.[10]. Values are compared between packages and we have checked that they agree well.

- cancellation of ultraviolet divergence, independence of the fictitious photon mass λ and the soft photon cut k_{cut} are also observed.

[2] At this moment, introduction of the soft photon factor is done by hand.

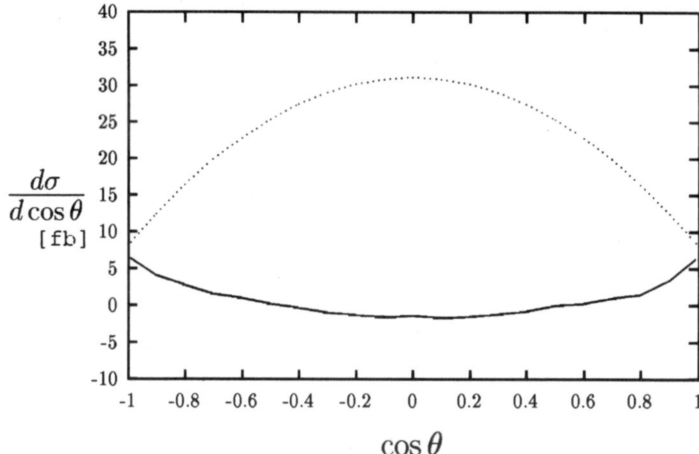

Figure 1: Differential cross section of $e^-e^+ \to HZ$ in the lowest order(dotted) and that in the one-loop level(solid).

5 Example: $e^-e^+ \to q\bar{q}\gamma$

Processes $2 \to 3$ are more complicated. For an example, we consider the QCD correction to $e^-e^+ \to q\bar{q}\gamma$ process. In this order, the diagrams are all QED-like, so that the fictitious gluon mass is enough for handling the infrared singularity. There appears no five-point function because a gluon does not couple to electrons.

In order to check the results, we prepared another three calculations for the matrix element and three kinds of integration of the Feynman integral. The first matrix element was calculated by hand with REDUCE, the second and third ones were obtained using helicity amplitude methods, here we call Amplitude-I and Amplitude-II, respectively. The method of Amplitude-I is the extension of the CHANEL[2], and we used the physical gauge for the emitted photon. The method of Amplitude-II is presented in this workshop[11]. We have two independent analytical calculations for the Feynman integrals, Analytic-I and Analytic-II. We also coped with the Feynman integral numerically(Numeric) according to the algorithm in Ref.[9].

Below we present the results obtained by various methods in the case of u-quark. Here, only QED diagrams are considered. We calculated at $W = 91.17$ GeV. We required the isolation of the photon from quark/anti-quark to be $10°$ and the minimum energy of the photon to be 1 GeV. The soft gluon cut was 0.1 GeV and the strong coupling constant was fixed as 0.12. We take the mass of u-quark to be 0.1 GeV. As seen from the Table-1 all the results are consistent within the errors which come from the 3-body phase space integration. The cross section of the hard gluon emission calculated by GRACE-T under the same cuts is 5.35 ± 0.01pb. Since the lowest order cross section is 0.852pb, the total QCD correction is about -23%. We examined the same calculations for the b-quark and obtained a good agreement, too.

	Matrix Element			
Feynman integral	Hand-Made	Amplitude-I	Amplitude-II	GRACE-L1
Analytic-I	-5.57 ± 0.01	-5.57 ± 0.01	-5.56 ± 0.01	-5.56 ± 0.01
Analytic-II	-5.54 ± 0.01	—	—	—
Numeric	-5.59 ± 0.05	—	—	—

Table 1 One-loop QCD corrected total cross sections for $q\bar{q}\gamma$ in *pb*.

In the computation of loop integrals numerically, we sometimes encounter a kind of numerical instability: The terms with higher powers of Feynman parameter in the numerator are unstable for some points in the phase space. This is caused by the fact that the matrix to reduce the numerator of a box diagram into scalar loop integrals is singular at those points. This is a new feature in the three-body final state in which computation becomes singular.

6 Conclusion

We have developed the system to compute one-loop diagrams automatically for the standard electroweak theory. While the system is only confined up to the four-point functions in the present version, the preparation of library to compute loop integrals makes it possible to compute diagrams including j-point functions for $j > 4$. The performance of GRACE–L1 is limited by the power of REDUCE. As is done in GRACE–T, we also plan to make a system, GRACE-Lx, which processes the loop in the amplitudes. Since the check of the system by use of gauge invariance is not easy in loop diagrams, the hand-made amplitudes for $e^-e^+ \to HZ$ written by FORM is vital in the present work. In general, the validity of the system can be achieved by comparison of a few different systems.

This work is supported in part by Ministry of Education, Science and Culture, Japan under Grant-in-Aid for International Scientific Research Program No.04044158, and in part by the Special Research Fund of Kogakuin University.

References

[1] MINAMI-TATEYA group, GRACE manual, KEK Report 92-19, 1993.

[2] H. Tanaka, Comput. Phys. Commn. **58**(1990)153.

[3] S. Kawabata, Comput. Phys. Commun. **41**(1986)127.

[4] T.Kaneko, Proceedings of the VII High Energy Physics and Quantum Field Theory, Sotchi, 1992;

T.Kaneko *et al.*, in "New Computing Techniques in Physics Research II", ed. D. Perret-Galix, p.659, 1992,World Scientific, Singapore.

[5] L.Kučera, Combinatorial Algorithms, Adam Hilger, Bristol, 1990.

[6] J.Fujimoto, M.Igarashi, N.Nakazawa, Y.Shimizu and K.Tobimatsu, Progr. Theor. Phys. Suppl., **100** (1990),1.

[7] J.Fleischer and F.Jegerlehner, Nucl.Phys. **B216** (1983), 469;

J.Küblbeck, H.Eck, and R.Mertig, in "New Computing Techniques in Physics Research II", ed. D. Perret-Galix, p.683, 1992,World Scientific, Singapore.

[8] J.A.M. Vermaseren, CAN, Kruislaan 413, 1098-SJ Amsterdam, ISBN 90-74116-01-9.

[9] J. Fujimoto, K.Kato, Y.Oyanagi and Y.Shimizu, Progr.Theor.Phys., **87**(1992), 1233.

[10] G.'tHooft and M.Veltman, Nucl.Phys. **B153** (1979), 365.

[11] J.X. Wang, talk given in this workshop.

PERTURBATIVE QCD WITH FEYNCALC

R. Mertig
Instituut-Lorentz, University of Leiden
P.O. Box 9506
2300 RA Leiden, The Netherlands

ABSTRACT

The application of *FeynCalc* to perturbative QCD calculations is described. Ingredients for the determination of the yet unknown higher order spin-dependent Altarelli-Parisi splitting functions are presented. As a calculational method the energy momentum tensor is coupled to the graviton field. The necessary Feynman rules are derived automatically by *FeynCalc*. The lowest order partonic structure functions and the (known) leading order Altarelli-Parisi splitting functions are correctly reproduced using the graviton as probing particle.

1. The development of the *FeynCalc* project

FeynCalc was designed and used as a tool for 1- and 2-loop radiative corrections in the Standard Model. For perturbative QCD calculations in the framework of dimensional regularizaion *FeynCalc* had to be extended and improved.

While *FeynCalc* is in general slower in speed than FORM [1], it offers more generality (including all usual *Mathematica* features). For example *FeynCalc* was used to check E.B. Zijlstra's FORM-programs involving different γ_5 prescriptions [2] and to show the violation of the Burkhardt-Cottingham sum rule in first order of α_s [3]. This involved the computation of the $\mathcal{O}(\alpha_s)$ QCD corrections to the transverse spin dependent structure function $g_2(x, Q^2)$.

In *FeynCalc*2.3 the old *FeynCalc*1.0 functions and objects related to $SU(3)$ have been replaced by $SU(N)$ ones, i.e., color factors can be calculated in terms of N (or, optionally, the Casimir operators $C_A = N$ and $C_F = (1 - N^2)/(2N)$.

Phase space integrals are only available for $2 \rightarrow 2$ DIS-processes. For the difficult $2 \rightarrow 3$ integrals a *FeynCalc*-link-program to existing FORM-programs [4] is projected. A general possibility to link *FeynCalc* to other programs like FORM or Schoonschip would be highly desirable and may be realized as a result of this conference. The compatability both to *FeynArts*1.0 and to the newest *FeynArts* version described in this proceedings is provided. The new *FeynArts* and *FeynCalc* versions can now also be used in one single *Mathematica* session.

2. Altarelli-Parisi splitting functions via the energy-momentum tensor

Since the Altarelli-Parisi (AP) splitting functions are process independent, the nature of the probing particle is irrelevant. For the objective of calculating the yet unknown next-to-leading order AP splitting functions P_{gg} and P_{gq} one may use the graviton field coupled to the energy-momentum tensor [5]. A detailed discussion on the next-to-leading order calculation will be given elsewhere.

Already the Feynman rules are rather involved. Therefore a function *FeynRule* was added to *FeynCalc* , enabling the fully automatic derivation of (Vertex-) Feynman rules from a given Lagrangian (see Appendix A). The new gravity-QCD-model-file for *FeynArts*1.0 could be generated to a large extend by a *FeynCalc* program, thus minimizing the danger of typing-errors.

The hadronic tensor with the graviton as probing particle reads

$$W_{\mu\nu\lambda\sigma} = \frac{1}{4\pi} \int d^4x \, e^{iqx} \langle p, s | [\Theta^{\mu\nu}(x), \Theta^{\lambda\sigma}(0)] | p, s \rangle \tag{1}$$

where $\Theta^{\mu\nu}$ is the energy-momentum tensor [6]. The partonic structure function $\mathcal{F}_p$ may be derived by contracting the hadronic tensor with a suitable projection operator. These projection factors can be determined by calculating the lowest order structure function $\delta(1-x)$ (see Appendix B).

At leading order the scattering process $\mathcal{G}+g \to g+g$ yields the Altarelli-Parisi splitting function P_{gg} via

$$\mathcal{F}_{gg} = \delta(1-x) + \frac{\alpha_s}{4\pi} \left(\frac{1}{\varepsilon} P_{gg}(x) + c_{gg}(x) + \varepsilon \right) a_{bb}(x) \tag{2}$$

where x is the Bjorken scaling variable and $\varepsilon = D-4$ from D-dimensional regularization. Squaring the 4 Born-level graphs for $\mathcal{G}+g \to g+g$ in D dimensions *FeynCalc* could reproduce the correct polarized and unpolarized P_{gg} (see Appendix C).

Similarly all other AP-splitting functions have been recalculated thus checking as well *FeynCalc* as the calculational approach using gravitons as probing particles.

3. Acknowledgements

It is a pleasure to thank Willy van Neerven for fruitful collaboration and WRI for providing a 2.3betaAlpha version of *Mathematica*.

4. References

1. J.A.M. Vermaseren, Symbolic manipulation with FORM, published by CAN, Kruislaan 413, 1098 SJ Amsterdam (1991), ISBN 90-74116-01-9.
2. E.B. Zijlstra, W.L. van Neerven, Phys.Lett. B297 (1992) 377.
3. R. Mertig and W.B. van Neerven, INLO-PUB-2/93.
4. E.B. Zijlstra, private communication.
5. W. van Neerven, private communication.
6. J.C. Collins, A. Duncan, S.D. Joglekar, Phys.Rev. D16 (1977), 438.

Appendix A

Automatic Feynman Rule derivation with FeynCalc

In this Mathematica-notebook the Lagrangian for the energy-momentum tensor coupled to the Graviton field is typed in as input (bold type). The output is typesetted automatically by Mathematica and the default setting $PrePrint = FeynCalcForm.
In the Feynman rules all momenta are understood to be incoming.

Load FeynCalc and define abbreviations. SUNT is the SU(N) T-matrix, SUNF is the structure constant f.

```
<< ~/HighEnergyPhysics/FeynCalc.m;
φ = QuantumField; P = Partial; γ = DiracMatrix; ℓ = LorentzIndex;
S = SUNIndex; T = SUNT; f = SUNF; gs = Subscripted[g[s]];
{p1, p2, p3, p4} = Array[Momentum[Subscripted[p[#]]]&, 4];
{l1, l2, l3, l4}  = Array[LorentzIndex[Subscripted[λ[#]]]&, 4];
{a1, a2, a3, a4}  = Array[SUNIndex[Subscripted[a[#]]]&, 4];
F[α_, β_][k_, m_, n_] := φ[P[α], A, ℓ[β], S[k]] -
    φ[P[β], A, ℓ[α], S[k]] - gs . f[m, n, k] . φ[A, ℓ[α], S[m]] . φ[A, ℓ[β], S[n]];
```

■ The Lagrangian for the energy-momentum tensor coupled to the gravity

The Lagrangian for massless QCD, without gauge fixing term.

```
LQCD = (-1/4 F[μ, ν][k, a, b] F[μ, ν][k, c, d] + I/2
      (φ[Bar[ψ]] . γ[μ] . φ[P[μ], ψ] - φ[P[μ], Bar[ψ]] . γ[μ] . φ[ψ]) +
      gs . φ[Bar[ψ]] . T[a] . γ[μ] . φ[ψ] φ[A, ℓ[μ], S[a]] + φ[P[μ], Bar[λ], S[a]]
      . (φ[P[μ], λ, S[a]] - gs . f[a, b, c] . φ[A, ℓ[μ], S[c]] . φ[λ, S[b]]));
```

Construct the energy-momentum tensor consisting out of gluon, quark and ghost fields (a la Collins, Duncan, Joglekar; without gauge fixing term, since we calculate a scattering process).

```
t1 = - MetricTensor[α, β] LQCD - F[α, σ][k, r, s] F[β, σ][k, z, w];
t2 = 1/4 I (φ[Bar[ψ]] . γ[α] . φ[P[β], ψ] -
      I gs . φ[Bar[ψ]] . γ[α] . T[a] . φ[A, ℓ[β], S[a]] . φ[ψ] -
      (φ[P[β], Bar[ψ]] + I gs . φ[Bar[ψ]] . T[a] . φ[A, ℓ[β], S[a]]) . γ[α] . φ[ψ]);
t3 = φ[P[α], Bar[λ], S[a]] (φ[P[β], λ, S[a]] - gs . f[a, b, c] φ[A, ℓ[β], S[c]] φ[λ, S[b]]);
```

Denote the graviton field by h. LG is the lagrangian for the graviton field coupled to the enerrgy-momentum tensor. G is the gravity coupling constant. For convenience we provide a factor i..

```
LG = I G . φ[h, ℓ[α], ℓ[β]] (t1 + t2 + t3 + ReplaceAll[ t2 + t3, {α :> β, β :> α}] )
```

$$i G\, h_{\alpha\beta}\Bigg(\frac{i}{4}(\overline{\psi}\,\gamma^{\alpha}\,\partial_{\beta}\psi - (i g_s\,\overline{\psi}\,T_a A^a_{\beta} + \partial_{\beta}\overline{\psi})\,\gamma^{\alpha}\,\psi + -i g_s\,\overline{\psi}\,\gamma^{\alpha}\,T_a A^a_{\beta}\,\psi) +$$

$$\frac{i}{4}(\overline{\psi}\,\gamma^{\beta}\,\partial_{\alpha}\psi - (i g_s\,\overline{\psi}\,T_a A^a_{\alpha} + \partial_{\alpha}\overline{\psi})\,\gamma^{\beta}\,\psi + -i g_s\,\overline{\psi}\,\gamma^{\beta}\,T_a A^a_{\alpha}\,\psi) -$$

$$(-g_s f_{rsk} A^r_{\alpha} A^s_{\sigma} + \partial_{\alpha} A^k_{\sigma} - \partial_{\sigma} A^k_{\alpha})\left(-g_s f_{zwk} A^z_{\beta} A^w_{\sigma} + \partial_{\beta} A^k_{\sigma} - \partial_{\sigma} A^k_{\beta}\right) +$$

$$\partial_{\beta}\,\overline{\lambda}^{a}\,(\partial_{\alpha}\lambda^a - g_s f_{abc}\lambda^b A^c_{\alpha}) + \partial_{\alpha}\,\overline{\lambda}^{a}\,(\partial_{\beta}\lambda^a - g_s f_{abc}\lambda^b A^c_{\beta}) -$$

$$g^{\alpha\beta}\Bigg(\partial_{\mu}\,\overline{\lambda}^{a}\left(-g_s f_{abc} A^c_{\mu}\lambda^b + \partial_{\mu}\lambda^a\right) + \frac{i}{2}(\overline{\psi}\,\gamma^{\mu}\,\partial_{\mu}\psi - \partial_{\mu}\overline{\psi}\,\gamma^{\mu}\,\psi) -$$

$$\frac{\left(-g_s f_{abk} A^a_{\mu} A^b_{\nu} + \partial_{\mu} A^k_{\nu} - \partial_{\nu} A^k_{\mu}\right)\left(-g_s f_{cdk} A^c_{\mu} A^d_{\nu} + \partial_{\mu} A^k_{\nu} - \partial_{\nu} A^k_{\mu}\right)}{4} +$$

$$g_s\,\overline{\psi}\,T_a\,\gamma^{\mu}\,\psi A^a_{\mu}\Bigg)\Bigg)$$

■ Calculation of the Feynman rules

□ Gluon - Gluon - Gluon

```
f3 = {φ[A, ℓ[α], S[a]][p1], φ[A, ℓ[μ], S[b]][p2], φ[A, ℓ[ν], S[c]][p3]}
```

$$\{A^a_{\alpha}(p_1), A^b_{\mu}(p_2), A^c_{\nu}(p_3)\}$$

```
g3 = FeynRule[LQCD, f3]
```

$$f_{abc}\,(g^{\alpha\mu}(p_1 - p_2)^{\nu} - g^{\alpha\nu}(p_1 - p_3)^{\mu} + g^{\mu\nu}(p_2 - p_3)^{\alpha})\, g_s$$

□ Gluon - Gluon - Gluon - Gluon

```
g4 = FeynRule[LQCD, Append[f3, φ[A, ℓ[ρ], S[d]][p4]]]
```

$$i(f_{abk} f_{cdk}\,(g^{\alpha\rho} g^{\mu\nu} - g^{\alpha\nu} g^{\mu\rho}) + f_{ack} f_{bdk}\,(g^{\alpha\rho} g^{\mu\nu} - g^{\alpha\mu} g^{\nu\rho}) + f_{adk} f_{bck}\,(g^{\alpha\nu} g^{\mu\rho} - g^{\alpha\mu} g^{\nu\rho}))\, {g_s}^2$$

□ Gluon - Antighost - Ghost

```
g5 = FeynRule[LQCD, {φ[A, ℓ[μ], a1][p1], φ[Bar[λ], a2][p2], φ[λ, a3][p3]}]
```

$$f_{a_1 a_2 a_3}\, p_2^{\mu}\, g_s$$

☐ **Graviton - Gluon - Gluon**

c3 = {φ[h, ℓ[α], ℓ[β]] [q], φ[A, l1, a1][p1], φ[A, l2, a2][p2]}

$\{h_{\alpha\beta}(q), A^{a_1}_{\lambda_1}(p_1), A^{a_2}_{\lambda_2}(p_2)\}$

Ggg = FeynRule[LG, c3]

$$-G\delta_{a_1 a_2}\Big((p_1\cdot p_2)g^{\alpha\lambda_2}g^{\beta\lambda_1}+(p_1\cdot p_2)g^{\alpha\lambda_1}g^{\beta\lambda_2}-(p_1\cdot p_2)g^{\alpha\beta}g^{\lambda_1\lambda_2}-g^{\beta\lambda_1}p_1^{\lambda_2}p_2^{\alpha}-$$
$$g^{\alpha\lambda_1}p_1^{\lambda_2}p_2^{\beta}+g^{\lambda_1\lambda_2}\big(p_1^{\beta}p_2^{\alpha}+p_1^{\alpha}p_2^{\beta}\big)-g^{\beta\lambda_2}p_1^{\alpha}p_2^{\lambda_1}-g^{\alpha\lambda_2}p_1^{\beta}p_2^{\lambda_1}+g^{\alpha\beta}p_1^{\lambda_2}p_2^{\lambda_1}\Big)$$

☐ **Graviton - Gluon - Gluon - Gluon**

c4 = Append[c3, φ[A, l3, a3][p3]]; Gggg = FeynRule[LG, c4]

$$iGf_{a_1a_2a_3}\,(g^{\beta\lambda_3}g^{\lambda_1\lambda_2}(p_1-p_2)^{\alpha}+g^{\alpha\lambda_3}g^{\lambda_1\lambda_2}(p_1-p_2)^{\beta}+g^{\alpha\lambda_2}g^{\beta\lambda_1}(p_1-p_2)^{\lambda_3}+$$
$$g^{\alpha\lambda_1}g^{\beta\lambda_2}(p_1-p_2)^{\lambda_3}-g^{\alpha\beta}g^{\lambda_1\lambda_2}(p_1-p_2)^{\lambda_3}-g^{\beta\lambda_2}g^{\lambda_1\lambda_3}(p_1-p_3)^{\alpha}-$$
$$g^{\alpha\lambda_2}g^{\lambda_1\lambda_3}(p_1-p_3)^{\beta}-g^{\alpha\lambda_3}g^{\beta\lambda_1}(p_1-p_3)^{\lambda_2}-g^{\alpha\lambda_1}g^{\beta\lambda_3}(p_1-p_3)^{\lambda_2}+$$
$$g^{\alpha\beta}g^{\lambda_1\lambda_3}(p_1-p_3)^{\lambda_2}+g^{\beta\lambda_1}g^{\lambda_2\lambda_3}(p_2-p_3)^{\alpha}+g^{\alpha\lambda_1}g^{\lambda_2\lambda_3}(p_2-p_3)^{\beta}+$$
$$g^{\alpha\lambda_3}g^{\beta\lambda_2}(p_2-p_3)^{\lambda_1}+g^{\alpha\lambda_2}g^{\beta\lambda_3}(p_2-p_3)^{\lambda_1}-g^{\alpha\beta}g^{\lambda_2\lambda_3}(p_2-p_3)^{\lambda_1})$$
$$g_s$$

☐ **Graviton - Gluon - Gluon - Gluon - Gluon**

c5 = Append[c4, φ[A, l4, a4][p4]]; Ggggg = FeynRule[LG, c5]

$$G\,(f_{ka_1a_2}f_{ka_3a_4}\,(g^{\alpha\lambda_2}\,(g^{\beta\lambda_4}g^{\lambda_1\lambda_3}-g^{\beta\lambda_3}g^{\lambda_1\lambda_4})+$$
$$g^{\alpha\lambda_4}\,(g^{\beta\lambda_2}g^{\lambda_1\lambda_3}-g^{\beta\lambda_1}g^{\lambda_2\lambda_3})-g^{\alpha\lambda_3}\,(g^{\beta\lambda_2}g^{\lambda_1\lambda_4}-g^{\beta\lambda_1}g^{\lambda_2\lambda_4})-$$
$$g^{\alpha\lambda_1}\,(g^{\beta\lambda_4}g^{\lambda_2\lambda_3}-g^{\beta\lambda_3}g^{\lambda_2\lambda_4})+g^{\alpha\beta}\,(g^{\lambda_1\lambda_4}g^{\lambda_2\lambda_3}-g^{\lambda_1\lambda_3}g^{\lambda_2\lambda_4}))+$$
$$f_{ka_1a_3}f_{ka_2a_4}\,(g^{\alpha\lambda_3}\,(g^{\beta\lambda_4}g^{\lambda_1\lambda_2}-g^{\beta\lambda_2}g^{\lambda_1\lambda_4})+$$
$$g^{\alpha\lambda_4}\,(g^{\beta\lambda_3}g^{\lambda_1\lambda_2}-g^{\beta\lambda_1}g^{\lambda_2\lambda_3})-g^{\alpha\lambda_2}\,(g^{\beta\lambda_3}g^{\lambda_1\lambda_4}-g^{\beta\lambda_1}g^{\lambda_3\lambda_4})-$$
$$g^{\alpha\lambda_1}\,(g^{\beta\lambda_4}g^{\lambda_2\lambda_3}-g^{\beta\lambda_2}g^{\lambda_3\lambda_4})+g^{\alpha\beta}\,(g^{\lambda_1\lambda_4}g^{\lambda_2\lambda_3}-g^{\lambda_1\lambda_2}g^{\lambda_3\lambda_4}))+$$
$$f_{ka_1a_4}f_{ka_2a_3}\,(g^{\alpha\lambda_4}\,(g^{\beta\lambda_3}g^{\lambda_1\lambda_2}-g^{\beta\lambda_2}g^{\lambda_1\lambda_3})+$$
$$g^{\alpha\lambda_3}\,(g^{\beta\lambda_4}g^{\lambda_1\lambda_2}-g^{\beta\lambda_1}g^{\lambda_2\lambda_4})-g^{\alpha\lambda_2}\,(g^{\beta\lambda_4}g^{\lambda_1\lambda_3}-g^{\beta\lambda_1}g^{\lambda_3\lambda_4})-$$
$$g^{\alpha\lambda_1}\,(g^{\beta\lambda_3}g^{\lambda_2\lambda_4}-g^{\beta\lambda_2}g^{\lambda_3\lambda_4})+g^{\alpha\beta}\,(g^{\lambda_1\lambda_3}g^{\lambda_2\lambda_4}-g^{\lambda_1\lambda_2}g^{\lambda_3\lambda_4})))$$
$$g_s^{\,2}$$

☐ **Graviton - Antiquark - Quark**

c6 = {φ[h, ℓ[α], ℓ[β]][q], φ[Bar[ψ]][p2], φ[ψ][p1]}

$\{h_{\alpha\beta}(q), \overline{\psi}(p_2), \psi(p_1)\}$

Gqq = FeynRule[LG, c6]

$$\frac{-G\big(-\gamma^{\beta}p_1^{\alpha}-\gamma^{\alpha}p_1^{\beta}+\gamma^{\beta}p_2^{\alpha}+\gamma^{\alpha}p_2^{\beta}+2\,g^{\alpha\beta}\,(p\!\!\!/_1-p\!\!\!/_2)\big)}{4}$$

☐ **Graviton - Antiquark - Quark - Gluon**

c7 = Append[c6, φ[A, ℓ[λ], S[a]][p3]]

$\{h_{\alpha\beta}(q), \overline{\psi}(p_2), \psi(p_1), A^{a}_{\lambda}(p_3)\}$

Gqqg = FeynRule[LG, c7]

$$\frac{G\,(2\gamma^{\lambda}g^{\alpha\beta}-\gamma^{\beta}g^{\alpha\lambda}-\gamma^{\alpha}g^{\beta\lambda})\,T_a\,g_s}{2}$$

☐ **Graviton - Antighost - Ghost**

c8 = {φ[h, ℓ[α], ℓ[β]][q], φ[Bar[λ], a2][p2], φ[λ, a1][p1]}

$\{h_{\alpha\beta}(q), \overline{\lambda}^{a_2}(p_2), \lambda^{a_1}(p_1)\}$

Guu = FeynRule[LG, c8]

$$G\delta_{a_1a_2}\big(-(p_1\cdot p_2)\,g^{\alpha\beta}+p_1^{\beta}p_2^{\alpha}+p_1^{\alpha}p_2^{\beta}\big)$$

☐ **Graviton - Antighost - Ghost - Gluon**

c9 = {φ[h, ℓ[α], ℓ[β]][q], φ[Bar[λ], a2][p2], φ[λ, a1][p1], φ[A, l3, a3][p3]}

$\{h_{\alpha\beta}(q), \overline{\lambda}^{a_2}(p_2), \lambda^{a_1}(p_1), A^{a_3}_{\lambda_3}(p_3)\}$

Gguu = FeynRule[LG, c9]

$$-iGf_{a_1a_2a_3}\big(g^{\beta\lambda_3}p_2^{\alpha}+g^{\alpha\lambda_3}p_2^{\beta}-g^{\alpha\beta}p_2^{\lambda_3}\big)g_s$$

■ Timing

The total time needed is

TimeUsed[]//FeynCalcForm

6.8 min

Appendix B

Lowest order parton structure functions

Unpolarized structure function

```
<< ~/HighEnergyPhysics/FeynCalc.m;
```

Get the amplitude for the Graviton-gluon-gluon vertex from a file produced by FeynArts1.0 (with the coupling constant G set to 1; the graviton indices GRmu and GRnu are replaced by mu and nu).

```
A = << ~/HighEnergyPhysics/work/grgg.amp; A =
    ReplaceAll[A, {GRmu :> μ, GRnu :> ν, li2 :> κ, li4 :> ξ, gi2 :> a, gi4 :> b}]
```

FeynAmpList[Process ->

{{GR[1], q, Sqrt[qq]}, {G[1], p, 0}} -> {{G[1], l, 0}}]

(FeynAmp($i\epsilon^{\kappa}(p)\epsilon^{\xi}(l)\delta_{ab}(-(l\cdot p)g^{\mu\xi}g^{\nu\kappa}-(l\cdot p)g^{\mu\kappa}g^{\nu\xi}+g^{\nu\xi}l^{\kappa}p^{\mu}-g^{\kappa\xi}l^{\nu}p^{\mu}+$

$g^{\mu\xi}l^{\kappa}p^{\nu}-g^{\kappa\xi}l^{\mu}p^{\nu}+g^{\nu\kappa}l^{\mu}p^{\xi}+g^{\mu\kappa}l^{\nu}p^{\xi}+g^{\mu\nu}((l\cdot p)g^{\kappa\xi}-l^{\kappa}p^{\xi}))$))

In front of the amplitude the information reads: Graviton(q) + Gluon(p) –> Gluon(l).
Furthermore q^2 = qq, p^2=0, l^2 = 0 is understood (and will be set later by SquareAmplitude).

```
ComplexIndex[μ] = λ; ComplexIndex[ν] = σ ; g = MetricTensor;
```

The projection factor multiplying the squared matrix element is : (qq = q^2; CA and CF are the Casimir operators of SU(N)).

$$a_g = \left(\frac{1}{2}\,\frac{1}{2\,\mathrm{CA\,CF}}\,\frac{2\,\mathrm{Pi}}{2\,\mathrm{ScalarProduct}[p,\,q]}\,\delta[1-x]\right.$$

$$\left.\frac{2}{\mathrm{qq}}\left(\frac{1}{4\,\mathrm{Pi}}\,\frac{1}{\left(1+\frac{\epsilon}{2}\right)}\,\frac{1}{\left(1+\frac{3}{4}\epsilon\right)}\right)g[\mu,\lambda]\;g[\nu,\sigma]\right)$$

$$\frac{-\delta(1-x)\,g^{\mu\lambda}g^{\nu\sigma}}{8\left(1+\frac{\epsilon}{2}\right)\left(1+\frac{3\epsilon}{4}\right)\mathrm{qq}\,(p\cdot q)\,C_A\,C_F}$$

This squares the amplitude A, performing polarization sums automatically, according to the information in the "Process -> ..." rule in A.

```
Fg = SquareAmplitude[
    A, Dimension -> D, SpinSumExternalMomentum -> 0,
      ExtraFactor -> ag , FinalSubstitutions -> {D -> (4 + ε)}]
```

$\delta(1-x)$

Polarized structure function

Calculate the polarized case without the function SquareAmplitude.

Introduce Lorentz indices alpha and beta for the gluons (amputate by hand).

```
B = ReplaceAll[A/Last/Last, {PolarizationVector[l, ξ,_] :> g[β, ξ],
      PolarizationVector[p, κ,_] :> g[α, κ], SUNDelta[_, _] :> 1}]
```

$$i\,g^{\alpha\kappa}g^{\beta\xi}(-(l\cdot p)g^{\mu\xi}g^{\nu\kappa}-(l\cdot p)g^{\mu\kappa}g^{\nu\xi}+g^{\nu\xi}l^{\kappa}p^{\mu}-g^{\kappa\xi}l^{\nu}p^{\mu}+$$
$$g^{\mu\xi}l^{\kappa}p^{\nu}-g^{\kappa\xi}l^{\mu}p^{\nu}+g^{\nu\kappa}l^{\mu}p^{\xi}+g^{\mu\kappa}l^{\nu}p^{\xi}+g^{\mu\nu}((l\cdot p)g^{\kappa\xi}-l^{\kappa}p^{\xi}))$$

```
ComplexIndex[α] = η ; ComplexIndex[β] = θ ; LC[x_, y_] :=
   Eps[LorentzIndex[x], LorentzIndex[y], Momentum[p], Momentum[q]];
```

The projection factor multiplying the squared matrix element is: given by:

$$b_g = \mathrm{ChangeDimension}\Big[$$

$$\left(\frac{1}{2}\,\frac{2\,\mathrm{Pi}}{2\,\mathrm{ScalarProduct}[p,\,q]}\,\delta[1-x]\right)\left(\frac{-1}{4\,\mathrm{Pi}}\,\frac{1}{\left(1+\frac{\epsilon}{2}\right)}\,\frac{(-32)}{(1+\epsilon)\,(4+\epsilon)}\,\frac{1}{\mathrm{qq}^3}\right)$$

$$\mathrm{LC}[\mu,\lambda]\;\mathrm{LC}[\alpha,\eta]\,g[\nu,\sigma]\,(-g[\beta,\theta]),\,D\Big]$$

$$\frac{8\,\delta(1-x)\,\epsilon_D^{\alpha\eta pq}\,\epsilon_D^{\mu\lambda pq}\,g_D^{\beta\theta}\,g_D^{\nu\sigma}}{\left(1+\frac{\epsilon}{2}\right)(1+\epsilon)(4+\epsilon)\,\mathrm{qq}^4}$$

```
B  = Contract[ ChangeDimension[B, D] ];
BC = ComplexConjugate[B];
w  = Collect2[Contract[B bg , EpsContract -> False], LorentzIndex];
Fg = Factor2[ReplaceAll[ Contract[Contract[w, BC]] , D -> (4 + ε)]]
```

$\delta(1-x)$

The reproduction of the correct lowest order structure function insures the correctness of the projecton factors.

Timing

```
TimeUsed[]/FCF
```

3.4 min

Appendix C

Below the *FeynCalc* listings for the determination of the polarized and unpolarized P_{gg} are given. Technical detail questions may be addressed to: rolfm@rulgm4.leidenuniv.nl

```
(* ************************************************************* *)
(* grggg.m *)
(* squaring Graviton + gluon --> gluon gluon *)

(* unpolarized case *)

(* load the 4 amplitudes produced by FeynArts1.0 *)
ammp = <<grggg.amp;
$VeryVerbose = 2;

(* averaging over initial quantum numbers:  1/2 *)
project = 1/2 /(2 CA CF) (* = 1/2 / (N^2-1) *);

ComplexIndex[GRmu] = GRla; ComplexIndex[GRnu] = GRsi;

(* the modified DIS-projection formula  *)
project = project (-1/(4 Pi)) 1/(1+epsilon/2) 1/(1+3/4 epsilon) (
           MetricTensor[GRmu, GRla, Dimension -> D]*
            MetricTensor[GRnu, GRsi, Dimension -> D]
                                                    );
(* the extra factor 2/q2 *)
project = project (2/q2);

SetOptions[SquareAmplitude, Factoring -> True,
                            ExtraFactor -> project ,
                            Dimension -> D
          ];

(* square it *)
rr = SquareAmplitude[ammp];

Write2["grggg.res", grggg = rr];

(* ************************************************************* *)

(* grgggps.m *)

(* perform the 2-->2 phase space integration for G + g--> g + g *)
<<psidis.m;
(*
 Process ->
{{GR[1], q, q2^(1/2)}, {G[1], p1, 0}} -> {{G[1], p2, 0}, {G[1], k, 0}}
*)
$VeryVerbose=2;
SetMandelstam[S, T, U, q,p1,-p2,-k, Sqrt[q2],0,0,0];
Get["grggg.res"];

ngrggg = PSDISIntegrate[grggg, T,U,{q,p1} -> {k,p2}, D];
ngrggg = Collect2[ngrggg /. D -> (4+epsilon), epsilon];
ngrggg = Series[ngrggg, {epsilon, 0, 1}]//Normal;
ngrggg = Collect2[ngrggg, epsilon];
ngrggg = Factor2[Map[#/.Plus->pluss&,ngrggg]]/.pluss->Plus
```

```
(* grgggpol.m     ( Graviton + gluon ---> gluon + gluon *)
(* polarized case*)

(* Process ->
{{GR[1], q, q2^(1/2)}, {G[1], p1, 0}} -> {{G[1], p2, 0}, {G[1], k, 0}} *)

$VeryVerbose = 2;
GG = GS = 1;
(* amputate *)
Pair[a_, Momentum[Polarization[p1,_,gli_SUNIndex]]] :=
  Pair[a, LorentzIndex[AL]] SUNDelta[SUNIndex[GLp1], gli];
Pair[a_, Momentum[Polarization[p2,_,gli_SUNIndex]]] :=
  Pair[a, LorentzIndex[MUp2]] SUNDelta[SUNIndex[GLp2], gli];
Pair[a_, Momentum[Polarization[k,_,gli_SUNIndex]]] :=
  Pair[a, LorentzIndex[MUk]] SUNDelta[SUNIndex[GLk], gli];
ComplexIndex[AL] = ALC;
ComplexIndex[MUp2] = MUp2C; ComplexIndex[MUk] = MUkC;
ComplexIndex[GRmu] = GRla;  ComplexIndex[GRnu] = GRsi;
ammp = <<grggg.amp;
ammp = Plus @@ Map[Last, ammp];
ammp = ammp /. Conjugate -> Identity;
ammp = PropagatorDenominatorExplicit[ammp];
ammp = ChangeDimension[ammp, D];
SetMandelstam[S, T, q2 - S - T, q, p1, -p2, -k, Sqrt[q2], 0, 0, 0];

(* averaging over initial quantum numbers:  1/2 *)
project = 1/2 /(2 CA CF) (* = 1/2 / (N^2-1) *);

(* identical particles in the final state *)
project = project / 2;
project =  project (-1/(4 Pi)) 1/(1+epsilon/2)  (
           LeviCivita[GRmu, GRla, mupe1, muq1, Dimension -> D]*
           LeviCivita[AL, ALC, mupe2, muq2, Dimension -> D]*
           FourVector[p1, mupe1] FourVector[q, muq1] *
           FourVector[p1, mupe2] FourVector[q, muq2] *
           MetricTensor[GRnu, GRsi, Dimension -> D] *
           PolarizationSum[MUp2, MUp2C(*, p2, n*), Dimension -> D] *
           PolarizationSum[MUk, MUkC(*, k, n*),    Dimension -> D]);
(* the extra factor *)
project = project (-32/((1 + epsilon)*(4 + epsilon)*q2*
                   (2 ScalarProduct[q, p1])^2 ));
(* start calculation ************************************** *)
project = Factor2[ Contract[project, EpsContract->False]];
   ammp = Contract[ ammp, Expanding -> False ];
    am = ExpandScalarProduct[ammp /. k -> (q + p1 - p2)];
    am = Collect2[am, LorentzIndex];
 cammp = ComplexConjugate[am];
nam = 0; lam = Length[am];
Print["lam = ",lam];
For[i = 1, i <= lam, i++, Write["stdout", i, "|"];
    nam = nam + Contract[am[[i]] project, Expanding -> False,
                         EpsContract -> False];
   ];

re = Contract[nam , cammp ];
Print["contdone"];
If[!FreeQ[re, LorentzIndex],
   re = Collect2[re, LorentzIndex, Factoring -> False];
   re = Contract[re]
  ];
rex = Map[Factor2, re]//Factor2
Print["rex = ",rex];
SetOptions[SUNSimplify, Factoring -> True];
rex = Factor2[ SUNSimplify[rex]/.D -> (4+epsilon)];
Write2["grgggpol.res", grgggpol  == rex];
```

DEVELOPMENT OF THE CompHEP PROJECT

A.E.Pukhov
Institute of Nuclear Physics, Moscow State University
119899 Moscow, Russia

1. Introduction

CompHEP program [1] is created for calculation of decay and high energy collision processes for elementary particles. It is a menu-driven system with text/graphical output of information and context help facility oriented on an user not experienced in programming. User input is checked for correctness. Notations used in CompHEP are very similar to those used in particle physics.

All computations are performed in tree approximation. Total number of 'in' and 'out' particles must not exceed 6. Squared diagrams method is used for calculation. Polarization of particles is not taken into account. However the left- and right-handed states are available for massless fermions.

CompHEP has four built-in models. One of them is the Standard Model (SU(3) x SU(2) x U(1)) for strong and electroweak interactions [2]. The Standard Model is available for calculations in physical (utitary) gauge and in 'tHooft - Feynman gauge.

User can change particle list, interaction vertices and model parameters.

Creation by the user of a new particle interaction model is possible.

CompHEP uses a specialized built-in symbolic package worked out specially for calculations in high energy physics. This package ensures fast calculation of squared matrix elements.

CompHEP generates FORTRAN code for further numerical calculations and REDUCE, FORM or MATHEMATICA codes for further analytical manipulations.

Project CompHEP was started in 1989.

The first versions of CompHEP were written on Turbo Pascal for IBM compatible PC. Now the programs are rewritten on C programming language and the installation on UNIX workstations is available. On UNIX workstations X11 library is used to support user-friendly interface.

CompHEP group now consists of the following programmers and physicists:

P.Baikov, E.Boos, M.Dubinin, V.Edneral, V.Ilyin, D.Kovalenko, A.Kryukov, A.Pukhov, S.Shichanin, A.Semenov.

2. CompHEP menus

CompHEP is a menu-driven system. Here we list the main menus that drive the process of squared matrix element calculation.

The first menu is

Menu 1
QED
Fermi Model
St.model (phys.gauge)
St.model (Feyn.gauge)
N E W M O D E L

This menu gives user a possibility to select a model of elementary particle interaction.

Here:

"QED" denotes Quantum Electrodynamics;

"Fermi" denotes QED with four-fermion weak interaction model. The interaction of fermion currents is implemented through auxiliary particles (W and Z bosons) having constant propagators;
"St.model (phys. gauge)" denotes the Standard Model with electroweak sector in physical (unitary) gauge and QCD in Feynman gauge;
"St.model (Feynman gauge) denotes the Standard Model with electroweak sector in 'tHooft - Feynman gauge and QCD in Feynman gauge;
"N E W M O D E L" is the option for creating of the new physical model by the user .

After the model selection user enters the following menu:

Menu 2

Enter process
View and Edit model
Delete changes

"Delete model" option removes the model created with the help of "N E W M O D E L" option from the previous menu, or reconstruct the built-in model if it had been changed with the help of "View and Edit model" option.

"View and Edit model" option allows user to view and change current physical model.

"Enter process" - option to enter the process from the keyboard. After the input of this command the list of particles will be shown contained in the model together with the corresponding notation conventions.

If the process input is correct, CompHEP constructs the corresponding Feynman diagrams and user enters the next menu.

Menu 3

Squaring
View diagrams

"View diagrams" - view graphic representation of constructed Feynman diagrams. Here user has a possibility to exclude some diagrams from further processing.

"Squaring" - generation of diagrams for squared S-matrix element.

When squaring is finished the main resulting menu is displayed:

Menu 3

View squared diagrams
Symbolic calculations
Write results
REDUCE program
Enter new process
Interface

Commands of main resulting menu:

"Symbolic calculations" - choose this option to start symbolic calculation for the squared diagrams generated. In the CompHEP system the calculation is done by the built-in symbolic manipulation package;

"Write results" - it is possible to write obtained analytical result for squared diagrams on hard disk in different formats, the corresponding files are placed in the subdirectory RESULTS. There are available FORTRAN, FORM, REDUCE and MATHEMATICA formats.

"REDUCE program" - this command generates for each squared diagram source code in REDUCE format for the following calculation of the matrix element squared by means of REDUCE. This is an alternative way to get symbolical results. This option may be used to check the build-in symbolic calculator. Physical model can be checked effectively with the help of this option. For this user can check by hand the REDUCE program which is direct realization of Feynman rules for squared diagrams.

"Enter new process" - return to the Menu 2 for input new collision or decay process.

"Interface" - it is possible to incorporate the CompHEP session with the work with another external package.

3. Structure of physical models in CompHEP

Physical model in CompHEP consists of four tables:
"Parameters" - list of parameters of the particles and vertices;
"Constraints" - shows the functional relations between the dependent parameters;
"Particles" - list of particles;
"Lagrangian" - list of vertices.

3.1. The table "Parameters"

All parameters used in the calculation must be described in "Parameters" table. This table consists of three fields: "Name", "Value" and "Comment". Example:

Name	Value	Comment
EE	0.313	electromagnetic coupling constant
GG	1.373	strong coupling constant
Me	0.511E-03	mass of the electron
SW	0.48	sin of the Weinberg angle
...	...	...

"Name" is the name of the parameter used for further model definition and symbolic results representation.

"Value" is numerical value of the parameter used for numerical calculations.

"Comment" is an commentary what the corresponding parameter means.

3.2. The table "Constraints"

Describes variables that are a function from parameters. For instance:

Name	Expression	Comment
CW	Sqrt(1-SW**2)	cos of the Weinberg angle
MW	MZ*CW	W boson mass
...	...	...

3.3. The table "Particles"

The "Particles" table consists of 8 fields:

Full name	P	aP	2*Spin	Mass	Width	Color	Aux
Electron	e1	E1	1	Me	0	1	
Neutrino	n1	N1	1	0	0	1	L
Z-boson	Z	Z	2	MZ	wZ	1	G
Gluon	G	G	2	0	0	8	G
Higgs	H	H	0	MH	wH	1	
...	...	...	...	...	...	...	...

"Full name" — full name of the particle. This field is only for user's convenience. Symbolic names of the particle and antiparticle used in the programs are stored in the next two fields.

"P", "aP" — these fields contain short names (not more than 2 symbols in each one) of the particle and antiparticle correspondingly. If the antiparticle is identical to particle the name of antiparticle must be the same.

"2*Spin" — double spin of the particle. There are available only 0, 1 and 2 values here (scalar, spinor and vector cases). This field defines the numerator of the corresponding propagator and in- and out- states.

"Mass" and "Width" — particle mass and width identificators. The values of these symbolic parameters have to be entered in the tables "Parameters" or "Constrains". There is possible to enter zero in these fields. In this case the corresponding parameter will not appear in the symbolic results.

"Color" — dimension of color group SU(3) representation: 3 for quarks, 8 for gluons, 1 for colorless particles. This field is used for squared diagram color weight evaluation.

"Aux" — special auxiliary field: "L", "R" are marks for left/right massless spinor particles, "G" is a mark for massive vector particle will be treated through gauge field in 'tHooft - Feynman gauge, "*" is a mark for massive vector particle will have constant propagator $\frac{1}{M^2}$. This option is used for implementation of four-fermion vertex from Fermi model of electroweak interaction.

Propagators of any particle are generated by CompHEP automatically. CompHEP uses the standard form for scalar, spinor and vector particle propagator [2]. However the field "Aux" provides user with some possibility to change propagators.

3.4. The table "Lagrangian"

This table is the list of vertices.

A1	A2	A3	A4	Factor	Lorentz part
U	u	G		GG	G(m3)
H	Z	Z		EE*MW/SW	m2.m3
H	H	Z	Z	(1/2)*(EE/(SW*CW))**2	m3.m4
E1	e1	Z		EE/(4*CW*SW)	G(m3)*((1-G5)-4*SW**2)
...	...	...	...	...	...

First four fields ("A1", "A2", "A3", "A4") contain the names of interacting particles. The last two fields "Factor" and "Lorentz part" define the vertex itself:

$$\frac{\delta^4 L}{\delta A_1(p1)\delta A_2(p2)\delta A_3(p3)\delta A_4(p4)} = Colorfactor * "Factor" * "Lorentzpart"$$

"Factor" is the scalar factor dependent only from model variables. It can not include summation of terms but only single term.

"Lorentz part" depends on model variables, Lorentz momenta (p1, p2, p3 and p4) and Lorentz indices (m1, m2, m3 and m4) corresponding particles listed in first four fields: p1 and m1 for first particle etc.

Here dot symbol "." is used for scalar product of Lorentz momenta ($p1.p2 = g_{\mu\nu}p1^\mu p2^\nu$) or indices ($p1.m1 = p1_{\mu_1}$, $m1.m2 = g_{\mu_1\mu_2}$); G denotes Dirac's gamma matrix ($G(m3) = \gamma^{\mu_3}$); G5 is gamma5 matrix ($G5 = \gamma^5$, $\gamma^5\gamma^5 = 1$, $\gamma^\mu\gamma^5 + \gamma^5\gamma^\mu = 0$).

Vertex does not include the QCD color structure. The corresponding factor is generated automatically during the symbolic calculations. It depends on particle's color weight. This agreement makes impossible to introduce in the table "Lagrangian" the 4-gluon vertex explicitly and display it on the screen. This vertex (and corresponding diagrams) is generated automatically in some special way during symbolic calculation of squared diagrams.

There is impossible now to introduce the imaginary unit i in the table "Lagrangian". The imaginary units appeared in vertices of standard Feynman rules typed in the textbooks are eliminated in CompHEP by special renormalization of propagators. This is possible to do if the Lagrangian is invariant under CP transformation. If this symmetry is violated (for example through the introducing of imaginary factors in Kobayashi - Maskawa mixing matrix there is no any renormalization which eliminates imaginary unit from the table "Lagrangian".

4. 'tHooft - Feynman gauge in CompHEP

In the case of physical (unitary) gauge the propagator for massive vector particles is

$$-\frac{g_{\mu_1\mu_2} - \frac{k_{\mu_1}k_{\mu_2}}{M^2}}{M^2 - k^2} = -(m1.m2 - k.m1 * k.m2/M * *2)/(M * *2 - k.k)$$

The term $\frac{k_{\mu_1}k_{\mu_2}}{M^2}$ leads to large (huge in some real cases) size of symbolic expressions for squared diagrams: these formulas will have nonunitary terms which cancel each other after summation over all squared diagrams of the process. However there is impossible to do this summation symbolically in physically interesting cases because of restricted memory. There more the numerical evaluation of matrix element is much preferable to do by the evaluation of separate squared diagrams but not of whole symbolic answer for summed up matrix element. In the last case the reducing to common denumerator of terms with different denumerators produces an expression with huge numerator which is not optimal for numerical evaluation.

The 'tHooft - Feynman gauge solves this problem in optimal way. In this case nonunitary terms will not be produced at all. However there are new vertices in the lagrangian in comparison with unitary gauge: vertices with goldstone partners of W and Z bosons (unphysical particles originated from gauge components of goldstone dublet field which provide masses of W and Z bosons) and Faddeev-Popov ghosts. In CompHEP goldstone partners are denoted as "W+.f", "W-.f" and "Z.f", and Faddeev-Popov ghosts and antighosts are denoted as ("W+.c", "W+.C"), ("W-.c", "W-.C"), ("Z.c", "Z.C") and ("A.c", "A.C"). In CompHEP the hermitian conjugation is defined in such a way that "W+.c" is antiparticle to "W-.c", "W+.C" is antiparticle to "W-.C", "Z.c" and "Z.C" (also "A.c" and "A.C") are antiparticles to their self. In the same time ".c" and ".C" ghosts have opposite fermion number.

Goldstone partners can appear in Feynman diagrams in the calculations of physical processes in 'tHooft - Feynman gauge even at tree level approximation. There more goldstone partners and Faddeev-Popov ghosts appear in the process of CompHEP calculation of squared diagrams because of Lorentz invariant procedure of summation over polarizations of in- and out-states of W, Z and A. The prescription consists of the replacement of the summation over physical polarizations on the summation over all (including unphysical states) polarizations. In unitary gauge this prescription produces correct answers. However in the case of 'tHooft-Feynman gauge there is necessary to generate additional diagrams with goldstone partners and Faddeev-Popov ghosts to cancel the contributions of unphysical polarizations. This prescription is Lorentz invariant procedure in comparison with summation over physical polarizations. In this prescription the line connected in- and out-states of bosons is replaced by the metric tensor $g_{\mu\nu}$

In addition, the original diagram generates diagrams with the same structure but with some lines (W, Z or A ones) replaced on goldstone partners and Faddeev - Popov ghosts. All possible diagrams generated by these replacements have to be taken into account. This prescription is based on the consideration of [3].

The essential point is that propagators for the goldstone partners and Faddeev - Popov ghosts are the same as for the corresponding vector boson for 'tHooft - Feynman gauge. So we can sum up the contributions of these diagrams without increasing significantly the final expression.

Effectively this means that number of squared diagrams is the same as in unitary gauge.

5. Run time estimations for numerical calculations

For $2 \to 3$ and $2 \to 4$ process types the numerical calculation by means of method Monte-Carlo is a standard way to get physical results. There is necessary large time for such calculations. So it is important to estimate number of arithmetic operations necessary for the evaluation of typical squared diagram.

In this section we estimate the upper limit for number of arithmetic operations necessary for one squared diagram. If we multiply this estimation on number of squared diagrams, than on the number of calls during Monte Carlo integration process and divide on CPU productivity then we get estimation for run time necessary for obtaining of the corresponding physical result.

Symbolic answer for squared diagram is rational function over scalar products of Lorentz momenta.

First note that there is no run time problems with denominator evaluation because of it is a product of propagators and we can evaluate it in this naturally optimized form.

Numerator for squared diagram is a polinomial which depends on scalar products, masses and other parameters. As a rule all model parameters and sum of energies of incoming particles (scalar product p1.p2) are not changed during Monte Carlo session: they can be evaluated only once for all points of phase space. So we can consider the numerator as a polinomial over scalar products (except of p1.p2) with constant coefficients evaluated once before the Monte Carlo session. As a result the number of kinematical variables is fixed: it equals to 1 for $2 \to 2$, and 4 for $2 \to 3$, and 8 for $2 \to 4$ types of processes. In the last case the number of independent kinematical variables is 7 really. However we can express the 8th variable through other seven ones by means nonlinear equation only. So for the numerical calculations we are forced to use 8 kinematical variables but not 7 ones.

In the case of 'tHooft - Feynman gauge it is possible to estimate the order of this polinomial. For any renormalizable gauge it equals to 2 for $2 \to 2$, and 4 for $2 \to 3$, and 8 for $2 \to 4$ types of processes.

The maximum number of terms in a polinomial is

$$N = \frac{(n+p)!}{n!p!},$$

where "n" is a number of variables and "p" is order of this polinomial. CompHEP FORTRAN output represents this polinomial in optimized form [1]. In this case the number of multiplication operations is not exceed number of summations. So the total number of arithmetic operations is restricted by $2N$. It equals to 70 for $2 \to 3$ and 990 for $2 \to 4$ types of reactions.

This work was supported by Russian Foundation of Fundamental Research (grant number 93-02-14428)

6. References

1. E. E. Boos et al. in *New computing techniques in physics research II*, ed. D. Perret-Gallix, Proc. of 2nd Int. Workshop AIHENP-92 (World Scientific, Singapore, 1992), p. 665.
   V. A. Ilyin, A. E. Pukhov and S. A. Shichanin. in *New computing techniques in physics research II*, ed. D. Perret-Gallix, Proc. of 2nd Int. Workshop AIHENP-92 (World Scientific, Singapore, 1992), p. 671.
2. C. Itzykson and J.-B. Zuber. *Quantum Field Theory* (McGraw - Hill, N. - Y., 1980)
3. *Phys. Rep.*, **129** (1985) 59.

CALCULATING VARIOUS QUANTITIES WITH MINCER: α_s^3 CORRECTION TO $\Gamma(Z^0 \to hadrons)$ and NNL APPROXIMATION TO DEEP INELASTIC STRUCTURE FUNCTIONS

S.A. LARIN[1], T. VAN RITBERGEN, J.A.M. VERMASEREN
NIKHEF-H, P.O. Box 41882, 1009 DB, Amsterdam, The Netherlands.

ABSTRACT

We give a resume´ of calculations that are presently performed by the authors with the improved MINCER program. These calculations include the α_s^3 correction to the decay width of $Z^0 \to$ *hadrons* and next-next-to-leading perturbative QCD corrections to the moments of structure functions of deep inelastic lepton-hadron scattering. Some field theory methods and their computer implementations are discussed.

1. Introduction

In the last decade, much effort has been put in testing the standard model of elementary particles. As we know it, so far this model has survived those tests. Since ongoing tests will be focused on increasing the experimental accuracy of some well-known processes (such as deep inelastic lepton hadron scattering and decay properties of electroweak bozons), theoretical predictions will have to match this accuracy. In practice, lacking exact non-perturbative methods this means that calculations involve many Feynman diagrams each containing many (i.e. 3,4) loops. Fortunately, the increase in symbolic manipulation power, combined with new field theory methods, makes some calculations feasible. One of the tools that facilitate multiloop calculations is the package MINCER [1] written for the system of symbolic manipulation FORM [2]. The first version of this package was written [3] for the system SCHOONSCHIP [4]. MINCER calculates massless 1,2 & 3-loop diagrams of the propagator type (i.e. two point functions) in dimensional regularization [5]. MINCER applies complicated recursion schemes that are derived using the method of integration by parts [6] and uses known results for basic integrals. In this short overview of some of our recent and ongoing projects MINCER plays an important role. Also new routines such as those for 1,2 & 3-loop massive vacuum diagrams were programmed.

2. $Z^0 \to$ hadrons

2.1. *generalities*

For the decay rate of a Z^0 bozon into hadrons, the quantity to be determined

[1] on leave from Institute for Nuclear Research (INR) of the Russian Academy of Sciences, Moscow 117312, Russia.

is the squared matrix element summed over all final (hadronic) states, $T^{\mu\nu}$. One can express this quantity as the imaginary part of a current correlator in the standard way

$$T^{\mu\nu} = \sum_h < 0|J^\nu|h >< h|J^\mu|0 > \; = \; 2Im\left(i\int d^4z e^{iq\cdot z} < 0|T(J^\nu(z)J^\mu(0))|0 >\right).$$

Here J^μ is the neutral weak quark current coupled to the Z bozon. The non-singlet contribution of the order α_s^3 (in the case of massless quarks) can be taken from the known calculation of the total cross section of e^+e^- annihilation into hadrons [7]. This result was confirmed in [8]. But the strongest confirmation of the correctness of this result comes from the paper [9] where the non-trivial connection between the result [7] and the α_s^3 approximation [10] to the deep inelastic Gross-Llewellyn Smith sum rule was established. In the following we will concentrate on the singlet contribution of the axial vector current to the correlator. The singlet contribution to the correlator is left over from the axial anomaly and is induced by the mass difference between top and bottom quarks. For treating the γ_5 matrix in dimensional regularization we use the prescription developed in [11] that is based on the original definition of γ_5 in [5].

2.2. α_s^2 correction

This correction was originally calculated in [12] and confirmed in [13]. In this paragraph we reproduce it to illustrate the techniques that we later apply to 4-loop diagrams. At the 3-loop level the only Feynman diagrams that contribute to the correlator are so called 'double triangle diagrams' with the triangles formed by bottom and top fermion loops. Corrections from these diagrams can be calculated by taking the bottom quark massless and the top quark mass large compared to the Z bozon mass. The diagram with two massive triangles is zero since the only physical cut is through two gluon propagators and Z^0 decay in two gluons is kinematically forbidden. We are therefore left with two diagrams to be calculated, one massive and one massless diagram. The massless diagram is of the MINCER topology LA and it's calculation and renormalization is straightforward.

We will now treat the expansion of the massive diagram. The general theory of asymptotic euclidean expansions was developed in [14, 15]. Practically we use the techniques developed in [16, 17]. We adopt the language of [16] and the definitions made therein. First we make renormalization of ultraviolet divergencies by means of the standard R-operation in the $\overline{MS}$ scheme. Then the expansion procedure consists of the following. We have to determine all *complete hard partitions* and make Taylor-expansions of the *hard subgraphs* of these partitions in their *external* momenta (which may be *internal* momenta of the whole Feynman diagram). After these Taylor expansions (truncated at a required order) the diagrams simplify significantly and we can perform the integrations over all internal momenta. Individual hard partitions may now contain artificial ultraviolet divergencies introduced by the Taylor expansions but after summation over all *complete hard partitions* the divergencies cancel to give the required finite result.

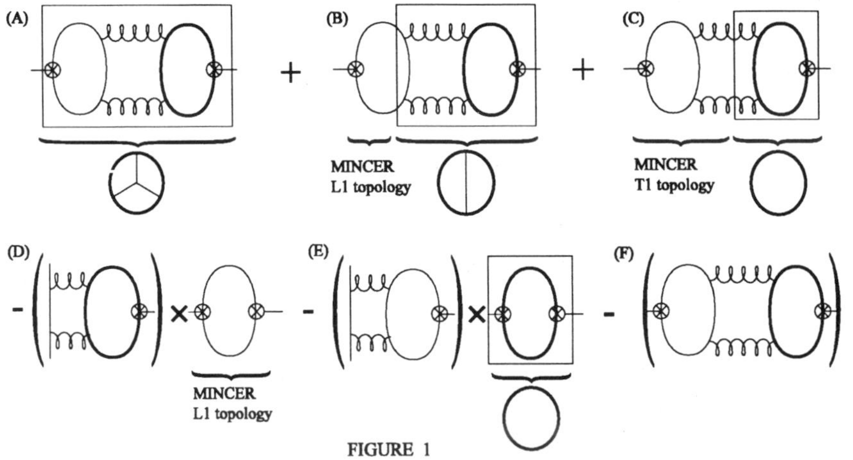

FIGURE 1

A diagrammatic representation of the ultraviolet R operation followed by the asymptotic expansion procedure, applied to the double triangle diagram is given in figure 1. Thick lines indicate top quark propagators, thin lines indicate the massless quark propagators and spiral lines - gluon propagators. Ultraviolet counterterms are indicated between round brackets. The hard subgraphs are surrounded by square boxes and these subgraphs are to be expanded in the external momenta of the box. The symbol ⊗ is used to indicate an axial vertex. After the Taylor expansions the hard subgraphs are reduced to massive vacuum integrals and the resulting topologies are indicated under the boxes. When the hard subgraphs are integrated out we are left with massless diagrams of the MINCER topologies that are also indicated. We have written efficient FORM procedures to perform the necessary massive vacuum integrals. The procedures include 3-loop topologies of the type Benz and non-planar and use recursions based on the recursion scheme of [18]. They are essential for the large mass expansion of 4-loop diagrams.

Making the Taylor expansions deep enough, we find the results for the contributions in figure 1. Summing the contributions, adding the massless double triangle diagram and taking the imaginary part by applying $Ln(-s - i\epsilon) = Ln(s) - i\pi$ (s is the squared 4-momentum of the Z bozon) we reproduced the large mass expansion of the known result [12].

2.3. α_s^3 correction

We will now show that the same method can be used to compute the α_s^3 correction. In contrast to the previous case where we could determine both the real and imaginary parts of a diagram, we are now able to compute only the necessary imaginary part of diagrams with the present available techniques. In order to find the imaginary part of a diagram we need to calculate only terms that contain logarithms of s. In dimensional regularization every massless integral receives a factor $(\frac{\mu^2}{s})^\epsilon = 1 + \epsilon Ln(\frac{\mu^2}{s}) + \cdots$. In contrast, vacuum massive integrals receive factors

$(\frac{\mu^2}{M^2})^\epsilon$ (M is the top quark mass) and do *not* give logarithms of s. This means that contributions without massless integrals do not have to be considered for our purpose.

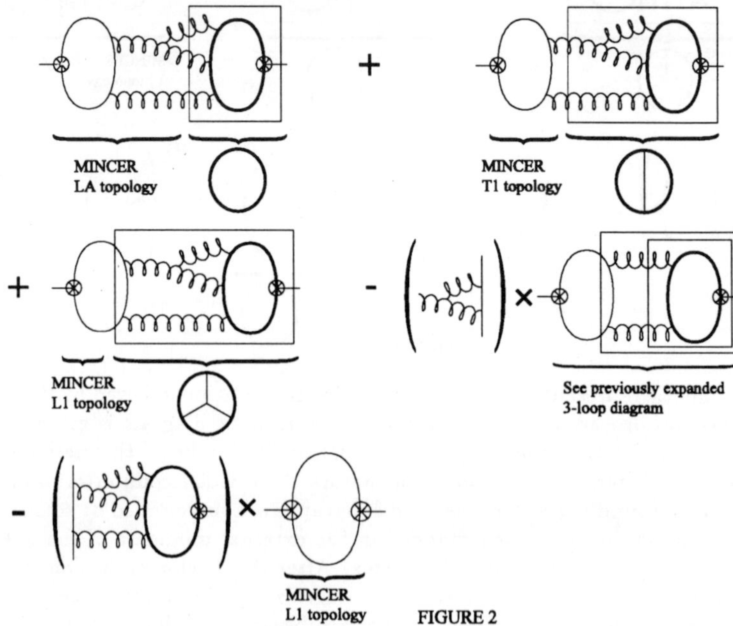

FIGURE 2

The expansion of one renormalized 4-loop diagram that contributes to the correlator is presented in figure 2. We have only indicated terms which contribute to the imaginary part. All 4-loop massive diagrams contributing to the correlator give after the large mass expansion topologies that can be calculated with either MINCER (the massless diagrams) or with the new procedures (the vacuum massive diagrams). The imaginary part of *massless* 4-loop diagrams are calculated with similar techniques that were used in the α_s^3 calculation of the total cross section of e^+e^- annihilation into hadrons [7].

3. The NNL approximation to deep inelastic structure functions

Another project which we perform with the help of MINCER is the calculation of next-next-to-leading (NNL) perturbative QCD corrections to the moments of the structure functions of deep inelastic lepton hadron scattering. The calculation is done by means of the 'method of projectors' [17]. This approach was previously successfully applied to the calculation of NNL corrections to several deep inelastic sum rules [19, 10] and to the lowest moment of the structure function F_L [20]. In the present calculation we apply this approach to the first 8 moments of the structure functions F_2 and F_L with a thoroughly optimized version of MINCER [21].

As a sample of the results of these calculations we present the answers for the anomalous dimension, γ_8 and coefficient function $C_{L,8}$ of the spin-8 non-singlet quark operator which are relevant for the 8th moment of the non-singlet structure functions F_L.

$$\gamma_8(a) =$$

$$\begin{aligned}
&+a(\tfrac{9883}{1260}C_F) \\
&+a^2(\tfrac{25870049}{762048}C_FC_A - \tfrac{36241943}{4762800}C_Fn_f - \tfrac{27040578211}{4000752000}C_F^2) \\
&+a^3(-\tfrac{19766}{315}C_FC_An_f\zeta_3 - \tfrac{1578915745223}{72013536000}C_FC_An_f + \tfrac{2510407}{132300}C_FC_A^2\zeta_3 + \tfrac{8101059985033}{41150592000}C_FC_A^2) \\
&+a^3(-\tfrac{38920977797}{18003384000}C_Fn_f^2 - \tfrac{2510407}{44100}C_F^2C_A\zeta_3 - \tfrac{3662576699059}{112021056000}C_F^2C_A + \tfrac{19766}{315}C_F^2n_f\zeta_3) \\
&+a^3(-\tfrac{91675209372043}{1680315840000}C_F^2n_f + \tfrac{2510407}{66150}C_F^3\zeta_3 - \tfrac{1093087100097437993}{6351593875200000}C_F^3)\ .
\end{aligned}$$

$$C_{L,8}(a) =$$

$$\begin{aligned}
&+aC_F\tfrac{4}{9} \\
&+a^2C_FC_A(\tfrac{14741729}{1190700} - \tfrac{16}{3}\zeta_3) \\
&+a^2C_Fn_f(-\tfrac{14234}{8505}) \\
&+a^2C_F^2(-\tfrac{21694349}{3572100} + \tfrac{32}{3}\zeta_3) \\
&+a^3C_FC_An_f(-\tfrac{23153083641529}{198037224000} + \tfrac{18459136}{363825}\zeta_3) \\
&+a^3C_FC_A^2(\tfrac{7653142193467}{18003384000} - \tfrac{9508318}{19845}\zeta_3 + \tfrac{2240}{9}\zeta_5) \\
&+a^3C_Fn_f^2(\tfrac{1435876}{229635}) \\
&+a^3C_F^2C_A(-\tfrac{7667007621800089}{27725211360000} + \tfrac{2476882549}{2182950}\zeta_3 - \tfrac{2720}{3}\zeta_5) \\
&+a^3C_F^2n_f(\tfrac{11844644404289}{198037224000} - \tfrac{914992}{10395}\zeta_3) \\
&+a^3C_F^3(-\tfrac{86167166469418457}{499053804480000} - \tfrac{111668693}{218295}\zeta_3 + \tfrac{7360}{9}\zeta_5) \\
&+a^3d_{abc}^2/n(-\tfrac{35555777437}{3000564000} - \tfrac{85378}{14175}\zeta_3 + \tfrac{160}{9}\zeta_5).
\end{aligned}$$

Here $a = \frac{g^2}{16\pi^2}$ is the QCD coupling, $C_F = 4/3$ and $C_A = 3$ are the Casimir operators of the defining and adjoint representation of the color group $SU(3)$, n_f is the number of active quark flavors, ζ is the Riemann zeta-function. $d_{abc}^2 = \frac{40}{3}$, d_{abc} are the symmetrical structure constants of the color group.

The calculation of the 8th non-singlet moment took the equivalent of more than 600 hours on a SGI workstation with a 100 MHZ MIPS 4400 chip.

4. Acknowledgements

We are grateful to M. Veltman for helpful discussions.

5. References

1. S.A. Larin, F.V. Tkachov, J.A.M. Vermaseren, preprint NIKHEF-H/91-18 (1991).
2. J.A.M. Vermaseren, *Symbolic Manipulation with Form* , Computer Algebra Nederland, Amsterdam, 1991.
3. S.G. Gorishny, S.A. Larin, F.V. Tkachov, INR preprint P-0330 (Moscow,1984); S.G. Gorishny, S.A. Larin, L.R. Surguladze, F.V. Tkachov, *Comp. Phys. Comm.* **55** (1989) 381.
4. M. Veltman, *SCHOONSCHIP*, CERN preprint, 1967; H. Strubbe, *Comp. Phys. Comm.* **8** (1974) 1.
5. G.'t Hooft, M.Veltman, *Nucl.Phys.* **B44** (1972) 189.
6. F.V. Tkachov, *Phys. Lett.* **100B** (1981) 65, *Teor. Mat. Fiz.* **56** (1983) 350; K.G. Chetyrkin, F.V. Tkachov, *Nucl. Phys.* **B192** (1981) 159.
7. S.G. Gorishny, A.L. Kataev, S.A Larin, *Phys. Lett.* **B212** (1988) 238, ibid. **B259** (1991) 144.
8. L.R.Surguladze, M.A. Samuel, *Phys. Rev. Lett.* **66** (1991) 560, erratum ibid, 2416.
9. D.J. Broadhurst, A.L. Kataev, preprint OUT-4102-45 , INR-820/93 (1993).
10. S.A. Larin and J.A.M. Vermaseren, *Phys. Lett.* **B259** (1991) 345.
11. S.A. Larin, *Phys. Lett.* **B303** (1993) 113.
12. B.A. Kniehl, J.H. Kuhn, *Nucl. Phys.* **B329** (1990) 547.
13. K.G. Chetyrkin, J.H. Kuhn, preprint TTTP93-8 (1993).
14. F.V. Tkachov, preprint INR P-358 (Moscow,1984), preprint FERMILAB-PUB-91/347-T (1991);
15. K.G. Chetyrkin, V.A. Smirnov, preprint INR G-518 (Moscow, 1987); K.G. Chetyrkin, *Teor.Mat.Fiz.* **76** (1988) 207.
16. S.G. Gorishny, *Nucl. Phys.* **B319** (1989) 633.
17. S.G. Gorishny, S.A. Larin, *Nucl. Phys.* **B283** (1987) 452.
18. D.J. Broadhurst, *Z. Phys.* **C54** (1992) 599.
19. S.A. Larin, F.V. Tkachov, J.A.M. Vermaseren, *Phys. Rev. Lett.* **66** (1991) 862.
20. S.A. Larin, F.V. Tkachov, J.A.M. Vermaseren, *Phys. Lett.* **B272** (1991) 121.
21. J.A.M. Vermaseren, these Proceedings.

THE CALCULATION OF ε'/ε BEYOND LEADING LOGARITHMS WITH THE HELP OF THE TRACER γ-ALGEBRA PACKAGE[†]

Markus E. Lautenbacher
Physics Department T31, Munich Technical University
D-85478 Garching, Germany

ABSTRACT

We present results of a recent 2-loop calculation of the effective $\Delta S = 1$ Hamiltonian at the next-to-leading order level. Together with an improved treatment of hadronic matrix elements these results are used to evaluate ε'/ε, the measure for direct $\mathcal{CP}$ violation in $K \to \pi\pi$, at the next-to-leading order. The TRACER γ-algebra system developed for the calculation is described shortly.

1. Introduction

For almost 30 years now $\mathcal{CP}$ violation in the K-meson system has been and still is an exciting field of studies. This is especially true for the determination of the measure for direct $\mathcal{CP}$ violation in $K \to \pi\pi$ decays, ε'/ε. Unfortunately, the latest experimental data [1,2]

$$\mathrm{Re}(\varepsilon'/\varepsilon) = \begin{cases} (23 \pm 6.5) \times 10^{-4} & \mathrm{NA31} \\ (7.4 \pm 6.0) \times 10^{-4} & \mathrm{E731} \end{cases} \tag{1}$$

are not yet fully conclusive and the theoretical side of the problem is plagued by uncertainties in the hadronic matrix elements and the role of a heavy top quark which can only be treated consistently at the next-to-leading order (NLO) level. In a series of recent papers [3–7] considerable progress has been achieved on the theoretical prediction of ε'/ε within the Standard Model. We will summarize some of this work below.

2. The Effective $\Delta S = 1$ Hamiltonian Beyond Leading Logarithms

After integrating out the heavy degrees of freedom of the electroweak theory the effective $\Delta S = 1$ Hamiltonian relevant for ε'/ε is given by

$$\mathcal{H}_{\mathrm{eff}}(\Delta S = 1) = \frac{G_F}{\sqrt{2}} V_{\mathrm{ud}} V_{\mathrm{us}}^* \sum_{i=1}^{10} \Big[z_i(\mu) + \tau\, y_i(\mu) \Big] \, Q_i(\mu) \,, \qquad \tau \equiv -\frac{V_{\mathrm{td}} V_{\mathrm{ts}}^*}{V_{\mathrm{ud}} V_{\mathrm{us}}^*} \tag{2}$$

with the $Q_i(\mu)$ being 4-fermion operators and $z_i(\mu)$, $y_i(\mu)$ the Wilson coefficients of the $\mathcal{CP}$ conserving and violating parts of $\mathcal{H}_{\mathrm{eff}}(\Delta S = 1)$, respectively. The most important among the current-current ($Q_{1,2}$), QCD penguin (Q_3-Q_6) and electroweak penguin operators (Q_7-Q_{10}) are

$$Q_2 = (\bar{s}u)_{\mathrm{V-A}} (\bar{u}d)_{\mathrm{V-A}} \;\; Q_6 = (\bar{s}_\alpha d_\beta)_{\mathrm{V-A}} \sum_q (\bar{q}_\beta q_\alpha)_{\mathrm{V+A}} \;\; Q_8 = \frac{3}{2} (\bar{s}_\alpha d_\beta)_{\mathrm{V-A}} \sum_q e_q \, (\bar{q}_\beta q_\alpha)_{\mathrm{V+A}} \,. \tag{3}$$

[†]Work done in collaboration with A.J. Buras and M. Jamin.

2.1. Calculation of Wilson Coefficients

The Wilson coefficients $C_i(\mu)$ satisfy the well known renormalization group (RG) equation

$$\left[\mu\frac{\partial}{\partial\mu} + \beta(g)\frac{\partial}{\partial g} - \hat{\gamma}^T(g^2,\alpha)\right]\vec{C}(\frac{M_W^2}{\mu^2}, g^2, \alpha) \equiv 0\,, \tag{4}$$

where the anomalous dimension matrix $\hat{\gamma}(g^2,\alpha)$ has the perturbative expansion

$$\hat{\gamma}(g^2,\alpha) = \frac{\alpha_s}{4\pi}\hat{\gamma}_s^{(0)} + \frac{\alpha_s^2}{(4\pi)^2}\hat{\gamma}_s^{(1)} + \frac{\alpha}{4\pi}\hat{\gamma}_e^{(0)} + \frac{\alpha\,\alpha_s}{(4\pi)^2}\hat{\gamma}_{se}^{(1)} + \cdots \tag{5}$$

The NLO matrices $\hat{\gamma}_s^{(1)}$ and $\hat{\gamma}_{se}^{(1)}$ in (5) were unknown until recently. They have been determined in refs. [3, 4] and confirmed in [7] by calculating the complete set of 28 current-current and 2×14 penguin 2-loop diagrams for $\Delta S = 1$ processes. A selection of these diagrams is shown in fig. 1. At 2-loop level scheme dependencies in $\hat{\gamma}$ enter the game and hence $\hat{\gamma}_s^{(1)}$ and $\hat{\gamma}_{se}^{(1)}$ have been calculated in the scheme with naively anticommuting γ_5 (NDR) and in the more elaborate 't Hooft-Veltman scheme (HV). The compatibility of $\hat{\gamma}$ obtained in both schemes has been shown successfully.

The solution of the RG equation (4) for the Wilson coefficients $C_i(\mu)$ is

$$\vec{C}(\mu) = \hat{U}(\mu, M_W, \alpha)\,\vec{C}(M_W) \tag{6}$$

with the RG evolution matrix from scale m_2 to m_1, $m_1 < m_2$ given by

$$\hat{U}(m_1, m_2, \alpha) \equiv T_g \exp\int_{g(m_2)}^{g(m_1)} dg'\,\frac{\hat{\gamma}^T(g'^2,\alpha)}{\beta(g')}\,. \tag{7}$$

From this it is easy to see that the corresponding RG evolution of the hadronic matrix elements from scale m_1 to scale m_2, $m_1 < m_2$ is determined by

$$\langle\vec{Q}(m_2)\rangle = \hat{U}^T(m_1, m_2, \alpha)\,\langle\vec{Q}(m_1)\rangle\,. \tag{8}$$

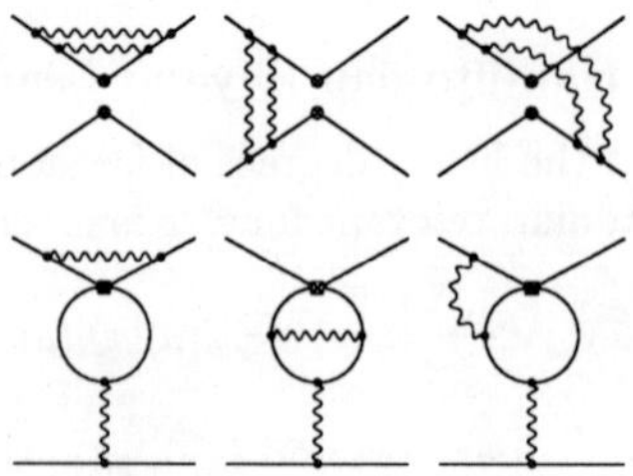

Figure 1: Some examples of 2-loop current-current and penguin diagrams in the effective theory contributing to the anomalous dimension matrices $\hat{\gamma}_s^{(1)}$ and $\hat{\gamma}_{se}^{(1)}$ of Eq. (5). The wavy lines denote either two gluons or a photon and a gluon, so that the whole diagram is of order $\mathcal{O}(\alpha_s^2)$ and $\mathcal{O}(\alpha\,\alpha_s)$, respectively.

2.2. Calculation of Hadronic Matrix Elements

Using the shorthand notations $\langle Q_i\rangle_I \equiv \langle(\pi\pi)_I\,|Q_i|\,K\rangle$, $I=0,2$ and $c \equiv \frac{G_F}{\sqrt{2}}\,V_{\rm ud}V_{\rm us}^*$ the matrix elements $\langle Q_{1,2}\rangle_I$ can be derived from a fit to $\mathcal{CP}$ conserving $K \to \pi\pi$ amplitudes [5]

$$\langle Q_1(m_c)\rangle_2 = \langle Q_2(m_c)\rangle_2 = \frac{{\rm Re}A_2}{c\,(z_1(m_c)+z_2(m_c))}\,, \tag{9}$$

$$\langle Q_1(m_c)\rangle_0 = \frac{{\rm Re}A_0}{c\,z_1(m_c)} - \frac{z_2(m_c)}{z_1(m_c)}\,\langle Q_2(m_c)\rangle_0\,, \tag{10}$$

with $\langle Q_2(m_c)\rangle_0$ parametrized by $B_2^{(1/2)}(m_c)$. In the NDR scheme one finds $B_{2,{\rm NDR}}^{(1/2)}(m_c) = 6.7 \pm 0.9$ [5]. The matrix elements $\langle Q_{9,10}\rangle_I$ can be expressed in terms of the elements $\langle Q_{1,2}\rangle_I$ given above. This approach of obtaining the $(V-A)\otimes(V-A)$ matrix elements from a fit to $\mathcal{CP}$ conserving $K \to \pi\pi$ decays has been introduced for the first time in ref. [5]. Among the remaining $(V-A)\otimes(V+A)$ matrix elements the by far most important ones are $\langle Q_6(m_c)\rangle_0$ and $\langle Q_8(m_c)\rangle_2$. They are parametrized by $B_6^{(1/2)}(m_c)$ and $B_8^{(3/2)}(m_c)$, respectively. With all $\langle Q_i\rangle_I$ now determined at scale $\mu = m_c$ the matrix elements at any other scale can be determined using Eq. (8).

2.3. Description of the TRACER *γ-Algebra System*

The calculation of the numerous 2-loop diagrams contributing to the $\Delta S = 1$ anomalous dimension matrix $\hat{\gamma}$ is rather tedious, especially in the HV scheme for γ_5, and hence indispensably calls for a computerized aid. As a result the TRACER γ-algebra system [8] was developed alongside with the work on refs. [3,4].

The TRACER system runs on top of the MATHEMATICA computer algebra system. It provides the user with means for algebraic manipulations as well as trace calculations for strings of γ-matrices. As a feature of particular interest, it offers not only the usual anticommuting γ_5-matrix, which is well known to be mathematically inconsistent in $D \neq 4$ space-time dimensions, but also the 't Hooft-Veltman version of γ_5 [9,10] which has no problems even in $D \neq 4$. In addition it provides a number of functions to tailor the output, e.g. a TeX interface.

In TRACER a string of γ-algebraic objects has the generic form

```
GammaTrace[1, expr, ...]   ≡  G[1, expr, ...]
```

where the notation for physical quantities can be looked up in the table below.

expr	Syntax	Physical Equivalent
`{mu}`	`G[1, {mu}]`	γ_μ γ-matrix
`G5`	`G[1, G5]`	γ_5 γ_5-matrix
`p`	`G[1, p]`	$\not{p}$ 4-momentum
`U`	`G[1, U]`	$\hat{1}$ unit matrix
`m U`	`G[1, m U]`	$m\,\hat{1}$ mass term

The scalar product is denoted as

```
S[expr1, expr2]  ≡  expr1.expr2
```

with the different physical meanings depending on the type of arguments explained again in the following table.

expr1	expr2	Syntax	Physical Equivalent	
p	q	p.q	$p \cdot q$	4-vector product
p	{mu}	p.{mu}	p_μ	4-vector component
{mu}	{nu}	{mu}.{nu}	$g_{\mu\nu}$	metric tensor

For more details the reader is referred to the literature [8]. TRACER is free upon request from the authors.

3. Phenomenological Analysis of ε'/ε Beyond Leading Logarithms

Combining Wilson coefficients and hadronic matrix elements determined in sects. 2.1 and 2.2, respectively, one obtains for the measure of direct $\mathcal{CP}$ violation

$$\frac{\varepsilon'}{\varepsilon} = \mathrm{Im}\lambda_t \left[P^{(1/2)} - P^{(3/2)} \right] , \tag{11}$$

$$\begin{aligned} P^{(1/2)} &= \sum P_i^{(1/2)} = r \sum y_i \langle Q_i \rangle_0 \left(1 - \Omega_{\eta+\eta'}\right) , \\ P^{(3/2)} &= \sum P_i^{(3/2)} = \frac{r}{\omega} \sum y_i \langle Q_i \rangle_2 , \end{aligned} \tag{12}$$

$$\lambda_i = V_{\mathrm{id}} V_{\mathrm{is}}^* , \qquad \omega = \frac{\mathrm{Re}A_2}{\mathrm{Re}A_0} , \qquad r = \frac{G_F \, \omega}{2\,|\varepsilon|\, \mathrm{Re}A_0} . \tag{13}$$

For the CKM parameters and the B_{K} parameter from K^0-$\bar{K}^0$ mixing we chose

$$|V_{\mathrm{cb}}| = \left\{ \begin{array}{l} 0.043 \pm 0.004 \\ 0.043 \pm 0.002 \end{array} \right. |V_{\mathrm{ub}}/V_{\mathrm{cb}}| = \left\{ \begin{array}{l} 0.10 \pm 0.03 \\ 0.10 \pm 0.01 \end{array} \right. B_{\mathrm{K}} = \left\{ \begin{array}{ll} 0.65 \pm 0.15 & \text{(range I)} \\ 0.65 \pm 0.05 & \text{(range II)} \end{array} \right. . \tag{14}$$

The resulting ranges for ε'/ε in the NDR scheme as a function of the QCD scale parameter $\Lambda_{\overline{\mathrm{MS}}}$ for a top quark mass of $m_t = 160\,\mathrm{GeV}$ and various choices for the hadronic parameters $B_6^{(1/2)}(m_c)$ and $B_8^{(3/2)}(m_c)$ are shown in fig. 2.

The numerical analysis of ε'/ε at the NLO in the NDR scheme shows that choosing parameters $\Lambda_{\overline{\mathrm{MS}}} = 300\,\mathrm{MeV}$, $m_t = 150\,\mathrm{GeV}$ and $B_6^{(1/2)}(m_c) = B_8^{(3/2)}(m_c) = 1$ one obtains

$$\varepsilon'/\varepsilon = \left\{ \begin{array}{ll} (4.9 \pm 2.0) \times 10^{-4} & 0 < \delta \le \frac{\pi}{2} \\ (3.4 \pm 1.7) \times 10^{-4} & \frac{\pi}{2} < \delta < \pi \end{array} \right. ,$$

which lies in the ball park of the E731 data given in Eq. (1). On the other hand, the NA31 result can be accommodated by $B_6^{(1/2)}(m_c) = 2$ and $B_8^{(3/2)}(m_c) = 1$

$$\varepsilon'/\varepsilon = \left\{ \begin{array}{ll} (16.9 \pm 5.7) \times 10^{-4} & 0 < \delta \le \frac{\pi}{2} \\ (11.8 \pm 5.0) \times 10^{-4} & \frac{\pi}{2} < \delta < \pi \end{array} \right. .$$

Hence, theory yet can give no definite answer as to which of the two experimental values is closer to the physical truth, although the lower values of E731 seem to be slightly more favored.

The detailed m_t dependence of ε'/ε at the NLO level is most easily studied with the help of the Penguin Box Expansion [11]. For this the reader is referred to ref. [12].

4. Conclusions

In summary, the calculation of the Wilson coefficients of $\Delta S = 1$ electroweak processes has been performed successfully at the NLO level. However, due to un-

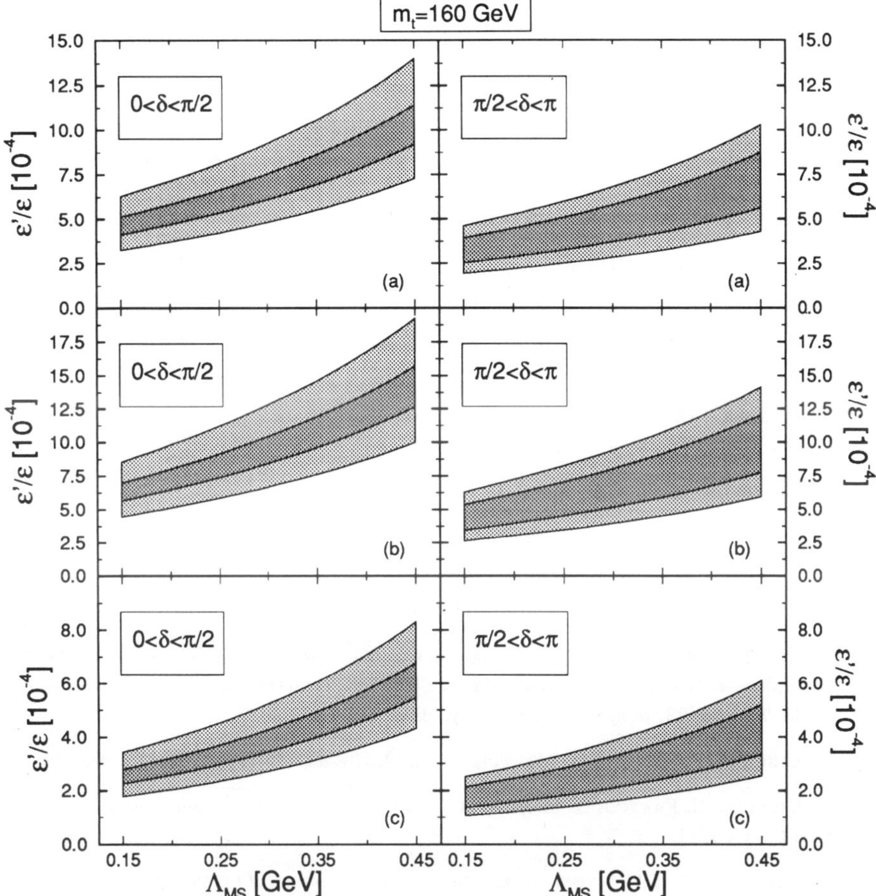

Figure 2: Ranges for ε'/ε in the NDR scheme as a function of $\Lambda_{\overline{\rm MS}}$ for $m_{\rm t} = 160\,{\rm GeV}$ and parameter ranges I (light gray) and II (dark gray) as of Eq. (14). The hadronic parameters are (a) $B_6^{(1/2)}(m_c) = B_8^{(3/2)}(m_c) = 1.5$, (b) $B_6^{(1/2)}(m_c) = 1.5$, $B_8^{(3/2)}(m_c) = 1$ and (c) $B_6^{(1/2)}(m_c) = B_8^{(3/2)}(m_c) = 1$.

certainties in hadronic matrix elements the theoretical prediction for ε'/ε still does not decide clearly for one of the competing experimental values for ε'/ε.

The NLO calculation presented here demonstrated that ε'/ε actually depends on just a few relevant parameters: First, fig. 2 clearly exhibits ε'/ε as sensitive function of the parameters

$$m_{\rm t}\,, \qquad \Lambda_{\overline{\rm MS}}\,, \qquad |V_{\rm cb}|\,, \qquad |V_{\rm ub}/V_{\rm cb}|\,.$$

Here improved accuracy is possible in principle by increased experimental efforts. Second, the figure also underlines the importance of the remaining hadronic parameters of ε'/ε

$$B_6^{(1/2)}, \qquad B_8^{(3/2)}.$$

However, these parameters lie on the purely theoretical side of the problem. Here progress clearly calls for new insights in the solution of non-perturbative QCD.

Acknowledgment

I gratefully acknowledge the most pleasant collaboration with Andrzej J. Buras and Matthias Jamin on ε'/ε in refs. [3–5] and on TRACER in [8].

References

[1] A. WAGNER (NA31), *talk presented at the European Physics Symposium on High Energy Physics, Marseille, July 1993* .

[2] L. K. GIBBONS *et al.*, *Phys. Rev. Lett.* **70** (1993) 1203.

[3] A. J. BURAS, M. JAMIN, M. E. LAUTENBACHER, and P. H. WEISZ, *Nucl. Phys.* **B400** (1993) 37.

[4] A. J. BURAS, M. JAMIN, and M. E. LAUTENBACHER, *Nucl. Phys.* **B400** (1993) 75.

[5] A. J. BURAS, M. JAMIN, and M. E. LAUTENBACHER, The Anatomy of ε'/ε Beyond Leading Logarithms with Improved Hadronic Matrix Elements, *Technical University Munich preprint,* **TUM-T31-35/92**, *to appear in* Nucl. Phys. **B** .

[6] M. CIUCHINI, E. FRANCO, G. MARTINELLI, and L. REINA, *Phys. Lett.* **B301** (1993) 263.

[7] M. CIUCHINI, E. FRANCO, G. MARTINELLI, and L. REINA, *Ecole Normale Superieure preprint* **LPTENS 93/11**.

[8] M. JAMIN and M. E. LAUTENBACHER, *Comp. Phys. Commun.* **74** (1993) 265.

[9] G. 'T HOOFT and M. VELTMAN, *Nucl. Phys.* **B44** (1972) 189.

[10] P. BREITENLOHNER and D. MAISON, *Commun. Math. Phys.* **52** (1977) 11, 39, 55.

[11] G. BUCHALLA, A. J. BURAS, and M. K. HARLANDER, *Nucl. Phys.* **B349** (1991) 1.

[12] A. J. BURAS and M. E. LAUTENBACHER, An Analytic Formula and an Upper Bound for ε'/ε in the Standard Model, *Technical University Munich preprint,* **TUM-T31-47/93**; *Max-Planck-Institut preprint,* **MPI-Ph/93-60** .

γ_5 Anomaly: Some Approaches in Dimensional Regularization

G.T.Gabadadze, A.A.Pivovarov

Institute for Nuclear Research of the Russian Academy of Sciences
Moscow 117312, Russia

Abstract

New methods for the calculation of the anomaly within the dimensional regularization are presented. The issue of normalization of the axial current within this schemes is discussed.

Introduction

In practical applications for many-loop calculations dimensional regularization is now most widely used. However the ingenious use of dimensional regularization for the calculation of the anomaly [1] seems to be cumbersome due to the presence of γ_5 matrix in corresponding expressions. The various schemes for the calculation of the anomaly with dimensional regularization have been elaborated (as a review see [2]). The only thing that requires the special consideration in these calculations is the accurate relative normalization of all operators entering the anomaly equation.

Last time a growing interest to the question of many-loop anomaly calculations has been stimulated by the axial current puzzle in supersymmetric theories (see [3] and refs. therein) and, more recently, by the proton spin problem (see e.g. [4]).

In this paper we will present two schemes of calculation of the axial anomaly within the framework of dimensional regularization. The first of them allows one to perform the many-loop calculations without introducing any additional properties for the γ_5 matrix. The second one is based on the expression for the anticommutator of γ_5 and γ_μ matrices and transparently demonstrates some properties of γ_5 matrix in d-dimensional space-time.

The discussion of the issue of normalization of the axial current in these two schemes is also presented.

1. The scheme-dependent corrections

In this section we discuss an explicit calculation of scheme-dependent two-loop contribution to the anomaly within the dimensional regularization. To be more concrete we study the $SU_c(3) \otimes U_{em}(1)$ model and concern about strong interaction corrections to the Abelian axial anomaly [5].

The three-point Green function which satisfies the conditions of Bose-symmetry in kinematics $k_1^2 = k_2^2$ where k_1 and k_2 are momenta of photons can be written down through three independent tensor structures and three invariant scalar functions F_i which we choose as follows

$$T^{\mu\nu\alpha}(k_1,k_2) = \int \langle 0|TJ^\mu(x)J^\nu(y)J^{5\alpha}(0)|0\rangle e^{ik_1x+ik_2y}dxdy = \epsilon^{\mu\nu\alpha\tau}q_\tau F_3(q^2,p^2)$$

$$+(q^\mu\epsilon^{\nu\alpha\rho\tau}p_\rho q_\tau + q^\nu\epsilon^{\mu\alpha\rho\tau}p_\rho q_\tau)F_1(q^2,p^2) + (p^\mu\epsilon^{\nu\alpha\rho\tau}p_\rho q_\tau - p^\nu\epsilon^{\mu\alpha\rho\tau}p_\rho q_\tau)F_2(q^2,p^2) \quad (1)$$

where $k_1 = p+q$, $k_2 = p-q$, J_μ is electromagnetic current and J^5_α is a third component of the Abelian axial isotriplet current, $J^5_\alpha = (\bar{u}\gamma_\alpha\gamma_5 u - \bar{d}\gamma_\alpha\gamma_5 d)/2$. The consequences of the vector current conservation, $\partial^\mu J_\mu = 0$, for invariant amplitudes F_i are

$$k_{1\mu}T^{\mu\nu\alpha}(k_1,k_2) = 0 = \epsilon^{\nu\alpha\rho\tau}p_\rho q_\tau(q^2F_1 + p^2F_2 + F_3), \quad (2)$$

$$k_{2\nu}T^{\mu\nu\alpha}(k_1,k_2) = 0 = -\epsilon^{\mu\alpha\rho\tau}p_\rho q_\tau(q^2F_1 + p^2F_2 + F_3), \quad (3)$$

while differentiation of the axial current in eq. (1) leads to the relation

$$p_\alpha T^{\mu\nu\alpha}(k_1,k_2) = \epsilon^{\mu\nu\rho\tau}p_\rho q_\tau F_3 \quad (4)$$

and the axial Ward identity would require $\epsilon^{\mu\nu\rho\tau}p_\rho q_\tau F_3 = 0$.

Since the quantity $F_3(p^2,q^2)$ is defined only up to a constant because of renormalization ambiguity according to the formal power counting, it is useful to introduce new dimensionless form factors $F_{1,2}(p^2,q^2) = f_{1,2}(p^2/q^2)/q^2$ and $F_3(p^2,q^2) = f_3(p^2/q^2)$. Then after taking the limit $p^2/q^2 \to 0$ in eqs. (2,3) we obtain

$$\lim_{q^2\to\infty} q^2F_1(q^2,p^2) = \lim_{z\to 0} f_1(z) = -f_3(0). \quad (5)$$

Eq. (5) is crucial for further consideration because it reduces the determination of the anomaly coefficient (4) to computation of the finite quantity F_1. First of all we briefly describe the general scheme of calculations. By symmetrising the tensor $T_{\mu\nu\alpha}$ over indices μ and ν we kill all terms except the invariant function F_1. The symmetrised amplitude is denoted by $S_{\mu\nu\alpha}$ and it does not contain singularity at small momentum p

$$S_{\mu\nu\alpha}(p,q) = \frac{1}{2}(T_{\mu\nu\alpha}(k_1,k_2) + T_{\nu\mu\alpha}(k_1,k_2)) = (q_\mu\varepsilon_{\nu\alpha\rho\tau} + q_\nu\varepsilon_{\mu\alpha\rho\tau})p^\rho q^\tau F_1(p^2,q^2). \quad (6)$$

Since to work with only one dimensional parameter q is more convenient for performing integration, we eliminate the small parameter p

$$D_{\nu\alpha\lambda} = (\frac{\partial}{\partial p^\lambda}(q^\mu S_{\mu\nu\alpha}))|_{p=0} = q^2F_1(q^2)\varepsilon_{\nu\alpha\lambda\tau}q^\tau.$$

The leading contribution to the quantity F_1 is finite due to formal power counting within the standard R-operation, its value does not depend on any regularization at all, and is rigorously fixed. Nevertheless some subgraphs can contain divergences at the intermediate stage of computation. In order to treat such cases we use dimensional regularization in d-dimensional space, $d = 4 - 2\epsilon$. Using only the cycle property of the trace operation one can write the expression for $D_{\nu\alpha\lambda}$ in any order of perturbation theory in the form

$$D_{\nu\alpha\lambda} = tr(\gamma_5 G_{\nu\alpha\lambda})$$

where $G_{\mu\nu\alpha}$ is represented as a set of Feynman diagrams. Then it is possible to expand the quantity $G_{\mu\nu\alpha}$ over the antisymmetric products of γ matrices Γ_n, which form a basis in the algebra of the Dirac's γ matrices, with finite coefficients $C^n_{\mu\nu\alpha}$

$$G_{\mu\nu\alpha} = \sum_{n=0}^{\infty} C^n_{\mu\nu\alpha}\Gamma_n.$$

All coefficients $C^n_{\mu\nu\alpha}$ are finite because of the finiteness of F_1 but some of them could contain divergences coming from subgraphs which eventually cancel each other. The anomaly is then given by $C^4_{\mu\nu\alpha}$ since only this coefficient survives in the limit $\epsilon \to 0$ after performing the trace operation. Taking into account these considerations one can write

$$C^4_{\mu\nu\alpha} = tr(G_{\mu\nu\alpha}\Gamma^4). \tag{7}$$

Eq. (7) does not contain γ_5 matrix and one can compute its right hand side within dimensional regularization. Let us consider now direct calculation. In the first order of perturbation theory one obtains from the triangle diagram

$$T_{\mu\nu\alpha}(p,q) = \int tr(\gamma^5(S(l+p+q)\gamma_\mu S(l)\gamma_\nu S(l-p+q) + (\mu \leftrightarrow \nu, q \leftrightarrow -q))\gamma_\alpha)dl \tag{8}$$

where $S(p) = \hat{p}^{-1}$ is the free fermion propagator. Making use the procedure explained above we get

$$\lim_{q^2\to\infty} q^2 F_1(q^2,p^2) = -F_3 = \pi^{-2}$$

that leads to the correct answer for the anomaly.

Now we turn over to calculation of strong interaction corrections to the three-point function given in eq. (8). The sum of all two-loop diagrams is finite due to Ward identity for the vector current and for divergent parts of the axial current. There is however an arbitrariness in definition of a finite part of the axial current in this order in the strong coupling constant $\bar{\alpha}_s$ since there are not any symmetry constraints to fix it in distinction from the vector current. In our case a chosen scheme of calculations defines the normalization of the axial current the anomalous derivative of which we calculate. For practical calculations of two-loop integrals it is convenient to use the Landau gauge. The leading order corrections to the propagator of massless fermion in Landau gauge are equal to zero and consequently instead of six diagrams one has to calculate only three diagrams with a vertex insertions.

Calculating these integrals within our scheme and then collecting all results together we get the following expression for the amplitude F_1

$$\lim_{q^2\to\infty} q^2 F_1 = \pi^{-2}(1 + C_F\frac{\bar{\alpha}_s}{\pi}) \tag{9}$$

where $\sum_{a=1}^{8} t^a t^a = C_F 1$.

2. The normalization of the axial current

Strictly speaking, to discuss the perturbative corrections to existent anomalies on the language of operator equations of the kind

$$\partial^\mu J^5_\mu = \frac{\alpha}{4\pi} F\tilde{F},$$

where $F_{\mu\nu}$ is the strength tensor of electromagnetic field, is possible only when the definitions of composite operators J^5_μ and $F\tilde{F}$ are rigorously fixed in corresponding order of perturbative expansion. One-loop or many-loop behavior of the operator equation depends on normalization of J^5_α and $F\tilde{F}$. With our choice of a model the operator $F\tilde{F}$ does not depend on $\bar{\alpha}_s$ to the considered order and we only have to elucidate which normalization is actually used for axial current in our scheme. In Landau gauge the expression for the axial vertex Λ^5_α in d-dimensional space-time is as follows

$$\Lambda^5_\alpha = \gamma_\alpha\gamma_5 - C_F\frac{\bar{\alpha}_s}{4\pi}(-6\gamma_\alpha\gamma_5 + \frac{1}{4\epsilon}\sigma_{\mu\nu}\gamma_\alpha\gamma_5\sigma^{\mu\nu} + ...) \tag{10}$$

The second term in parenthesis is inevitable since no commutation rule for γ_5 matrix has been specified in d-dimensional space-time. Without γ_5 matrix in this expression the quantity in parenthesis is equal to zero but in our situation we can not make such a simplification. Hence we conclude that within dimensional regularization there is a mixing between an ordinary operator of the axial current and a new operator $\bar{q}\sigma_{\mu\nu}\gamma_\alpha\gamma_5\sigma^{\mu\nu}q$ which appears in d-dimensional space-time. Just this operator generates the finite corrections to the anomaly. All these observations are eventually equivalent to the assertion that the axial current in four dimensional space-time has a normalization which differs from one for the vector current (the quantity in parenthesis in eq. (10) is equal to zero for the vector current). This difference causes the appearence of corrections to the anomaly in the second order of perturbation theory expansion.

Now it is instractive to compare the normalization of the axial current in two different schemes of computations – in the scheme of dimensional regularization and within the regularization with higher covariant derivatives, for instance. As is well known the latter method of regularization provides the normalization of the axial current that satisfies the Adler-Bardeen theorem [6]. For this purpose we consider an operator product expansion for T-product of two vector currents sandwiched between the vacuum and the pion state. The leading contribution reads

$$i\int\langle 0|TJ_\mu(x)J_\nu(0)|\pi\rangle e^{iqx}dx = \frac{2i\varepsilon_{\mu\nu\alpha\beta}q^\alpha}{q^2}C(\bar{\alpha}_s)\langle 0|J_5^\beta|\pi\rangle + ...$$

and the ellipsis denotes contributions of operators with higher dimension. Calculating a coefficient $C(\bar{\alpha}_s)$ within two schemes we obtain the following result for the matrix elements of the axial current

$$\frac{\langle q|J_5^\beta|q\rangle^{dim}}{\langle q|J_5^\beta|q\rangle^{hd}} = 1 + C_F\frac{\bar{\alpha}_s}{\pi},$$

where $|q\rangle$ is a quark state in the Fock space and the marks "*dim*" and "*hd*" denote dimensional regularization and regularization with higher derivatives respectively. This result is in an exact agreement with our previous calculations. In other words it means that corrections to the expression for the anomaly in this case appear due to a specific normalization of J_5^β that is the immediate consequence of dimensional regularization. To restore the Adler-Bardeen normalization of the axial current one must make the finite multiplicative renormalization of its matrix elements in the corresponding order of perturbation theory. So we can write the relation between Adler-Bardeen current and the axial current normalized within the dimensional regularization

$$(J^5_\mu)^{Ren}_{AB} = (1 - \frac{\alpha_s}{\pi}C_F)(J^5_\mu)^{Ren}_{dim.\ reg}. \tag{11}$$

It is important to point out that the experimentally observable quantity – the amplitude of the decay $\pi^0 \to 2\gamma$ – does not depend on any special choice of normalization condition of the axial current [7].

3. Anticommutator of γ_5 and γ_α matrices

As we have already shown the anomaly calculation within the above scheme leads to the normalization of the axial current that differs from that of the vector current at the one-loop level. The same result takes place within the t'Hooft-Veltman [8] prescription as well [9]. The following questions might be interesting:

1) Which properties of the γ_5 matrix in d-dimensional space-time are responsible for the appearance of these corrections?

2) Is it possible to perform all calculations using a self-consistent definition of the anticommutator of the γ_5 and γ_α matrices?

Now we try to investigate these questions.

One of possible definitions of γ_5 matrix can be [8]

$$\gamma_5 = \frac{1}{4!}\varepsilon^d_{\mu\nu\alpha\beta}\gamma^\mu\gamma^\nu\gamma^\alpha\gamma^\beta \tag{12}$$

where $\varepsilon^d_{\mu\nu\alpha\beta}$ is a completely antisymmetric tensor. Taking into account the general tensor structure of the anticommutator $[\gamma_5, \gamma_\alpha]_+$ one can write it as

$$\gamma_\alpha\gamma_5 + \gamma_5\gamma_\alpha \sim \Gamma_{\alpha\mu\nu\rho\lambda}\varepsilon^d_{\mu\nu\rho\lambda}, \tag{13}$$

where the quantity $\Gamma_{\alpha\mu\nu\rho\lambda}$ is antisymmetric under the permutations of all indexes.

All scalar expressions including this tensor should vanish in the limit $\epsilon \to 0$.

Now it is convenient to rewrite (13) in the following form

$$\gamma_\alpha\gamma_5 + \gamma_5\gamma_\alpha$$

$$= \frac{1}{4!}(\gamma^\mu\gamma^\nu\gamma^\alpha\gamma^\rho\gamma^\lambda + \gamma^\lambda\gamma^\rho\gamma^\alpha\gamma^\nu\gamma^\mu)\varepsilon^d_{\mu\nu\rho\lambda} \equiv B_\alpha, \tag{14}$$

which, as we see, is just the consequence of the definition (12). Some properties of the quantity B_α are:

$$\gamma^\mu B_\mu = B_\mu\gamma^\mu = -4\epsilon\gamma_5, \tag{15a}$$

$$tr B_\mu\gamma_\alpha\gamma_\beta\gamma_\rho = 0. \tag{15b}$$

As is well known, the suggestion that the anticommutator in eq. (14) is equal to zero in the d-dimensional space-time results in the contradiction

$$(4-d) tr\gamma_5\gamma_\mu\gamma_\nu\gamma_\alpha\gamma_\beta = 0. \tag{16}$$

Using the correct expression (14) and performing the same manipulations one gets a formula

$$2(4-d) tr\gamma_5\gamma_\mu\gamma_\nu\gamma_\alpha\gamma_\beta = -16\epsilon\varepsilon^d_{\mu\nu\alpha\beta},$$

which agrees with the definition (12).

Let us turn over the calculation of the anomaly within this approach. As we stressed the use of eq. (14) with vanishing right hand side leads to the eq. (16) that reveals no anomaly. So, we conclude that, within the correct approach, the anomaly is determined by the divergent integral containing B_μ. Indeed, writing the expression for the ordinary triangle diagram we find

$$p^\alpha T_{\alpha\mu\nu} = \frac{1}{2}(k_1 + k_2)^\alpha T_{\alpha\mu\nu} =$$

$$e^2 \int \frac{d^d l}{(2\pi)^d} trS(l-k_2)\hat{p}\gamma_5 S(l+k_1)\gamma_\mu S(l)\gamma_\nu + cross - symmetric\ \ term$$

$$= e^2 \int \frac{d^d l}{(2\pi)^d} (trS(l-k_2)(l+k_1)^\alpha B_\alpha S(l+k_1)\gamma_\mu S(l)\gamma_\nu$$

$$-trS(l-k_2)\gamma_5\gamma_\mu S(l)\gamma_\nu - tr\gamma_5 S(l+k_1)\gamma_\mu S(l)\gamma_\nu) + cross - symmetric\ \ term$$

The second and the third integrals in the r.h.s of this expression are equal to zero and anomaly is determined by the integral containing B_μ. Such a calculations seems more difficult than the ordinary one because the expression for B_μ contains five gamma matrices. However, it is crucial for the computational simplification to point out that the tensor

$$M_{\alpha\mu\nu\rho\lambda\tau} = tr\gamma_\alpha\gamma_\mu\gamma_\nu\gamma_\rho\gamma_\lambda B_\tau$$

is antisymmetric with respect to the permutations of the first five indexes and the expression for it can be written without any calculation of traces

$$12\ tr\gamma_\alpha\gamma_\mu\gamma_\beta\gamma_\nu\gamma_\lambda B_\tau = g_{\lambda\tau}\varepsilon^d_{\alpha\mu\beta\nu} - g_{\alpha\tau}\varepsilon^d_{\lambda\mu\beta\nu} - g_{\mu\tau}\varepsilon^d_{\alpha\lambda\beta\nu}$$

$$-g_{\beta\tau}\varepsilon^d_{\alpha\mu\lambda\nu} - g_{\nu\tau}\varepsilon^d_{\alpha\mu\beta\lambda}.$$

Using this expression it is easy to obtain the final result

$$q^\alpha T_{\alpha\mu\nu} = \frac{\alpha}{2\pi}\varepsilon^d_{\mu\nu\alpha\beta}k_1^\alpha k_2^\beta,$$

which is the well known expression for the anomaly. Let us now turn to discussion of the issue of normalization of the axial current. In accordance with the Adler-Bardeen theorem, if the axial current is normalized as the vector one, the higher order corrections to the one-loop expression for the anomaly disappear. Hence we would like to have the following relation between the axial and vector vertexes

$$(\Lambda^5_\mu)^{Ren} = (\Lambda_\mu)^{Ren}\gamma_5. \tag{17}$$

In order to satisfy eq. (17) it is necessary to perform the permutations of γ_5 matrix with two other γ matrices in the expression for the axial-vertex diagram. As a result of this operation we have additional terms. For the last term in eq. (10) one can write

$$\sigma_{\mu\nu}\gamma_\alpha\gamma_5\sigma^{\mu\nu} = \sigma_{\mu\nu}\gamma_\alpha\sigma^{\mu\nu}\gamma_5 - 64(\epsilon\gamma_\alpha\gamma_5 + B_\alpha).$$

The first term in the r.h.s. of this equation guaranties the desired normalization (17) and the second one should be subtracted within the framework of R – operation. This is equivalent to the finite multiplicative renormalization in eq. (11).

Conclusions

We formulate new schemes for the calculation of the anomaly by dimensional regularization.

The first of them does not require the additional commutation relations for γ_5 matrix in d-dimensional space-time. This scheme is applicable to any order of perturbation theory that makes it especially attractive.

It has been shown that within our method the normalization of the axial current differs from that for the vector one. In order to restore the conditions of the Adler-Bardeen theorem one should make a finite multiplicative renormalization of the axial current.

The second scheme is immediately based on the expression for the anticommutator $[\gamma_5, \gamma_\alpha]_+$. It allows one to perform very simple and instructive calculation for the one-loop anomaly and derive in the leading order the renormalized axial current which coincides with the Adler-Bardeen one.

Acknowledgements

The authors are indebted to V.A. Matveev for support and encouragement and to K.G. Chetyrkin, A.L. Kataev, S.A. Larin and V.A. Rubakov for useful discussions.

One of us (G.G.) thanks Prof. K-H. Becks for invitation at Oberammergau workshop.

References

[1] S. Adler, Phys. Rev. 177(1969)2426;
J.S. Bell, R. Jackiw, Nuovo Cimento A60(1969)47.

[2] J.C. Collins, Renormalization, Cambridge University Press, 1987;
M. Bos, PREPRINT UCLA/92/TEP/41 (1992).

[3] A.A. Anselm, A.A. Johansen, ZhETP 96(1989)1181, Yad.Fiz. 52(1990)882.

[4] G.M. Shore, G. Veneziano, Nucl. Phys. B381(1992)23.

[5] G.T. Gabadadze, A.A. Pivovarov, Pisma JETP 54(1991)305.

[6] S. Adler, W. Bardeen, Phys. Rev. 182(1969)1517.

[7] G.T. Gabadadze, A.A. Pivovarov, Yad.Fiz. 56(1993)257.

[8] G. t'Hooft, M. Veltman, Nucl. Phys. B44(1972)189.

[9] S.A. Larin, Phys. Lett. B303(1993)113;
K.G. Chetyrkin, J.H. Kuhn, PREPRINT TTP 93-7, 1993.

PROGRAMMING OF THE γ_5-MATRIX IN DIMENSIONAL REGULARIZATION IN MULTILOOP CALCULATIONS

S.A. LARIN[1]
NIKHEF-H, P.O. Box 41882, 1009 DB, Amsterdam, The Netherlands.

ABSTRACT

The technique for efficient programming of the γ_5 matrix in dimensional regularization in multiloop computation of Feynman diagrams is elaborated. The calculation of the three-loop anomalous dimension of the singlet axial current is presented as an example of the application of this technique.

1. Introduction

At present, dimensional regularization [1] and the minimal subtraction (MS) scheme [2] are indispensable tools for multiloop calculations of Feynman diagrams in Quantum Field Theory. The majority of *practical* multiloop calculations of physical quantities was done during last years using the technique of dimensional regularization. But everybody who performed multiloop calculations involving the γ_5-matrix agrees, probably, that the problem how to deal with this matrix in dimensional regularization consumes a lot of time and efforts. The following approaches to this problem were used in *practical* calculations: the prescriptions [3, 4] based on the original definition by 't Hooft and Veltman [1], the prescription keeping the four-dimensional anticommutation relation for γ_5 in D-dimensions [5] and dimensional reduction [6]. Here $D = 4 - 2\epsilon$, and ϵ is the famous parameter of dimensional regularization.

The most known and widely used technique to deal with γ_5 is the formalization [4] of the original definition [1]. In this technique one splits D-dimensional γ-matrices γ_μ into 4-dimensional $\overline{\gamma}_\mu$ and $(D-4)$-dimensional $\hat{\gamma}_\mu$-matrices: $\gamma_\mu = \overline{\gamma}_\mu + \hat{\gamma}_\mu$. Then one postulates that γ_5 anticommutes, as usual, with four-dimensional $\overline{\gamma}_\mu$- matrices and commutes with $\hat{\gamma}_\mu$- matrices. This is a self-consistent scheme which allows in principle to perform all calculations involving γ_5 in dimensional regularization. Unfortunately splitting γ-matrices into 4-dimensional and $(D-4)$-dimensional objects is a serious complication and can produce undesirable difficulties during programming and computing in multiloop calculations.

But it is possible to use the original definition [1] without splitting γ-matrices. This drastically simplifies practical multiloop computations. The corresponding technique was developed for the case of the non-singlet axial current in [7, 8] and extended for the case of the singlet axial current in [9].

[1] on leave from Institute for Nuclear Research (INR) of the Russian Academy of Sciences, Moscow 117312, Russia.

The effectiveness of this approach is confirmed by its use in such advanced calculations as the calculation of the deep inelastic structure function F_3 at the two-loop QCD level [10] and the calculation of the heavy top quark limit of the singlet axial current [11]. We would like to stress here that the approach is also used in the calculation of the α_s^3 correction to the hadronic decay width of the Z-bozon [12].

In the present contribution we discuss this technique following in part the paper [9] and stress its efficiency for multiloop computations. As the example of the application we present the calculation of the three-loop anomalous dimension of the singlet axial current, based on the calculation of the axial anomaly relation [13] at the three-loop level.

All calculations are performed within massless perturbative QCD. The computations are done by means of the program Mincer [14] written for the system of symbolic manipulations Form [15]. This program provides unique possibilities for performing multiloop calculations, see also [12].

2. The non-singlet axial current: the non-anomalous case

Let us first consider the case of the non-singlet axial current:

$$J_\mu^{5a}(x) = \overline{\psi}(x)\gamma_\mu\gamma_5 t^a \psi(x), \tag{1}$$

where ψ is a quark field and t^a is a generator of a flavor group.

In our opinion, the most practical definition of γ_5 for multiloop calculations in dimensional regularization (and the only one known to be self-consistent) is the original definition due to 't Hooft and Veltman [1]:

$$\gamma_5 = i\frac{1}{4!}\varepsilon_{\nu_1\nu_2\nu_3\nu_4}\gamma_{\nu_1}\gamma_{\nu_2}\gamma_{\nu_3}\gamma_{\nu_4}, \tag{2}$$

here the Levi-Civita ε-tensor is unavoidably a four-dimensional object and should be taken outside the R-operation where any object can be safely considered as a four-dimensional one; the indices $\nu_1 \ldots \nu_4$ are D-dimensional inside the R-operation as all other indices within dimensional regularization. From this definition one can unambiguously derive [9] the following definition of the non-singlet axial current originally proposed in [3]:

$$J_\mu^{5a} = i\frac{1}{3!}\varepsilon_{\mu\nu_1\nu_2\nu_3}\overline{\psi}\gamma_{\nu_1}\gamma_{\nu_2}\gamma_{\nu_3}t^a\psi. \tag{3}$$

The definition (3) is compact and saves computational time enormously so we will use it in what follows.

The anticommutativity of γ_5 is violated by the definition (2), so the standard properties of the axial current and Ward identities which are valid in four-dimensional regularization are also violated. In particular the renormalization constant Z_{MS}^{ns} of the non-singlet axial current in the MS-scheme is not equal to one any more. It was calculated in the three-loop approximation in QCD in [8]; we remind here the two-loop expression:

$$Z_{MS}^{ns} \quad = \quad 1 + a^2\frac{1}{\epsilon}(\frac{22}{3}C_F C_A - \frac{4}{3}C_F n_f), \tag{4}$$

where we use the notation $a = \frac{\alpha_s}{4\pi} = \frac{g^2}{16\pi^2}$ for the strong coupling constant, C_F and C_A are the Casimir operators of the defining and the adjoint representations of the color group and n_f is the number of quarks flavors.

To restore the renormalization invariance of the non-singlet axial current (i.e. to nullify its anomalous dimension) one should perform [16] an extra finite renormalization or in other words to introduce the extra finite renormalization constant $Z_5^{ns}(a)$. So the correct renormalized non-singlet axial current is:

$$(J_\mu^{5a})_R = Z_5^{ns}\ Z_{MS}^{ns}\ i\frac{1}{3!}\varepsilon_{\mu\nu_1\nu_2\nu_3}\overline{\psi}_B\gamma_{\nu_1}\gamma_{\nu_2}\gamma_{\nu_3}t^a\psi_B, \tag{5}$$

where $\psi_B = Z_2^{\frac{1}{2}}\psi$ is a bare quark field. The full anomalous dimension can be now nullified:

$$\gamma_J^{ns}(a) = \mu^2\frac{d}{d\mu^2}\log(Z_5^{ns}\ Z_{MS}^{ns}) = 0. \tag{6}$$

One can use the recipe of [7] and find Z_5^{ns} from the requirement that the renormalized axial and vector vertices coincide:

$$Z_5^{ns}R_{MS} < \overline{\psi}\ J_\mu^{5a}(0)\ \psi >= R_{MS} < \overline{\psi}\ J_\mu^a(0)\ \psi > \gamma_5, \tag{7}$$

where $J_\mu^a(x) = \overline{\psi}(x)\gamma_\mu\gamma_5 t^a\psi(x)$ is the vector current. This relation means that anticommutativity of the γ_5-matrix is effectively restored, so the standard Ward identities are also restored. This prescription for Z_5^{ns} automatically ensures zero anomalous dimension (6) because the anomalous dimension of the vector vertex is naturally zero.

The first impression is that the calculation of the axial vertex is rather cumbersome within adopted prescription (3) for the axial current: taking the ε-tensor outside R-operation seems to leave inside R-operation three extra uncontracted indices ν_1, ν_2, ν_3. But it is always possible to contract first a quantity under consideration with an extra ε-tensor to produce a scalar object. In our case we first multiply the axial vertex with $\gamma_\mu\gamma_5 = i\frac{1}{3!}\varepsilon_{\mu\rho_1\rho_2\rho_3}\gamma_{\rho_1}\gamma_{\rho_2}\gamma_{\rho_3}$ and take the trace. Now outside the R_{MS}-operation the product of four-dimensional ε-tensors can be represented in a standard way as the determinant of the four-dimensional metric tensors:

$$\varepsilon_{\mu\nu_1\nu_2\nu_3}\varepsilon_{\mu\rho_1\rho_2\rho_3} = g_{\nu_1\rho_1}(g_{\nu_2\rho_2}g_{\nu_3\rho_3} - g_{\nu_2\rho_3}g_{\nu_3\rho_2}) + \dots \tag{8}$$

Then outside the R_{MS}-operation these four-dimensional metric tensors can be safely considered as D-dimensional ones (which will add only inessential $O(\epsilon)$-terms to the renormalized axial vertex). These D-dimensional metric tensors $g_{\mu_i\rho_j}$ can be safely taken inside the R-operation. So one obtains finally a scalar expression inside the R-operation containing only D-dimensional objects which makes the practical calculations straightforward.

Calculating the axial vertex and the vector vertex in eq.(7) one finds Z_5^{ns}. The three-loop approximation for Z_5^{ns} was obtained in [8]; we recall here the two-loop expression:

$$Z_5^{ns} = 1 + a(-4C_F) + a^2(22C_F^2 - \frac{107}{9}C_FC_A + \frac{2}{9}C_Fn_f). \tag{9}$$

So to calculate any *physical* quantity involving the non-singlet axial current one can use the prescription (5) with Z^{ns}_{MS} and Z^{ns}_5 given in (4) and (9). This prescription has all typical features, of the approach developed in [4], where D-dimensional indices split into 4-dimensional and $(D-4)$-dimensional indices. But in our approach all indices during the calculations are D-dimensional which avoids in practical calculations all complications connected with splitting indices.

In the case of several γ_5 matrices in one fermion line one can naively anticommute them (because of the validity of eq.(7)) as in the prescription with anticommuting γ_5 [5] and use the standard property $\gamma_5^2 = 1$. So e.g. the spectral density (i.e. the imaginary part) of the correlator of two axial non-singlet currents coincides in the approach under consideration with one of two vector non-singlet currents.

3. The singlet axial current: the axial anomaly

Let us consider now the case of the singlet axial current which we define in the analogy with the non-singlet current (3):

$$J^5_\mu = \overline{\psi}\gamma_\mu\gamma_5\psi = i\frac{1}{3!}\varepsilon_{\mu\nu_1\nu_2\nu_3}\overline{\psi}\gamma_{\nu_1}\gamma_{\nu_2}\gamma_{\nu_3}\psi. \tag{10}$$

It is known that the singlet axial current is nontrivially renormalized because of the axial anomaly and the renormalization constant of the singlet axial current is nontrivial at the two-loop level [13, 17]. To receive the correct renormalized singlet axial current we need, as in the previous cases, to introduce the finite renormalization constant Z^s_5 in addition to the standard ultraviolet renormalization constant Z^s_{MS} within the MS-scheme:

$$(J^5_\mu)_R = Z^s_5 Z^s_{MS}(J^5_\mu)_B = Z^s_5 Z^s_{MS} i\frac{1}{3!}\varepsilon_{\mu\nu_1\nu_2\nu_3}\overline{\psi}_B\gamma_{\nu_1}\gamma_{\nu_2}\gamma_{\nu_3}\psi_B. \tag{11}$$

We have calculated within the MS-scheme the renormalization constant Z^s_{MS} of the singlet current at the three-loop level:

$$Z^s_{MS} = 1 + a^2[C_F C_A(\frac{22}{3\epsilon}) + C_F n_f(\frac{5}{3\epsilon})] + a^3[C_F C_A n_f(-\frac{22}{27\epsilon^2} + \frac{149}{81\epsilon})$$
$$+C_F C_A^2(-\frac{484}{27\epsilon^2} + \frac{3578}{81\epsilon}) + C_F n_f^2(\frac{20}{27\epsilon^2} + \frac{26}{81\epsilon}) + C_F^2 C_A(-\frac{308}{9\epsilon}) + C_F^2 n_f(-\frac{22}{9\epsilon})]. \tag{12}$$

To fix Z^s_5 one can require within dimensional regularization the conservation of the one-loop character of the operator relation for the axial anomaly which is valid in four-dimensional regularization:

$$(\partial_\mu J^5_\mu)_R = a\frac{n_f}{2}(G\tilde{G})_R, \tag{13}$$

where $G\tilde{G} = \varepsilon_{\mu\nu\lambda\rho}G^a_{\mu\nu}G^a_{\lambda\rho}$ and $G^a_{\mu\nu} = \partial_\mu A^a_\nu - \partial_\nu A^a_\mu + g f^{abc}A^b_\mu A^c_\nu$ is the gluonic field strength tensor.

To calculate Z^s_5 one have to calculate matrix elements of the l.h.s. and r.h.s. of the anomaly operator equation (13) between gluon states (so the famous anomalous triangle one-loop diagrams plus higher loops appear in the l.h.s.). In total it was necessary to compute about one hundred three-loop diagrams for the l.h.s. Some typical three-loop diagrams to be calculated are shown in fig.1.

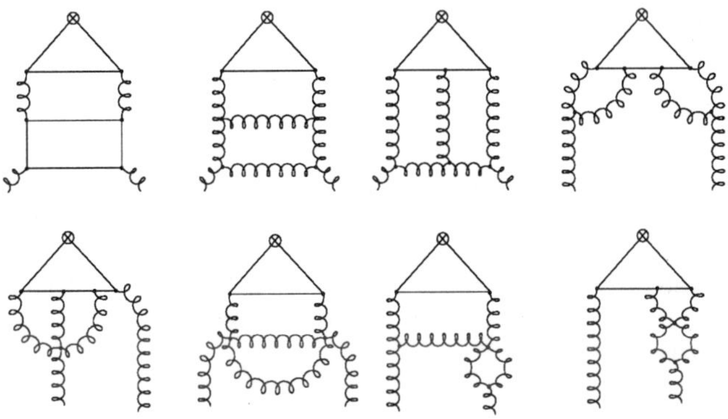

figure 1

Some typical two-loop diagrams contributing to the r.h.s can be obtained from the diagrams of fig.1 by shrinking the upper fermion loop into a point.

We can obtain the desired finite constant Z_5^s by restoring the one-loop character of the anomaly relation. The result is:

$$Z_5^s = 1 + a(-4C_F) + a^2(22C_F^2 - \frac{107}{9}C_AC_F + \frac{31}{18}C_Fn_f). \tag{14}$$

Thus both sides of the axial anomaly relation receive the non-trivial higher order corrections if one considers their matrix elements. But these corrections are matched and the one-loop character of the operator equation (which is valid in four-dimensional regularization) can be preserved in dimensional regularization.

Now we can calculate the full anomalous dimension of the singlet current in $O(a^3)$ approximation:

$$\begin{aligned} \gamma_J^s(a) &= \mu^2 \frac{d\log(Z_5^s Z_{MS}^s)}{d\mu^2} \\ &= +a^2(-6C_Fn_f) + a^3(-\frac{142}{3}C_FC_An_f + \frac{4}{3}C_Fn_f^2 + 18C_F^2n_f). \end{aligned} \tag{15}$$

The first term agrees with the calculation [17].

Thus in all calculations of *physical* quantities involving the singlet axial current one can apply the calculational power of dimensional regularization by using for the singlet axial current the prescription (11) with corresponding renormalization constants given in (12) and (14).

The generalization of the considered γ_5-prescription for the massive case is straightforward. Within the MS-scheme the ultraviolet renormalization constants do not depend on masses [18]. The same is valid for the finite renormalization constants Z_5, so the obtained finite constants can be applied also in the massive case.

Thus we should conclude that we have the efficient universal prescription to deal with the γ_5 matrix in dimensional regularization. The efficiency is ensured by keeping all indices during calculations as D-dimensional. It gives the possibility to use directly for multiloop computations with γ_5 such elaborated programs as the program Mincer [14] for the symbolic manipulation system Form [15].

4. Acknowledgements

I am grateful to D.J. Broadhurst, K.G. Chetyrkin, G.T. Gabadadze, V.A. Ilyin, A.L. Kataev, W.L. van Neerven, T. van Ritbergen and J.A.M. Vermaseren for helpful discussions.

5. References

1. G.'t Hooft, M.Veltman, *Nucl.Phys.* **B44** (1972) 189;
2. G.'t Hooft, *Nucl. Phys.* **B61** (1973) 455.
3. D.A. Akyeampong, R. Delborgo, *Nuov. Cim.* **17A** (1973) 578.
4. P. Breitenlohner, D. Maison, *Comm. Math. Phys.* **52** (1977) 11.
5. M. Chanowitz, M. Furman, I. Hinchliffe, *Nucl. Phys.* **B159** (1979) 225.
6. W. Siegel, *Phys. Lett.* **B84** (1979) 193.
7. S.G. Gorishny, S.A. Larin, *Phys. Lett.* **B172** (1986) 109.
8. S.A. Larin, J.A.M. Vermaseren, *Phys. Lett.* **B259** (1991) 345.
9. S.A. Larin, *Phys. Lett.* **B303** (1993) 113.
10. W.L. van Neerven, E.B. Zijlstra, *Phys. Lett.* **B297** (1992) 377.
11. K.G. Chetyrkin, J.H. Kuhn, Preprint TTP93-8 (1993), Karlsruhe.
12. S.A. Larin, T. van Ritbergen, J.A.M. Vermaseren, these Proceedings.
13. S.L. Adler, *Phys.Rev.* **177** (1969) 2426; J.S. Bell, R. Jackiw, *Nuov. Cim.* **60A** (1969) 47; S.L. Adler, W. Bardeen, *Phys. Rev.* **182** (1969) 1517.
14. S.A. Larin, F.V. Tkachov, J.A.M. Vermaseren, Preprint NIKHEF-H/91-18 (1991).
15. J.A.M. Vermaseren, *Symbolic Manipulation with Form* (Computer Algebra Nederland, Amsterdam, 1991).
16. T.L. Trueman, *Phys.Lett.* **B88** (1979) 331.
17. J. Kodaira, *Nucl. Phys.* **B165** (1980) 129.
18. J.C. Collins, *Nucl. Phys.* **B80** (1974) 341.

NUMERICAL APPROACH TO HIGH ORDER RADIATIVE CORRECTIONS IN QUANTUM ELECTRODYNAMICS

TOICHIRO KINOSHITA
Newman Laboratory, Cornell University
Ithaca, NY 14853, U. S. A.

and

Theory Group, KEK
Tsukuba, Ibaraki 305, Japan

ABSTRACT

A systematic procedure for high precision numerical evaluation of high order QED amplitudes is described. It is based on an algebraic manipulation program for converting momentum space integrals into those over Feynman parameters and an adaptive-iterative Monte Carlo numerical integration routine VEGAS. Problems encountered in applying this scheme to the evaluation of α^3 and α^4 corrections to the lepton anomalies and the $\alpha^2(Z\alpha)$ correction to the hyperfine structure of th e muonium ground state are discussed.

1. Introduction

In applications of quantum electrodynamics (QED), one routinely runs into multi-dimensional momentum space integrals. Although substantial progress has been made recently on analytic techniques, most integrals have thus far been inaccessible to analytic evaluation except for some low order ones or in the massless lim it. Where analytical evaluation is not yet available, numerical integration is the only way to obtain results relevant to experiment. Thus, it is important to establish a systematic procedure for high precision numerical evaluation of high order radiative corrections in QED and related processes.

Of course there are many ways to achieve such a goal. The approach I have chosen and applied to the evaluation of α^3 and α^4 radiative corrections to the lepton anomalies [1] and the $\alpha^2(Z\alpha)$ hyperfine structure of the mu onium ground state [2] consists of two steps:

1. Convert a Feynman amplitude defined in terms of momentum variables into an integral over Feynman parameters. This can be achieved using programs for algebraic manipulation.

2. Evaluate this integral by an adaptive-iterative Monte-Carlo integration routine, such as VEGAS [3]. This has thus far been the only effective method for evaluating such an integral, which may be huge and multi-dimensional in general.

In this talk I will discuss how this scheme has actually worked for these problems. More detailed discussion can be found in Refs. [1] and [2].

2. Automatic conversion of momentum space integrals into Feynman-parametric integrals

Conversion of a momentum space Feynman amplitude into an integral over Feynman parameters can be carried out most easily by writing a generic algebraic manipulation program. To accomplish this, I have written such programs using REDUCE, SCHOONSCHIP, and FORM at various occasions.

In my formulation, a significant simplification has been achieved using the analogy of Feynman diagram with electric network (following Bjorken and Landau) and expressing the integrand in terms of so-called Kirchhoff currents in addition to Feynman parameters (which play the role of electric resistance). These auxiliary functions themselves can be expressed in terms of Feynman parameters in a simple and well-defined manner. By stressing common features of related diagrams through parametric functions defined directly from the topology of the diagram, one can reduce redundancy of effort in writing programs for different d iagrams and obtain means of cross-checking related diagrams. This is particularly useful when one has to deal with a large number of Feynman diagrams with similar structure.

Such a formulation would be deficient, however, if it does not provide a rule for locating and subtracting ultraviolet (UV) and infrared (IR) divergences. We have developed an algebraic scheme based on UV and IR power counting, which provides a systematic way for constructing UV and IR subtraction terms. This is relatively straightforward for UV divergences. However, execution of this scheme for IR divergences is more subtle and requires an in-depth analysis of divergence mechanism. Consequently the treatment of IR divergence tends to be more specific to individual diagram and is difficult to make it fully automatic.

The IR divergence for the lepton anomalous magnetic moments is only logarithmic, and can be dealt with most easily by constructing certain intermediate on-shell renormalization procedure. On the other hand, IR divergences in the muonium hyperfine structure behaves as $\lambda^{-1/2}$, where λ is the photon mass. Thus far, this case has been handled only by evaluating the integrals for a number of small values of λ and extrapolating the result to zero photon mass.

3. Numerical integration routine VEGAS

Detailed description of VEGAS can be found in Ref. [3]. Let me sketch it briefly using the one-dimensional integral

$$I = \int_0^1 dx \; f(x) \tag{1}$$

as an example. Suppose N points x are randomly chosen from a set of points distributed in the interval [0, 1] with probability density $p(x)$. Then the integral I

is approximated by

$$S^{(1)} = \frac{1}{N}\sum_{x}\frac{f(x)}{p(x)}, \tag{2}$$

which converges to I as $N \to \infty$, where $p(x)$ is normalized to unity. The quantity $S^{(1)}$ will fluctuate about the true value I as different sets of N random points are chosen. The variance of this fluctuation is given by

$$\sigma^2 = \frac{1}{N}\left(\int_0^1 dx\frac{f^2(x)}{p(x)} - \left[\int_0^1 dx\ f(x)\right]^2\right) \simeq \frac{S^{(2)} - (S^{(1)})^2}{N-1}, \tag{3}$$

where

$$S^{(2)} = \frac{1}{N}\sum_{x}\left(\frac{f(x)}{p(x)}\right)^2. \tag{4}$$

In VEGAS, rather than dealing with the density function $p(x)$ of an arbitrary shape, one deals with a step function with Q steps such that the probability of a random number being chosen from any g iven step is a constant, equal to $1/Q$ for all steps $(0 = x_0 \le \cdots \le x_Q = 1, \Delta x_i = x_i - x_{i-1})$:

$$p(x) = \frac{1}{Q\Delta x_i}, \quad x_i - \Delta x_i \le x < x_i, \quad i = 1, \cdots, Q, \tag{5}$$

where

$$\sum_{i=1}^{Q}\Delta x_i = 1. \tag{6}$$

The probability distribution is tailored to a particular integrand simply by adjusting the incremental size Δx_i. This is done in the following way: Given N integrand evaluations, the probability distribution is refined by subdividing each increment Δx_i into $m_i + 1$ sub-increments, where

$$m_i = K\frac{\bar{f}_i\Delta x_i}{\sum_j \bar{f}_j\Delta x_j} \tag{7}$$

and

$$\bar{f}_i \equiv \sum_{x\in x_i-\Delta x_i}^{x_i} |f(x)| \propto \frac{1}{\Delta x_i}\int_{x_i-\Delta x_i}^{x_i} dx\ |f(x)|. \tag{8}$$

Thus each increment is subdivided into as many as $(K+1)$ sub-increments, and its contribution to the weight function increases in proportion to its contribution to the integral of $|f(x)|$, which i s the expected behavior of the optimal variance:

$$p(x) = \frac{|f(x)|}{\int dx|f(x)|}. \tag{9}$$

In order to restore the number of increments to its original value $(= Q)$, groups of the new increments are amalgamated into larger increments, the number of sub-increments in each group being constan t (to preserve the relative increment density). The net effect is to alter the increment sizes, while keeping the total number constant, so that the smallest increments occur where $|f(x)|$ is largest. The new grid is further refined in subsequent iterations until the optimal grid has been obtained.

The values $\bar{f}_i$ are discarded after each iteration to ease storage limitation. However, a cumulative estimate of the integral and its error can be made that utilizes every evaluation of the integrand:

$$\bar{I} = \sigma_I^2 \sum_i \frac{I_i}{\sigma_i^2} \quad with \quad \sigma_I = \left(\sum_i \frac{1}{\sigma_i^2} \right)^{-1/2} . \tag{10}$$

Here I_i and σ_i are the integral and standard deviation evaluated in the i-th iteration using Eqs. (2) and (3). The quality of the result of integration can be estimated by

$$\chi^2 \simeq \sum_i \frac{(I_i - \bar{I})^2}{\sigma_i^2}. \tag{11}$$

This algorithm can be readily generalized to integrals of arbitrary dimensionality.

The number of iterations and the number of integrand samplings per iteration needed clearly depend upon the complexity of the integrand and the accuracy being sought. Usually, initial few iterations give large errors, but they are essential for establishing a relatively stable grid structure. Once the (approximately) optimal grid has been found, the uncertainty (σ_I) in the subsequent iterations become roughly proportional to $N^{-1/2}$, N being the number of integrand evaluations .

4. Problems in high-precision evaluation of Feynman integrals

In this section I will discuss some salient features of the numerical work specific to higher-order QED calculations. In particular, it is important to understand thoroughly the nature of error estimates generated by VEGAS in order to assure the credibility of numerical results.

Errors encountered in these calculations arise primarily from the following three features of our integrands:

(a) In general our integrands are singular on some boundaries of the integration domain, whether renormalized or not.

(b) In our scheme, renormalization is performed numerically, which means mutual cancellation of ∞'s at every singular point on a boundary of the integration domain.

(c) The sheer size of the integrand, particularly in the order α^4, makes it difficult to accumulate an adequate number of integrand samplings for high precision work.

At first sight, the feature (a) seems to indicate that it is hopeless to obtain a reliable result for such a singular integrand. Actually, our integral is multidimensional and the size of the integration measure at singular points is so small that the integral itself is mathematically well defined and convergent once it is ren ormalized. Nevertheless, the singular nature of the integrand may be a cause for concern in the sense that the interval adaptation by VEGAS might not reach the optimal stage (9) as rapidly as one wou ld like. In some cases, it may even run away after just a few iterations. This problem can be alleviated, however, by "stretching" the integration variables. Suppose VEGAS finds after a few iterations that the integrand samplings are highly concentrated at one end of the integration domain, say $x_1 = 0$, of the axis x_1 of the hypercube. In such a case, if one maps x_1 into x_1' as

$$x_1' = x_1^{a_1}, \tag{12}$$

where a_1 is some real number greater than 1, the domain near $x_1 = 0$ is stretched out and random samplings in the x_1' variable give more attention to the region near $x_1 = 0$ from the beginn ing of iteration. Also, the Jacobian $a_1 x^{a_1 - 1}$ of this transformation helps reduce the peaking of the integrand. Similarly, the singularity at $x_1 = 1$ can be weakened by the stretching

$$x_1' = 1 - (1 - x_1)^{b_1}, \tag{13}$$

where b_1 is a real number greater than 1. One advantage of such a simple-minded mapping is that it may be applied independently to all variables $x_1, \ldots, x_n$. By an appropriate choice of $a_1, \ldots, a_n$ and/or $b_1, \ldots, b_n$, one can cure the run-away problem and accelerate the speed of convergence of the iteration considerably. Note that all these stretchings are one-to-one mappings of a unit hypercube onto another.

In practice, based on the result of several iterations by VEGAS, one decides to stretch some axes by choosing appropriate trial values of stretch parameters a_i and b_i, and then run several more iterations to see what happens to the structure of the integrand along each axis. One then readjusts the stretch parameters and tries again. In this manner one may be able to find, after a few trials, a set of stretch parameters that gives a reasonable speed of convergence, although this method will never be perfect in view of the complex structure of the integrand.

It will be obvious from this argument that there is a great deal of freedom in the choice of integration variables even if one restricts oneself to the unit hypercube. Note that the criterion of square integrability, which seems to be necessary for mathematically reliable estimates of variance, is actually dependent on the choice of integration variable. For the type of integrals we are interested, we can always make an appropriate stretch mapping so that the resulting integral is square integrable with respect to the new measure.

For some problems, one may have difficulty in finding a satisfactory stretching. In such a case, a simple alternative approach is to map the interval (0, 1) onto a slightly smaller interval (δ, 1 - δ), $0 \leq \delta \ll 1$, for all axes, which removes the troublesome region (as well as some harmless regions) from the integration domain. This will introduce an error of order $\delta^{1/2}$. Of course, the result must be monitored closely by varying δ.

Let us now discuss feature (b). In our approach, the integrand, whether renormalized or not, actually diverges on some boundaries of the integration domain. In cases that require no renormalization, the integral converges in spite of this singularity because it is compensated by a small enough integration measure. When renormalization is required, the integral is made convergent as a consequence of point-by-point cancellation of divergences by carefully tailored counter-terms.

All this would pose no problem if each step of computation were carried out with infinite precision. In reality we have to perform calculations in finite precision, typically in double precision arithmetics. The intended cancellation of divergences may fail occasionally because cancelling terms have no more than 12 or 13 significant digits (in double precision) and their difference tends to be dominated n ear a singularity by roundoff errors which may cause undesirable fluctuations. The difficulty is compounded further by register overflow. To the extent that this takes place in very small parts of the integration domain, it does not cause appreciable error in most integrals since the error is proportional to the fractional volume of th e integration domain involved. However, in some cases where this is a serious problem, it is found that the error can be reduced to a manageable level by adopting quadruple precision arithmetic, although it slows down the computati on substantially. In practice quadruple precision is needed only in the neighborhood of some singularities. Thus one may evaluate the bulk of the integral in double precision, resorting to quadruple precision only where it is absolutely needed. In this manner we have been able to reduce the impact of computation slowdown considerably.

Finally, let us discuss feature (c). The huge size of our integrands (particularly in the case of eighth-order magnetic moment calculation) means that it requires a large amount of computing time in order to accumulate adequate sampling statistics. Some of these integrands consist of more than 20,000 complicated rational functions, resulting in FORTRAN source code of 300 to 500 kilobytes in size. The bulkiness prevents us from performing a sufficient number of integrand samplings per iteration except on the fastest computers. This also makes it very time-consuming to experiment with alternate choices of variable stretching which are often required to test the reliability of numerical results. Indeed, this, combined with the unavailability (until mid 1980's) of adequate computing power, has been the main reason why it was not possible for a long time to calculate the eighth-order term of the electron anomalous magnetic moment with reasonable precision.

5. Acknowledgments

This research is supported in part by the U. S. National Science Foundation. The author would like to thank the hospitality of the National Laboratory for High Energy Physics (KEK), Japan, where he is visiting on sabbatical leave from Cornell University.

6. References

1. T. Kinoshita, in *Quantum Electrodynamics*, ed. by T. Kinoshita (World Scientific, Singapore, 1990), p. 218.
2. T. Kinoshita and M. Nio, to be published.
3. G. P. Lepage, *J. Comput. Phys.* **27** (1978) 192.

PROGRESS WITH MULTI-LOOP DIAGRAMS SINCE AI-92

D. J. BROADHURST
Physics Department, Open University, Milton Keynes MK7 6AA, UK

ABSTRACT

Progress with multi-loop diagrams, since the AI-92 workshop, is reported in the following 7 areas: summation of infinite series of ladder diagrams; large-N_f expansions in QED and QCD; 5-loop calculations in ϕ^4 theory; 4-loop QED β-function and muon-anomaly contributions; connections between deep-inelastic and annihilation processes at $O(\alpha_s^3)$; heavy-quark effective theory; 2-loop diagrams with masses.

I shall report on 7 developments, [1]–[7], in multi-loop perturbation theory, indicating, where appropriate, related contributions to this workshop, AI-93, where further references may be obtained. At AI-92 [8], I discussed the algebraic computability of diagrams; thanks to the activities of many workers in the last 18 months, the domain of calculability has been extended, considerably.

1. Summation of infinite series of ladder diagrams

As reported by Davydychev, in these proceedings, results are now available for massless L-loop diagrams contributing to 3-point and 4-point functions in $\lambda\phi^3/3!$ theory [9]. Previously, only 2-point ladder diagrams were known for arbitrary L [10]. In [1] it was shown that all these L-loop results are obtainable, via conformal transformations, from the x-space propagator

$$P_L(x_0, x_{L+1}) \equiv \left[\prod_{k=1}^{L} \int \frac{\mathrm{d}^4 x_k}{\pi^2 x_k^2}\right] \prod_{m=0}^{L} \frac{1}{(x_{m+1} - x_m)^2}, \tag{1}$$

representing spatial propagation, from x_0 to x_{L+1}, with L interactions, traced back to the origin, and given by the generating function

$$\sum_{L=1}^{\infty} (-g^2)^L P_L(x, y) = \int_0^{\infty} \frac{\mathrm{d}t}{(\mathrm{e}^t x - \mathrm{e}^{-t} y)^2} \frac{\mathrm{d}}{\mathrm{d}t} J_0\left(4gt\left[1 + \frac{1}{2t}\log\frac{x^2}{y^2}\right]^{1/2}\right), \tag{2}$$

which allows analytic continuation to the strong-coupling regime. In the case of the 2-point function, the infinite-coupling limit exhibits a remarkable feature: it is given by *minus* the missing *zero*-loop term, which perturbation theory both defines (by continuation to $L = 0$) and lacks. Specifically, the ladder contributions to the derivative of the self-energy, for time-like $p^2 = w^2$, give the infinite-coupling limit

$$\sum_{L=1}^{\infty} \frac{1-L}{2} \binom{2L}{L} \zeta(2L-1)(-\alpha)^L \to \frac{-\zeta(-1)}{2} = \frac{1}{24}, \tag{3}$$

as $\alpha \equiv (\lambda/4\pi w)^2 \to \infty$. Note that a infinite series of diagrams, each of which is infinite as $\alpha \to \infty$, yields a finite constant, familiar to string theorists. The strong-coupling limit of 3- and 4-point ladder sums is currently under study.

2. Large-N_f expansions of QED and QCD Green functions

As reported by Gracey, in these proceedings, all-orders results are now available for anomalous dimensions [11] in 4-dimensional gauge theories, in the limit $N_f \to \infty$. These are obtained by evaluation of skeleton contributions to the Dyson equations, in the neighbourhood of the large-N_f critical coupling, $\alpha^* \propto \varepsilon/N_f$, in $D \equiv 4-2\varepsilon$ dimensions. Experience with the Gross-Neveu model, near $D=2$, suggests the possibility of reaping further rewards in QED and QCD. Here I report results of the first attempt to extract large-N_f results for renormalized Green-functions [2], as opposed to anomalous dimensions.

For the $O(1/N_f)$ L-loop contributions to the Gell-Mann–Low function, with $[L-1]$ electron loops, the result

$$\Psi_L^{[L-1]} = \frac{(L-1)!}{(-3)^{L-1}}\left[-2L+4-\frac{L+4}{2^L}+\frac{16}{L-1}\sum_{L/2>s>0} s\left(1-2^{-2s}\right)\left(1-2^{2s-L}\right)\zeta(2s+1)\right], \quad (4)$$

was obtained in [2] and exploited in QCD studies [5]. An algorithm was found to compute, algebraically, the corresponding coefficients of the on-shell β-function of QED, allowing one to reconstruct the $O(1/N_f)$ photon propagator, to arbitrarily high order in α, and hence to find $O(1/N_f^2)$ contributions to the muon anomaly, ignoring terms of order m_e/m_μ. Thus the 4-loop analytical results of [12] and the corresponding 5-loop analysis of [13] were extended to 20 loops, analytically, and to 100 loops, to high precision. The behaviour of these contributions at large orders was studied in the on-shell, $\overline{\text{MS}}$, and MOM renormalization schemes. Extensions to QCD Green functions, and to higher orders in $1/N_f$, are under study.

3. 5-loop propagator in ϕ^4 theory

As reported by Chetyrkin, in these proceedings, a method [14] is fully articulated for the algebraic computation of 3-loop massless 2-point diagrams, giving 4-loop anomalous dimensions by the device of infra-red rearrangement, after the subtraction of subdivergences. Beyond that level, inspirational methods have been used to obtain 5-loop anomalous dimensions in ϕ^4 theory, with results recently corrected in [15]. Here I report further progress [3], yielding the $\overline{\text{MS}}$-renormalized 5-loop propagator of massless ϕ^4 theory, whose contributions include those from the following diagrams.

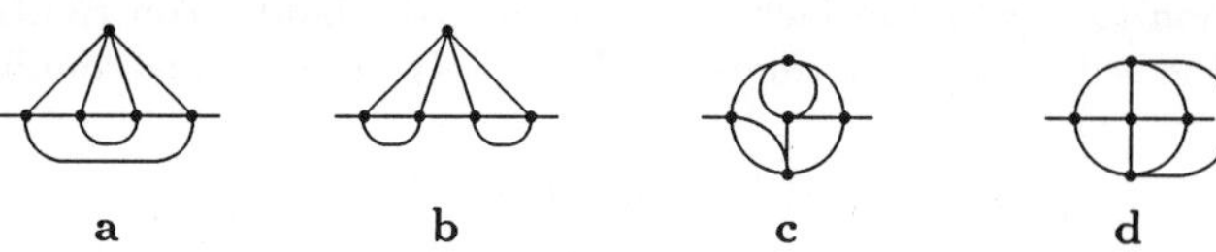

The essential idea is to exploit subdivergences in determining the Laurent expansions of the *un*-subtracted diagrams. For example, diagram (b) reduces to a 3-loop diagram, with two propagators raised to the power ε. Denoting the corresponding 3-loop diagram with propagators raised to the powers $\alpha \pm \beta$ by $F(\alpha, \beta)$, one may obtain its residues at $\alpha \in \{-3\varepsilon/2, -\varepsilon \pm \beta, -2\varepsilon \pm \beta\}$, from 2-loop diagrams that determine the subdivergences of the diagram and its overall divergence, for arbitrary ε. With $\alpha = \pm\beta$, $F(\alpha, \beta)$ is trivially reducible to a 2-loop diagram; with $\alpha = -\varepsilon$, a conformal transformation reduces it to a 2-loop diagram. From these 8 special cases, one obtains $F(\alpha, \beta)$, through $O(\varepsilon^2)$, for any α and β of $O(\varepsilon)$. Setting $\alpha = \varepsilon$ and $\beta = 0$, one obtains the Laurent expansion *and* the finite part of diagram (b).

Diagrams (a) and (c) also yield to this method. Diagram (d), whose $O(1/\varepsilon)$ term proved so troublesome in [14], may be determined from cases reducible to 2-loop diagrams and to a single finite 3-loop diagram, with the familiar value $20\zeta(5)$. Combining diagrams (a) to (d) with the remaining simpler diagrams, which are all reducible to 2-loop form, one confirms the revised 5-loop anomalous dimension, given in [15], and obtains, for the first time, a renormalized 5-loop Green function, as opposed to an anomalous dimension. The success of this approach suggests that the 6-loop anomalous field dimension may be calculable.

4. 4-loop QED β-function and muon-anomaly contributions

As reported by Kinoshita, in these proceedings, there exists a systematic method for obtaining numerical results for the QED contributions to the anomalous magnetic moments of the electron and muon, up to 4 loops, by Monte Carlo integration over Feynman parameters. At AI-92 [8], Kataev reported on successful analytical checks of certain 4-loop muon-anomaly contributions, due to insertions of 3-loop vacuum polarization diagrams with 2 electron loops. Here I report on analytical checks of the corresponding diagrams with only one electron loop, where discrepancies have been discovered [4] and resolved [17].

To check the corresponding numerical results of [18], one needs the asymptotic 3-loop photon propagator in the on-shell scheme, obtainable from the $\overline{\text{MS}}$ methods of the INR group [19] and the on-shell (OS) methods of [20]–[23]. Combining these, one finds significant disagreement with numerical integrations in [18]. Accordingly, the $\overline{\text{MS}}$ and OS results were rederived, ***ab initio***, using REDUCE and FORM, respectively, confirming a 25σ discrepancy [4]. Numerical re-evaluation [17], using VEGAS instead of RIWIAD, resulted in a shift by 27σ, giving final agreement, at a level consistent with a reasonable coefficient of the m_e/m_μ term, which cannot be determined by renormalization-group analysis.

An interesting by-product of this ASTEC collaboration is the analytical determination of the 4-loop coefficient of the QED β-function in the OS scheme, which differs from those of the $\overline{\text{MS}}$ and MOM schemes by less than 10%, showing that scheme-independent contributions dominate at 4 loops. Most notable is the contribution $\beta_4^{[1]} = -46$, from 4-loop diagrams with only one electron loop [19]. Confirmation of its value has recently come from the following unexpected source.

5. Connections between deep-inelastic and annihilation processes

As reported by Larin, in these proceedings, results have been obtained [24] for the 3-loop corrections to the Bjorken polarized and Gross–Llewellyn-Smith (GLS) sum rules of deep-inelastic scattering (DIS), in the $\overline{\rm MS}$ scheme. They were obtained by correcting the chiral-symmetry-breaking defects of the D-dimensional 't Hooft–Veltman γ_5 prescription, by a renormalization (with *no* anomalous dimension) of the antisymmetric tensor current $\overline{\psi}_i(\gamma_\kappa\gamma_\lambda\gamma_\mu - \gamma_\mu\gamma_\lambda\gamma_\kappa)\psi_j$, corresponding, for $D = 4$, to contracting the axial current A^ν with $\varepsilon_{\kappa\lambda\mu\nu}$. Here I report the discovery [5] of 7 intricate relations between these DIS results and those of INR group [19] for e^+e^- annihilation, namely for the 4-loop Adler function of the vector current, V_μ. The combined effect of these 7 relations is very easy to state: the product of the DIS and e^+e^- radiative corrections is of the form $(A + B\alpha_s)\beta(\alpha_s) + O(\alpha_s^4)$, for an arbitrary colour gauge theory, where A and B are linear and quadratic in the colour factors. This factorization of the β-function reduces from 11 to 4 the number of independent colour factors in the product. The 7 requisite relations between the complicated ζ-function-dependent coefficients of [19, 24] are wondrous to behold and reduce, effectively to zero, the probability that either set of results is in error.

In [5] this new connection was construed as deriving from Crewther's arguments in [25], according to which the conformal structure of the leading term in the OPE for the 3-point function $A_\kappa(x)V_\lambda(y)V_\mu(z)$ (giving the anomaly [26] that determines π_0 decay) is fixed by the product of 2-point OPE terms, with coefficients measured in DIS and annihilation. This relation is trivially satisfied in the quark-parton model, and hence at a trivial UV-stable fixed point. Though the process of renormalization breaks conformal invariance, by the introduction of a scale, Crewther's assumptions ought to hold at *any* fixed point, $\beta(\alpha^*) = 0$, where both the scale dependence and also the anomalous dimension of $\varepsilon_{\mu\nu\alpha\beta}G^{\mu\nu}G^{\alpha\beta}$ vanish, ensuring the essentially one-loop nature of the anomaly [26]. A fuller elucidation of this discovery is clearly required. In the meantime, one may rest assured that the previously unchecked term in the 4-loop QED function, namely $\beta_4^{[1]} = -46$, is correct, since it is obtained from the *reciprocal* of the scheme-independent radiative corrections to the 3-loop GLS sum rule in quenched QED, giving (with $a \equiv \alpha/4\pi$)

$$\beta^{[1]}_{\rm QED}(a) = \frac{\frac{4}{3}a^2}{1 - 3a + \frac{21}{2}a^2 - \frac{3}{2}a^3 + O(a^4)} = \tfrac{4}{3}a^2 + 4a^3 - 2a^4 - 46a^5 + O(a^6)\,. \tag{5}$$

6. Heavy-quark effective theory at 2 and 3 loops

Whilst considerable refinements to HQET [27] have been made recently, three obstacles to the computation and exploitation of multi-loop HQET diagrams persist [8, 23]: (i) the absence of an infra-red rearrangement procedure to enable the extraction of 3-loop anomalous dimensions from the finite parts of 2-loop 2-point diagrams; (ii) the fearsome complexity of polylogarithms in radiative effects involving a change of heavy-quark velocity; (iii) the fact that some leading-order HQET radiative corrections are huge [6].

Since Andrey Grozin, my HQET collaborator, was not able to attend this workshop, I offer the following observations here: (i) When 3-loop anomalous dimensions and 2-loop finite effects are formally commensurate in conventional QCD, the latter tend to dominate numerically [20], though the former are easy enough to derive. An algorithm is still required to relate the 3-loop anomalous dimension of the heavy-light mesonic current to the finite parts of 2-loop diagrams, obtainable from [28]. By contrast, 3-loop effects in the correlators of heavy-light-light baryonic currents appear easier to tackle. (ii) The only viable algebraic approaches to heavy-heavy radiative corrections appear to involve expansions at small velocity transfer [29], or asymptotic studies at large velocity transfer [30]. Otherwise one must labour to evaluate polylogs whose complexity is at least as great as those that derive from the mass ratios of multi-flavour QED [21]. (iii) The radiative corrections of HQET derive from quantum effects within a finite virtuality of the mass-shell of an indefinitely heavy quark. One should therefore not be surprised to find finite HQET effects with the same character as the finite differences between OS and $\overline{\mathrm{MS}}$ renormalization in QCD [20, 21], namely with large [6] coefficients. There is no non-relativistic guidance on how to sum them in the heavy-light sector; the problem is worse than trying to sum final-state radiative corrections to the Kurie plot of a high-Z nucleus, at highly-relativistic electron energies, where the Coulomb wavefunction is inapplicable. Fast brown muck is nasty.

7. 2-loop diagrams with masses

Finally I comment briefly on massive 2-loop diagrams, which are also considered in the talks by Fleischer and by Davydychev, where numerical approximations are obtained from small- and/or large-momentum expansions. Here I remark that all the 4-dimensional results of [31] have recently [7] been obtained as special cases of general D-dimensional results, obtained by algebraic methods in all the mass cases needed for the 2-point functions of QED and heavy-quark QCD. As might be expected, D-dimensional methods [7] are much easier than 4-dimensional ones [31]: the dilogarithms of the latter result from simple hypergeometric series for the former, of the form ${}_pF_q$, with $p = q+1$. Thanks to the exact results, accelerated convergence of expansions is easily investigated, with dramatic improvements being achieved by the use of Padé approximants, whose utility might be expected, since the 2-point functions are of Stieltjes-type. If, however, one needs 2-point integrals with a variety of internal masses, as in the electro-weak sector, the two-dimensional integral of logarithms, given by Kreimer [32], is convenient for Monte Carlo integration. There is however a faster alternative, when accuracy is needed: one-dimensional quadrature, with an integrand rapidly computable in terms of logarithms and elliptic integrals [31], the latter being speedily found by the process of the arithmetic-geometric mean. In the single-mass case, however, *analytical* applications of [7] are in progress, to obtain 2-loop gluon-condensate contributions in QCD sum rules and 2-loop coupling-constant evolution across heavy-quark thresholds.

Acknowledgements I am most grateful to my collaborators: Jochem Fleischer, Norman Gray, Andrey Grozin, Andrei Kataev, Karl Schilcher, and Oleg Tarasov.

1. D J Broadhurst, *Phys. Lett.* **B307** (1993) 132.
2. D J Broadhurst, *Z. Phys.* **C58** (1993) 339.
3. D J Broadhurst, OUT–4102–46 (October 1993).
4. D J Broadhurst, A L Kataev and O V Tarasov, *Phys. Lett.* **B298** (1993) 445.
5. D J Broadhurst and A L Kataev, *Phys. Lett.* **B315** (1993) 179.
6. D J Broadhurst and A G Grozin, *Phys. Lett.* **B274** (1992) 421.
7. D J Broadhurst, J Fleischer and O V Tarasov, *Z. Phys.* **C60** (1993) 287.
8. D J Broadhurst, in *New Computing Techniques in Physics Research II*, ed. D Perret-Gallix (World Scientific, 1992), p 579; A L Kataev, *ibid* p 631.
9. N I Ussyukina and A I Davydychev, *Phys. Lett.* **B305** (1993) 136.
10. V V Belokurov and N I Ussyukina, *J. Phys.* **A16** (1983) 2811.
11. J A Gracey, *Mod. Phys. Lett.* **A7** (1992) 1945; LTH–316 (1993).
12. T Kinoshita et al, *Phys. Lett.* **B254** (1991) 235; **B260** (1991) 193.
13. A L Kataev, *Phys. Lett.* **B284** (1992) 401.
14. K G Chetyrkin and F V Tkachov, *Nucl. Phys.* **B192** (1981) 159.
15. H Kleinert et al, *Phys. Lett.* **B272** (1991) 39.
16. K G Chetyrkin, A L Kataev and F V Tkachov, *Phys. Lett.* **B99** (1981) 147.
17. T Kinoshita, *Phys. Rev.* **D47** (1993) 5013.
18. T Kinoshita, B Nizic and Y Okamoto, *Phys. Rev.* **D41** (1990) 593.
19. S G Gorishny et al, *Comput. Phys. Commun.* **55** (1989) 381; *Phys. Lett.* **B256** (1991) 81; **B259** (1991) 144; **B273** (1991) 141; **B275** (1992) 512 (Erratum).
20. N Gray, D J Broadhurst, W Grafe and K Schilcher, *Z. Phys.* **C48** (1990) 673.
21. D J Broadhurst, N Gray and K Schilcher, *Z. Phys.* **C52** (1991) 111.
22. D J Broadhurst, *Z. Phys.* **C54** (1992) 599.
23. D J Broadhurst, On-shell renormalization in QCD and QED, in *Quarks 92* (World Scientific, 1993) in press.
24. S A Larin and J A M Vermaseren, *Phys. Lett.* **B259** (1991) 345.
25. R J Crewther, *Phys. Rev. Lett.* **28** (1972) 1421.
26. S L Adler and W Bardeen, *Phys. Rev.* **182** (1969) 1517.
27. A G Grozin, BUDKER–92–97 (1992).
28. D J Broadhurst and A G Grozin, *Phys. Lett.* **B267** (1991) 105.
29. E Bagan and P Ball, *Phys. Lett.* **B301** (1993) 249.
30. A G Grozin, Novosibirsk preprint (October 1993).
31. D J Broadhurst, *Z. Phys.* **C47** (1990) 115.
32. D Kreimer, *Phys. Lett.* **B273** (1991) 277.

Automatic reduction of loop amplitudes

Jian-Xiong WANG[1] **and Yoshimitsu SHIMIZU**

National Laboratory for High Energy Physics

Oho 1-1, Tsukuba, Ibaraki 305, Japan

Abstract

In order to develop an automatic system to calculate loop-diagrams in renormalizable theories, we have completed the main three parts of the system FDC[2]. These are written on REDUCE and RLISP.

---

[1]on leave of absence from Institute of Theoretical Physics, Academia Sinica

[2]Feynman Diagram Calculation

1 Introduction

Even we start with some simple ideas to solve a physical problem, we often find that the perturbative calculation to realize them needs very lengthy, complicated and tedious work. This is particularly true in gauge theories or standard model. Thus it is a common recognition that a system of computer programs is desirable which can help us to perform complicated works in an automatic way. To develop such programs is recently becoming a very interesting field as experiments require higher order perturbative calculations.

There are a lot of works[1] on this subject, but, the present situation is far from complete because of the complexity of the whole problem. We have to overcome many technical problems such as calculating tree Feynman diagrams, one-loop diagrams, further two-loop, $\cdots$, removing infrared and ultraviolet divergences, utilizing gauge invariance, imposing renormalization conditions, and so on. Thus if we want to solve all of these problems in an automatic way, we have to do a number of things.

Listing up the main issues required to complete the program for automatic calculation of amplitudes including loop-diagrams, we have the following items:

1. Build the physics model, i.e. construct the Lagrangian and quantize it.
2. Generation of all Feynman diagrams and amplitudes for a given process.
3. Analytic calculation of amplitudes of these diagrams and generation of the expression of the total squared amplitude
4. Feynman parameter integration [2].
5. Phase space integration.

If all these steps are performed by hand, one can find many tricks to simplify the results. For an automatic calculation system, applicable to general processes, it is quite difficult to introduce these tricks. However we can still expect that some systematic procedures can be introduced to simplify the calculations, such as symmetry of external identical bosons, gauge invariance and better method to evaluate amplitude squared.

It is obvious that we could not to computerize anything more before step 1. The physical model must be given by physicists prior to step 1. In this work, we try to realize a system from this starting point, i.e, step 1, towards the final results, but now on the middle way, just step 1, 2, 3 have been constructed. This system is written in REDUCE[3], some procedures work in REDUCE Algebraic mode and others in REDUCE Symbolic mode.

The paper is organized as follows. In Section 2, we give a description about the package FDC1 for Step 1. Section 3 is devoted to the package FDC2 which generate Feynman diagrams for a given process. In section 4, we present FDC3 which is used to perform analytic calculation of each diagram and generate total squared amplitude. Finally, the summary and conclusions are presented in Section 5.

2 To build the physical model

First of all, the physical model must be built in. It can be done by just introducing all the Feynman rules, counterterms and constant obtained from a handbook of the physical model. This is not a so hard task. But it is very unconvenient for a user to input new gauge or new model, such as SU(5) or suppersymmetry models. When a physical model is constructed, to quantize it and give all Feynman rules and counterterms by hand needs quit a lot of work, through there is a standard procedure to do it.

For this propose, we have constructed a package FDC1. It makes the input Of the Physical model easy. The following is the input data for Standard Model.

```
mode:='((gauge (1 u  1 b g2 ksi (gfix1 a ksi))
              (2 su 2 a g1 ksi (gfix2 b ksi)))
        (scalar ((h 2 h1 h2) (potential vv h) v0))
        (Fermions ((el 2 (eu 1 1) (eed 1 1)))
                  ((er 1 (eed 1 r)) (yukawa ff1 el h er))
                  ((ql 2 (qu 1 1) (qd 1 1) ) )
                  ((qdr 1 (qd 1 r) ) (yukawa ff2 ql h qdr) )
                  ((qur 1 (qu 1 r)) (yukawa ff3 ql h qur))))$

phymass:='((w wm) (z zm) (p 0) (h0 hm) (eed (fm eed))
           (eu 0) (qd (fm qd)) (qu (fm qu)))$
pchalist:='((a w z) (b p) (h1 x2) (h2 h0 x3))$
constl:='(g1 g2 v0 nu2 lad1)$
constflf:='(fc cos sin y)$
h1:=i*x2(1)$
h2:=(v0+h0(0)-i*x3(0))/2**0.5$
gfix1:=pd(a,v)+ksi*hmix(a,v)$
gfix2:=pd(b,v)+ksi*hmix(b,v)$
vv:=nu2*dager(h)*h-lad1*(dager(h)*h)**2$
ff1:=fc(eed)*dager(el)*h*er$
ff2:=fc(qd)*dager(el)*h*er$
ff3:=fc(qu)*dager(ql)*i*tp(1,2,2,2)*cconj(h)*qur$
ycharge:='((1 (h 1)))$
;end$
```

user do not needed to worry about ghost fields, which will appear when the gauge fields are quantized. In the input it is easy to change the gauge fixing terms to another gauge.

The output of this part is the list of all Feynman rules and the expression of counterterms which depends on the scheme (choice of a set of independent parameters as input quantities). The output are of two types, one is built in the system as the physical model. Another appeares in REDUCE algebraic mode, which is easy to be understood by user.

3 Generation of Feynman diagrams and amplitudes

The package FDC2 generate Feynman diagrams for a given process. The program is based on a naive way to construct diagrams, i.e, when a matrix element is calculated, we have:

$$S_{fi} = \ldots < 0|\psi(x_1)\psi(x_2)\ldots\psi(x_m)\sum_n \frac{i^n}{n!}\int dy_1^4\mathcal{L}(y_1)\ldots\int dy_n^4\mathcal{L}(y_n)|0> \ldots$$

We then use the Wick's theorem to reduce it and keep only the connected parts. Of course, all the terms which are equivalent should be generated only once. We do not consider about topology of diagram, so in some sense, there is no concept of diagram in the program. Instead, in each step where the Wick's theorem is used. the symmetry among points is searched so that we can keep only one out of many equivalent ways to apply the theorem. The symmetry factor is calculated based on this symmetry countering. In this sense, the way in our package is quite different from that in GRACE or FeynArts, they consider and generate all the topology at first.

If the package is applied to ϕ^4 theory, then each generated term is corresponding to each topology created in GRACE or FeynArts. In the case of ϕ^n theory, all the topologies of the theory with 3-point vertex, 4-point vertex, ... n-point vertex will be generated.

This package is designed to generate Feynman diagrams in any order of perturbation. The number of generated diagrams is countered in two ways, i.e, the number of diagrams which need to be numerically calculated and the number of diagrams which need to be analytically calculated. The second number must less or equal to the the first one. There may be a fermion-loop in a diagram in which many fermions appear. In this case the diagram should be analytically calculated once. In an another case, some internal particles will appear, which can be either Z-boson or photon, then these two diagrams need an analytically calculation once.

We have checked partly the generated amplitudes of quark selfenergy up to 4-loop in $SU(3) \times SU(2) \times U(1)$ Standard Model.

The output of this package is written in RLISP which will be used as the input for the next step.

4 Analytic calculation of each diagram

The package FDC3 is used to make an analytic calculation of loop diagrams. It is the largest package among these three package. In fact, it contains two separated packages FDC31 and FDC32 whose details are presented in another paper[4]. This package was tested and checked with other calculations for the case of QCD correction to $e^+e^- \rightarrow q\bar{q}\gamma$.

FDC31

In order to make the calculation efficient, we have to simplify the amplitudes. For this purpose we constructed FDC31. In which, a general method to factorize the calculation of loop amplitudes is presented and realized in REDUCE and RLISP environment.

In the method, before performing calculation of Feynman diagrams, tensor and gamma matrix bases and a general form of the total amplitude are generated, gauge invariance condition is applied to reduce the number of the independent coefficients of the bases. Also the symmetry group of identical particles is generated and used to simplify the expression of the total amplitude. Based on the above simplified amplitude, its square is calculated. By introducing the tensor and gamma matrix bases, we have avoided the problem, the square increasing of calculation as the number of diagrams increases. This problem appears when calculating the amplitude square directly. Moreover, the method utilizes the gauge invariance and symmetry on analytic level, therefore, most of the important cancellations will take place in analytic level, that makes numerical calculation easier. This part generates REDUCE source code for the next step and FORTRAN source code used in the final numerical calculation of cross section (except coefficients of the bases, see FDC32).

The gamma matrix bases chosen in our procedure is different from the sixteen gamma matrices since generally the later leads to more complicate general results. The method to construct the gamma matrix bases is limited to the case with one fermion line. We will present different way to deal with the case with more than one fermion lines in a forthcoming paper.

Our method, to simplify calculation and to get compact analytic results, makes the program rather lengthy and complicated, but it will save much CPU when in evaluating numerical results.

FDC31

The role of this part is to perform Feynman parameterization and loop momentum integration for loop diagram. All the divergent terms are separated from finite part and all the counterterms can be determined to the given perturbative order once the renormalization scheme is fixed. Relations due to the symmetry of the lagrangian imposed on counterterms are used to check the calculation. An example of two-loop diagram is presented to show that the produced counterterms are consistent with those obtained in other works. In this step only the one-particle irreducible diagrams are calculated. All the reducible diagrams are expressed in terms of known irreducible diagrams.

The amplitudes obtained above are projected into the bases generated in FDC1, and the contributions to the coefficients of the bases are determined. The contribution from each diagram to the total amplitude is therefore simply added to the coefficient of each base. The outputs of this part are REDUCE source code and FORTRAN source code. The latter is used to calculate the coefficients numerically.

5 Summary and Conclusion

We have constructed first 3 packages towards building a system FDC which is expected to perform perturbative calculation in an automatic way. The first and second have not yet been connected. When the connection is completed, one can get all the FORTRAN source programs necessary to calculate amplitude, except the scalar Feynman parameter integral library. We like to emphasize that in order to develop a practicable system which makes away with large scale calculation, we have to make a lot of study to simplify and compact the calculation in this package.

Acknowledgements

The author would like to thank Dr. J. Fujimoto and Prof. K. Kato for many discussion, to Prof. T. Kaneko and Dr. T. Ishikawa for helpful discussions especially on LISP and RLISP, to Dr. H. Tanaka for discussion about the interface of the system to CHANEL. One of us (J. X. W) is indebted to the Japan Society for the Promotion of Science for financial support.

References

[1] T. Kaneko, S. Kawabata and Shimizu, Comput. Phys. Commun. 43 (1987) 279. H. Tanaka, Comput. Phys. Commun. 58 (1990) 153; H. Tanaka, T. Kaneko and Y. Shimizu, Comput. Phys. Commun. 64 (1991) 149. E.E. Boos *et al.*, in "*New Computing Techniques in Physics Research*", ed. D. Perret-Gallix and W. Wojcik, p.573,1990, Edition du CNRS Paris; J. Kublbeck, M. Bohm and A. Denner, Comput. Phys. Commun. 60 (1990) 165; R. Mertig, M. Bohm and A. Denner, Comput. Phys. Commun. 64 (1991) 345.

[2] G. J. Oldenborgh, Comput. Phys. Commun. 66 (1991) 1. J. Fujimoto, Y. Shimizu, K.Kato and Y. Oyanagi, Prog. Theor. Phys. (1992), 1233 Chao-Hsi Chang and Jian-Xiong Wang, Comput. Phys. Commun. 69 (1992) 330.

[3] REDUCE User's Manual, version 3.0, A. C. Hearn et al.; Standard LISP REPORT, J.B. Marti et al..

[4] J. X. Wang, preprint KEK-TH-305, will appear in Computer Phys. Commun..

POSSIBLE AND IMPOSSIBLE IN MULTILOOP RENORMALIZATION GROUP

K.G.CHETYRKIN
Institute for Nuclear Research of the Russian Academy of Sciences, Moscow, 117312, Russia

ABSTRACT

The talk briefly discusses the current state of the art in doing multiloop renormalization group calculations in a completely analytic way. It stresses on the problem of reliability of available and future results and focuses on what can and should be done to facilitate calculations and minimize the chance of making errors via a proper use of computer algebra methods.

1. Introduction

The method of the renormalization group (RG) (see e.g. the book[1]) is of vital importance in modern quantum field theory. It may suffice to recall that the famous idea of the asymptotic freedom bases on the RG concept of the running coupling constant. The RG functions — β-functions and various anomalous dimensions — serve as coefficients in the RG equations and are expressed in terms of Feynman Integrals (FI's). The complexity of these integrals grows drastically with the number of loops.

Yet the last decade or so has seen successful stories of solving truly multiloop physically important RG problems. It has become possible due to significant breakthroughs in understanding the interplay between UV and IR structures of FI's, in creating analytical tools specially tuned for calculating RG functions as well as developing computer algebra programs and algorithms to cope with vast proliferation of diagrams in higher orders of perturbation theory.

In the same time the fraction of the published results which prove to be wrong and had been corrected afterwards increases significantly when one goes from 3-loop problems to 4-loops ones, not speaking about the 5-loop level. These facts call for attention and taking special measures against their repetitions in future.

In the talk I will discuss analytical calculational methods available along with some distinguished results obtained with their help. I will try in particular to analyze why incorrect results were obtained and what can and should, in my opinion, be done to improve the reliability of multiloop RG calculations.

I will not dwell on any specific technical details and also will completely skip over numerous fruitful applications of the ideas and methods developed in doing multiloop RG calculations to other physical problems (see in this connection[2,3,4]).

2. The current status and limitations of analytical methods

Let me begin by noting that all the analytical methods to be discussed can effectively be put into action only within a "calculation friendly" environment provided by dimensional regularization[5] (DR) and minimal subtraction (MS-) scheme[6]. The importance of the environment for multiloop calculations is difficult to overestimate[(1)]. It should also be stressed that from a purely calculational point of view the most valuable feature of dimensionally regularized FI's is their similarity with the usual absolutely convergent integrals. As a result almost all standard operations with dimensionally regularized FI's can be mathematically justified.

2.1. Infrared rearrangement and R^*-operation

The naturalness and convenience of the MS-scheme for RG calculations comes from the following remarkable statement[7]:

Theorem 1. *Any UV counterterm for any Feynman integral and, consequently, any RG function in arbitrary minimally renormalized model is a polynomial in momenta and masses.*

This observation was effectively employed by A. Vladimirov[8] to simplify considerably the calculation of the RG functions. The method was further developed and named Infrared ReaRrangement (IRR) in[9]. It essentially amounts to an appropriate transformation of the IR structure of FI's by setting zero some external momenta and masses (in some cases after some differentiation is performed with respect to the latter). As a result the calculation of UV counterterms is much simplified by reducing the problem to evaluating massless propagator-type FI's (so-called p-integrals). The method of IRR was ultimately refined and freed from unessential qualifications with inventing a so-called R^* -operation[10,11]. The following statement was proven[11]

Theorem 2. *Any (L+1)-loop UV counterterm for any Feynman integral may be expressed in terms of pole and finite parts of some appropriately constructed p-integrals.*

It should be noted that the R^* -operation is absolutely essential for the proof of Theorem 2, though in most (but not in all!) practical cases one could proceed without it. However, such a practice heavily limits the possibilities of checking calculations by choosing different ways of IRR. Next, Theorem 2 is a key statement for multiloop RG calculations as it reduces the general task of evaluation of (L+1)-loop UV counterterms to a well-defined and distinctly posed purely mathematical problem: the calculation of L-loop p-integrals. In the following we shall refer to the latter as the L-loop Problem.

2.2. p-integrals

While 1-loop Problem is done with ease already at the 2-loop level the evaluation of the master two-loop diagram (see Fig. 1a) is not absolutely trivial. The 2-loop Problem was solved after inventing and developing the Gegenbauer polynomial technique in x-space[9] (GPTX). *In principle* GTPX is also applicable to compute

[(1)]Of course, in computing, say, 1-loop β- function one could use a variety of approaches; however already at 2-loop level it is almost impossible to avoid a use of DR unless one is playing with a toy, usually scalar, model.

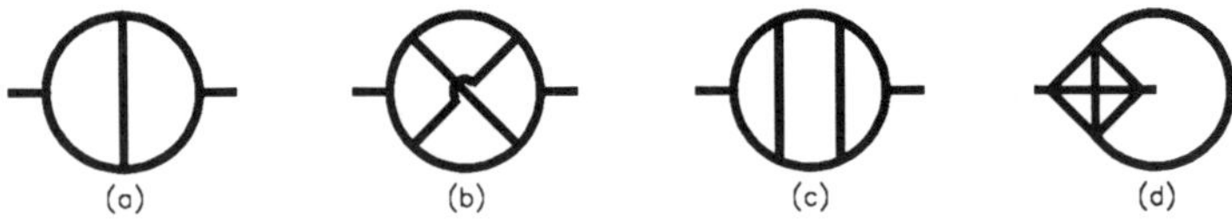

Figure 1: (a) the master 2-loop diagram; (b),(c) the master 3-loop non-planar and planar scalar diagrams; a 5-loop scalar diagram contributing to the field anomalous dimension of the φ^4 theory.

some quite non-trivial 3-loop p-integrals[(2)]. However, in practice calculations quickly get clumsy, especially for diagrams with nominators.

The main breakthrough at the 3-loop level happened with elaborating the method of integration by parts[12,13] of dimensionally regularized integrals. All (about a dozen) topologically different 3-loop p-integrals were neatly analyzed in[13] and a concrete calculational algorithm was suggested for every topology. As a result the algorithm of integration by parts for 3-loop p-integrals (AIP3) was developed and the following statement was obtained (by primitive p- integrals below we understand the p-integrals which may be performed by repeated application of the 1-loop integration formulas):

Theorem 3. *Any 3-loop p-integral may be identically transformed into a sum of primitive 1-loop p-integrals and two basic 3-loop p-integrals pictured in Fig. 1(b,c).*

Thus Theorems 2 and 3 solve *at least in principle* the task of analytical evaluation of any RG function in any model at the 4-loop level. It goes without saying that the solution of concrete RG problems is almost impossible without intensive use of the computer algebra methods. In addition to the old problem of taking long traces of Dirac γ matrixes AIP3, when applied even to a single 3-loop p-integral, generically produces dozens or even hundreds of terms. At the moment there exist essentially three different packages which implement AIP3. They are written in SCHOONSCHIP[14] (the works[15]) and in FORM[16] (the works[17,18]).

However, in genuine 4-loop calculations the reduction to p-integrals derscribed by Theorem 2 is far from being trivial and also includes a lot of purely algebraic manipulations. Typical steps here are

- to produce all the relevant diagrams with correct group and symmetrical factors;
- to reduce the initial FI to logarithmically divergent ones via a proper differentiation with respect masses and external momenta;
- to identify UV and IR divergent subgraphs of the resulting integral;
- to perform in a recursive way the R- and R^* to operations on the subgraphs.

[(2)] For example, the basic scalar non-planar 3-loop diagram of Fig. 1b was first done via GPTX.

In spite of its quite routine nature such a work is highly error-prone and time-consuming. At the moment I do not know about any algebraic package which would be able to do all three steps without human help.

2.3. Prospects

What about the 4-loop Problem? To my opinion, the chances to solve it — that is to expicitly construct a calculational algorithm for any of hundreds of toplogically different 4-loop p-integrals — are extremely small, at least in the foreseeable future. The main difficulty here is not the total amount of time and efforts to be invested but rather the absence of any general idea how it could be done in principle. Indeed, even for a simple φ^4 model the 5-loop RG calculations could be done analytically[19] up to the very end only due to the fact that for every genuine 4-loop scalar FI a special, sometimes quite involved trick was invented. For instance, the result for the diagram (d) of Fig. 1 was obtained in[13] only after as many as six seemingly ad hoc identities had been found and then combined. One can hardly hope that all non-scalar 4-loop p-integarls abounding in gauge theories could be treated in a similar way[(3)].

3. Some distinguished results and their reliability

In order to be definite I will consider only three distinguished problems, viz.

1. Calculation of the β-function in QCD[21,22] (3-loop level);
2. The calculation of the strong interaction corrections to the photon anomalous dimension in QCD or, equivalently, the evaluation of the famous R ratio in massless QCD[23,24] (4-loop level);
3. The calculation of the β-function in scalar φ^4 model[19] (5-loop level).

As for the reliability of these results I would like to say the following.

- The 3-loop result for the QCD β-function is very reliable indeed as it was obtained by two independent groups[21,22] which where using different methods and different algebraic computer programs.
- The initial solution of the second problem obtained in[25] was later essentially corrected by two (not quite independent! See the discussion in[3,26]) groups[23,24]. Moreover, the groups were using very similar algebraic computer programs. To my opinion, the existing results of works[23,24] should be considered just reliable. It means that a recalculation made by a really independent group using really different algebraic programs and employing a general covariant gauge would be strongly welcomed. It should be also noted here that a successful implicit test of the results of the works[23,24] has been recently performed in[27].

[(3)] Of course, some restricted classes of FI's can in some cases be computed in very high if not all numbers of loops[2].

- A similar situation exists with the third problem. Here a recent recalculation[19] revealed a few errors in the results obtained in[20], most of them of trivial nature, such as typos in the way of renormalizing away UV and IR subdivergencies. Again it would be nice to have a really independent recalculation.

4. What can and should be done to avoid errors in future

The first result for $R(s)$ was wrong due to a bug in a computer program. Once the bug is removed the result should be correct *provided* no errors are made in doing IRR. However, in the case of the β-function in the ϕ^4 model the errors were made in the process of doing IRR for hundreds of diagrams. It means that recalculations of both results would again require vast human efforts spent on doing something which may well be done by a computer!

Thus the problem of creating an algebraic program capable to cope with the combinatorics of R and R^* operations is on the agenda. Such a package should have the facilities to do the following:

1. generate all relevant diagrams and calcuclate their symmetrical and group weights;
2. identify divergent subgraphs;
3. use the IRR procedure in order to express the UV countertem for every diagram in terms of p-integrals;
4. evalute the resulting p-integrals.

It seems probable that the package will use different algebraic languages for implementing different steps. For instance, the step 4 needs a system like FORM or SCHOONSCHIP, while the first three steps are more conveniently to be programmed with MATHEMATICA[28]. In fact, already existing packages such as FeynArts[29], FeynCalc[30] and MINCER[17] may well be used as a starting point in creating the new package.

Once the program is written it can be very reliably debugged by repeating the existing calculations. But its main applications will be to perform new calculations at the 3- and 4-loop levels in different field-theoretical models.

5. Conclusions

I begin with the pessimistic prediction that the 4-loop Problem will remain unsolved in this century. Unfortunately, this also means that the corrections of order α_s^4 to the total cross section of e^+e^- annihilation into hadrons and to the tau-lepton semihadronic decay rate will presumably remain unknown for a long time. On the other hand there are a lot of tasks which can and I hope will be done in the near future. They include in particular:

- Creating a "universal" package for multiloop RG calculations; it should work automatically starting from a given Lagrangian and a given set vertex functions and ending with the relevant renormalization constants and RG functions.
- Recalculation with the help of the package of previously done multiloop RG problems.
- Computation of RG functions of the standard model and various gauge and supersymmetric models at the four loop level.

Acknowledgments

I am indebted to A. A. Vladimirov, D. I. Kazakov and D. V. Shirkov for numerous discussions which ignited my interest to multiloop calculations many years ago. I appreciate the warm hospitality of Institute of Theoretical Particle Physics at the Karlsruhe University where this work was finished. I thank Mrs. M. Frasure and A. Kwiatkowski for their help in preparing the manuscript. A stipendium from the HERAUES-stiftung is gratefully acknowledged.

References

1. N. N. Bogoliubov and D. V. Shirkov, *Introduction to the theory of the quantized fields,* 4th supplemented edition (Nauka, Moscow, 1984).
2. D. J. Broadhurst, in: *New Computing Techniques in Physics Research II,* ed. D. Perret-Gallix (World Scientific, 1992) 579.
3. A. L. Kataev, in: *New Computing Techniques in Physics Research II,* ed. D. Perret-Gallix (World Scientific, 1992).
4. D. J. Broadhurst, A. L. Kataev and O. V. Tarasov, *Phys. Lett.* **B298** (1993) 445.
5. G.'t Hooft and M. Veltman, *Nucl. Phys.* **B44** (1972) 189;
   C. G. Bollini and J. J. Giambiagi, *Phys. Let.* **40B** (1972) 566;
   G. M. Cicuta and E. Montaldi, *Nuovo Cim. Lett.* **4** (1972) 329;
   J. F. Ashmore, *Nuovo Cim. Lett.* **4** (1972) 289.
6. G. 't Hooft, *Nucl. Phys.* **B61** (1973) 455 .
7. J. C. Collins, *Nucl. Phys.* **B92** (1975) 477.
8. A. A. Vladimirov, *Teor. Mat. Fiz.* **43** (1980) 210.
9. K. G. Chetyrkin, A. L. Kataev and F. V. Tkachov, *Nucl. Phys.* **B174** (1980) 345; Preprint *INR P-0170* (1980) unpublished.
10. K. G. Chetyrkin and F. V. Tkachov, *Phys. Lett.* **B114** (1982) 340.
11. K. G. Chetyrkin and V. A. Smirnov, *Phys. Lett.* **B144** (1984) 419.
12. F. V. Tkachov, *Phys. Lett.***B100** (1981) 65
13. K. G. Chetyrkin and F. V. Tkachov, *Nucl. Phys.***B192** (1981) 159.

14. M. Veltman, *SCHOONSCHIP, A CDC 6600 program for symbolic evaluation of algebraic expressions, CERN Report* 1967 (unpublished).

15. S. G. Gorishny, S. A. Larin and F. V. Tkachov, Preprint *INR P-0330* (1984) (unpublished);
S. G. Gorishny, S. A. Larin, L. R. Surguladze and F. V. Tkachov, *Comput. Phys. Commun.* **55** (1989) 381;
L. R. Surguladze, Preprint *INR P-0643* (1989) (unpublished.)

16. J. A. M. Vermaseren, *Symbolic Manipulation with FORM, Version 2, CAN, Amsterdam*, 1991.

17. S. A. Larin, F. V. Tkachov, J. A. M. Vermaseren, *Preprint NIKHEF-H*/91-18 (1991).

18. L. R. Surguladze, FERMILAB-PUB-92/191 -T (1992).

19. H. Kleinert, J. Neu, V. Schulte- Frohlinde, K. G. Chetyrkin and S. A. Larin *Phys. Lett.* **B272** (1992) 39;
D. I. Kazakov, *Phys.Lett.* **B 133** (1983) 406.

20. K. G. Chetyrkin, S. G. Gorishny, S. A. Larin and F. V. Tkachov, *Phys. Lett.* **B132** (1983) 351; Preprint INR P-0453, Moscow, 1986.

21. O. V. Tarasov, A. A. Vladimirov, A. Yu. Zharkov, *Phys. Lett.* **B93** (1980)429.

22. S. A. Larin and J. A. M. Vermaseren, *Phys. Lett.* **B303** (1993) 334.

23. S. G. Gorishny, A. L. Kataev and S. A. Larin, *Phys. Lett.* **259 (1991) 144.**

24. L. R. Surguladze, M. A. Samuel, *Phys. Rev. Lett.* **66** (1991) 560, 2416 (erratum);
L. R. Surguladze, M. A. Samuel, *Phys. Rev.* **D44** (1991) 1602.

25. S. G. Gorishny, A. L. Kataev, S. A. Larin, *Phys. Lett.* **B212** (1988) 238.

26. A. L. Kataev, *University of Michigan report UM-TH-91-04* (1991) (unpublished).

27. D. J. Broadhurst and A. L. Kataev, *Phys. Lett.* **B315** (1993) 179.

28. S. Wolfram, *Mathematica : A System for Doing Mathematics by Computer* (Addison-Wesley , New York, 1988).

29. J. Kublbeck, M. Bohm and A. Denner, *Comput. Phys. Commun.* **60** (1990) 165.

30. R. Mertig, M. Bohm and A. Denner, *Comput. Phys. Commun.* **64** (1991) 345.

PROGRESS IN EVALUATING SOME COMPLICATED TYPES OF FEYNMAN DIAGRAMS WITH TWO (AND MORE) LOOPS

A. I. DAVYDYCHEV
Institute for Nuclear Physics, Moscow State University,
119899, Moscow, Russia
and
Department of Physics, University of Bergen,
Allégaten 55, N-5007 Bergen, Norway

ABSTRACT

Problems occurring in physically important non-trivial examples of loop calculations are discussed. A procedure of deriving expansions of two-loop self-energy diagrams with different masses is constructed. The cases of small and large external momentum are considered. The coefficients of the expansions are calculated analytically. Comparison with numerical results shows good agreement below the first threshold and above the last threshold of the diagram. An approach to evaluating three- and four-point ladder diagrams with massless internal particles and off-shell external momenta is presented. Exact results for an arbitrary number of loops are obtained. Evaluating non-planar diagrams is also discussed.

In the present talk I would like to discuss some recent results for loop diagrams obtained in collaboration with J.B. Tausk [1,2], V.A. Smirnov [2] and N.I. Ussyukina [3,4,5]. The papers [1,2] are devoted to the examination of massive two-loop self-energy diagrams, whilst refs. [3,4,5] deal with some three- and four-point diagrams with massless internal particles and off-shell external momenta.

1. Two-loop massive self-energy diagrams

It is known that any two-loop self-energy diagram (see Fig. 1) can be reduced to scalar integrals [6]. Moreover, in the case when the powers of denominators are integer, the diagram in Fig. 1b can be reduced to Fig. 1a by the decomposition of

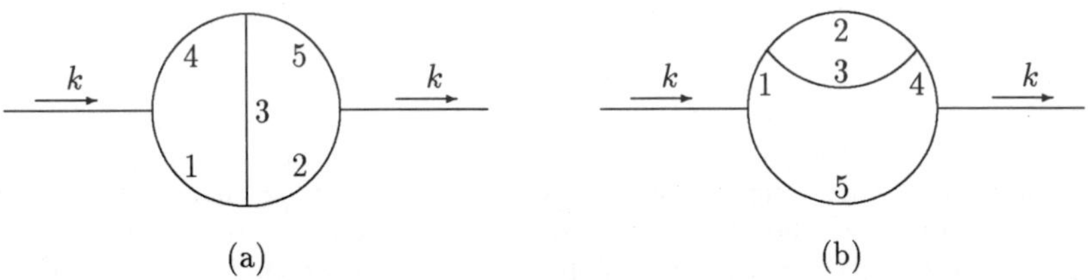

Figure 1: Two-loop self-energy diagrams

the first and the fourth denominators. So, we can express all two-loop two-point contributions in terms of the scalar integrals corresponding to Fig. 1a,

$$J(\{\nu_i\};\{m_i\};k) = \int\int \frac{\mathrm{d}^n p\, \mathrm{d}^n q}{D_1^{\nu_1} D_2^{\nu_2} D_3^{\nu_3} D_4^{\nu_4} D_5^{\nu_5}}, \tag{1}$$

where $D_i = (p_i^2 - m_i^2 + i0)$ are massive denominators, ν_i are the powers of these denominators, $n = 4-2\varepsilon$ is the space-time dimension [7], and p_i are constructed from the external momentum k and the loop integration momenta p and q (with due account of the momentum conservation).

For some special cases the integrals (1) can be evaluated exactly [8,9] and expessed in terms of the polylogarithms Li_N [10] with $N \leq 3$. On the other hand, the problem of evaluating such diagrams when all the internal lines are massive is more complicated, and exact expressions are not known. For example, in ref. [11] the result was presented in terms of a two-fold integral representation.

In the papers [1,2] we studied expansions of massive two-loop two-point diagrams for the case of small [1] and large [2] values of the external momentum squared. Note that in general the diagram in Fig. 1a has four physical thresholds (with respect to k^2), corresponding to different cuts of the diagram; namely,

$$(m_1+m_4)^2,\ (m_2+m_5)^2,\ (m_1+m_3+m_5)^2,\ (m_2+m_3+m_4)^2.$$

The cases considered correspond to the situations when we are either below the lowest threshold [1] or above the highest one [2].

For small values of k^2, the expansion of the integral (1) is a usual Taylor expansion (we consider the general case, when the lowest thershold is not equal to zero). Coefficients of the expansion can be represented in terms of two-loop massive "vacuum" diagrams. Moreover, for the case when all powers of denominators are integer, they can be reduced to vacuum diagrams, each of them depending no more than on three different masses (see Fig. 2). These diagrams with higher (integer)

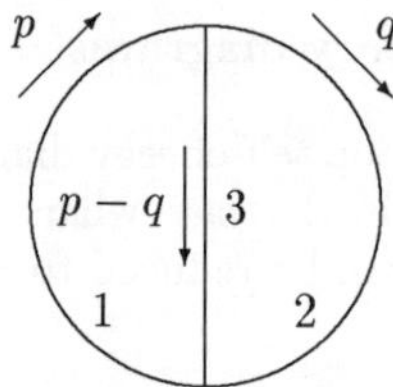

Figure 2: The vacuum two-loop diagram

powers of denominators (occurring in the coefficients of the expansion) can be evaluated by use of the integration-by-parts technique [12] (see also [13]), and all of them (up to finite in ε parts) can be represented in terms of dilogarithms or Clausen's function [10]. By use of the REDUCE system [14], we constructed an algorithm of analytical evaluation of the coefficients of the expansion. We compared the results for some diagrams with a numerical calculation based on the integral representation [11], and we found a nice agreement in the region below the lowest threshold.

The case when k^2 is larger than the highest threshold of the diagram is more complicated, because the corresponding expansion is not a usual Taylor expansion, but also contains logarithms and squared logarithms of $-k^2$ (in four dimensions) yielding an imaginary part when the momentum is time-like. To obtain this expansion, we applied a general mathematical theorem on asymptotic expansions of Feynman integrals in the limit of large external momentum. Expansions of this kind were presented in refs. [15,16]. For our case (1), the asymptotic expansion theorem gives

$$J_\Gamma \underset{k^2 \to \infty}{\sim} \sum_\gamma J_{\Gamma/\gamma} \circ \mathcal{T}_{\{m_i\};\{q_i\}} J_\gamma, \tag{2}$$

where Γ is the main graph (see Fig.1a), γ are subgraphs involved in the asymptotic expansion (see below), Γ/γ is the reduced graph obtained from Γ by shrinking the subgraph γ to a single point, J_γ denotes the dimensionally-regularized Feynman integral corresponding to a graph γ (for example, J_Γ is given by (1)), $\mathcal{T}_{\{m_i\};\{q_i\}}$ is the operator of Taylor expansion of the integrand in masses and "small" momenta q_i (that are "external" for the given subgraph γ, but do not contain the "large" external momentum k), and the symbol "$\circ$" means that the resulting polynomial in these momenta should be inserted into the numerator of the integrand of $J_{\Gamma/\gamma}$. It is implied that the operator $\mathcal{T}$ acts on the integrands before the loop integrations are performed.

In our case (see Fig. 1a), the sum (2) goes over all subgraphs γ that become one-particle irreducible when we connect the two vertices with external momentum k by a line. These subgraphs (there are five different types of them) are shown in Fig. 3 (dotted lines correspond to the lines that do not belong to γ).

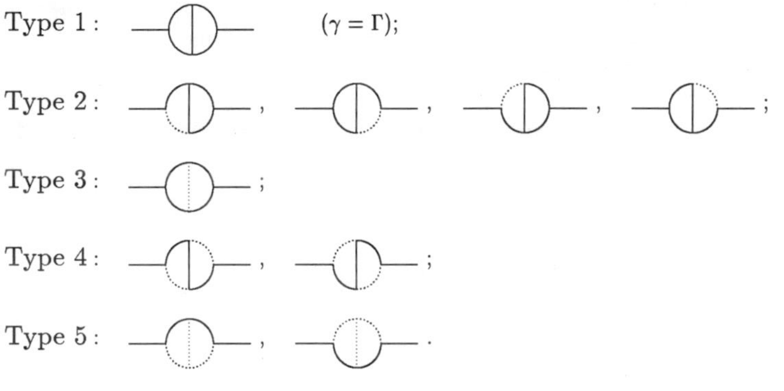

Type 1 : ($\gamma = \Gamma$);

Type 2 : , , , ;

Type 3 : ;

Type 4 : , ;

Type 5 : , .

Figure 3: The subgraphs γ contributing to the large k^2 expansion

The reduced graphs Γ/γ correspond to the dotted lines and can be obtained by shrinking all solid lines to a point. In such a way, we obtain that for the second and third type (see Fig. 3) $J_{\Gamma/\gamma}$ corresponds to a massive tadpole, for the fourth type we obtain a product of two massive tadpoles, while for the fifth type we get a two-loop massive vacuum integral with three internal lines, corresponding to Fig. 2. All other integrals (occurring in separate terms of the expansion) can also be evaluated.

We have realized this algorithm also by use of the REDUCE system [14]. The analytical expressions for the coefficients contain powers and logarithms of the masses and the external momentum squared, and also the function of masses corresponding to the two-loop vacuum diagram (Fig. 2) that can be expressed in terms of dilogarithms. For some special cases occurring in the Standard Model, we compared our expansion with the result of the numerical integration [11], and we found good agreement in the region above the highest threshold. The asymptotic value as $k^2 \to \infty$ is, of course, $-6\zeta(3)\pi^4/k^2$ (in the case of unit powers of denominators).

Recently some of these algorithms have been used for the renormalization group analysis of hadronic decays of a charged Higgs boson [17].

2. Three- and four-point massless diagrams

The methods considered in Section 1 can be also applied to massive diagrams with larger number of external lines. When constructing the expansion for large values of the external momentum invariants, we shall need results for the corresponding diagrams with massless internal particles. The examination of such diagrams is also important for calculation of some types of radiative corrections and for analysis of some theoretical models.

In papers [3,5] we considered two-loop three- and four-point contributions with off-shell external momenta. We have also considered three- and four-point ladder contributions with an arbitrary number of loops [4]. We used the following tools: (i) the Feynman parametric representation, (ii) the "uniqueness" conditions [18], (iii) Mellin–Barnes contour integrals and (iv) Fourier transform to the coordinate space. We considered only scalar diagrams (corresponding to the massless ϕ^3 theory), because expressions occurring in realistic calculations can be reduced to such scalar integrals.

For the two-loop three-point diagram $C^{(2)}$ (see Fig. 4a), we obtained a simple

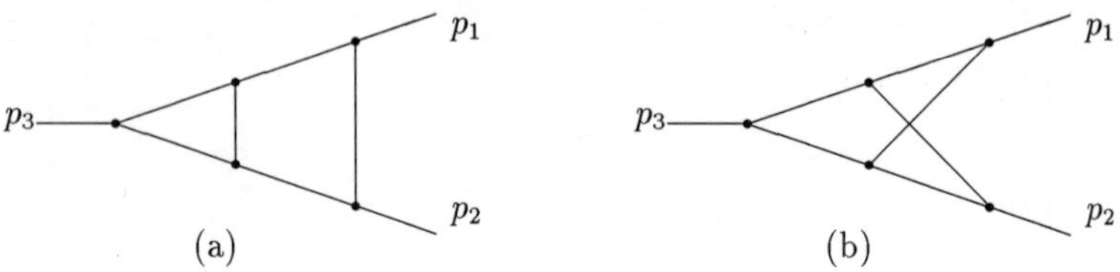

Figure 4: The planar (a) and non-planar (b) two-loop three-point diagrams

integral representation,

$$C^{(2)}(p_1^2,p_2^2,p_3^2) = -\frac{1}{2}\left(\frac{i\pi^2}{p_3^2}\right)^2 \int_0^1 \frac{d\xi\ \ln\xi}{y\xi^2+(1-x-y)\xi+x}\left(\ln\frac{y}{x}+\ln\xi\right)\left(\ln\frac{y}{x}+2\ln\xi\right) \tag{3}$$

with $x \equiv p_1^2/p_3^2$, $y \equiv p_2^2/p_3^2$. The integral (3) can be easily calculated in terms of polylogarithms Li_N [10] with $N \leq 4$.

For the non-planar ("crossed") two-loop three-point diagram $\widetilde{C}^{(2)}(p_1^2,p_2^2,p_3^2)$ (see Fig. 4b), we found that

$$\widetilde{C}^{(2)}(p_1^2,p_2^2,p_3^2) = \left(C^{(1)}(p_1^2,p_2^2,p_3^2)\right)^2, \tag{4}$$

where $C^{(1)}$ is a function corresponding to one-loop triangle diagram,

$$C^{(1)}(p_1^2,p_2^2,p_3^2) = -\frac{i\pi^2}{p_3^2}\int_0^1 \frac{d\xi}{y\xi^2+(1-x-y)\xi+x}\left(\ln\frac{y}{x}+2\ln\xi\right). \tag{5}$$

Since the integral (5) can be evaluated in terms of dilogarithms Li_2, the result for $\widetilde{C}^{(2)}$ contains products of dilogarithms.

We examined a four-point ladder diagram $D^{(2)}$ ("double box", see Fig. 5),

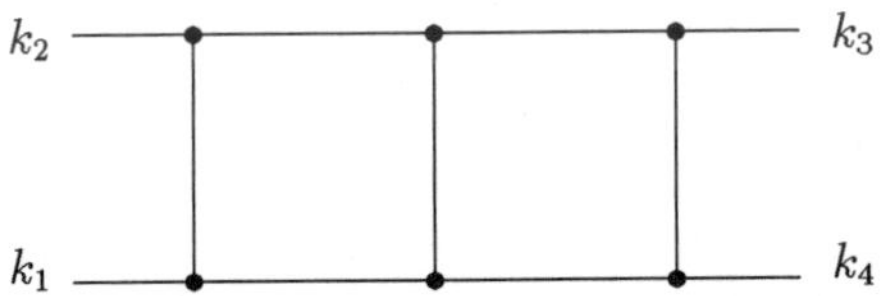

Figure 5: The "double box" diagram

and showed that the corresponding function can be reduced to $C^{(2)}$, namely,

$$D^{(2)}(k_1^2,k_2^2,k_3^2,k_4^2,s,t) = t\,C^{(2)}(k_1^2\,k_3^2,\ k_2^2\,k_4^2,\ s\,t), \tag{6}$$

where $s \equiv (k_1+k_2)^2$ and $t \equiv (k_2+k_3)^2$ are usual Mandelstam variables. Therefore, the "double box" can also be evaluated in terms of Li_N with $N \leq 4$.

We managed to generalize some of the results for one- and two-loop diagrams to the case of arbitrary number of loops [4]. We considered L-loop three-point ladder diagrams $C^{(L)}$ (Fig. 6a), and we obtained the following generalization of the formulae (5), (3):

$$C^{(L)}(p_1^2,p_2^2,p_3^2) = -\frac{1}{L!\,(L-1)!}\left(\frac{i\pi^2}{p_3^2}\right)^L$$

$$\times\int_0^1 \frac{d\xi}{y\xi^2+(1-x-y)\xi+x}\ \ln^{L-1}\xi\left(\ln\frac{y}{x}+\ln\xi\right)^{L-1}\left(\ln\frac{y}{x}+2\ln\xi\right). \tag{7}$$

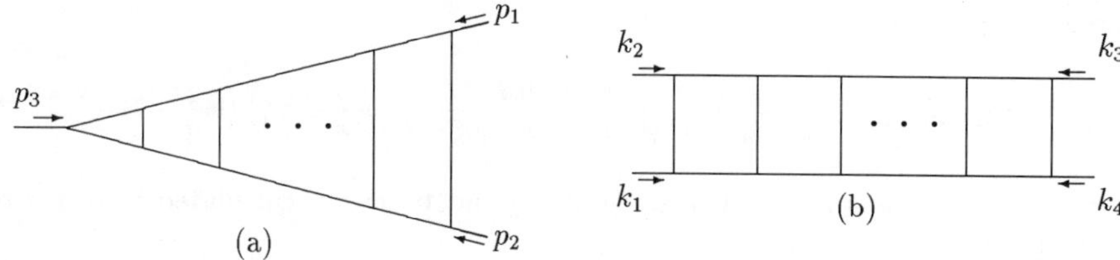

Figure 6: The three-point (a) and four-point (b) L-loop diagrams

This integral can be calculated in terms of polylogarithms Li_N with $N \le 2L$.

Moreover, for the four-point L-loop ladder diagram $D^{(L)}$ (shown in Fig. 6b) we found that

$$D^{(L)}(k_1^2, k_2^2, k_3^2, k_4^2, s, t) = t^{L-1}\, C^{(L)}(k_1^2 k_3^2, k_2^2 k_4^2, st). \tag{8}$$

This formula generalizes eq. (6) and holds also for one-loop and zero-loop (tree) cases. Thus, the four-point L-loop diagram can be also expressed in terms of Li_N with $N \le 2L$. Note that another derivation of the formula (8) was presented in ref. [19].

We have checked the results for L-loop diagrams by reducing them to the $(L+1)$-loop two-point functions (by means of additional integration), and we found complete agreement with known results for such two-point diagrams [20].

Acknowledgements

I would like to thank the AIHENP-93 Organizing Committee for their help, and the International Science Foundation (New York) for financial support. Research was supported, in part, by the Research Council of Norway.

References

1. A. I. Davydychev and J. B. Tausk, *Nucl. Phys.* **B397** (1993) 123.
2. A. I. Davydychev, V. A. Smirnov and J. B. Tausk, Leiden preprint INLO-PUB–5/93 (1993); *Nucl. Phys.* **B**, to appear.
3. N. I. Ussyukina and A. I. Davydychev, *Phys. Lett.* **B298** (1993) 363.
4. N. I. Ussyukina and A. I. Davydychev, *Phys. Lett.* **B305** (1993) 136.
5. N. I. Ussyukina and A. I. Davydychev, *Yad. Fiz.* **56** (No. 11) (1993) 172.
6. G. Weiglein, R. Scharf and M. Böhm, Würzburg preprint (1993).
7. G. 'tHooft and M. Veltman, *Nucl. Phys.* **B44** (1972) 189;
   C. G. Bollini and J. J. Giambiagi, *Nuovo Cim.* **12B** (1972) 20.
8. G. Källén and A. Sabry, *Dan. Mat. Fys. Medd.* **29** (1955) No.17;
   J. Schwinger, *Particles, Sources and Fields*, Vol.2 (Addison-Wesley, 1973);

D. J. Broadhurst, *Phys. Lett.* **B101** (1981) 423; *Z. Phys.* **C47** (1990) 115;
T. H. Chang, K. J. F. Gaemers and W. L. van Neerven, *Nucl. Phys.* **B202** (1982) 407;
A. Djouadi, *Nuovo Cim.* **100A** (1988) 357;
B. A. Kniehl, *Nucl. Phys.* **B347** (1990) 86;
D. J. Broadhurst, J. Fleischer and O. V. Tarasov, *Z.Phys.* **C60** (1993) 287.

9. R. Scharf and J. B. Tausk, Leiden preprint INLO-PUB–7/93 (1993); *Nucl. Phys.* **B**, to appear.
10. L. Lewin, *Polylogarithms and Accociated Functions* (North-Holland, Amsterdam, 1981).
11. D. Kreimer, *Phys. Lett.* **B273** (1991) 277.
12. F. V. Tkachov, *Phys. Lett.* **B100** (1981) 65;
K. G. Chetyrkin and F. V. Tkachov, *Nucl.Phys.* **B192** (1981) 159.
13. A. I. Davydychev, *J. Phys.* **A25** 5587.
14. A. C. Hearn, *REDUCE User's Manual*, RAND publication CP78 (Santa Monica, 1987).
15. F. V. Tkachov, Moscow preprint INR P-358 (1984);
G. B. Pivovarov and F. V. Tkachov, Moscow preprints INR P-0370, П-459 (1984);
K. G. Chetyrkin and V. A. Smirnov, Moscow preprint INR G-518 (1987);
K. G. Chetyrkin, *Teor. Mat. Fiz.* **75** (1988) 26; **76** (1988) 207; Munich preprint MPI-PAE/PTh 13/91 (1991);
S. G. Gorishny, *Nucl. Phys.* **B319** (1989) 633.
V. A. Smirnov, *Commun. Math. Phys.* **134** (1990) 109.
16. V. A. Smirnov, *Renormalization and Asymptotic Expansions* (Birkhäuser, Basel, 1991).
17. A. Czarnecki and A. I. Davydychev, Edmonton preprint Alberta–Thy-38-93 (1993).
18. A. N. Vasil'ev, Yu. M. Pis'mak and Yu. R. Honkonen, *Teor. Mat. Fiz.* **47** (1981) 291;
N. I. Ussyukina, *Teor. Mat. Fiz.* **54** (1983) 124;
D. I. Kazakov, *Phys. Lett.* **B133** (1983) 406;
D. I. Kazakov and A. V. Kotikov, *Teor. Mat. Fiz.* **73** (1987) 348.
19. D. J. Broadhurst, *Phys. Lett.* **B307** (1993) 132.
20. V. V. Belokurov and N. I. Ussyukina, *J. Phys.* **A16** (1983) 2811;
D. J. Broadhurst, *Phys. Lett.* **B164** (1985) 356.

DIFFERENTIAL EQUATION METHOD. THE CALCULATION OF N-POINT DIAGRAMS

A.V.KOTIKOV
Laboratory of Particle Physics, JINR
141980, Dubna, Moscow region, Russia

ABSTRACT

A new method of massive Feynman diagrams calculation is presented. It is provides a fairly simple procedure of obtaining the result without the D-space (for the dimensional regularization) integrals calculation. Some diagrams are calculated as an illustration of this method capacities.

Here we present a method which, using the rule of integration by parts[1-3], allows to obtain the result for the massive diagrams without the calculation of the complicate D-space (in case of the dimensional regularization) Feynman integrals. Throughout the article we use the following notation. The use of dimensional regularization is assumed. All the calculations are performed in momentum space of dimension $D = 4 - 2\epsilon$. The dotted and solid lines of a diagram correspond to the massless and massive (for simplicity euclidean) propagators, respectively,

$$\frac{1}{[p^2]^\alpha} = \overset{\alpha}{\bullet - - - \bullet} \quad \overrightarrow{p} \quad , \qquad \frac{1}{[p^2+m^2]^\alpha} = \underset{m^2}{\overset{\alpha}{\bullet\!\!-\!\!\!-\!\!\!-\!\!\bullet}} \quad \overrightarrow{p}$$

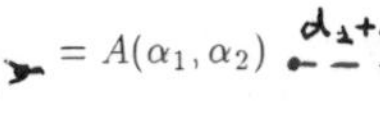

α and m are called the index and mass of this line, respectively. Further (except specially described places) all solid lines contain the same mass m . Index of a line is equal to 1 and masses do not marked.

Let us consider the rules of the Feynman diagrams calculation.

Rule 1. Massless loops are integrated due to the graphical identity

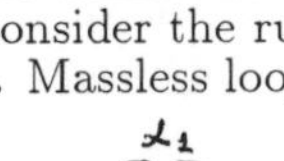

$$\text{(massless loop with indices } \alpha_1, \alpha_2) = A(\alpha_1,\alpha_2) \overset{\alpha_1+\alpha_2-D/2}{\bullet - - - - \bullet} ,$$

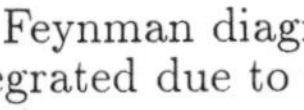

where

$$A(\alpha_1,\alpha_2) = \pi^{D/2}\frac{a(\alpha_1)a(\alpha_2)}{a(\alpha_1+\alpha_2-D/2)}, \quad a(\alpha) = \frac{\Gamma(D/2-\alpha)}{\Gamma(\alpha)},$$

Γ is the Euler Γ – function

Rule 2. Massive tadpoles are integrated due to the graphical identity

$$\text{(massive tadpole with indices } \alpha_1, \alpha_2) = B(\alpha_1,\alpha_2)\frac{1}{(m^2)^{\alpha_1+\alpha_2-D/2}},$$

where

$$B(\alpha_1,\alpha_2) = \pi^{D/2}\frac{\Gamma(D/2-\alpha_2)\Gamma(\alpha_1+\alpha_2-D/2)}{\Gamma(\alpha)\Gamma(D/2-1)}$$

Rule 3. For a triangle the following recurrent relation is valid (here the line with index α_i contains also the mass m_i)

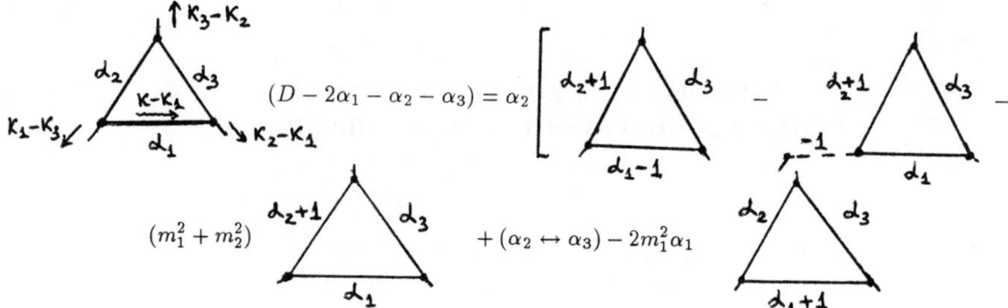

Pay attention to the fact, that the lower line of the triangle is distinguished and two other lines are similar. The lower line is called "distinguished line".

The rule 3, i.e. the rule of integration by parts, can be obtained by means of multiplication of the subintegral magnitude by the number $D = \frac{d}{dq_\mu}(q-p)^\mu$ and using the relationship $\int d^D q \text{ div } (\ ) = 0$ for the regularized Feynman integrals.

The key idea of this method: by application of the rules 1-3 the results of the different massive FDC are obtained without the calculation of the complicated D-space integrals. Notice that all information about D-space Feynman integrals structure is already contained in the rules 1 and 2 for massless loop and massive tadpole, respectively.

Consider a few specific examples.

A. Propagator-type diagrams. (see also[4])

We divide all diagrams into tree types

- the master diagrams which can't be obtained as loops and chains combination,
- the diagrams, which are the combination of loops and chains
- the diagrams which contain only loops.

1. As the master diagram example we consider

$$= I_1$$

Applying sequentially the rule 3 to the right triangle with the vertical and sidelong distinguished lines, we get

$$I_1(D-4) = 2\,[\ \ - \ \ - m^2 \ \ - m^2 \ \ ]$$

$$I_1(D-4) = \ \ - m^2 \ \ - q^2$$

Combining these two eq.s, we obtain

$$I_1(D-4)\left(1 - 2\frac{m^2}{q^2}\right) = 2[\ \ - \ \ ] - 2\frac{m^2}{q^2} \ \ - 2m^2\left(1 - \frac{m^2}{q^2}\right)$$

The last diagram in the right hand said (r.h.s.) is the differential with respect to m^2 of the initial diagram I_1. Hence, the r.h.s. has the form

$$2f_1 + 2m^2(1 - m^2/q^2)\frac{d}{dm^2}I_1,$$

where the function f_1 does not contain master diagrams.

Thus,we get differential equation with respect to m^2 (DE) for the initial diagram. The result for I_1 is

$$I_1(t) = \Big[\frac{t^2}{(1-t)(1-2t)}\Big]^\epsilon \Big(C_1 + \int_0^t \frac{d\psi f_1(\psi)}{\psi^{2\epsilon}(1-\psi)^{1-\epsilon}(1-2\psi)^{1-\epsilon}}\Big),\ (t = \frac{q^2}{q^2+m^2})$$

The constant C_1 can be obtained by comparison with massless limit $t = 1$.

2. As the example of the second type of diagrams we consider the diagram

$$= I_2$$

Applying the rule 3 to the loop with the massless distinguished line, we have

$$I_2(D-3) = \quad - \quad$$

3. The examples of the third type of diagrams.

a). A simple loop with one massive line.

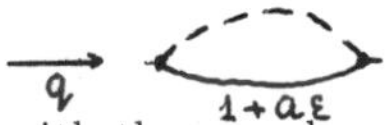

$$= I_3$$

Applying the rule 3 with the massless distinguished line, we get

$$I_3(1-(a+2)\epsilon) = (1+a\epsilon)\,[\quad -(q^2+m^2) \quad]$$

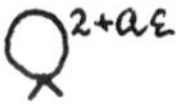

By analogy with subsect.1 we have

$$(q^2)^{(a+1)\epsilon} I_3(t) = (1/t)^{1-(a+2)\epsilon} \int_0^t \frac{d\psi}{\psi^\epsilon(1-\psi)^{(a+1)\epsilon}} (1+a\epsilon)B(2+a\epsilon, 0)$$

Notice that this is sometimes more convenient form of the diagram I_3

$$= \int_0^1 \frac{d\psi}{\psi^{\alpha+1-D/2}(1-\psi)^\epsilon}\ \alpha B(\alpha+1, 0) \qquad (1)$$

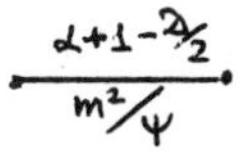

It is clear to see that the rule 3 allows us to reduce L-loop diagram to the $(L-1)$-loop diagram with one propagator, which contains the integration parameter in its mass.

b). A simple loop with two massive lines.

$$= I_4$$

By analogy with the previous subsection we get

$$I_4 = (4^\epsilon/2)B(2,0) \int_0^1 \frac{d\psi}{\psi^\epsilon(1-\psi)^{1/2}} \qquad (2)$$

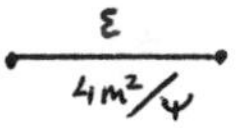

c). The two-loop diagram

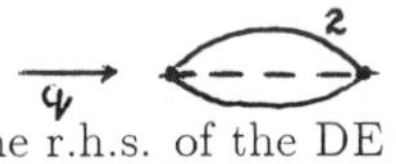

$$= I_5,$$

which appears in the r.h.s. of the DE for many complicated diagrams.

Differentiating eq.(2) with respect to the mass and applying this result along with the eq.(1) to the diagram I_5, we get

$$I_5 = 4^\epsilon \Gamma(1+\epsilon) \int_0^1 \frac{d\psi}{\psi^{1+\epsilon}(1-\psi)^{1/2}} \quad =$$

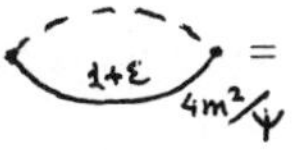

$$4^{\epsilon}\Gamma(2\epsilon)\int_0^1 \frac{d\psi}{\psi^{1+\epsilon}(1-\psi)^{1/2}}\int_0^1 \frac{d\sigma\sigma^{-\epsilon}}{[q^2(1-\psi)\sigma+4m^2]^{2\epsilon}}$$

After the integration we have

$$I_5 = \frac{\Gamma^2(1+\epsilon)}{(1-\epsilon)(m^2)^{2\epsilon}}\Big[\frac{1}{2\epsilon^2} - 1 - p\ln(\frac{1+\sqrt{p}}{1-\sqrt{p}}) - \frac{1+p}{4p}(\ln(\frac{1+\sqrt{p}}{1-\sqrt{p}}))^2 + O(\epsilon)\Big],\ (p=\frac{q^2}{q^2+4m^2})$$

Thus, the usage of the rule of the integration by parts allows us to obtain the DE for the massive diagrams with r.h.s. contains more simple diagrams.

B. Vertex-type diagrams.

4. Consider one-loop vertex-type diagram with one massive line.

$$= I_6$$

Applying the rule 3 with massive and one massless distinguished lines, we have

$$(D-4)I_6 = \quad -[(p-q)^2+m^2] \quad + (p\leftrightarrow k) - 2m^2 \qquad (3)$$

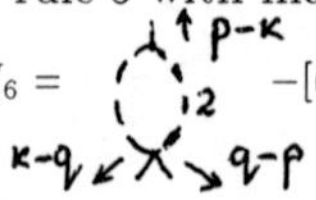

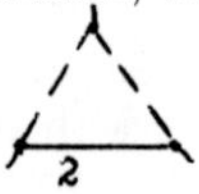

$$(D-4)I_6 = \quad -[(p-q)^2+m^2] \quad + \quad -(p-k)^2 \qquad (4)$$

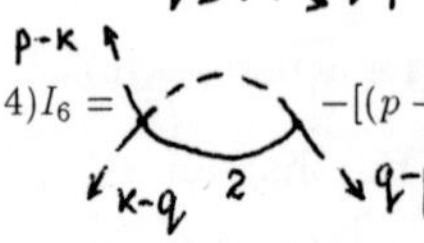

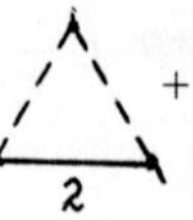

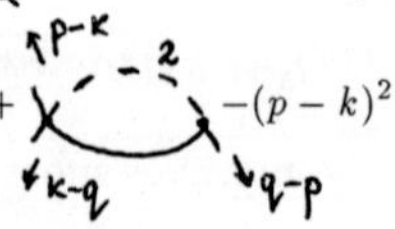

Combining three eq.s (3), (4) and (4) with substitution $(p\leftrightarrow k)$ and putting without the lack of generality $q=0$, we get

$$(D-4)I_6\psi_6 = \Phi_6 - 2\phi_6\frac{d}{dm^2}I_6$$

Here $\psi_6 = 1-(k^2+p^2+2m^2)/(k-p)^2$, $\phi_6 = m^2+(k^2+m^2)(p^2+m^2)/(k-p)^2$ and $\Phi_6 = f_2(p-k) - [(k^2+m^2)f_1(k)+(p^2+m^2)f_1(p)]/(p-k)^2$, where

$$f_2(p) = 2 \qquad , f_1(p) = \qquad + $$

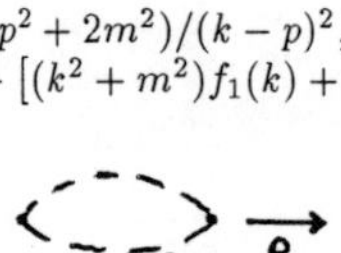

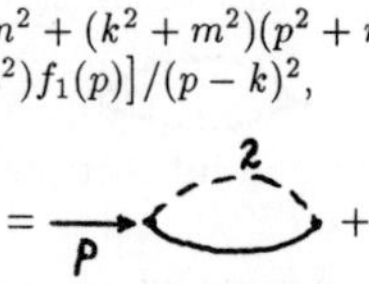

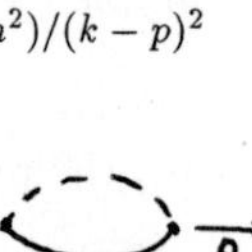

Defining the variable $x = m^2/(p-k)^2$, we get for the initial diagram in the symmetrical point $k^2=p^2=(k-p)^2=\mu^2$

$$\phi_3(x) = (1+x+x^2)\mu^2$$

$$I_6(x) - I_6(0) = [J_1(x,0) - 2J_1(x,1)]$$

where

$$J_1(x,a) = \int_0^x \frac{d\psi}{1+\psi+\psi^2}\ln(\psi+a)$$

The values of diagram $I_6(0)$ is found in the papers[6,5].

5. Here we consider the diagram,

$$= I_7,$$

which is contained in the ghost-gluon- ghost vertex-type function in QCD.

Applying the rule 3 to diagram I_7 with massive distinguished lines we get

$$I_7(D-3) = \quad - \quad - 2m^2 \left[2 \quad - 2 \quad + \quad \right] \tag{5}$$

The first and third diagrams in r.h.s. of eq.(5) can be expressed by the rule 2 in the form

$$-4m^2 \quad = \frac{\Gamma(1+\epsilon)}{(m^2)^\epsilon}(1/\epsilon + 2)I_6(0)$$

These diagrams contain ultraviolet infinity of the initial diagram. All other diagrams in r.h.s. of eq.(5) are ultraviolet-finite. Using eq.(2) we have that three other diagrams from r.h.s. of eq.(5) give the following contribution

$$-\int_0^1 \frac{d\tau}{\tau}\sqrt{1-\tau}\, I_6(4x/\tau)$$

The full contribution of the initial two-loop diagram can be given in the form

$$I_7(x) = \frac{\Gamma(1+\epsilon)}{(m^2)^\epsilon}\left[(1/\epsilon + 2)I_6(0) - \int_0^1 \frac{d\tau}{\tau}\sqrt{1-\tau}\, I_6(4x/\tau)\right]$$

Thus, the usage of the rule of integration by parts allows us to obtain the DE for the massive vertex-type diagrams, with r.h.s. containing the propagator-type diagrams only.

C. N-point diagrams (see also[7])

Consider of N- point diagram (here the line with momentum $k - k_i$ contains also the mass m_i)

$$= V(N)$$

Applying the rule 3 with the distinguished line with momentum $k - k_1$, we get

$$V(N)(D-N-1) = V_2^{(1)}(N-1) - \left[(k_1 - k_2)^2 + m_1^2 + m_2^2 \right] V_2(N) +$$
$$(2 \leftrightarrow 3) + \ldots + (N-1 \leftrightarrow N) - 2m_1^2 V_1(N)$$

Hereafter the upper symbol (in bracket) marks the initial diagram line, which is cancelled by the application of the rule 3. The lower one marks the line, which have index 2. Applying also the rule 3 with the other distinguished lines, we get the matrix equation

$$\hat{A}\vec{V}(N) = \vec{C}, \left(\vec{V}(N) = \begin{bmatrix} V_1 \\ \vdots \\ V_N \end{bmatrix}\right) \tag{6}$$

where $C_i = \sum_{j \neq i} V_j^{(i)}(N-1) - (D-N-1)V(N)$

$$a_{ii} = 2m_i^2; \quad a_{ij}(i \neq j) = (k_i - k_j)^2 + m_i^2 + m_j^2$$

The solution of the eq.(6) is the standart one

$$V_i(N) = \frac{\det \hat{A}_i}{\det \hat{A}},$$

where the matrix $\hat{A}_i$ equal to one $\hat{A}$ with respect elements a_{ij} , which are replaced by elements C_j. We have

$$-\det \hat{A} \, \frac{d}{da}\vec{V}(N) = \sum_1^N \det A_i, \;\; C_i = -\frac{d}{da}V^{(i)}(N-1) - (D-N-1)V(N) \tag{7}$$

It is clear to see that the eq.s (7) allow us to reduce the N-point diagram to $(N-1)$-point diagrams.

Resume. **The usage of the rule of the integration by parts allows us to obtain the DE for the massive diagrams with r.h.s. contains more simple diagrams. By using this procedure several times one can get the loops with the several number of lines. These loops are integrated either by means of eqs. (1) and (2) decreasing step by step the number of loops or by method of Feynman parameters in the more complicated case.**

The basic difference from the massless case is the necessity to integrate the final result with respect to the mass the needed number of times. Thus, this method (at least in principle) is as powerful in the calculation of massive Feynman diagrams as the rule of integration by parts in the massless case.

The essential property of the massive case is the continuous application of the rule of integration by parts to both complicated and more simple diagrams. In the massless case the rule 3 is used only with master diagrams.

The essential difference from the ordinary methods (for example, Feynman parameters method) of the calculation of massive Feynman diagrams is the appearance of the complicated functions in the final result only. Hence, these functions do not get in the way in the process of the calculation.

References.

1. A.N.Vassiliev, Y.M.Pis'mak and Y.P.Honkonen, *Teor.Mat.Fiz.* **47** (1981) 291.
2. K.G.Chetyrkin and F.V.Tkachov, *Nucl.Phys.* **B192** (1981) 159.
3. D.I.Kazakov, *Phys.Lett.* **133B** (1983) 406.
4. A.V.Kotikov, *Phys.Lett.* **254B** (1991) 158.
5. A.V.Kotikov, *Phys.Lett.* **259B** (1991) 314.
6. E.E.Boos and A.I.Davydychev, *Vestn.Mosk.Univ.* **28** (1987) 6.
7. A.V.Kotikov, *Phys.Lett.* **267B** (1991) 123.

UNSTABLE PARTICLES AND GAUGE INVARIANCE

Geert Jan van Oldenborgh
Paul Scherrer Institute, CH-5232 Villigen PSI, Switzerland

ABSTRACT

When computing the properties of reactions involving unstable charged particles care has to be taken to use a gauge invariant amplitude. In this talk we present methods to (automatically) obtain such an amplitude, both at the tree level and in one-loop calculations, using only a minimal number of diagrams. The numerical difference with gauge variant methods commonly used will be discussed for the case of tree level W pair production.

1. Introduction

Many of the cross sections one would like to evaluate with the methods described elsewhere in these proceedings involve the production and decay of unstable particles. The W and Z bosons both have a sizable width Γ ($\Gamma/m \approx 1/40$), and the same may be true for the top quark and Higgs boson (depending on their masses). For these particles the well-known narrow width approximation will not be sufficient for many purposes. Including off-shell effects, however, involves a resummation of some of the higher order graphs, which means that care has to be taken to keep the matrix element gauge invariant. This last point is not only of academic interest as gauge breaking terms are often much larger numerically than the gauge invariant result. Therefore even suppression factors Γ/m may not be enough to avoid inaccuracies.

The Dyson resummation of higher order graphs in the propagator, which leads to a finite width, is clearly only needed when the unstable particle can kinematically be on its mass shell. Indeed, for an unstable particle with space-like momentum $p^2 < 0$ the imaginary part of the self energy is zero, hence no finite width should be used. One can therefore not use a complex mass in the Lagrangian, even disregarding the problems with the standard model mass generation mechanism which would follow (for instance a complex value for $\sin^2\theta_W$). It is a purely kinematical problem which particles should be given a finite width and which not.

The amplitude can thus be divided into a set of *resonant* diagrams, which contain a divergence within the allowed phase space as an unstable particle is taken on-shell, and *non-resonant* graphs without such divergences. In principle the contribution of the non-resonant diagrams is suppressed with respect to the resonant ones by $\Gamma/m \propto \alpha$. We first discuss ways to only use the resonant tree graphs to get the dominant behaviour, next increasing the accuracy of the result by including the non-resonant tree graphs and the resonant one-loop graphs. A more detailed discussion will be given in a forthcoming publication.[1] The tree level results are based on Ref..[2]

2. Resonant Tree Graphs

In this section some schemes will be given to deal with the resonant tree graphs. For simplicity these are first discussed for a single unstable particle with momentum p and mass m; the extension to more particles will be given later later. We assume that the diagram generator has already sorted out which diagrams ones are resonant and which are not. Without resummation the full (gauge invariant) amplitude for this process therefore has the form

$$\mathcal{M} = \frac{R(p^2)}{p^2 - m^2} + N(p^2) \ . \tag{1}$$

The essential variable here is the virtuality p^2 of the unstable particle. The other kinematical variables are assumed to be CMS angles, which are independent of p^2. Unfortunately, this amplitude gives an infinite cross section when the integration region includes the pole at $p^2 = m^2$.

The simplest approximation one can make is to completely factorize the process in production and decay: the narrow width approximation. This fails to take into account terms of $\mathcal{O}(\Gamma)$; the corresponding dimensionless parameter will usually be Γ/m (about 1/40 for the W), but can be much larger near the threshold for production of the unstable particles.

The most important of these $\mathcal{O}(\Gamma)$ corrections are included by simply using the on-shell expression for the matrix element, but treating the kinematics and unstable particle propagator off-shell. This off-shell propagator is derived by resumming the one-loop corrections to the propagator, giving rise to a simple Breit-Wigner propagator $1/(p^2 - m^2 + im\Gamma)$ with Γ defined by the relation $m\Gamma = \mathrm{Im}\Pi(m^2)$ in terms of the self energy Π evaluated at the real mass. We refer to this procedure as the "resonant" scheme. This procedure obviously respects gauge invariance and will serve as the base line to compare the others with.

The next step is usually to also evaluate the matrix element off-shell, we refer to this procedure as the "off-shell" method. This also gives an answer below threshold. In the case of a charged resonance however, this procedure violates $U(1)_{em}$ gauge invariance, even if the width is given by the physical (on-shell) quantity, which is gauge invariant. The reason is that the original resonant graphs are not the only graphs which lead to the particular final state: non-resonant graphs have to be included.

3. Non-resonant Tree Graphs

One reason to include the non-resonant graphs is to conserve gauge invariance. Another is that they can be sizable due to some other near-divergence, such as large collinear logarithms. The "naive" way to include these non-resonant graphs is to just add them to the resonant terms with a complex mass. However, as the amplitude was gauge invariant for $\Gamma = 0$ it follows that it must be gauge variant for $\Gamma > 0$ (if the resonant graphs are not separately gauge invariant).

An obviously gauge invariant procedure to include the non-resonant graphs was introduced by Zeppenfeld *et al.*[3] The price to save gauge invariance in this "overall" scheme is an incorrect treatment of the non-resonant contribution close to mass shell.

name	kinematics	matrix element	gauge invariant?	treatment non-resonant terms	threshold behaviour
narrow width	on-shell	$R(m^2)$	yes	none	undefined below threshold
resonant	off-shell	$\dfrac{R(m^2)}{p^2 - m^2 + im\Gamma}$	yes	none	zero below threshold
off-shell	off-shell	$\dfrac{R(p^2)}{p^2 - m^2 + im\Gamma}$	no	none	wrong below threshold
naive	off-shell	$\dfrac{R(p^2)}{p^2 - m^2 + im\Gamma} + N(p^2)$	no	ok	ok
overall	off-shell	$\dfrac{R(p^2)}{p^2 - m^2 + im\Gamma} + \dfrac{p^2 - m^2}{p^2 - m^2 + im\Gamma} N(p^2)$	yes	wrong	ok
pole-scheme	off-shell	$\dfrac{R(m^2)}{p^2 - m^2 + im\Gamma} + \dfrac{R(p^2) - R(m^2)}{p^2 - m^2} + N(p^2)$	yes	ok	wrong around threshold

Table 1. The six different schemes to treat unstable particles at tree level.

However, it can be argued that the difference is of higher order in this region of phase space. We thus have a prescription which is correct to leading order both on resonance and away from it, but these contributions are formally of different order in α.

Another gauge invariant way to include the off-shell effects is to systematically separate orders in Γ.[4] As the residue at the pole $p^2 = m^2$ is gauge invariant one can add the finite width in the first term without breaking gauge invariance. This corresponds to adding and resumming only the gauge invariant part of the propagator corrections. In this scheme the cross section is given as the sum of the resonant cross section plus $\mathcal{O}(\Gamma)$ corrections, which are both gauge invariant. It is thus a natural starting point for higher order corrections.

However, this "polescheme" also has some undesirable properties. The first (resonant) term has a discontinuity when the threshold for the production of the unstable particle is crossed. Approaching from below, one even encounters the original (non-resummed) singularity in the propagator. The accuracy of this scheme is thus doubtful around threshold.

The difference between the last three schemes is of order Γ^2 for total cross sections and singly differential distributions, but of order Γ for doubly and higher differential distributions. This follows from the occurrence of a Levi-Civita tensor in the $\mathcal{O}(\Gamma)$ terms. A summary of all schemes is given in table 1.

The extensions to the case where there are more unstable particles are largely straightforward. There now appears a hierarchy of n-fold resonant diagrams. The

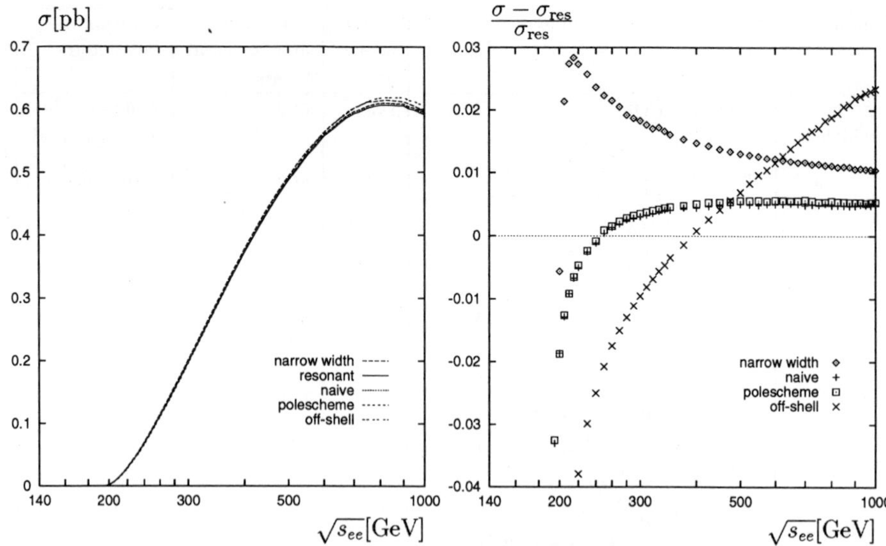

Fig. 1. Cross sections for $\gamma\gamma \to \ell^- \bar{\nu}_\ell W^+$ in the different schemes defined in the text with $p_\perp(\ell) > 0.02\sqrt{s_{ee}}$

first three methods just use the maximally resonant ones. In the "polescheme" one only needs these and the ones where only one particle is non-resonant; the "overall" scheme uses all diagrams. The threshold behaviour of the 'polescheme" now is worse than before. However, in this area the worth of any fixed-order calculation is doubtful because of possible bound state effects.

The numerical influence on the total cross section is illustrated in Fig. 1, where the total cross section for $\gamma\gamma \to \ell^- \bar{\nu}_\ell W^+$ is plotted, with the W^+ considered stable and a photon beam derived from backscattered electrons.[5] The physically relevant off-shell effects will be twice as large. To avoid the collinear divergence in the final state a minimum transverse momentum $p_\perp(\ell) > 0.02\sqrt{s}$ is demanded; this should also give an indication of detector acceptance. It can be seen that the narrow width approximation misses some important effects. The graph for the offshell scheme is misleading, in that we used the formula $\sum_i \epsilon_i^\mu \epsilon_i^\nu = -g^{\mu\nu}$ which is not valid when the photon couples to a non-conserved current. It does show that one has to be careful when breaking gauge invariance. One should also note that the differences between the various schemes are expected to be much larger in doubly differential distributions.

4. Resonant One-loop Corrections

When computing the $\mathcal{O}(\alpha)$ corrections one is normally only interested in the resonant part. We will thus have to identify which diagrams contribute to this, and then to find a gauge invariant way to isolate just the resonant part. A generalization of the "polescheme" is proposed for this last step. In the case of a neutral resonance (like the Z boson) the isolation step is not necessary, as the resonant diagrams already form a gauge invariant subset; the only bone of contention left in this case amounts to the proper definition of the mass of the unstable particle.

There are three classes of resonant diagrams. The first one contains the corrections to the resonant propagator. A gauge invariant part of these are resummed to give the finite width; the rest (vanishing at $p^2 = m^2$) has to be treated perturbatively. The second group, the factorizable diagrams, comprises the corrections to either production or decay of an unstable particle; these diagrams obviously retain the original resonant propagators. Added to this, charged resonances will give rise to logarithms $\log(p^2 - m^2)$ as the virtuality and finite width of the unstable particle regulate the infrared divergence. The third class of non-factorizable diagrams is given by the set of infrared divergent diagrams with the photon (or gluon) spanning one or more resonant propagators. Simple power counting arguments show that no other non-factorizable diagrams can give a resonant contribution. One would suspect that even the contributions from these diagrams may cancel against the corresponding soft Bremsstrahlung diagrams. As the hard Bremsstrahlung is easily seen to be non-resonant this would imply the intuitively satisfying result that initial-final and final-final state interaction vanish in the narrow width approximation.

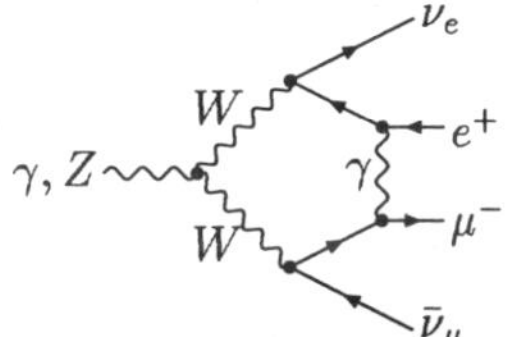

Fig. 2. A double resonant five point function occurring in W pair production

However, this is not the case. As an example consider the doubly resonant five point function shown in Fig. 2. Using the Schouten identity one can decompose it in five four point functions as

$$E_0 = \frac{-1}{2\Delta_{5s}} \left(\Delta^{\gamma}_{ps} D_0(\gamma) + \Delta^{\mu}_{ps} D_0(\mu) + \Delta^{W^-}_{ps} D_0(W^-) + \Delta^{W^+}_{ps} D_0(W^+) + \Delta^{e}_{ps} D_0(e) \right) \tag{2}$$

where $D_0(p)$ denotes the scalar four point function with the propagator corresponding to particle p missing and the Δ's denote determinants of external and internal momenta.[6] The overall factor $-1/2\Delta_{5s}$ is quadratic in the virtualities $p_{\pm}^2 - m^2$ of the W bosons and thus gives rise to the double resonance. The $D_0(\gamma)$ and Δ^{γ}_{ps} are both finite, whereas for the other combinations the determinant is linear in the virtuality and the four point function linearly divergent.

The corresponding scalar Bremsstrahlung integral can be decomposed similarly, except that the equivalent of the $D_0(\gamma)$ never appears. The other four terms exactly match the corresponding terms in the virtual diagram, leaving this one doubly resonant term. There is no symmetry which would cause it to disappear, even when integrated over phase space, and there are no other graphs which will give rise to the same four point function. It therefore is a double resonant term of order α which is not suppressed by additional factors Γ/m. The failing cancellations can be traced back to the fact that, in the limit $\Gamma \ll \lambda$ (λ is an infrared regulator, normally the photon mass) the linear and quadratic infrared divergences do *not* cancel. Only a numerical study can decide whether the non-factorizable diagrams give rise to measurable effects.

The diagrams we have selected as resonant diagrams in general do not form a gauge invariant subset. One can select a gauge invariant part by taking the residue at the pole, as was done in the previous sections. Unfortunately, this residue is ill-defined due to the occurrence of the divergent logarithms. (These also do not cancel between the virtual and Bremsstrahlung graphs.) These logarithms only occur in the scalar functions however. As the coefficients of the scalar functions are gauge invariant, one can evaluate these any way one likes.

The following recipe therefore serves as the extension of the "polescheme" for one-loop (and soft Bremsstrahlung) diagrams:

- The reduction of the diagram to scalar functions should be done on-shell (if the unstable particle is charged).
- The scalar functions can be evaluated in any convenient non-divergent way; on-shell or off-shell. This also allows one to take into account the finite width in non-divergent but non-analytical logarithms like the Coulomb singularity and threshold effects.

This schemes have been implemented in the FF library.[7]

5. Conclusion

In this talk we have presented a scheme to systematically include off-shell effects in amplitudes including unstable particles, explicitly keeping gauge invariance. Unfortunately it breaks down near the threshold for the production of the unstable particles. In this region, however, the occurrence of bound state effects makes any fixed-order perturbation series suspect. The numerical difference of this procedure with the (gauge variant) ones commonly used are small in the tree level total cross section, at least in the case of W pair production and the unitary gauge. Larger differences are expected for doubly differential cross sections.

6. Acknowledgements

The results presented here are based on work done together with Daniel Wyler and Andre Aeppli in the context of the one-loop corrections to W pair production.

References

1. A. Aeppli, G. J. van Oldenborgh and D. Wyler. (to appear).
2. A. Aeppli, F. Cuypers and G. J. van Oldenborgh. Phys. Lett. **B314** (1993) 413.
3. D. Zeppenfeld, J. A. M. Vermaseren and U. Baur. Nucl. Phys. **B375** (1992) 3.
4. R. G. Stuart. Phys. Lett. **B262** (1991) 113, Phys. Lett. **B272** (1991) 353.
5. I. F. Ginzburg, Kotkin G. L., V. G. Serbo and V. I. Telnov. Nucl. Instr. Meth. **205** (1983) 47.
6. G. J. van Oldenborgh. One-loop Calculations with Massive Particles. Ph.D. thesis, Universiteit van Amsterdam, 1991.
7. G. J. van Oldenborgh. Comp. Phys. Comm. **66** (1991) 1.

APPLICATION OF PADÉ APPROXIMANTS TO THE CALCULATION OF FEYNMAN DIAGRAMS[1]

J. FLEISCHER
Fakultät für Physik, Universität Bielefeld
D-33615 Bielefeld, Germany

ABSTRACT

Taylor expansions of two-point functions in terms of the external momentum squared have been evaluated in their region of analyticity very successfully by accelerating their convergence with Padé approximants (see Ref. [1]). In the present work it is demonstrated that the method works extremely well also for three point functions and it is conjectured that indeed it works for arbitrary n-point functions below thresholds. Beyond that, making use of a method proposed in Ref. [2], it is shown how to perform the analytic continuation to kinematical regions above thresholds on the cut. In this manner we obtain a method which should be particularly useful for massive higher loop diagrams in all kinematical regions. Numerical examples are given.

1. Expansion of three-point functions in terms of external momenta squared

Here we have two independent external momenta in d dimensions. The general expansion of (any loop) scalar 3-point function with its momentum space representation $C(p_1,p_2)$ reads

$$C(p_1,p_2) = \sum_{l,m,n=0}^{\infty} a_{lmn}(p_1^2)^l(p_2^2)^m(p_1\cdot p_2)^n = \sum_{L=0}^{\infty}\sum_{l+m+n=L} a_{lmn}(p_1^2)^l(p_2^2)^m(p_1\cdot p_2)^n, \quad (1)$$

where the coefficions a_{lmn} are to be determined from the given diagram. They are obtained by applying the following differential operators $\Box_{ij} = \frac{\partial}{\partial p_{i\mu}}\frac{\partial}{\partial p_j^{\mu}}$ several times to both sides of (1):

$$\Box_{11}(p_1^2)^l(p_2^2)^m(p_1\cdot p_2)^n$$
$$= (p_2^2)^m\{2l(d+2(l-1)+2n)\cdot(p_1^2)^{l-1}(p_1\cdot p_2)^n + n(n-1)(p_1^2)^l p_2^2(p_1\cdot p_2)^{n-2}\}, \quad (2)$$

similarly for $\Box_{22}$ and

$$\Box_{12}(p_1^2)^l(p_2^2)^m(p_1\cdot p_2)^n$$
$$= \{n(2(l+m)+n+d+1)(p_1^2)^l(p_2^2)^m(p_1\cdot p_2)^{n-1} + 4l\cdot m(p_1^2)^{l-1}(p_2^2)^{m-1}(p_1\cdot p_2)^{n+1}\}. \quad (3)$$

[1]report on work in collaboration with O.V. Tarasov

This procedure results in a system of linear equations for the a_{lmn}. For fixed L (see equation (1)) we obtain a system of $(L+1)(L+2)/2$ equations of which, however, maximally $[L/2]+1$ couple ($[x]$ standing here for the largest integer $\leq x$). These linear equations are easily solved with REDUCE, e.g., for arbitrary d.

For the purpose of demonstrating the method, I confine myself to the case $p_1^2 = p_2^2 = 0$, which is e.g. physically realized in the case of the Higgs decay into two photons ($H \rightarrow \gamma\gamma$) with p_1 and p_2 the momenta of the photons. In this case only the coefficients a_{00n} are needed. They are each obtained from a "maximally coupled" system of $[n/2]+1$ linear equations. Solving these systems of equations we obtain a sequence of differential operators which project out from the r.h.s. of (1) the coefficients a_{00n}, of which I give in the following the first few ($Df_0 = 1$):

$$\begin{aligned} Df_1 &= \frac{1}{d}\Box_{12}, \qquad Df_2 = -\frac{1}{2(d-1)d(d+2)}\{\Box_{11}\Box_{22} - d\Box_{12}^2\}, \\ Df_3 &= -\frac{1}{2(d-1)d(d+2)(d+4)}\{\Box_{11}\Box_{22}\Box_{12} - \frac{1}{3}(d+2)\Box_{12}^3\}, \\ Df_4 &= \frac{1}{8(d-1)d(d+1)(d+2)(d+4)(d+6)} \\ &\quad \left\{\Box_{11}^2\Box_{22}^2 - 2(d+2)\Box_{11}\Box_{22}\Box_{12}^2 + \frac{1}{3}(d+2)(d+4)\Box_{12}^4\right\}, \cdots . \end{aligned} \tag{4}$$

Applying the operator Df_n to the (scalar) momentum space integral $C(p_1, p_2)$ and putting the external momenta equal to zero, yields the expansion coefficients a_{00n}.

It is instructive to first look at the one-loop integral ($d = 4$ since the integral is finite):

$$C_0(m_1, m_2, m_3; p_1, p_2) = \frac{1}{i\pi^2}\int \frac{d^4k}{((k+p_1)^2 - m_1^2)((k+p_2)^2 - m_2^2)(k^2 - m_3^2)}. \tag{5}$$

Performing the above procedure, it is seen that for $d = 4$ the coefficients a_{00n} can be written as

$$i\pi^2 a_{00n} = \frac{2^n}{n+1}\int d^4k \frac{(k^2)^n}{c_1^n c_2^n} \cdot \frac{1}{c_1 c_2 c_3} \tag{6}$$

with $c_i = k^2 - m_i^2 (i = 1, 2, 3)$. For $m_3 \neq 0$ we have

$$a_{000} = \frac{1}{i\pi^2}\int \frac{d^4k}{c_1 \cdot c_2 \cdot c_3} = C_0(m_1, m_2, m_3; 0, 0) = \frac{1}{m_3^2} g(r, s), \tag{7}$$

where $g(r,s) = \frac{1}{r-s}\left(\frac{r \ln r}{1-r} - \frac{s \ln s}{1-s}\right)$ and $r = \frac{m_1^2}{m_3^2}, s = \frac{m_2^2}{m_3^2}$. The function $g(r,s)$ can now be considered as *generating function* of the scalar one-loop vertex since all higher negative powers of the $c_i's$ can be obtained by differentiation with respect to m_i^2: writing $k^2/c_1 = 1 + \frac{m_1^2}{c_1}$, we finally have

$$a_{00n} = \frac{2^n}{(n+1)!}\frac{1}{(m_3^2)^{n+1}}\frac{\partial^n}{\partial s^n}\sum_{\nu=0}^{n}\binom{n}{\nu}\frac{r^\nu}{\nu!}\frac{\partial^\nu}{\partial r^\nu} g(r, s), \tag{8}$$

which can easily be evaluated with the help of a formulae manipulating program.

Similarly we proceed in the two-loop case: starting e.g. from the scalar integral

$$C(m_1,\cdots,m_6;p_1,p_2)$$
$$= \frac{1}{(i\pi^2)^2}\int\frac{d^4k_1d^4k_2}{((k_1+p_1)^2-m_1^2)((k_1+p_2)^2-m_2^2)((k_2+p_1)^2-m_3^2)((k_2+p_2)^2-m_4^2)(k_2^2-m_5^2)((k_1-k_2)^2-m_6^2)} \tag{9}$$

in analogy to (5) we now have with obvious abbrevations

$$(i\pi^2)^2 a_{00n} = \frac{2^n}{n+1}\int d^4k_1 d^4k_2 F_n\cdot\frac{1}{c_1\cdot c_2\cdot c_3\cdot c_4\cdot c_5\cdot c_6}, \tag{10}$$

where

$$F_n = \sum_{\nu=0}^{n} c_1^{-(n-\nu)}c_3^{-\nu}\sum_{\nu'=0}^{n} c_2^{-(n-\nu')}c_4^{-\nu'}\cdot A^n_{\nu\nu'}(k_1,k_2), \tag{11}$$

and

$$A^n_{\nu\nu'}(k_1,k_2) = \sum_{0\le 2\mu\le\nu+\nu'\le n+\mu} a^{n\mu}_{\nu\nu'}(k_1^2)^{n-(\nu+\nu')+\mu}(k_2^2)^{\mu}(k_1\cdot k_2)^{\nu+\nu'-2\mu}, \tag{12}$$

$a^{n\mu}_{\nu\nu'}$ being rational numbers with the properties

$$a^{n\mu}_{\nu\nu'} = a^{n\mu}_{\nu'\nu} \qquad \text{and} \qquad \sum_{\mu} a^{n\mu}_{\nu\nu'} = 1, \tag{13}$$

the first property being equivalent to $A^n_{\nu\nu'}(k_1,k_2) = A^n_{\nu'\nu}(k_1,k_2)$. Besides that we also have the nontrivial symmetry property

$$A^n_{\nu\nu'}(k_1,k_2) = A^n_{n-\nu',n-\nu}(k_2,k_1) \tag{14}$$

such that the essentially different $A^n_{\nu\nu'}$ are the following ones:

$$A^n_{\nu\nu'}, \quad \text{with} \quad \nu\le\nu' \quad \text{and} \quad \nu+\nu' = 2,\cdots,n \quad \left(0\le\mu\le\left[\frac{\nu+\nu'}{2}\right], \quad \text{see (12)}\right)$$

while we have explicitely $A^n_{00} = (k_1^2)^n$ and $A^n_{01} = (k_1^2)^{n-1}(k_1\cdot k_2)$ or $a^{n0}_{00} = a^{n0}_{01} = 1$ (see (13)). Finally we observe $a^{n0}_{02} = 4/3$ independent of n. Some of the other higher coefficients are given explicitely:

$$\begin{array}{llll}
n=2: & a^{20}_{11} = 1/3. & & \\
n=3: & a^{30}_{11} = 4/9; & a^{30}_{03} = 2, & a^{30}_{12} = 2/9. \\
n=4: & a^{40}_{11} = 1/2; & a^{40}_{03} = 2, & a^{40}_{12} = 1/3; \\
 & a^{40}_{04} = 16/5, & a^{41}_{04} = -12/5; & a^{40}_{13} = 1/5, \\
 & a^{41}_{13} = 11/5; & a^{40}_{22} = 4/45, & a^{41}_{22} = 22/45 \quad \text{etc.}
\end{array}$$

In the two-loop case the "generating functions", i.e. the diagrams with all external momenta equal to zero, were calculated in Ref. [3]. Let us consider as an example $m_6 = 0$ (gluon exchange) and all other masses $= m_t$ (top mass). In order to properly generate by differentiation the higher negative powers of c_1, c_2 and c_2, c_4, respectively, we have to introduce even in this case different masses corresponding to $c_i's$ with different integration momenta and take the limit $m_1, m_2 \to m_t$ at the end. In the notation of Ref. [3] we thus obtain

$$\begin{aligned} \int \frac{d^4k_1 d^4k_2}{(k_1^2 - m_1^2)^2 (k_2^2 - m_2^2)^3 (k_1 - k_2)^2} &= \ < 2m_1 | 3m_2 | 0 > \\ = \quad \frac{1}{2} \frac{\partial^2}{\partial (m_2^2)^2} < 2m_1 | m_2 | 0 > &= \ (i\pi^2)^2 \frac{1}{2} \frac{1}{(m_1^2)^2} \frac{\partial^2}{\partial a^2} f(a, 0) \end{aligned} \tag{15}$$

with $f(a,0) = Sp(1) - Sp(a) - lna \cdot ln(1-a)$ and $a = m_2^2/m_1^2$ ($Sp(x) = -\int_0^1 \frac{dt}{t} ln(1 - xt)$). After differentiation we finally have as generating function for the diagram under consideration

$$g(a) = \frac{1}{2} \frac{1}{(m_1^2)^2} \left\{ \frac{1}{a} \frac{1}{1-a} + \frac{lna}{(1-a)^2} \right\}. \tag{16}$$

For the two-loop case under consideration, as can be seen from the above ((11) and following), it does not seem to be possible to write down the analog of (8) in a compact form as for the one-loop vertex. In the general mass case, however, it should be possible to write a procedure in any formulae manipulating language to obtain by differentiation of the corresponding generating function the coefficients a_{00n} (and others). In the present special case, namely that of equal masses, it was possible to obtain the higher coefficients by direct differentiation and reduction to Γ−functions. The above procedure was only used to verify the lowest ones. In the general mass case, however, the described method will become indispensable. In table 1 we show numerical results below the threshold, i.e. for $q^2 < 4m_t^2$. As can be seen, even very close to the threshold and also for large spacelike q^2, the summation of the Taylor series by means of Padé approximants in terms of the "ϵ-algorithm" [4] produces excellently convergent sequences.

2. Analytic continuation on the cut

Of course it is not satisfying to be confined with ones method to the kinematical region below thresholds. One way to use the above expansion for the calculation above threshold is analytic continuation, provided the expansion coefficients are known with high enough precision. This is, however, no limitation if one uses modern formulae manipulating systems. It may even turn out that adding Feynman diagrams is best performed on the level of these coefficients (if not already on the level of the generating functions).

Table 1: diagonal Padé approximants for various momenta squared below threshold

q^2/m^2	[10/10]	[11/11]	$Ref.[5], method I$
3.99	12.09	12.07	
3.9	5.9866	5.9863	5.990 ± 0.01
3.0	1.57522559708	1.57522559707	1.576 ± 0.002
−8.0	0.15944823388526	0.15944823388521	0.1595 ± 0.0001
−40.0	0.033534605	0.033534597	

Assume, the following Taylor expansion of a three-point function is given:

$$C(p_1, p_2) = \sum_{m=0}^{\infty} a_m y^m \equiv f(y) \tag{17}$$

and the function on the r.h.s. has a cut for $y \geq 1$. In the above case of $H \to \gamma\gamma$ one would introduce $y = z \equiv \frac{q^2}{4m_t^2}$ with $q^2 = (p_1 - p_2)^2$ as new variable. The method proposed in Ref. [2] consists in setting up a Taylor series in a new variable, x, related to y by the Ricatti differential equation

$$y'(x) = y^2 - 1, \tag{18}$$

a solution of which is $y(x) = -(e^x - e^{-x})/(e^x + e^{-x})$, or by inversion

$$x(y) = \frac{1}{2} ln \frac{-1/y + 1}{-1/y - 1}, \tag{19}$$

i.e. a mapping of the whole y-plane on the strip with $-\pi/2 \leq Im(x) \leq \pi/2$ in the $x-$plane. In particular the region of convergence of the series (17), i.e. $1 \geq y \geq -1$ is mapped on the real x-axis $-\infty \leq x \leq +\infty$ with $x(0) = 0$. The (upper side of the) cut $(1 \leq y \leq +\infty)$ is mapped into $-\infty \leq Re(x) \leq 0$ and $Im(x) = -\pi/2$. Furthermore the mapping (19) has the nice property

$$Re(x(z > 1)) = x(\frac{1}{z} < 1) < 0. \tag{20}$$

Thus we obtain on the cut an exact representation for the function $f(y)$ by expanding in a Taylor series in x around $x_C \equiv x(1/z)$. Using (18) we finally have

$$f(z > 1) = \sum_{n=0}^{\infty} \frac{\left\{(y^2 - 1)\frac{d}{dy}\right\}^n f(y)\,|_{y=1/z<1}}{n!} (-i\frac{\pi}{2})^n, \tag{21}$$

which is convergent when no singular point of $f(y(x))$ lies inside the circle with center x_C and radius $\pi/2$. In particular all coefficients of the series (21) are convergent if the expansion (17) for $f(y)$ is used since the derivatives are taken at points $0 \leq y = \frac{1}{z} = \frac{4m_t^2}{q^2} < 1$ if $z > 1$ on the cut, i.e. they are taken in the region of analyticity of the original series. The term $n = 0$ in (21) is just the original series, but now taken for $y = 1/z < 1$, which was dealt with in part 1! What concerns the higher terms in (21), it turns out to be of great use near threshold to factor out an overall (y^2-1) and apply Padé approximations to the remaining coefficient-series (sub-series). Observe: if in (17) the coefficients $a_0, \cdots, a_M$ are known, then in the sub-series of (21) only coefficients $a_0^{(n)}, \cdots, a_{M-n}^{(n)}$ are completely known and only these are taken into account.

Numerical results are given in table 2. In Ref. [5] the diagram $H \rightarrow \gamma\gamma$ was calculated by means of Monte-Carlo integration with different ways (called method I and II, respectively) of dealing with the singularities. The method of analytic continuation yields results consistent with their method II. In general this comparison indicates that our results with the total number of 23 coefficients in the original series (17) can be assumed to have an error $\lesssim 1\%$. Increasing the number of coefficients, one can obtain any desired precision.

It is important at this point to stress the methodic difference in obtaining the results close to the threshold and far away. In the latter case $z = 4m_t^2/q^2$ becomes quickly small and the sub-series converge well. One needs, however, relatively many terms in the summation over n in (21). The situation is different close to the threshold (see also table 2): for the summation over n relatively few terms are needed but the sub-series are badly convergent - they can, however, be summed quite accurately with Padé approximants (see the $\{M,N\}_P$'s in the Padé-column of table 2). Very close to the threshold, however, the Padé's for the higher (than indicated) sub-series become (drastically) unstable at a certain order and yield consequently wrong results if taken to that order. Since also the contributions from the unaccelerated sums become quickly nonsensical (less and less terms of a badly convergent series), I have confined myself in table 2 to those approximants where the highest order Padé's calculable from the set of complete coefficients known in the sub-series are still stable. Generally speaking these will be different for the real and imaginary parts. In this empirical manner excellent results are also obtained close to the threshold.

For the very high q^2, as was mentioned above, the sub-series converge well. Since, however, in the main series (21), very many terms are contributing, in this case also the convergence of the main series must be accelerated by means of Padé's (applied to real and imaginary part seperately). The result for $m_H^2/m_t^2 = 40$ obtained in this manner (see last row of table 2) indicates that a considerable improvement is thus achieved. Indeed, the partial sums change so slowly in this case that the series seems to have reached convergence – only comparison with Ref. [5] indicates that in fact the acceleration is proper.

As a final remark I want to stress the following general property of the proposed

Table 2: Results for the two-loop $H\gamma\gamma$ diagram above threshold. $\{M, N\}$ means complete sums with M terms in (17) and N terms in (21) and the index P stands for acceleration by Padé's in the sub-series

m_H^2/m_t^2	*sums*		*sums*		*Padé's*		*Ref.* [5],	*methodII*
	Re	*Im*	*Re*	*Im*	*Re*	*Im*	*Re*	*Im*
	$\{20,11\}$	$\{20,11\}$	$\{22,13\}$	$\{22,13\}$	$\{22,2\}_P$	$\{20,5\}_P$		
4.05	3.79	1.36	3.91	1.55	5.15	10.51	5.17 ± 0.02	10.47 ± 0.02
					$\{22,6\}_P$	$\{22,5\}_P$		
4.1	2.665	2.31	2.704	2.62	2.660	9.08	2.65 ± 0.01	9.08 ± 0.02
					$\{22,6\}_P$	$\{22,9\}_P$		
4.2	1.00	3.76	1.00	3.78	0.499	7.417	0.51 ± 0.01	7.40 ± 0.01
					$\{22,8\}_P$	$\{22,9\}_P$		
4.5	−1.18	3.83	−1.23	4.05	−1.428	4.782	-1.44 ± 0.01	4.80 ± 0.01
					$\{22,10\}_P$	$\{22,11\}_P$		
5.0	−1.951	2.643	−1.966	2.704	−1.987	2.758	-2.03 ± 0.03	2.761 ± 0.005
6.0	−1.762	1.115	−1.772	1.123			-1.78 ± 0.03	1.128 ± 0.004
7.0	−1.404	0.481	−1.412	0.479			-1.41 ± 0.02	0.480 ± 0.003
8.0	−1.121	0.187	−1.125	0.179			-1.14 ± 0.02	0.183 ± 0.002
10.0	−0.758	−0.0429	−0.751	−0.0534			-0.76 ± 0.01	-0.055 ± 0.002
	$\{20,17\}$	$\{20,17\}$	$\{22,19\}$	$\{22,19\}$	[5/5] *on* (21) :			
40.0	−0.0615	−0.0702	−0.0586	−0.0676	−0.0476	−0.0622	-0.0463 ± 0.0003	-0.0639 ± 0.0002

method: may it take enormous effort to calculate the expansion coefficients of the original series (17), they can, however, be calculated analytically once for ever as a function of the mass ratios and following the demonstrated procedure, the diagrams can then be calculated in a very simple manner in the whole kinematical region of interest.

Acknowledgements

I am grateful to Junpei Fujimoto for providing the numerical values of Ref. [5] for comparison.

References

[1] D.J. Broadhurst, J. Fleischer and O.V. Tarasov, *Z.f.Physik*, **C60** (1993) 287.

[2] J. Fleischer, *Nucl. Physics*, **B37** (1972) 59.

[3] J. van der Bij and M. Veltman, *Nucl. Phys.* **B231** (1984) 205.

[4] D. Shanks, *J. Math. Phys.* **34** (1955); P. Wynn, *Math. Comp.* **15** (1961) 151; G.A. Baker, P. Graves-Morris, Padé approximants, in *Encyl.of math.and its appl.*, Vol. **13**, pp Addison-Wesley (1981).

[5] J. Fujimoto, Y. Shimizu, K. Kato and Y. Oyanagi, *KEK preprint* 92-213 (March 1993).

A CLOSED ANALYTICAL FORMULA FOR TWO-LOOP MASSIVE TADPOLES WITH ARBITRARY TENSOR NUMERATORS

K.G.CHETYRKIN
Institute for Nuclear Research of the Russian Academy of Sciences, Moscow, 117312, Russia

ABSTRACT

Using the integration by parts method we derive a closed analytical expression for the result of the integration of an arbitrary dimensionally regulated tadpole diagram composed of a massless propagator and two massive ones, each raised into an arbitrary power, and including an arbitrary tensor numerator. We also briefly discuss the implementation of the formula in the algebraic manipulation language of FORM.

1. Introduction

Feynman integrals (FI) with complicated tensor numerators are usually difficult to work with. Even in the cases when they are known to be analytically calculable *in principle*, in practical terms their evaluation often implies a tedious, time-consuming and error-prone labour of reducing the problem to the calculation of a host of properly constructed scalar integrals. Moreover, the number of the latter integrals grows very fast as the numerator's structure gets more complicated.

There exist only a few examples when the task is completely solved. A good example is provided with the so-called *p-integrals*, that is completely massless Feynman integrals depending on only one external momentum. Here the explicit result is known for one-loop integrals[1], while, say, a given three-loop tensor p-integral can be done by employing a rather cumbersome and time-costly method of harmonic projections[2].

In this talk we discuss another useful class of Feynman integrals — integrals without external momenta at all but comprising massive lines[(1)] as well as massless ones. They will be referred to as *m-integrals*. Such integrals naturally appear in many problems where the mass m may be treated as a "heavy" one, much larger than all other mass scales involved.

In the one-loop case m-integrals are rather trivial and we shall concentrate on two-loop m-integrals pictured in Fig. 1. 2-loop m-integrals with only one massive line (Fig. 1a) may be reduced to 1-loop case after firstly integrating the 1-loop p-subintegral. 2-loop m-integrals with more than 1 massive lines (Fig. 1b,c) are not so easy to do. We show how the use of the integration by parts method leads to a

(1) It is understood that all the massive propagators depend on one and the same mass m.

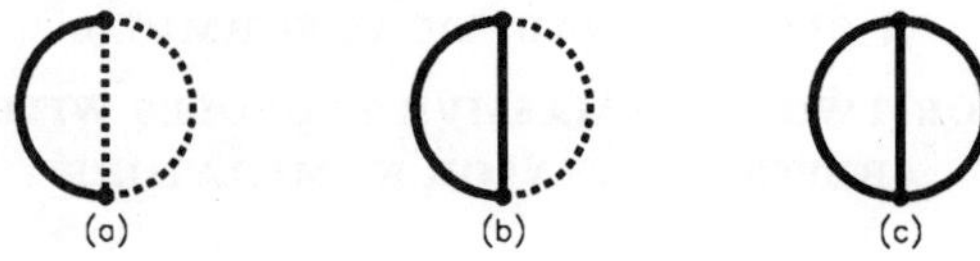

Figure 1: Diffferent cases of 2-loop m-integrals: dashed lines are massless; solid lines have mass m.

simple and general result for arbitrary (not necessarily scalar) two-loop m-integral with two massive and one massless line (see Fig. 1b). In principle, our method allows also to reduce a tensor integral with 3 massive lines of Fig. 1c to a similar scalar integral. The latter can probably be done (at least in some particular cases) again through integration by parts[3]. But to the best of our knowledge no explicit integration formula for this integral exists if the powers of all three propagators are arbitrary.

2. Setting the problem

We begin with a bit more complicated FI of the same topology as shown in Fig. 1b but with a non-zero external momentum q. The corresponding analytical expression reads (in Euclidean space)

$$\frac{1}{(\pi^2)^{2-\epsilon}}\int\frac{\mathrm{d}^D\ell_1\,\mathrm{d}^D\ell_2\;\mathcal{P}(p)}{(p_1^2+m^2)^\alpha(p_2^2+m^2)^\beta(p_3^2)^\gamma} \tag{1}$$

where ℓ_1 and ℓ_2 are the loop momenta, $p=\{p_1,p_2,p_3\}$ are the propagator momenta and $\mathcal{P}()$ is a tensor nominator. We shall deal with three possibilities of expressing the momenta p in terms of the loop momenta and the external momentum, viz.,

$$p_1=\ell_1+q,\quad p_2=\ell_2,\quad p_3=-(\ell_1+\ell_2), \tag{2}$$

$$p_1=\ell_1,\quad p_2=\ell_2+q,\quad p_3=-(\ell_1+\ell_2), \tag{3}$$

$$p_1=\ell_1,\quad p_2=\ell_2,\quad p_3=q-(\ell_1+\ell_2). \tag{4}$$

In a particular case of $\mathcal{P}()\equiv 1,\ \ q=0$ the result for Eq. (1) is known[4]

$$(\text{Eq. (1) with } \mathcal{P}(p)\equiv 1, \text{ and } q=0)=(m^2)^{D-\alpha-\beta-\gamma}\frac{\Gamma(D/2-\gamma)}{\Gamma(\alpha)\Gamma(\beta)\Gamma(D/2)}M(\alpha,\beta,\gamma)$$

with

$$M(\alpha,\beta,\gamma)=\frac{\Gamma(\alpha+\gamma-D/2)\Gamma(\beta+\gamma-D/2)\Gamma(\alpha+\beta+\gamma-D)}{\Gamma(\alpha+\beta+2\gamma-D)}. \tag{5}$$

Our aim is to generalize this result on the case of arbitrary tensor polynomial

$\mathcal{P}(p)$, still keeping $q=0$. To simplify the formulas we shall set $m=1$ below.

3. Solution through the integration by parts method

As is well known within dimensional regularization the value of the FI Eq. (1) does not depend on specifying the propagator momenta and all three choices (2-4) are completely equivalent. The fact can be conveniently expressed through some differential identities. The essence of the integration by part method for dimensionally regulated Feynman integrals consists of the use of such identities in order to simplify integrals to be computed[1].

Let us try to apply the method in our case. It is convenient to write eq. Eq. (1) in a condensed form as follows

$$\text{eq. Eq. (1)} = \int \mathcal{P}(p) I(\alpha,\beta,\gamma) \tag{6}$$

and to introduce four differential operators acting on Eq. (6)

$$\tilde{p}_i^\mu = \frac{1}{2}\frac{\partial}{\partial p_i^\mu}, \quad i=1,2,3 \text{ and } \tilde{q}^\mu = \frac{1}{2}\frac{\partial}{\partial q^\mu}.$$

Explicitly, one has

$$\int I(\alpha,\beta,\gamma) = \frac{\Gamma(D/2-\gamma)}{\Gamma(\alpha)\Gamma(\beta)\Gamma(D/2)} M(\alpha,\beta,\gamma),$$

$$\begin{aligned} \tilde{p}_i^\mu \left(\mathcal{P}(p) I(\alpha,\beta,\gamma)\right) &= \left(\frac{1}{2}\frac{\partial}{\partial p_i^\mu}\mathcal{P}(p)\right) I(\alpha,\beta,\gamma) - \alpha_i \mathcal{P}(p) I(\alpha+\delta_{1i},\beta+\delta_{2i},\gamma+\delta_{3i}) p_i^\mu, \\ \tilde{q}^\mu \left(\mathcal{P}(p) I(\alpha,\beta,\gamma)\right) &= \left(\frac{1}{2}\frac{\partial}{\partial p_3^\mu}\mathcal{P}(p)\right) I(\alpha,\beta,\gamma) - \gamma \mathcal{P}(p) I(\alpha,\beta,\gamma+1) p_3^\mu, \end{aligned} \tag{7}$$

with $\alpha_1=\alpha, \alpha_2=\beta, \alpha_3=\gamma$. The equivalence of the three momentum patterns (2-4) may now be expressed as a chain of identities

$$\int \tilde{p}_1^\mu \mathcal{P}(p) I(\alpha,\beta,\gamma) = \int \tilde{p}_2^\mu \mathcal{P}(p) I(\alpha,\beta,\gamma) = \int \tilde{p}_3^\mu \mathcal{P}(p) I(\alpha,\beta,\gamma) = \int \tilde{q}^\mu \mathcal{P}(p) I(\alpha,\beta,\gamma). \tag{8}$$

These equalities allow us to evaluate immediately an integral of the form

$$\int \tilde{\mathcal{P}}_{2n}(\tilde{p}) I(\alpha,\beta,\gamma)\Big|_{q=0} = \frac{\Gamma(D/2-\gamma)}{\Gamma(\alpha)\Gamma(\beta)\Gamma(D/2)} \left\{ \left(\frac{\Box_{\tilde{q}}}{4}\right)^n \tilde{\mathcal{P}}(\tilde{q})_{2n} \right\} \frac{(-)^n (\gamma)_n}{n!(2-\epsilon)_n} M(\alpha,\beta,\gamma+n), \tag{9}$$

with the Pochhammer symbol $(a)_n = \dfrac{\Gamma(a+n)}{\Gamma(a)}$, $D=4-2\epsilon$ and $\tilde{\mathcal{P}}_{2n}(\tilde{p})$ being an arbitrary tensor in $\tilde{p}=\{\tilde{p}_1,\tilde{p}_2,\tilde{p}_3\}$ of rank $2n$. Indeed, in view of Eq. (8) we may freely replace $\tilde{\mathcal{P}}_{2n}(\tilde{p})$ by $\tilde{\mathcal{P}}_{2n}(\tilde{q}) \equiv \tilde{\mathcal{P}}_{2n}(\tilde{p}_1=\tilde{q},\tilde{p}_2=\tilde{q},\tilde{p}_3=\tilde{q})$. Now the result Eq. (9) comes from three simple observations:

(i) The integral $\int I(\alpha,\beta,\gamma)$ is a scalar function of q and thus only the scalar component of the polynomial $\tilde{\mathcal{P}}_{2n}(\tilde{q})$ (that is proportional to $\tilde{q}^{2n}$) will survive after setting $\tilde{q}=0$ in the very end of the calculation.

(ii) $\left(\frac{\Box_{\tilde{q}}}{4}\right)^n (\tilde{q})^n = n!(2-\epsilon)_n$.

(iii) $\left(\frac{\Box_{\tilde{q}}}{4}\right)^n \frac{1}{\tilde{q}^{2\gamma}} = (\gamma)_n(\gamma - 1 + \epsilon)_n \frac{1}{(\tilde{q}^2)^{\gamma+n}} = (-)^n(\gamma)_n(2 - \epsilon - \gamma - n)_n \frac{1}{(\tilde{q}^2)^{\gamma+n}}$.

Thus we are left with the task of finding a representation of the initial integral Eq. (1) as a linear combinations of integrals of the form displayed in Eq. (9). The problem is solved by the use of the following identity

$$\mathcal{P}_n(q)f(q^2) = \sum_{\sigma=0}^{[n/2]} \frac{(-)^\sigma}{4^\sigma} \left\{\Box_{\tilde{q}}^\sigma \mathcal{P}_n(\tilde{q})\right\} f^{(-n+\sigma)}(q^2) \tag{10}$$

where $f(x)$ is an arbitrary smooth function of x and $f^{(n)}(x)$ is defined in such a way that

$$\frac{d}{dx} f^{(n)}(x) \equiv f^{(n+1)}(x). \tag{11}$$

It should be clear now that once the initial integral Eq. (1) has been expressed as a linear combinations of integrals of the form Eq. (9) it may be done without any problem. Indeed, without essential loss of generality we may assume that the polynomial $\mathcal{P}(p) = \mathcal{P}_{n_1,n_2}(p)$ does not depend on p_3 and meets the following homogeneity equation:

$$\mathcal{P}_{n_1,n_2}(\lambda_1 p_1, \lambda_2 p_2) \equiv \lambda_1^{n_1}\lambda_2^{n_2}\mathcal{P}_{n_1,n_2}(p_1,p_2).$$

Now a direct application of Eq. (8), Eq. (9) and Eq. (10) gives

$$\int \mathcal{P}_{n_1,n_2}(p_1,p_2) I(\alpha,\beta,\gamma) =$$
$$\frac{\Gamma(D/2-\gamma)}{\Gamma(\alpha)\Gamma(\beta)\Gamma(D/2)} \sum_{\sigma_3=0}^{[(n_1+n_2)/2]} \sum_{\sigma_2=0}^{[n_2/2]} \sum_{\sigma_1=0}^{[n_1/2]} \frac{(\gamma)_{\sigma_3}}{\sigma_3!(2-\epsilon)_{\sigma_3}} M(\alpha+\sigma_1-n_1, \beta+\sigma_2-n_2, \gamma+\sigma_3) \tag{12}$$
$$\frac{(-)^{(n_1+n_2+\sigma_3)}}{4^{(\sigma_1+\sigma_2+\sigma_3)}} \left\{\Box_{p_3}^{\sigma_3} \left\{\Box_{p_2}^{\sigma_2}\Box_{p_1}^{\sigma_1}\mathcal{P}_{n_1,n_2}(p_1,p_2)\right\}\big|_{p_1=p_2=p_3}\right\}\big|_{p_3=0}$$

which is the formula we wanted.

The algebraic structure of Eq. (12) is very similar to that of the corresponding formula for 1-loop p-integrals in[1]. This observation has allowed us to perform a simple algebraic programming of Eq. (12) in FORM[5] by closely following the routine ONE.PRC from the package MINCER[2].

Acknowledgments

I appreciate the warm hospitality of Institute of Theoretical Particle Physics at the Karlsruhe University where this work has been finished. I thank Mrs. M. Frasure and A. Kwiatkowski for their help in preparing the manuscript. A stipendium from the HERAUES-stiftung is gratefully acknowledged.

References

1. F. V. Tkachov, *Phys. Lett.* **B100** (1981) 65;
   K. G. Chetyrkin and F. V. Tkachov, *Nucl. Phys.* **B192** (1981) 159.

2. S. A. Larin, F. V. Tkachov, J. A. M. Vermaseren, *Preprint NIKHEF-H*/91-18 (1991).

3. A. I. Davydychev and J. B. Tausk, *Nucl. Phys.* **B397** (1993) 123;
A. I. Davydychev, V. A. Smirinov and J. B. Tausk, *Preprint INLO-PUB*-5/93.

4. M. Veltman, *SCHOONSCHIP, A CDC 6600 program for symbolic evaluation of algebraic expressions, CERN Report* 1967 (unpublished).

5. J. A. M. Vermaseren, *Symbolic Manipulation with FORM*, Version 2, CAN, Amsterdam, 1991.

HANDLING PARTICLE MULTIPLETS AND FERMION NUMBER VIOLATION IN FEYNMAN DIAGRAM AND AMPLITUDE GENERATION

HAGEN ECK and JOSEF KÜBLBECK
Physikalisches Institut, Universität Würzburg, Am Hubland
97074 Würzburg, FRG

ABSTRACT

Looking at the great variety of models under current investigation in high energy physics it is desirable to have a flexible computer algebra program available that can handle non-standard Feynman rules. It should treat models with many particles efficiently, handle fermion number violating couplings and provide the possibility to work in different representations of the Lorentz group. Here we want to explain how this is done in *FeynArts 2.0*. Appropriate subsequent packages like *FeynCalc* [1] or *FeynWeyl* [2] do the computeralgebraic evaluation so that one arrives at an expression that can be evaluated numerically.

1. Introduction

Meanwhile there exist a lot of computer algebra packages that can create all Feynman graphs and amplitudes for a certain process from a given model. However, each of these packages was originally designed for a certain set of purposes and has its limitations. When applying them to elaborate models like supersymmetric extensions of the standard model or models with $SU(N)$ gauge groups problems may appear. Our aim was to write a new version of *FeynArts* that can deal with the following complications:

- Fermion number is not conserved in all interactions. This requires unique algorithmic Feynman rules for fermion violating couplings that should also yield the correct signs from the permutation of the external particles.
- Big models involve multiplets with many particles but the available computer power is limited. Usually one will work on the multiplet level, which means to use indexed fields. On the other hand it may be convenient to have all explicit graphs available for reasons of checking or of comparing with other models. If the number of graphs on the multiplet level is still big it even may be favorable to do the algebraic computation with generic graphs i.e. graphs that consist of basic field *types* (scalars, fermions, ghosts, vector bosons) only.
- Sometimes it is necessary to do calculations with different representations of the Lorentz group even in the same model. E.g. for tree calculations not only with massless external particles [3] the Weyl-van der Waerden-formalism is well approved. Changes of the representation should be possible without having to enter a completely new model.

The generation of Feynman diagrams and amplitudes involves 3 basic steps, namely *(1)* generation of topologies, *(2)* insertion of particles (fields) and *(3)* translation of diagrams into amplitudes. How *FeynArts 1.0* performs the generation of topologies including the correct combinatorial factors, creates Feynman graphs from a given set of couplings and external particles, and forms the amplitudes for models without fermion number violation is described in [4] and [5]. In the following sections we want to show how *FeynArts 2.0* deals with fermion number violation and different *levels of insertion.* Of course this concerns only steps *(2)* and *(3)*.

2. Treating Fermion Number Violating Fields

2.1. The Problem

For Dirac fermions it is straightforward to implement the well-known algorithm of multiplying the Grassman valued couplings along the inverse direction of the fermion flow. Supersymmetric gauge theories involve interactions that violate fermion number conservation, so that it is no longer appropriate to speak of a definite fermion flow. The problem becomes most evident for Majorana fermions. The fact that Majorana fields are self-conjugate implies that there exist Wick contractions of these fields which are different from those of Dirac fields. Together with the anticommutativity of fermion fields this obscures the determination of the relative sign of interfering Feynman graphs (RSIF).

For a computer algebra program we want simple algorithmic Feynman rules for fermion violating interactions which yield the amplitude including the correct RSIF. It is furthermore desirable to have them close to the Feynman rules for Dirac particles. In particular we wanted to avoid charge conjugation matrices so that the rules how to treat these matrices in programs that do the analytic evaluations are obsolete. Feynman rules for Majorana fermions were given in Refs. [6, 7, 8, 9]. However, those Feynman rules seem to be either unnecessarily complicated or do not allow the determination of the relative sign of interfering Feynman graphs directly from the diagrams.

2.2. The Solution

We have derived Feynman rules for Majorana (and Dirac) fermions which are close to the rules for Dirac fermions, avoiding charge conjugation matrices. Since the fermion *number* flow is violated we introduce a continuous fermion flow, i.e. an (arbitrary) orientation of each fermion line. Based on this fermion flow we can form chains of Dirac matrices and determine the RSIF as one does for Dirac fermions. We only need the well-known Dirac propagator for all fermions, and we have to introduce in general two Feynman rules for each vertex (one for the two possible orientations), which is still less than in the usual treatment. These two expressions are closely related: When switching the orientation we have to exchange left and right handed parts. For vector and axial vector couplings we get a minus sign in addition. We refer to this as 'flipping the vertex'. The explicit Feynman rules and their derivation can be found in [10] and [11].

In the model files of *FeynArts* for every vertex only one analytic expression has to be entered, the one for the coupling *or* the one for its flipped counterpart. E.g. for a neutralino-electron-selectron vertex we would have to enter

$$Coupling(\widetilde{\chi^0}, \bar{e}, \tilde{e}) = ie(c_- \omega_- + c_+ \omega_+)$$

or

$$Coupling(\bar{e}, \widetilde{\chi^0}, \tilde{e}) = ie(c_+ \omega_- + c_- \omega_+),$$

where ω_- and ω_+ are the left- and right-handed projection operators, respectively. (We have fixed some convention in which order we write down the fermions in *Coupling*.) So the algorithm has to perfom the following things: chains of coupling matrices of a given fermion line are formed as usual now following an arbitrary orientation. When a vertex is hit with fermions in the opposite order as in the model file, c_- and c_+ are flipped and a minus sign is provided for vector and axial vector couplings. All sign ambiguities have disappeared and the RSIF is determined exactly as in the case of Dirac fermions.

3. Generating Feynman Diagrams with Field Classes

3.1. General Remarks

In models with many fields it is favorable to treat the particles as indexed fields. This allows to enter couplings of the models in matrix notation and reduces the huge number of explicit graphs to a smaller number of graphs where the particles carry representation indices. However, for some applications it may be desirable to have the full set of graphs and amplitudes created by the program.

As the used models respect space time translational symmetry we always have a small number of generic fields and couplings (fields and couplings with distinct Lorentz structures). The other group symmetries vary of course but the Lorentz structure is always present and and its evaluation usually is the tough part of the calculation. (It contains contraction of indices, evaluation of traces, tensor integral decompositions, etc.) Therefore, for the calculation of the Lorentz part it is in many cases more economic to calculate the amplitudes on the generic level. For the final evaluation of course all values which the model allows for the generic coupling constants must be inserted. This requires a strict decoupling of the Lorentz structure of the generic graphs from the non-Lorentz parts of the couplings. On the other hand if there are only few explicit graphs per generic graph the explicit couplings may have a simpler structure than the generic one (in its generality), so that it is convenient to have the amplitudes (besides the graphs) also on different levels.

We want to create the Feynman graphs and amplitudes on three levels. The new algorithms of *FeynArts 2.0* comprise appropriate sets of particles in classes of particles, which can (but need not) be identical with the multiplets.

3.2. Generic Fields, Classes and Particles

The three insertion levels on which Feynman graphs are generated are: `Generic`, `Classes` and `Particles`.

a) The `Generic` fields are fermionic fields (`F`), scalar fields (`S`), ghost fields (`U`) and vector boson fields (`V`)[1]. In renormalizable theories there exist exactly 11 couplings among these fields.

b) Field `Classes` are user-defined sets of fields, carrying one or more additional indices. Classes are identified by their class numbers that can be chosen by the user (e.g. `F[1], V[3]`,...).

c) `Particles` are the members of the classes. They carry additional indices which may be unspecified. `F[2, {3}]` denotes the 3rd member of the second fermion class whereas `V[4, {ColorIndex[3]}]` denotes an arbitrary member of the 4th vector boson class carrying an unspecified index named `ColorIndex` with counter '3'.

The generic propagators and couplings in a certain Lorentz representation are provided in a generic model file. If the coupling is a sum, each of its terms is provided with a coupling matrix which has to be specified in the individual model files. E.g. the generic file in Dirac representation contains the fermion vector boson coupling:

$$\mathit{Coupling}(\texttt{F[i], F[j], V[k,}\mu\texttt{]}) := c_-(\texttt{F[i], F[j], V[k]})\gamma_\mu\omega_- + c_+(\texttt{F[i], F[j], V[k]})\gamma_\mu\omega_+$$

Thus for the QED vertex of a lepton `F[2]` (2nd fermion class) to a photon `V[1]` (1st vector boson class) we only have to specify the c_- and c_+:

$$c_-(\texttt{F[2], -F[2], V[1]}) = ie\,; \qquad c_+(\texttt{F[2], -F[2], V[1]}) = ie$$

In addition to the couplings of the classes the individual model file contains the definition of each class. For a class of three leptons it reads:

```
F[2] == { SelfConjugate -> False,
          Indices -> {Index[Generation]},
          Mass -> MLE,
          PropagatorLabel -> ComposedChar[{"e", Index[Generation]} ],
          PropagatorType -> Straight,
          PropagatorArrow -> Forward },
```

together with the index range definition

```
IndexRange[ Index[Generation] ] = {1,2,3}.
```

This fixes that the three leptons are not self-conjugate fields, have a mass of `MLE` (which will carry a `Generation` index for the different class members) and appear as a straight line with an arrow in a Feynman graph.

QED, QCD, the electroweak standard model (SM) and a yet unchecked SUSY model are already entered in *FeynArts 2.0*, others can be written by the user.

3.3. Creating Feynman Diagrams

After specification of the wanted set of topologies and of the external fields,

[1] We have not included a graviton field.

the creation of Feynman diagrams is done in three steps corresponding to the three levels of insertion.

a) `Generic` graphs: the algorithm for the insertion of fields into topologies of *FeynArts 1.0* is used. All propagators of the topology that do not carry a field are successively filled with generic fields (`F`, `S`, `U` or `V`) so that the arising vertices form a subset of the 11 generic ones.

b) `Classes` insertion: For every propagator containing a generic field the vertices at the end of this propagator are checked for possible replacements of this generic field by one of the corresponding classes. Each of the two vertices allows for a certain set of replacements. Thus for the propagator the intersection of these possible replacements is allowed. This procedure finds all possible 'classes graphs' belonging to a generic graph.

c) `Particles` insertion: For every class contained in a Feynman graph we now insert all members of the classes to find the explicit graphs, i.e. we form the outer product over the ranges of all unspecified indices in a graph.

Besides the advantages of the level concept mentioned above, this procedure has the additional benefit of minimizing intermediate expression swell. During steps *a)* and *b)* we have to consider only a small number of fields in every iteration: four generic fields in step *a)* and the classes of the corresponding generic type for each generic field in step *b)*.

We have also included QED, QCD and the SM in a form in which each class contains only one particle. This grants compatibility with *FeynArts 1.0.*

3.4. Creating Feynman Amplitudes

The translation of a Feynman diagram to a Feynman amplitude is performed on the generic level of the diagram. The algorithm for this procedure is the algorithm of *FeynArts 1.0* with the additional feature of handling models that do not conserve the fermion number as described in section 2.

Thus the result of this translation is a `Generic` amplitude. If the user wants amplitudes on the `Classes` or `Particles` level a list of possible replacements of the generic couplings by the desired ones is appended to the generic amplitude. Therefore it is possible to perform the calculation of the Lorentz part of the amplitude without concerning the concrete couplings of this diagram. The results of this calculation can be input — together with the list of concrete couplings — to a numeric program that computes all concrete diagrams by looping over the list of couplings.

For users who want to deal with the `Classes` amplitudes or with the `Particles` amplitudes of the process we provide a routine that changes a `Generic` amplitude to a set of `Classes` or `Particle` amplitudes.

3.5. Amplitudes in different Lorentz representations

The strict decoupling of the Lorentz structure of the generic graphs from the non-Lorentz parts of the couplings in the coupling files has a further advantage:

Not only has the Lorentz structure of the couplings to be entered only once for all models in a certain representation. Also once the coupling matrices of an individual model are entered they can be run on top of any generic model. In other words: the user has the possibility to change the representation of the Lorentz group in which he calculates without having to touch the concrete couplings.

We provide generic models in the Lorentz representation and the Weyl-van der Waerden-representation.

4. Conclusion

On the basis of the algorithms of *FeynArts 1.0* we have developed a version that provides enhanced possibiblities to cope with non standard models.

The package, *FeynArts 2.0*, is able to handle fermion number violating couplings. The algorithm is based on Feynman rules which are close to those for Dirac fermions and provide the correct relative sign of interfering Feynman graphs automatically.

Three levels of Feynman graphs and amplitudes are created. This minimizes intermediate expression swell, allows one to choose the most economic level for algebraic evaluation and provides a clear handling of models with a great number of particles (e.g. in big multiplets).

1. R. Mertig, M. Böhm and A. Denner, *Comp. Phys. Comm.***64** (1991) 345.
2. T. Grund, program package *FeynWeyl*, in preparation.
3. S. Dittmaier, "Weyl-van der Waerden-formalism for helicity amplitudes of massive particles, University of Würzburg preprint 1992.
4. J. Küblbeck, M. Böhm and A. Denner, *Comp. Phys. Comm.***60** (1990) 165.
5. H. Eck and J. Küblbeck, "Computeralgebraic Generation of Feynman Graphs and Amplitudes" in *New Computing Techniques in Physics Research II*, ed. D. Perret-Gallix, World Scientific, Singapore (1992).
6. S.K. Jones and C.H. Llewellyn Smith, *Nucl. Phys.* **B217** (1983) 145.
7. H.E. Haber and G.I. Kane, *Phys. Rep.* **117** (1985) 76.
8. E.I. Gates and K.L. Kowalski, *Phys. Rev.* **D37** (1988) 938.
9. J. Gluza and M. Zrałek, *Phys. Rev.* **D45** (1992) 1693.
10. A. Denner, H. Eck, O. Hahn and J. Küblbeck, *Nucl. Phys.* **B387** (1992) 467.
11. A. Denner, H. Eck, O. Hahn and J. Küblbeck, *Phys. Lett.* **B291** (1992) 278.

Lazy Functional Programming and Manipulation of Perturbation Series

Jerzy Karczmarczuk

Department of Computer Science, University of Caen, Esplanade de la Paix
14032 Caen, France karczma@univ-caen.fr

ABSTRACT

This talk is an introductory tutorial on the techniques of *lazy functional programming* and their possible application to formal manipulations in theoretical physics, in particular to perturbation expansions in field theories. We will *not* present another ready-to-use package in the style of FeynArts[1]. Our ambition is rather methodological: to popularize the powerful semantics of polymorphic, higher-order functions, and the usage of lazy, "virtually infinite" data structures. The examples will be written in Gofer[2] — a portable dialect of the programming language Haskell[3]. We will stress upon the algorithmic properties of such mathematical objects as functions or series, rather than on the syntactic (structural) manipulation of data. We show on a simplified model how to apply in a clean and legible way the functional techniques to generate Feynman graphs. The presented approach is by no means a substitution for the standard computer algebra techniques, but rather a complement.

0. Motivation

Everybody knows that the path between a mathematical formula and the related computer algorithm might be quite long. The situation is especially difficult in the case of the perturbative approach to nonlinear systems, for example in the quantum field theory, but also in other domains, such as the accelerator physics[4].

In principle, the automatic generation of Feynman diagrams from the basic rules should be easy. We take a theory, for simplicity the $\varphi^3+\varphi^4$ model, we write the recursive Dyson-Schwinger equation for the connected Green functions generating functional:

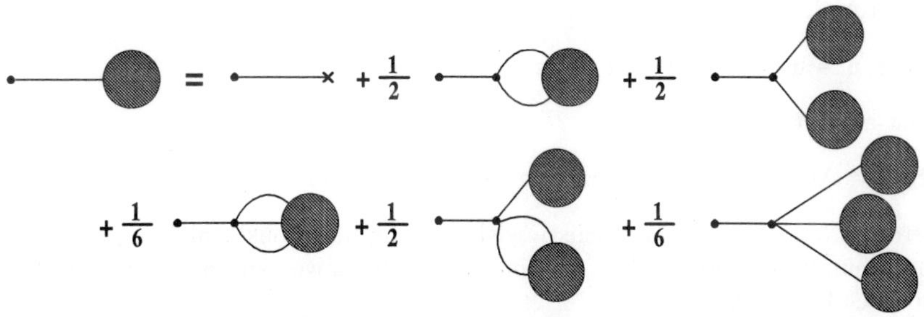

$$= \quad + \frac{1}{2} \quad + \frac{1}{2} \quad + \frac{1}{6} \quad + \frac{1}{2} \quad + \frac{1}{6}$$

and, if our favourite programming language can cope with the algebra of vertices and propagators, the task is solved, the Green functions with more legs are just the derivatives of the left hand side. Of course, the actual computation of the graphs, all the kinematics etc., is another story.

We know however that in order to transform the recurrence above into a program, one usually has to reformulate everything, so that higher graphs are expressed through the simpler ones. The simplicity of the D-S equation is due to the fact that its solution represents the *full infinite series.* Computers don't like infinities. Or is it just a prejudice?

We would like to propose a technique based on the *lazy functional programming* paradigms which permits to generate such series as above *directly* from the D-S equation. We will show how to deal effectively with the infinite data structures and infinite iterative processes, provided the infinities are only potential, "virtual", and all the *final* results of the program are finite.

1. What is Functional Programming

1.1. Syntax and Style

In our examples we will use Haskell[3, 5] — a *pure* functional language[6, 7, 8]. Its syntax is simple, with very few keywords, and some layout conventions which will not be discussed here. Perhaps the only non-obvious convention which deserves some comment at this point is the *normal*, left-to-right evaluation order, unless the operator precedences tell otherwise. The form denoting the application of the function `f` to its arguments `x`, `y` and `z` — `f x y z` is read as `((f x) y) z`.

A program in Haskell is a set of function definitions, and a top-level call of a function which may be called "main". Interpretive versions of Haskell, such as Gofer permit also a more conversational style of work.

Bodies of functions are expressions: applications of functions to expressions. There are no assignments, no side-effects. Even the I/O is side-effect free, it is just the processing of some specific *lazy* data streams: strings of characters and/or the operating systems directives. There are no iterative control structures such as loops. But Haskell is *not* just a subset of "C'. Here functions are "first-class data", objects which may be anonymous, which can be passed as arguments and returned as results of function calls. The following definition of a function which computes the factorial:

```
fact n = if n==0 then 1 else n*fact(n-1)
```

is just a shorthand for:

```
fact = \n -> if n==0 then 1 else n*fact(n-1)
```

where the backslash is the Asciistic vision of the notorious lambda used in Lisp to denote functional forms. Even without assignments we can have local variables in functions, for example:

```
cosh x = 0.5*(y + 1.0/y)
         where y = exp x
```

as this can be translated internally into functional form such as

```
cosh x = (\y -> 0.5*(y+1.0/y)) (exp x)
```

etc. This translation is in fact more complicated, as all the definitions may be potentially recursive.

The language is equipped with the conditionals and with more general `case` forms, but they are less omnipresent than in "C", thanks to the possibility of using *pattern* parameters in the definitions:

```
fact 0 = 1
fact n = n*fact (n-1)
```

The second clause is activated when the first one fails, i.e. when the value of the argument is different from zero. The compiler transforms automatically such patterns into the case analysis, but the user may write programs in the "rule-based" style, as in Prolog, or Mathematica. The compiler performs also another very important action: *the type inference.* Haskell is a strongly typed language. Since the definitions are static, without mutable assignments, there is (almost) no need for explicit typing: the function `fact` takes an integer and returns an integer, this is obvious from both clauses. Analogously, the compiler will find and may inform us that `cosh :: Float -> Float` which is the standard way of presenting the type of a functional object. This remark should convince the reader that functional languages are not toys, and that their compilation is fairly efficient.

What is the type of the function `sqr x = x*x`? The argument `x` may be integer, or floating (or a matrix, or ...). The only information available is that `x` can be multiplied, so the compiler constructs an *overloaded* function of type:

```
sqr :: Num a => a -> a
```

which means that x belongs to a *class* `Num` of *all* numbers, defined in the *standard prelude* loaded during the initialization of the compiler. This prelude may be modified: the author's private Gofer prelude defines the multiplication operator within the class `Monoid` and the addition in `Ring`. These classes include numbers, polynomials, power series, etc.

The word "class" may suggest some common points with the object-oriented programming, its philosophy and terminology being, however, different. Here the notion of class is applicable to *types.* Inside a class we declare a set of operations, and then we specify that a given data type is an *instance* of this class. Here is a simplified example:

```
class Monoid a where
   (*) :: a->a->a
   ...
infix 6 :+                          -- (operator; means "+i*")
data Complex = Float :+ Float
```

```
instance Monoid Complex where
   (x1:+y1)*(x2:+y2) = (x1*x2-y1*y2):+(x1*y2+x2*y1)
   ...
```

The recognition of the two different (*) is static. However, a polymorphic function, as sqr is compiled with a hidden parameter — a dictionary, whose entry: (*) is provided by the appropriate instance definition. The binding is dynamic, less efficient than static, but the program might be *much* shorter.

1.2. Datatypes and Higher-order Functions

Haskell permits the construction of user specific datatypes: named records and unions. The field constructors may be identifiers, and also user-defined operators. Here is the definition of a binary tree:

```
infix 5 :^:          -- Will denote the internal node of the tree
data Tree a = Leaf a | (Tree a) :^: (Tree a)
```

Note that Tree is a *polymorphic, recursive* datatype: its leaves may belong to any (but all to the same) type. Among the predefined polymorphic types we have lists or *streams* whose constructor is a right-associative operator denoted by the colon: (:). A list of three elements: x, y and z may be denoted as [x,y,z] or as x:y:z:[].

We have also anonymous, heterogeneous records: tuples, denoted as (x,y), (x,y,z) etc.

We give here just one example of general list processing: the mapping-filtering functional which transforms all the elements of a list by the application of a given function f, and rejects those which do not fulfil the predicate p (two separate functions, map and filter are predefined in the standard prelude):

```
mapfilter f p []      = []
mapfilter f p (x:xq) | p x        = f x : rest
                     | otherwise = rest
                     where rest = mapfilter f p xq
```

Exercise: how to transform a list by sequential application of two functions, f and g to its elements? Of course, we may write map g (map f l) but a "canonical" way consists in using the composition of functions: map (g . f) l where the composition operator (.) is predefined as: (f . g) x = f (g x).

We see that the composition operator is truly polymorphic: we don't know anything about the argument x. The composition of functions facilitates the definitions of functions in abstraction of their arguments. The function odd may be defined as not . even, and instead of defining the inequality as:

```
x /= y = not (x == y)
```

we can write: `(/=) = ((.) not) . (==)` (admittedly a slight perversion...)

Function manipulations belong to the standard programming style in Haskell. The construction `(map f)` is not a syntax error, but a partial application, or a *section*: a function which accepts one argument, and then applies the mapping. The form `(1 +)` is a section which adds 1 to something, and `(/ 2)` divides something by 2.

*1.3. Data Constructors and Functions are **Lazy***

The examples above serve only to introduce the syntax. The most interesting semantic property of Haskell is its *non-strictness* or *laziness* which means that the arguments of a function are evaluated *only when needed*, and that the data constructors do not evaluate their arguments either. A lazy data structure is a functional object, a *thunk*, which, *when accessed*, produces the appropriate value or set of values. If the datatype is recursive, such as lists or trees, we may effectively operate upon infinite data structures. The infinite list `[1,1,1,...]` we define as `ones = 1 : ones`. We may also define `ones` as `repeat 1`, where we use the predefined function `repeat x = x : repeat x`. This is usually implemented as a compact, cyclic structure, realisable also in strict languages:

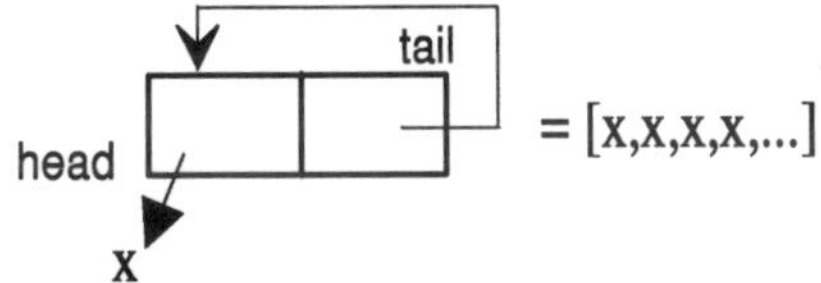

But the following is more difficult to code in Fortran or "C":

```
allIntegs = enumFrom 0
      where enumFrom n = n : enumFrom (n+1)
```

which returns the list `[0,1,2,3,...]`. Such infinite list may be effectively processed, as only a finite part of it exists physically in the memory — the tail is hidden within a thunk. For example, the result of `map (* 2) allIntegs` is equal to `[0,2,4,6,...]`. The arithmetic sequences and the results of mapping, eventually filtered, have in Haskell a compact notation: `allIntegs` can be denoted as `[0..]`, and all the odd integers whose predecessors are not divisible by 5 can be defined using the compact list comprehension notation for the mapping and/or filtering as

```
[n+1 | n <- [1,3..], n `rem` 5 /= 0]
```

The standard way of getting out of the infinite data stream is to take its finite initial segment, for example: `take 100 allIntegs` where the function `take` is predefined as

```
take 0 _ = []
take n (x:xq) = x : (take (n-1) xq)
```

The character '_' denotes an irrelevant, dummy variable used in patterns. In the next section we will show how easy is the manipulation of infinite series with this technique. The user doesn't need to worry about the truncations and the related control structures. In order to prevent the system from generating superfluous terms, it suffices not to look at them!

Here is the Haskell version of the Eratosthenes sieve. The idea is to build a sequence starting at 2, whose tail is constructed from all the integers, filtered by the elimination of all the multiples of the head. The tail undergoes the same process, and without further comment we present the generator of *all* primes:

```
primes = sieve 2 where
         sieve n = n: filter (\x ->x 'rem' n /= 0) (sieve (n+1))
```

which may be rephrased also as

```
primes = map head (iterate sieve [2..]) where
         sieve (p:xs) = [ x | x<-xs, x 'rem' p /= 0 ]
```

where `head :: [a] -> a` is defined as `head (x:_) = x`. The functional `iterate` shows another side of lazy, infinite streams: the possibility of representing an *infinite process* — in fact, we remove from `[2..]` all the multiples of 2, then the multiples of 3, then of 5, etc. We don't need to iterate *ad infinitum*, but if we don't know how many primes will be needed, it may be useful to keep them all. This example shows a typical lazy functional technique to avoid the excessive memory load during the generation of huge data streams — the filtering should be combined with the generation process. In such a way we may, for example, eliminate the isomorphic structures during the generation of a long sequence of graphs. The function `iterate` defined as `iterate f x = x : iterate f (fx)` returns the list: `[x, f x, f (f x), f (f (f x)), ...]`. Its applications will be shown later.

1.4. Streams and Nondeterministic Computations

Suppose that a given algorithmic problem has many solutions (or none). Such situation arises often in the domain of artificial intelligence, and in combinatorics. The fact that the semantics of Prolog is nondeterministic, that we can generate through backtracking a potentially infinite sequence of solutions, is one of the reasons of its popularity. Can we implement this nondeterminism? The idea is to generate the collection of *all* solutions, and the laziness prevents the program from crashing. Here is the program which computes the collection of all subsets of a given set (a list). We shall use the `Tree` datatype, whose nodes should be interpreted as alternatives: left branch *or* the right one. The algorithm is quite trivial: Either you accept the head of the list, or not; then, do the same with the tail.

```
allcombs [] = Leaf []
allcombs (x:q) =
  let z = allcombs q     in     z :^: (map (x :) z)
```

(Note the usage of the overloaded function `map` for structures other than lists.) This example will work for a list of 200 elements, but it is better not to try to print the full result, lazy data structures should be processed element by element and discarded.

2. Power Series Manipulation

2.1. Basic Arithmetics

Everybody knows how to process the power series. But again, the effective coding of a formula might not be entirely trivial. The examples below show however, that it might be trivial as well ...

An univariate power series $u_0 + u_1 x + u_2 x^2 + u_3 x^3 + \ldots$ will be represented by the list `[u0,u1,u2,...]`. Adding or subtracting such objects may be defined as:

```
u + v = zipWith (+) u v
u - v = zipWith (-) u v
```

where `zipWith` is a simple list combinator:

```
zipWith op (u0:uq) (v0:vq) = (u0 `op` v0) : zipWith op uq vq
```

The multiplication is defined recursively. If $U = U_0 + x \cdot \overline{U}$, $\quad V = V_0 + x \cdot \overline{V}$, then $U \cdot V = U_0 \cdot V_0 + x \cdot (U \cdot \overline{V} + V_0 \cdot \overline{U})$. In Haskell:

```
u@(u0:uq) * (v0:vq) = (u0*v0) : (v0 `cmul` uq + vq*u)
```

This example works, because the head of the solution is obtained immediately, and in order to get the first element of the tail, we need only the head of the recursive call of the defined operator. We have introduced the function `cmul` which multiplies the series by a constant: `cmul c l = map (c *) l`, and usable syntactically also as an operator: ``c `cmul` series``, because the automatic conversion procedures in Haskell are rudimentary. This will probably change in future. We have shown also how to destructure a named parameter `u` with the form `@`.

The division $W = U/V$ is also defined recursively. We have $U = W \cdot V$, or $U_0 + x \cdot \overline{U} = V_0 \cdot W_0 + x \cdot (W_0 \cdot \overline{V} + V \cdot \overline{W})$ Rearranging the equation we get the definition:

```
(u0:uq)/v@(v0:vq) = w0 : (uq - w0 `cmul` vq)/v
                    where w0 = u0/v0
```

Other manipulations of power series are also easy to code[9]. If $W = \log U$, then $W' = U'/U$.

```
log u@(u0:_) = integ (log u) (diff u / u)
```

where

```
diff (_:uq) = zipWith (gener (*)) [1..] uq
integ c u = c : zipWith (gener (/)) u [1..]
```

The function gener is a simple combinator of a binary operator and the type conversion function:

```
gener op x y = op x (fromInteger y)
```

(The polymorphic function fromInteger is necessary to ensure the correct type of the operand; the elements of the series may belong to any field.) For the exponential $W = \exp U$ we have $W' = W \cdot U'$, or $W = \int W \cdot U'$:

```
exp u@(u0:_) = w where w0= exp u0
                       w = integ w0 (diff u * w)
```

A power with an arbitrary real (or complex, etc.) exponent uses the identities: $W = U^\alpha$, $W' \cdot U = \alpha \cdot W \cdot U'$, from which we get $W = \int \alpha \cdot W \cdot U'/U$, and obviously $W_0 = U_0^\alpha$. The reader may write himself the one-line Haskell code.

If the series fulfils some complicated non-linear equation, we can try to obtain its solution through the Newton algorithm. For example, if $W = \sqrt{U}$, we apply the iterative formula: $W_{(n+1)} = \frac{1}{2}\left(W_{(n)} + U/W_{(n)}\right)$. It is easy to push this iteration to infinity using the functional iterate. We obtain a stream of streams. The final step consists in extracting the first element from the first stream, next two elements from the second, next four from the third, etc., knowing that the number of correct terms doubles with every iteration. With the aid of two standard functions: (take n list) and (drop n list) which return respectively the n-prefix and the n-tail of the list we have

```
sqrtS y = xtr 1 (iterate (\x -> half (x+y/x)) (fromInteger 1))
   where  xtr n (x:q) = (take n (drop (n-1) x) : (xtr (2*n) q)
```

The last example in this section is a list whose n-th element is the number of all the partitions (the number of Young diagrams) of the number n. Its generating power series is:

$$Z(x) = \prod_{n=1}^{\infty} \frac{1}{(1 - x^n)}.$$

Once more we see an infinite iteration problem. We could use the iterate functional, but this time we will try to implement directly the *open recursion* in the spirit of the generator of [1..]. Our stream is equal to the limit as $n \to \infty$ of

$$Z_n = Z_{n-1} \cdot S_n, \quad \text{where} \quad Z_1 = 1; \quad S_n = \frac{1}{(1 - x^n)}$$

We reformulate this as follows:

$$Z = F_1, \quad \text{where} \quad F_n = S_n \cdot S_{n+1} \cdot S_{n+2} \cdot \ldots = S_n \cdot F_{n+1}.$$

We know that each F_n begins with 1 followed by $n-1$ zeros, so defining W_n through $F_n = 1 + x^n W_n$ we get

$$W_n = 1 + xW_{n+1} + x^n W_n$$

and, denoting by (++) the concatenation of lists, we have:

```
partgen = 1:(w 1) where
    w n = 1:(zipWith (+) (w (n+1)) ((copy (n-1) 0)++(w n)))
```

which gives us: `[1,1,2,3,5,7,11,15,22,30,42,56,77,101,135,...]`.

2.2. More Advanced Operations on Power Series

We begin with the algorithm for the composition of series. Let $U(x) = U_0 + U_1x + U_2x^2 + \ldots; V = V_1x + V_2x^2 + \ldots$ Find $W = U(V)$. Note the absence of the constant term in V; otherwise we would have to compute explicitly the infinite sum. The solution is "just" the infinite Hörner scheme:

$$U(V) = U_0 + x(V_1 + V_2x + \ldots) \times \Big(U_1 + x(V_1 + V_2x + \ldots) \times (U_2 + x(\ldots))\Big)$$

or

```
comp u v@(_:vq) = cv u where
                      cv u@(u0:uq) = u0:(vq * cv uq)
```

The reversion of a power series is the solution of the following problem. Given

$$z = t + V_2t^2 + V_3t^3 + \ldots, \quad \text{find} \quad t = z + W_2z^2 + W_3z^3 + \ldots$$

The solution presented in the book of Knuth[9] is quite long. Using streams we can readily reduce it to the composition problem. It is easy to see that if we define the stream p by the equation $t = z(1 - zp)$, p fulfils the following identity:

$$p = (1 - zp)^2 \left(V_2 + V_3z(1 - zp) + V_4z^2(1 - zp)^2 + \ldots\right)$$

```
serrev v@(_:_:vt) = t            -- We know that V = [0, 1, V2,...]
   where  t = (fromInteger 0) : m
          m = (fromInteger 1) : negate (sqr m)*(comp vt t)
```

We end this section by a very short program which computes the diagonal Padé approximants for arbitrary power series. One of the popular ways of generating the rational approximations of real numbers is the continuous fraction expansion. We can apply the same technique also to the power series (whose coefficients must belong to a field, otherwise the division is not defined). Consider the following expansion:

$$u_0 + u_1x + u_2x^2 + u_3x^3 = u_0 + \cfrac{w_1x}{1 + \cfrac{w_2x}{1 + \cfrac{w_3x}{1 + \cdots}}}$$

In order to transform this expansion into a rational function we truncate it (by not expanding further), and then we fold it back:

```
pade 0 (u0:_) = (cmul u0 (fromInteger 1),(fromInteger 1))
pade m (u0:us@(u1:uq)) = (cmul u0 p + cmul u1 (mulByX q),p)
        where (p,q) = pade (m-1) (cmul u1 (fromInteger 1) / us)
              mulByX s = (fromInteger 0) : s
```

Of course, a "professional" program should verify that the linear term of the series does not vanish, otherwise the expansion starts with its successor, but this is just a technical detail.

3. Functional Calculus and Feynman Diagrams

3.1. Green Functions and Their Generating Functionals

The generation of diagrams is cumbersome. Even before their evaluation, regularization etc., finding all topologies with correct symmetry and combinatorial factors is a challenge[1]. In theory the process is well defined: if we know the Langrangian of the model, it suffices to apply the Wick theorem. The most compact representation of the generating functional for the Green functions $G_{i_1,i_2,\ldots i_n}$:

$$Z[J] = \sum_{m}^{\infty} \frac{1}{m!} G_{i_1,i_2,\ldots i_n} J_{i_1} J_{i_2} \ldots J_{i_n}$$

where i_k represent all the indices, discrete and continuous, is the Feynman path integral formula, but for the algebraic manipulation it is better to rewrite $Z[J]$ in terms of functional derivatives. As we need only the derivatives of Z, we write the Dyson-Schwinger equation:

$$\frac{\partial S}{\partial \phi_i}\left[\frac{\delta}{\delta J}\right] Z[J] = -J_i Z[J]$$

for the full generating functional, or

$$\frac{\partial S}{\partial \phi_i}\left[\frac{\delta}{\delta J} + \frac{\partial W}{\partial \phi}\right] = -J_i$$

for the connected generating functional $W = \log Z$. In the last formula the differentiation operators act inside the (polynomial) action S, commute to the right and disappear. For the ϕ^3 theory: $S_I = \frac{1}{3!} g\phi^3$ we will have

$$W_i' = \Delta_{ij} J_j + \frac{1}{2!}\Delta_{ikg}\left[W_{kk}'' + W_k' W_k'\right]$$

where X_k denotes $\partial X/\partial J_k$. This formula, despite its simplicity displays the main problem with the generation algorithm, and the *algorithm* is the only thing that we are here interested in: the formula above is an open recursive definition, without termination condition. *This* difficulty presents itself even in the 0-dimensional case, where J is a scalar. Denoting W' by φ and normalizing Δ to 1 we have

$$\varphi = J + \frac{1}{2} g\left(\varphi' + \varphi^2\right)$$

The solution in Haskell is based on the following reasoning:

We treat φ as a series in g, whose coefficients are some (unbound) polynomials in J. *We should know how to manipulate them.* In a reasonable theory this means the handling of indices, recognition of isomorphic monomials, etc. Lazy programming will not replace the standard algebraic manipulation. Here the functions of J are just univariate series, so we need only to know how to push inside the series the differentiation of its coefficients, and the multiplication by a constant. These are linear operations, so after having defined:

```
df s = map diff s
cm c s = map (cmul c) s
```

we have our solution (in the field of rational numbers `num%den`:

```
phi = phi0 : cm (1%2) ((df phi) + phi*phi)
      where phi0 = mulByX (fromInteger 1)   -- just: series(x) = x
```

If we want, for example, to compute the propagator, we have to *transpose* our series, treat this as a function of J, whose coefficients are series in g.

```
wprime = transps phi where
         transps s = (map head 1) : (transps (map tail s))
d2 = head (diff wp)
```

And if we `(take 10 d2)` we find a small mistake in the book of Cvitanović[10]... The correct answer is

$$d_2 = 1 + g^2 + \frac{25}{8} g^4 + 15 g^6 + \frac{12155}{128} g^8 + \ldots$$

A very compact Feynman diagram generator for more reasonable theories is the subject of a forthcoming paper by the author.

4. Conclusions

We have shown how to harness the unbound computational problems such as infinite lists, open recursion, unbound iterations or cascading alternatives with the aid of the lazy evaluation mechanism. This technique seems to be particularly useful for the manipulation of the perturbation series. The generation of Feynman graphs becomes much easier. This technique is not meant to replace the standard computer algebra manipulations such as the tensor algebra, but simplifies their application. In several cases it is possible to avoid the memory overflow by the intermediate expression swell (especially if one uses filtering already during the generation stage), and — what is even more important for the design — lazy algorithms offer a dramatic simplification of the control structures in the program. Laziness is not always a sin!

Appendix: How to obtain Haskell

Actually there exist four implementations of Haskell (including Gofer), and a fifth one (Bristol) in preparation, all of them available via anonymous FTP. The archives mentioned below contain the binaries for most Unix platforms together with the sources, examples and documentation. Gofer runs also on PC and MacIntosh computers. Connect to one of the following sites and see the directory `/pub/haskell` and its subdirectories.

```
nebula@cs.yale.edu
ftp.dcs.glasgow.ac.uk
ftp.cs.chalmers.se,
```

References

1. R. Mertig, J. Kühlbeck *FeynArts and FeynCalc: Computeralgebraic Generation and Calculation of Feynman Diagrams for Radiative Corrections* (Proceedings, New Computer Techniques in Physics Research, ed. D. Perret-Gallix, Lyon-Villeurbane, 1990).
2. M. P. Jones, *Gofer Functional Programming Environment* (Yale, 1992).
3. P. Hudak, P. Wadler (Eds.) *Report on the Functional Programming Language Haskell*, (Yale Technical Reports, 1988).
4. B. Autin, *Symbolic Approach to Perturbed Non-linear Dynamical Systems*, (these Proceedings, Oberammergau, 1993).
5. P. Hudak, *Conception, Evolution, and Application of Functional Programming Languages*, ACM Computing Surveys **21** (1989) 360.
6. William H. Burge, *Recursive Programming Techniques* (Addison-Wesley, 1975).
7. P. Henderson, *Functional Programming, Application and Implementation*, (Prentice-Hall, 1980).
8. R. Bird, P.Wadler, *Introduction to Functional Programming*, (Prentice Hall, 1988).
9. D. E. Knuth, *The Art of Computer Programming, (Vol. 2: Seminumerical Algorithms)*, (Addison-Wesley, 1981).
10. P. Cvitanović, *Field Theory* (Nordita Lecture Notes, København, 1983).

A NEW METHOD FOR COMPUTING ONE-LOOP INTEGRALS

J. FRANZKOWSKI
Institut für Physik, Johannes Gutenberg-Universität
D-55099 Mainz, Germany

ABSTRACT

A new program package for calculating one-loop Feynman integrals is presented, based on a new method avoiding Feynman parametrization and the contraction due to Passarino and Veltman (cf. [9] and [10]). The package is calculating one-, two- and three-point functions both algebraically and numerically to all tensor cases. This program is written as a package for Maple.

1. Introduction

The program package was developed in collaboration with L. Brücher, D. Kreimer and R. Stemler. Our approach is completely different from other procedures and packages. In our method the loop momentum integrations are performed directly, using the residue theorem and introducing $\mathcal{R}$ functions (cf. [5]), a class of special functions with some nice and useful features. We get some recursion relations for the involved $\mathcal{R}$ functions (cf. [1] or [2]), which allow a fast and memory efficient evaluation of the integrals under consideration.

The program is designed to reduce the $\mathcal{R}$ functions to a set of fundamental functions which are rewritten in terms of logarithms and dilogarithms. Up to now the different one-, two- and three-point functions are programmed as procedures in Maple [11]. Therefore our results are available numerically as well as algebraically. We tested our procedures using Maple V on a VAX 4000-90 workstation.

2. Description of the method

A detailed description of the method may be found in [1] or [2]. Here I would like to give a brief introductory overview only. Of course the calculation is based on dimensional regularization. Instead of introducing Feynman parameters the integral is split up into orthogonal and parallel space (cf. [6]), so that for example the two-point integral takes on the form

$$\begin{aligned} &B^{(p_0 p_1)}(q, m_1, m_2) \\ &= \frac{2\pi^{\frac{D-1}{2}}}{\Gamma(\frac{D-1}{2})} \int_{-\infty}^{\infty} dl_{\|} \int_{0}^{\infty} dl_{\perp}\, l_{\perp}^{D-2} \frac{(l_{\|})^{p_0}\,(l_{\perp})^{p_1}}{[(l_{\|}+q)^2 - l_{\perp}^2 - m_1^2 + i\varrho]\,[l_{\|}^2 - l_{\perp}^2 - m_2^2 + i\varrho]} \end{aligned} \tag{1}$$

This represents the most general case, depending on one external momentum q and two arbitrary masses m_1 and m_2. The splitting of space requires a distinction between $l_{\|}$ and $l_{\perp}$ which we define as

$$\begin{aligned} l_{\|} &= \frac{l \cdot q}{\sqrt{q^2}} \\ l_{\perp} &= \sqrt{l_{\|}^2 - l^2} \end{aligned} \tag{2}$$

In this notation $l_{\|}$ describes the component of l which is parallel to the external momentum q, whereas $l_{\perp}$ represents the orthogonal complement.

The integrals (1) can be reduced to a sum of $\mathcal{R}$ functions by using (4) with the help of partial fraction and residue theorem. The complete procedure is described in detail in [2] and [1]. The result in the two-point-case may be written according to [3]:

$$\begin{aligned} & B^{(p_0 p_1)}(q, m_1, m_2) \\ &= \frac{i\pi^{\frac{3}{2}-\varepsilon}}{\Gamma(\frac{3}{2}-\varepsilon)} \Gamma(\tfrac{1}{2}+\varepsilon-\tfrac{p_1}{2}) \Gamma(\tfrac{3}{2}-\varepsilon+\tfrac{p_1}{2}) \frac{(-1)^{\frac{p_1}{2}} (e^{-i\pi})^{-\varepsilon}}{q(-2\varepsilon+p_1+1)} \\ & \Big[(-1)^{p_0-[p_0]} (\tfrac{q}{2}+A)^{1+p_0-[p_0]} B(\varepsilon - \tfrac{p_1+[p_0]}{2}, \tfrac{[p_0]+1}{2}) \\ & \qquad \mathcal{R}_{-\varepsilon+\frac{p_1+[p_0]}{2}} \left(-\tfrac{p_1+1}{2}+\varepsilon, 1; -m_1^2+i\varrho, -(\tfrac{q}{2}+A)^2\right) \\ & + \sum_{i=0}^{p_0} \binom{p_0}{i} (-q)^{p_0-i} (\tfrac{q}{2}-A)^{1+i-[i]} B(\varepsilon - \tfrac{p_1+[i]}{2}, \tfrac{[i]+1}{2}) \\ & \qquad \mathcal{R}_{-\varepsilon+\frac{p_1+[i]}{2}} \left(-\tfrac{p_1+1}{2}+\varepsilon, 1; -m_2^2+i\varrho, -(\tfrac{q}{2}-A)^2\right)\Big] \end{aligned}$$

where $\varepsilon = \frac{4-D}{2}$ represents the usual dimensional regularization parameter. We used the abbreviations

$$A = \frac{m_2^2 - m_1^2}{2q}$$

$$[a] = \begin{cases} a, & a \text{ even} \\ a+1, & a \text{ odd} \end{cases}$$

The result of the three-point integral is obtained by a quite similar procedure and may be found in [2] or [4].

3. Evaluation and expansion of the $\mathcal{R}$ functions

After having expressed all possible integrals in terms of $\mathcal{R}$ functions we get easy and fast computable results by applying some recurrence relations for the $\mathcal{R}$ functions under consideration.

We achieve these relations by first reducing the number of different $\mathcal{R}$ functions to a set of fundamental $\mathcal{R}$ functions. This is possible due to the already mentioned features of the $\mathcal{R}$ functions, especially the relation (5) and (6). The reduction takes place in two steps. First of all we increase the parameter of the $\mathcal{R}$ function to get it close to 0 or 1. This can easily be achieved by using formula (5). In the second step we reduce the index of the $\mathcal{R}$ function by virtue of a convenient formula derived from (6). In the case of the two point-function it has the form:

$$\begin{aligned} \mathcal{R}_t(b_1, b_2; z_1, z_2) &= \frac{-1}{t+\beta-1} [((t-1)z_1 z_2)\mathcal{R}_{t-2}(b_1, b_2; z_1, z_2) \\ &\quad + ((1-t-b_1)z_1 + (1-t-b_2)z_2)\, \mathcal{R}_{t-1}(b_1, b_2; z_1, z_2)] \end{aligned} \tag{3}$$

Both steps mentioned above may be performed easily by computer, due to the recursive definition of the formulae (5) and (6).

Up to now we have acquired a result which is exact to all powers of ε, though we are interested only in an expansion in powers of ε. Therefore we now substitute the expansion for the fundamental $\mathcal{R}$ functions in terms of ε, resulting in simply computable functions like logarithms and dilogarithms. This is a rather cumbersome task but has to be performed only once for each type of N-point function. The procedure, especially its subtleties concerning analytical continuation, is described in detail in [7]. For completeness we here just cite the expansion for the scalar two-point case

$$\begin{aligned} \mathcal{R}_{-\varepsilon}(-\tfrac{1}{2}+\varepsilon, 1, z_1, z_2) &= \frac{\pi i}{B(\varepsilon, \frac{1}{2})} \frac{(z_2-z_1)^{\frac{1}{2}-\varepsilon}}{z_2^{\frac{1}{2}}} \\ &+ \tfrac{1}{2} z_1^{-\varepsilon} \left(\frac{z_1}{z_2}\right)^{1-\varepsilon} \left[\left(1+\sqrt{1-\frac{z_1}{z_2}}\right)^{-1+2\varepsilon} + \left(1-\sqrt{1-\frac{z_1}{z_2}}\right)^{-1+2\varepsilon}\right] + \mathcal{O}(\varepsilon^2) \end{aligned}$$

The three-point integrals are calculated in a quite similar way (cf. [2] or [4]).

4. The one-loop functions

Our aim is to calculate one-loop integrals both algebraically and numerically. Therefore the algorithm is implemented in Maple [11], a 'language' for symbolic computations. To make the implementation for any user most comfortable, the procedures are bound to a package, which may be read from the Maple system during run-time. The user-interface consists of three elementary functions

- OneLoop1Pt(p, m)
- OneLoop2Pt(p_0, p_1, q, m_1, m_2)
- OneLoop3Pt$(p_0, p_1, p_2, q_1, q_{20}, q_{21}, m_1, m_2, m_3)$

The functions directly correspond to the following integrals

- One-point function:

$$A^{(p)}(m) = \int d^D l \frac{(l)^p}{[l^2 - m^2 + i\varrho]}$$

- Two-point function:

$$B^{(p_0 p_1)}(q, m_1, m_2)$$
$$= \frac{2\pi^{\frac{D-1}{2}}}{\Gamma(\frac{D-1}{2})} \int_{-\infty}^{\infty} dl_{\|} \int_{0}^{\infty} dl_{\perp}\, l_{\perp}^{D-2} \frac{(l_{\|})^{p_0} (l_{\perp})^{p_1}}{[(l_{\|} + q)^2 - l_{\perp}^2 - m_1^2 + i\varrho]\,[l_{\|}^2 - l_{\perp}^2 - m_2^2 + i\varrho]}$$

- Three-point function:

$$C^{(p_0 p_1 p_2)}(q_1, q_2, m_1, m_2, m_3)$$
$$= \frac{2\pi^{\frac{D-2}{2}}}{\Gamma(\frac{D-2}{2})} \int_{-\infty}^{\infty} dl_{0\|} \int_{-\infty}^{\infty} dl_{1\|} \int_{0}^{\infty} dl_{\perp}\, l_{\perp}^{D-3} \frac{(l_{0\|})^{p_0} (l_{1\|})^{p_1} (l_{\perp})^{p_2}}{[(l_{0\|} + q_1)^2 - l_{1\|}^2 - l_{\perp}^2 - m_1^2 + i\varrho]}$$
$$\frac{1}{[(l_{0\|} + q_{20})^2 - (l_{1\|} + q_{21})^2 - l_{\perp}^2 - m_2^2 + i\varrho]\,[l_{0\|}^2 - l_{1\|}^2 - l_{\perp}^2 - m_3^2 + i\varrho]}$$

Again we follow our notation which distinguishes between parallel and orthogonal space:

$$l_{0\|} = \frac{l \cdot q_1}{\sqrt{q_1^2}}$$
$$l_{1\|} = -\frac{l \cdot q'_2}{\sqrt{q'^2_2}}; \quad q'_2 = q_2 - \frac{q_1 \cdot q_2}{q_1^2} q_1$$
$$l_{\perp} = \sqrt{l_{0\|}^2 - l_{1\|}^2 - l^2}$$
$$q_{20} = \frac{q_1 \cdot q_2}{\sqrt{q_1^2}}$$
$$q_{21} = \sqrt{q_{20}^2 - q_2^2}$$

The masses are denoted by m_i and the definitions of (2) are of course still valid. In this notation the indices p_i represent the powers of the different components of the loop momentum l and should not be mixed with Lorentz indices of the corresponding tensor. This notation is perhaps not very often used, but it corresponds directly to this method of integration. The usual notation may be recovered by an additional algorithm described in the next section.

The results of the one-loop functions are Laurent expansions in terms of the ultraviolet regulator ε. The output of the described procedures consists of a list where the significant coefficients of this expansion are denoted.

input	output
OneLoop1Pt(p, m)	$[\varepsilon^{-1}$-*term*, ε^{0}-*term*$]$
OneLoop2Pt(p_0, p_1, q, m_1, m_2)	$[\varepsilon^{-1}$-*term*, ε^{0}-*term*$]$
OneLoop3Pt$(p_0, p_1, p_2, q_1,$ $q_{20}, q_{21}, m_1, m_2, m_3)$	$[\varepsilon^{-2}$-*term*, ε^{-1}-*term*, ε^{0}-*term*, $\{abb.\}]$

In the output "ε^{-1}-term" is meant to represent the divergent part (the coefficient of $\frac{1}{\varepsilon}$) whereas "ε^{0}-term" is describing the finite part. Of course the ε^{-2}-term has to vanish in any case, but is left in order to check the accuracy of the program in numerical calculations. In the three-point case "abb." is written for some abbreviations which are introduced to get a more compact result.

5. The Passarino-Veltman coefficients

To receive a more familiar representation of the output a procedure is added which extracts the Passarino-Veltmann coefficients from our programs

- PassVelt1Pt(M, m)
- PassVelt2Pt(M, q, m_1, m_2)
- PassVelt3Pt$(M, q_1, q_{20}, q_{21}, m_1, m_2, m_3)$

for the different n-point cases, where M represents the tensor rank.

6. Conclusion and further projects

Our aim was to introduce a program package that gives any computer user the possibilty to calculate one-loop particle processes without much effort in manpower and hardware.

As a further project we plan to extend this package to the four- and five-point function.

Appendix

Conventions

$D = 4 - 2\varepsilon$	dimension in the dimensional regularization method
$\mathcal{R}_t(b; z)$	$\mathcal{R}$ function, t is called index,
	$b = (b_1, \ldots, b_n)$ parameters, $z = (z_1, \ldots, z_n)$ arguments (cf. [5])
$\beta = \sum_{i=1}^n b_i$	sum of parameters of an $\mathcal{R}$ function

Some fundamental properties of $\mathcal{R}$ functions

- formula for integrals

$$\int_0^\infty x^{\alpha-1} \prod_{i=1}^k (z_i + w_i x)^{-b_i}\, dx = B(\beta - \alpha, \alpha)\, \mathcal{R}_{\alpha-\beta}(b, \frac{z}{w}) \prod_{i=1}^k w_i^{-b_i} \tag{4}$$

- increasing of parameters

$$\beta\, \mathcal{R}_t(b, z) \;=\; (\beta + t)\, \mathcal{R}_t(b + e_i, z) \;-\; t z_i\, \mathcal{R}_{t-1}(b + e_i, z) \tag{5}$$

- associated functions (k is the number of parameters)

$$\sum_{i=0}^k A_i^{(k)}(t, b, z) \mathcal{R}_{-t-i}(b, z) \;=\; 0 \tag{6}$$

with

$$A_i^{(k)}(t, b, z) = \frac{(t, i)(\beta - t - k, k - i)}{t(\beta - t - k)} \left(t + i - \sum_{j=1}^k b_j z_j \frac{\partial}{\partial z_j} \right) E_i^{(k)}(z) \tag{7}$$

$E_i^{(k)}(z)$ is the elementary symmetric function, gained from

$$(1 + x z_1) \cdots (1 + x z_k) = \sum_{i=0}^k x^i E_i^{(k)}(z) \tag{8}$$

References

[1] D. Kreimer, *Dissertation* (Univ. Mainz, 1992).

[2] J. Franzkowski, *Diploma thesis* (Univ. Mainz, 1992).

[3] D. Kreimer, *Z. Phys.* **C54** (1992) 667.

[4] D. Kreimer, *Int. J. Mod. Phys.* **A8** (1993) 1797.

[5] B. C. Carlson, *Special Functions of Applied Mathematics* (Academic Press, 1977).

[6] J. Collins, *Renormalization* (Cambridge University press, 1984).

[7] L. Brücher, J. Franzkowski and D. Kreimer, *Univ. Mainz preprint* MZ-TH 93-11.

[8] L. Brücher, J. Franzkowski and D. Kreimer, *Univ. Mainz preprint* MZ-TH 93-14.

[9] G. 't Hooft and M. Veltman, *Nucl. Phys.* **B153** (1979) 365.

[10] G. Passarino and M. Veltman, *Nucl. Phys.* **B160** (1979) 151.

[11] B. W. Char, K. O. Geddes, G. H. Gonnet, B. L. Leong, M. B. Monagan and S. M. Watt, *Maple V* (Springer, 1991).

TWO-LOOP INTEGRALS FOR ARBITRARY MASSES

ULRICH KILIAN
Institut für Physik, Johannes Gutenberg Universität
D 55099 Mainz, Germany

and

DIRK KREIMER
Physics Dept., University of Tasmania
GPO Box 252C, Hobart 7001, Australia

ABSTRACT

In this talk we present a method which is designed to simplify the calculation of massive feynman diagrams. We restrict ourselves to the two-loop two-point and three-point functions, where we discuss the general mass cases. The calculation of the Feynman integrals is achieved by a systematic evaluation via partial fractions and shift operations. We avoid the use of parametrization formulas. Our aim will be to give the simplest integral representations possible for the generic cases.

1. The two-loop two-point function

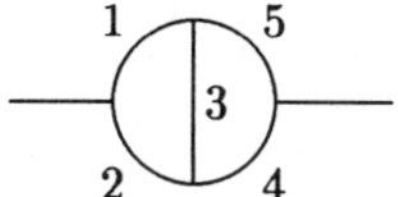

Fig.1: The master two-loop two-point function; the labels denote the propagators

In the first part of the talk we present a new calculation[1] of the so-called normalized master two-loop two-point function

$$I(q^2) = -\frac{q^2}{\pi^4} \int d^4l_1 \int d^4l_2 \prod_{i=1}^{5} \frac{1}{P_i} \tag{1}$$

with inverse propagators

$$\begin{aligned} P_1 &= l_1^2 - m_1^2 + i\eta \,, \\ P_2 &= (l_1+q)^2 - m_2^2 + i\eta \,, \\ P_3 &= (l_1+l_2)^2 - m_3^2 + i\eta \,, \\ P_4 &= (l_2-q)^2 - m_4^2 + i\eta \,, \\ P_5 &= l_2^2 - m_5^2 + i\eta \,. \end{aligned} \tag{2}$$

The most general case is the one with five different masses. In the following we develop an expression which allows its calculation in every case, i.e. on-shell, off-shell and with arbitrary masses.

The starting-point of the calculation is the separation of space into a parallel space spanned by the linear independant exterior momenta and an orthogonal space which is the orthogonal complement of the parallel space. For the two-loop two-point integral the parallel space is a one-dimensional space which we choose to be the zero-component of Minkowski space. We express the integral in the restframe of the exterior momentum $q \equiv q_0$ in which the loop momenta satisfy

$$\begin{aligned} l_1 &\equiv (l_{10}, \vec{l}_2), \quad l_2 \equiv (l_{20}, \vec{l}_2), \\ l_{1\perp} &= |\vec{l}_1|, \quad l_{2\perp} = |\vec{l}_2|, \\ \vec{l}_1 \cdot \vec{l}_2 &= l_{1\perp} l_{2\perp} z. \end{aligned} \tag{3}$$

The inverse propagators are now defined by

$$\begin{aligned} P_1 &= l_{10}^2 - l_{1\perp}^2 - m_1^2 + i\eta\,, \\ P_2 &= (l_{10} + q_0)^2 - l_{1\perp}^2 - m_2^2 + i\eta\,, \\ P_3 &= (l_{10} + l_{20})^2 - l_{1\perp}^2 - l_{2\perp}^2 - 2 l_{1\perp} l_{2\perp} z - m_3^2 + i\eta\,, \\ P_4 &= (l_{20} - q_0)^2 - l_{2\perp}^2 - m_4^2 + i\eta\,, \\ P_5 &= l_{20}^2 - l_{2\perp}^2 - m_5^2 + i\eta\,. \end{aligned} \tag{4}$$

The space-like $\vec{l}_1, \vec{l}_2$ span the orthogonal space, which is integrated first. This integration involves one non-trivial angular integration over z. All together we are left with a five-dimensional integration

$$I(q^2) = -\frac{8q^2}{\pi^2} \int_{-\infty}^{\infty} dl_{10} \int_{-\infty}^{\infty} dl_{20} \int_0^{\infty} dl_{1\perp} \int_0^{\infty} dl_{2\perp} l_{1\perp}^2 l_{2\perp}^2 \int_{-1}^{1} dz \prod_{i=1}^{5} \frac{1}{P_i}. \tag{5}$$

All integrations are well defined for $\eta \neq 0$.

The z integration needs only the propagator $1/P_3$, which has the simple form $1/(a+bz)$. The result of this integration is

$$\log[(l_{10}+l_{20})^2 - m_3^2 + i\eta - (l_{1\perp}+l_{2\perp})^2] - \log[(l_{10}+l_{20})^2 - m_3^2 + i\eta - (l_{1\perp}-l_{2\perp})^2], \tag{6}$$

which can be expanded

$$\begin{aligned} \log[w_3^2 - (l_{1\perp} + l_{2\perp})^2] &= \log[w_3 + l_{1\perp} + l_{2\perp}] + \log[w_3 - l_{1\perp} - l_{2\perp}], \\ \log[w_3^2 - (l_{1\perp} - l_{2\perp})^2] &= \log[w_3 + l_{1\perp} - l_{2\perp}] + \log[w_3 - l_{1\perp} + l_{2\perp}], \end{aligned} \tag{7}$$

where we have defined $w_3 = \sqrt{(l_{10}+l_{20})^2 - m_3^2 + i\eta}$. The next step of the calculation uses the even behaviour of the integrand under the transformations $l_{1\perp} \to -l_{1\perp}, l_{2\perp} \to -l_{2\perp}$ which enables us to extend the integration intervals to $[-\infty, \infty]$.

So we can use the residue theorem to do the $l_{1\perp}, l_{2\perp}$ integration. The crucial point is that each of the four logarithms in Eqs. (7) has its cut in the upper *or* in the lower complex $l_{1\perp}$ (respectively $l_{2\perp}$) half-plane making the integrand meromorphic in the closed complementary half-plane.

The poles of the inverse propagators are given by

$$\begin{aligned} w_1 &= \sqrt{l_{10}^2 - m_1^2 + i\eta}, \\ w_2 &= \sqrt{(l_{10} + q_0)^2 - m_2^2 + i\eta} \end{aligned} \tag{8}$$

for $P_{1,2}$ and

$$\begin{aligned} w_4 &= \sqrt{(l_{20} - q_0)^2 - m_4^2 + i\eta} \\ w_5 &= \sqrt{l_{20}^2 - m_5^2 + i\eta} \end{aligned} \tag{9}$$

for $P_{4,5}$. The integrand decreases for $l_{1\perp} \to \infty, l_{2\perp} \to \infty$ rapidly enough, so that we can apply the residue theorem. Introducing the dimensionless variables $x = l_{10}/q, y = l_{20}/q$ we end with the simple result

$$I(q^2) = 4\int_{-\infty}^{\infty} dx \int_{-\infty}^{\infty} dy \frac{1}{(w_1^2 - w_2^2)(w_4^2 - w_5^2)} [L_{325} - L_{315} + L_{314} - L_{324}], \tag{10}$$

where

$$L_{ijk} = \log(w_i + w_j + w_k). \tag{11}$$

The analytic continuation $\eta \to 0$ turns out to be well defined. The four logarithms can be summarized in one:

$$L_{325} - L_{315} + L_{314} - L_{324} = \log\left(\frac{\{325\}\{314\}}{\{315\}\{324\}}\right), \tag{12}$$

where $\{ijk\} = (w_i + w_j + w_k)$. As an immediate consequence one sees that the integral exists because the argument of the logarithm in Eq. (12) tends to one for large $|x|$ or $|y|$. We also note that at the location of the poles of the integrand the argument of the logarithm tends to one, so we can conclude the integral to exist as an improper Riemann integral.

We ended up with a simple integral representation for the master two-loop two-point function which is valid for all cases. Analytical and numerical evaluations can easily be performed. We confirmed our result by such tests. This result turned out to be in excellent agreement with methods based on asymptotic expansions[2] and was used for the calculation of self-energies of γ-, W- and Z-bosons recently[3].

Our handling of divergent tensor integrals makes use of the fact that the powercounting of a divergent integral becomes better if the massless integral (or zero exterior momenta) is substracted. We give an example for the two-loop two-point function,

$$I^{(2,0,0,0)} = \int d^D l d^D k \frac{l_0^2 k_0^0 l_\perp^0 k_\perp^0 z^0}{P_1 P_2 P_3 P_4 P_5}, \tag{13}$$

which is logarithmic divergent in l and convergent in k. Now we subtract

$$\begin{aligned} I^{(2,0,0,0)} &= \left[(I^{(2,0,0,0)} - L^{(2,0,0,0)}) - (\tilde{I}^{(2,0,0,0)} - \tilde{L}^{(2,0,0,0)}) \right]_4 \\ &\quad + L_D^{(2,0,0,0)} + \tilde{I}_D^{(2,0,0,0)} - \tilde{L}_D^{(2,0,0,0)} \end{aligned} \tag{14}$$

where

$$\begin{aligned} L_*^{(2,0,0,0)} &= \int d^*l d^*k \frac{l_0^2}{\tilde{P}_1 \tilde{\bar{P}}_2 \bar{P}_3|_{k=0} P_4 P_5} \\ \tilde{L}_*^{(2,0,0,0)} &= \int d^*l d^*k \frac{l_0^2}{\tilde{P}_1 \tilde{\bar{P}}_2 \bar{P}_3|_{k=0} \tilde{P}_4 \tilde{P}_5} \\ \tilde{I}_*^{(2,0,0,0)} &= \int d^*l d^*k \frac{l_0^2}{\tilde{P}_1 \tilde{P}_2 \tilde{P}_3 \tilde{P}_4 \tilde{P}_5}. \end{aligned} \tag{15}$$

We used the definitions

$$\tilde{P}_i = P_i|_{m_i^2=0}\,, \; \bar{P}_i = P_i|_{q=0}\,, \; \tilde{\bar{P}}_i = P_i|_{q=0, m_i^2=0}. \tag{16}$$

The first four integrals in Eq. (14) are now convergent and can be calculated numerically using the two-fold integral representation presented in this talk. The substracted integrals are known special cases of two-loop integrals. This method is equivalent to the techniques used in the above mentioned calculation of boson self-energies[3]. We will investigate a generalization to arbitrary tensor cases in the future.

2. The two-loop three-point function

We now turn to the two-loop three-point function[4]. We restrict ourselves to the planar diagram. It turns out as an advantage of our method that the crossed diagram is not much more difficult than the planar diagram.

Fig.2: The planar and the crossed diagrams of the two-loop three-point function

In the case of the three-point function the parallel space is a two-dimensional space. We choose for our calculation the restframe of q_1, so we have

$$\begin{aligned} q_1 &\equiv q_{10}, \quad q_2 \equiv (q_{20}, q_{21}) \\ l &\equiv (l_0, l_1, \vec{l}_\perp), \quad k \equiv (k_0, k_1, \vec{k}_\perp) \\ l_\perp &= |\vec{l}_\perp|, \quad k_\perp = |\vec{k}_\perp|. \end{aligned} \tag{17}$$

The explicit expressions for the inverse propagators are

$$\begin{aligned}
P_1 &= (l_0+q_{10})^2 - l_1^2 - l_\perp^2 - m_1^2 + i\eta \\
P_2 &= (l_0-q_{20})^2 - (l_1-q_{21})^2 - l_\perp^2 - m_2^2 + i\eta \\
P_3 &= (l_0+k_0)^2 - (l_1+k_1)^2 - l_\perp^2 - k_\perp^2 - 2l_\perp k_\perp z - m_3^2 + i\eta \\
P_4 &= (k_0-q_{10})^2 - k_1^2 - k_\perp^2 - m_4^2 + i\eta \\
P_5 &= (k_0+q_{20})^2 - (k_1+q_{21})^2 - k_\perp^2 - m_5^2 + i\eta \\
P_6 &= k_0^2 - k_1^2 - k_\perp^2 - m_6^2 + i\eta.
\end{aligned} \tag{18}$$

The scalar product of the loop momenta involves a non-trivial angular integration. It has the form

$$\int_{-1}^{1} dz \frac{1}{\sqrt{1-z^2}} \frac{1}{a+bz} = \pi \frac{1}{\sqrt{a^2-b^2}}, \tag{19}$$

where

$$\begin{aligned}
a &= (l_0+k_0)^2 - (l_1+k_1)^2 - l_\perp^2 - k_\perp^2 - m_3^2 + i\eta \\
b &= 2l_\perp k_\perp.
\end{aligned} \tag{20}$$

This allows us to make the shifts $l_0 \to l_0 + l_1$ and $k_0 \to k_0 + k_1$ which locates both the cuts in l_1 and k_1 of the square-root in Eq. (19) in either the upper or lower half-plane, depending on the sign of (l_0+k_0). Furthermore, all propagators become linear in l_0 and k_0, respectively:

$$\begin{aligned}
P_1 &= l_0^2 + q_{10}^2 + 2l_1(l_0+q_{10}) + 2l_0 q_{10} - l_\perp^2 - m_1^2 + i\eta \\
P_2 &= l_0^2 + q_{20}^2 - q_{21}^2 + 2l_1(l_0 - q_{20} + q_{21}) - 2l_0 q_{20} - l_\perp^2 - m_2^2 + i\eta \\
P_4 &= k_0^2 + q_{10}^2 + 2k_1(k_0 - q_{10}) - 2k_0 q_{10} - k_\perp^2 - m_4^2 + i\eta \\
P_5 &= k_0^2 + q_{20}^2 - q_{21}^2 + 2k_1(k_0 + q_{20} - q_{21}) - 2k_0 q_{20} - k_\perp^2 - m_5^2 + i\eta \\
P_6 &= k_0^2 + 2k_1 k_0 - k_\perp^2 - m_6^2 + i\eta.
\end{aligned} \tag{21}$$

Before applying the residue theorem, we partial fractionize the above product of propagators

$$\begin{aligned}
\frac{1}{P_1P_2P_4P_5P_6} &= \frac{1}{P_1-P_2}\left[\frac{1}{P_2} - \frac{1}{P_1}\right] \times \\
&\times \left[\frac{1}{(P_4-P_5)(P_4-P_6)P_4} + \frac{1}{(P_5-P_6)(P_5-P_4)P_5} \right. \\
&\left. + \frac{1}{(P_6-P_5)(P_6-P_4)P_6}\right].
\end{aligned} \tag{22}$$

All together we have 36 residues but the poles on the real axis of the differences $P_i - P_j$ don't contribute. The poles which do not lie on the real axis contribute, if they are located in the interior of the integration contour Γ, and this in turn is determined by the sign of $(l_0 + k_0)$. The location of the poles is fixed by the signs of the expressions coming together with l_1 and k_1 in Eq. (21).

Having done the l_1 and k_1 integrations with the help of the residue theorem, one ends with expressions of the following form for the $l_\perp^2$ and $k_\perp^2$ dependance,

$$\int_0^\infty dl_\perp^2 dk_\perp^2 \frac{[\mathrm{Poly}(l_\perp^2, k_\perp^2)]^{-1/2}}{(a_l l_\perp^2 + b_l)(a_k k_\perp^2 + b_k)}, \tag{23}$$

where $\mathrm{Poly}(l_\perp^2, k_\perp^2)$ is a polynomial in $l_\perp^2$ and $k_\perp^2$ of second degree

$$\mathrm{Poly}(x, y) = a(x - x_+(y))(x - x_-(y)). \tag{24}$$

The $l_\perp^2$ integration can be done by using the integral representation of a $\mathcal{R}$-function (for the definition and some properties of $\mathcal{R}$-functions consult the talk of J. Franzkowski in this workshop)

$$\int_0^\infty dl_\perp^2 dk_\perp^2 \frac{[\mathrm{Poly}(l_\perp^2, k_\perp^2)]^{-1/2}}{(a_l l_\perp^2 + b_l)(a_k k_\perp^2 + b_k)} = \frac{1}{\sqrt{a a_l}} \mathcal{R}_{-1}(1/2, 1/2, 1; -x_+(k_\perp^2), -x_-(k_\perp^2), b_l/a_l). \tag{25}$$

After a transformation this result can be expressed in terms of logarithms and square roots,

$$\mathcal{R}_{-1}(1/2, 1/2, 1; z_+, z_-, z_0) = 2\frac{\log(u_+) - \log(u_-)}{u_+ - u_-}, \tag{26}$$

where[1]

$$\begin{aligned} u_+ &= z_0 + \sqrt{z_+ z_-} + \sqrt{z_0^2 - z_0(z_+ + z_-) + z_+ z_-} \\ u_- &= z_0 + \sqrt{z_+ z_-} - \sqrt{z_0^2 - z_0(z_+ + z_-) + z_+ z_-}. \end{aligned} \tag{27}$$

We end with a representation as a threefold integral for the generic case of six different masses. The crossed function can be expressed, with partial fractions and shifting of momenta, in terms of the same function as in the planar case. Only the number of terms increases. We checked the result numerically for some simple mass cases. Numerical results for more interesting situations for both cases will be presented elsewhere[5].

We have derived integral representations valid for planar and non-planar massive two-loop two- and three-point functions. The results turn out to have a systematic structure, which makes them manageable for numerical evaluations.

3. References

1. D. Kreimer, *Phys. Lett.* **B273** (1991) 277.

[1]Note that in ref. 4 one have to replace Eq. (9) by the corrected formula Eq. (27) of this paper.

2. A.I. Davydychev, J.B. Tausk, *Nucl. Phys.* **B397** (1993) 123.
   A.I. Davydychev, V.A. Smirnov, J.B. Tausk, *Leiden preprint* **INLO-PUB-5/93** (1993).
3. J.B. Tausk, *Ph.D. thesis* **Leiden** (1993).
4. D. Kreimer, *Phys. Lett.* **B292** (1992) 341.
5. A. Czarnecki, U.Kilian and D. Kreimer, in preparation.

REDUCE USAGE FOR CALCULATIONS IN CHIRAL MESONIC MODELS

ALEXANDER V. LANYOV
Particle Physics Laboratory, Joint Institute for Nuclear Research
Head Post Office, P.O.Box 79, Moscow 101000 Russia

ABSTRACT

We describe a REDUCE system for a wide range of calculations, encountered during the work in chiral mesonic models. Specific operations for this area have been introduced by symbolic-mode programming. Some simple facilities have also been provided for a convenient practical work with large expressions and their physical analysis.

Computer algebra systems (CAS), such as REDUCE or FORM, have long been used in high-energy physics, and, as a rule, in the sphere of calculations of high-order perturbation theory. Besides this "classical" usage field, CAS can also be used for the time-consuming operations in the mesonic models, where the largest difficulty usually consists in calculating the amplitude of a given process by a given Lagrangian, which can have up to dozens or hundreds terms. It is also highly desirable to have some means to obtain such Lagrangians and currents, consisting of various mesonic operators, and to work with them.

We used CAS REDUCE for implementation since it is widely used by physicists and is available at many computers. It is a "completely open" system, giving a possibility to access the internal data structures directly and to define new operations and data types by means of symbolic-mode programming. At first we have written what we called a "package of procedures" which can be called by a user to perform some operations [1]. Then, as the package was developed during the usage, it began to tend to become closer to a "user package" really extending a range of operations and data types in REDUCE. During implementation of these operations, it has been found that symbolic-mode programming is required or at least is "natural" for this purpose. The code size is approximately 120 procedures in the symbolic mode of REDUCE 3.4 (over 2000 lines).

The package has been designed and used for work in the field of chiral Lagrangians describing the properties and interactions of the lowest mesonic nonets (pseudoscalar, scalar, vector and axial-vector mesons) and photons (see e.g. Ref. [2] for the papers describing the physical models).

The usage of this package made it possible to perform calculations of bosonized mesonic Lagrangian and weak and electromagnetic currents taking into account higher-order terms and explicit chiral symmetry breaking by current quarks masses.

Then calculations of amplitudes of various processes were performed taking into account the following effects:

- p^4-contribution to strong interaction Lagrangian and the corresponding contributions to weak currents (additional higher-order terms to usual chiral p^2-Lagrangian);
- renormalization of Weinberg-Salam weak interaction by hard gluons leading to Shifman-Vainstein-Zakharov type Lagrangian (additional currents contributing to the weak nonleptonic transitions);
- pole diagrams;
- explicit chiral symmetry breaking by current quarks masses;
- violation of isotopic symmetry leading to $(\pi–\eta–\eta')$-mixing;
- meson rescattering effects (meson one-loop diagrams).

It is important to take these effects into account, e.g. for predictions of effects of direct CP violation in $K^{\pm} \to 3\pi$ decays [3].

The following "physical" features have been implemented:

- representation of mesonic operators with necessary arguments (momentum and polarization) with different types of derivatives over space-time 4-vectors ("short" ones and also "long" ones, containing various gauge bosons);
- calculation of amplitude of a given process by a given Lagrangian;
- means for variation over some tensor structure which can be a non-commutative one and also have some dummy indices (by means of this procedure weak and electromagnetic currents are obtained);
- means for working with physical expressions, including transforming them to a unique "standard" form with an account of additional symmetries: with respect to cyclic permutations under the trace sign ($\mathrm{tr}\, ABC \equiv \mathrm{tr}\, BCA$, if A, B, C are non-commuting operators) or redesignation of dummy indices ($A_{\mu\nu} B_\mu C_\nu \equiv A_{\nu\mu} B_\nu C_\mu$);[1]
- some transformation tools between different REDUCE notations were required, since at different stages of calculations different notations are "natural";
- means for testing the results on some symmetry properties, e.g. Hermitian or internal-parity conjugation;
- definitions specific for our model (matrices, particle properties, physical constants, kinematics of decays, etc.);

[1] However, the combination of these two freedoms actually sometimes still cause problems in transforming physically identical expressions to a unique "standard" form.

- some other utilities (for calculations with 4-vectors, sorting procedures, etc.).

The following "system" features have been implemented for REDUCE as a rather primitive effort to add some desirable possibilities to it:

- tools for automatic execution of several REDUCE programs with some parameters;
- some tools for "careful calculations" (heave use of substitutions at all stages of calculation of some expression in order not to cause memory overflow, e.g. when multiplying several large expressions while there are active substitutions which can reduce some intermediate expressions);
- LaTeX output of large expressions resulting in our calculations (on the basis of RLFI package from REDUCE library);
- some primitive database for work with terms in an expression.

Using this package, it is possible to write procedures for the calculation of amplitudes of strong, weak and electromagnetic Lagrangians in chiral mesonic models and also of currents and Lagrangians. In fact, all stages of calculation in our model can be performed in this way.

The experience of the package usage proved that it provides a convenient computing environment for work in the chiral mesonic models.

I express my gratitude to my co-authours A. A. Bel'kov and A. Schaale for the support of this work and useful comments. I thank V. P. Gerdt, J. Heinrich and Yu. F. Pirogov for useful discussions.

1. A. A. Bel'kov and A. V. Lanyov, Preprint JINR E11-91-162, (Dubna, 1991); *Proc. ISSAC'91*, July 15-17, 1991, Bonn, Germany, ed. S. M. Watt, (ACM Press, N.Y., 1991), p. 454.
2. S. Weinberg, *Physica*, **96A** (1979) 327;
J. Gasser and H. Leutwyler, *Ann. Phys.* **158** (1984) 142; *Nucl. Phys.* **B250** (1985) 465;
A. A. Bel'kov, G. Bohm, D. Ebert, A. V. Lanyov and A. Schaale, *Int. J. Mod. Phys.* **A7** (1992) 4757; **A8** (1993) 1313.
3. A. A. Bel'kov, G. Bohm, D. Ebert, A. V. Lanyov and A. Schaale, *Phys. Lett.* **B300** (1993) 283.

LARGE N_f CRITICAL EXPONENT APPROACH TO COMPUTING PERTURBATIVE RGE FUNCTIONS

J.A. GRACEY
DAMTP, University of Liverpool, P.O. Box 147, Liverpool, L69 3BX, UK

ABSTRACT

We review the large N_f self consistency programme for computing critical exponents in d-dimensions from which one can deduce the $\overline{\text{MS}}$ coefficients of the perturbative RGE functions to all orders in the coupling and several orders in $1/N_f$ for four dimensional gauge theories. The latest results for QED and QCD are presented.

1. Introduction

The renormalization group equation, (RGE), governs the quantum properties of renormalizable quantum field theories and it depends on fundamental functions such as the β-function and mass anomalous dimension, γ_m. Conventionally, they are calculated order by order in powers of the perturbative coupling constant by renormalizing Green's functions via some subtraction prescription like $\overline{\text{MS}}$, which is a mass independent scheme. For complicated theories such as QED and QCD their perturbation series is known to three orders[1,2,3] although the 4-loop QED $\overline{\text{MS}}$ β-function has recently been found[4]. These higher order calculations were carried out mainly by using computer algebra packages which substantially reduced the amount of (real) time used to obtain results. With the current demand for even higher order results, it is now crucial that independent checks for such calculations be developed to ensure their credibility. The large N_f critical exponent method we present here, which is applicable to theories with N_f fundamental fields, achieves this and importantly always provides new information on the perturbation series to all orders in the coupling constant at successive orders in $1/N_f$. Instead of computing the renormalization constants one calculates the d-dimensional critical exponents of the Green's function in the neighbourhood of the non-trivial zero of the d-dimensional β-function where the propagators have a simple scaling or conformal structure[5]. At this critical point the RGE simplifies to the extent that the anomalous part of the critical exponent, which arises for example because of radiative corrections, has a very simple relationship to the corresponding RGE function. Therefore, for theories which are large N expandable, one can deduce the numerical coefficients of the coupling constant of the respective RGE function. Moreover, the methods[5] and the masslessness of the integrals means that one can probe theories to several orders beyond the leading one.

2. The method

It is best to give an indication of the power of the method by outlining the computation of the electron anomalous dimension in QED at leading order[6], $O(1/N_f)$. First, the location of the critical point in whose neighbourhood we will determine the exponent is given by $\beta(g_c) = 0$ where $g_c \neq 0$ in d-dimensions and to deduce information on the $(4-2\epsilon)$-dimensional perturbation series, it is given to four loops in $\overline{\text{MS}}$ from[4]

$$\begin{aligned}\beta(g) &= (d-4)g + \frac{2N_f}{3}g^2 + \frac{N_f}{2}g^3 - \frac{N_f(22N_f+9)}{144}g^4 \\ &- \frac{N_f}{64}\left[\frac{616N_f^2}{243} + \left(\frac{416\zeta(3)}{9} - \frac{380}{27}\right)N_f + 23\right]g^5 + O(g^6)\end{aligned} \tag{1}$$

as

$$g_c \sim \frac{3\epsilon}{N_f} - \frac{27\epsilon^2}{4N_f^2} + \frac{99\epsilon^3}{16N_f^2} + \frac{77\epsilon^4}{16N_f^2} \tag{2}$$

where $g = (e/2\pi)^2$ and $d = 4 - 2\epsilon$. The form of the lagrangian for the critical point analysis is[6]

$$L = i\bar{\psi}^i \partial\!\!\!/ \psi^i + A_\mu \bar{\psi}^i \gamma^\mu \psi^i - \frac{(F_{\mu\nu})^2}{4e^2} \tag{3}$$

where ψ^i is the electron field, $1 \leq i \leq N_f$, A_μ is the photon field and $F_{\mu\nu} = \partial_\mu A_\nu - \partial_\nu A_\mu$. The electron coupling constant has been scaled into the photon field to enable us to use the method of uniqueness[7] at $O(1/N_f^2)$. Near g_c the fields obey asymptotic scaling and we take the following conformal forms for the propagators in coordinate space as $x^2 \to 0$ as[8]

$$\psi(x) \sim \frac{A x\!\!\!/}{(x^2)^\alpha} \quad , \quad A_{\mu\nu}(x) \sim \frac{B}{(x^2)^\beta}\left[\eta_{\mu\nu} + \frac{2\beta}{(2\mu-2\beta-1)}\frac{x_\mu x_\nu}{x^2}\right] \tag{4}$$

in the Landau gauge where A and B are x-independent amplitudes and α and β are their dimensions. From dimensional analysis, we define

$$\alpha = \mu + \tfrac{1}{2}\eta \quad , \quad \beta = 1 - \eta - \chi \tag{5}$$

where $d = 2\mu$, η is the electron anomalous dimension and χ is the 3-vertex anomalous dimension. From the critical RGE $\eta = \gamma_2(g_c)$, where $\gamma_2(g)$ is the renormalization group function corresponding to the electron wave function renormalization. So by calculating η and postulating the most general structure for $\gamma_2(g)$, where the coefficient at $O(g^n)$ is a polynomial in N_f of degree $(n-1)$, one can deduce the coefficients of $\gamma(g)$ by setting $g = g_c$ from (2) and expanding it in powers of ϵ to compare with the ϵ expansion of η.

To find η one solves the skeleton Dyson equation of the ψ and A_μ fields, truncated to $O(1/N_f)$, in the critical region where (4) are satisfied and the relevant graphs are given in Fig. 1. The scaling behaviour of the 2-point functions ψ^{-1} and

$A^{-1}_{\mu\nu}$ are deduced from (4) by inverting in momentum space, with the latter restricted to the transverse subspace[5,6], which is the physical region. Thus, Fig. 1 becomes

$$0 = \frac{(\alpha-1)a(\alpha-\mu-1)}{(\mu-\alpha+1)a(\alpha-1)\pi^{2\mu}} + \frac{2(2\mu-1)(\beta-\mu+1)z}{(2\mu-2\beta-1)} \tag{6}$$

$$0 = \frac{(2\mu-2\beta-1)a(\beta-\mu)}{2(\mu-\beta)a(\beta)\pi^{2\mu}} - \frac{8(\alpha-1)z}{(2\alpha-1)} \tag{7}$$

where $a(\alpha) = \Gamma(\mu-\alpha)/\Gamma(\alpha)$, $z = A^2B$ and only the transverse part of the photon equation is relevant. Eliminating z and expanding $\eta = \sum_{i=1}^{\infty} \eta_i/N_f^i$ then

$$\eta_1 = \frac{(2\mu-1)(\mu-2)\Gamma(2\mu)}{4\Gamma^2(\mu)\Gamma(\mu+1)\Gamma(2-\mu)} \tag{8}$$

and the ϵ-expansion of (8) is in agreement with the 3-loop electron wave function renormalization[9].

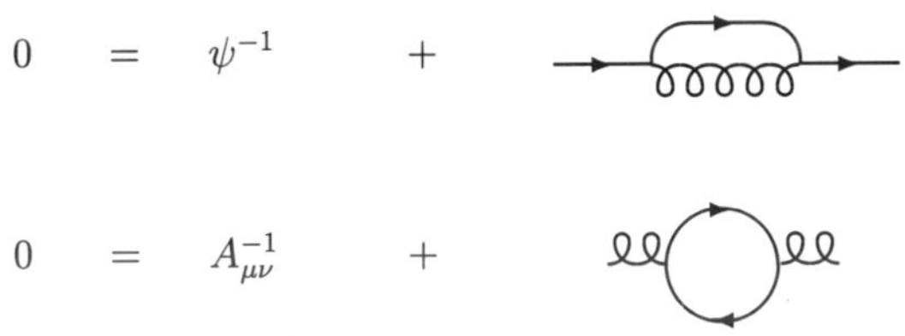

$0 = \psi^{-1} +$

$0 = A^{-1}_{\mu\nu} +$

Figure 1. Leading order skeleton Dyson equations.

3. Results

The calculation of Eq. (8) has been extended[8] to $O(1/N_f^2)$ which involved computing the higher order corrections to Eqs. (6) and (7). As the method uses propagators with non-zero anomalous dimensions this means that corrections with dressed propagators only are considered which implies that there is only one (vertex) correction to each of the one loop graphs of Fig. 1. They are computed using integration by parts and uniqueness[7] and lead to[8]

$$\eta_2 = \frac{\eta_1^2}{2}\left[\frac{6\mu(\mu-1)\hat{\Theta}(\mu)}{(2\mu-1)(\mu-2)} + \frac{3}{\mu} + \frac{6}{(\mu-1)} - \frac{2}{3(\mu-2)^2} - \frac{56}{9(2\mu-1)} - \frac{35}{9(\mu-2)}\right] \tag{9}$$

where $\hat{\Theta}(\mu) = \psi'(\mu-1) - \psi'(1)$ and $\psi(x)$ is the logarithmic derivative of the Γ-function. Again (9) agrees with the 3-loop result[9] and further implies that to fourth order

$$\gamma_2(g) = -\frac{(4N_f+3)}{16}g^2 + \left[\frac{5N_f^2}{72} + \frac{3N_f}{32} + \frac{3}{64}\right]g^3 + \left[\frac{35N_f^3}{1296} + \left(\frac{1}{27} - \frac{\zeta(3)}{4}\right)N_f^2 + a_1N_f + a_2\right]g^4 \tag{10}$$

in the Landau gauge where a_i are unknown. (In QED, $\gamma_2(g)$ is gauge dependent but it is has been shown that the only appearance of the gauge parameter is at one loop. The higher order coefficients are independent of the gauge choice[10].)

The β-function can also be deduced within the formalism by considering the following corrections to (4)[11],

$$\begin{aligned} \psi(x) &\sim \frac{A \not{x}}{(x^2)^\alpha} \left[1 + (x^2)^\lambda A'\right] \\ A_{\mu\nu}(x) &\sim \frac{B}{(x^2)^\beta} \left[\eta_{\mu\nu} + \frac{2\beta}{(2\mu - 2\beta - 1)} \frac{x_\mu x_\nu}{x^2} \right. \\ & \left. + \; (x^2)^\lambda B' \left(\eta_{\mu\nu} + \frac{2(\beta - \lambda)}{(2\mu - 2\beta + 2\lambda - 1)} \frac{x_\mu x_\nu}{x^2} \right) \right] \end{aligned} \tag{11}$$

and computing $\lambda = -\frac{1}{2}\beta'(g_c)$ where A' and B' are additional amplitudes and $\lambda = \mu - 2 + O(1/N_f)$ from (1). Substituting these into the graphs of Fig. 1 together with a 2-loop correction to the photon Dyson equation the solution of the appropriate consistency equation yields[11]

$$\lambda = \mu - 2 - \frac{(2\mu - 3)(\mu - 3)\eta_1}{N_f} + O\left(\frac{1}{N_f^2}\right) \tag{12}$$

which is gauge independent.

Several points are in order. First, Eq. (12) is in agreement with the original large N_f result[12] which was performed by explicitly calculating the photon bubble sum and renormalizing the Green's function by absorbing the simple poles in ϵ minimally to deduce $\beta(g)$. The exact agreement of Eq. (12) with Ref. 12 means that the large N_f exponent programme produces results which encode $\overline{\text{MS}}$ information. Secondly, it offers the possibility of going beyond $O(1/N_f)$ to deduce $O(1/N_f^2)$ corrections. An algorithm has been conceived to achieve this[11] which requires the computation of several 3 and 4 loop corrections to the photon 2-point Function. This will allow one to check the 4-loop result[4] and deduce higher order coefficients. The algorithm has been tested out successfully in the simpler 4-fermi model[13].

Another gauge independent quantity which is important for the QED RGE is the electron mass anomalous dimension. Its $O(1/N_f)$ part was computed in Ref. 12 and the $O(1/N_f^2)$ correction has recently beed found[14] by calculating the critical exponent corresponding to the renormalization of the composite operator $\bar{\psi}\psi$, as

$$\gamma_m(g_c) = -\frac{2\eta_1}{(\mu - 2)N_f} - \frac{6\eta_1^2}{(\mu - 2)^2(2\mu - 1)N_f^2} \left[\mu(\mu - 1)\hat{\Theta}(\mu) + \frac{1}{\mu} - \frac{\mu}{(\mu - 1)} + \frac{2}{3}(2\mu - 3) \right] \tag{13}$$

which agrees with the 3-loop result of Ref. 9. Moreover, in the same way that we can write down part of the 4-loop structure of $\gamma_2(g)$, we have that

$$\begin{aligned} \gamma_m(g) &= -\frac{3}{2}g + \left[\frac{10N_f}{3} - \frac{3}{2}\right]\frac{g^2}{8} + \left[\frac{140N_f^2}{27} + (46 - 48\zeta(3)) - \frac{129}{2}\right]\frac{g^3}{32} \\ &+ \left[\left(\frac{83}{1296} - \frac{\zeta(3)}{9}\right)N_f^3 + \left(\frac{5\zeta(3)}{4} - \frac{3\zeta(4)}{4} - \frac{19}{216}\right)N_f^2 + b_1 N_f + b_2\right] g^4 + O(g^5) \end{aligned} \tag{14}$$

where b_i are unknown.

Finally, we have extended some of the leading order QED results to QCD, in the Landau gauge, by considering the scaling behaviour of 2 and 3 point functions and following the same strategy as for QED. For instance, we have found the quark and gluon anomalous dimensions at $O(1/N_f)$ are

$$\eta_1 = \frac{C_2(R)}{T(R)}\eta_1^{\text{QED}} \quad , \quad \eta_1 + \chi_1 = -\frac{C_2(G)\eta_1^{\text{QED}}}{2(\mu-2)T(R)} \tag{15}$$

where η_1^{QED} is given by Eq. (8), $T(R)$, $C_2(R)$ and $C_2(G)$ are the usual Casimirs of $SU(N_c)$ and the second result of Eq. (15) corresponds to the QED Ward identity in the abelian limit. If one writes the leading $O(1/N_f)$ part of the RGE function corresponding to the renormalization of the quark-quark-gluon vertex, $\gamma_1(g)$, as

$$\gamma_1(g) = \frac{3}{8}C_2(G)g + \sum_{n=2}^{\infty} c_n[T(R)N_f]^{n-1}g^n \tag{16}$$

then Eq. (15) implies

$$\begin{aligned}
c_2 &= -\left[\frac{1}{4}C_2(R) + \frac{5}{48}C_2(G)\right] \quad , \quad c_3 = \frac{1}{27}\left[\frac{15}{8}C_2(R) - \frac{35}{32}C_2(G)\right] \\
c_4 &= \frac{1}{81}\left[\frac{35}{16}C_2(R) + \left(\frac{9}{4}\zeta(3) - \frac{83}{64}\right)C_2(G)\right] \\
c_5 &= -\frac{1}{216}\left[\left(4\zeta(3) - \frac{83}{36}\right)C_2(R) - \left(3\zeta(4) - \frac{5}{3}\zeta(3) - \frac{65}{48}\right)C_2(G)\right]
\end{aligned} \tag{17}$$

The first two coefficients, c_2 and c_3, are in agreement with those given in the recent three loop calculation[4] of $\gamma_1(g)$ when that result is restricted to the Landau gauge whilst c_4 and c_5 are new coefficients at four and five loops respectively. (In fact the structure of $\gamma_1(g)$ is such that the covariant gauge parameter does not appear in leading $O(1/N_f)$ part to three loops which is true to higher orders too.)

4. Conclusions

We have demonstrated that there is useful information encoded within the critical exponents calculated in the large N_f self consistency method to the extent that one can deduce part of the perturbative series of the RGE functions of 4-dimensional gauge theories to all orders. Whilst this allows one to have a useful independent way of checking the results already determined it offers, more importantly, a way of reducing the effort needed to carry out high order work. For example, in Eqs. (10) and (14) the unknown coefficients a_i and b_i can be deduced by isolating from all the contributing 4-loop Feynman graphs those with only one or no electron loops and calculating their relevant pole piece for the renormalization constant. This will reduce computing time substantially.

References

1. D.J. Gross & F. Wilczek, *Phys. Rev. Lett.* **30** (1974), 1343; H.D. Politzer, *Phys. Rev. Lett.* 30 (1974), 1346; W.E. Caswell, *Phys. Rev. Lett.* **33** (1974), 244; D.R.T. Jones, *Nucl. Phys.* **B75** (1974), 531; E.S. Egorian & O.V. Tarasov, *Theor. Math. Phys.* **41** (1979), 26.
2. O.V. Tarasov, A.A. Vladimirov & A.Yu. Zharkov, *Phys. Lett.* **93B** (1980), 429.
3. S.A. Larin & J.A.M. Vermaseren, *Phys. Lett.* **B303** (1993), 334.
4. S.G. Gorishny, A.L. Kataev, S.A. Larin & S.R. Surguladze, *Phys. Lett.* **B256** (1991), 81.
5. A.N. Vasil'ev, Yu.M. Pis'mak & J.R. Honkonen, *Theor. Math. Phys.* **46** (1981), 157; **47** (1981), 291.
6. J.A. Gracey, *J. Phys.* **A24** (1991), L431.
7. M. d'Eramo, L. Peliti & G. Parisi, *Lett. Nuovo Cim.* **2** (1971), 878.
8. J.A. Gracey, *Mod. Phys. Lett.* **A7** (1992), 1945.
9. O.V. Tarasov, *JINR* P2-82-900.
10. B. Zumino, *J. Math. Phys.* **1** (1960), 1.
11. J.A. Gracey, *Int. J. Mod. Phys.* **A8** (1993), 2465.
12. A. Palanques-Mestre & P. Pascual, *Commun. Math. Phys.* **95** (1984), 277.
13. J.A. Gracey, LTH-312 to appear in *Int. J. Mod. Phys. A.*
14. J.A. Gracey, LTH-316 to appear in *Phys. Lett. B.*

THE OPTIMIZATION OF A HUGE FORM PROGRAM

S.A. LARIN[1], T. van RITBERGEN and J.A.M. VERMASEREN
NIKHEF-H
P.O.Box 41882, 1009DB Amsterdam

ABSTRACT

We show some techniques and tricks that improve the performance of the FORM version of MINCER program.

The computation of massless propagator-like three loop diagrams was first undertaken[1] with a program that was written for the program SCHOONSCHIP[2], running on the CDC at Dubna. It was named MINCER[3]. This program has been used to compute the α_S^3 contribution in the vector channel of the reaction $e^+e^- \to hadrons$. After this MINCER was translated[4] to the language of FORM[5] and several sum rules and moments of structure functions were calculated with it. In order to do these calculations much attention was given to the optimization of the program. Recently we computed higher moments of the structure function F_L in three loop order and to obtain these moments it turned out that the MINCER program needed many extra optimizations and improvements in order to be able to deal with such a gigantic task. We will describe some of these optimizations. These could also be used to speed up other programs.

First a short description of the MINCER program as it is implemented in FORM. The diagrams to be computed are all in a single file. With the folding structure of FORM it is possible to select a single diagram from this very rudimentary database. Each diagram has an instruction with it that specifies its topology. Based on the topology the proper treatment of the diagram is selected from other 'databases' that contain information about all different topologies. The diagrams are first treated according to the process under consideration, as it may be necessary to use projections, take traces, etc. The most complicated projection is the harmonic projection which we will encounter below. After this we eliminate as many dotproducts between different vectors as possible after which the terms are ready for integration. This is done by recursive application of relations that are obtained by partial integration. This scheme can be quite complex. After a number of these partial integrations we reach the point that one of the integrals can be done by means of a 'simple' substitution. Then we repeat the dotproduct rewriting and partial integrations at the two loop level. Etc.

The first example is a hybrid between careful programming and an adaptation to FORM itself. One of the problems in Symbolic Manipulation is the intermediate expression swell. Often the size of an intermediate result can be more restricting than the amount of CPU time. One place in which we encountered such problems is the harmonic projection. A harmonic tensor in the vector Q is a symmetric and traceless tensor. An example of a harmonic tensor with two indices is $H^{(2)}_{\mu\nu} = g_{\mu\nu}/D - Q_\mu Q_\nu/Q.Q$ in which D is the dimension of the indices μ and ν.

[1]on leave from the Institute for Nuclear Research (INR) of the Russian Academy of Sciences, Moscow 117312

When computing higher moments of structure functions one expands in powers of P (the hadron momentum) and the moments are the coefficients of terms in $Q.P^n$ (or $Q.P^{n+2}$ for the $P_\mu P_\nu$ projection in the vector current). The loop integral is expanded in powers of P and the integral itself gives a harmonic tensor in Q. To project onto this tensor we can replace the tensor $P_{\mu_1}\ldots P_{\mu_n}$ by an n-th order harmonic tensor in Q. The problem with this tensor is that in its general form it contains an enormous number of terms. For an 8-th order tensor we have 764 terms and the 10-th order tensor has 9496 terms. This is when the D in the denominators are not expanded in terms of ϵ, because that would make things worse. The projection should take place in a very early stage, because the 3-loop algorithms cannot deal with the momentum P. Early attempts to implement these projections failed miserably. We can however design a practical scheme based on some simple observations: In any 3-loop diagram of the propagator type (which we get at $P = 0$) there are only 4 independent momenta, and $Q^{\mu_n} H^{(n)}_{\mu_1\ldots\mu_n} = H^{(n-1)}_{\mu_1\ldots\mu_{n-1}}$. Hence we have to compute only contractions of harmonic tensors with three different vectors. The art is now to do this in such a way that no term is generated twice. This required the introduction of a new function:

```
id  P.P = 0;
id  Q.P = Q.Q;
ToTensor,T,P;
id  T(??) = sum_(j,0,nargs_(..),2,Q.Q^(j/2)*
      acc(hfac(nargs_(..),j))*distrib_(1,j,dd,T,..));
ToVector,T,Q;
id  dd(??) = dd_(..);
id  n = acc(4-2*ep);
```

The commands ToTensor collects all indices of occurrences of a vector P into the tensor `T`. Or: $P_{\mu_1}\ldots P_{\mu_n} = T_{\mu_1\ldots\mu_n}$. The command ToVector is the opposite operation. These commands can be worked around but they make the programming much easier. The statement that substitutes `T` makes the split of the indices over Kronecker delta's and indices of the vectors Q. The function `hfac` is a table that contains the coefficients, expanded in ϵ. We keep the power series in ϵ always inside a function `acc` that is treated as a polynomial coefficient. The new function is `dd_`. For an even number of indices it produces a symmetric tensor that consists of combinations of Kronecker delta's. Hence

```
dd_(m1,m2) = d_(m1,m2);
dd_(m1,m2,m3,m4) = d_(m1,m2)*d_(m3,m4)
      +d_(m1,m3)*d_(m2,m4)+d_(m1,m4)*d_(m2,m3);
```

This is not special as it is possible to generate these terms with the `distrib_` function. The special properties have to do with equal arguments. When several indices of the `dd_` function are contracted with identical vectors the output will greatly simplify. The evaluation of the function has this built in with the proper combinatorics. The evaluation of `dd_(p1,p1,p1,p1,p1,p2,p2,p2,p2,p2)` gives only 3 terms, while there would be 945 terms if all vectors would be different. Because the `distrib_` function also generates no double terms (if the arguments are properly symmetrized) the above program generates no terms twice. With this routine the harmonic projections ceased to be the dominant part of the program, both in speed and in memory requirements.

The second example concerns the treatment of dotproducts when computing such complicated diagrams. In principle we evaluate traces, make the harmonic projection if necessary and then rewrite the dotproducts that the integration algorithms cannot deal with to squares and one dotproduct that involves two unequal vectors. When computing second moments the harmonic projection was relatively trivial and the major size of the intermediate expressions was obtained during the rewriting of the dotproducts. Already the rewriting into an independent set of vectors at a very early stage (just after the trace and before the expansion in P) improves things dramatically. But this is not sufficient. The eventual scheme relies on a very careful order of the operations:

- Make the projection in the vector current (multiply either by $\delta_{\mu\nu}$ or by $P_\mu P_\nu$).
- Apply Ward identities. They have the tendency to reduce the amount of work later. To facilitate their application the three-gluon vertices are written in a special form that maximizes the chance that Ward identities can be used.
- Use symmetries of the diagram.
- Reduce the number of momenta in the numerator to an independent set that is suitable for this diagram. There are various choices here and they can be specified in the database with the diagrams.
- Take traces.
- Make the expansion in P.
- Reduce the momenta again (the denominators may have given dependent momenta again).
- Apply symmetries again.
- Make the harmonic projection.
- Rewrite the mixed dotproducts that are left.

The final rewriting can be done in many different ways. It should be done statement by statement, each followed by a `.sort` instruction. The order of the statements can make a difference in the amount of disk space needed. Hence it took some work to determine the proper permutation of the substitutions. After this has been done we have reached an irreducible object in the sense that if we keep integration separate from the rest there is no way to economize on the expression as it has a unique form.

What we learn from the above is that the order of operations can be very crucial with respect to the size and the time needed for a program. Also the placement of the `.sort` instructions has a big influence on the practical aspects of a computation. If for instance all dotproducts are substituted in a single module, we would never be capable to compute an 8-th moment. The way this went in the program the integration phase has become the most time and space consuming phase, due to all these tricks. The taking of the traces is actually a very insignificant step and the harmonic projection and the dotproduct substitutions are far more expensive (the dotproducts take more time than the harmonic projection). Actually for the dotproducts to use less disk space we need another trick, which we call the AB-trick (AB stands for AntiBracket). To understand it we need to know a little bit about what FORM does when it generates a large expression.

During the processing of an expression terms are generated one by one and each is offered to the sorting routine as soon as it is complete. The sorting routine puts it in a buffer which is called the 'small' buffer. If this buffer happens to be full, its contents are sorted and the sorted results are placed in the 'large' buffer. This buffer can contain a number of these sorted 'patches'. If the large buffer is full the 'patches' are merged and the result is written to an intermediate file. This file can contain as many 'patches' as needed, unless the disk space is insufficient. Once the last term has been treated this way, the 'patches' in the disk file have to be merged to give the final result. When the size of the final result is called x, the intermediate sort file will have a size of at least x, but it may be many times larger. The AB-trick is meant to give the sort file a size that is only a little bit larger than x. This is done by making sure that terms that differ only by a coefficient are generated close to each other, so that they meet each other already during the sorting that takes place inside the real memory. To do this one should group those terms that can generate identical terms already in the previous sort. We give an example. Assume that we have the replacement $Q \cdot p_6 \to (p_1 \cdot p_1 - Q \cdot Q - p_6 \cdot p_6)/2$ then we code this as

```
AB  Q.Q,p1.p1,p6.p6,Q.p6;
.sort
id  Q.p6 = p1.p1/2-Q.Q/2-p6.p6/2;
```

after which we can put the AB statement that prepares for the next substitution. The AB statement puts everything outside brackets, except for the objects that are specified. When (in this case) terms have different objects outside the brackets, they cannot possibly cancel after the substitution, and hence (with the exception of boundary effects due to a finite buffer size) all cancellations can take place inside memory. In one of the substitutions with the trick we had that the final result took 78.8 Mbytes and the intermediate file took 81.5 Mbytes, while in a case where the trick was not applied (almost all dotproducts were involved in the substitution) we had a final result of 39.1 Mbytes while the intermediate file was 136.8 Mbytes. The total need for disk space in FORM is usually the sum of these numbers.

Of course the reduction identities due to partial integration have to be programmed very carefully too, but we will not consider them here. The next very special step is the first loop integral. It is at this point that the Mincer program makes the largest space and time demands. Let us see what is happening here. Assume that a ladder diagram is reduced to diagrams that have a one loop subintegral that can be done. Pictorially we have:

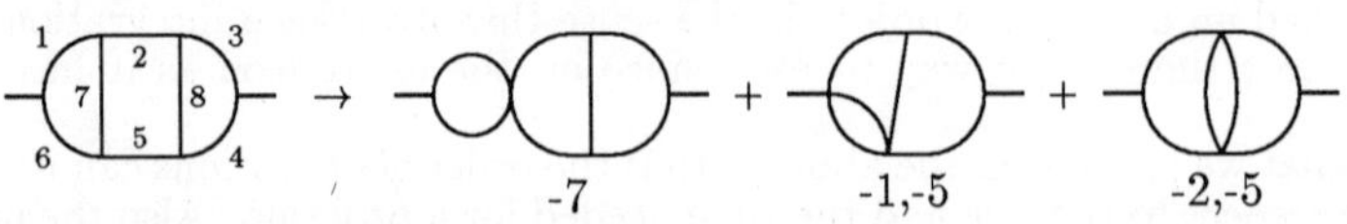

1 2 3 7 8 6 5 4 → + +

-7 -1,-5 -2,-5

The negative numbers refer to the momenta that are missing in the denominators. Of course it is still possible that some of these momenta occur in the numerator (terms generated during the rewriting of the momenta, or descending from the one mixed dotproduct that could not be reduced). These have to be eliminated. So, suppose that in the second integral on the right hand side we replace $p_1 \cdot p_1 \to p_2 \cdot p_2 + 2p_2 \cdot p_7 + p_7 \cdot p_7$, then the last term may kill a $p_7 \cdot p_7$ in the numerator and it can happen that the resulting integral is also missing the seventh momentum in the denominator. This gives an easier integral and by carefully rewriting the integrals, using symmetries, we can obtain as small a set of integrals as possible. This is quite important as one can see in the sequel.

Once the one loop integrals have been brought to a standard form they have to be replaced by a formula that contains a harmonic projection and an object G that stands for a combination of 6 Pochhammer symbols. The harmonic projection does not generate terms that cancel in a significant way and hence a sort at this point would be quite pernicious. It would make a much heavier demand on disk space than would be the case otherwise. The function G should be substituted first. The integrals that occur most frequently have been tabulated, and the others are substituted by inserting the Pochhammer symbols which have also been tabulated. These combined tables are quite large. Each table element is an expansion in $\epsilon = 2 - D/2$ to order ϵ^6. This means that for those integrals for which the function G has not been tabulated we have to multiply 6 of these expansions. Hence we can economize greatly by using the 'Keep brackets' command:

```
B   p8,p9,int3,x,FQ,naar;
.sort
Keep Brackets;
```

We first take all objects that have to deal with the integral outside brackets. The vector p_9 is the integration momentum and p_8 is the other momentum in the loop. FQ is a tensor, used for the contractions with p_9, `int3` fulfills the role of $d^D p_8$ and x and `naar` indicate some special cases. Then we sort and tell FORM to do the instructions till the next `.sort` only on the part outside the brackets. This means that FORM will multiply the inside of the brackets with the outside only after all work in this module has been done. The result is that the multiplication of the expansions in ϵ is done only once, even though there may be many terms inside the brackets. Because there is also a polynomial in ϵ for each term inside the brackets this means that when the G is not tabulated we have $5+n$ polynomial multiplications rather than $6n$ multiplications when there are n terms inside the bracket. The above work on the one loop integral improved the execution speed by about a factor 2 and gave also a significant reduction of the required disk space.

Below we will give some parameters for our most complicated diagram. This is given by:

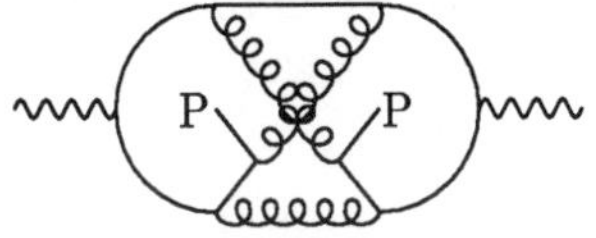

The diagram d14c4

For the $n = 6$ moment in the $P_\mu P_\nu$ projection we obtained during various stages of the project:

Date	seconds	Mbytes
28-3	152345	166.9
22-4	108772	183.0
26-4	68526	146.7
28-4	56483	129.1
1-5	50296	114.6
2-5	47748	104.3
4-5	43315	101.3
5-5	33179	83.3

Timings and space on a 25 MHz NeXT-station

Only after this we could start to compute the $n = 8$ moments. As a function of the moment we have

Moment	seconds	Mbytes
2	359	0.6
4	4693	13.6
6	33179	83.3
8	213000	408.2

Timings and space on/converted to a 25 MHz NeXT-station

The time for the full calculation (both projections) for the $n = 6$ moment was 110–120 hours on a 100MHz SGI Challenge computer and for the $n = 8$ moment it was more than 600 hours. The most complicated diagrams had to be run on another computer because this machine did not have enough free disk space. This seems to be a general problem for big calculations. If we extrapolate it seems that the $n = 10$ calculation would need about 3000 hours and 2 Gbytes of disk space. Currently we do not have those resources.

1. S.G.Gorishny, A.L.Kataev and S.A.Larin, *Phys. Lett.* **B212** (1988) 238, ibid. B259 (1991) 144; L.R.Surguladze and M.A.Samuel, *Phys.Rev.Lett.* **66** (1991) 560, erratum ibid, 2416.
2. M.Veltman, SCHOONSCHIP, CERN preprint, 1967; H.Strubbe, *Comp. Phys. Commun.* **8** (1974) 1.
3. S.G. Gorishny, S.A. Larin and F.V. Tkachov, INR preprint P-0330 (Moscow, 1984); S.G. Gorishny, S.A. Larin, L.R. Surguladze and F.V. Tkachov, *Comp. Phys. Comm.* **55** (1989) 381.
4. S.A. Larin, F.V. Tkachov and J.A.M. Vermaseren, preprint NIKHEF-H/91-18.
5. J.A.M. Vermaseren, *Symbolic Manipulation with FORM*, published by CAN (Computer Algebra Nederland), Kruislaan 413, 1098 SJ Amsterdam, 1991, ISBN 90-74116-01-9. An earlier version of FORM can be obtained for free by means of anonymous ftp from nikhefh.nikhef.nl. Version 1 is however insufficient to run the program described in this publication.

SYMBOLIC COMPUTATION OF THE HEAT KERNEL EXPANSION ON CURVED MANIFOLDS

V.P. GUSYNIN
Institute for Theoretical Physics, 252143 Kiev, Ukraine

and

V.V. KORNYAK
Institute of Mathematics, 252601 Kiev, Ukraine

ABSTRACT

An algorithm based on covariant (intrinsic) pseudodifferential calculus for computing DeWitt-Seeley-Gilkey coefficients of elliptic operators on compact curved manifold with torsion is considered. To implement this algorithm a program package for algebraic and tensor manipulations has been written in C language. Some computer results are presented.

1. Introduction

The coefficients in the asymptotic expansion of the diagonal heat kernel elements are of fundamental importance in theoretical physics, analysis, spectral geometry and topology of manifolds. They determine one-loop divergences of the effective action, the axial and trace anomalies [1-5] in quantum field theory, high-temperature expansion of the partition function in statistical physics, the indices of elliptic operators [6,7], geometric and topological properties of manifolds [6-11].

The asymptotic expansion we are interested in, reads [6,7,12-14]

$$\langle x|e^{-tA}|x\rangle \sim \sum_{m\geq 0} \mathrm{E_m(x|A)t^{\frac{m-n}{2r}}}, \qquad \mathrm{t} \to 0_+ \,, \tag{1}$$

where A is a positive elliptic differential operator of the order $2r$ acting, in general, on the bundle of k-tensors with the values in a vector space V, the base M being the compact closed n-dimensional manifold. The expansion coefficients $E_m(x|A)$ are endomorphisms of the fibre at x and prove to be the local covariant quantities of given dimensionality constructed from the coefficients of the operator A, curvatures, torsion and their derivatives. The E_m's are called DeWitt-Seeley-Gilkey (DWSG) coefficients though many people contributed into their investigation beginning from Hadamard (see [15] for historical background of the question).

The typical examples of operators A used in physics and mathematics are

$$-\Box + X, \tag{2}$$

$$\Box^2 + V^{ij} D_i D_j + N^i D_i + X, \tag{3}$$

$$-g^{ab}\Box + aD^a D^b + X^{ab}. \tag{4}$$

Here $\Box \equiv D_i D^i$ is a covariant generalization of the Laplace operator, D_i is the covariant derivative involving affine and bundle connections, V, N, X are tensors and a is a scalar parameter (bundle indices are assumed implicitly). Operators (2), (3) with leading term being a power of the Laplacian are called *minimal* operators whereas operator (4) is an example of the *non-minimal* one. Up to now the most advanced results have been obtained for second-order operator (2): see DeWitt [1] for E_2, E_4, Sakai [16] and Gilkey [14] for E_6, Avramidi [17] and Amsterdamski et al. [18] for E_8 (E_m vanish for m odd). Progress to higher orders m has been rather slow because of combinatorial explosion with increasing m. Thus we may expect that computer will help to get further progress in this field.

2. Algorithm of Calculation

To compute E_m several algorithms were elaborated [1,14,19–25]. Some of them are based on particular properties of considered operator and are not universal. There are also computer implementations of these algorithms based on general purpose computer algebra systems (for example, in [26] *Mathematica* and in [27] REDUCE and FORM were used). Here we consider the computer implementation of an algorithm based on Widom [28] covariant generalization of pseudodifferential calculus (so-called intrinsic symbolic calculus). This algorithm has been developed in [29–33] and permits calculation of DWSG coefficients for operators of general type and of arbitrary order on curved manifolds (including torsion), while being explicitly covariant with respect to gauge and general coordinate transformations. We describe it here briefly.

Let us use the integral representation of the heat operator $exp(-tA)$ in the Dunford operator calculus by the formula

$$e^{-tA} = \int_C \frac{id\lambda}{2\pi} e^{-t\lambda}(A-\lambda)^{-1},$$

where the contour C goes counterclockwise around the spectrum of an operator A. Then the matrix elements of the resolvent $(A-\lambda)^{-1}$ take the following representation

$$G(x,x',\lambda) \equiv \langle x|\frac{1}{A-\lambda}|x'\rangle = \int \frac{d^n k}{(2\pi)^n \sqrt{g(x')}} e^{il(x,x',k)} \sigma(x,x',k;\lambda),$$

where $l(x,x',k)$ is a phase function, $\sigma(x,x',k;\lambda)$ is an amplitude, $g(x) = det g_{ab}(x)$. We choose the phase $l(x,x',k)$ to be a covariant generalization [28,34] to curved manifolds of the flat space phase $k_a(x-x')^a$. The generalization of the linearity with respect to x is the requirement for higher symmetrized covariant derivatives to vanish at the point $x = x'$, i.e.

$$[D_a l] = k_a, \qquad \{D_{a_1} \ldots D_{a_m}\} l \Big|_{x=x'} = [\{D_{a_1} \ldots D_{a_m}\} l] = 0, \quad m > 1. \tag{5}$$

Here the curly and square brackets mean symmetrizing in all indices and transition to coincidence limit ($x = x'$) respectively. The resolvent of operator A satisfies the equation $(A-\lambda)G = 1$ which leads to an equation for the amplitude:

$$(A(x, D_a + iD_a l) - \lambda)\sigma(x,x',k;\lambda) = I(x,x'). \tag{6}$$

The transport function $I(x,x')$(having bundle and Lorentz indices) is defined by the conditions

$$[I] = E, \qquad [\{D_{a_1}\dots D_{a_m}\}I] = 0, \quad m \geq 1, \tag{7}$$

where E is the unit matrix. The functions $l(x,x',k)$ and $I(x,x')$ introduced by means of Eqs.(5),(7) play an important role in the intrinsic symbolic calculus [28,34]. They are universal functions reflecting purely geometric properties of the manifold. Since we are interested in the diagonal matrix values of operators it is sufficient to consider the coincidence limits for nonsymmetrized covariant derivatives $[D_{a_1}\dots D_{a_m}l]$ and $[D_{a_1}\dots D_{a_m}I]$. These all are polynomials in the torsion T^a_{bc}, curvature tensors R^a_{bcd}, W_{ab} of the given connection and their covariant derivatives and are obtained directly from (5),(7) by reducing all terms to a unified index ordering with the help of the Ricci identity for the commutator of covariant derivatives

$$[D_a,D_b]\varphi^{b_1\dots b_l}_{a_1\dots a_k} = \sum_{i=1}^{l} R^{b_i}_{jab}\varphi^{b_1\dots b_{i-1}jb_{i+1}\dots b_l}_{a_1\dots a_k} - \sum_{i=1}^{k} R^{j}_{a_iab}\varphi^{b_1\dots b_l}_{a_1\dots a_{i-1}ja_{i+1}\dots a_k} + T^j_{ab}D_j\varphi^{b_1\dots b_l}_{a_1\dots a_k} + W_{ab}\varphi^{b_1\dots b_l}_{a_1\dots a_k}.$$

For example, for lowest coincidence limits we have

$$[D_a l] = k_a, \qquad [D_aD_bl] = \frac{1}{2}k_iT^i_{ab},$$

$$[D_aD_bD_cl] = k_i(-\frac{1}{3}R^i_{cab} - \frac{1}{3}R^i_{bac} + \frac{1}{2}D_aT^i_{bc} + \frac{1}{6}D_bT^i_{ac} + \frac{1}{6}D_cT^i_{ab} + \frac{1}{4}T^i_{aj}T^j_{bc} + \frac{1}{12}T^i_{jb}T^j_{ac} + \frac{1}{12}T^i_{jc}T^j_{ab}),$$

$$[D_aI] = 0, [D_aD_bI] = \frac{1}{2}W_{ab}, [D_aD_bD_cI] = \frac{1}{3}D_aW_{bc} + \frac{1}{3}D_cW_{ba} + \frac{1}{12}T^i_{ac}W_{bi} + \frac{1}{12}T^i_{bc}W_{ai} + \frac{1}{4}T^i_{ab}W_{ic}.$$

The number of terms in Eqs.(5),(7) is equal to $m!$ so the complexity increases very rapidly with the order m and calculation by hand becomes impractical.

At last to generate expansion (1) we expand the amplitude σ in degrees of homogenity of $k : \sigma = \sum_{m=1}^{\infty}\sigma_m(x,x',k;\lambda)$. This allows one to obtain the recursion equations for σ_m from Eq.(6).Solving these equations we can express DWSG coefficients by integrals of $[\sigma_m]$[29–31]

$$E_m(x|A) = \int \frac{d^nk}{(2\pi)^n\sqrt{g}}\int_C \frac{id\lambda}{2\pi}e^{-\lambda}[\sigma_m](x,k,\lambda) \equiv J([\sigma_m]).$$

The integration over λ and k is performed by using formulae [31–33]

$$J\Big(\frac{k^{2p}k_{a_1}\dots k_{a_{2s}}}{(k^{2r}-\lambda)^l[(1-a)k^{2r}-\lambda]^m}\Big) = g_{\{a_1\dots a_{2s}\}}\frac{\Gamma(n/2+s+p/r)}{(4\pi)^{n/2}2^s r\Gamma(n/2+s)\Gamma(l+m)}F(m,p+s+\frac{n}{2r};l+m;a).$$

Here $F(a,b;c;z)$ is the Gauss hypergeometric function and $g_{\{a_1\dots a_{2s}\}}$ is the symmetrized sum of metric tensor products. (The odd coefficients E_{2m+1} are always zeros because $[\sigma_{2m+1}]$ contain odd powers of k. The hypergeometric functions can be reduced to elementary ones.) Substituting the expressions for coincidence limits of l and I functions into E_m we get the final result in terms of geometric quantities.

During computation various tensor simplifications must be carried out using, for example, the above mentioned Ricci identity and the Bianchi and cyclic identities

$$D_aR^b_{cde} + D_dR^b_{cea} + D_eR^b_{cad} + T^i_{ad}R^b_{cei} + T^i_{de}R^b_{cai} + T^i_{ea}R^b_{cdi} = 0,$$

$$D_a W_{bc} + D_b W_{ca} + D_c W_{ab} + W_{ai} T^i_{bc} + W_{bi} T^i_{ca} + W_{ci} T^i_{ab} = 0,$$

$$R^a_{bcd} + R^a_{cdb} + R^a_{dbc} + D_b T^a_{cd} + D_c T^a_{db} + D_d T^a_{bc} + T^a_{bi} T^i_{cd} + T^a_{ci} T^i_{db} + T^a_{di} T^i_{bc} = 0.$$

Also, the symmetry properties of tensors $R^a_{bcd}, T^a_{bc}, W_{ab}$ have to be taken into account.

3. Outline of the Program Package

The package is written in the C language with the help of the *Borland C++ 2.0* system. Its total length is about 8000 lines. It contains different functions including tensorial and algebraic simplifications, differentiation, integration, substitutions, two-dimensional output, processing input information and so on. To calculate DWSG coefficients these functions are gathered into two programs DWSGCOEF and COLIM. Internal structures for tensor and algebraic expressions are lists. Tensor multiindices are stored in hash tables to minimize space and time expenses. A single index is a byte containing the following information: whether the index is free or dummy, whether it is lower or upper, ordinal number of index. A list node contains address of multiindex in hash table (for tensor node) and some other information (type of object, sign of object, pointers to function arguments etc.)

The COLIM program computes coincidence limits of l or I functions and writes them to the disk. Once computed and stored on the disk the coincidence limits can be used in many calculations for different operators A.

The DWSGCOEF program computes E_m coefficients executing the following steps:

- Reading input information (operator, order m, etc.)
- Computation of a set of asymptotic operators needed for constructing recursion equations.
- Computation of σ_m.
- Taking the coincidence limit $[\sigma_m]$.
- Integration of $[\sigma_m]$ over wave vector k and spectral parameter λ to obtain the coefficient E_m.
- Reading the coincidence limits of l and I functions from disk and substituting them into E_m.
- Output of final result.

The following session file has been produced by 33MHz AT 386 computer under DOS when calculating E_4 for minimal operator (2)

```
The program calculates De Witt - Seeley - Gilkey coefficients
for differential operators on curved manifolds.
Version of 12th June 1993. V. V. Kornyak
```

```
Enter required ordinal number of DWSG coefficient
4
Include torsion ?
Enter Y or N
n
Include bundle curvature ?
Enter Y or N
y
Is there file with your input data ?
Enter Y or N
y
Enter name of file containing input information
mini.in
Input data:
Additional tensors: X;
Operator: -[] + X;
```

Set of asymptotic operators:

(0) $A_0 = - \lambda + D_i 1 D^i 1$

(1) $A_1 = - i D_i D^i 1 - 2 i D_i 1 D^i$

(2) $A_2 = - D_i D^i + X$

DWSG coefficients:

(0) $E_0 = (4 \pi)^{- 1/2 n}$

(2) $E_2 = (4 \pi)^{- 1/2 n} (1/6 R - X)$

(4) $E_4 = (4 \pi)^{- 1/2 n} (- 13/180 D_i D^i R + 1/72 R^2 - 1/6 R X + 1/5 D_i D_j R^{ij} + 1/90 D_i D_j R^{ji} - 1/180 R_{ij} R^{ij} - 1/36 R_{ij} W^{ij} - 2/27 R_{ijkl} R^{ijkl} + 103/1080 R_{ijkl} R^{ikjl} - 23/360 R_{ijkl} R^{iljk} + 1/12 D_i D_j W^{ij} + 1/12 W_{ij} W^{ij} - 1/6 D_i D^i X + 1/2 X^2)$

```
Time: 29.29 sec
Number of nodes: 6386  Node space: 38316 bytes
Number of indices: 3883
Total space: 42199 bytes
```

4. Conclusion

The calculation of E_2 for nonminimal operator (4) requires 56944 bytes of memory and run time 3 min 15 sec. Analogous computation in [26] uses two programs written in C and *Mathematica*. The C program is very fast whereas *Mathematica* computation (dealing with combining terms) took two days of operation of a SUN 3/60 workstation. So we conclude that general purpose computer algebra systems are of little use for scientific problems of "world record" level. Such problems seem to require the writing of special programs [35]. We have computed also the coefficients E_0, E_2, E_4 for operators (2), (3) and their generalizations (with and without torsion). The inclusion of the torsion increases the computational efforts considerably. All these computations required no more time than several tens of seconds and no more space than several tens or hundreds Kbytes. Certainly higher coefficients require more powerful computer system than AT 386 under DOS (i.e. without 640K memory limit). As we saw each term E_m in series (1) is a linear combination of the monomials of definite order formed from tensors by multiplication and index-pair contraction. Thus we need to specify a basis of independent terms which should be somehow "natural" with respect to some criteria. This problem has been solved in [36] by applying the representation theory of the symmetric, general linear and orthogonal groups. Of course, the problem has no unique answer and different bases are useful for different purposes. Having generated any basis the next important task is to develope an algorithm for reducing the answer for E_m to chosen basis. But that is for the future.

5. Acknowledgements

We would like to thank Profs. A.G. Zagorodny and A.U. Klimyk for the opportunity to use their computers without which this work could not be done. We are very grateful to Prof. S.A.Fulling for his interest in this work and reading the manuscript.

6. References

1. B.DeWitt, *Dynamical Theory of Groups and Fields* (Gordon and Breach, New York, 1965).
2. N.D.Birrell and P.C.W.Davies, *Quantum Fields in Curved Space* (Cambridge U.P., Cambridge, 1982).
3. L.S.Brown, *Phys.Rev.* **D15** (1977) 1469.
4. J.S.Dowker, *J.Phys.* **A11** (1978) 347.
5. S.M.Christensen and M.J.Duff, *Nucl.Phys.* **B154** (1979) 301.
6. M.A.Atiyah, R.Bott and V.K.Patodi, *Invent.Math.* **19** (1973) 279.

7. P.B.Gilkey, *The Index Theorem and the Heat Equation* (Publish or Perish, Boston, 1974).
8. S.Minakshisundaram, *J.Indian Math.Soc.* **17** (1953) 37.
9. S.Minakshisundaram and A.Pleijel, *Can.J.Math.* **1** (1949) 242.
10. M.Kac, *Amer.Math.Monthly* **73** (1966) 1.
11. M.P.McKean and I.M.Singer, *J.Diff.Geom.* **1** (1967) 43.
12. R.T.Seeley, *Proc.Symp.Pure Math.*, Amer.Math.Soc. **10** (1967) 288.
13. P.Greiner, *Arch.Rat.Mech.Anal.* **41** (1971) 163.
14. P.B.Gilkey, *J.Diff.Geom.* **10** (1975) 601.
15. R.Schimming, *Calculation of the Heat Kernel Coefficients*, Preprint of Ernst-Moritz-Arndt- Universität, 1993.
16. T.Sakai, *Tôhoku Math.J.* **23** (1971) 589.
17. I.G.Avramidi, *Nucl.Phys.* **B355** (1991) 712.
18. P.Amsterdamski, A.L.Berkin and D.J.O'Connor, *Class.Quant.Grav.* **6** (1989) 1981.
19. T.S.Bunch and L Parker, *Phys.Rev.* **D20** (1979) 2499.
20. R.Endo, *Progr.Theor.Phys.* **71** (1984) 1366.
21. A.O.Barvinsky and G.A.Vilkovisky, *Phys.Repts.* **119** (1985) 1.
22. R.I.Nepomechie, *Phys.Rev.* **D31** (1985) 3291.
23. H.W.Lee and P.Y.Pac, *Phys.Rev.* **D33** (1986) 1012:
24. H.W.Lee, P.Y.Pac and H.K.Shin, *Phys.Rev.* **D35** (1987) 2440.
25. S.A.Fulling and G.Kennedy, *Trans.Amer.Math.Soc.* **310** (1988) 583.
26. S.A.Fulling, in *Proceedings of Third International Colloquium on Differential Equations*, ed. D.Bainov and V.Covachev (VSP Intern. Science Publ., 1993).
27. A.A.Bel'kov, A.V.Lanyov, D.Ebert and A.Schaale, *Heat-Kernel Calculation of the Quark Determinant and Computer Algebra*, Preprint DESY 92-153, 1992.
28. H.Widom, *Bull.Sci.Math.* **104** (1980) 19.
29. V.P.Gusynin, *Phys.Lett.* **B225** (1989) 233.
30. V.P.Gusynin, *Nucl.Phys.* **B333** (1990) 296.
31. V.P.Gusynin, *Ukrainian Math.Journ.* **43** (1991) 1154.
32. V.P.Gusynin and E.V.Gorbar, *Phys.Lett.* **B270** (1991) 29.
33. V.P.Gusynin, E.V.Gorbar and V.V.Romankov, *Nucl.Phys.* **B362** (1991) 449.
34. L.Drager, *On the Intrinsic Symbol Calculus for Pseudodifferential Operators on Manifolds*, Dissertation,1978.
35. S.A.Fulling, *J. Symbolic Comput.* **9** (1990) 73.
36. S.A.Fulling, R.C.King, B.G.Wybourne and C.J.Cummins, *Class.Quant.Grav.* **9** (1992) 1151.

The effects of the QCD corrections to $\Gamma(H^0 \rightarrow b\bar{b})$

Andrei L. Kataev
Institute for Nuclear Research of the Academy of Sciences of Russia, Moscow 117312, Russia
and **Victor T. Kim**
St.Petersburg Nuclear Physics Institute of the Academy of Sciences of Russia, Gatchina 188350, Russia

Abstract

We consider in detail various theoretical uncertainties of the perturbative predictions for the decay width of $H^0 \rightarrow b\bar{b}$ process in the region 60 $GeV < M_H \leq 2M_W$. We calculate the order $O(\alpha_s^2)$-contributions to the expression for $\Gamma_{Hb\bar{b}}$ through the pole b-quark mass m_b and demonstrate that they are important for the elimination of the numerical difference between the corresponding expression and the renormalization-group (RG) improved one through the running b-quark mass. In the case of m_b=4.8 GeV and $\Lambda_{\overline{MS}}^{(5)}$=250 (150) MeV we observe the identical cancellation of the next-to-leading order RG-improved corrections to $\Gamma_{Hb\bar{b}}$ at M_{ch}=65 (130) GeV and the indications to the asymptotic explosion of the corresponding next-to-next-to-leading parametrization. We are considering the hypothesis, which might be related to this observation. The theoretical importance of the order $O(m_b^2/M_H^2)$ massive corrections to $\Gamma_{Hb\bar{b}}$ for modeling of threshold effects is demonstrated.

1. Among the most intriguing modern theoretical problems are the investigations of the properties of still non-discovered Higgs particles. The special attention is paid nowadays to the searches of a Standard Model Higgs boson. Various aspects of "Higgs hunting" were discussed in detail from the theoretical [1],[2] and the phenomenological [2],[3] points of view. The existing experimental lower bound on a Standard Model Higgs boson mass is $M_H > 60$ GeV at 95% confidence level (see e.g. [4]). This lower bound came from the analysis of the LEP data taking into account of the theoretical expression for $H^0 \rightarrow q\bar{q}$ decay rate with the QCD corrections of order $O(\alpha_s)$ [5]. One of the main decay channels of a Higgs boson in the intermediate mass range 60 $GeV < M_H < 2M_W$ is the decay to the $b\bar{b}$ final states with the coupling constant being proportional to the b-quark mass. In this mass range the theoretical uncertainties of $\Gamma_{Hb\bar{b}}=\Gamma(H^0 \to b\bar{b})$ are closely related to the theoretical uncertainties of the branching ratio of the process $H^0 \to \gamma\gamma$ which is known as one of the most typical reactions for the searches of not too heavy Higgs bosons.

The massless QCD corrections to $\Gamma(H^0 \to hadrons)$ were considered at the next-to-next-to-leading order (NNLO) using the concept of the running b-quark mass $\overline{m}_b(M_H)$ [6],[7] and taking into account the 3-loop NNLO corrections to the QCD β-function [8], and to the anomalous mass dimension function [9]. In the process of the consideration of phenomenological consequences of the results of ref.[7] it was noticed that the variation of the running b-quark mass from the b-quark mass-shell (namely from the $\overline{m}_b(m_b)$-value) to the M_H-scale (namely to the $\overline{m}_b(M_H)$-value) in the leading order (LO) of perturbation theory results in the negative corrections, which diminish the value of the corresponding Born expression for $\Gamma_{Hb\bar{b}}$ by over 50% [3].

However, the running mass is not the unique way of defining mass parameters. Indeed, in total analogy with the physical mass of electron, in QCD one can also define the pole quark mass (for the general consideration of the corresponding renormalization group equations see ref.[10]). The expression for the decay width $H^0 \to q\bar{q}$ ($H^0 \to l^-l^+$) has been explicitly calculated in terms of the pole quark masses m_c and m_b in refs.[5], [11] at the $O(\alpha_s)$ ($O(\alpha)$)-level. The presented in ref.[2] numerical studies of these results did not reveal the effect of the 50% reduction of the Born approximation. Therefore, it is important to understand the origin of the observed in ref.[2] puzzle of the differences between various parametrizations of the QCD results for $\Gamma_{Hb\bar{b}}$ in the experimentally interesting region of M_H values.

Using the 2-loop relation between the running and the pole quark masses [12] and the results of refs.[7] we calculate the expression for $\Gamma_{Hb\bar{b}}$ at the α_s^2-level in terms of the pole quark mass in two different forms. The first one will contain the $ln(M_H^2/m_b^2)$-contributions explicitly, while in the second one they will be summed up through the renormalization group (RG) technique.

Our results demonstrate that for the RG-nonimproved expression for $\Gamma_{Hb\bar{b}}$ calculated by us α_s^2-contribution produce the negative correction which is responsible for the elimination of the numerical difference between various parametrizations of the QCD results for $\Gamma_{Hb\bar{b}}$.

We also present the more detailed analysis of the RG-improved expression for $\Gamma_{Hb\bar{b}}$ taking into account the effects, neglected in the course of estimates of ref.[3]. Among them are the corrections responsible for the relation between the running mass $\overline{m}_b(m_b)$ and the pole mass m_b at the 1-loop and 2-loop levels and the order $O(m_b^2/M_H^2)$-corrections to $\Gamma_{Hb\bar{b}}$. The characteristic of the proposed parametrization is discussed, namely the cancellation of the next-to-leading order (NLO) corrections at the scales M_{ch}=65 (130) GeV for the values of $\Lambda^{(5)}_{\overline{MS}}$= 250 (150) MeV and the indications to the asymptotic explosion of the corresponding NNLO approximations. The importance of taking into account of the order $O(m_b^2/M_H^2)$-corrections for modeling the threshold effects is demonstrated.

2. We start from technical considerations. At the Born level of perturbation theory one has

$$\Gamma_{Hb\bar{b}} = \Gamma(H^0 \to b\bar{b}) = \Gamma_0^{(b)}\beta^3 \tag{1}$$

where $\Gamma_0^{(b)} = \frac{3\sqrt{2}}{8\pi}G_F M_H m_b^2$ and $\beta = \sqrt{1-\frac{4m_b^2}{M_H^2}}$. The corresponding $O(\alpha_s)$ leading order (LO) approximation of the decay width reads [5, 11]

$$\Gamma_{Hb\bar{b}} = \Gamma_0^{(b)}\beta^3\Big[1+\delta\frac{\alpha_s(M_H)}{\pi}\Big] \tag{2}$$

where

$$\delta = \frac{4}{3}\Big[\frac{A(\beta)}{\beta}+\frac{3+34\beta^2-13\beta^4}{16\beta^3}\ln\frac{1+\beta}{1-\beta}+\frac{3(-1+7\beta^2)}{8\beta^2}\Big] \tag{3}$$

with $A(\beta) = (1+\beta^2)\Big[4Li_2\Big(\frac{1-\beta}{1+\beta}\Big)+2Li_2\Big(-\frac{1-\beta}{1+\beta}\Big)-3\ln\frac{2}{1+\beta}\ln\frac{1+\beta}{1-\beta}-2\ln\frac{1+\beta}{1-\beta}\ln\beta\Big] - 3\beta\ln\frac{4}{1-\beta^2}-4\beta\ln\beta$. Here $Li_2(x) = \int_0^x(dt/t)\ln(1-t)$ is the Spence function. In the limit $m_b^2/M_H^2 \to 0$ (or $\beta \to 1$) the expression for the α_s-correction to $\Gamma_{Hb\bar{b}}$ appears to be infinite, namely it contains the logarithmic singularities

$$\Gamma_{Hb\bar{b}} = \Gamma_0^b\Big[[1+\Big(3-2ln(x)\Big)\frac{\alpha_s(M_H)}{\pi}]-6\frac{m_b^2}{M_H^2}[1+\Big(\frac{4}{3}-4ln(x)\Big)\frac{\alpha_s(M_H)}{\pi}]+O(\frac{m_b^4}{M_H^4})\Big] \tag{4}$$

where $x = M_H^2/m_b^2$.

We will now consider the expression for $\Gamma_{Hb\bar{b}}$ in terms of $\overline{m}_b = \overline{m}_b(M_H)$ taking into account the order $O(\overline{m}_b^2/M_H^2)$-contributions. Since $\overline{m}_b(M_H)$ is defined through the running coupling constant $\alpha_s(M_H)$ already in the lowest order of perturbation theory, we will identify the Born RG-improved expression with the LO approximation, which can be defined as

$$\Gamma_{Hb\bar{b}} = \Gamma_0^{(b)}\frac{\overline{m}_b^2}{m_b^2}\Big(1-\frac{4\overline{m}_b^2}{M_H^2}\Big)^{\frac{3}{2}} \tag{5}$$

where in the LO of perturbation theory one has $\overline{m}_b(M_H)=(\alpha_s(M_H)/\alpha_s(m_b))^{12/23}m_b$. At the NNLO level of perturbation theory the expanded in $O(\overline{m}_b^2/M_H^2)$ -terms expression for $\Gamma_{Hb\bar{b}}$ can be presented in the following form

$$\Gamma_{Hb\bar{b}} = \Gamma_0^{(b)}\frac{\overline{m}_b^2}{m_b^2}\Big[[1+\Delta\overline{\Gamma}_1+\Delta\overline{\Gamma}_2] - 6\frac{\overline{m}_b^2}{M_H^2}[1+\Delta\overline{\Gamma}_1^{(m)}+\Delta\overline{\Gamma}_2^{(m)}] + O\Big(\frac{\overline{m}_b^4}{M_H^4}\Big)\Big] \quad (6)$$

Here $\Delta\overline{\Gamma}_1$ ($\Delta\overline{\Gamma}_1^{(m)}$) and $\Delta\overline{\Gamma}_2$ ($\Delta\overline{\Gamma}_2^{(m)}$) are the NLO and NNLO QCD corrections to the "coefficient functions" of the $H^0 \to b\bar{b}$ decay width $\Gamma_{Hb\bar{b}}$. The general analytic expressions for $\Delta\overline{\Gamma}_1$ and $\Delta\overline{\Gamma}_2$ were obtained in ref. [7] in the $\overline{MS}$-scheme with the help of the SCHOONSCHIP analytical system. In the case of QCD with $f = 5$ numbers of flavours the results read [7]

$$\Delta\overline{\Gamma}_1 = \frac{17}{3}\frac{\alpha_s(M_H)}{\pi} \quad (7)$$

$$\Delta\overline{\Gamma}_2 = 29.14\Big(\frac{\alpha_s(M_H)}{\pi}\Big)^2 \quad (8)$$

The result for $\Delta\overline{\Gamma}_1$ coincides with the one obtained previously [13]. The expression for $\Delta\overline{\Gamma}_2$ was confirmed recently [14] using the FORM analytical system.

Let us determine the NLO α_s correction to the order $O(\overline{m}_b^2/M_H^2)$ -contributions using the results of calculations of ref. [15]. In the $\overline{MS}$-scheme it reads

$$\Delta\overline{\Gamma}_1^{(m)} = 5C_F\frac{\alpha_s(M_H)}{\pi} = (C_F = 4/3) = \frac{20}{3}\frac{\alpha_s(M_H)}{\pi} \quad (9)$$

The corresponding NNLO α_s^2-term is still unknown. In principle, it can be obtained after careful analysis of the results of the calculations of the 3-loop order $O(\overline{m}^2)$- corrections to the 2-point function of the scalar quark currents in the eucledian region [16] and the inclusion of the effects of the analytical continuation to the physical region (π^2-terms). However, varying this correction between $\Delta\overline{\Gamma}_2^{(m)}=0$ and $\Delta\overline{\Gamma}_2^{(m)}=40(\alpha_s(M_H)/\pi)^2$ we will demonstrate that this term is important only in the case if one is interested in the detailed studies of the threshold behaviour of the corresponding correction to the physical quantity.

3. Let us now express the corresponding approximation for $\Gamma_{Hb\bar{b}}$ of eq.(6) with the radiative corrections defined by eqs.(7,8,9) through the b-quark pole mass. We will use the obtained in ref.[12] 2-loop relation between the normalized on the mass-shell running b-quark mass in the $\overline{MS}$-scheme and the pole b-quark mass. This relation was also confirmed recently [14]. In the rough estimates presented in ref.[3] the corrections of order α_s and α_s^2 in the results of ref.[12] were neglected. However, in view of the sizeable value of $\alpha_s(m_b)$ it is desirable to take these corrections into account in the detailed studies. Using the results of calculations of ref.[12] one can get the following 2-loop relation between the $\overline{MS}$-scheme running and pole b-quark masses :

$$\overline{m}_b^2(M_H) = m_b^2\Big[1 - \Big(\frac{8}{3} + 2\ln(x)\Big)\frac{\alpha_s(M_H)}{\pi} - \tag{10}$$

$$-\Big(26.89 - 2.08\sum_{f=u}^{c}(1-\frac{m_f}{m_b}) - \frac{1}{12}\ln^2(x) + \frac{245}{36}\ln^2(x)\Big)\Big(\frac{\alpha_s(M_H)}{\pi}\Big)^2\Big]$$

This expression was also used in the course of derivation of the $O(\alpha_s^2 m_b^2/M_Z^2)$ corrections to the axial Z-boson decay rate [17] with taking into account the results of ref.[7]. We will apply now eq.(6,10) for the determination of the higher order corrections to the expression of $\Gamma_{Hb\bar{b}}$ through the pole mass m_b. The corresponding expression has the following form

$$\Gamma_{Hb\bar{b}} = \Gamma_0^{(b)}\Big[[1 + \Delta\Gamma_1 + \Delta\Gamma_2] - 6\frac{m_b^2}{M_H^2}[1 + \Delta\Gamma_1^{(m)} + \Delta\Gamma_2^{(m)}] + O\Big(\frac{m_b^2}{M_H^4}\Big)\Big] \tag{11}$$

where

$$\Delta\Gamma_1 = \Big(3 - 2\ln(x)\Big)\frac{\alpha_s(M_H)}{\pi} \tag{12}$$

$$\Delta\Gamma_2 = \Big(-4.54 - 2.08\sum_{f=u}^{c}\frac{m_f}{m_b} - 18.14\ln(x) + 0.08\ln^2(x)\Big)\Big(\frac{\alpha_s(M_H)}{\pi}\Big)^2 \tag{13}$$

$$\Delta\Gamma_1^{(m)} = \Big(\frac{4}{3} - 4\ln(x)\Big)\frac{\alpha_s(M_H)}{\pi} \tag{14}$$

$$\Delta\Gamma_2^{(m)} = \Delta\overline{\Gamma}_2^{(m)} - \Big(65.6 + 4.16\sum_{f=u}^{c}\frac{m_f}{m_b} + 29.6\ln(x) - 4.17\ln^2(x)\Big)\Big(\frac{\alpha_s(M_H)}{\pi}\Big)^2 \tag{15}$$

The results for $\Delta\Gamma_1$ and $\Delta\Gamma_1^{(m)}$ are in agreement with the ones obtained from the complete LO massive dependence of $\Gamma_{Hb\bar{b}}$ after expanding it in powers of m_b^2/M_H^2 (see eq.(4)). The expressions for the order α_s^2 corrections are new.

4. In order to understand the relative value of different QCD corrections to $\Gamma_{Hb\bar{b}}$ discussed and derived in Secs.2,3 we plot at Figs.1-3 the expression for the ratio $R_{Hb\bar{b}} = \Gamma_{Hb\bar{b}}/\Gamma_0^{(b)}$ ($\Gamma_0^{(b)} = 3\sqrt{2}/(8\pi)G_F M_H m_b^2$) as the functions of M_H for the values of $\Lambda_{\overline{MS}}^{(5)} = 150$ MeV and $m_b = 4.8$ GeV without and with order m_b^2/M_H^2-corrections.

Fig.1 shows the dependence of $R_{Hb\bar{b}}$ from M_H in the case of the pole quark mass parametrization of $\Gamma_{Hb\bar{b}}$ including the $O(m_b^2/M_H^2)$-corrections. The curves (c) compare

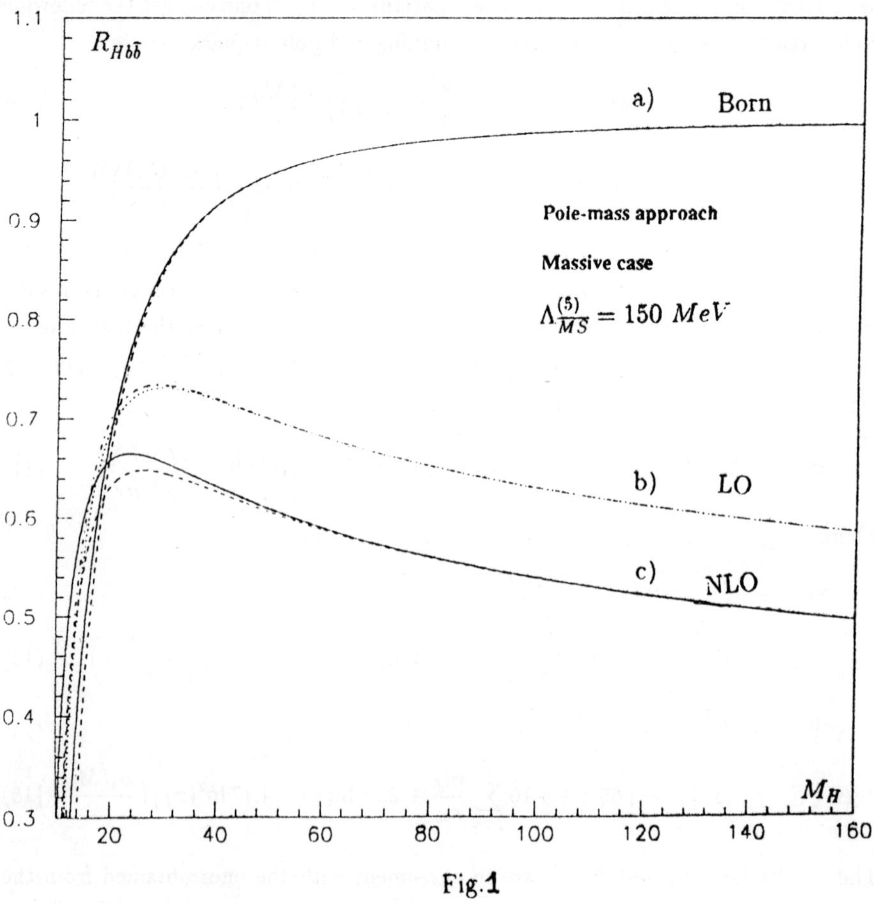

Fig.1: The pole-mass parametrization of the different approximations of $R_{Hb\bar{b}}$ with the massive-dependent corrections ($\Lambda^{(5)}_{\overline{MS}} = 150\ MeV$).

present order $O(\alpha_s^2 m_b^2/M_H^2)$ term $\Delta\overline{\Gamma}_2^{(m)}$ (compared solid curves (c) at Figs. 1,2,3 with $\Delta\overline{\Gamma}_2^{(m)} = 0$ and the corresponding dashed curves (c) with $\Delta\overline{\Gamma}_2^{(m)} = 40(\alpha_s(M_H)/\pi)^2$) demonstrate that this term is not very important in the rough analysis of the threshold behavior of the corresponding 2-loop approximations. However, in order to perform the more detailed theoretical analysis of this topic it is desirable to calculate this term explicitly.

Taking into account the discussed above QCD uncertainties and the current uncertainties of the value of the parameter $\Lambda_{\overline{MS}}^{(5)}$ (for the detailed discussions see ref.[21]) we obtain the following theoretical estimate for $\Gamma_{Hb\bar{b}}$ in the intermediate region of M_H values $60\ GeV < M_H < 160\ GeV$

$$\Gamma_{Hb\bar{b}} = \Big(0.55 \div 0.45\Big)\frac{3\sqrt{2}}{8\pi}G_F M_H m_b^2 \tag{16}$$

which support the rough estimates presented in [3].

5. In order to demonstrate the theoretical ambiguities of the analysed approximations of $\Gamma_{Hb\bar{b}}$ and of $R_{Hb\bar{b}}$-ratio we consider their variations due to the variation of the parameter $\Lambda_{\overline{MS}}^{(5)}$ from the used in the previous section value $\Lambda_{\overline{MS}}^{(5)} = 150\ MeV$ (which is in agreement with the existing at present QCD analysis of experimental data for the deep-inelastic processes [21, 22]) to $\Lambda_{\overline{MS}}^{(5)} = 250\ MeV$. This value was proposed as the current most careful world average in the extensive review of the available theoretical and experimental results [21] and has the error bars $\pm 100\ MeV$. The typical example of the behaviour of $R_{Hb\bar{b}}$ for $\Lambda_{\overline{MS}}^{(5)} = 250\ MeV$ is demonstrated at Fig.3 where the RG-improved massive dependence is presented with the similar notations as at Fig.2. The corresponding curves lie over 5% lower of the corresponding curves of Fig.2.

Notice that for the model case of m_b=4.8 GeV and $\Lambda_{\overline{MS}}^{(5)}$=250 MeV the most special feature of the proposed RG-improved expression of $\Gamma_{Hb\bar{b}}$, namely the cancellation of the NLO corrections, demonstrates itself at the scale $M = 65\ GeV$ (compare with the characteristic scale M=130 GeV for $\Lambda_{\overline{MS}}^{(5)}$=150 MeV).

It was noticed in the works of ref.[23],[24] that the characteristic scales of the nullification of the NLO QCD corrections to different correlators can be related to the characteristic scales of the bound states in the corresponding channels.

Moreover, the further studies of the effects of the NNLO QCD corrections for the correlators of vector currents revealed the special feature of the perturbative approximations, namely the asymptotic explosion of the corresponding series at the typical scale of resonances (ρ-meson) (see [25]).

In view of these facts we are proposing the hypothesis that the cancellation of NLO corrections in the discussed above RG-improved parametrization of the heavy quark (b-quark) dependent contributions and the indications to the asymptotic explosion of the corresponding NNLO parametrizations (see Figs.2,3) might be related to the indications of the existence of the certain characteristic mass scale M_{ch}, which considerably depends from the value of the parameter $\Lambda^{(5)}_{\overline{MS}}$ and less significantly from the value of on-shell b-quark mass. In the case of m_b=4.8 GeV the values of this mass scale are

$$M_{ch} \approx 65\ GeV\ for\ \Lambda^{(5)}_{\overline{MS}} = 250\ MeV \tag{17}$$

$$M_{ch} \approx 130\ GeV\ for\ \Lambda^{(5)}_{\overline{MS}} = 150\ MeV \tag{18}$$

The possible meaning of these mass scales is not yet clear to us.

It should be noted that in the region 60 $GeV < M_H <$ 160 GeV all above made conclusions are not affected by the 1-loop electroweak (EW) corrections, calculated in ref.[11]. Indeed, for the analyzed values of a Higgs mass these negative EW corrections are negligible (less then -1%) [11]. They start to play the significant role in the region $M_H >$ 250 GeV [11], not considered in our work. Another source of theoretical uncertainties in the $\Gamma_{Hb\bar{b}}$-expression comes from the higher order EW-corrections (which due to the smallness of the 1-loop EW effects in the analyzed region of M_H-values should be small) and from the mixed EW-QCD contributions. In the more detailed preliminary version of our work [26] we have demonstrated that the part of these corrections, namely the order $O(\alpha\alpha_s)$ and $O(\alpha^2)$ corrections to $\Gamma_{Hb\bar{b}}$, as analytically calculated by us [26], are small. Note, that in our considerations of the order $O(\alpha_s^2)$ QCD massive-dependent corrections to $\Gamma_{Hb\bar{b}}$ we did not take into account the 3-loop light-by-light-type diagrams, analogous to the ones, considered in refs.[27],[28] in the case of the axial contribution to the $Z^0 \to b\bar{b}$ decay, which in the letter case give some additional α_s^2-contributions. In our case we also expect the appearance of the massive corrections of the similar origin. However, we think that these not yet calculated contributions will not change significantly all results and conclusions.

This work was partly supported by the Russian Fund of the Fundamental Research, Grant N 93-02-14428 and is based on the results pre-printed in refs.[26, 29].

References

[1] M. Veltman, *The Higgs System*, Instituut-Lorentz preprint (1989).

[2] J. F. Gunion, H. E. Haber, G. L. Kane and S. Dawson, *The Higgs Hunter's Guide*, Addison-Wesley Publishing Co. (1990).

[3] Z. Kunszt and W. J. Stirling, Proc. ECFA LHC Workshop, Aachen 4-9 October 1990, CERN 90-10, vol.2, p.428;
R. Kleiss, Z. Kunszt and W. J. Stirling, *Phys. Lett.* **B253** (1991) 269.

[4] J.-F. Grivaz, preprint LAL 92-59, Orsay, 1992.

[5] E. Braaten and J. P. Leveille, *Phys. Rev.* **D22** (1980) 715;
M. Drees and K. Hikasa, *Phys. Rev.* **D41** (1990) 1547; *Phys. Lett.* **B240** (1990) 455; ibid. **B262** (1991) 497 (Err.).
A. I. Davydychev and M. N. Dubinin, in Proc. XIV Workshop "Problems on High Energy Physics and Field Theory" (Protvino, Russia, July 1991), Nauka, Moscow, 1992, p.294.

[6] S. G. Gorishny, A. L. Kataev and S. A. Larin, *Sov. J. Nucl. Phys.* **40** (1984) 329.

[7] S. G. Gorishny, A. L. Kataev, S. A. Larin and L. R. Surguladze, *Mod. Phys. Lett.* **A5** (1990) 2703; *Phys. Rev.* **D43** (1991) 1633.

[8] O. V. Tarasov, A. A. Vladimirov and A. Yu. Zharkov, *Phys. Lett.* **B93** (1980) 429.

[9] O. V. Tarasov, Joint Institute for Nuclear Research preprint JINR **P2-82-900** (1982) (unpublished).

[10] D. V. Shirkov, *Nucl. Phys.* **B371** (1992) 467.

[11] D. Yu. Bardin, B. M. Vilensky and P. Kh. Khristov, *Sov. J. Nucl. Phys.* **53** (1991) 152;
A. Dabelstein and W. Hollik, *Z. Phys.* **C53** (1992) 507.
B. A. Kniehl, *Nucl. Phys.* **B376** (1992) 3.

[12] N. Gray, D. J. Broadhurst, W. Grafe and K. Schilcher, *Z. Phys.* **C48** (1990) 673.

[13] N. Sakai, *Phys. Rev.* **D22** (1980) 2220;.
T. Inami and T. Kubota, *Nucl. Phys.* **B179** (1981) 171.

[14] O. V. Tarasov, private communications.

[15] L. R. Surguladze and F. V. Tkachov, *Nucl. Phys.* **B331** (1990) 35.

[16] L. R. Surguladze, Institute for Nuclear Research preprint INR П-0644 (1989), unpublished.

[17] K. G. Chetyrkin, J. H. Kühn and A. Kwiatkowski, *Phys. Lett.* **B282** (1992) 221.
K. G. Chetyrkin, A. Kwiatkowski and J. H. Kühn, talk presented by A. Kwiatkowski at the International Europhysics Conference on HEP, Marseille, June 1993; preprint University Karlsruhe TTP93-15, June 1993.

[18] A. I. Davydychev and J. B. Tausk, *Nucl. Phys.* **B397** (1993) 123.

[19] A. I Davydychev, V. A. Smirnov and J. B. Tausk, Instituut-Lorentz preprint INLO-PUB-5/93 (1993), submitted to Nucl. Phys.

[20] D. J. Broadhurst, J. Fleischer and O. V. Tarasov, *Z. Phys.* **C60** (1993) 287.

[21] G. Altarelli, rapporteur talk at the Conference " QCD-20 Years Later", Aachen, June 1992, CERN preprint CERN-TH.6623/92 (1992).

[22] J. Chýla and A. L. Kataev, *Phys. Lett.* **B297** (1992) 385.

[23] A. L. Kataev, N. V. Krasnikov and A. A. Pivovarov, *Nucl. Phys.* **B198** (1982) 508;
A. A. Pivovarov and E. N. Popov, *Phys. Lett.* **B205** (1988) 79;
K. G. Chetyrkin, S. G. Gorishny, A. L. Kataev, S. A. Larin and F. V. Tkachov, *Phys. Lett.* **116B** (1982) 455;
S. G. Gorishny, A. L. Kataev and S. A. Larin, *Theor. Math. Phys.* **70** (1987) 262.

[24] K. G. Chetyrkin, A. Yu. Ignatiev, V. A. Matveev, M. E. Shaposhnikov and A. N. Tavkhelidze, *Phys. Lett.* **117B** (1982) 251.

[25] J. Chýla, A. L. Kataev and S. A. Larin, *Phys. Lett.* **B267** (1991) 269.

[26] A. L. Kataev and V. T. Kim, Annecy preprint ENSLAPP-A-407/92, November 1992.

[27] B. A. Kniehl and J. H. Kühn, *Phys. Lett.* **B224** (1989) 229; *Nucl. Phys.* **B329** (1990) 547.

[28] K. G. Chetyrkin and A. Kwiatkowski, *Phys. Lett.* **B305** (1993) 285.

[29] A. L. Kataev and V. T. Kim, preprint Institute for Nuclear Research, INR-826/93, Moscow, 1993.

FOURIER ANALYSIS OF THE MULTI-DIMENSIONAL DISTRIBUTION OF EVENTS

Sergei A. SADOVSKY[1]
Institute for High Energy Physics
Protvino, 142284, Moscow Region, Russia

and

Andrei E. SOBOL
Physical Department, Moscow State University,
Vorob'evy Gory, Moscow, 119899, Russia

ABSTRACT

The package for the finite Fourier parametrization of the multi-dimensional distribution of events is described. The parametrization procedure is based on the maximum entropy principle. Few statistics of events distribution, so-called E, χ_1^2 and χ_L^2-statistics, is used to control the convergence of the procedure. The case of the exactly positive density of events as well as the essentially different case of zeroed density in some subregions of the definition region are considered. Two-dimensional parametrizations of Monte-Carlo event distributions are discussed as simplest examples of the package application.

1. Introduction

Analytic representation of event distributions is a topical problem in high energy physics. For instance, it is important for parameters estimation, physical hypotheses testing or distributions smoothing. In the low dimensional space of measurements ($m \leq 2$) the distributions are usually presented by one- and two-dimensional histograms and its analytic forms can be easily found using the well known procedures of fitting, parametrization or smoothing [1,2]. But the difficulties found in such approach increase exponentially with m limiting the method application by small values of m, say $m \leq 3$.

In the present paper the parametrization of multi-dimensional distributions is considered in terms of generalized Fourier series. The parametrization procedure is based on the maximum entropy principle. It is free from disadvantages of the traditional approaches. The mathematical grounds of the procedure are described in details elsewhere [3,4]. Here we shall discuss the main features of this procedure using the Monte-Carlo (MC) detection efficiency as a useful and flexible example.

2. Problem Formulation and Mathematical Grounds of Package

The MC efficiency is characterized by the event set which satisfies all detec-

[1]The author gratefully acknowledges the International Science Foundation in Washington for financial support of his participation in the Third International Workshop on Software Engineering, Artificial Intelligence and Expert Systems for High Energy and Nuclear Physics.

tion conditions of an experiment. We assume that from the set $E_o\{i\}$ of N_o initial MC events[2] the subset $E\{i\}$ $(i = 1,2,...N)$ satisfies all cuts of the experiment, from the trigger and up to kinematical and physical ones. And each event is described by $m-$dimensional vector of measured variables $\mathbf{x}$ defined, for simplicity, on the hypercube $\Delta_m : -1 \leq x_k \leq 1,\ k = 1,2,...m$. We further assume here without loss of generality that initial MC events are distributed uniformly in the hypercube Δ_m.

The detection efficiency ε as a function of vector $\mathbf{x}$, can be presented by generalized Fourier series

$$\varepsilon(\mathbf{x}) = \sum_{\lambda \geq 0} \varepsilon_\lambda f_\lambda(\mathbf{x}) \tag{1}$$

over a full system of the orthonormal functions $f_\lambda(\mathbf{x})$ on the hypercube Δ_m. Estimations of the expansion coefficients $\tilde{\varepsilon}_\lambda$ on the subset $E\{i\}$ of events as well as estimations of its dispersions $\tilde{\sigma}^2_\lambda$ are given [4,5], respectively, by formulas[3]:

$$\tilde{\varepsilon}_\lambda = \frac{1}{N} \sum_i f_\lambda(\mathbf{x}_i), \tag{2}$$

$$\tilde{\sigma}^2_\lambda = \frac{1}{N^2} \sum_i (f_\lambda(x_i) - \tilde{\varepsilon}_\lambda)^2. \tag{3}$$

The other important characteristic of the coefficient ε_λ is the noise component $\tilde{n}_\lambda$ which is defined as the ratio

$$\tilde{n}_\lambda = \tilde{\sigma}_\lambda / \tilde{\varepsilon}_\lambda. \tag{4}$$

If one truncates the sum (1) by the finite number of terms L in the appropriate way (say, by terms with $\tilde{n}_\lambda \leq 1$), then eq.(1-4) practically solves the problem of the efficiency parametrization in comparatively simple cases.

The main reasons, which restrict application of the Fourier parametrization of distributions in multi-dimensional space, are the problems of the terms significance and convergence of the parametrization. Solutions of these problems have been proposed in [3] by introducing few statistics of the events distribution $E\{i\}$ which depend also on the parametrization (1).

The first statistics is the entropy of the parametrization (1) found:

$$\tilde{h}(\varepsilon) = -\frac{1}{N} \sum_i \ln \varepsilon(\mathbf{x}_i). \tag{5}$$

According to the maximum entropy principle [3], the sum (1) should be restricted by such a term set at which the entropy achieves the maximum. If one orders the expansion coefficients ε_λ in the increasing noise component n_λ the entropy becomes practically monotone decreasing function of the terms number L outside some transition region, see below. And thus, the maximum entropy principle requires the sum (1) limiting by the minimal number of n_λ-ordered terms at all other equal conditions.

---

[2] In notations we follow the paper [3].

[3] We can consider that function $\varepsilon(\mathbf{x})$ is normalized to unit, otherwise the scale factor N/N_o restores the true values of ε_λ coefficients.

The second statistics is the so called [3] $E(\varepsilon)$-statistics,

$$E(\varepsilon) = \frac{1}{N}\sum_i \frac{1}{\varepsilon(\mathbf{x}_i)}. \tag{6}$$

At sufficiently large L $E(\varepsilon)$ is an estimation of the volume of subregions in the hypercube, where the efficiency is exactly positive. In the case of everywhere positive efficiency this statistics is the estimation of the hypercube volume V_m itself. Therefore, it can be used to distinguish the last case from the essentially different case of zero efficiency in some subregions in the hypercube[4].

In the case of everywhere positive efficiency the deviation of $E(\varepsilon)$ from the value of V_m being normalized to its own dispersion $D(\varepsilon) = 1/N^2 \sum_i (1/\varepsilon(\mathbf{x}_i) - E(\varepsilon))^2$,

$$\chi_1^2(\varepsilon) = (V_m - E(\varepsilon))^2/D(\varepsilon), \tag{7}$$

can be used as the agreement criteria of the parametrization found and the set of events $E\{i\}$. Together with the above defined $h(\varepsilon)$- and $E(\varepsilon)$-statistics the $\chi_1^2(\varepsilon)$-statistics provides an efficient tool to control Fourier parametrization of the everywhere positive distributions of events.

In the case of zeroed efficiency in subregions of Δ_m an additional analysis is required, and two other statistics should be introduced. First one is the ε_{min}-statistics [3] equal to minimal value of $\varepsilon(\mathbf{x})$ in Δ_m. In the case of exactly positive density of events this statistics should be positive at sufficiently large L, and thus it can be used for the case identification together with $E(\varepsilon)$-statistics. If ε_{min} is negative at large L, then regions with zeroed density in Δ_m should exist. In this case the ε_{min}-statistics characterized the precision ε_o achieved during the efficiency calculation on the bases of the given set of events $E\{i\}$. The value $\tilde{\varepsilon}_o = -\varepsilon_{min}$ can be used then as its empirical estimation. Due to limited number of events in the $E\{i\}$-set we can not distinguish the sufficiently small density from the exactly zeroed one in the same region. Therefore, the value of $\tilde{\varepsilon}_o$ depends on the real density behavior in Δ_m, the event number in the $E\{i\}$-set, and the parametrization found.

The other statistics, $\chi_L^2(\varepsilon)$, is dedicated to replace the $\chi_1^2(\varepsilon)$ in the case under consideration, because the last one becomes useless now. The $\chi_L^2(\varepsilon)$ is defined by

$$\chi_L^2 = \sum_{\lambda=1}^{L} (a_\lambda - E_\lambda(\varepsilon))^2/[D_\lambda(\varepsilon) + d_\lambda(\varepsilon)], \tag{8}$$

where $E_\lambda(\varepsilon) = 1/N \cdot \sum_i f_\lambda(\mathbf{x}_i)/\varepsilon(\mathbf{x}_i)$, $D_\lambda(\varepsilon) = 1/N^2 \sum_i [f_\lambda(\mathbf{x}_i)/\varepsilon(\mathbf{x}_i) - E_\lambda(\varepsilon)]^2$, $a_\lambda = \int_\Delta f_\lambda(\mathbf{x}) d^m\mathbf{x}$ and $d_\lambda(\varepsilon)$ is dispersion of numerical estimation of a_λ-value (the summations and integration are extended here over the region, where $\varepsilon(\mathbf{x}) > \varepsilon_o$).

But the above defined estimation of the ε_o is too crude to be used in quantitative criteria like (8). A more precise estimation can be obtained by variation ε_o near the value $\tilde{\varepsilon}_o$ (with subsequent interpolation) to satisfy the equation

$$E(\varepsilon) = \int_{\varepsilon(\mathbf{x})>\varepsilon_o} d^m x, \tag{9}$$

[4] $S(\varepsilon)$-statistics, which was introduced in [3] to distinguish these cases, is not effective in practice because the events number in $E\{i\}$-set is fixed or can not be varied significantly to study asymptotic behavior of $S(\varepsilon)$.

where for $E(\varepsilon)$ calculation the summation in (6) is extended over the region $\varepsilon(\mathbf{x}) > \varepsilon_o$.

3. Examples of Package Application

The statistics discussed above allow to carry out an efficient analysis of event sets in multi-dimensional spaces. But here we consider the package application for parametrization of MC event sets in two-dimensional space only. These simple examples allow us to compare the method discussed and traditional methods directly and to simplify visualization of distributions. The examples of the package application in three- and four-dimensional spaces can be found elsewhere [3,4]. The maximum dimension of the space, where the package had been successfully applied, is equal to eight [6]. In such high-dimensional space the special methods of the terms number reduction should be used, details see in [3,4].

3.1. Exactly Positive Distribution

We start from the exactly positive distribution. As example we consider here the function defined on the unit sphere S_2 $(-1 \leq \cos\theta \leq 1,\ 0 \leq \phi \leq 2\pi)$:

$$\varepsilon_p(\Omega) = A \cdot [\sin^2\theta \cdot e^{-(2\phi^2 - \cos\theta - 0.5)^2} + 0.1], \tag{10}$$

where A is a normalization constant, fig.1a. Further we fixed the sample $E_p\{i\}$ with 50 000 MC events (see, fig.1b) and employed the package described above for the event analysis. The result is presented in fig.1c. Fourier parametrization was carried out in terms of spherical harmonics: $Y_\lambda(\Omega) = ReY_{lm}(\cos\theta, \phi),\ ImY_{lm}(\cos\theta, \phi)$.

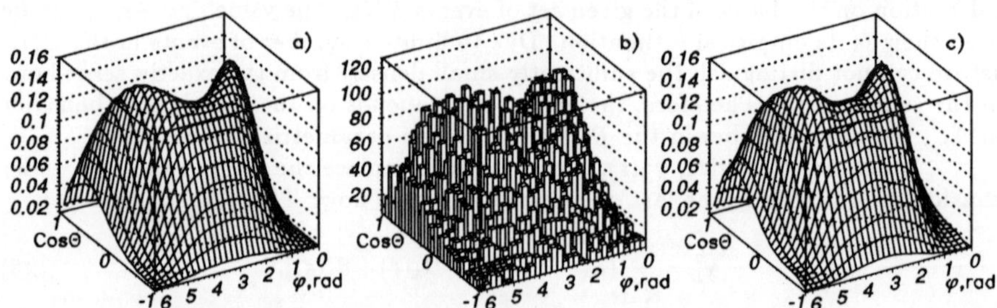

Figure 1. The original density of events $\varepsilon_p(\Omega) = A \cdot [\sin^2\theta \cdot e^{-(2\phi^2 - \cos\theta - 0.5)^2} + 0.1]$ **(a)**; the sample of 50 000 MC events distributed according to the density $\varepsilon_p(\Omega)$ **(b)**; Fourier restoration of the events density by spherical harmonics, $L = 13$ **(c)**.

The procedure of the package application was as follows:

- the maximum order of the coefficients L_{max} was chosen from heuristic point of view (100 in the example discussed);

- the expansion coefficients (2) and their dispersions (3) were evaluated and ordered with the increasing noise (4), fig.2a, that corresponds to monotone decreasing entropy (5) as a function of terms number L (fig.2b) out of the transition region $(L > 6)$;

- the statistics ε_{min}, fig.2b, and ε_o were evaluated as functions of L;
- the entropy, $E(\varepsilon)$ and $\chi_1^2(\varepsilon)$ statistics as well as the corresponding Confidence Level (C.L.) $P(\chi_1^2)$ were calculated over the event set, where $\varepsilon(\Omega_i) > \varepsilon_o$, fig.2b.

Figure 2 allows us to choose the optimal terms number $L = 13$, at which $E(\varepsilon)/4\pi = 0.998$ and $\chi_1^2(\varepsilon) = 0.253$ corresponds to 62% of C.L. It is interesting to compare the last values with traditional ones obtained with help of 2-dimensional histogram with 30×30 bins: $\chi_{900}^2/900 = 1.073$, i.e. 6.3% C.L. For original distribution (10) the same values are equal to: $\chi_{900}^2/900 = 1.014$ and 38% C.L.

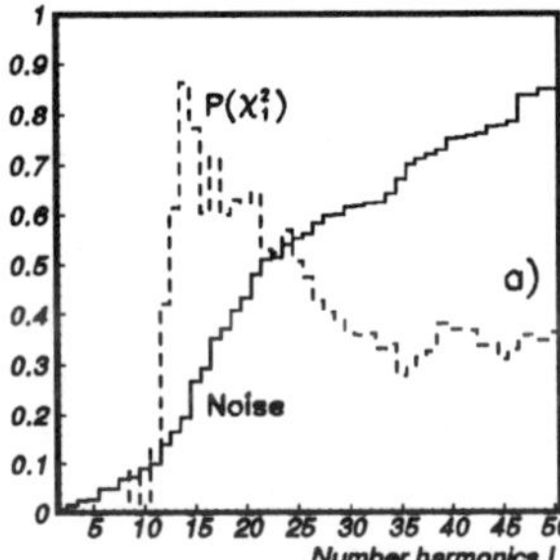

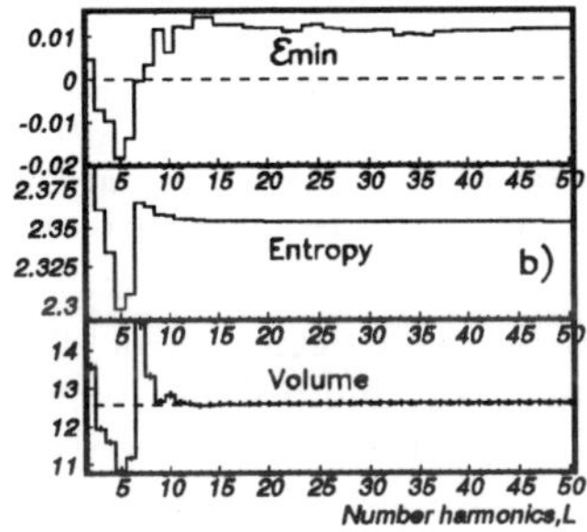

Figure 2. The noise of coefficients (solid line) and the confidence level (dashed line) corresponding to χ_1^2 as function of L (a); ε_{min}, Entropy and $E(\varepsilon)$-statistics (Volume) as L-functions, the dashed line at the bottom shows the value of $V_2 = 4\pi$ (b).

3.2. Distribution Zeroed in Subregions of Δ_m

As an example of zeroed distribution in Δ_2-subregions we consider here the distribution $\varepsilon_o(\Omega)$ which is equal to $A \cdot [\sin^2\theta \cdot e^{-(2\phi^2 - \cos\theta - 0.5)^2} - 0.1]$ if the last function is positive and equal to zero in the opposite case (compare with function (10)). Further on we fixed as before the sample $E_o\{i\}$ with 50 000 MC events and employed the Fourier parametrization package. After n_λ-ordering of the coefficients the L-dependence of ε_{min}-, $E(\varepsilon)$- and $h(\varepsilon)$-statistics are shown in fig.3(a,b). First it is clearly seen, that $E(\varepsilon)$ is at least 20% below $V_2 = 4\pi$, fig.3b, and so there should be subregions in S_2 where the distribution density is negligible. And second, the transition region continues up to $L \simeq 14$, because all statistics show strong variation at smaller values of L. Furthermore, the entropy becomes practically monotone decreasing function of L outside this region. And we conclude, that suitable values of L should be greater than 14.

The variation of $P(\chi_L^2)$ with the increasing L confirms our conclusion, see fig.3b. This statistics allows to fix the region $19 \div 21$ as one of optimal values of L. We choose $L = 20$ at which $P(\chi_L^2) = 62\%$, finally. The distribution obtained is shown in fig.3c. Traditional method based on 2-dimensional histogram with 30×30 bins gives 22% C.L. For original distribution the same value is equal to 65%.

4. Conclusion

The main idea of the method [3] realized in the package discussed above is to

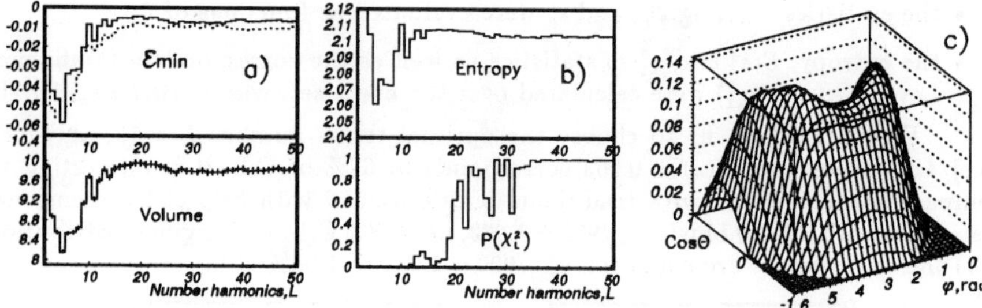

Figure 3. Statistics for Fourier parametrization of 50 000 $E_o\{i\}$-events distributed according to the probability density $\varepsilon_o(\Omega)$ (see text): **a)** ε_{min} (solid line) and $-\varepsilon_o$ (dashed lined) at the top, E_2-statistics at the bottom; **b)** at the top the Entropy and at the bottom the confidence level $P(\chi^2_L)$ corresponding to χ^2_L. **c)** The restoration of the events distribution density by 20 spherical harmonics.

propose statistics which can control effectively convergence of the parametrization without using of any multi-dimensional distributions of events. The computation procedure does not use matrix operations, and memory requirement increases linearly with the term number in expansion (1). The package uses mainly computer operations which can be easily vectorized. A freedom in choosing orthonormal functions can be used further for significant reduction of the terms number, see [3,4]. Calculation of orthonormal functions can be followed by a recourse method.

All listed features make the package application effective for the event distribution analysis in multi-dimensional spaces.

In conclusion the authors would like to thank Yu.D.Prokoshkin for fruitful discussions.

References

1. R. Brun and D. Lienart, *HBOOK User Guide*, CERN Program Library, 1987.
2. R. Brun, M. Hansroul, H. Wind, *MUDIFI, Preprint CERN/DD/EE 80-1*, CERN Program Library, E999, 1985.
3. S. A. Sadovsky, A. A. Kondashov, V. D. Samoylenko, *Proceedings of XXVI International Conference of High Energy Physics*, (August 6-12, Dallas, 1992), p. 1791, AIP Conference Proceedings NO. 272.
4. A. A. Kondashov, S. A. Sadovsky, and V. D. Samoylenko, *Preprint IHEP 91-167*, Protvino, 1991.
5. W. T. Eadie et al. *Statistical Methods in Experimental Physics*, (North-Holland, Amsterdam-London, 1971), p. 40.
6. V. D. Samoylenko, *Private communication.*

GSL: A PORTABLE STANDARD LISP INTERPRETER

A. M. RAPORTIRENKO
Laboratory of Computing Techniques and Automation
Joint Institute for Nuclear Research, 141980 Dubna, Russia

ABSTRACT

This paper describes the first release of the *GNU* Standard Lisp interpreter *(GSL)*. This dialect of *LISP* language has been created as a part of the large computer algebra system *REDUCE*. The interpreter allows dynamically load compiled *LISP* programs and programs written in other high level languages (such as *C*, *FORTRAN* etc.). That is very useful to mix the numeric and symbolic calculations. *GSL* is free software. The interpreter is written in *C*, and uses some *GNU* library packages.

Introduction

The dialect of *LISP* language, Standard Lisp has been created as a part of the large computer algebra system *REDUCE*[3]. Presently it is possible to say that this dialect of *LISP* implements the minimal set of functions which is big enough for effective programming of computer algebra problems.

At present, there are two portable implementations of the Standard Lisp interpreter — *PSL*[10] and *CSL*[11]. The necessity in developing of the new portable interpreter *GSL* was caused, in general, by the following factors:

- to have not commercial, portable, easily extendable *LISP* interpreter which implement the Standard Lisp dialect of the *LISP* language;
- to have abilities for dynamic loading not only compiled *LISP* modules, but also modules written in languages such as *C*, *FORTRAN*, C^{++} etc.;
- to achieve high efficiency in the algebraic objects transformations by their including in the basic set of the interpreter objects.

In the interpreter developing only software of the Free Software Foundation was used. The interpreter written in *C* with use the extensions of the *GNU C*[4] compiler.

1. The basic set of the interpreter objects

Before start the description of the basic set of the interpreter's data types it is necessary to mention the pointer structure.

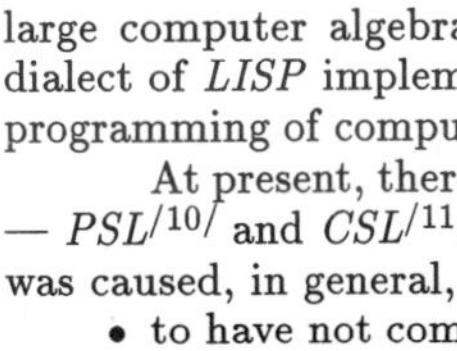

g	type	addr

addr_size (width of the addr field); *sizeof(long)* (width of the whole pointer)

The *GSL* pointer occupies one lisp cell (the machine word) and contains two fields—the type field *type* and the address field *addr*. The width of the address field *addr_size* is not fixed and may be changed before compilation of the interpreter. The type field contains code corresponding to the type of the lisp object address field point to. The higher bit *g* of the pointer is used by the garbage collector to mark lisp object.

The basic data set of the interpreter are formed by[1]: *integers, floatings, dotted pairs, identifiers, strings, vectors and codes.*

Below the brief description of the objects corresponding to the basic data types is given.

1.1. Integers

The representation of an integer depends on its value and may be as follows:

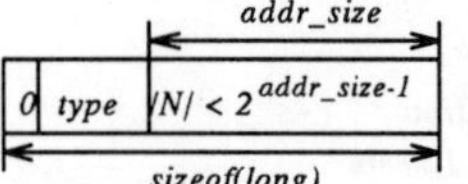

If the absolute value of the integer is less than 2^{addr_size-1}, it is placed directly to the address field of the pointer, as shown in figure. The *type* field of the pointer contains a code which corresponds to such a "small integer". Standard twos complement representation of the signed value is used. If the absolute value of the integer is not less than 2^{addr_size-1} and is less than $2^{sizeof(longlong)-2}$, then it is represented as a longlong integer and occupies two 32–bit words.

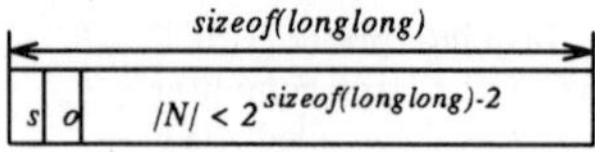

Standard twos complement representation of the signed value is also used in this case. The bit s is the sign bit, and the bit o is used to fix overflow conditions in the arithmetic operations.

To represent the integer N, the absolute value of which is not less than $2^{sizeof(longlong)-2}$, the decomposition over base $b = 2^{sizeof(long)}$, $N = \sum_{i=0}^{n} c_i b^i, c_i < b$ is used. The structure of the arbitrary precision integer is shown in figure. The unsigned value of N is represented as array of coefficients c_i. The number of machine words which are allocated for coefficients is placed into the field *alloc*. The absolute value of the field *size* is the number of 32–bit machine words which are used to store coefficients. The sign of N is the same as the sign of the value in the field *size*. The arbitrary precision integer arithmetic in *GSL* is based on the *GNU* multiple precision arithmetic library *gmp*.[9]

1.2. Floatings

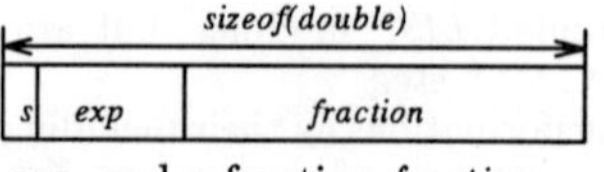

Floating–point numbers in *GSL* are represented as double precision values of the *C* language, and in *IEEE* format have three stored components: a sign s, an biased exponent *exp*, and a fraction *fraction*.

1.3. Identifiers

An identifier in *GSL* is constructed from four pointers, as shown in figure. The field $addr_{name}$ contains the address of the string which represents the print name of the identifier. The pair of fields $type_v$ and $addr_{value}$ forms the pointer to the value of the identifier. The pointer to the identifier property list is constructed from fields $type_p$ and $addr_{plist}$. It points to *nil* if the identifier does not have properties and flags, or points to a list of the form: $(flag_1 \cdots flag_k\ (indicator_1.property_1) \cdots (indicator_l.property_l))$. In this case the identifier has been flagged with flags $flag_1 \cdots flag_k$ and it has properties $property_1 \cdots property_l$ with indicators $indicator_1 \cdots indicator_l$, receptively. The field $type_a$ contains the code which determines the identifier usage. Its value corresponds to one of the following cases:

global – identifier has been declared as *global*. It may not appear as a parameter in lambda expressions or as *prog* form variable;

fluid – identifier has been declared as *fluid*. It may appear as a parameter in lambda expressions or as a *prog* form variable. In this case its previous value is temporarily saved. The parameter value or *nil* is bound to the identifier. After exit, its previous value is restored;
void – identifier does not have value. If it is used as a parameter in lambda expressions or as a *prog* form variable, then it temporary declared as fluid. After exit, its value become undefined.
expr, fexpr, macro – identifier has been defined as a function of type *expr*, *fexpr* or *macro*. The field $addr_{value}$ points to the lambda expression.
subr, fsubr, msubr – identifier has been defined as the function of type *subr*, *fsubr* or *msubr*. The field $addr_{value}$ points to the function–pointer. The functions of this types are a compiled analog of functions with type *expr*, *fexpr* or *macro*, receptively.

If identifier is entered to the *oblist*, it belongs to one of the 512 sublists called "buckets". The identifiers in a "bucket" are connected through their $addr_{next}$ fields.

Bit t is used to mark the function which can be traced.

1.4. Strings

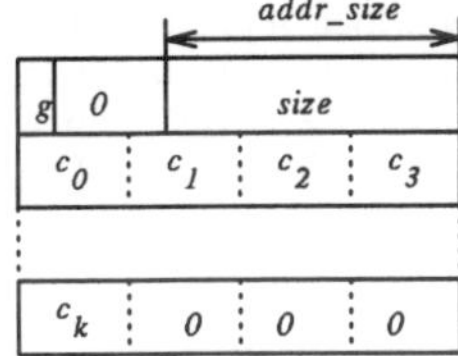

A string is a lisp object which is used mainly to represent the print names of the identifiers and file names of the streams. It consist of auxiliary field *size* and the zero terminated character array with characters c_i. The string size in *GSL* is not limited, and the number of characters in the string are placed to the field *size*. Bit g is also used by the garbage collector.

1.5. Vectors

A vector is a complex structure in which an integer index is used to access random values in the vector. It contains two auxiliary fields *alloc*, *size* and array of pointers. The field *alloc* contains the initial vector size. The current size of the vector is placed to the field *size*. When vector has been created, both fields *alloc* and *size* contain the same value. While executing it is possible to decrease value in the field *size*. Then the released cells become accessible after the garbage collection. Each vector component is formed by the pair of fields $type_i$, $addr_i$. Bits g are used to mark vector and its components.

1.6. Function–pointers

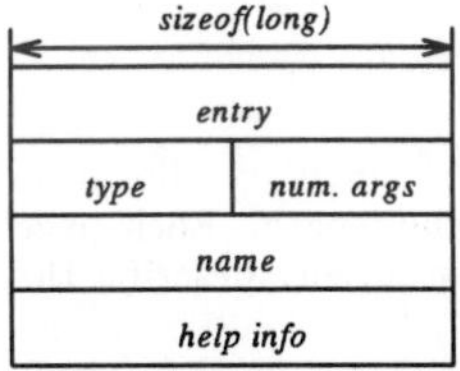

A function–pointer is created for each function of the interpreter and for each compiled lisp function. It contains information which is necessary to provide correct access to the function. The field *entry* contains entry point address of the function. The code of the function type *(subr*, *fsubr* or *msubr)* is placed into field *type*. The field *num.args* contains number of arguments of the function. Cells *name* and *help info* point to the *C*–language strings. The former represents the function–pointer name†, and the latter contains short help info for this function.

---

† In general function–pointer has the same name as identifier to which function is bounded.

1.7. Dotted–pairs

		addr_size
g	$type_a$	$addr_a$
g	$type_d$	$addr_d$

A dotted–pair is represented as a pair of pointers. It forms its *car* and *cdr* portions. The *car* portion contains fields $type_a, addr_a$, and the *cdr* portion — $type_d, addr_d$. Bits g are used by the garbage collector to mark the dotted–pair itself as well as its components.

2. The auxiliary objects

In the previous section the basic set of the interpreter objects has been briefly described. In this section we describe another set of objects which are specific ones for internal use in *GSL*. At present time this set are formed by: *streams* and *bit–scales*.

2.1. Streams

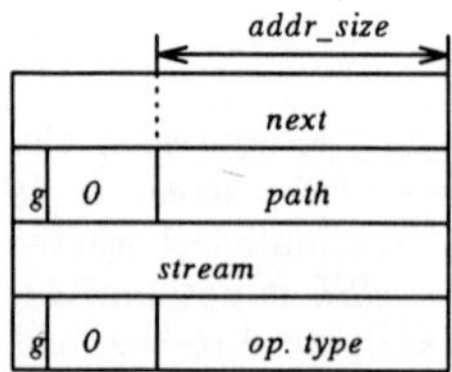

This object is created by the *LISP* function *open*. The stream is constructed from four 32-bit machine words, as shown in figure. The field *path* contains address of the string which represents the file name. The value returned by *C* function *fopen* is placed in the filed *stream*. The operation type code (read or write), which is performed on the file, is contained in the field *op.type*. The field *next* is used to create a chain of the currently selected files. The chain mechanism is used when the *LISP* function *rds (wrs)* is called, the address of the stream of the previous selected *read (write)* file is placed to the field *next* of the currently selected file. When stream in the head of the chain is closed, it will be removed from the chain and *read (write)* operation will be performed for the next file of the chain.

2.2. Bit–scales

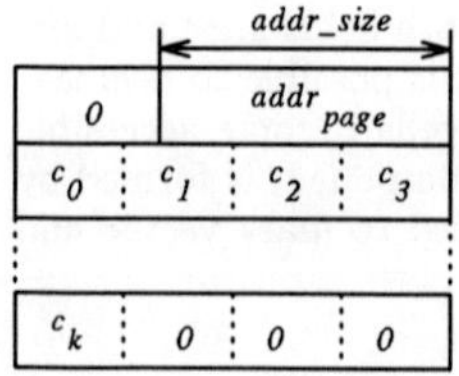

The bit–scale is created for each memory page which contains objects without the garbage collector bit $g^{\dagger}$. It contains the field $addr_{page}$ and the character array. The address of the memory page is placed to the field $addr_{page}$. Each bit of the character c_i is the garbage collector bit for some object of the memory page.

3. Memory allocation

Memory allocation in *GSL* is based on the *GNU Emacs* [5, 6] memory allocation strategy. The memory for the objects is allocated by pages of some length. Each page contains objects of the same type. In the following subsections we briefly describe the memory allocation algorithms.

3.1. Memory allocation for fixed length objects

† At present, *floatings* and longlong *integers* belong to such objects.

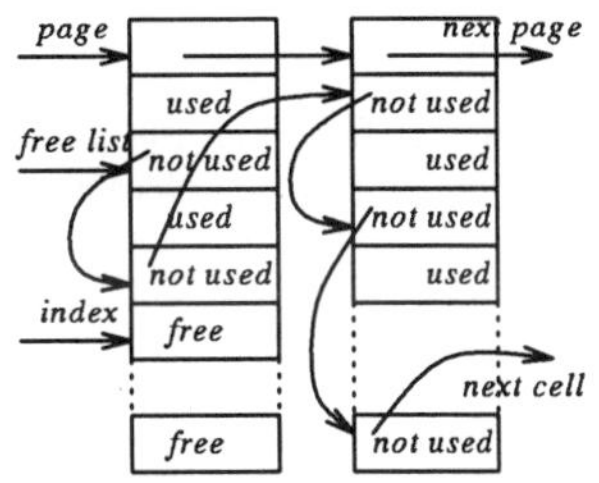

The memory page of the fixed length objects† is represented as an array, each element of which is an object of a given type. All pages with the objects of the same type are linked together in the list *page*, as shown in figure. Unused objects of the same type from all pages also are linked in the list *free list*. The first page of *page* may be partially filled. In this case the variable *index* points to the first free cell of the page. When the request to create new object is issued, it is made an attempt to receive an object from *free list*. If it is empty, the value of the variable *index* is analyzed. If the first page is not filled, the object is received from it. If no free objects of the given type have been found, then the request to allocate new page is issued. The allocated page is appended to the list of pages and the object is received from it.

3.2. Memory allocation for variable length objects

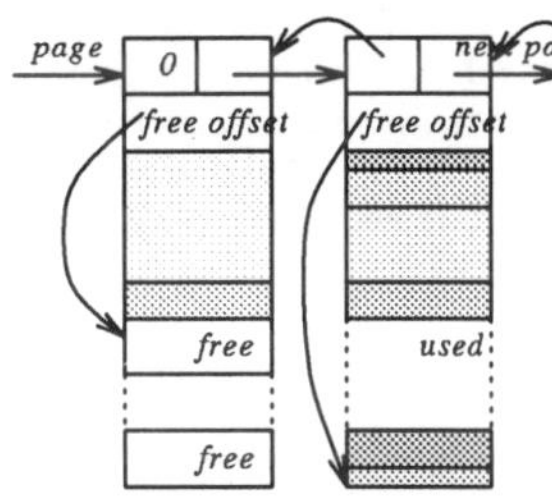

All pages with the variable length objects‡ of the same type are linked together in the double linked list *page*, as shown in figure. The cell *free offset* points to the first unused byte of the page. When request to create new object has received, the value of the variable *free offset* of the first page is analyzed. If the page has enough space for the new object, it is resided on it. If not, then the request to allocate new page is issued. The allocated page is appended to the list of pages and the object is created on the new page.

3.3. Memory allocation for big variable length objects

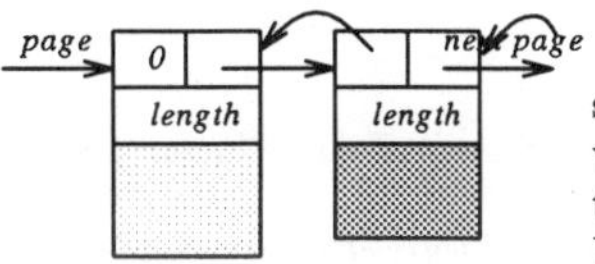

The variable length objects whose size is greater than page size belong to the big variable length objects. All pages with big variable length objects of the same type are linked together in the double linked list *page*, as shown in figure. Each page contains only one element, and the field *length* contains its size. If new object has created, its page is appended to the list of pages.

4. Garbage collection

The garbage collector of the *GSL* is based on the garbage collector of the *GNU Emacs*/5, 6/. When the interpreted function is evaluated, the garbage collector may be invoked only from *eval* and *apply* functions. The value of the global variable *gc-cons-threshold* gives amount of the memory which may be allocated between two calls to the garbage collector.

The first pass marks all the used objects. After that, all unused fixed length objects are linked together into corresponding free lists. All used variable length objects

† To the fixed length objects belong: longlong *integers*, *floatings*, *dotted-pairs*, *identifiers*, *function-pointers* and *bit-scales*.

‡ To the variable length objects belong: big *integers*, *strings* and *vectors*, the size of each is not greater then page size

are moved to use space of the unmarked objects. All cleared memory pages are returned to the system. All pages with unused a big variable length objects are returned to the system. Moreover, if big object has been resized, and its current size is less than the page size, it is moved to the variable length objects of the same type and its page is released.

5. Compilation and dynamic loading

The compilation of the *LISP* programs is provided in several steps. At the first step the source module *file.sl* is processed by the program *sl2lap*, which forms two files *file.l* and *file.lap*. The file *file.l* contains *LISP* expressions which can not be compiled. The *lap* codes of the *LISP* compiler[2] are written to the file *file.lap*. At the next step the program *lap2c* generates *C* programs and writes it to the files *file.c* and *file.h*. To compile *file.c* *GNU C*[4] compiler is to be used. The compiled *LISP* modules may be later combined into library files. The dynamic loader of the *GSL* is based on the *GNU* library package *dld*.[7, 8] The named relocatable object or the library file may be loaded into memory. But there is a difference in its loading. Specifically, if the named file is a relocatable object file, it is completely loaded into the memory. If it is a library file, then only modules defining an unresolved external reference are loaded. The set of standard libraries can be predefined at the dynamic loading stage to resolve externals references. The *unexec*[5, 6] facility is allowed to dump *GSL's* memory into executable file. It is especially useful for generation of large systems, such as *REDUCE*[3].

Acknowledgments

This paper could not be done without the *GNU* software has become available to the public due to Richard Stallman and others. I would like to acknowledge the encouragement and assistance of Vladimir Gerdt.

References

1. J. B. Marti, A. C. Hearn, M. L. Griss, C. Griss, *The Standard Lisp Report, SIGPLAN Notices, ACM*, 14, (10), 48-68, (1979).
2. M. L. Griss, A. C. Hearn, *A Portable LISP Compiler, Software–Practice and experience*, 11, (6), 541-605, (1981).
3. A. C. Hearn, *REDUCE User's Manual, Version 3.4*, RAND, Santa Monica, 1991.
4. R. M. Stallman, *Using and Porting GNU CC*, Free Software Foundation, Cambridge, 1993.
5. R. M. Stallman, *GNU Emacs manual*, Free Software Foundation, Cambridge, 1993.
6. B. Lewis, D. LaLiberte, *GNU Emacs Lisp Reference Manual*, Free Software Foundation, Cambridge, 1993.
7. W. Wilson Ho, *A Dynamic Link/Unlink Editor*, Free Software Foundation, Cambridge, 1991.
8. W. Wilson Ho, R. A. Olsson, *An Approach to Genuine Dynamic Linking, Software–Practice and experience*, 21, (4), 375-390, (1991).
9. T. Granlund, *The GNU Multiple Precision Arithmetic Library*, Free Software Foundation, Cambridge, 1993.
10. M. L. Griss, E. Bensen, G. Q. Maguire Jnr., *PSL: A Portable LISP System*, Proc. ACM Sym. Lisp and Functional Programming, 88-97, (1982).
11. A. C. Norman, *Codemist Standard Lisp (CSL) technical overview and details*, Codemist Ltd., 1991.

Symbolic approach to perturbed non-linear dynamical systems

Bruno Autin

CERN, 1211 Geneva 23 (Switzerland)

Abstract

A dynamical system is, in its most general form, described by a system of six coupled first order differential equations which cannot be solved rigorously. An approximate solution is nevertheless possible and fruitful using a perturbation method which can be decomposed in two steps: the conversion of the original system into sub-systems characteristic of the order at which the perturbation is treated then the recursive integration of the various sub-systems. The method of time-dependent perturbations has been adopted.

Introduction

Dynamical systems have been among the first topics to be treated by symbolic methods when the description of the motion of the moon by Delaunay in the XIX century has been checked, and found incorrect, with the first version of Macsyma. The problems of celestial mechanics can be transposed to particle accelerators [1] and especially to storage rings where the forces are no longer gravitational but electro-magnetic. More important than the difference in the nature of the forces is the difference in the time scales of the revolution times. The earth takes one year to rotate around the sun whereas a particle completes one turn of machine in a time of the order of 1 to 100 μ-seconds so that an instability of the motion may show up much faster in terrestrial machines than in cosmic systems. Dramatic disruptions appear for instance when a collider is tuned near the so-called beam-beam limit in which the non-linear perturbation of one beam is due to the self field of the opposite beam. On the other hand, relatively slow and thus more controlable blow-up's of the beam envelope occur when the non linearities have their origin in the guiding fields of the electro-magnets. The first type of instability is considered as a fundamental limitation, the only way to cope with it is to sacrifice the luminosity of a collider by operating the machine in a regime where the build-up time of the instability is long enough. The second type of instability can, to some extent, be prevented by a suitable arrangement of the non-linear fields so that they compensate each other globally. The rules of compensation can only be drawn from an analytical theory, most often a first order perturbation theory, which leads to a straightforward integration. However, with the advent of new machines, such as very bright synchrotron light sources or superconducting colliders, higher order perturbations become necessary. The integration, albeit possible, is much more delicate than for the first order theory because it has to include all the correlations between the fields. The core of the present study is devoted to the generation of the sub-systems associated with each iteration in a time-dependent perturbation theory. The solution to the systems is given in the form of integrals which are not evaluated.

Perturbation of a differential system

The mechanism of the perturbation technique will first be explained for a single differential equation then generalized to a set of differential equations. Let us consider the first order equation which describes the evolution of the action J of a dynamical system with respect to a variable s:

$$dJ \,/\, ds = f(J)$$

In its ground state, the system is submitted to a certain field of forces F such that the action is a constant of the motion and has the value J_0. The right hand side of the equation is then zero. Under the effect of new fields assumed to be small with respect to F, the action J_0 is modified and can be expanded in the form

$$J = J_0 + J_1 + J_2 + \ \ldots$$

where J_1, J_2, ... are the first, second, ... order perturbations of J. The iterative solution of the differential equation at the order n is obtained by substituting to J at the right hand side its expression at the order n-1. J_1 is thus obtained by integrating

$$dJ_1 \,/\, ds = f\left(J_0\right)$$

and the integral form is

$$J_1(L) = \int_0^L J_0(s)\, ds$$

At the second order, the differential equation becomes

$$d\left(J_1 + J_2\right) \,/\, ds = f\left(J_0 + J_1\right)$$

which is reduced to

$$dJ_2 \,/\, ds = \left(df \,/\, dJ\right)_0 \, J_1(s)$$

The right hand side is known since the derivative of f with respect to J is taken for J_0 and J_1 is the result of the previous iteration. The solution to the equation is thus

$$J_2(L) = \int_0^L \left(dF \,/\, dJ\right)_0 \, J_1(s)\, ds$$

There is at this point a confusion to be avoided in the notations. J_1 is an integral and the previous result is therefore a double integral whose two variables need be distinguished. The detailed form of the solution will therefore be written

$$J_2(L) = \int_0^L \left(dF / dJ\right)_0 ds_1 \int_0^{s_1} J_1\left(s_0\right) ds_0$$

At the third order, the need for a truncation of the right hand side appears so that the full expansion

$$d\left(J_1 + J_2 + J_3\right) / ds = f\left(J_0 + J_1 + J_2\right) =$$
$$f\left(J_0\right) + \left(df / dJ\right)_0 \left(J_1 + J_2\right) + \left(1 / 2\right)\left(d^2 f / dJ^2\right)_0 \left(J_1 + J_2\right)^2$$

is reduced to

$$d\left(J_3\right) / ds = \left(df / dJ\right)_0 J_2 + \left(1 / 2\right)\left(d^2 f / dJ^2\right)_0 J_1^2$$

since the terms in $J_1 J_2$ and J_2^2 are of third and fourth order respectively and the integration gives:

$$J_3(L) = \int_0^L \left(\left(dF / dJ\right)_0 J_2(s) + \left(1 / 2\right)\left(d^2 F / dJ^2\right)_0 J_1(s)^2\right) ds$$

The first contribution is a triple integral and the second one a double integral whose detailed forms are

$$J_3(L) = \int_0^L \left(dF / dJ\right)_0 ds_2 \int_0^{s_2} \left(dF / dJ\right)_0 ds_1 \int_0^{s_1} J_1\left(s_0\right) ds_0 +$$
$$\left(1 / 2\right) \int_0^L \left(d^2 F / dJ^2\right)_0 ds_2 \left(\int_0^{s_2} J_1\left(s_0\right) ds_0\right)^2$$

The multiple integrals are the signature of the correlations between the perturbing fields for which a detailed discussion at the second order is given in [2]. Once the perturbation has been illustrated for a single differential equation, its extension to a system of differential equations is done by replacing the scalar J by a vector **J**.

Symbolic treatment of the perturbation

As soon as the calculation deals with a high order of iteration, the method becomes untractable by hand and one has to resort to a symbolic technique which is composed of two parts: the transformation of the exact but unsolvable system into an approximate solvable system at any order and the integration of the approximate systems starting at the lowest order. The integration is outside the scope of this paper but the perturbation calculation is organized to be suitable for a further integration.

As it has been seen in the previous section , the initial system is split into as many sub-systems as there are iterations, say n. The program thus produces a list of the right hand sides of the sub-systems. If the initial set contains a single equation, the list is a n-vector, if it is made of m equations, the output list is a (n*m)-matrix. The integration of these lists provide J_1, J_2, ... or $J_1, J_2, \ldots$

The code is written with *Mathematica* and sets the perturbation in a recursive form as it has been presented so that the expression of the n-th iteration contains the non evaluated results of the n-1 previous iterations. Each approximation of J is denoted J[i] where i is the iteration number. The evaluation is made by means of substitution rules collected in a single list:

$$\{J[1]- > rhs1, J[2]- > rhs2, \cdots\}$$

The right hand sides result from a Taylor expansion of the initial right hand side to the order n of the iteration and the expansion is truncated to eliminate spurious higher order terms. The recognition of the order of a special term uses the argument of J and the power of the term in the series expansion; if the sum of the argument and of the exponent exceeds n, the term is rejected.

At the iteration n+1, the name s of the variable is updated to sn so that s is converted to the sequence of values (s0, s1, s2, ...) in the same way as J is split into (J[1], J[2], ...). The purpose of this re-labeling of the variable is to obtain expressions ready for multiple integrations.

Description of the code

The program of perturbation of differential equations is part of a package of non linear dynamics called NLDynamics. Its name is ODEPerturbation and it is called with the command

ODEPerturbation[{equations},iteration]

The "equations" are logical equalities of the type

$$J'[s] == f[s, J]$$

where "'" means the derivative of J with respect to s. "iteration" is an integer denoting the wanted iteration. This syntax is remarkably simple as compared with the *Mathematica* command for differential equations

DSolve[{equations},{functions},variable]

It automatically detects the variable and the functions by deciphering the left hand sides. The perturbation of a single differential equation is thus given by

$$\textbf{ODEPerturbation}\left[\{j'[s] = f[s,j]\},2\right] = \{f(s0,j0), j(1)f^{(0,1)}(s1,j0)\}$$

The result is a list of two elements which are the result j(1) and j(2) of the two iterations

$$f(s0, j0)$$

and

$$j(1)\, f^{(0,1)}(s1,j0)$$

As f acts on two variables s and j, $f^{(0,1)}$ is the *Mathematica* notation for the partial derivative of f with respect to the second variable j. The variable s is labled s0 at the first iteration so that, after substitution of j(1), there is no ambiguity to integrate j(2) first with respect first to s1 from 0 to s0 then to s0 from 0 to some fixed limit say L.

Application to a dynamical system

A dynamical system is in the most general case described in the so called *phase-space* by three actions J_x, J_y, J_z and three angles ϕ_x, ϕ_y, ϕ_z corresponding to the coordinates (x, y, z) of the real space. The system of differential equations is derived from the potential V of the non-linear field through the Hamilton's equations [3]:

$$dJ \,/\, ds = -\partial V \,/\, \partial\phi$$
$$d\phi \,/\, ds = \partial V \,/\, \partial J$$

In particle accelerators, the functions of the right hand sides can be decomposed into a product of three functions which depend either on the action or on the angles or even on the variable only. One of the most crucial points is the determination of the transverse motion to determine as a basic parameter as the aperture of the vacuum chamber. The system to solve is then made of four coupled equations

$$dJ_x \,/\, ds = f_1(s)\, g_1\left(s, J_x, J_y\right) h_1\left(s, \phi_x, \phi_y\right)$$
$$dJ_y \,/\, ds = f_2(s)\, g_2\left(s, J_x, J_y\right) h_2\left(s, \phi_x, \phi_y\right)$$
$$d\phi_x \,/\, ds = g_3\left(s, J_x, J_y\right) h_3\left(s, \phi_x, \phi_y\right)$$
$$d\phi_y \,/\, ds = g_4\left(s, J_x, J_y\right) h_4\left(s, \phi_x, \phi_y\right)$$

If the entire system is called *sys*, its perturbative solution at the second order is given by

ODEPerturbation[sys,2]

which returns:

$$
\begin{aligned}
&\{\{f1(s0)\,g1(s0,j_{x0},j_{y0})\,h1(s0,\phi_{x0},\phi_{y0}),\\
&\quad f1(s1)\,h1(s1,\phi_{x0},\phi_{y0})\left(j_y(1)\,g1^{(0,0,1)}(s1,j_{x0},j_{y0})+j_x(1)\,g1^{(0,1,0)}(s1,j_{x0},j_{y0}\right)+\\
&\quad f1(s1)\,g1(s1,j_{x0},j_{y0})\left(\phi_y(1)\,h1^{(0,0,1)}(s1,\phi_{x0},\phi_{y0})+\phi_x(1)\,h1^{(0,1,0)}(s1,\phi_{x0},\phi_{y0})\right)\},\\
&\{f2(s0)\,g2(s0,j_{x0},j_{y0})\,h2(s0,\phi_{x0},\phi_{y0}),\\
&\quad f2(s1)\,h2(s1,\phi_{x0},\phi_{y0})\left(j_y(1)\,g2^{(0,0,1)}(s1,j_{x0},j_{y0})+j_x(1)\,g2^{(0,1,0)}(s1,j_{x0},j_{y0})\right)+\\
&\quad f2(s1)\,g2(s1,x0,j_{y0})\left(\phi_y(1)\,h2^{(0,0,1)}(s1,\phi_{x0},\phi_{y0})+\phi_x(1)\,h2^{(0,1,0)}(s1,\phi_{x0},\phi_{y0})\right)\},\\
&\{g3(s0,j_{x0},j_{y0})\,h3(s0,\phi_{x0},\phi_{y0}),\\
&\quad h3(s1,\phi_{x0},\phi_{y0})\left(j_y(1)\,g3^{(0,0,1)}(s1,x0,j_{y0})+j_x(1)\,g3^{(0,1,0)}(s1,j_{x0},j_{y0})\right)+\\
&\quad g3(s1,j_{x0},j_{y0})\left(\phi_y(1)\,h3^{(0,0,1)}(s1,\phi_{x0},\phi_{y0})+\phi_x(1)\,h3^{(0,1,0)}(s1,\phi_{x0},\phi_{y0})\right)\},\\
&\{g4(s0,j_{x0},j_{y0})\,h4(s0,\phi_{x0},\phi_{y0}),\\
&\quad h4(s1,\phi_{x0},\phi_{y0})\left(j_y(1)\,g4^{(0,0,1)}(s1,j_{x0},j_{y0})+j_x(1)\,g4^{(0,1,0)}(s1,x0,j_{y0})\right)+\\
&\quad g4(s1,j_{x0},j_{y0})\left(\phi_y(1)\,h4^{(0,0,1)}(s1,\phi_{x0},\phi_{y0})+\phi_x(1)\,h4^{(0,1,0)}(s1,\phi_{x0},\phi_{y0})\right)\}\}
\end{aligned}
$$

At this level, it is easy to check the result at a glance but, as soon as the order increases, the number of terms grows in an an inflationary way and the interest of a symbolic code becomes obvious.

Acknowledgments

Fabrice Ajalbert from the University of Clermont-Ferrand made a major contribution to the development of the symbolic code.

References

[1] *Nonlinear Dynamics Aspects of Particle Accelerators*, Edited by J.M. Jowett, M. Month and S. Turner, Lecture Notes in Physics, 247, Springer-Verlag (1985)

[2] B. Autin, *Non-linear Betatron Oscillations*, CERN 90-04 (1990)

[3] H. Goldstein, *Classical Mechanics*. Addison Wesley (1980)

LIST OF PARTICIPANTS

AGTERHUIS Eric
Dep. of Physics and Astronomy
Utrecht University
P.O. Box 80.000
NL-3508 TA Utrecht
NETHERLANDS
agter@fys.ruu.nl

ANDREE H.M.A.
Inst. for Theor. Physics
Rijksuniv. Utrecht
P.O.Box 80006
Princetonplein 5
NL-3508 TA Utrecht
NETHERLANDS
taal@fys.ruu.nl

AUTIN Bruno
CERN
European Organization
for Nuclear Research
PS-Division
CH-1211 Genève 23
SWITZERLAND
autin@cernvm.cern.ch

BALKE A.C.
Dep. of Physics and Astronomy
Utrecht University
Post Box 80.000
NL-3508 TA Utrecht
NETHERLANDS

BARILLERE Renaud
CERN ECP-PT
CH-1211 Geneva 23
SWITZERLAND
renaud_barillere@macmail.cern.ch

BECKS Karl-Heinz
Physik Dept.
Univ. of Wuppertal
Gauss Str. 20
D-42097 Wuppertal
GERMANY
becks@physik.uni-wuppertal.d400.de

BEHRENS Ulf
ZEUS Collaboration
Deutsches Elektronen-Synchrotron
Notkestr. 85
D-22603 Hamburg
GERMANY
behrens@vxdesy.desy.de

BENHAR Omar
INFN
Sezione Santá
Viale Regina Elena, 299
I-00161 Rome
ITALY
VAXSAN::BENHAR
BENHAR@ISS.INFN.IT

BERGMANN Andreas
Department of Informatics
Rheinische Friedr. Wilhelms
Univ. Bonn
Römerstr. 164
D-53117 Bonn
GERMANY
andreas@uran.informatik.uni-bonn.de

BLOM Jo
Physics Department
Rijksuniv. Utrecht
P.O.Box 80000
Princetonplein 5
NL-3508 TA Utrecht
NETHERLANDS
blom @fys.ruu.nl

BORMANN Christian
Inst. f. Kernphysik
Johann-Wolfgang-Goethe Univ.
August Euler-Str. 6
D-60486 Frankfurt am Main
GERMANY
bormann@c3000.ikf.physik.uni-frankfu
rt.de

BOTELLA Bernard
Dassault Electronique
55 Quai Marcel Dassault
F-92214 St. CLOUD
FRANCE
botella@dassault-elec.fr

BRANCHINI Paolo
INFN
Sezione Sanità
Viale Regina Elena, 299
I-00161 Rome
ITALY
vaxsan::branchini
branchini@iss.infn.it

BRAND Susanne
Inst. f. Experimentalphysik I
Ruhr-Univ. Bochum
Postfach 10 21 48
Universitätsstrasse 150
D-44780 Bochum
GERMANY
sbrand@tau.ep1.ruhr-uni-bochum.de

BROADHURST David
Physics Department
The Open University
Walton Hall
Milton Keynes MK7 6AA
UNITED KINGDOM
d.broadhurst@open.ac.uk

BRUN René
European Organization
for Nuclear Research
CH-1211 Genève 23
SWITZERLAND
brun@cernvm.cern.ch

BURTON Jackie
SSCL
2550 Beckleymeade av.
Dallas Tx 75237
U.S.A.
Burton@poobah.ssc.gov

CALL Darrel
Theor. Div.
Los Alamos National Lab.
Los Alamos NM 87545
U.S.A
70400.1033@compuserve.com

CHETYRKIN Konstantin G.
Institute of Nuclear Research
of Russian Academy of Sciences
Profsoyuznaja Str. 7 a
Moscow 117312
RUSSIA
chet@inucres.msk.su

CORMELL Laird R.
SSCL MS 2003
2550 Beckleymeade av.
Dallas Tx 75237
U.S.A.
cormell@sscvx1.ssc.gov

D'ANTONE Ignazio
Istituto Nazionale Fisica Nucleare
Univ. degli Studi di Bologna
Via Irnerio 46
I-40126 Bologna
ITALY
dantone@bologna.infn.it

DAHM Jochen
Fachbereich Physik
Bergische Univ.
Gesamthochschule Wuppertal
Gauss Str. 20
D-42097 Wuppertal
GERMANY
dahm@physik.uni-wuppertal.dbp.de

DAVYDYCHEV Andrei
Division of Theoretical Physics
Institute of Nuclear Physics
Moscow State University
119899 Moscow
RUSSIA
davyd@vsfys1.fi.uib.no

DE LAAT C.A.T.M.
Physics Department
Rijksuniv. Utrecht
P.O.Box 80000
Princetonplein 5
NL-3508 TA Utrecht
NETHERLANDS
delaat@fys.ruu.nl

DECHANT Bernd
GSI
Ges. f. Schwerionenforschung mbH
Postfach 110 552
Planckstr. 1
D-64220 Darmstadt
GERMANY
dechant@v6000a.gsi.de

DEGENER Thomas
EP1
Ruhr-Univ. Bochum
Postfach 10 21 48
Universitätsstrasse 150
D-44780 Bochum
GERMANY
thomas@tau.ep1.ruhr-uni-bochum.de

DENBY Bruce
MS 318
Fermi Nat. Accelerator Lab.
P.O.Box 500
Wilson Rd.
Batavia IL 60510
U.S.A
denby@fnal.fnal.gov

DÖBBELING Hans
GSI
Ges. f. Schwerionenforschung mbH
Postfach 110 552
Planckstr. 1
D-64220 Darmstadt
GERMANY
dob@hp9a.gsi.de

DUVAL Pierre-Yves
CCMP - IN2P3
Faculte des Sciences de Luminy
Case 907
70 route Léon Lachamp
F-13288 Marseille Cedex 9
FRANCE
duval@marvax.in2p3.fr

ECK Hagen
Inst. f. Theor. Physik I
Univ. Würzburg
Am Hubland
D-97074 Würzburg
GERMANY

FINE Valery
Joint Inst. for Nuclear Research
Dubna, Head Post Office
P.O.Box 79
101 000 Moscow
RUSSIA
fine@main1.jinr.dubna.su
fine@vxcern.ch

FLEISCHER Jochen
Fakultät f. Physik
Univ. Bielefeld
Postfach 86 40
Universitätsstrasse
D-33501 Bielefeld 1
GERMANY
fleischer@physik.uni-bielefeld.de

FORTNER Michael
Physics Department
Northern Illinois University
De Kalb, IL 60015
U.S.A
fortner@fnal.fnal.gov

FRANZKOWSKI Johannes
Johannes Gutenberg Universität
Institut fuer Physik
D-55099 Mainz
GERMANY
franzkowski@vipmza.physik.uni-mainz.de

FUCHS Paul
Laboratoire National
SATURNE
F-91191 Gif-sur-Yvette Cedex
FRANCE
FUCHS@lnsbm2.cern.ch
33608::fuchs

GABADADZE Grigori T.
Institute fir Nuclear Research
Academy of Sciences
117 234 Moskva
RUSSIA
kataev@inucres.msk.su

GHESQUIERE Claude
L.P.C.
College de France
11 Pl. M. Berthelot
F-75005 Paris
FRANCE
ghesq@frcpn11.in2p3.fr

GILS Hans Jürgen
Kernforschungszentrum
Karlsruhe GmbH
Institut für Kernphysik III
Postfach 3640
D-76021 Karlsruhe
GERMANY
gils@ik3dc1.kfk.de

GRACEY John
DAMTP
Univ. of Liverpool
P.O. Box 147
Liverpool L69 3BX
UNITED KINGDOM
jag@s-a.amtp.liv.ac.uk

GRUBER Andreas
Max-Planck-Inst.
f. Physik und Astrophysik
Föhringer Ring 6
D-80805 München
GERMANY
aeg@iws104.mppmu.mpg.de

HAGGE Lars
ZEUS Collaboration
Deutsches Elektronen-Synchrotron
Notkestr. 85
D-22603 Hamburg
GERMANY
hagge@vxdesy.desy.de

HAHN Siegfried
Fachbereich Physik
Bergische Univ. Wuppertal
Wuppertal
Gauss Str. 20
42097 Wuppertal
GERMANY
hahns@vxcern.cern.ch

HAVEMAN Jan
Dept. of Physics & Astronomy
Utrecht University
P.O.Box 80000
NL-3508 TA Utrecht
NETHERLANDS
haaf@fys.ruu.nl

HEMKER Andreas
Fachbereich Physik
Bergische Univ. Wuppertal
Gauss Str. 20
D-42097 Wuppertal
GERMANY
hemker@physik.uni-wuppertal.dbp.de

ILYIN Vyacheslav A.
Symbolic Computation Lab.
Moscow State Univ.
Leninskie Gory
Nucl. Physics Inst.
119 899 Moskva
RUSSIA
ilyin@v1.ph.qmw.ac.uk

IVANOV Victor
Joint Inst. for Nuclear Research
Dubna, Head Post Office
P.O.Box 79
101 000 Moscow Dubna
RUSSIA
victor@lcta21.jinr.dubna.su

JONES Robert
CERN
ECP/DS
CH-1211 Geneva 23
SWITZERLAND
jones@online.cern.ch

KAMMERLOCHER Hans
Computer Center
Deutsches Elektronen-Synchrotron
Notkestr. 85
D-22603 Hamburg
GERMANY
hans@x4u.desy.de

KANEKO Toshiaki
Faculty of general education
Meiji Gakuin University
Totsuka
Yokohama 244
JAPAN
kaneko@minami.kek.jp

KARCZMARCZUK Jerzy
Lab. d'Informatique
Univ. de Caen
Esplanade de la Paix
F-14032 Caen Cedex
FRANCE
karczma@univ-caen51.fr

KARIMÄKI Veikko
CERN
European Organization
for Nuclear Research
ECP-Division
CH-1211 Genève 23
SWITZERLAND
karimaki@cernvm.cern.ch

KATAEV Andrey
Acad. of Sciences of the USSR
Inst. for Nuclear Research
60th October Anniversary
Prospect 7A
117 312 Moskva
RUSSIA
kataev@inucres.msk.su

KATO Kiyoshi
Dept. of Physics
Kogakuin Univ.
1-24-2 Nishishinjuku
Shinjuku
Tokyo 160
JAPAN
katok@jpnku01.bitnet

KIESLING Christian
Max-Planck-Inst.
f. Physik und Astrophysik
Postfach 40 12 12
Föhringer Ring 6
8000 München 40
GERMANY
cmk@dmumpiwh.binet

KILIAN Ulrich
Inst. f. Physik
Johannes-Gutenberg-Univ.
Postfach 39 80
Staudinger Weg 7
D-55099 Mainz
GERMANY
kilian@vipmza.physik.uni-mainz.de

KINOSHITA Toichiro
KEK
Nat.Lab. for High Energy Phys.
1-1 Oho
Tsukuba 305
Ibaraki-Ken 305
JAPAN
tk@crnlnuc.bitnet

KISEL Ivan V.
Lab. of Nuclear Problems
Joint Institut for Nuclear Research
Head Post Office
P.O. Box 79
101000 Moscow
RUSSIA
kisel@main2.jinr.dubna.su

KOLANOSKI Hermann
Institut f. Physik
Univ. Dortmund
D-44221 Dortmund
GERMANY
hermann@physik.uni-dortmund.de

KORNYAK Vladimir V.
Institute of Mathematics
Tereschenskaya str. 3
Kiev-4, 252601
UKRAINIA
vgusynin@gluk.apc.org
mathkiex@sovam.com

KOTIKOV Anatoly V.
Lab. of Particle Physics
Joint Inst. for Nuclear Research
Dubna, Head Post Office
P.O.Box 79
101 000 Moskva Dubna
RUSSIA
kotikov@lshe7.jinr.dubna.su

KRAMER Helmut
Inst. f. Kernphysik
Univ. Mainz
Becherweg 45
D-55099 Mainz
GERMANY
kramer@asterix.kph.uni-mainz.de

KRÄMER-FUHRMANN
GMD
National Res. Center f.Comp. Science
Schloss Birlinghoven
D-53757 Sankt Augustin
GERMANY
kraemer-fuhrmann@gmd.de

KRÄMERKÄMPER Thomas
Lehrstuhl f. Exp. Phys.
Univ. Dortmund
Inst. für Physik
Postfach 50 05 00
D-4600 Dortmund 50
GERMANY
kraemerk@physik.uni-dortmund.de

KUNZE Marcel
Inst. f. Experimentalphysik
Ruhr-Univ. Bochum
Postfach 10 21 48
Universitätsstrasse 150
4630 Bochum 1
GERMANY
marcel@cernvm.cern.ch
marcel@tau.epl.ruhr-uni-bochum.de

KURIHARA Yoshimasa
KEK
Oho 1-1
Tsukuba
Ibaraki-ken
Tsukuba
JAPAN
kato@jpnku01.bitnet

KURSAWE Frank
Lehrstuhl f. Informatk
Univ. Dortmund
Postfach 50 05 00
D-44221 Dortmund
GERMANY
kursawe@ls11.informatik.uni-dortmund.de

LADES Martin
Inst. f. Neuroinformatik, ND03
Ruhr-Univ. Bochum
Universitätsstrasse 150
D-44780 Bochum
GERMANY
martin@neuroinformatik.ruhr-uni-bochum.de

LANGE Jens S.
Inst. f. Experimentalphysik
Ruhr-Univ. Bochum
Universitätsstrasse 150
D-4630 Bochum 1
GERMANY
soeren@tau.ep1.ruhr-uni-bochum.de

LANSKE Dankfried
III.Physikal. Inst. (A)
RWTH Aachen
Physikzentrum
Sommerfeldstr. 28
D-52056 Aachen
GERMANY
lanske@acv3a1.cern.ch

LARIN Sergey A.
Institut for Nuclear Research
60 October Prospect, 7A
Moscow 117312
RUSSIA
larin@inucres.msk.su

LAUTENBACHER Markus
Physics Dept. T30
Tech.Univ. München
James-Franck-Strasse
8046 Garching bei München
GERMANY
lauten@feynman.t30.physik.tu-muenchen.de

LE FOLL Joseph
DAPNIA Bat. 141
CEN Saclay
F-91191 Gif-sur-Yvette Cedex
FRANCE
lefoll@frsac12.bitnet

LEBRETON Evelyne
Phys. Nucl. et Hautes Energies
Universite Pierre et Marie Curie
Tour 33-RDC
4 place Jussieu
F-75005 Paris
FRANCE
evelyne@frcpn11.in2p3.fr

LEE Jungil
Department of Physics
Seoul National University
Seoul, 151-742
KOREA
jungil@phyy.snu.ac.kr

LEGRAND Iosif
CERN
European Organization
for Nuclear Research
ECP-Division
CH-1211 Genève 23
SWITZERLAND
ilegrand@cernvm.cern.ch

LINDSEY Clark
Royal Institut of Technology
Department of Physics
Frescativägen
S-10405 Stockholm
SWEDEN
lindsey@msivax.physto.se

LITTMANN Enno
Technische Fakultät
Univ. Bielefeld
Universitätsstrasse
D-33615 Bielefeld
GERMANY
littmann@techfak.uni-bielefeld.de

MANDRIOLI Gianni
Istituto Nazionale Fisica Nucleare
Univ. degli Studi di Bologna
Via Irnerio 46
I-40126 Bologna
ITALY
mandrioli@bologna.infn.it

MARTIN Liliane
L.P.C.
College de France
11 Pl. M. Berthelot
F-75005 Paris
FRANCE
lila@frcpn11.in2p3.fr

McFARLANE Kenneth W.
SSC Laboratory, MS 2005
2550 Beckleymeade Ave
Dallas, TX 75115
U.S.A.
mcfarlan@diamond.gem.ssc.gov

MERTIG Rolf
Lorentz Inst. for Theor. Physics
Rijksuniv. Leiden
P.O. Box 9506
Nieuwsteeg 18
NL-2300 RA Leiden
NETHERLANDS
rolfm@rulgm4.LeidenUniv.nl

MICHELOTTO Michele
INFN
Padova
Via F. Marzolo 8
I-35131 Padova
ITALY
michelotto@mvxpd5.pd.infn.it
vaxfpd::michelotto

MOECK Juergen
Max-Planck-Inst.
f. Physik und Astrophysik
Föhringer Ring 6
D-80805 München
GERMANY
jum@iws104.mppmu.mpg.de

NEMRY Myriam
Inst. de Physique Nucléaire
Univ. Catholique de Louvain
Dépt. de Physique
Chemin du Cyclotron 2
B-1348 Louvain-la-Neuve
BELGIUM
Nemry@sisc.ucl.ac.be

NIITTYLAHTI Jarkko
CERN
European Organization
for Nuclear Research
CH-1211 Genève 23
SWITZERLAND
jarkko@cernvm.cern.ch

NILSSON Sigward
Stockholm University
Fysikum
Vanadisvägen 9
S-11346 Stockholm
SWEDEN
sn@vana.physto.se

NOWAK Jörg P.
Fachbereich Physik
Bergische Univ. Wuppertal
Wuppertal
Gauss Str. 20
D-42097 Wuppertal
GERMANY
nowak@physik.uni-wuppertal.dbp.de

O'NEALE Stephen
School of Physics
Univ. of Birmingham
P.O.Box 363
Chancellor's Court
Birmingham B15 2TT
UNITED KINGDOM
oneale@cernvm

ODORICO Roberto
Dipart. di Fisica
Univ. degli Studi di Bologna
Via Irnerio 46
I-40126 Bologna
ITALY
vxcern::odorico
odorica@bologna,infn.it

OFFERMANN Verena
Physics Department
Univ. of Wuppertal
Gauss Str. 20
D-42097 Wuppertal 1
GERMANY
offermann@physik.uni-wuppertal.d400.de

OLDENBORGH van, Geert Jan
Theory Group
Paul Scherrer Institute
CH-5232 Villingen PSI
SWITZERLAND
gj@csun.psi.ch

OLIN Art
Tri-Univ. Meson Facility
TRIUMF
4004 Wesbrook Mall
Vancouver BC
CANADA
triumfcl

PALICHIK Vladimir
LCTA
Joint Inst. for Nuclear Research
Dubna, Head Post Office
P.O.Box 79
141980 Moskva Dubna
RUSSIA
palichik@main1.jinr.dubna.su

PERRET-GALLIX Denis
L.A.P.P.
Lab. de Phys. des Particules
B.P. 110
Ch. de Bellevue
F-74941 Annecy le vieux Cedex
FRANCE
perretg@cernvm.cern.ch

PUKHOV Alexander E.
Symbolic Computation Lab.
Moscow State Univ.
Research Inst. for Nucl. Phys.
119 899 Moskva
RUSSIA
pukhov@compnet.msu.su

RAPORTIRENKO Anatoly
Joint Inst. for Nuclear Research
P.O.Box 79
Moscow, 101000
RUSSIA
ram@jinr.dubna.su

REMIDDI Ettore
Dipart. di Fisica
Univ. degli Studi di Bologna
Via Irnerio 46
I-40126 Bologna
ITALY
remiddi@bo.infn.it
vaxbo::remiddi

RIBARICS Pal
Max-Planck-Inst.
f. Physik und Astrophysik
Postfach 40 12 12
Föhringer Ring 6
D-80805 München
GERMANY
par@dmumpiwh.bitnet

RITBERGEN van, Timo
Nat. Inst. for Nucl. Phys.-H.E.-
NIKHEF-H
P.O.Box 41882
NL-1009DB Amsterdam
NETHERLANDS
t46@nikhefh.nikhef.nl

SADOVSKY SergeI
Inst. for High Energy Physics
142 284 Protvino (Moscow Region)
RUSSIA
sadovsky@mx.ihep.su

SCHMIDT, Manfred
University of Wuppertal
D-42097 Wuppertal
GERMANY

SEIXAS Jose Manoel de
COPPE/UFRJ
Programa de Engenharia Eletrica
Caixa Postal 68504
21945-970 Rio de Janeiro
BRAZIL
seixasj@vxcern.cern.ch
seixas@lpc.ufrj.br

SHIMIZU Yoshimitsu
KEK
Nat.Lab. for High Energy Phys.
1-1 Oho
Tsukuba 305
Ibaraki-Ken 305
JAPAN
shimiz@minami.kek.jp

SHIRKOV Dmitrij
Bogoliubov Theor. Lab.
Joint Inst. for Nuclear Research
Dubna, Head Post Office
141980 Dubna
RUSSIA
shirkovd@th-head.jinr.dubna.su

SOMERS Francis
Inst. de Physique Nucléaire
Univ. Catholique de Louvain
Dépt. de Physique
Chemin du Cyclotron 2
B-1348 Louvain-la-Neuve
BELGIUM
somers@sisc.ucl.ac.be

SPAHN Martin
Max-Planck-Institut fuer Kernphysik
Saupfercheckweg 1
D-69117 Heidelberg
GERMANY
spahn@e3.mpi-hd.mpg.de

SZAFRAN Sylviane
Lab. de Physique Corpusculaire
Collège de France
11 place Marcelin-Berthelot
F-75005 Paris Cedex 05
FRANCE
sylviane@frcpn11.in2p3.fr

VANDERKELEN Mark
Department for Mathematical Physics
and Astronomy
University of Ghent
Krijgslaan 281 / S9
B-9000 Ghent
BELGIUM
vdkelen@steevast.rug.ac.be

VELTMAN Martinus
Univ. of Michigan
Harrison M. Randall Lab.
Physics Dept.
Ann Arbor MI 48109-1120
U.S.A
veltman@mich.physics.lsa.umich.edu

VERMASEREN Jozef Antoon
NIKHEF-H
Kruislaan 409
P.O. BOX 41882
NL-1009 DB Amsterdam
NETHERLANDS
t68@nikhefh.nikef.nl

VERMEULEN Jos
Nat. Inst. for Nucl. Phys.- H.E.
NIKHEF-H
Kruislaan 409
P.O.Box 41882
NL-1009 DB Amsterdam
NETHERLANDS
i73@nikhefh.nikhef.nl

WANG Jian-Xiong
Theory Group
Insitute of High Energy Physics
Academia Sinica
100039, Beijing
P.R. CHINA
jxwang(p)bepc3@scs.slac.stanford.edu

WILDNER Elena
European Organization
for Nuclear Research
CH-1211 Genève 23
SWITZERLAND
mal@cernvm.cern.ch